Applied Discrete Structures

Useful for

- UG and PG students of Mathematics, Computer Science and Engineering.
- GATE and many other Entrance and Competitive Examinations.

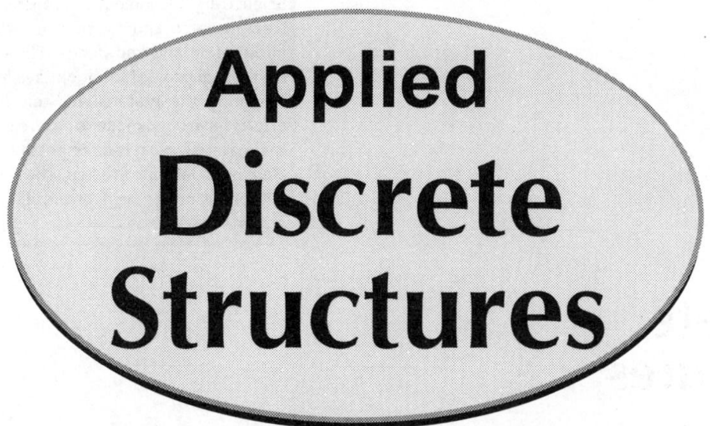

Applied Discrete Structures

Useful for

- UG and PG students of Mathematics, Computer Science and Engineering.
- GATE and many other Entrance and Competitive Examinations.

Dr. SUDHIR KUMAR PUNDIR
M.Sc., M.Phil, NET, Ph.D.
Head
Department of Mathematics
S.D. (P.G.) College,
Muzaffarnagar (U.P.)

CBS

CBS Publishers & Distributors Pvt. Ltd.

New Delhi • Bengaluru • Chennai • Kochi • Kolkata • Mumbai
Hyderabad • Nagpur • Patna • Pune • Uttarakhand • Jharkhand

Applied
Discrete Structures

ISBN: 978-93-86827-59-3

First Edition: 2019

Published by **Satish Kumar Jain** and produced by **Varun Jain** for
CBS Publishers & Distributors Pvt. Ltd.,
4819/XI Prahlad Street, 24 Ansari Road, Daryaganj, New Delhi - 110002
delhi@cbspd.com, cbspubs@airtelmail.in • www.cbspd.com
Ph.: 23289259, 23266861, 23266867 • Fax: 011-23243014
Corporate Office: 204 FIE, Industrial Area, Patparganj, Delhi - 110 092
Ph: 49344934 • Fax: 011-49344935
E-mail: publishing@cbspd.com • publicity@cbspd.com

Branches:
- ***Bengaluru:*** 2975, 17th Cross, K.R. Road, Bansankari 2nd Stage, Bengaluru - 70 • Ph: +91-80-26771678/79 • Fax: +91-80-26771680 E-mail: cbsbng@gmail.com, bangalore@cbspd.com
- ***Chennai:*** No. 7, Subbaraya Street, Shenoy Nagar, Chennai - 600030 Ph: +91-44-26681266, 26680620 • Fax: +91-44-42032115 E-mail: chennai@cbspd.com
- ***Kochi:*** Ashana House, 39/1904, A.M. Thomas Road, Valanjambalam, Ernakulum, Kochi • Ph: +91-484-4059061-65 Fax: +91-484-4059065 • E-mail: cochin@cbspd.com
- ***Kolkata:*** 6-B, Ground Floor, Rameshwar Shaw Road, Kolkata - 700014 Ph: +91-33-22891126/7/8 • E-mail: kolkata@cbspd.com
- ***Mumbai:*** 83-C, Dr. E. Moses Road, Worli, Mumbai - 400018 Ph: +91-9833017933, 022-24902340/41 • E-mail: mumbai@cbspd.com

Representatives:
- Hyderabad: 0-9885175004
- Nagpur: 0-9021734563
- Patna: 0-9334159340
- Pune: 0-9623451994
- Jharkhand: 0-9811541605
- Uttarakhand: 0-9716462459

Printed at:
India Binding House, Noida, UP (India)

The book entitled **'Applied DISCRETE STRUCTURES'** meet the needs of engineering and science students of U.G. and P.G. levels. Besides, it will also be very useful for students preparing for various competitive examinations.

The contents of the book are designed from curricula offered by various universities across the country. In each chapter of the book an ample amount of theory is given which is supported by solved examples along with their answers.

I express my gratitude to the authors and publishers of various books I consulted during the preparation of the book.

I wish to sincerely thank **Sh. S.K. Jain** and **Sh. Varun Jain**, Managing Director, CBS publishers and distributors, New Delhi for their encouragement and help in bringing out this publication in a present nice from.

My special thanks to **Sh. B.M. Singh, Sh. Sunil Dutt, Sh. Puneet Verma** and entire team of CBS publishers, New Delhi whose encouragement and unstinted support enable me to complete this book. Mr. Kanti Prasad, M/s Balaji Computers also deserve special mention for nice type setting.

I must also record my appreciation due to my wife **Dr. Rimple**, daughter **Rijuta**, son **Shrish** for their understanding and love during the long period that I have taken to complete this book.

Above all I am thankful to Almighty God without whose grace nothing is possible for any one.

Readers are welcomed to point out errors, if any and send their valuable suggestions for improving the quality of the book.

Dr. SUDHIR KUMAR PUNDIR

email : skpundir05@yahoo.co.in

CONTENTS

Chapter 1

Introduction

- Number system
- Sets and its type
- Operation on sets
- ordered pairs
- Relation
- Functions
- Types of relation and function
- Some particular functions
- Algebra of functions
- Recursively defined function
- Functions for computer science.

1.1 INTRODUCTION

The theory of sets was developed by German mathematician George Cantor. The concept of sets appears in all mathematics. This chapter introduces the notations and terminology of set theory. Classical set theory, also termed as crisp set theory, is fundamental to the study of pure mathematics.

1.2 NUMBER SYSTEM

The number system plays a key role in Mathematics. The real number system **R** is one of the most important and beautiful mathematical system. There are different ways of introducing the real number system, but the most common way is to start with Peano's Axioms for the natural numbers. The axioms for natural numbers, discovered by the Italian Mathematician Peano are :

(i) 1 is a natural number.

(ii) Each natural number n has a successor $(n + 1)$.

(iii) Two natural numbers are equal if their successors are equal.

(iv) Except 1, each natural number is a successor of natural number.

(v) Any set of natural numbers which contains 1 and the successor of every natural number $(k + 1)$ whenever it contains k in the set N of natural numbers.

- Axiom (*v*) is commonly known as the axiom of induction or principle of finite induction.
- The above axioms completely define the set of natural numbers.

Definition : *The numbers* 1, 2, 3,... *are called natural numbers.* We represent the set of natural numbers by **N**.

i.e., $\mathbf{N} = \{1, 2, 3, \ldots\}$

The Peano's axioms can be used to extend the set **N** of natural numbers to another large system, known as the set of integers.

Definition : *The numbers..., –3, –2, –1, 0, 1, 2, 3, ... are called integers.* We represent the set of integers by **Z**.

i.e., $\mathbf{Z} = \{...... 3, -2, -1, 0, 1, 2, 3, ...\}$

Integers can be used to define the rational numbers.

Definition : *Any number of the form p/q, where p, q* $\in \mathbf{Z}$, $q \neq 0$ *and p, q have no common factor (except ±1) is called a rational number.*

The set of rational numbers is denoted by **Q**.

$$\therefore \qquad \mathbf{Q} = \left\{\frac{p}{q};\ p, q \in \mathbf{Z},\ q \neq 0\right\}$$

☛ The set of rational numbers consists of integers and fractions.

Definition : *Any number which is not rational, is called an Irrational number.* For example, $\sqrt{2}$, $\sqrt{3}$ etc. It should be noted that every rational number can be expressed as a terminating or recurring decimal whereas every irrational number can be expressed as a non-terminating infinite decimal.

1.2.1 Real Number

A number which is either rational or irrational is called a real number. The set of real number is denoted by **R**.

1.2.2 Integral Powers of a Real Number

Let $\mathbf{a} \in \mathbf{R}$, and n be any positive integer then we can define $a^n = a \,.\, a \,.\, a \,...\, n$ times.

In particular $a^1 = a$

$a^2 = a.a$

$a^3 = a.a.a = a^2.a$ and so on.

Also, if n is any negative integer, then we have $x^{-n} = (x^n)^{-1} = (x^{-1})^n$

1.2.3 Positive and Negative Real Number

(i) A real number a is called positive, if $a > 0$ and the set of all positive real numbers, denoted by $\mathbf{R}^+$, is given by $\mathbf{R}^+ = \{x : x \in \mathbf{R}, x > 0\}$

(ii) A real number a is called negative if $a < 0$ and the set of all negative real numbers, denoted by $\mathbf{R}^-$, is given by $\mathbf{R}^- = \{x : x \in \mathbf{R}, x < 0\}$

1.3 INTERVAL

A subset S of **R** *is called an interval if* $a, b \in S$, $x \in \mathbf{R}$ *such that* $a < x < b$ *implies* $x \in S$.

There are following four type of intervals

(i) a ○———○ $b \Rightarrow \;]a, b[= \{x : a < x < b\}$

(i) a ●———● $b \Rightarrow \; [a, b] = \{x : a \leq x \leq b\}$

(iii) a ○———● $b \Rightarrow \;]a, b] = \{x : a < x \leq b\}$

(iv) a ●———○ $b \Rightarrow \; [a, b[= \{x : a \leq x < b\}$

☛ The set $]a, b[$ in which the end points are not included, is called an open interval.

☛ The set $[a, b]$, also contains both its end points, is called a closed interval.

☛ The sets $[a, b[$ and $]a, b]$ are called half open (or half closed) intervals or semi-open (or semi-closed) as they contain only one end point.

Apart from the four type of intervals listed above, there are a few more types : These are

(i) $]a, \infty[= \{x : a < x\}$ (open right ray)
(ii) $[a, \infty[= \{x : a \leq x\}$ (closed right ray)
(iii) $]-\infty, b[= \{x : x < b\}$ (open left ray)
(iv) $]\infty, b] = \{x : x \leq b\}$ (closed left ray)
(v) $]-\infty, \infty[= \mathbf{R}$ (open interval)

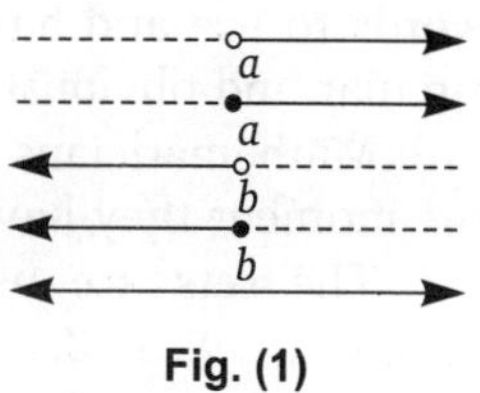

Fig. (1)

- If S is any interval and if c and d are two elements of S, then all numbers lying between c and d are also elements of S.
- The proper use of a bracket, for example, parenthesis for open and square brackets for closed and end points, it self specifies the interval. As such, to emphasize the nature of an interval, we shall drop the used 'description' and shall simply express the interval by using the appropriate brackets.

1.3.1 Length of an Interval

The number $b - a$ is called length of the intervals $]a, b[$, $[a, b[$, $]a, b]$ *and* $[a,b]$. If the length of the interval is finite, the interval is said to be finite and if the length is infinite, then it is known as infinite interval.

1.3.2 Absolute value of a Real Number

The absolute value of a real number a denoted by $|a|$ is the real number a, $-a$ or 0 according as a is positive, negative or zero, *i.e.*,

$$|a| = \begin{cases} a & \text{if } a \geq 0 \\ -a & \text{if } a < 0 \end{cases}$$

From the above definition, it is clear that

(i) $|a| = \max\{a - a\}$ (ii) $-|a| = \min\{a - a\}$ (iii) $|a| \geq a \geq -|a|$

1.3.3 Some Useful Results

(i) $|xy| = |x|.|y|$
(ii) $|x + y| \leq |x| + |y|$
(iii) $|x - y| \geq ||x| - |y||$
(iv) $|x - y| \leq |x| + |y|$
(v) $\left|\dfrac{x}{y}\right| = \dfrac{|x|}{|y|}$
(vi) If $\in > 0$, then $|x - y| < \varepsilon \Leftrightarrow y - \in < x < y + \in$

1.4 CONCEPT OF SETS

The theory of sets is one of the most important tools of pure mathematics. Pure mathematics is the study of sets equipped with assigned structures, known as mathematical systems. In this section, we shall study some fundamental concepts of set theory.

Definition : *'A set is a well defined collection of objects'.*

The objects of a set are called the elements or members of that set and their membership is defined by certain conditions.

The basic concept used can be defined. Suppose, for example, one defines the set by "A set is well defined collection of objects" one naturally asks what is meant by a collection. Perhaps, then one defines "A collection is an aggregate of things". What then is an aggregate? How our language is finite, so after sometime we will run out of new words to use and have to repeat some words already questioned. The definition is then circular and obviously worthless.

Mathematicians realize that there must be some undefined or primitive concepts at the moment they have agreed that set shall be such a primitive concept.

The sets are usually denoted by the capital letters of English alphabets : Say *A, B, C, …, X, Y, Z.*

For example :

(i) The collection of the letters *a, b, c, d,…*

(ii) The collection of all natural numbers denoted by **N.**

(iii) The students of M.Sc., Mathematics in C.C.S. University, Meerut.

(iv) The collection of vowels in English alphabet. This set containing only five elements, namely, *a, e, i, o, u*.

(v) The collection of all states in Indian union.

If *S* is a set, an object *a* in the collection *S* is called an element of *S*. This fact is expressed in symbol as $a \in S$ (read as *a* is in S or *a* belongs to S). If *a* is not in *S*, we write $a \notin S$. For example, $4 \in \mathbf{R}$, the set of real numbers, but $\sqrt{-2} \notin \mathbf{R}$.

Here, Greek letter $\in$ denotes 'belongs to'. It is the abbreviation of the Greek word meaning 'is'.

- By the term 'well defined' we mean that we are given a collection of objects, with certain definite property, so that we are able to determine whether a given object belongs to our collection or not. Thus, every collection of objects is not a set.
- Set and aggregate both have the same meaning.
- The elements of a set must be distinguished from one another. The collection of sand particles does not form a set.
- The collection of rich persons of a city is not a set. However the collection of those persons of city whose wealth exceeds, a fixed amount, say rupees fifty thousands, is a set.
- The order is not preserved in case of a set, whereas order is necessarily preserved in case of sequence. That is to say, each of the sets {1, 2, 3}, {3, 2, 1}, {1, 3, 2} denotes the same set.
- The repetition of an element does not change the nature of a set, *i.e.*, each of the sets {1, 2, 3}, {1, 2, 2, 3}, {1, 3, 3, 2} denotes the same sets.

1.4.1 Representation of a Set

There are two ways of representing a set :

(i) Roster or tabulation method

(ii) Set-builder or rule method

Roster Method : In this method, the elements of the set are listed within braces, and separated by comma.

For example :

(i) A = {1, 2, 3, 4, 5, 6}

(ii) The set of vowels of English alphabet may be represent as {*a, e, i, o, u*}.

(iii) The set of a natural numbers from 1 to 100 may be written as $\mathbf{N} = \{1, 2, 3, \ldots, 100\}$. We use three dots in the middle to include the missing elements.

(iv) The set of positive integers, which is a non-ending set may be written as $\mathbf{Z}^+ = \{1, 2, 3, 4, 5, \ldots\}$. The three dots in the end means that the elements continue in the same manner.

(v) The set of prime numbers is written as $P = \{2, 3, 5, 7, 11, 13, 17, 19, \ldots\}$

Set-Builder Method : In this method, we first try to find a property which characterizes the elements of a set, that is, a property P, which all the elements of the set possess and which no other objects possess. Then, we describe the set as $\{x : x$ has property $P\}$

This is to be read as "the set of all x such that x has property P".

For example :

(i) The set of all integers can be written as $\mathbf{Z} = \{x : x \text{ is an integer}\}$

(ii) The set $A = \{1, 2, 3, 4, 5\}$ can be written as $A = \{x \in \mathbf{N} : x \leq 5\}$

(iii) The set of complex numbers can be written as $\mathbf{C} = \{a + ib : a, b \in \mathbf{R}\}$.

(iv) The set $A = \{1, 8, 27, \ldots\}$ can be written as $A = \{x^3 : x \in \mathbf{Z}^+\}$.

ILLUSTRATIVE EXAMPLES

1. *Use the Roster method to identify each set :*

(a) *the set of possible integers greater than 8 and less than 14.*

(b) *the set of numbers whose elements are the first five positive odd integers.*

(c) *the set of even positive integers.*

(d) *the set of even positive integers that are divisible by 10.*

(e) *the set of all vowels in English alphabets which precedes r.*

Solution : (a) $\{9, 10, 11, 12, 13\}$ (b) $\{1, 3, 5, 7, 9\}$

(c) $\{2, 4, 6, 8, 10, \ldots\}$ (d) $\{10, 20, 30, 40, 50, \ldots\}$

(e) $\{a, e, i, o\}$

2. *Use the set-builder method, identify the following sets :*

(a) $A = \{1, 3, 5, 7, 9, \ldots\}$ (b) $B = \left\{1, \frac{1}{4}, \frac{1}{9}, \frac{1}{16}, \frac{1}{25}, \ldots\right\}$

(c) $C = \{0, 1, 2, 3, \ldots\}$ (d) $D = \left\{\frac{1}{2}, \frac{2}{3}, \frac{3}{4}, \frac{4}{5}, \ldots\right\}$

Solution : (a) The set of odd positive integers.

(b) Here, elements of the set B are the reciprocals of the squares of the natural numbers.

So, the set $B = \left\{\frac{1}{n^2} : n \in \mathbf{N}\right\}$

(c) The set of whole numbers.

(d) Here, each element in the given set has the denominator one more than the numerator. Hence

$$D = \left\{x : x = \frac{n}{n+1}; n \in \mathbf{N}\right\}$$

3. *Write the set* $\left\{\frac{1}{2}, \frac{2}{5}, \frac{3}{10}, \frac{5}{26}, \ldots\right\}$ *in the set-builder form.*

Solution : We observe that each element in the given set has the denominator one more than the square of the numerator. Also, the numerator begins with 1. Hence, in the set builder form, the given set can be written as

$$\left\{x : x = \frac{n}{n^2+1}, n \in \mathbf{N}\right\}$$

Exercise 1.1

1. Which of the following collections are sets?
   (i) All mathematics students in your college.
   (ii) All poor hockey players in a college.
   (iii) All odd numbers less than 20.
   (iv) The collection of good teachers in your college.
   (v) All successful and rich people in your city.
   (vi) The people in your immediate family (father, mother, sister., brother).
2. Write the members of each of following sets by the Roster method.
   (i) $\{x : x$ is odd whole number less than $14\}$
   (ii) $\{x : x^2 < 36$ and $x \in \mathbf{N}\}$
   (iii) $\{x :$ squares of all whole numbers less than $8\}$
   (iv) $\{x : x$ is a prime number, $10 < x < 20\}$
   (v) $\{x : x$ is a composite number less than $20\}$
   (vi) $\{x : x < x\}$
3. Rewrite the followings sets using set builder method.
   (i) $A = \{2, 4, 6, 8, \ldots\}$
   (ii) $B = \left\{1, \frac{1}{2}, \frac{1}{3}, \frac{1}{4}, \ldots\right\}$
   (iii) $C = \{0, 3, 6, 9, 12, \ldots\}$
   (iv) $D = \{0, 4, 6, 8, 10, \ldots\}$
4. List the elements of the following sets.
   (i) $A = \{x : x^2 \le 16 : x \in \mathbf{Z}\}$
   (ii) $B = \{x : 1 \le x \le 5$ and $x \in \mathbf{N}\}$
   (iii) $C = \{x : x \in \mathbf{N}$ and x is a factor of $15\}$
   (iv) $D = \{x : x$ is a month of year having 31 days$\}$
   (v) $E = \{x : x \in \mathbf{Z}$ and $3x - 2 = 3\}$
   (vi) $F = \{x : x$ is an integer lying between $-\frac{1}{2}$ and $\frac{1}{2}\}$
5. Use the appropriate symbols $\in$ or $\notin$ to fill in the blanks below :
   (i) 12 ... the set of all numbers dividing 84.
   (ii) K... the set of all vowels of the English alphabets.
   (iii) $\frac{1}{2}$ ... the set of natural number.
   (iv) India ... the set of members of UNO.
   (v) $\sqrt{2}$ ... the set of rational numbers.
   (vi) 15 ... the set of multiples of 3.

ANSWERS

1. (i), (iii), (vi)

2.(i) $\{1, 3, 5, 7, 9, 11, 13\}$ (ii) $\{1, 2, 3, 4, 5\}$ (iii) $\{0, 1, 4, 9, 16, 25, 36, 49\}$
(iv) $\{11, 13, 17, 19\}$ (v) $\{1, 4, 6, 8, 9, 10, 12, 14, 15, 16, 18\}$ (vi) ϕ

3.(i) $A = \{x : x = 2n : n \in \mathbf{N}\}$ (ii) $\{1/n : n \in \mathbf{N}\}$
(iii) $\{x : x = 3n, n$ is the whole number$\}$ (iv) $\{x : x = 2n, n$ is the whole number$\}$

4.(i) $\{-4, -3, -2, -1, 0, 1, 2, 3, 4\}$ (ii) $\{1, 2, 3, 4, 5\}$ (iii) $\{3, 5\}$
(iv) {Jan, March, May, July, August, October, December} (v) ϕ (vi) 0

5.(i) $\in$ (ii) $\notin$ (iii) $\notin$ (iv) $\in$ (v) $\notin$ (iv) $\in$

1.5 TYPES OF SETS

(i) Empty Set : A set containing no elements is called empty set and is denoted by the symbol ϕ.

For example :

(i) $\phi = \{x : x$ is a negative integer whose square is $-1\}$
(ii) $\phi = \{x : x$ is a natural number lying between 2 and 3$\}$
(iii) $\phi =$ {the set of such persons, who never die}
(iv) $\phi = \{x : x$ is a real number, $x^2 < 0\}$
(v) $\phi = \{x : x$ is an even prime number greater than five$\}$

(vi) ϕ = {the set of real numbers which are solution of equation $x^2 + 1 = 0$}

(vii) ϕ = {x : x is a straight line passing through three distinct points on a circle}

- The empty set is also known as **null set** or **void set**.
- In Roster method, the empty set is denoted by {}.
- To describe the null set, we can use any property, which is not true for any element.
- It is wrong to use the expression 'an empty' or 'a null set' as there is one and only one empty set through, it may have many descriptions. We shall always call 'The empty or the null set'.
- A set consisting of at least one element is called a non-empty or non-void set.
- {ϕ} is not a null set.

(ii) Singleton Set : *Set containing only one element is a Singleton set.* The set {a} is a singleton set.

- {0} is not a null set, since it contains 0 as its member. It is a Singleton set.
- A room containing only one man is not same thing as a man. In a similar way, the singleton set {*a*} is not the same thing as the element *a*.

(iii) Finite set : *A set is said to be finite if it consists of only finite number of elements.* Here,the process of counting the different elements comes to an end.
For example :
(i) Set of natural numbers less than 50.
(ii) Set of all persons in a city.
(iii) Set of English alphabets.
(iv) Set of all persons on the earth.

(iv) Infinite Set : *A set which is not finite, i.e.,* it contains infinite number of elements. Here, process of counting the different elements never comes to an end.
For example :
(i) Set of natural number **N** = {1,2,3,...}
(ii) Set of all points of plane. (iii) Set of all even integers.
(iv) Set of rational numbers lying between two integers.

(v) Equal Sets : *Two sets are said to be equal if they contain exactly the same elements.*
For example :

A = {x : x is a letter in the word 'Area'}, *i.e.,* $A = \{a, r, e\}$

and $B = \{y$: y is a letter in the word 'ear'$\}$, *i.e.,* $B = \{a, r, e\}$

Here A and B are equal sets.

1.5.1 Cardinal Number of a Set

The number of distinct elements contained in a finite set A is called cardinal number of A and is denoted by n(A).

1.5.2 Equivalent Sets

Two finite sets are said to be equivalent if they have the same cardinal number.

- Equivalent sets are not always equal but equal sets are always equivalent.
- The number of distinct elements in a finite set is also called the order of the set. If the order of a set is zero, the set is empty.
- If the order of a set is one, the set is singleton.
- The order of an infinite set is never defined.

1.6 SUBSET

Let A and B be two sets. The set A is said to be a subset of the set B if every element of A is also an element of B. Symbolically, we write $A \subseteq B$.

When A is a subset of B, it means that 'A is contained in B' or 'B contains A'. Here, B is called super set of A and is written as $B \supset A$.

- Every set is a subset of itself.
- Empty set is a subset of every set.
- If A is not a subset of B, we write $A \not\subseteq B$.
- An element can not be a subset of a set, only a set can be subset of a set.

1.6.1 Proper Subset

We know that for A to be a subset of B all that is needed is that every element of A is in B. It is possible that every element of B may or may not be in A. If it so happens that every element of B is also in A, then we will have $B \subset A$. Obviously, then A and B are the same set, so that we have $A \subset B$ and $B \subset A \Leftrightarrow A = B$.

If every element of *A* is in *B*, but every element of *B* is not in *A*, *i.e.*, if $A \subset B$ and $B \not\subset A$, then *A* is said to be a proper subset of *B*.

For Example :

(i) {a, b} is a proper subset of {a, b, c}.

(ii) Set of natural numbers **N** is a proper subset of set **Z** of integer.

- Here, it follows that every element of *A* is an element of *B* and *B* contains at least one element which does not belong to *A*.
- If the subset is not proper, it is called **improper subset.** $A \subseteq A$ and $\phi \subseteq A$ are improper subsets.

1.6.2 Number of Subsets of a Set

If A is a set that contains n distinct elements. Let $0 < r \leq n$. If we consider those subsets of A that have r elements each, then we know that the number of ways in which r elements can be chosen out of n elements is ${}^nC_r$. Therefore, the number of subsets of A having r elements each is ${}^nC_r$.

Hence, the total number of subsets of A is equal to

$${}^nC_0 + {}^nC_1 + {}^nC_2 + \ldots + {}^nC_n = (1+1)^n = 2^n$$

For example :

(i) If a set *A* has one element, then it has $2^1 = 2$ subsets.

(ii) If a set *A* has two elements, then it has $2^2 = 4$ subsets.

- The number of proper subsets of a set with n elements is 2^{n-1}.
- The collection of all possible subsets of a given set A is called Power set. It is denoted by $P(A)$. For example : If $A = \{1, 2, 3\}$ then the power set $P(A) = \{\phi, \{1\}, \{2\}, \{3\}, \{1,2\}, \{1,3\}, \{2,3\}, \{1,2,3\}\}$.
- $P(\phi) = \{\phi\}$
- The power set of any given set is always non-empty.

1.7 UNIVERSAL SET

In any discussion, we are given particular set and we consider different subsets of the given set. This given set is called Universal Set. It is denoted by U.

For Example :

(i) The universal set is of real numbers **R,** while considering the set of natural numbers, whole numbers, integers and rational numbers.

(ii) The set of alphabets is the universal set from which the letters of any word may be chosen to form a set.

(iii) In geometry, we discuss set of lines, triangles and circles, then the universal set is the plane, in which the lines, triangles and circles lie.

- Universal set is a super set of each of the given sets.
- The universal set is not unique.

1.7.1 Complement of a Set

Let U be the universal set and the set $A \subseteq U$. Complement of set A with respect to the universal set U is the set of all those elements of U which are not the elements of A and is denoted by A' or A^c, $A' = \{x : x \in U \text{ and } x \notin A\}$

For Example :

(i) If $U = \{1, 2, 3, 4, 5, 6, 7, 8, 9, 11\}$ and $A = \{1, 2, 3\}$ then

$$A' = \{4, 5, 6, 7, 8, 9, 11\}$$

- Complement of the universal set is the null set ϕ and vice versa.
- $(A')' = A$
- If $A \subseteq B$, then $B' \subseteq A'$.
- $x \in A' \Leftrightarrow x \notin A$.

ILLUSTRATIVE EXAMPLES

1. *Let $A = \{1, 2, 3\}$, then find $P(A)$.*

Solution : Since $A = \{1, 2, 3\}$ then, $P(A) = \{\phi, \{1\}, \{2\}, \{3\}, \{1, 2\}, \{1, 3\}, \{2, 3\}, \{1, 2, 3\}\}$

2. *Let $A = \{a, b, c, d\}$, $B = \{a, b, c\}$ and $C = \{b, d\}$, find all sets X such that*

(i) $X \subset B$ and $X \subset C$ *(ii) $X \subset A$ and $X \not\subset B$.*

Solution : (i) Here, we have $P(B) = \{\phi, \{a\}, \{b\}, \{c\}, \{a, b\}, \{a, c\}, \{b, c\}, \{a, b, c\}\}$ and $P(C) = \{\phi, \{b\}, \{d\}, \{b, d\}\}$, then $X \subset B$ and $X \subset C$ implies $x \in P(B)$ and $x \in P(C)$

$X = \{\phi, \{b\}\}$

(ii) Here, we have, $X \subset A$ and $X \not\subset B$, which implies that $X \in P(A)$ and $X \in P(B)$

Therefore $X = \{\{d\}, \{a, b, d\}, \{b, c, d\}, \{a, c, d\}, \{a, d\}, \{b, d\}, \{c, d\}, \{a, b, c, d\}\}$

3. *Write down all the subsets of the following sets.*

(i) $\{a\}$ *(ii) $\{a, b\}$* *(iii) $\{a, b, c\}$* *(iv) ϕ*

Solution : (i) Let $A = \{a\}$. Since A contains only one element, therefore, the total number of subsets is $2^1 = 2$, which are given by ϕ and $\{a\}$.

(ii) Here, total number of subsets = $2^2 = 4$, which are given by ϕ, $\{a\}$, $\{b\}$, $\{a, b\}$

(iii) Here, total number of subsets = $2^3 = 8$, given by

$$\phi, \{a\}, \{b\}, \{c\}, \{a, b\}, \{a, c\}, \{b, c\}, \{a, b, c\}$$

(iv) Since ϕ contains no element therefore the number of subsets = $2^0 = 1$. The only subset is ϕ.

4. *Which of the following sets are empty. Also, give the reason.*

(i) $A = \{x : x \neq x, x \text{ is a real number}\}$ *(ii)* $B = \{x : x + 4 = 4\}$

(iii) $C = \{x : x^3 - 3 = 0 \text{ and } x \text{ is rational number}\}$

Solution : (i) Here, $A = \{x : x \neq x, x \text{ is a real number}\}$. Since $x \neq x$ is not true

$\Rightarrow \quad A = \phi$

(ii) $B = \{x : x + 4 = 4\} = \{x : x = 0\} = \{0\}$

$\Rightarrow \quad B$ has one element 0, therefore $B \neq \phi$.

(iii) Since there is no rational number whose square is 3, so $x^2 - 3 = 0$ is not satisfied for any rational numbers. Therefore, C is an empty set.

5. *Which of the following sets are finite and which are infinite*

(i) The set of natural numbers divisible by 2.

(ii) The set of natural numbers less than 8.

(iii) The set of integers whose square is even.

(iv) The set of integers greater than –18.

(v) The set of lines passing through a point.

(vi) The set of points of a plane at a fixed distance from a given point in the plane.

(vii) The set of points common to two given parallel lines.

(viii) The set of the roots of a polynomial of nth degree.

Solution : (i) The given set is {2, 4, 6, 8,...}. It has an infinite number of elements, therefore it is an infinite set.

(ii) The given set is {1, 2, 3, 4, 5, 6, 7}. It has seven elements, *i.e.,* finite number of elements. Hence, it is a finite set.

(iii) The given set is {...,–8, –6, –4, –2, 0, 2, 4, 6, 8,...}. It has infinite number of elements, therefore it is an infinite set.

(iv) Here, the given set is $\{-17, -16, \ldots, 0, 1, 2, \ldots\}$. It has infinite number of elements therefore, it is an infinite set.

(v) Since infinite number of lines can pass through a fixed point, therefore the given set is an infinite set.

(vi) Since the points in a plane at a fixed distance from a given point in the plane lie on a circle with the given point as center and the number of points on a circle is infinite. Therefore, the given set is an infinite set.

(vii) Since two parallel lines cannot meet anywhere, therefore, the set of points common to two given parallel lines is empty, therefore the given set cannot be infinite. Hence, it is a finite set.

(viii) Since, a polynomial of nth degree always have almost n roots. Therefore, the given set is always a finite set.

6. *Which of the following sets are equivalent ϕ, $\{0\}$ and $\{\phi\}$.*

Solution : Since ϕ has no element. Also, $\{0\}$ and $\{\phi\}$, each contains one element namely 0 and ϕ respectively. Hence, $\{0\}$ and $\{\phi\}$ are equivalent.

7. *Which of the following sets are equal* $A = \{1, 2, 3\}$, $B = \{2, 3, 4\}$, $C = \{3, 2, 1\}$, $D = \{2, 3, 5\}$

Solution : Since $1 \in A$ but $1 \notin B$, therefore $A \neq B$. A and C have exactly the same element, therefore $A = C$.

Also, $1 \in C$ but $1 \notin D \quad \Rightarrow \quad C \neq D$

$4 \in B$ but $4 \notin C \quad \Rightarrow \quad B \neq C$

$4 \in B$ but $4 \notin C \quad \Rightarrow \quad B \neq C$

$1 \in A$ but $1 \notin D \quad \Rightarrow \quad A \neq D$

Hence, only A and C are equal sets.

Exercise 1.2

1. Fill in the blanks :
   (i) A set which contains no elements is called......set.
   (ii) If $A = \{1, 2, 3\}$ and $B = \{3, 2, 1\}$ then they are said to be......
   (iii) If $A = \{a, b, c\}$ and $B = \{c, d, e\}$ then they are said to be......
   (iv) If every element of a set B is also an element of A, then B is said to be......of A.
   (v) The empty set is a......of every set.
   (vi) Every set is a.........of itself.
   (vii) The set **Z** of integers is a......of set of natural numbers **N**.
2. Which of the following sets are equal?
   (i) $A = \{1, 2, 3\}$
   (ii) $B = \{1, 2, 2, 3\}$
   (iii) $C = (x \in \mathbf{R} : x^3 - 6x^2 + 11x - 6 = 0)$
3. Which of the following sets are equivalent to the set $\{4, 7, 11, 17, 20\}$
   (i) $\{5, 1, 2, 3, 4\}$
   (ii) {all odd numbers less then 10}
   (iii) {the months of a year of 30 days}
   (iv) {all the prime numbers which lie between 10 and 25}
4. Which of the following sets are finite and which are infinite?
   (i) $\{x \in \mathbf{N} : x > 10\}$
   (ii) $\{x \in \mathbf{N} : x < 100\}$
   (iii) $\{x \in \mathbf{R} : 1 \le x \le 2\}$
   (iv) Set of vowels in English alphabets.
   (v) The set of prime numbers less than 100.
   (vi) The set of multiple of 8.
5. Which of the following statements are true? Give the reason.
   (i) For any two sets A and B either $A \subseteq B$ or $B \subseteq A$.
   (ii) Every subset of a finite set is finite.
   (iii) A subset of an infinite set may be finite.
   (iv) Every set has a proper subset.
   (v) A set containing n elements have 2^n subsets.
   (vi) If $A = \{1, 2, 3, 4, 5, 6\}$ and $B =$ {whole numbers less than 6}, then $A = B$.
   (vii) The empty set has no proper subset.
6. Examine which of the following sets are empty :
   (i) The set of tigers in your class.
   (ii) The set of triangles having three equal sides.
   (iii) The set of all numbers which, when added to zero, yield sum greater than the original.
   (iv) The set of odd numbers which are divisible by 2.
   (v) The set of men, who never die.
7. Which of the following statements are true?
   (i) If $x \in A$ and $A \subset B$, then $x \in B$
   (ii) If $A \subset B$ and $B \subset C$ then $A \subset C$
   (iii) If $A \not\subset B$ and $B \not\subset C$, then $A \not\subset C$
   (iv) If $x \in A$ and $A \not\subset B$, then $x \in B$
   (v) If $A \subset B$ and $x \notin B$, then $x \notin A$.
8. Are the following sets, *i.e.*, (A and B) are equal.
   (i) $A = \{x : x$ is a letter of the word 'LITTLE'$\}$
   $B = \{x : x$ is a letter of the word 'TITLE'$\}$
   (ii) $A = \{x : x$ is a letter in the word 'FOLLOW'$\}$
   $B = \{x : x$ is a letter in the word 'WOLF'$\}$
   (iii) $A = \{x : x$ is a letter in the word 'LOYAL'$\}$
   $B = \{x : x$ is letter in the word 'ALLOY'$\}$
9. Write down all possible subsets of each of the following sets.
   (i) $\{a\}$ (ii) $\{0, 1\}$ (iii) $\{a, b, c\}$
   (iv) $\{1, \{1\}\}$ (v) ϕ
10. Which of the following statements are true :
   (i) $\{a, \phi\} \in \{a, \{a, \phi\}\}$
   (ii) If $A \in B$ and $B \subseteq C$, then $A \subseteq C$
   (iii) If $A \in B$ and $B \subseteq C$, then $A \subseteq C$
   (iv) If $A \subset B$ and $B \in C$, then $A \in C$
   (v) If $A \subseteq B$ and $B \in C$, then $A \subseteq C$.

ANSWERS

1.(i) Empty (ii) equal (iii) equivalent (iv) subset (v) subset (vi) subset (viii) super set

2. $A = B = C$ **3.** (i), (ii), (iv)

4. (ii), (iv), (v) are finite sets and (i), (iii), (vi) are infinte

5.(i) F (ii) T (iii) T (iv) F (v) T (vi) F (vii) T

6. (i), (iii), (iv), (v) **7.** (i), (ii), (v)

8. (i) Equal, (ii) Equal, (iii) Equal

9.(i) ϕ, $\{a\}$; (ii) ϕ, $\{0\}$, $\{1\}$, $\{0, 1\}$; (iii) ϕ, $\{a\}$, $\{b\}$, $\{c\}$, $\{a, b\}$, $\{b, c\}$, $\{a, c\}$, $\{a, b, c\}$;
(iv) $\{1\}$; $\{1\}$, $\{\{1\}\}$, $\{1, \{1\}\}$; (v) ϕ, $\{\phi\}$;

10. (i), (ii), (iii), (iv), (v)

1.8 VENN DIAGRAMS

A set can be represented by closed figures like circles, triangles, rectangles, etc. The point in the interior of the figure represents the elements of the set. Such a representation is called a Venn diagram. In Venn diagram, the universal set is usually represented by a rectangular region and its subset by closed bounded regions inside the rectangular region. For example, if A is a subset of B, *i.e.*, $A \subset B$. This is as shown in Figure 2.

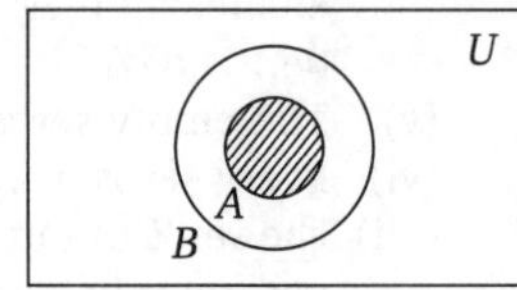

Fig. (2)

- The diagrams drawn to represent sets are called Venn diagram or Venn-Euler diagrams, after the name of British mathematician **Venn.**
- If A and B are two sets, which are not equal, but have common elements, then to represents A and B, we draw two intersecting circles.
- Two disjoint sets are represented by two non-intersecting circles.
- Venn diagrams are to be used for clarity and are no substitute for precise proof.

1.9 OPERATIONS ON SETS

1.9.1 Union and Intersection Operations

(i) Union of Two Sets

Let A and B be two sets. Then Union of A and B, denoted by $A \cup B$ is the set of those elements, which either belongs to A or B or to both A and B.

It should be noted that the common elements are to be taken only once.

Symbolically : $A \cup B = \{x : x \in A \text{ or } x \in B\}$

It is shown in the adjoining Figure (3).

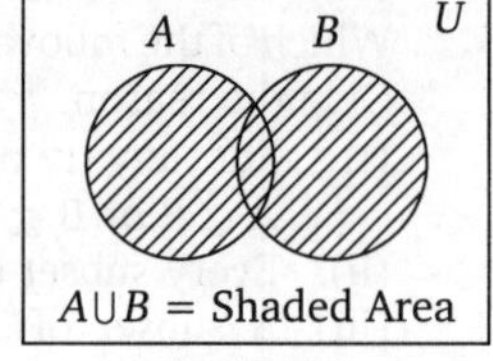

Fig. (3)

For Example :

(i) Let $A = \{3, 4, 5, 6, 7\}$ and $B = \{5, 6, 7, 8, 9\}$
Then $A \cup B = \{3, 4, 5, 6, 7, 8, 9\}$

(ii) Let $A = \{x : x = 2n, n = 1, 2, 3, \ldots\} = \{2, 4, 6, 8, \ldots\}$
and $B = \{x : x = 3n, n = 1, 2, 3, \ldots\} = \{3, 6, 9, 12, \ldots\}$
Then $A \cup B = \{x : x \text{ is multiple of 2 or a multiple of 3}\}$
$= \{2, 3, 4, 6, 8, 10, 12, \ldots\}$

(iii) Let A = set of even natural numbers $= \{2, 4, 6, 8, \ldots\}$
and B = set of natural numbers $= \{1, 2, 3, 4, 5, \ldots\}$
Then $A \cup B = \{1, 2, 3, 4, \ldots\}$

- $x \in (A \cup B) \Leftrightarrow x \in A$ or $x \in B$.
- $x \notin (A \cup B) \Leftrightarrow x \notin A$ and $x \notin B$.

- $A \cup B = B \cup A$, *i.e.*, union of sets is commutative.
- $A \cup A' = U$ and $A \cup U = U$.
- $A \cup \phi = A$
- If $A, B, C, D, \ldots, Z$ is a finite family of sets, then their union is denoted by $A \cup B \cup C \cup D \ldots \cup Z$.
- $(A \cup B) \cup C = A \cup (B \cup C)$, *i.e.*, a union of sets is associative.

(ii) Intersection of Two sets

Let A and B be two sets. Then intersection of A and B, denoted by $A \cap B$ is the set of all those elements,which belongs to both A and B.

Symbolically : $A \cap B = \{x : x \in A \text{ and } x \in B\}$

It is shown in the adjoining Figure 4.

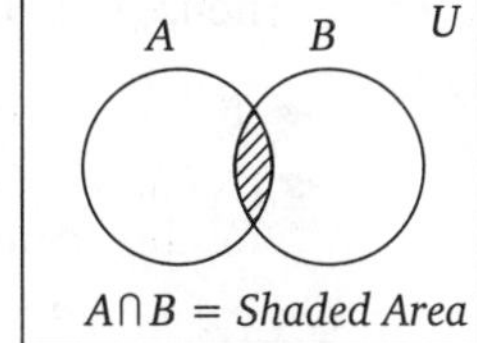

Fig. (4)

For Example :

(i) Let $A = \{2, 4, 6, 8, 10\}$ and $B = \{1, 2, 3, 4, 5\}$
Then $A \cap B = \{2, 4\}$

(ii) If $A = \{x : x = 3n, n \in \mathbf{Z}\}$
and $B = \{x : x = 4n, n \in \mathbf{Z}\}$
Then $A \cap B = \{x : x \text{ is multiple of 3 and } x \text{ is a multiple of 4}\}$
$= \{x : x \text{ is a multiple of 3 and 4 both}\} = \{x : x = 12n, n \in \mathbf{Z}\}$

- $x \in (A \cap B) \Leftrightarrow x \in A$ and $x \in B$.
- $x \notin (A \cap B) \Leftrightarrow x \notin A$ or $x \notin B$.
- $A = A \cap A$, *i.e.*, intersection of sets is idempotent.
- $A \cap \phi = \phi$
- $A \cap U = A$, where $\cup$ is a universal set.
- $A \cap B = B \cap A$, *i.e.*, intersection of sets is commutative.
- $(A \cap B) \cap C = A \cap (B \cap C)$, *i.e.*, intersection of sets is associative.
- If $A, B, C, \ldots, Z$ is a finite family of sets, then their intersection is denoted by $A \cap B \cap C \ldots \cap Z$.

(iii) Distributive Property of Union and Intersection

(i) $A \cup (B \cap C) = (A \cup B) \cap (A \cup C)$ (ii) $A \cap (B \cup C) = (A \cap B) \cup (A \cap C)$

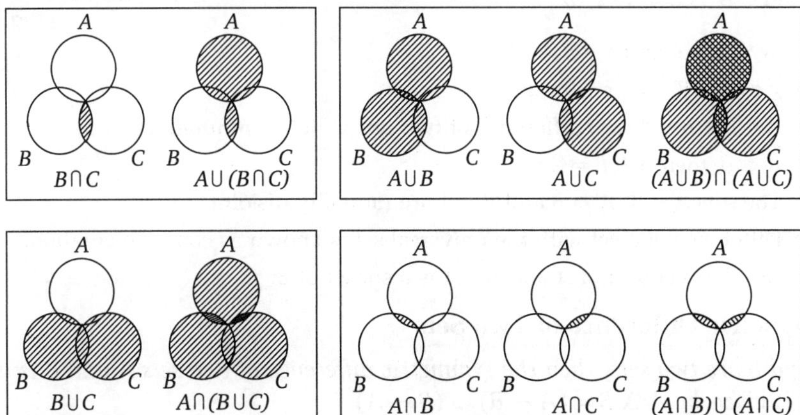

Fig. (5)

1.9.2 Disjoint Sets

When two sets have no common elements, they are called disjoint sets. Thus, if $A \cap B = \phi$, then A and B are disjoint. It is shown in the adjoining Figure 6.

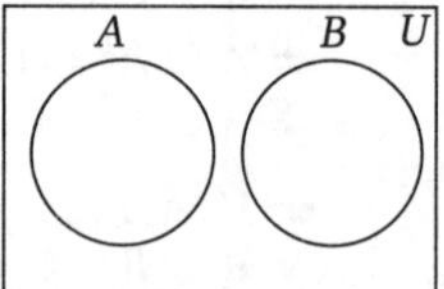

Fig. (6)

For Example :

(i) If $A = \{2, 4, 6, 8\}$ and $B = \{1, 3, 5, 7, 9\}$
Then, $A \cap B = \phi$

(ii) If A = Boys in school
B = Girls in school
Then, $A \cap B = \phi$

- If $A \cap B \neq \phi$, then A and B are said to be intersecting or overlapping sets.
- A family of sets is said to be pairwise disjoint family of sets if and only if any two sets of this family are disjoint. For example, classes of A_2, A_3, A_5 and A_7 defined as
$A_2 = \{2, 2^2, 2^3, ...\}$; $A_3 = \{3, 3^2, 3^3, ...\}$; $A_5 = \{5, 5^2, 5^3, ...\}$
and $A_7 = \{7, 7^2, 7^3, ...\}$ are pairwise disjoint
- $\phi \cap A = \phi$, *i.e.*, null set is disjoint from every subset.

1.9.3 Difference of Two Sets

If A and B are two sets, then the set of all elements which belong to A but do not belong to B is called the difference of sets A and B and is denoted by A ~ B. The set of all elements which belong to B but do not belong to A is called the difference of sets B and A and is denoted by $B \sim A$.

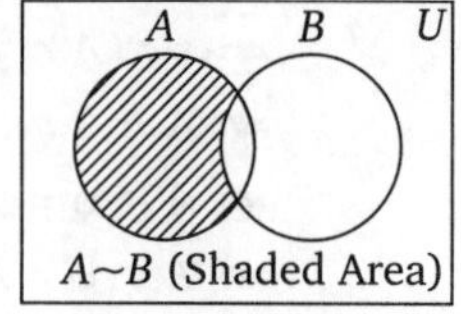

A~B (Shaded Area)

Fig. (7)

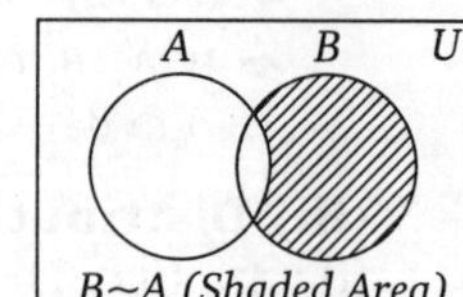

B~A (Shaded Area)

Fig. (8)

Therefore,

$$A \sim B = \{x : x \in A \text{ and } x \notin B\} = A \cap B'$$
and $$B \sim A = \{x : x \notin A \text{ and } x \in B\} = B \cap A'$$

For Example :

(i) Let $A = \{1, 2, 3, 4, 5\}$
and $B = \{-1, 0, 1, 2\}$
Then, $A \sim B = \{3, 4, 5\}$
and $B \sim A = \{-1, 0\}$

- $x \in (A \sim B) \Leftrightarrow x \in A$ and $x \notin B$.
- $x \notin (A \sim B) \Leftrightarrow x \notin A$ or $x \in B$.
- $A \sim B \neq B \sim A$, *i.e.*, difference of two sets is not commutative.
- $A \subset B$, then $A \sim B = \phi$.
- The sets $A \sim B$, $A \cap B$ and $B \sim A$ are mutually disjoint.
- Difference of a set with the universal set is known as complementation.
- $A \sim B$ is a subset of A and $B \sim A$ is a subset of B.

1.9.4 Symmetric Difference of Two Sets

If A and B are two sets, then the symmetric difference of two sets A and B is denoted by $A \Delta B$ and is given by $A \Delta B = (A \sim B) \cup (B \sim A)$

Symbolically : $A \Delta B = \{x : (x \in A \text{ and } x \notin B) \text{ or } (x \in B \text{ and } x \notin A)\}$

For example :

(i) If $A = \{1, 2, 3, 4, 5, 6, 7, 8\}$ and $B = \{1, 3, 5, 6, 7, 8, 9\}$

Then $A \sim B = \{2, 4\}$ and $B \sim A = \{9\}$
and $A \Delta B = \{2, 4, 9\}$

Equivalent Sets : Two finite sets A and B are equivalent if their cardinal numbers are same, *i.e.*, $n(A) = n(B)$.

A B U

BDA (Shaded Area)

Fig. (9)

1.9.5 Law of Excluded Middle and Law of Contradiction

Two special properties of set operations are known as the excluded middle axioms and law of contradiction. The excluded middle axioms are very important because they are the only set operations described here that are not valid for both classical sets and fuzzy sets. Let A be any subset of universal set X. Then, we define

(i) Axiom of the excluded middle : $A \cup \bar{A} = X$
(ii) Axiom of the contradiction : $A \cap \bar{A} = \phi$

Theorem 1.

(i) $A \cup \phi = A$ *(ii)* $A \cap \phi = \phi$ *(iii)* $A \cup A = A$
(iv) $A \cap A = A$ *(v)* $A \cup B = B \cup A$ *(vi)* $A \cap B = B \cap A$

Proof :

(i) Let x be an arbitrary element of $A \cup \phi$.
i.e., $x \in A \cup \phi$
Then, by definition $x \in A \cup B \Leftrightarrow x \in A$ or $x \in B$
i.e., $x \in A \cup \phi \Leftrightarrow x \in A$ or $x \in \phi$ ($\because \phi$ is a null set $\Rightarrow x \notin \phi$)
$\Leftrightarrow x \in A$
Therefore, $A \cup \phi = A$

(ii) Let x be an arbitrary element of $A \cap \phi$
$x \in A \cap \phi \Leftrightarrow x \in A$ and $x \in \phi$ ($\because \phi$ is a null set)
Therefore, $A \cap \phi = \phi$.

(iii) Let x be an arbitrary element of $A \cup A$
$x \in A \cup A \Leftrightarrow x \in A$ or $x \in A$ (Repeated statement)
$\Leftrightarrow x \in A$
Therefore, $A \cup A = A$

(iv) Let x be an arbitrary element of $A \cap A$
$x \in A \cap A \Leftrightarrow x \in A$ or $x \in A$ (Repeated statement)
$\Leftrightarrow x \in A$
Therefore, $A \cap A = A$

(v) Let x be an arbitrary element of $A \cup B$
$x \in A \cup B \Leftrightarrow x \in A$ or $x \in B$ (Writing in reverse order)
$\Leftrightarrow x \in B$ or $x \in A \Leftrightarrow x \in B \cup A$
Therefore, $A \cup B = B \cup A$

(vi) Let x be an arbitrary element of $A \cap B$
$x \in A \cap B \Leftrightarrow x \in A$ and $x \in B$ (Writing in reverse order)
$\Leftrightarrow x \in B$ and $x \in A \Leftrightarrow x \in B \cap A$
Therefore, $A \cap B = B \cap A$

Theorem 2.

(i) $A \cup (B \cup C) = (A \cup B) \cup C$ *(ii)* $A \cap (B \cap C) = (A \cap B) \cap C$
(iii) $A \cup (B \cap C) = (A \cup B) \cap (A \cup C)$ *(iv)* $A \cap (B \cup C) = (A \cap B) \cup (A \cap C)$

Proof :

(i) Let x be an arbitrary element of $A \cup (B \cup C)$, then $x \in A \cup (B \cup C)$

$\Leftrightarrow x \in A$ or $x \in (B \cup C)$ $\Leftrightarrow$ $x \in A$ or $(x \in B$ or $x \in C)$
$\Leftrightarrow$ $(x \in A$ or $x \in B)$ or $x \in C$ (By associativity)
$\Leftrightarrow$ $x \in (A \cup B)$ or $x \in C$ $\Leftrightarrow$ $x \in (A \cup B) \cup C$
Therefore, $A \cup (B \cup C) = (A \cup B) \cup C$

(ii) Let x be an arbitrary element of $A \cap (B \cap C)$, then $x \in A \cap (B \cap C)$
$\Leftrightarrow$ $x \in A$ and $x \in (B \cap C)$ $\Leftrightarrow$ $x \in A$ and $x \in (B \in C)$
$\Leftrightarrow$ $(x \in A$ and $x \in B)$ and $x \in C$ (By associativity)
$\Leftrightarrow$ $x \in (A \cap B)$ and $x \in C$ $\Leftrightarrow$ $x \in (A \cap B) \cap C$
Therefore, $A \cap (B \cap C) = (A \cap B) \cap C$

(iii) Let x be an arbitrary element of $A \cup (B \cap C)$, then $x \in A \cup (B \cap C)$
$\Leftrightarrow$ $x \in A$ or $x \in (B \cap C)$ $\Leftrightarrow$ $x \in A$ or $(x \in B$ and $x \in C)$
$\Leftrightarrow$ $(x \in A$ or $x \in B)$ and $(x \in A$ or $x \in C)$
$\Leftrightarrow$ $x \in (A \cup B)$ and $(x \in A \cup C)$ $\Leftrightarrow$ $x \in (A \cup B) \cap (A \cup C)$
Therefore, $A \cup (B \cap C) = (A \cup B) \cap (A \cup C)$.

(iv) Let x be an arbitrary element of $A \cup (B \cup C)$, then $x \in A \cap (B \cup C)$
$\Leftrightarrow$ $x \in A$ and $x \in (B \cup C)$ $\Leftrightarrow$ $x \in A$ and $(x \in B$ or $x \in C)$
$\Leftrightarrow$ $(x \in A$ and $x \in B)$ or $(x \in A$ and $x \in C)$ $\Leftrightarrow$ $x \in (A \cap B)$ or $(x \in A \cap C)$
Therefore, $A \cap (B \cup C) = (A \cap B) \cup (A \cap C)$

Theorem 3.
(i) $(A')' = A$ *(ii)* $A \cup A' = U$, *where U is the universal set.*
(iii) $A \cap A' = \phi$ *(iv)* $(A \cup B)' = A' \cap B'$ *(De' Morgan's Law)*
(v) $(A \cap B)' = A' \cup B'$ *(De' Morgan's Law)*

Proof :

(i) Let x be an arbitrary element of $(A')'$
$x \in (A')'$ $\Leftrightarrow$ $x \notin A'$ $\Leftrightarrow$ $x \in A$
Therefore, $(A')' = A$

(ii) Let x be an arbitrary element of $(A \cup A')$
$x \in (A \cup A')$ $\Leftrightarrow$ $x \in A$ or $x \in A'$ $\Leftrightarrow$ $x \in A$ or $x \in U - A$
$\Leftrightarrow$ $x \in A$ or $(x \in U, x \notin A)$ $\Leftrightarrow$ $x \in U$
Therefore, $A \cup A' = U$

(iii) The x be an arbitrary element of $(A \cap A')$
$x \in (A \cap A')$ $\Leftrightarrow$ $x \in A$ or $x \in A'$ $\Leftrightarrow$ $x \in A$ and $x \notin A'$
Therefore, $A \cap A' = \phi$

(iv) The x be an arbitrary element of $(A \cup B)'$
$x \in (A \cup B)'$ $\Leftrightarrow$ $x \notin (A \cup B)$ $\Leftrightarrow$ $x \notin A$ and $x \in B$
$\Leftrightarrow$ $x \in A'$ and $x \in B'$ $\Leftrightarrow$ $x \in A' \cap B'$
Therefore, $(A \cup B)' = A' \cap B'$

(v) Let x be an arbitrary element of $(A \cap B)'$
$x \in (A \cap B)'$ $\Leftrightarrow$ $x \notin (A \cap B)$ $\Leftrightarrow$ $x \notin A$ or $x \notin B$
$\Leftrightarrow$ $x \in A'$ or $x \in B'$ $\Leftrightarrow$ $x \in A' \cap B'$
Therefore, $(A \cap B)' = (A' \cup B')$

ILLUSTRATIVE EXAMPLES

1. *Show that (i)* $A \subset (A \cup B)$, *(ii)* $(A \cap B) \subset A$.

Solution : (i) Let $x \in A$ be arbitrary then $x \in A$, certainly but may or may not belong to B.

$\Rightarrow$ $x \in A \cup B$

Therefore, $x \in A \Rightarrow x \in A \cup B$ gives $A \subset A \cup B$.

(ii) Let $x \in A \cap B$ where x is arbitrary.

$x \in A \cap B \Rightarrow x \in A$ and $x \in B$.

In particular, $x \in A \cap B \Rightarrow x \in A$

Therefore, $A \cap B \subset A$.

☛ Similarly we can show that (i) $B \subset (A \cup B)$ and (ii) $A \cap B \subset B$.

2. *Let A and B be two sets, if $A \cap X = B \cap X = \phi$ and $A \cup X = B \cup X$ for some set X, prove that $A = B$.*

Solution : Given that $A \cup X = B \cup X$

$\Rightarrow A \cap (A \cup X) = A \cap (B \cup X)$ (taking intersection by A on both sides)

$\Rightarrow A = A \cap (B \cup X)$ $(\because A \cap (A \cup X) = A)$

$\Rightarrow A = (A \cap B) \cup (A \cap X)$ (By distributive law)

$\Rightarrow A = (A \cap B) \cup \phi \Rightarrow A = A \cap B$

$\Rightarrow A \subset (A \cap B) \Rightarrow A \subset B$ ...(1)

Again consider, $A \cup X = B \cup X$

$\Rightarrow B \cap (A \cup X) = B \cap (B \cup X)$ (By taking intersection with B)

$\Rightarrow B \cap (A \cup X) = B \Rightarrow (B \cap A) \cup (B \cap X)$ (By distributive law)

$\Rightarrow (B \cap A) \cup \phi$ (Given $B \cap X = \phi$)

$\Rightarrow (B \cap A) = B$ $(\because A \cap B = B \cap A)$

$\Rightarrow A \cap B = B \Rightarrow B \subset A \cap B \Rightarrow B \subset A$

Hence, (1) and (2) gives $A \subset B$ and $B \subset A$ ...(1)

$\Rightarrow A = B$

3. *For any two sets A and B, show that*

(i) $P(A \cap B) = P(A) \cap P(B)$ *(ii) $P(A) \cup P(B) \subset P(A \cup B)$*

Solution. (i) Let $X \in P(A \cap B) \Rightarrow x \subset A \cap B$

$\Rightarrow x \subset A$ and $x \subset B \Rightarrow X \in P(A)$ and $X \in P(B)$

$\Rightarrow x \in P(A) \cap P(B)$

Therefore, $P(A \cap B) \subset P(A) \cap P(B)$ ...(1)

Now, let $X \in P(A) \cap P(B) \Rightarrow X \in P(A)$ and $X \in P(B)$

$\Rightarrow x \subset A$ and $x \subset B \Rightarrow x \subset A \cap B \Rightarrow x \in P(A \cap B)$

Therefore, $P(A) \cap P(B) \subset P(A \cap B)$ ...(2)

From (1) and (2), we conclude that

$P(A \cap B) \subset P(A) \cap P(B)$ and $P(A) \cap P(B) \subset P(A \cap B)$ which gives

$$P(A \cap B) = P(A) \cap P(B)$$

(ii) Let $X \in P(A) \cup P(B) \Rightarrow X \in P(A)$ or $X \in P(B)$

$\Rightarrow x \subset A$ or $x \subset B \Rightarrow x \subset A \cup B \Rightarrow X \in P(A \cup B)$

Therefore, $P(A) \cup P(B) \subset P(A \cup B)$.

☛ Converse of the result (ii) is not necessarily true. For example, let $A = \{1,2\}$ and $B = \{4,5,6\}$, then we find that $x = \{1, 2, 3,5\}$ which is a subset of $A \cup B$. Therefore, $X \in P(A \cup B)$. But $X \notin P(A), X \notin P(B)$. So,

$$X \notin P(A) \cup P(B) \Rightarrow P(A \cup B) \not\subset P(A) \cup P(B).$$

☛ If A and B are any two sets, then

(i) $A - B = A \cap B'$ (ii) $A - B = A \Leftrightarrow A \cap B = \phi$

(iii) $(A - B) \cup B = A \cup B$ (iv) $A \subset B \Leftrightarrow B' \subset A'$

(v) $(A - B) \cup (B - A) = (A \cup B) - (A \cap B)$

☛ If A and B are any two sets, then
(i) $A-(B\cap C)=(A-B)\cup(A-C)$ (ii) $A-(B\cup C)=(A-B)\cap(A-C)$
(iii) $A\cap(B-C)=(A\cap B)=(A\cap C)$

Exercise 1.3

1. Let $A=\{a,b\}$, $B=\{a,b,c\}$. Is $A\subset B$. Find $A\cup B$ and $A\cap B$.
2. If $A=\{1,2,3,4\}$, $B=\{2,4,6,8\}$, $C=\{3,4,5,6\}$ and universal set $U=\{1,2,3,4,\ldots,9\}$. Verify that $A\cap(B\cup C)=(A\cap B)\cup(A\cap C)$.
3. If A, B, C are subsets of a set X, then show that $A\in B$ and $B\in C\Rightarrow A\in C$.
4. Find the union of the following sets :
   (i) $A=\{x:x$ is an even integer$\}$, $B=\{x:x$ is an odd integer$\}$
   (ii) $A=\{x:x$ is a multiple of $2\}$, $B=\{x:x$ is a multiple of $3\}$
   (iii) $A=\{x:x$ is a rational number$\}$, $B=\{x:x$ is an irrational number$\}$
   (iv) $A=\{x:x$ is a negative integer$\}$, $B=\{x:x$ is a non-negative integer$\}$
5. Find the intersection of the following sets,
   (i) $A=\{x:x$ is an even integer$\}$, $B=\{x:x$ is an odd integer$\}$
   (ii) $A=\{x:x$ is a rational number$\}$, $B=\{x:x$ is an irrational number$\}$
   (iii) $A=\{x:x$ is a multiple of $5\}$, $B=\{x:x$ is a multiple of $2\}$
   (iv) $A=\{x:x$ is a rational number$\}$. $B=\{x:x$ is a real number$\}$
6. If $A=\{1,2,3,4\}$, $B=\{2,4,6,8\}$ and $C=\{3,4,5,6\}$, find
   (i) $(A\cup B)\cap C$ (ii) $A\cup(B\cap C)$
7. Write T for true and F for false statement.
   (i) $A\in(A\cup B)$ (T/F)
   (ii) $(A\cup B)\in B$ (T/F)
   (iii) $(A\cap B)\in A$ (T/F)
   (iv) $A\cup A=A$ and $A\cap A=A$ (T/F)
   (v) If $A\cap B=\phi$, then $A\cap\phi=B$ (T/F)
   (vi) For any sets A and B, $A\cup B=B\cup A$ (T/F)
   (vii) If A and B are disjoint sets, then intersection of their union and their intersection is the null set. (T/F)
   (viii) If A is the proper subset of U, then the union of A and A' is U. (T/F)
   (ix) $U'=\phi$ and $\phi'=U$ (T/F)
   (x) $(A\cup B)'=A'\cap B'$ (T/F)
   (xi) $A\cap A'$ is always empty (T/F)
   (xii) $(A\cap B)'=A'\cup B'$ (T/F)
8. If $A=\{1,2,3,4,5,6,7,8\}$ and $B=\{1,3,5,6,7,8,9\}$, then show that $A\,\Delta\,B=\{2,4,9\}$
9. Let $A=\{x:x\in\mathbf{N}\}$,
   $B=\{x:x=2n:n\in\mathbf{N}\}$,
   $C=\{x:x=2n-1,n\in\mathbf{N}\}$
   and $D=\{x:x$ is a prime natural number$\}$. Find
   (i) $A\cap B$ (ii) $A\cap C$
   (iii) $A\cap D$ (iv) $B\cap C$
   (v) $B\cap D$ (vi) $C\cap D$
10. For any two sets A and B, prove that $P(A)=P(B)$ implies that $A=B$
11. For any two sets A and B, show that
    (i) $A\cup(A\cap B)=A$
    (ii) $A\cap(A\cup B)=A$
    (iii) $(A\cup B)\cap(A\cap B')=A$
    (iv) $A'\cup B=U\Rightarrow A\subset B$
    (v) $A\subset B\Leftrightarrow B'\subset A'$
    (vi) $B\subset B\subset A\Leftrightarrow A\cap B=B$
12. Let $A=\{1,2,3,4\}$, $B=\{2,3,4,5\}$ and $C=\{4,5,6,7\}$. Verify that
    (i) $A\cup(B\cap C)=(A\cup B)\cap(A\cup C)$
    (ii) $A\cap(B\cup C)=(A\cap B)\cup(A\cap C)$
    (iii) $A\cap(B-C)=(A\cap B)-(A\cap C)$
    (iv) $A-(B\cup C)=(A-B)\cap(A-C)$
    (v) $A-(B\cap C)=(A-B)\cup(A-C)$
    (vi) $A\cap(B\,\Delta\,C)=(A\cap B)\,\Delta\,(A\cap C)$
13. Show that
    (i) If a sets has only even element, then it has 2 subsets.
    (ii) If $B\subset A$ and B has one element less than that of A, show that A has twice as many subset as B has.
    (iii) A set with 2 elements has 2^2 subsets, a set with 3 elements has 2^3 subsets and so on.
14. If $x=\{4^n-3n-1:n\in\mathbf{N}\}$ and $Y=\{9\,(n-1):n\in\mathbf{N}\}$, show that $X\subset Y$.
15. Show that $A-B$, $A\cap B$ and $B-A$ are pair-wise disjoint.
16. Show that $A\cup B\subseteq A\cap B$ implies that $A=B$.

ANSWERS

1. Yes, $\{a, b, c\}$, $\{a, b\}$;

4.(i) $A \cup B = \{x : x \text{ is a non-zero integer}\}$ (ii) $A \cup B = \{x : x \text{ is a multiple of 2 or 3}\}$
(iii) $A \cup B = \{x : x \text{ is a real number})$ (iv) $A \cup B = \{x : x \text{ is an integer}\}$

5.(i) ϕ (ii) ϕ (iii) ϕ (iv) $\{x : x \text{ is a rational number}\}$

6.(i) $\{3, 4, 6\}$ (ii) $\{1, 2, 3, 4, 6\}$

7.(i) T (ii) F (iii) T (iv) T (v) F (vi) T (vii) T
(viii) T (ix) T (x) T (xi) T (xii) T

9.(i) B (ii) C (iii) D (iv) ϕ (v) 2 (vi) $D - \{2\}$

1.10 SOME RESULTS ON VENN DIAGRAMS

In n is a finite set, then $n(A)$ = No. of elements in the set A.

The following results may be remembered for direct application.

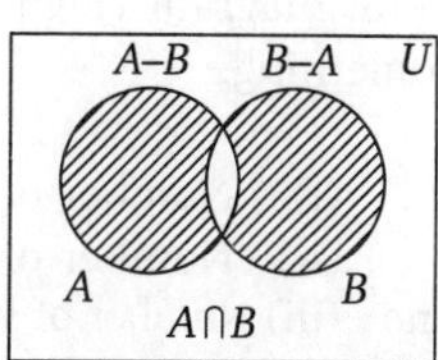

Fig. (10)

(i) $n(A \cup B) = n(A) + n(B) - n(A \cap B)$

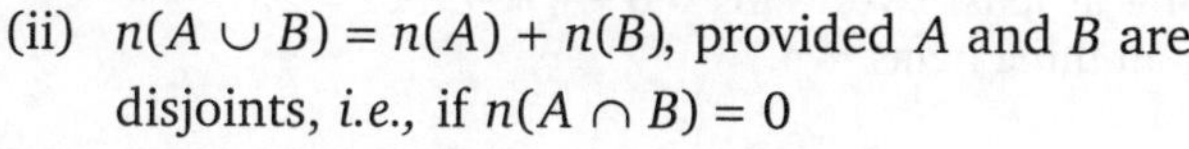

(ii) $n(A \cup B) = n(A) + n(B)$, provided A and B are disjoints, *i.e.*, if $n(A \cap B) = 0$

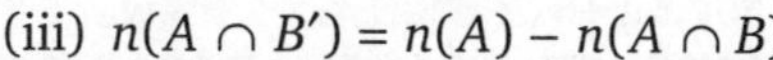

(iii) $n(A \cap B') = n(A) - n(A \cap B)$

(iv) $n(B \cap A') = n(B) - n(A \cap B)$

(v) $n(A \cup B) = n(A \cap B') + n(B \cap A') + n\,(A \cap B)$

(vi) $n(A \,\Delta\, B) = n(A) + n(B) - 2n\,(A \cap B)$

(vii) $n(A' \cup B') = n[(A \cap B)'] = n\,(U) - n(A \cap B)$

(viii) $n(A' \cap B') = n[(A \cup B)'] = n\,(U) - n\,(A \cup B)$

(ix) $n(A - B) = n(A) - n(A \cap B) \Rightarrow n(A - B) + n\,(A \cap B) = n(A)$

(x) $n(A \cup B \cup C) = n(A) + n(B) + n(C) - n(A \cap B) - n(B \cap C) - n(A \cap C) + n(A \cap B \cap C)$

ILLUSTRATIVE EXAMPLES

1. *In a group of athletic teams in a school, 21 are in the basket ball, 26 in the hockey team and 29 in the football team. If 14 play hockey and basket ball, 12 play football and basket ball, 15 play hockey and football and 8 play all the three games. Find (i) How many players are there in all (ii) How many play football only.*

Solution. Let A, B and C denote the set of players, who play basket ball, hockey and football respectively. Then, according to question, we have

$$n(A) = 21,\ n(B) = 26,\ n(C) = 29$$

$$n(A \cap B) = 14,\ n\,(A \cap C) = 12,\ n\,(B \cap C) = 15 \text{ and } (A \cap B \cap C) = 8$$

Therefore, $n(A \cup B \cup C) = [n(A) + n(B) + n(C) + n(A \cap B \cap C)] - [n(A \cap B) + n(A \cap C) + n(B \cap C)]$

$$= [21 + 26 + 29 + 8] - [14 + 12 + 15] = 43$$

Hence, the total number of players is 43. Now, the number of players playing football only is $[29 - (7 + 8 + 4)] = 10$.

2. *In a canteen, out of 123 students, 42 students buy ice cream, 36 buy buns and 10 buy cakes, 15 students buy ice-cream and buns, 10 ice-cream and cakes, 4 cakes and buns but not ice-cream and 11 buy ice-cream and buns but no cakes. Draw Venn diagram to illustrate the above information and find (i) How many students buy nothing at all (ii) How many students buy at least two items. (iii) How many students buy all three items.*

Solution : Define the sets A, B and C such that

A = Set of students who buy cakes

B = Set of students who buy ice-cream

C = Set of students who buy buns

According to question, we have.

$$n(A) = 10;\ n(B) = 42;\ n(C) = 36;\ n(B \cap C) = 15;$$

$$n(A \cap B) = 10;\ n[(A \cap C) - B] = 4$$

$$n[(B \cap C) - A] = 11 \text{ and } n[A - B \cup C] = 10$$

Now we have $n(B \cup C) = n(B) + n(C) - n(B \cap C)$

$$= 42 + 36 - 15 = 63$$

$$n(B \cup C) - n(B) = 63 - 42 = 21$$

and $\quad n(B \cup C) - n(C) = 63 - 36 = 27$

The above distribution of the students can be illustrated by Venn diagram (Figure 11). Now, total number of students buying something.

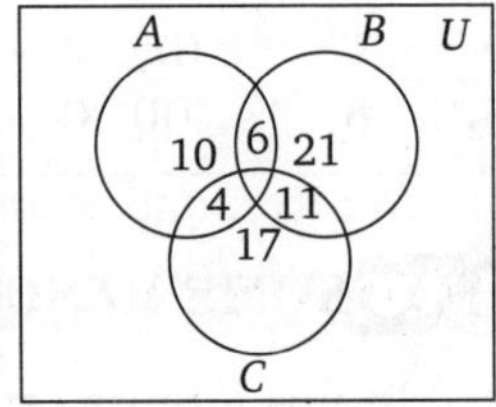

Fig. (11)

$$= 10 + 6 + 21 + 4 + 4 + 11 + 17 = 73$$

(i) Number of students who did not buy anything = 123 – 73 = 50

(ii) Number of students buying at least two items = 6 + 4 + 4 + 11 = 25

and (iii) Number of students buying all three items = 4.

Exercise 1.4

1. Out of 80 students who secured first class marks in Mathematics or in Physics, 50 obtained first class marks in Mathematics, 10 in both Physics and Mathematics. How many students secured first class marks in Physics only ?

2. The Mathematics club in a school held an open house on three afternoons. 115, 110 and 135 students attended both the first, second and third afternoons respectively. 25 attended just the first day, 30 attended both the first and second days, 80 attended both the first and third days, and 60 attended both the second and third days. How many attended (i) all three days (ii) just the second day (iii) just the third day ?

3. In a school of 250 pupils, 100 are girls, and 200 pupils stay at school for lunch. If 40 girls go home for lunch. Find the number of boys who go home for lunch.

4. In a class of 150 students, the following results were obtained in a certain examination. 45 students failed in Maths; 50 students failed in Physics, 48 students failed in Chemistry, 35 failed in both Maths and Chemistry, 25 failed in all the three subjects. Find the number of students who have failed in at least one subject.

ANSWERS

1. 30 **2.** 20, 30, 15 **3.** 10 **4.** 71

1.11 ORDERED PAIR

Sometimes, there are situations in which order is very important. Some results may be affected by order and others are not.

Definition : *An ordered pair is a pair of entries whose components occur in a specific order. It is written by listing the two components in the specific order, separating them by a comma and enclosing the pair in parentheses.*

Symbolically : If A and B are two sets, then by ordered pair of elements, we must mean a pair $(a, b) : a \in A, b \in B$ in that order.

- It may be noted that (a, b) is not the same as $\{a, b\}$. The former denotes an ordered pair whereas the latter denotes a set.
- $(a, b) \neq (b, a)$ unless $a = b$.
- Ordered pair may have the same first and second components, *i.e.,* two elements of an ordered pair need not be distinct.
- Two ordered pairs are said to be equal when both the first components are equal and also their second components are equal.

1.11.1 Cartesian Product of Two Sets

The set of all ordered pairs of elements (a, b), $a \in A$, $b \in B$ is called the cartesian product of two sets A and B. It is denoted by $A \times B$.

Symbolically : $A \times B = \{(a, b) : a \in A, b \in B\}$.

For Example :

If $A = \{2, 3\}$ and $B = \{4, 5, 6\}$, then $A \times B = \{(2, 4), (2, 5), (2, 6), (3, 4), (3, 5), (3, 6)\}$

- $A \times B = \phi \Leftrightarrow A = \phi$ or $B = \phi$
- If A and B are finite sets, then $n(A \times B) = n(A).n(B)$
- If either A or B are infinite sets, then $A \times B$ is an infinite set.

1.11.2 Ordered Triplet

If A, B, C are three sets, then by ordered triple product of elements, we mean a triplet (a, b, c): $a \in A$, $b \in B$, $c \in C$ in that order.

This is also called **ordered 3-tuple.**

The set of all ordered triplets $(a, b, c) : a \in A, b \in B, c \in C$ is also called the Cartesian triple product of three sets A, B and C and and is denoted by $(A \times B \times C)$.

Symbolically : $A \times B \times C = \{(a, b, c) : a \in A, b \in B, c \in C\}$.

- In general, the cartesian product of n sets $A_1, A_2, ..., A_n$ is an ordered n tuples $(a_1, a_2, ..., a_n)$, where $a_1 \in A_1$, $a_2 \in A_2, ..., a_n \in A_n$. It is denoted by $A_1 \times A_2 ... \times A_n$ or briefly by $\prod_{i=1}^{n} A_i$ where $\prod$ stands for the product.

ILLUSTRATIVE EXAMPLES

1. *If $A = \{1, 2\}$ and $B = \{a, b, c\}$, find the value of $A \times B$, $B \times A$, $A \times A$, $B \times B$.*

Solution : We have $A = \{1, 2\}$ and $B = \{a, b, c\}$.

Therefore,

$$A \times B = \{(1, a), (1, b), (1, c), (2, a), (2, b), (2, c)\}$$
$$B \times A = \{(a, 1), (a, 2), (b, 1), (b, 2), (c, 1), (c, 2)\}$$
$$A \times A = \{(1, 1), (1, 2), (2, 1), (2, 2)\}$$
$$B \times B = \{(a, a), (a, b), (a, c), (b, a), (b, b), (b, c), (c, a), (c, b), (c, c)\}.$$

2. *If $A = \{1, 2, 3\}$, $B = \{a, b, c, d\}$ and $C = \{-1, -2\}$, find $A \times B$, $B \times A$ and $C \times (B \cup C)$.*

Solution. Given that $A = \{1, 2, 3\}$, $B = \{a, b, c, d\}$ and $C = \{-1, -2\}$.

Therefore, $A \times B = \{(1, a), (1, b), (1, c), (1, d), (2, a), (2, b), (2, c), (2, d), (3, a), (3, b), (3, c), (3, d)\}$

$$B \times A = \{(a, 1), (b, 1), (c, 1), (d, 1), (a, 2), (b, 2), (c, 2), (d, 2), (a, 3), (b, 3), (c, 3), (d, 3)\}$$

Also, $B \cup C = \{a, b, c, d, -1, -2\}$

Therefore, $C \times (B \cup C) = \{(-1, a), (-1, b), (-1, c), (-1, d), (-1, -1), (-1, -2), (-2, a), (-2, b), (-2, c), (-2, d), (-2, -1), (-2, -2)\}$

3. *Find the values of a and b if* $(4a - 2, b + 4) = (2a, 4)$.

Solution. Since we know that two ordered pairs (a_1, b_1) and (a_2, b_2) are said to be equal if $a_1 = a_1$ and $b_1 = b_2$. Therefore, for the equality of two given ordered pairs, we have $4a - 2 = 2a$ and $b + 4 = 4$

Therefore, $4a - 2a = 2 \quad \Rightarrow \quad a = 1$ and $b + 4 = 4 \Rightarrow b = 0$.

4. *If* $A = \{1, 2, 3, 4\}$ *and* $B = \{4, 5\}$, *represent* $A \times B$, $B \times A$ *and* $B \times B$ *pictorially and find their values.*

Solution : Given $A = \{1, 2, 3, 4\}$ and $B = \{4, 5\}$

$\therefore \quad A \times B = \{(1, 4), (1, 5), (2, 4), (2, 5), (3, 4), (3, 5), (4, 4), (4, 5)\}$

$B \times A = \{(4, 1), (5, 1), (4, 2), (5, 2), (4, 3), (5, 3), (4, 4), (5, 4)\}$

and $B \times B = \{(4, 4), (4, 5), (5, 4), (5, 5)\}$

Pictorially, $A \times B$ and $B \times A$ can be represented as shown in Figure 12.

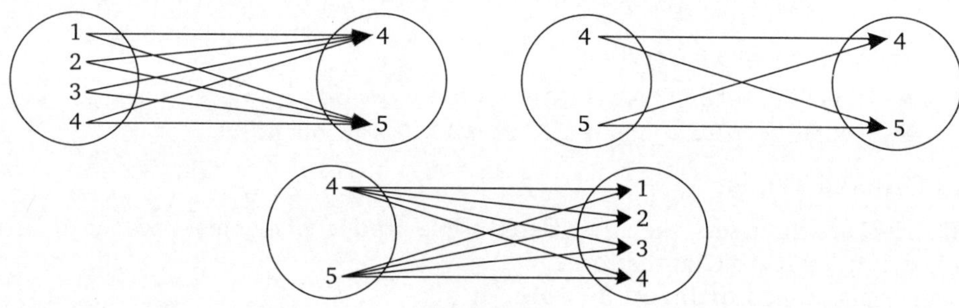

Fig. (12)

5. *Let* $A = \{1, 2, 3, 4\}$ *and* $B = \{5, 7, 9\}$. *Determine (i)* $A \times B$, *(ii)* $B \times A$. *Also represent* $A \times B$ *and* $B \times A$ *graphically.*

Solution : (i) Given $A = \{1, 2, 3, 4\}$ and $B = \{5, 7, 9)$. Then,

$A \times B = \{(1, 5), (1, 7), (1, 9), (2, 5), (2, 7), (2, 9), (3, 5), (3, 7), (3, 9), (4, 5), (4, 7), (4, 9)\}$

Graphically, it can be represented as shown in Figure 13.

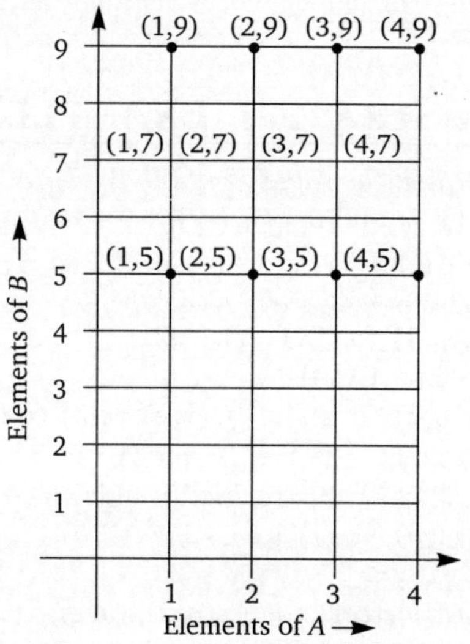

Fig. (13)

Now, $B \times A = \{(5,1),(5,2),(5,3),(5,4),(7,1),$
$(7,2),(7,3),(7,4),(9,1),(9,2),(9,3),(9,4)\}$

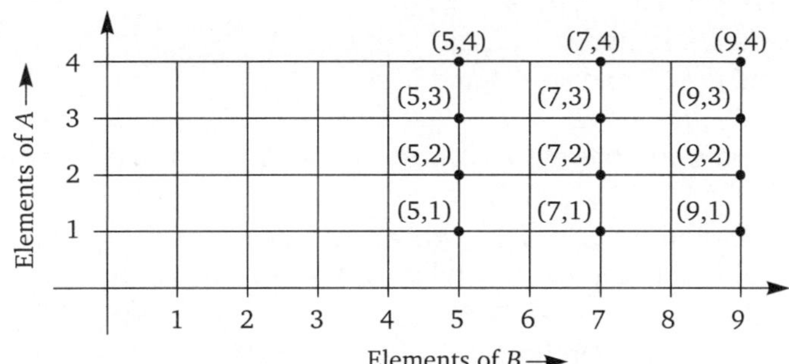

Fig. (14) : $B \times A$

Graphically, it can be represented as shown in Figure 14.

$$A \times (B \cap C) \subset (A \times B) \cap (A \times C) \quad \ldots(1)$$

Conversely.

If $(x, y) \in (A \times B) \cap (A \times C)$

Then, $(x, y) \in A \times B$ and $(x, y) \in A \times C$

$\Rightarrow \quad x \in A, y \in B$ and $x \in A, y \in C \quad \Rightarrow \quad x \in A, y \in B$ and $y \in C$

$\Rightarrow \quad x \in A$ and $y \in (B \cap C) \quad \Rightarrow \quad (x, y) \in A \times (B \cap C)$

But (x, y) is arbitrary, therefore

$$(A \times B) \cap (A \times C) \subseteq A \times (B \cap C) \quad \ldots(2)$$

From (1) and (2), we conclude that

$$A \times (B \cap C) = (A \times B) \cap (A \times C)$$

(ii) $(x, y) \in A \times (B \cup C)$

Then, $x \in A$ and $y \in (B \cup C)$

$\Rightarrow \quad x \in A$ and $y \in B$ or $y \in C \quad \Rightarrow \quad (x \in A$ and $y \in B)$ or $(x \in A$ and $y \in C)$

$\Rightarrow \quad \{(x, y) \in (A \times B)\}$ or $\{(x, y) \in (A \times C)\}$

$\Rightarrow \quad (x, y) \in (A \times B) \cup (A \times C)$

Since (x, y) is arbitrary, therefore

$$A \times (B \cup C) \subseteq (A \times B) \cup (A \times C) \quad \ldots(1)$$

Conversely.

If $(x, y) \in (A \times B) \cup (A \times C)$

Then, $(x, y) \in (A \times B)$ or $(x, y) \in (A \times C)$

$\Rightarrow \quad (x \in A$ and $y \in B)$ or $(x \in A$ and $y \in C)$

$\Rightarrow \quad x \in A$ and $(y \in B$ or $y \in C) \quad \Rightarrow \quad (x, y) \in A \times (B \cup C)$

Since (x, y) is arbitrary, therefore

$$(A \times B) \cup (A \times C) \subseteq A \times (B \cup C) \quad \ldots(2)$$

From (1) and (2), we conclude that $A \times (B \cup C) = (A \times B) \cup (A \times C)$.

Theorem 1. *For any three subsets A, B and C, we have*

(i) $A \times (B \cap C) = (A \times B) \cap (A \times C)$ *(ii)* $A \times (B \cup C) = (A \times B) \cup (A \times C)$

Proof : (i) If $(x, y) \in A \times (B \cap C)$

Then, $x \in A$ and $y \in (B \cap C)$

$\Rightarrow \quad x \in A$ and $y \in B$ and $y \in C \quad \Rightarrow \quad x \in A, y \in B$ and $x \in A, y \in C$

$\Rightarrow \quad (x, y) \in A \times B$ and $(x, y) \in (A \times C) \Rightarrow (x, y) \in (A \times B) \cap (A \times C)$

But (x, y) is arbitrary, therefore

$C \times (B \cup C) = \{(-1, a), (-1, b), (-1, c), (-1, d), (-1, -1),$
$(-1, -2), (-2, a), (-2, b) (-2, c), (-2, d), (-2, -1), (-2, -2)\}.$

Theorem 2. *For any sets A, B, C, D we have $(A \times B) \cap (C \times D) = (A \cap C) \times (B \cap D)$.*

Proof : If $(a, b) \in (A \times B) \cap (C \times D)$, then

$\Rightarrow$ $(a, b) \in (A \times B)$ and $(a, b) \in (C \times D)$

$\Rightarrow$ $(a \in A$ and $b \in B)$ and $(a \in C$ and $(b \in D)$

$\Rightarrow$ $(a \in A$ and $a \in C)$ and $(b \in B$ and $b \in D)$

$\Rightarrow$ $a \in (A \cap C)$ and $b \in (B \cap D)$ $\Rightarrow$ $(a, b) \in (A \cap C) \times (B \cap D)$

Since (a, b) is arbitrary, therefore $(A \times B) \cap (C \times D) \subseteq (A \cap C) \times (B \cap D)$ ...(1)

Now, let $(a, b) \in (A \cap C) \times (B \cap D)$

$\Rightarrow$ $a \in (A \cap C)$ and $b \in (B \cap D)$ $\Rightarrow$ $(a \in A$ and $a \in C)$ and $(b \in B$ and $b \in D)$

$\Rightarrow$ $(a \in A$ and $b \in B)$ and $(a \in C$ and $b \in D)$

$\Rightarrow$ $(a, b) \in (A \times B) \cap (C \times D)$

Since (a, b) is arbitrary, therefore $(A \cap C) \times (B \cap D) \subseteq (A \times B) \cap (C \times D)$ ...(2)

From (1) and (2), we conclude that $(A \times B) \cap (C \times D) = (A \cap C) \times (B \cap D)$.

- Here, it is evident that $(A \times B) \cap (B \times A) = (A \cap B) \times (B \cap A)$.
- $A \times (B' \cup C')' = A \times (B \cap C) = (A \times B) \cap (A \times C)$.
- $A \times (B' \cap C')' = A \times (B \cup C) = (A \times B) \cup (A \times C)$.

Theorem 3. *If A and B are two non-empty sets having n elements in common, then $A \times B$ and $B \times A$ have n^2 elements in common.*

Proof : We know that $(A \times B) \cap (C \times D) = (A \cap C) \times (B \cap D)$

$\Rightarrow$ $(A \times B) \cap (B \times A) = (A \cap B) \times (B \cap A)$

$\Rightarrow$ $(A \times B) \cap (B \times A) = (A \cap B) \times (A \cap B)$

Since $(A \times B)$ has n elements, therefore $(A \cap B) \times (B \cap A)$ has n^2 elements.

$\Rightarrow$ $(A \times B) \cap (B \times A) = (A \cap B) \times (B \cap A)$ has n^2 elements.

Hence, $(A \times B)$ and $(B \times A)$ have n^2 elements in common.

- (i) For any three sets, A, B, C; $A \times (B - C) = (A \times B) - (A \times C)$
- (ii) If A and B are any two non-empty sets, then $A \times B = B \times A$ iff $A = B$.
- (iii) If $A \subseteq B$, then $A \times A \subseteq (A \times B) \cap (B \times A)$
- (iv) If $A \subseteq B$, then $A \times C \subseteq B \times C$ for any set C.
- (v) If $A \subseteq B$ and $C \subseteq D$, then $A \times C \subseteq B \times D$.
- (vi) $A \times B = A \times C \Rightarrow B = C$.

Exercise 1.5

1. If $A = \{a, b, c\}$, $B = \{d\}$, $C = \{2\}$, then verify that
   (i) $A \times (B \cup C) = (A \times B) \cup (A \times C)$
   (ii) $A \times (B \cap C) = (A \times B) \cap (A \times C)$
   (iii) $A \times (B - C) = (A \times B) - (A \times C)$
   (iv) $(A \cap B) \times C = (A \times C) \cap (B \times C)$
2. If $A = \{2, 3\}$, $B = \{1, 2, 3\}$, $C = \{2, 3, 4\}$, show that $A \times A = (B \times B) \cap (C \times C)$.
3. If $A = \{1, 2, 3\}$, $B = \{4, 5\}$ and $C = \{1, 2, 3, 4, 5\}$, then show that $(C \times B) - (A \times B) = B \times B$.
4. The ordered pairs $(2, 7)$, $(4, 8)$ and $(5, 9)$ are among nine elements of the set $A \times B$. Determine the other six elements of $A \times B$.
5. Let $A = \{2, 3, 5, 7\}$, $B = \{1, 12, 13, 15\}$. How many elements are there in $A \times B$? In $B \times A$? Is $A \times B = B \times A$? Is $n(A \times B) = n(B \times A)$?
6. Let A and B be two sets. Show that the sets $A \times B$ and $B \times A$ have an element in common if and only if the sets A and B have an element in common.
7. Some elements of $A \times B$ are (a, x), (a, y), (d, z). If $A : \{a, b, c, d\}$, find the remaining elements of $A \times B$ such that $n(A \times B)$ is least.
8. If A and B are two sets having 3 elements in common. If $n(A) = 5$, $n(B) = 4$, find $n(A \times B)$ and $n\{(A \times B) \cap (B \times A)\}$.

9. The ordered pairs (1, 1), (2, 2) and (3, 3) are among the elements in the set $A \times B$. If A and B have 3 elements each, how many elements in all does the set $A \times B$ have? Also find the remaining elements.

10. If A and B are two sets such that $n(A) = 3$ and $n(B) = 2$. If $(x, 1), (y, 2), (z, 1)$ are in $A \times B$, find A and B, where x, y, z are distinct.

11. Write 'T' for true and 'F' for false statement :

(a) If $A = (a, b)$ and $B = (b, a)$, then $A \times B = \{(a, b), (b, a)\}$ (T/F)

(b) $\{(a, x), (a, y), (b, x), (b, y)\}$ is product set. (T/F)

(c) If $n(A) = x$ and $n(B) = y$ and $A \cap B = \phi$, then $n(A \times B) = xy$ (T/F)

(d) If A and B are non-empty sets, then $A \times B$ is a non-empty set of ordered pairs (x, y) such that $x \in A$ and $y \in A$. (T/F)

12. (a) If $A = \{1, 2, 3\}$, $B = \{4, 5\}$ and $C = \{1, 2, 3, 4, 5\}$. Find

(i) $A \times B$,

(ii) $C \times B$,

(iii) $B \times B$

(b) If $A = \{1, 2, 3, 4\}$ and $B = \{5, 7, 9\}$,

find $(A \times B) \cap (A \cap B)$.

ANSWERS

4. (2, 8), (2, 9), (4, 7), (4, 9), (5, 7), (5, 8) **5.** 16, 16, No, Yes

7. (a, y), $(a, 2)$, (b, x), (b, y), (b, z), (c, x), (c, z), (d, x), (d, y) **8.** 20, 9

9. 9, (1, 2), (1, 3), (2, 1), (2, 3), (3, 1), (3, 2)

10. (i) $A = \{x, y, z\}$, $B = \{1, 2\}$, (ii)(a) F (b) T, (c) T (d) F

12. (a) (i) $A \times B = (1, 4), (1, 5), (2, 4), (2, 5), (3, 4), (3, 5)$

(ii) $C \times B = \{(1, 4), (1, 5), (2, 4), (2, 5), (3, 4), (3, 5), (4, 4), (4, 5), (5, 4), (5, 5)\}$

(iii) $B \times B = \{(4, 4), (4, 5), (5, 5)\}$ (b) ϕ

1.12 RELATION

Let us take two sets of natural numbers N_1 and N_2. We define R as a relation between them such that N_1 is a square of N_2. Then we can write $1R1$, $2R4$, $3R9$,...

In terms of ordered pair, we can write

$$R = \{(1, 1), (2, 4), (3, 9), (4, 16), \ldots\} = (\{x, y) : x, y \in \mathbf{N} \text{ and } y = x^2\}$$

The relation from set **N** to **N** is a subset of $\mathbf{N} \times \mathbf{N}$ such that $(x, y) \in \mathbf{R}$ iff $y = x^2$.

Definition : Let A and B be two sets. Then a relation R from A to B is a subset of $A \times B$.

Symbolically : R is a relation from A to $B \Leftrightarrow R \subseteq A \times B$.

- If R is a relation from A to B, then A is called the domain and B the range of R.
- If R is a relation from a non-empty set A to a non-empty set B and if $(a, b) \in R$, then we write aRb, read as "a is related to b by the relation R." On the other hand, if $(a, b) \notin R$, we write a R b and say that 'a is not related to b by the relation R'.
- In particular, any subset $A \times A$ defined a relation in A, known as **Binary relation**.

For Example :

If $a, b \in \mathbf{N}$ and R is defined as "a is divisor of b" then R is relation on **N.** The subset $\mathbf{N} \times \mathbf{N}$, which corresponds to the relation R is $S = \{(n, r) : n \in N, r \in \mathbf{N}\}$ Here, it is clear that $(1, 3), (2, 4), (3, 9), (4, 8), (4, 4)$, are in S, whereas (2, 3), (4, 5), (5, 6) are not in S.

If R is a relation from set $A = \{1, 2, 3\}$ to the set $B = \{-1, -2\}$ defined by $x + y = 0$, then $R = \{(1, -1), (2, -2)\}$ Here, domain of R is $\{1, 2\}$ and Range $= \{-1, -2\}$.

If $A = \{a, b, c, d, e\}$ and $B = \{f, g, h, i\}$ and let $R = \{(a, g), (a, i), (d, h), (e, f)\}$ be a relation from A to B, then Domain of $R = \{a, d, e\}$ and Range of $R = \{g, i, h, f\}$

If $a, b \in \mathbf{R}$ the set of real numbers and $\mathbf{R}$ is "$|a - b|$ is a rational number" then R is a relation on R. The subset S of $\mathbf{R} \times \mathbf{R}$ which corresponds to the relation is

$$S = \{(a, b + a) : a \in \mathbf{R}, b \in \mathbf{Q}\}$$

It is observed that $\left(1, 2\frac{1}{2}\right), \left(\pi, 1 - \frac{1}{2}\right)$ belongs to S, while $(\sqrt{2}, \pi + \sqrt{2}) \notin S$.

If $A = \{2, 3, 4\}$ and $B = \{a, b, c\}$, then $R = \{(2, b), (3, c), (2, a), (4, a)\}$ being a subset of $A \times B$, is a relation from $A \times B$. Here $(2, b), (3, c), (2, a), (4, a) \in R$, so we may write $2Rb$, $3Rc$, $2Ra$, $4Ra$. But $(3, b) \notin R$ therefore, $3\ R\ b$. If $a, b \in \mathbf{N}$ and R is defined by "$a - b$ is divisible by a number $n \in \mathbf{N}$", then R is a relation on $\mathbf{N}$.

The subset S of $\mathbf{N} \times \mathbf{N}$ corresponding to the relation by

$$S = \{n, n + rm : n \in \mathbf{N}, r \in \mathbf{N}\}$$

Here, $m = 3$, $(2, 8), (5, 11) \in S$ $[\because 2 - 8 = 6$, which is divisible by 3]

while $(3, 8) \in S$ $[\because 3 - 8 = 5$, which is not divisible by 3]

1.12.1 Total number of relations

Let A and B be two non-empty finite sets consisting p and q elements respectively, then $A \times B$ consists of pq ordered pairs. Therefore, total number of subset of $A \times B$ is 2^{pq}.

- For a non-empty set A, $\phi \subseteq A \times A$, therefore it is a relation on A, this relation is called **void** or **empty** relation on A.
- The void relation ϕ and the universal relation $A \times B$ are called trivial relation from A to B.
- The void and universal relation on set A respectively the smallest and the largest relation on A.

Identity Relation. Let A be a set. The identity relation on A is the relation $I_A = \{(x, x) : x \in A\}$ on A.

For example : If $A = \{a, b, c\}$, then the relation $I_A = \{(a, a), (b, b), (c, c)\}$ is the identity relation. $R = \{(a, a), (b, b)\}$ is not an identity relation as $(c, c) \notin R$.

1.11.2 Inverse of a relation

Let A, B be two non-empty sets and R be a relation from a set A to B and let (x, y), number of the subset D of $A \times B$ corresponding to the relation R from A to B.

To the relation R from the set A to the set B, there corresponds a relation from the set B to the set A called the inverse of the relation, denoted by R^{-1} such that the subset $B \times A$ corresponding to the relation R^{-1} is $= \{(y, x) : (x, y) \in D\}$.

i.e., $yR^{-1}x \Leftrightarrow xRy$

For example :

(i) Let $A = \{a, b, c\}$ and $B = \{1, 2, 3\}$ be two sets and let $R = \{(a, 1), (a, 2), (b, 1), (b, 2)\}$ be a relation from A to B then $R^{-1} = \{(1, a), (2, a), (1, b), (2, b)\}$.

(ii) If $A = \{1, 2, 3\}$, $B = \{5, 6, 7\}$ and let $R = \{(1, 5), (2, 5), (2, 7)\}$ be a relation from A to B.

Then $R^{-1} = \{(5, 1), (5, 2), (7, 2)\}$ which is a relation from B to A.

also, Domain $(R) = \{1, 2\} =$ Range (R^{-1})

and, Range $(R) = \{5, 7\} =$ Domain (R^{-1})

(iii) The inverse of the relation 'is less than' In **R** "is greater than".

- It may be noted that sometimes, the inverse of a relation coincides with the relation itself.
- The inverse of the relation "perpendicular to" in the set of straight lines coincides with itself.

1.13 CLASSIFICATION OF RELATIONS

(a) Reflexive Relation. Let R be a relation on a set A.

"A relation R is said to be reflexive if $(x, x) \in R \ \forall \ x \in A$"

i.e., $x \, R \, x \ \forall \ x \in A$.

For Example :

In a set of integers, a relation R defined by $x \, R \, y$ iff $x - y$ is divisible by 4, then R is a reflexive relation because $x - x = 0$ which is a divisible by 4.

The universal relation on a non-empty set A is reflexive.

The relation "is less then" *i.e.,* '<' in the set of natural number, is not reflexive, because no member have the relation is less than to itself.

The relation "is a factor of" in the set of rational number is reflexive, since every rational number is a factor of itself.

The relation "is less than or equal to" *i.e.,* $\leq$ is in the set of natural number is reflexive.

$$n \leq n \ \forall \ n \in \mathbf{N}$$

(b) Symmetric Relation. A relation R on a set A is said to be symmetric if

$$(y, x) \in R \text{ whenever } (x, y) \in R \ \forall x, y \in R$$

i.e., $x \, R \, y \Leftrightarrow y \, R \, x \ \forall x, y \in R$

For Example :

Let l_1, l_2 be two lines such that l_1 is perpendicular to l_2, *i.e.,* $l_1 \perp l_2$. Then $l_1 \perp l_2 \Rightarrow l_2 \perp l_1$. Therefore the relation $\perp$ is said to be symmetric.

The identity and the universal relation on a non-empty set are symmetric relations.

Consider the set **N** of natural numbers and the relation 'is less than'. This relation is not symmetric. Since if $2 < 3$ then $3 \nless 2$.

Let $A = \{1, 2, 3\}$ and relations R_1 and R_2 defined by

$$R_1 = \{(1, 2) \ (1, 3), \ (3, 1) \ (2, 1)\} \text{ and } R_2 = \{(1, 2) \ (2, 3), \ (3, 1)\}$$

Then R_1 is a symmetric relation, but R_2 is not symmetric.

(c) Transitive Relation. A relation R on a set A is said to be transitive iff $(x, y) \in R$ and $(y, z) \in R$

$\Rightarrow \ (x, z) \in R \ \forall x, y, z \in A$ *i.e.,* $x \, R \, y, y \, R \, z \Rightarrow x \, R \, z$.

For Example :

Let a, b, c be three numbers such that a is a factor of b and b is a factor of c, then obviously a is a factor of c. Therefore, 'is a factor of' is a transitive relation.

If l_1, l_2, l_3 are three lines such that $l_1 \perp l_2$ and $l_2 \perp l_3$ then it is obvious that l_1 is parallel to l_3. Therefore the relation $\perp$ is not transitive.

The identity and universal relations on a non-empty set are transitive.

Let l_1, l_2, l_3 be three straight lines, such that l_1 is parallel to l_2 and l_2 is parallel to l_3 then it is clear that l_1 is parallel to l_3. Therefore, 'is parallel to' is a transitive relation.

(d) Anti-symmetric Relation. A relation R on a non-empty set A is said to be an anti symmetric relation iff $(x, y) \in R$ and $(y, x) \in R \Rightarrow x = y \ \forall \ x, y \in R$

- The identity relation R on a set A is an Anti-symmetric relation.
- If $(x,y) \in R$ and $(y,x) \notin R$, then it may be noted that $x = y$.
- The universal relation on a set A containing at least two elements is not Anti-symmetric.

1.13.1 Equivalence Relations

A relation R on a set E is said to be equivalence if it is
(i) Reflexive, (ii) Symmetric and (iii) Transitive.

For Example :

In a set of integers, a relation R is defined by $x\ R\ y$ if and only if $x - y$ is divisible by 4. Then R is an equivalence relation. Since

(a) For $x\ R\ x, x - x = 0$ is divisible by 4. Therefore, it is reflexive.

(b) For $x\ R\ y$. Let $x - y = 4m$ so $y - x = 4m$, which is also divisible by 4. Therefore, it is symmetric.

(c) For $x\ R\ y$, let $x - y = 4m$; for $y\ R\ z$, let $y - z = 4n$. By adding these two equations, we get $x - z = -4(m + n)$, which is divisible by 4. Therefore it is transitive.

Let R be a relation on the set of all lines in a plane L defined by $(l_1, l_2) \in R$, if and only if line l_1 is parallel to l_2, then R is an equivalence relation because

(a) For each line $l \in L$, we have l is parallel to $l \quad \Rightarrow \quad l\ R\ l \Rightarrow R$ is reflexive.

(b) Let $l_1, l_2 \in L$ such that $(l_1, l_2) \in R$, then
$\Rightarrow \quad (l_1, l_2) \in R \Rightarrow l_1$ is parallel to $l_2 \Rightarrow l$ is symmetric.

(c) Let $l_1, l_2, l_3 \in L$ such that (l_1, l_2) and $(l_2, l_3) \in R$, then obviously $(l_1, l_3) \in R$ because if l_1 is parallel to l_2 and l_2 is parallel to l_3, then l_3 should be parallel to l_1.

1.13.2 Congruence Modulo 'm'

Let m be an arbitrary but fixed integer. If $x - y$ is divisible by m, then two integers x and y are said to be congruence modulo m of one another.

Symbolically : $x \equiv y \pmod{m} \Leftrightarrow x - y$ is divisible by m.

For Example : $32 \equiv 2 \pmod{3}$, as $32 - 2 = 30$ which is divisible by 3.

1.13.3 Composition of relations

Let R_1 and R_2 be two relations from sets A to B and B to C respectively, then we can define a relation $R_1 \circ R_2$ from A to C, such that $(x, z) \in R_1 \circ R_2$ if and only if there exist $y \in Y$ such that $(x, y) \in R_1$ and $(y, z) \in R_2$.

This relation is called composition of R_1 and R_2.

- $R_1 \circ R_2 \neq R_2 \circ R_1$.
- $(R_2 \circ R_1)^{-1} = R_1^{-1} \circ R_2^{-1}$.

For Example : Let A, B, C be three sets such that

$$A = \{-1, -2\},\ B = \{p, q, r\} \text{ and } C = \{\alpha, \beta, \gamma\}$$

Also, $R_1 = \{(-1, p), (-1, r), (-2, q)\}$ is a relation from A and B and

$R_2 = \{(p, \alpha), (q, \beta), (r, \gamma)\}$ is a relation from set B to C.

Then, $R_2 \circ R_1$ is a relation from A to C given by $R_2 \circ R_1 = \{(-1, \alpha), (-1, \gamma), (-z, \beta)\}$.

Theorem 4. *The intersection of two equivalence relations on a set is an equivalence relation.*

Proof : Let R_1, R_2 be two equivalence relations on a set A. To show $(R_1 \cap R_2)$ also an equivalence relation.

(i) Let $a \in A$ and a is arbitrary.

Since R_1 and R_2 both are reflexive on A.

$\therefore \quad (a,a) \in R_1$ and $(a,a) \in R_2 \Rightarrow (a,a) \in R_1 \cap R_2$.

Therefore, $(R_1 \cap R_2)$ is a reflexive.

(ii) Let $a, b \in A$ such that $(a,b) \in (R_1 \cap R_2)$

$$(a,b) \in R_1 \cap R_2 \Rightarrow (a,b) \in R_1 \text{ and } (a,b) \in R_2.$$

Also, R_1 and R_2 both are symmetric on A.

Therefore, $(b,a) \in R_1$ and $(b,a) \in R_2 \Rightarrow (b,a) \in R_1 \cap R_2 \Rightarrow (R_1 \cap R_2)$ is symmetric on A.

(iii) Let $a, b, c \in A$ such that $(a,b) \in R_1 \cap R_2$, $(b,c) \in R_1 \cap R_2$

Then, $(a,b) \in R_1 \cap R_2$ and $(b,c) \in R_1 \cap R_2$

$\Rightarrow \{(a,b) \in R_1$ and $(a,b) \in R_2\}$ and $\{(b,c) \in R_1$ and $(b,c) \in R_2\}$

$\Rightarrow \{(a,b) \in R_1, (b,c) \in R_1\}$ and $\{(a,b) \in R_2, (b,c) \in R_2\}$

$\Rightarrow (a,c) \in R_1$ and $(a,c) \in R_2$ $\quad [\because R_1$ and R_2 both are transitive$]$

$\Rightarrow (a,c) \in R_1 \cap R_2$

Therefore, $(R_1 \cap R_2)$ is transitive on A.

From (i), (ii) and (iii), we have that $R_1 \cap R_2$ is reflexive, symmetric and transitive, and hence $R_1 \cap R_2$ is an equivalence relation.

☛ The union of two equivalence relations on a set is not necessarily an equivalence relation.

Theorem 5. *If R is an equivalence relation, then R^{-1} is also an equivalence relation.*

Proof : Let R be an equivalence relation on a set A. Then by definition of relation on a set, we have $R \subseteq A \times A \Rightarrow R^{-1} \subseteq A \times A$

Therefore, R^{-1} is a relation on A.

Now, to show R^{-1} is an equivalence relation.

(i) Let $a \in A$, then $(a,a) \in R$ $\quad (\because R$ is an equivalence relation$)$

$\Rightarrow (a,a) \in R^{-1}$

Thus, $(a,a) \in R^{-1} \ \forall \ a \in R \quad \Rightarrow \quad R^{-1}$ is reflexive on A.

(ii) Let $(a,b) \in R^{-1}$, then $(a,b) \in R^{-1} \Rightarrow (b,a) \in R$

$\Rightarrow (a,b) \in R$ $\quad (\because R$ is symmetric$)$

$\Rightarrow (b,a) \in R^{-1}$

Therefore R^{-1} is symmetric.

(iii) Let $(a,b) \in R^{-1}$ and $(b,c) \in R^{-1}$, then $(a,b) \in R^{-1} \Rightarrow (b,a) \in R$

and $(b,c) \in R^{-1} \Rightarrow (c,b) \in R$

Now, $(c,b) \in R$ and $(b,a) \in R$

$\Rightarrow (c,a) \in R$ $\quad (\because R$ is transitive$)$

$\Rightarrow (a,c) \in R^{-1}$

Therefore, R^{-1} is transitive.

From (i), (ii) and (iii), we conclude that R^{-1} is an equivalence relation.

ILLUSTRATIVE EXAMPLES

1. *Let* **Z** *be the set of integers. Define a relation R on* **Z** *such that x R y holds if and only if x – y is divisible by 5, $x \in$* **Z**, *$y \in$* **Z**. *Show that R is an equivalence relation.*

Solution : (i) For each $x \in \mathbf{Z}$, $x - x = 0$ and 0 is divisible by 5.

Therefore, for all $x \in \mathbf{Z}$, $x R x \Rightarrow x$ is reflexive.

(ii) Let $x R y \Rightarrow x - y$ is divisible by 5.

$\Rightarrow$ $y - x = [-(x - y)]$ is divisible by 5.

Thus $xRy = yRx$

Therefore, R is symmetric.

(iii) Let us suppose xRy and yRz, then $(x - y)$ and $(y - z)$ are both divisible by 5. Hence, 5 is also a divisor of $(x - y) + (y - z)$.

$\Rightarrow$ 5 is a divisor of $(x - z)$

Therefore, xRy, $yRz \Rightarrow xRz$. $\Rightarrow$ R is transitive.

From (i), (ii) and (iii), we conclude that R is an equivalence relation.

2. *Let* $\mathbf{N} \times \mathbf{N}$ *be the set of ordered pairs of natural numbers. Also, let R be the relation in* $\mathbf{N} \times \mathbf{N}$, *defined by (a, b) R (c, d) if and only if a + d = b + c. Show that R is an equivalence relation.*

Solution : (i) For all $(a, b) \in \mathbf{N} \times \mathbf{N}$, we have $a + b = b + a$ *i.e.*, $(a, b) R (b, a)$

Therefore, R is reflexive.

(ii) Let $(a, b) R (c, d)$, then, by definition of R

$$(a + d) = (b + c) \quad \text{or} \quad (c + b) = (d + a)$$

$$(c, d) R (a, b) \Rightarrow R \text{ is symmetric.}$$

(iii) Let us suppose $(a, b) R (c, d)$ and $(c, d) R (e, f)$, then

$$a + d = b + c \text{ and } c + f = d + e$$

$\Rightarrow$ $(a + d) + (c + f) = (b + c) + (d + e)$ $\Rightarrow$ $a + f = b + e$ $\Rightarrow$ $(a, b) R (e, f)$

Therefore, R is transitive.

Hence, from (i), (ii) and (iii), we conclude that R is an equivalence relation.

3. *If R is the relation for natural numbers defined by x + 4y = 20. Find the domain and range.*

Solution : Let $x + 4y = 20$. $\Rightarrow$ $y = \dfrac{20 - x}{4}$

For $x = 4$, $y = 4$ and for $x = 8$, $y = 3$.

For $x = 16$, $y = 1$ and for $x = 12$, $y = 2$.

Therefore, Domain = $\{4, 8, 12, 16\}$ and range = $\{4, 3, 2, 1\}$

4. *A relation R defined on the set of integers* **Z**, *as follows* $(x, y) \in R \Leftrightarrow x^2 + y^2 = 25$

Express R and R^{-1} *as the sets of ordered pairs and hence find their respective domains.*

Solution. Since $(x, y) \in R \Leftrightarrow x^2 + y^2 = 25$ $\Leftrightarrow$ $y = \pm\sqrt{25 - x^2}$

If $x = 0 \Rightarrow y = \pm 5$.

Therefore, $(0, 5) \in R$ and $(0, -5) \in R$

Now, $x = \pm 3$ $\Rightarrow$ $y = \sqrt{25 - 9} = \pm 4$

$$(3, 4) \in R, (-3, 4) \in R, (3, -4) \in R \text{ and } (-3, -4) \in R$$

$$x = \pm 4 \Rightarrow y = \pm 3.$$

Therefore, $(4,3) \in R$, $(-4, 3) \in R$, $(4, -3) \in R$ and $(-4, -3) \in R$

$$x = \pm 5 \Rightarrow y = \sqrt{25 - 25} = 0 \quad \therefore \quad (5, 0) \in R \text{ and } (-5, 0) \in R$$

Here, it is clear that for any other integral value of x, y is not an integer.

Therefore, $R = \{(0, 5), (0, -5), (3, 4), (-3, 4), (3, -4), (-3, -4), (4, 3), (-4, 3), (4, -3)$
$(-4, -3), (5, 0), (-5, 0)\}$

and $R^{-1} = \{(5,0), (-5,0), (4,3), (4,-3), (-4,3), (-4,-3), (3,4), (3,-4), (-3,4)$
$(-3,-4), (0,5), (0,-5)\}$

Also, Domain $(R) = \{(0, 3, -3, 4, -4, 5, -5)\}$ = domain of (R^{-1}).

5. *Consider the set $A = \{a, b, c\}$. Give an example of a relation R on A which is*
(i) Reflexive and symmetric but not transitive.
(ii) Symmetric and transitive, but not reflexive.
(iii) Reflexive and transitive, but not symmetric.

Solution : (i) Given $A = \{a, b, c\}$

Let $R = \{(a,a),(a,b),(b,a),(b,c),(c,b),(b,b),(c,c)\}$ on A.

Clearly, R is reflexive and symmetric but not transitive

(ii) Let $R = \{(a,a),(a,b),(b,a),(b,b)\}$ on A.

Here, R is symmetric and transitive but not reflexive.

(iii) Let $R = \{(a,a),(b,b),(c,c),(a,b)\}$ on A.

Here, R is reflexive, transitive but not symmetric.

6. *If R is a relation in $\mathbf{N} \times \mathbf{N}$, show that the relation R is defined by $(a, b) R (c, d)$ if and only if $ad = bc$ is an equivalence relation.*

Solution : (i) Since $ab = ba \ \forall \ a, b, \in \mathbf{N}$.

Therefore, $(a, b) R (a, b) \ \forall \ a, b, \in \mathbf{N} \Rightarrow R$ is reflexive.

(ii) We have $(a, b) R (c, d)$ iff $ad = bc \ \forall \ a, b, c, d \in \mathbf{N}$.

Now, $(c, d) R (a, b)$ iff $cb = da \ \forall \ a, b, c, d \in \mathbf{N} \Rightarrow R$ is symmetric.

(iii) We have $(a, b) R (c, d)$ iff $ad = bc \ \forall \ a, b, c, d \in \mathbf{N}$.

Therefore, $(a, b) R (c, d), (c, d) R (e, f) \Rightarrow (a, b) R (e, f) \ \forall \ a, b, c, d \in \mathbf{N}$

Using $(a, d), (c, f) = (b, c)\ (d, e)$

$\Rightarrow \quad (a, f) = (b, e) \quad \Rightarrow \quad R$ is transitive.

Hence, from (i), (ii) and (iii), we conclude that R is an equivalence relation.

7. *Let R_1 and R_2 be two relations on a set A, where $A = \{1, 2, 3, 5\}$ such that*

$$R_1 = \{(1,1), (1,2), (1,5), (2,1), (2,5)\}$$

and $$R_2 = \{(3,3), (3,2), (2,3), (1,2), (2,1)\}$$

Then, which of the following statement is false

(i) $R_1 \cup R_2$ is symmetric *(ii) $R_1 \cap R_2$ is transitive*
(iii) $R_1 \cap R_2$ is symmetric *(iv) $R_1 \cup R_2$ is transitive.*

Solution : (i) As $(1,2) \in R_1$, also $(2,1) \in R_1$, therefore, it is symmetric and as $(1,2) \in R_2$, also $(2,1) \in R_2 \Rightarrow R_2$ is symmetric.

Now, $R_1 \cup R_2 = \{(1,1),(1,2),(1,5),(2,1),(2,5),(3,3),(3,2),(2,3)\}$

In $R_1 \cup R_2$, as $(1,2) \in R_1 \cup R_2$, also $(2,1) \in R_1 \cup R_2 \Rightarrow R_1 \cup R_2$ is symmetric.

Therefore, (i) is true.

(ii) We have, $R_1 \cap R_2 = \{(1,2),(2,1)\}$

$\Rightarrow \quad (1,1)$ should also belong to $R_1 \cap R_2$.

But in this case $(1,1) \notin R_1 \cap R_2$ is not transitive.

Therefore, (ii) is false.

(iii) We have, $R_1 \cap R_2 = \{(1,2),(2,1)\}$

$(1,2) \in R_1 \cap R_2$ and also $(2,1) \in R_1 \cap R_2$.

$\Rightarrow \quad R_1 \cap R_2$ is symmetric.

Therefore, (iii) is true.

(iv) In $R_1 \cup R_2$, $(1,2) \in R_1 \cup R_2$ and $(2,5) \in R_1 \cup R_2$, also $(1,5) \in R_1 \cup R_2$.

$\Rightarrow \quad R_1 \cup R_2$ is transitive. Therefore, (iv) is true.

8. *If A be the set of all triangles in a plane and $R = \{(a, b) : \Delta a = \Delta b\}$, i.e., $aRb \Leftrightarrow$ Area of triangle a = Area of triangle b, then show that R is an equivalence relation.*

Solution : (i) Since, for all $a \in A$, we have $\Delta a = \Delta b$

Therefore, $aRa \Rightarrow R$ is reflexive.

(ii) For any $a, b \in A$, we have $(a, b) \in R \quad \Rightarrow \quad \Delta a = \Delta b$

$\Rightarrow \quad \Delta b = \Delta a \quad \Rightarrow \quad (b, a) \in R$

Therefore, $(b, a) \in R$, i.e., $bRa \Rightarrow R$ is symmetric.

(iii) For all $a, b, c \in A$, we have $(a, b) \in R$, $(b, c) \in R$

$\Rightarrow \quad \Delta a = \Delta b$ and $\Delta b = \Delta c \quad \Rightarrow \quad \Delta a = \Delta c \quad \Rightarrow \quad (a, c) \in R$

Therefore, R is transitive.

Hence, from (i), (ii) and (iii), we conclude that R is an equivalence relation.

9. *If **Z** be a set of non-zero integers and a relation R defined by $xRy \Rightarrow x^y = y^x \ \forall x, y \in \mathbf{Z}$, then show that R is not an equivalence relation on **Z**.*

Solution : (i) Let $x \in \mathbf{Z}$, then $x^x = x^x, \forall x \in \mathbf{Z} \quad \Rightarrow \quad xRx, \forall x \in \mathbf{Z}$

Therefore, R is reflexive.

(ii) Let $x, y \in \mathbf{Z}$ such that xRy, i.e., $x^y = y^x$

$$x^y = y^x \Rightarrow y^x = x^y$$

Therefore, $xRy \Rightarrow yRx, \forall x, y \in \mathbf{Z}$

$\Rightarrow \quad R$ is symmetric.

(iii) Let $x, y, z \in \mathbf{Z}$ such that xRy and yRz

i.e., $x^y = y^x$ and $y^z = z^y$ which does not give $x^z = z^x$

$\Rightarrow \quad R$ is not transitive.

Hence, from (iii), we conclude that R is not an equivalence relation.

10. *Let $A = \mathbf{R} \times \mathbf{R}$ (**R** is the set of real numbers) and define the following relation on $A : (a, b)\, R\, (c, d)$ iff $a^2 + b^2 = c^2 + d^2$*

(i) Verify that (A, R) is an equivalence relation.

(ii) Describe geometrically what the equivalence classes are for this reason.

Solution : (i) We have $(a, b)\, R\, (c, d) \Rightarrow a^2 + b^2 = c^2 + d^2$

$\Rightarrow \quad c^2 + d^2 = a^2 + b^2 \quad \Rightarrow \quad (c, d)\, R\, (a, b)$ ...(1)

$\Rightarrow \quad R$ is symmetric.

Now, $(a, b)\, R\, (c, d)$ and $(c, d)\, R\, (x, y) \Rightarrow a^2 + b^2 = c^2 + d^2$

and $c^2 + d^2 = x^2 + y^2 \quad \Rightarrow \quad a^2 + b^2 = x^2 + y^2$

$\Rightarrow \quad (a, b)\, R\, (x, y)$ ...(2)

$\Rightarrow \quad R$ is transitive.

Again, $(a, b)\, R\, (a, b) \Leftrightarrow a^2 + b^2 = a^2 + b^2 \quad \Rightarrow \quad R$ is reflexive. ...(3)

Hence, from (1), (2) and (3), we conclude that R is an equivalence relation.

(ii) For any point (a, b), the sum $a^2 + b^2$ is the square of the distance from the origin. The equivalence classes are, therefore, the set of points in the place which have the same distance from the origin. Hence, the equivalence classes are concentric circles centered on the origin.

11. *Let R be the binary relation defined as $R = \{(a, b) \in R^2 : a - b \le 3\}$*

Determine whether R is reflexive, symmetric, anti symmetric and transitive.

Solution : We have $(a, b) \in R^2 : a - b \le 3$

$\Rightarrow \quad (a, a) \in R^2 : a - a \le 3$, i.e., $0 \le 3$, which is true. So, R is reflexive.

In a similar way, we can easily show that R is neither symmetric, anti symmetric nor transitive.

1.13.4 Relations other than equivalence

Given a relation R in a set X. Then R is

(i) non-reflexive if $\exists x$, such that $(x, x) \notin R$.

(ii) anti reflexive or irreflexive if $i_x \cap R = \phi$ (where i_x is the identity relation on x or $\forall n \in X : (x, x) \notin R$

(iii) non-symmetrical if for some $(x, y) \in R$, we have $(y, x) \notin R$

(iv) anti-symmetric if $R \cap R^{-1} = i$, *i.e.*, $(x, y) \in R$ and $(y, x) \in R \Rightarrow x = y$.

(v) asymmetric if $R \cap R^{-1} = \phi$, *i.e.*, $(x, y) \in R \Rightarrow (y, x) \notin R$

(vi) non-transitive if $R \circ R \notin R$.

(vii) anti-transitive if $(R \circ R) \cap R = \phi$.

(viii) a reflexive and symmetric, but not transitive relation is called a tolerance relation.

(ix) a non-symmetric transitive relation is called an ordered relation.

(x) a reflexive, anti-symmetric and transitive relation is called partial-ordered, relation.

Exercise 1.2

1. If R is the relation 'is less than' from $A = \{1, 2, 3, 4, 5\}$ to $B = \{1, 4, 5\}$, find the set of ordered pairs corresponding to R. Also find R^{-1}.

2. A relation R defined from a set $A = \{2, 3, 4, 5\}$ to a set $B = \{3, 6, 7, 10\}$ as follows :
$(x, y) \in R \Rightarrow x$ divides y. Write R as a set of ordered pairs and determine the domain and range of R. Also find R^{-1}.

3. Find the domain and range of $A = \{1, 2, 3, 4, 5, 6\}$ when the relations are defined as
(i) xR_1y if and only if $x - y > 0$
(ii) xR_2y if and only if $x + y < 0$

4. Two sets A and B are given by $A = \{1, 2, 8, 9\}$ and $B = \{2, 3, 4, 6, 7\}$ and if R is the relation from A to B given by $\{(1, 2), (1, 3), (2, 4), (2, 6)\}$, then which of the following statement is true?
(i) Domain (R) = Range (R^{-1}) and Range (R) = Domain (R^{-1})
(ii) Domain (R) = Domain (R^{-1}) and Range (R) = Range (R^{-1})
(iii) Domain (R) = Range (R^{-1}) and Range (R) = Domain (R^{-1})
(iv) Domain (R) = Range (R)

5. If R is a relation on a set A, then which of the following statement is not true?
(i) if R is reflexive then R^{-1} is reflexive
(ii) if R is symmetric then R^{-1} is symmetric
(iii) if R is transitive, then R^{-1} is transitive
(iv) None of these.

6. Find the domain and range of the following relations :
(i) $R = \{(x + 1, x + 5) : x \in \{0, 1, 2, 3, 4, 5\}$
(ii) $R = \{(x, x^3) : x$ is a prime number, less than $10\}$
(iii) $R = \{(a, b) : a \in \mathbf{N}, a < 5, b = 4\}$
(iv) $R = \{(a, b) : b = |a - l|, \quad a \in \mathbf{Z}$, and $|a| \le 3\}$

7. Let R_1 be the relation defined on the set of reals $\mathbf{R}$ such as $(a, b) \in R_1$ if and only if $1 + ab > 0$ for all $a, b, \in \mathbf{R}$. Show that R_1 is reflexive, symmetric but not transitive.

8. Let R be a relation on $\mathbf{N} \times \mathbf{N}$, defined by $(a, b)\, R\, (c, d)$ if and only if $ad\,(b + c) = bc\,(a + d)$. Show that R is an equivalence relation.

9. Show that the relation 'congruence modulo m' on the set of integers is an equivalence relation.

10. Let R_1 be a relation on the set of reals defined by
$R_1 = \{(a, b) \in R \times R : a^2 + b^2 = 1\}$. Show that R_1 is not an equivalence relation on R.

11 In a set L of all straight lines in a plane, discuss which of the following two relations are equivalence relations L.
(i) $R_1 = \{(x, y) : x, y \in L$ and x is parallel to $y\}$
(ii) $R_2 = \{(x, y) : x, y \in L$ and x is perpendicular to $y\}$.

12. Show that the relation $R = \{(a, b) : a - b =$ even integer $\forall\ a, b \in \mathbf{Z}\}$, *i.e.*, $aRb \Leftrightarrow a - b =$ even integer, is an equivalence relation.

13. Show that the relation R in $\mathbf{N}$, the set of natural numbers, defined by xRy if $x^2 - 4xy + 3y^2 = 0$, $(x, y \in \mathbf{N})$ is reflexive, not symmetric and not transitive.

14. For the given relation R on a set S, determine which are equivalence relations :
(i) S is the set of all rational numbers, aRb if and only if $a = b$.
(ii) S is the set of all real numbers iff
(a) $|a| = |b|$,
(b) $a \geq b$
(iii) S is the set of all triangles in a plane, aRb iff a is congruent to b.
(iv) S is the set of all triangles in a plane, aRb iff a and b have equal perimeters.

15. An integer m is said to be related to another integer n if m is a multiple of n. Show that the relation is reflexive and transitive but not symmetric.

16. Let R be a relation defined on the set of natural numbers $\mathbf{N}$ as $R = \{(x, y) : x, y \in \mathbf{N}, 2x + y = 41\}$. Find the domain and range of R.

17. Let O be the origin. Define a relation between two points P and Q in a plane if $PO = OQ$. Show that the relation is an equivalence relation.

18. Given the relation $R = \{(1, 2), (2, 3)\}$ on the set of natural numbers $\mathbf{N}$, add a minimum of ordered pairs so that the enlarged relation is symmetric, transitive and reflexive.

19. Let $\mathbf{N}$ denote the set of all natural numbers and R be the relation on $\mathbf{N} \times \mathbf{N}$ defined by $(a, b)\, R\, (c, d) \Leftrightarrow ad\,(b + c) = bc\,(a + d)$. Show that R is an equivalence relation.

20. Show that the relation, which is symmetric and transitive, is not necessarily reflexive.

ANSWERS

1. $aRb = \{(1, 4), (1, 5), (2, 4), (2, 5), (3, 4), (3, 5), (4, 5)\}$,
$R^{-1} = \{(4, 1), (5, 1), (4, 2), (5, 2), (4, 3), (5, 3), (5, 4)\}$

2. Domain $(R) = \{2, 3, 5\}$, Range $(R) = \{3, 6, 10\}$, $R^{-1} = \{(6, 2), (10, 2), (3, 3), (6, 3), (10, 5)\}$

3. (i) $\{2, 3, 4, 5, 6\}, \{1, 2, 3, 4, 5\}$, (ii) ϕ, ϕ **4.** (iii) **5.** (iv)

6. (i) Domain $(R) = \{1, 2, 3, 4, 5, 6\}$, Range $(R) = \{5, 6, 7, 8, 9, 10\}$
(ii) Domain $(R) = \{2, 3, 5, 7\}$, Range $(R) = \{8, 27, 125, 243\}$
(iii) Doman $(R) = \{1, 2, 3, 4\}$, Range $(R) = \{4\}$
(iv) Domain $(R) = \{0, -1, -2, -3, 1, 2, 3\}$, Range $(R) = \{1, 2, 3, 4, 0, 1, 2\}$

11. $R_1 =$ Equivalence relation, $R_2 =$ Not equivalence

14. (i), (ii)

16. Domain $(R) = \{1, 2, \ldots, 19, 20\}$, Range $(R) = \{39, 37, 35, \ldots, 5, 3, 1\}$

18. $\{(1, 2), (2, 1), (2, 3), (3, 2), (1, 3), (3, 1), (1, 1), (2, 2), (3, 3), (4, 4), \ldots\}$

1.14 FUNCTIONS

Definition : *Let A and B be two sets, then the rule or correspondence, which associates each element of A to a unique element of B, is called a function from set A to set B.*

If a general element of set A is denoted by x, and of set B is denoted by y, then we say that y is a function of x if, for every $x \in A$, one and only one value of $y \in B$ can be determined.

Symbolically : If f is a function from a set A to a set B, then we write $f : A \to B$, read as f is a function from A to B or f maps A to B.

1.14.1 Range and Domain of a Function

Let an element $y \in B$ be corresponded by an element $x \in A$, then y is called the image of x and is denoted by $f(x)$. Here, x is defined as the pre-image of y.

The set A is called the domain and the set B is called the co-domain of the function f.

The set of all f-images of the element of A, is called image set or the range of f and is denoted by

$$f(A) \quad \text{or} \quad \{f(x) : x \in A\}$$

Evidently, $f(A) \subseteq B$.

Thus, a mapping $f : A \to B$ is the set of ordered pairs $\{(a, b) : a \in A,\ b \in B\}$

So that no two ordered pairs have the same finite element.

$$f = \{(a, b) : a \in A,\ b \in B,\ b = f(x)\ \forall\, a \in A\}$$

For example : Let $A = \{-2, -1, 0, 1, 2\}$ and B is the set of natural numbers for every $x \in A$, $f(x) \in B$ and $f(x) = x^2$.

Here, A is the domain and B is the co-domain.

$f(a)$ is the value of the function $f(x)$, when x takes the value a, *i.e.*, when x is replaced by a.

The elements of the co-domain which is equal to $f(x)$ form the range.

When $x = -2$, $f(-2) = (-2)^2 = 4$.

When $x = -1$, $f(-1) = 1$

When $x = 0$, $f(0) = 0$

When $x = 1$, $f(1) = 1$

When $x = 2$, $f(2) = 4$.

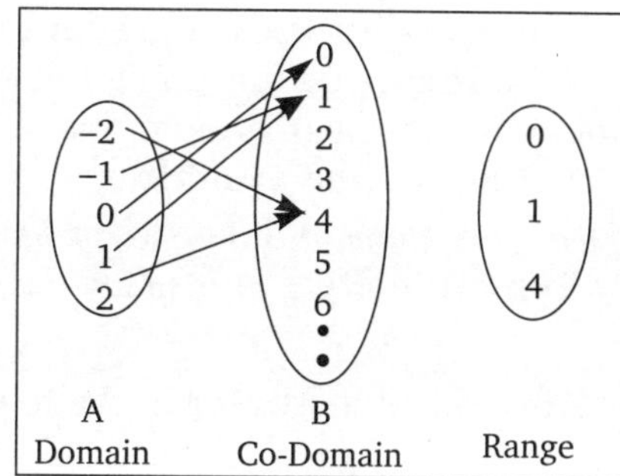

Fig. (15)

- If $f : A \to B$, then a single element in A can not have more than one image in B. However, two or more elements in A may have the same images in B.
- Every element in A must have its image in B, but every element in B may not have its pre-image in A.
- To each element x in A, there exists a unique element y in B such that $y = f(x)$.
- The unique element y of B is called the value of f at x (the image of f under x), and written as $y = f(x)$.
- The range of f consist of those elements in B which appear as the image of at least one element in A.

  In other words, we can say range of a function is the image of its domain.
- Range is a subset of co-domain.

1.15 TYPES OF FUNCTIONS

(a) One-one Function : *A function f from A to B, i.e., $f : A \to B$ is said to be one-one (or injective) iff distinct elements of A have distinct images.*

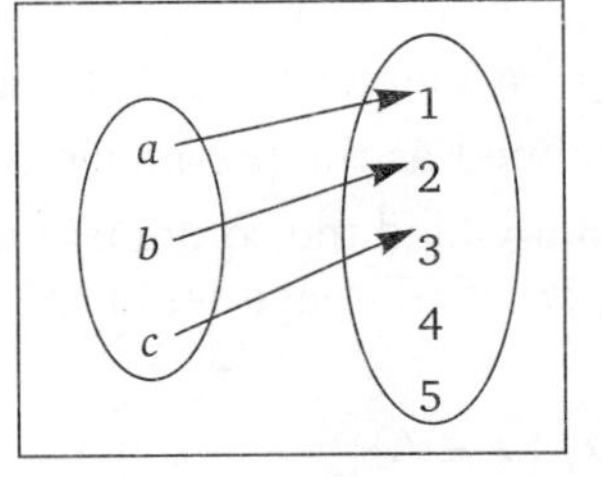

Fig. (16)

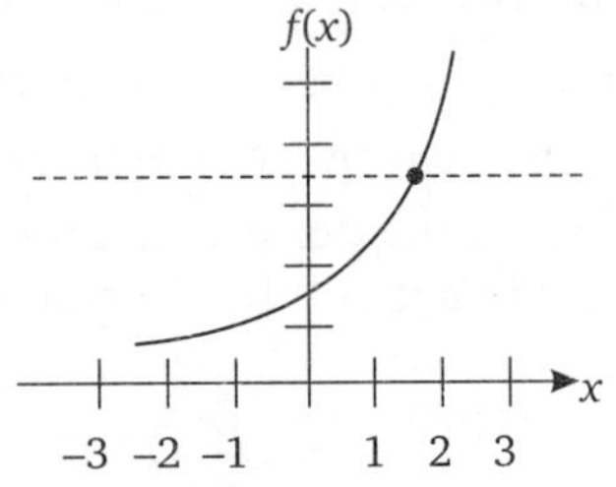

Fig. (17)

Symbolically : f is one-one if for $x_1, x_2 \in A$, we have

$$x_1 \neq x_2 \quad \Rightarrow \quad f(x_1) \neq f(x_2)\ \forall x_1, x_2 \in A$$

or $$f(x_1) = f(x_2) \quad \Rightarrow \quad x_1 = x_2\ \forall x_1, x_2 \in A.$$

It is alo called **Univalent function.**

Graphically, a function is one-one if and only if no line parallel to x-axis meets the graph of the function in more than one point.

(b) Many-one Function : *A function $f : A \to B$ is called many-one, if at least one element of co-domain B has two or more than two pre-images in domain A.*

Symbolically : f is many-one if for $x_1, x_2 \in A$, we have

$$x_1 \neq x_2 \Rightarrow f(x_1) = f(x_2)$$

This can be illustrated in the following figures.

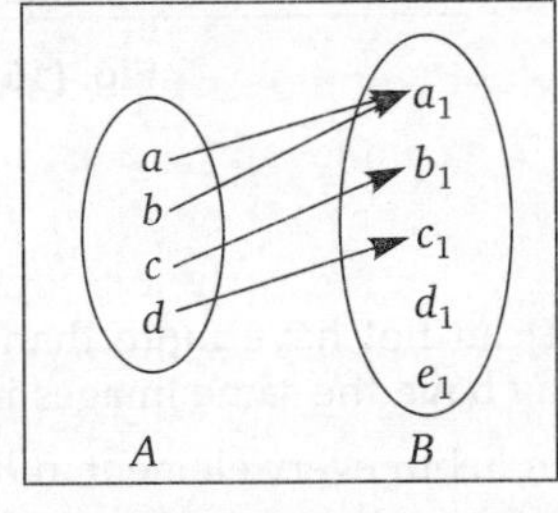

Fig. (18)

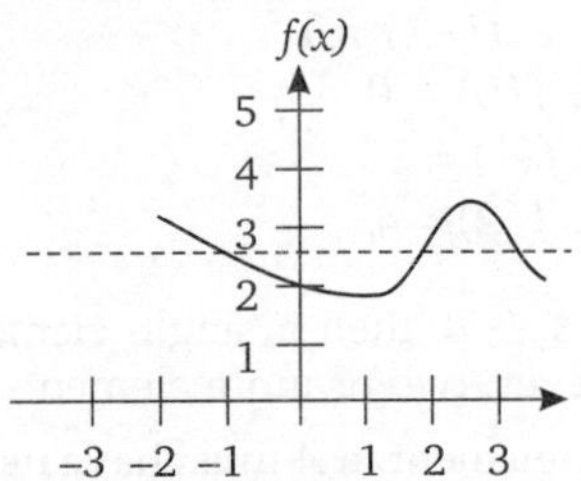

Fig. (19)

Graphically, a function is many-one if and only if a line parallel to x-axis meets the graph of the function in more than one point.

☛ One-many function does not exist.

(c) Onto Function : *A function $f: A \to B$ is called an onto function, if there is no element of B which is not an image of some element of A, i.e.,* every element of B appears as the image of at least one element of A. This is illustrated in Figure 20.

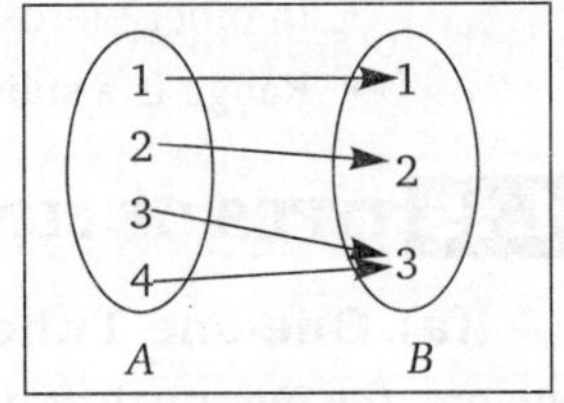

Fig. (20) : Onto Function

☛ In an onto function, Range = Co-domain.

☛ Onto function is also called subjective.

(d) Into Function : *A function $f : A \to B$ is called an into function, if there is at least one element of set B which has no pre-image in the set A.* This is illustrated in Figure 21.

☛ In an into function, Range $\subset$ Co-domain.

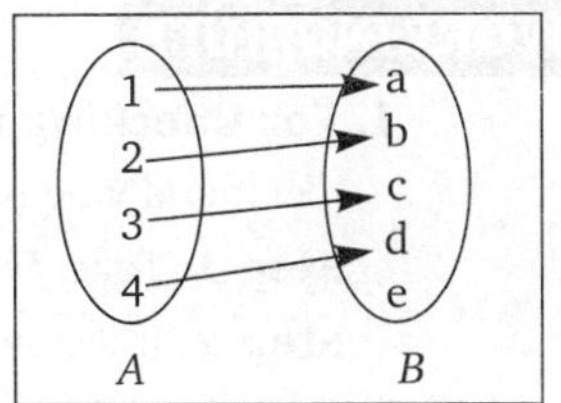

Fig. (21) : Into function

(e) One-one Into Function : *A function $f : A \to B$ is called a one-one into function, if it is both one-one and into function, i.e.,* the different points in A are joined to different points in B and there are some points in B which are not joined to any point in A. This is illustrated in Figure 22.

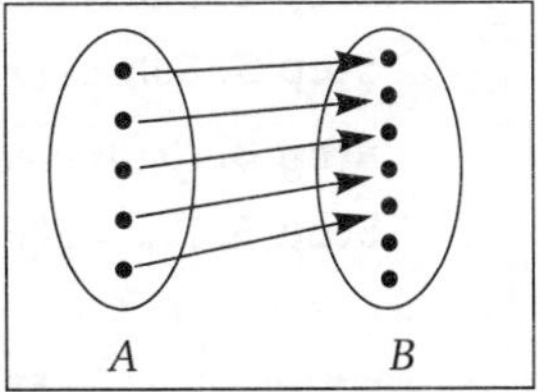

Fig. (22) : One-One Onto Function

Symbolically : One-one into function is defined as

(i) Range $\subset$ Co-domain.

(ii) $f(x_1) \neq f(x_2)$

$\Rightarrow \quad x_1 \neq x_2$.

(f) One-one Onto Function : *A function $f : A \to B$ is both one-one and onto, i.e., the different points in A are joined to different points in B and no point in B is left vacant.* This is illustrated in Figure 23.

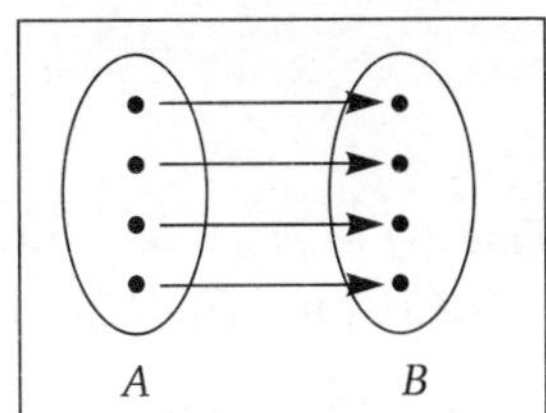

Fig. (23) : One-One Into Function

☛ One-one onto mapping is also known as bijective or one-to-one.

☛ For a one-one onto function

Range = Co-domain, and $x_1 \neq x_2 \Rightarrow f(x_1) \neq f(x_2)$

or $f(x_1) = f(x_2) \Rightarrow x_1 = x_2$.

(g) Many-One Into Function : *A function $f : A \to B$ which is both many-one and into function is called a many-one into function, i.e.,* two or more points in A are joined to some points in B and there are some points in B which are not joined to any point in A. Therefore, for many-one into function.

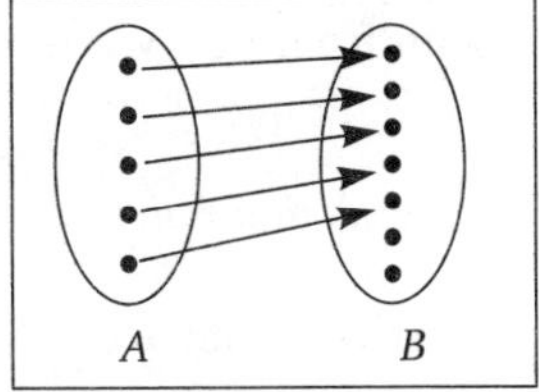

Fig. (24) : Many-One Into Function

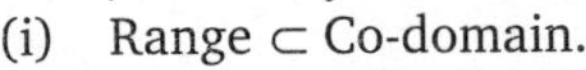

(i) Range $\subset$ Co-domain.

(ii) $x_1 \neq x_2 \Rightarrow f(x_1) = f(x_2)$.

(h) Many-One onto Function : *If function $f : A \to B$ is both many-one and onto function is called a many-one onto function, i.e.,* in B one point is joined to at least one point in A and two or more points in A are joined to some points in B. Therefore, for many-one onto function

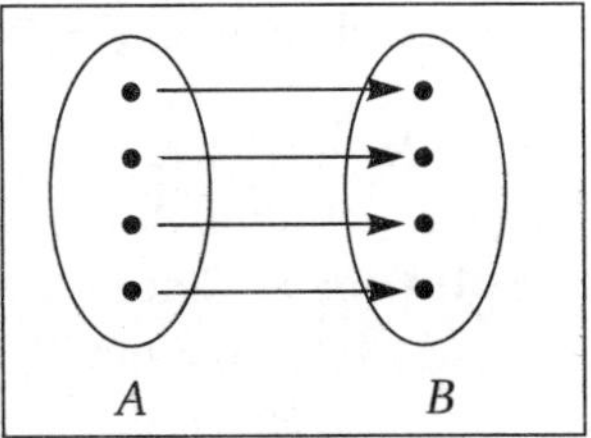

Fig. (25) : Many-One Onto Function

(i) Range = Co-domain.

(ii) $x_1 \neq x_2 \quad \Rightarrow \quad f(x_1) = f(x_2)$.

WORKING RULE

1. For Checking the Injectivity (One-one) of the Function

Let x and y be two arbitrary elements in the domain of f.

Step 1. Take $f(x) = f(y)$.

Step 2. If we get $x = y$, after solving $f(x) = f(y)$. Then, $f : A \to B$ is one-one.

2. For checking the surjectivity (Onto) of a Function.

Step1. Take an arbitrary element y in the co-domain.

Step 2. Put $f(x) = y$.

Step 3. Solve $f(x) = y$ for x and obtain x in terms of y.

Step 4. Get the equation of the form $x = g(y)$.

Step 5. If $x = g(y)$ belongs to domain f, for all values of y, then f is onto.

ILLUSTRATIVE EXAMPLES

1. *Let $f : \mathbf{R} \to \mathbf{R}$ be a function defined by*

$$f(x) = \begin{cases} 3x - 1 & \text{when } x > 3 \\ x^2 - 2 & \text{when } -2 \le x \le 3 \\ 2x + 3 & \text{when } x < -2 \end{cases}$$

Find (i) $f(2)$, (ii) $f(4)$, (iii) $f(-1)$, (iv) $f(-3)$

Solution : (i) $f(2) = (2)^2 - 2 = 4 - 2 = 2$.

(ii) $f(4) = 3(4) - 1 = 12 - 1 = 11$ (iii) $f(-1) = (-1)^2 - 2 = 1 - 2 = -1$

(iv) $f(-3) = 2(-3) + 3 = -6 + 3 = -3$

2. *For the function $y = +\sqrt{x}$, say whether it is a function or not. If it is a function, find its domain and range.*

Solution : Here we have $y = +\sqrt{x}$ ...(1)

Since y is real if $x \ge 0$ and is unique and finite for each $x \ge 0$.

Then (1) is a function with domain $[0, \infty[$.

Again from (1), $y \ge 0 \;\forall\; x \ge 0$.

Hence, Range $= [0, \infty[$.

3. *Find the domain of* $f(x) = \dfrac{x^3 - x^2 + 4x + 7}{3x + 11}$

Solution : Since f is defined for all real values of x except when $3x + 11 = 0$.

i.e., when, $x = \dfrac{-11}{3}$

Hence, domain of $f = \mathbf{R} - \left\{\dfrac{-11}{3}\right\}$

4. *Let $f : \mathbf{N} - \{1\} \to \mathbf{N}$ be defined by $f(n) =$ the highest prime factor of n. Show that f is neither one-one nor onto. Also, find the range f.*

Solution : Since we have

$f(6) =$ the highest prime factor of $6 = 3$

$f(9) =$ the highest prime factor of $9 = 3$

$f(12) =$ the highest prime factor of $12 = 3$

Therefore, f is a many-one function.

Clearly, image of any $n \in \mathbf{N} - \{1\}$ is the largest prime number that divides n. So the range of f consists of prime number only. Consequently, range of $f \neq \mathbf{N}$ (Co-domain)

$\Rightarrow$ f is not onto function.

Hence, f is neither one-one nor onto. The range of f is the set of all prime numbers.

5. *Let $A = \{1, 2\}$. Find all one-to-one function from A to A.*

Solution : Let $f : A \to A$ be a one-one function.

Then, for $f(1)$, there are two choices, *i.e.,* 1 or 2.

Let us first suppose $f(1) = 1$.

As $f : A \to A$ is one-one, $f(2) = 2$.

Therefore, we have $f(1) = 1$, $f(2) = 2$.

Now, let $f(1) = 2$

Since, $f : A \to A$ is one-one, therefore $f(2) = 1$.

Therefore, we have $f(1) = 2$ and $f(2) = 1$.

Hence, we have two one-one function say f and g from A to A given by $f(1) = 1$, $f(2) = 2$ and $f(2) = 1$ and $f(1) = 2$.

6. *Let $A = \{x \in \mathbf{R} : -1 \leq x \leq 1\} = B$. Show that $f : A \to B$ given by $f(x) = x|x|$ is one-one and onto.*

Solution : Let x, y be any two elements in A, then

$$x \neq y \Rightarrow x|x| \neq y|y| \Rightarrow f(x) \neq f(y).$$

Therefore, f is one-one.

Since, range of $f = f(A) = B$, so $f : A \to B$ is onto mapping. Hence, f is one-one and onto.

7. *Find the domain and range of the function.*

$$f(x) = -\sqrt{-5 - 6x - x^2}$$

Solution : Given that, $f(x) = -\sqrt{-5 - 6x - x^2}$

For f to be real, $-5 - 6x - x^2 \geq 0 \quad \Rightarrow \quad x^2 + 6x + 5 \leq 0$

$\Rightarrow \quad x^2 + 6x \leq -5 \quad \Rightarrow \quad x^2 + 6x + 9 \leq -5 + 9 \quad \Rightarrow \quad (x+3)^2 \leq 4$

$\Rightarrow \quad |x+3|^2 \leq 4 \quad \Rightarrow \quad |x+3| \leq 2 \quad \Rightarrow \quad -2 \leq x + 3 \leq 2$

$\Rightarrow \quad -2 - 3 \leq x \leq 2 - 3 \quad \Rightarrow \quad -5 \leq x \leq -1$

Therefore, domain of $f(x) = [-5, -1]$

To find the range of $f(x)$, put $y = f(x)$.

Therefore, $y = -\sqrt{-5 - 6x - x^2}$, $y \leq 0$.

Therefore, $y^2 = -5 - 6x - x^2 \quad \Rightarrow \quad x^2 + 6x + (y^2 + 5) = 0$

For real x, discriminant ≥ 0. $\quad$ *i.e.,* $(6)^2 - 4 \times 1 \times (y^2 + 5) \geq 0$

$\Rightarrow \quad 36 - 4y^2 - 20 \geq 0 \quad \Rightarrow \quad -4y^2 \geq -16$

$\Rightarrow \quad y^2 \geq 4 \quad \Rightarrow \quad |y|^2 \leq 4 \quad \Rightarrow \quad |y| \leq 2 \quad$ *i.e.,* $-2 \leq y \leq 2$

But $y \leq 0$, therefore, $-2 \leq y \leq 0$.

Hence, Range of $f = [-2, 0]$.

8. *For a finite set A, if $f : A \to A$ is a one-one function, show that f is onto.*

Solution : Let $A = \{a_1, a_2, \ldots, a_n\}$ be a finite set.

Since $f : A \to A$ is one-one function, therefore $f(a_1), f(a_2), \ldots, f(a_n)$ are distinct elements of the set A, but A has only n elements. Therefore,

$$A = \{f(a_1), f(a_2), \ldots, f(a_n)\}$$

$\Rightarrow$ Co-domain = Range

Hence, every element in A (co-domain) has its pre-image in the domain A.

$\Rightarrow$ $f : A \to A$ is onto.

☛ For a finite set A, if $f : A \to A$ is onto function, then f is one-one.

9. *If $f : \mathbf{R} \to \mathbf{R}$ be a function defined by $f(x) = 4x^3 - 7$, show that the function f is bijective.*

Solution : Given that $f(x) = 4x^3 - 7$; $x \in \mathbf{R}$

***f* is one-one :** Let $x_1, x_2 \in \mathbf{R}$

Now, $f(x_1) = f(x_2) \quad \Rightarrow \quad 4x_1^3 - 7 = 4x_2^3 - 7 \quad \Rightarrow \quad 4x_1^3 = 4x_2^3$

$\Rightarrow \quad x_1^3 = x_2^3 \quad \Rightarrow \quad x_1^3 - x_2^3 = 0 \quad \Rightarrow \quad (x_1 - x_2)(x_1^2 + x_1x_2 + x_2^2) = 0$

$$\Rightarrow \quad (x_1 - x_2)\left[\left(x_1 + \frac{x_2}{2}\right)^2 + \frac{3x_2^2}{4}\right] \qquad \left\{\because \left(x_1 + \frac{x_2}{2}\right)^2 + \frac{3x_2^2}{4} \neq 0\right\}$$

$\Rightarrow \quad (x_1 - x_2) = 0 \quad \Rightarrow \quad x_1 = x_2.$

Therefore, f is one-one.

***f* is onto :** Let $c \in \mathbf{R}$

$$f(x) = c \quad \Rightarrow \quad 4x^3 - 7 = c \quad \Rightarrow \quad x = \left(\frac{c+7}{4}\right)^{1/3}$$

$$\text{Now, } \left(\frac{c+7}{4}\right)^{1/3} \in \mathbf{R} \text{ and } f\left[\left(\frac{c+7}{4}\right)^{1/3}\right] = 4\left[\left(\frac{c+7}{4}\right)^{1/3}\right]^3 - 7 = c + 7 - 7 = c$$

which implies that c is the image of $\left(\frac{c+7}{4}\right)^{1/3}$.

Therefore, f is onto. Hence, f is bijective function.

10. *Let A and B be two sets. Prove that $f : A \times B \to B \times A$ defined by $f(a, b) = (b, a)$ is one-one and onto.*

Solution : *f* is one-one : Let (a_1, b_1) and $(a_2, b_2) \in A \times B$ such that

$$f(a_1, b_1) = f(a_1, b_2)$$

$\Rightarrow \quad (b_1, a_1) = (b_2, a_2) \quad \Rightarrow \quad b_1 = b_2$ and $a_1 = a_2$

Therefore, $(a_1, b_1) = (a_2, b_2)$

Thus, $f(a_1, b_1) = f(a_2, b_2) \quad \Rightarrow \quad (a_1, b_1) = (a_2, b_2) \ \forall \ (a_1, b_1), (a_2, b_2) \in A \times B$

$\Rightarrow \quad f$ is one-one.

***f* is onto :** Let $(b, a) \in B \times A$, such that $b \in B$ and $a \in A$.

$\Rightarrow \quad (a, b) \in A \times B$

Therefore, for all $(b, a) \in B \times A$, there exist $(a, b) \in A \times B$ such that $f(a, b) = (b, a)$

$\Rightarrow \quad f$ is onto. Hence, f is one-one and onto.

Exercise 1.5

1. Let $A = \{-2, -1, 0, 1, 2\}$ and $f : A \to \mathbf{Z}$ given by $f(x) = x^2 - 2x - 3$. Find
   (i) the range of f,
   (ii) pre-image of 6, –3 and 5.
2. Find the domain and range of the following function $f(x) = \sqrt{(x-1)(3-x)}$.
3. Find the range of the following function $f(x) = \dfrac{1}{(2x-3)(x+1)}$.
4. Find the domain and range of the following functions :
   (i) $f(x) = \dfrac{x^2 - 1}{x - 1}$ (ii) $y = -|x|$
   (iii) $f(x) = \dfrac{|x-1|}{x-1}$ (iv) $y = \sqrt{x-3}$
5. If $A = \{-1, 0, 2, 5, 6, 11\}$, $B = \{-2, -1, 0, 18, 28, 108\}$ and $f(x) = x^2 - x - 2$, find $f(A)$.
6. Let A be the set of two positive integers. Let $f : A \to \mathbf{Z}^+$, set of positive integers be defined by $f(n) = p$, where p is the highest prime factor of n. If range of $f = \{3\}$, find A.

7. Find the domain for which the function $f(x) = 2x^2 - 1$ and $g(x) = 1 - 3x$ are equal.

8. Let $f : \mathbf{R} \to \mathbf{R}$ and $f_2 : \mathbf{C} \to \mathbf{C}$ be two functions defined as $f_1(x) = x^3$ and $f_2(x) = x^3$. Show that they are not equal.

9. Let $A = \{p, q, r, s\}$ and $B = \{1, 2, 3\}$. Which of the following relations from A to B not a function?
   (i) $R_1 = [(p, 1), (q, 2), (r, 1), (s, 2)\}$
   (ii) $R_2 = \{(p, 1), (q, 1), (r, 1), (s, 1)\}$
   (iii) $R_3 = \{(p, 1), (q, 2), (p, 2), (s, 3)\}$
   (iv) $R_4 = \{(p, 2), (q, 3), (r, 2), (s, 2)\}$

10. Write the following relations as sets of ordered pairs and find which of them are functions :
   (i) $\{(x, y) : y = 3x, x \in \{1, 2, 3\}, y \in \{3, 6, 9, 12\}\}$
   (ii) $\{(x, y) : y > x + 1, x = 1, 2$ and $y = 2, 4, 6\}$
   (iii) $\{(x, y) : x + y = 3, x, y \in \{0, 1, 2, 3\}\}$

11. Express the following functions as sets of ordered pairs, and find their range
   (i) $f_1 : A \to \mathbf{R} : f_1(x) = x^2 + 1$ where $A = \{-1, 0, 2, 4\}$
   (ii) $f_2 : A \to \mathbf{N} : f_2(x) = 2x$, where $A = \{x : x \in \mathbf{N}, x \le 10\}$

12. Let $f : \mathbf{R} \to \mathbf{R}$ be a function such that $f(x) = 2^x$. Determine
   (i) Range of f
   (ii) $\{x : f(x) = 1\}$
   (iii) Whether $f(x + y) = f(x).\, f(y)$ holds

13. Let $f : \mathbf{R}^+ \to \mathbf{R}$, be a function such that $f(x) = \log x$. Determine
   (i) the image set of domain of f (ii) $\{x : f(x) = -2\}$
   (iii) Whether $f(x\, y) = f(x) + f(y)$ holds

14. Give an example of a map which is
   (i) one-to-one but not onto
   (ii) not one to one, but onto
   (iii) neither one-to-one nor onto.

ANSWERS

1.(i) $f(A) = (-4, -3, 0, 5)$, (ii) $\phi, \{1, 2\}, -2$ 2. Domain $= [1, 3]$, Range $= [-1, 1]$

3. $]-\infty, \frac{-8}{25}] \cup [0, \infty[$

4.(i) $\mathbf{R} - \{1\}, R - \{2\}$,   (ii) $\mathbf{R} : \mathbf{R} - \mathbf{R}^+$,   (iii) $\mathbf{R} - \{1\}, \{-1, 1\}$,
   (iv) $[3, \infty[, [0, \infty]$

5. $f(A) = \{1, -2, 18, 28, 108\}$ 6. $A = \{3, 6\}$ or $(3, 9)$ or $\{3, 12\}$ etc.

7. $\{-2, 1/2\}$ 9. (iii)

10.(i) $\{(1, 3), (2, 6), (3, 9)\}$, function,   (ii) $\{(1, 4), (1, 6), (3, 4), (3, 6)\}$, not function
   (iii) $\{(0, 3), (1, 2), (2, 1), (3, 0)\}$, function

11.(i) $f_1 = (x, f(x) : x \in A) = \{(-1, 2), (0, 1), (2, 5), (4, 17)\}$
   (ii) $f_2 = \{(x, g(x)) : c \in A) = \{(1, 2), (2, 4), (3, 6), \ldots, (10, 20)\}$

12.(i) Range of $f = \mathbf{R}^+$, the set of positive real numbers, (ii) $(x : f(x) = 1) = \{0\}$,
   (iii) $f(x + y) = f(x).\, f(y)$ holds for all $x, y \in \mathbf{R}$

14.(i) $n \to n^2 : \mathbf{N} \to \mathbf{N}$ (ii) $n \to |n| : \mathbf{Z} \to \mathbf{N} \cup \{0\}$
   (iii) $n \to |n|^2 : \mathbf{Z} \to \mathbf{N} \cup \{0\}$

1.16 SOME PARTICULAR FUNCTIONS

(a) Even Function : *A function $f : A \to B$ is said to be an even function if $f(-x) = f(x)$ for all $x \in A$.*

For Example :

$$f(x) = x^2 + 1 \quad \Rightarrow \quad f(-x) = x^2 + 1$$

Therefore, $f(x) = f(-x)$.

Here, $f(1) = 1^2 + 1 = 2$, $f(-1) = (-1)^2 + 1 = 2$

The graph of an even function is symmetric about the function axis.

In figure 26, for the function $y = f(x)$, if we take any positive value of x, *e.g.*, $x = a$, $y = f(a)$. For the same value of x, but in the negative axis, *i.e.*, $x = -a$, we get $y = f(-a)$. We can observe that, $f(a) = f(-a)$, *i.e.*, for both positive and negative values of x, we get a positive value of y. Thus, $y = f(x)$ is an even function.

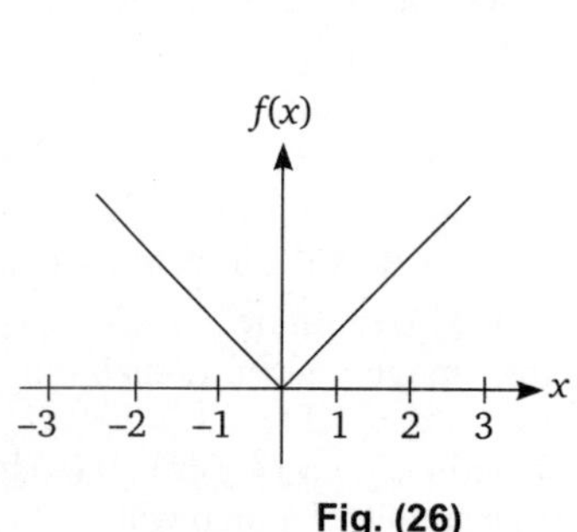

Fig. (26)

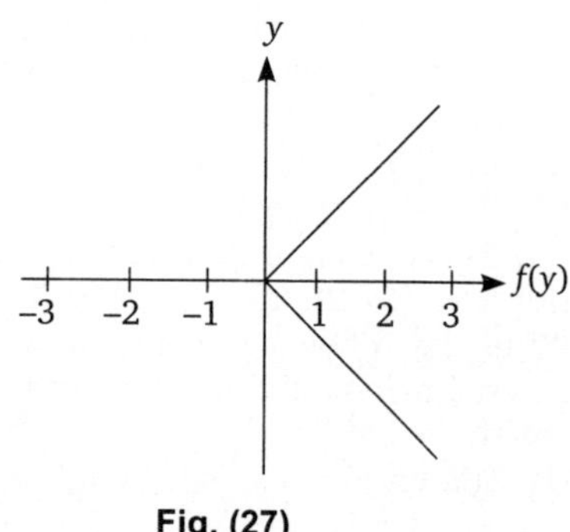

Fig. (27)

Since argument holds true if we get a negative value of y for both negative and positive values of x, *i.e.*, $y = -|f(x)|$.

Similarly, in Figure 18, x is a function of y and the same argument holds for $x = f(y)$.

(b) Odd Function : *A function $f : A \to B$ is said to be an odd function if $f(-x) = f(x)$ for all $x \in A$.*

For example : $f(x) = x^3$

$\Rightarrow \quad f(-x) = (-x)^3 = -x^3 = -f(x)$.

The graph of an odd function is a double reflection, first in the function axis and then in the other axis. For the function $y = f(x)$, if we take any positive value of x, *e.g.*, $x = a$, $y = f(a)$. For the same value of x, but in the negative axis, *i.e.*, $x = -a$, we get $y = f(-a)$. In the graph, we observe that $f(-a) = -f(a)$.

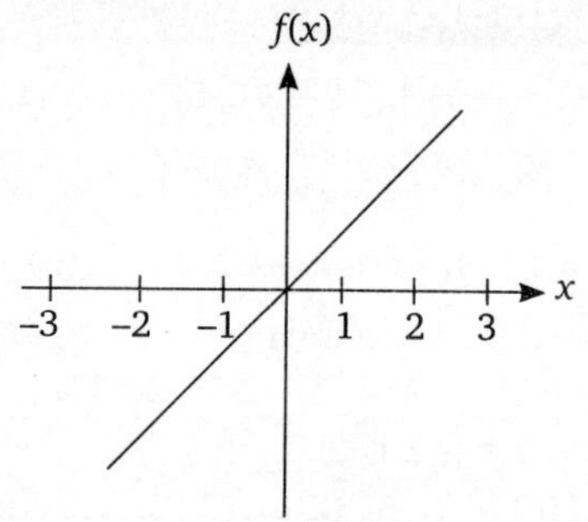

Fig. (28)

☛ There are functions, which are neither even nor odd, *e.g.*, $f(x) = x^3 + x^2 + x + 5$.

(c) Inverse Function : *Let $f : A \to B$ be a one-one onto function, then a function $g : B \to A$ which associates each element $y \in B$ to a unique element $x \in A$ such that $f(x) = y$ is called the inverse of f, i.e., $f(x) = y \Leftrightarrow g(y) = x$*

It is denoted by f^{-1}

$$f^{-1}(y) = x \Leftrightarrow f(x) = y$$

Fig. (29)

For Example : $f(x) = 2x$, where $x \in A$

$$f^{-1}(x) = x/2, \text{ where } x \in B$$

☛ The inverse function $f^{-1} : B \to A$ is defined only when f is one-one and onto. Further, if inverse function f^{-1} exists, then it is also one-one onto.

☛ The inverse relation of function is symmetric.

(d) Constant Function : *A function $f : A \to B$ is said to be constant function if each element of domain is associated with a single element of the co-domain B.* This can be illustrated in Figure 21. Graphically, a constant function will be a straight line parallel to either of two axes.

For Example : $f(x) = 3$, *i.e.*, $y = 3$.

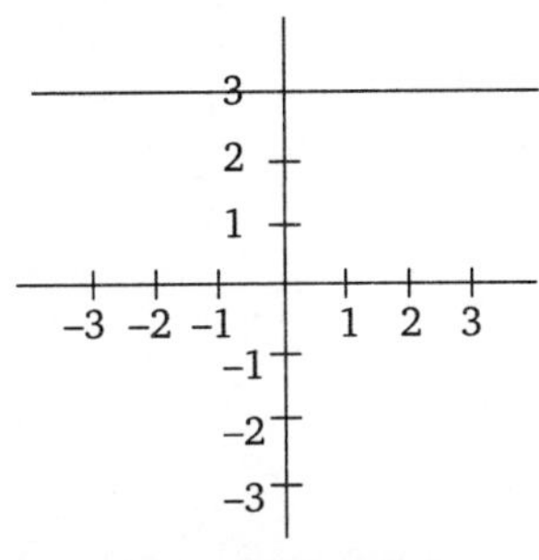

Fig. (30)

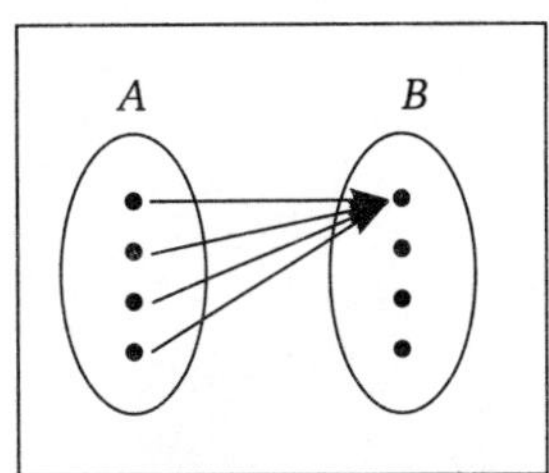

Fig. (31)

(e) Step Function : *A function $f : A \to B$ is said to be step function if $f(x) = [x]$, where $[x]$ denotes the greatest integer less than or equal to x for all $x \in A$.*

For Example : If $f(x) = [x]$, $f([2.46]) = 2$, $f([3]) = 3$; $f([-2.29]) = -3$.

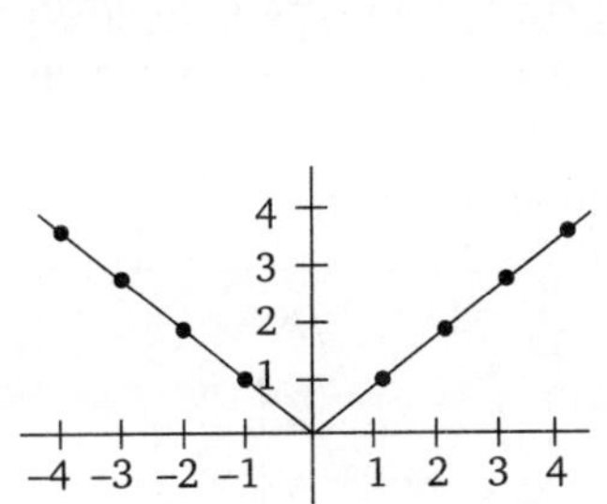

Fig. (32) : Step function

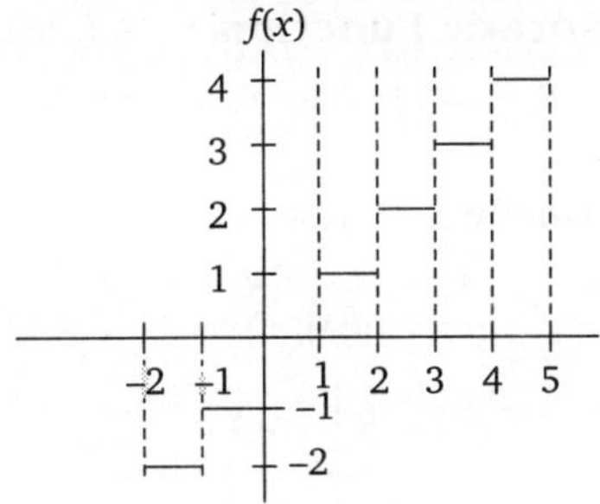

Fig. (33) : Modulus function

Graphically, it can be represented as follows

$f(x) = 0$, when $0 \le x < 1$
$f(x) = 1$, when $1 \le x < 2$
$f(x) = -1$, when $-1 \le x < 0$

☛ Step function is also known as integer function.

(f) Modulus Function : *A function $f : A \to B$ is said to be modulus function if $f(x) = |x|$, i.e., $f(x)$ takes only the magnitude of x.*

Symbolically : $f(x) = \begin{cases} -x & \text{if } x < 0 \\ x & \text{if } x \ge 0 \end{cases}$

The graph of the modulus function, $f(x) = |x|$ is shown in Figure 33.

$$f(1) = f(-1) = 1;\ f(2) = f(-2) = 2;\ f(3) = f(-3) = 3$$

☛ Modulus function is always an even function.

(g) Linear Function : *A function $f : A \to B$ of the form $f(x) = ax + b$, where $a, b \in \mathbf{R}$ is called a linear function.*

The graph of a linear function is always a straight line.

For Example : Graph of $f(x) = 4x - 2$ is shown in figure 34.

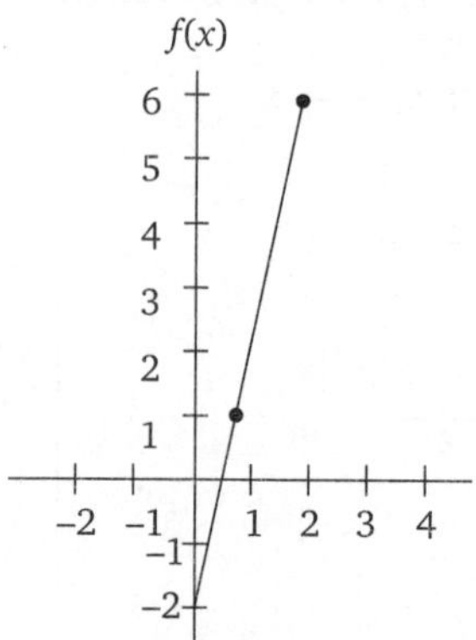

Fig. (34) :
Linear Function

x	1	0	2
$f(x)$	2	– 2	6

(h) Quadratic Function : *A function $f : A \to B$ is called a quadratic function if it is of the form $y = ax^2 + bx + c$, where $a, b, c \in \mathbf{R}$, $a \neq 0$.* The graph of such a function is called parabola.

For Example : Graph of $f(x) = 2x^2$ is shown in Figure 35.

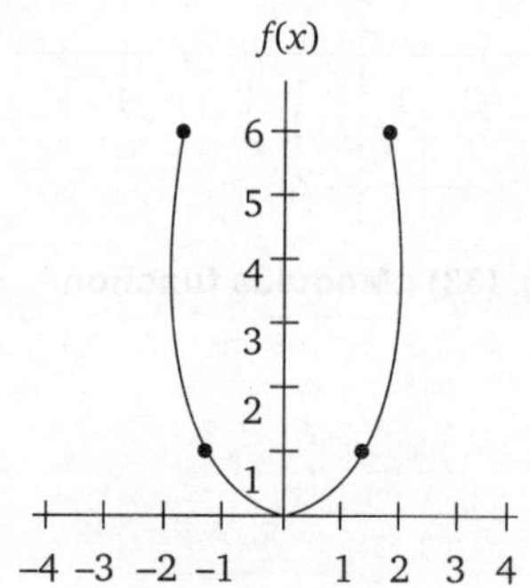

Fig. (35) :
Quadratic Function

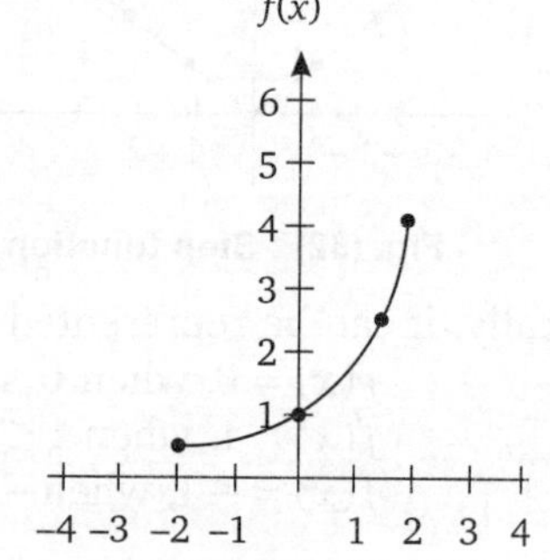

Fig. (36) :
Exponential Function

x	– 2	1	– 1	2
$f(x)$	8	2	2	8

(i) Exponential Function : A function $f : A \to B$ of the form $f(x) = a^x$, where $a > 0$ and $x \in \mathbf{R}$ is called an exponential function.

The exponential function a^x will never be negative for all $a > 0$ and $x \in \mathbf{R}$.

Therefore, domain of $f(x)$ is $\mathbf{R}$ and range of $f(x)$ is the set of positive real numbers.

For Example : Graph of the exponential function $f(x) = 2^x$ is shown in Figure 36.

x	-2	-1	0	1	2	∞	$-\infty$
$f(x)$	$1/4$	$1/2$	1	2	4	∞	0

(j) Logarithmic Function : *For any $a > 0$, $a \neq 1$, a function $f(x)$ defined by*

$$f(x) = \log_a x \ \forall x > 0$$

is called logarithmic function.

By definition of logarithms, we have

$$a^y = x \iff \log_a x = y$$

Here, it is clear that $x > 0$ for all $y \in \mathbf{R}$ and $a > 0$, $a \neq 1$.

So, $f(x)$ is defined for all $x > 0$. Thus, domain of $f(x)$ is the set of positive real numbers.

The graph of the logarithmic function is shown in Figure 37.

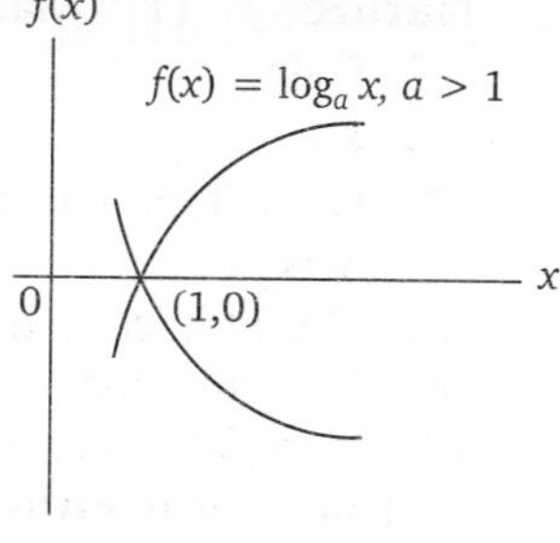

Fig. (37)

(k) Trigonometric functions

(i) *Sine Function*

Definition : $y = \sin x$

Domain : $x \in \mathbf{R}$

Range : $y \in [-1, 1]$

Nature : (1) Many to one, one to one in $x \in \left[-\frac{\pi}{2}, \frac{\pi}{2}\right]$

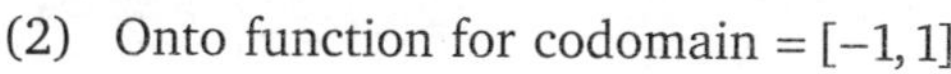

(2) Onto function for codomain $= [-1, 1]$

(3) Monotonically increasing for all

$$x \in \left(2n\pi - \frac{\pi}{2}, 2n\pi + \frac{\pi}{2}\right)$$

(4) Monotonically decreasing for all $x \in \left(2n\pi + \frac{\pi}{2}, 2n\pi + \frac{3\pi}{2}\right)$

(5) Odd function

(6) Periodic with period 2π.

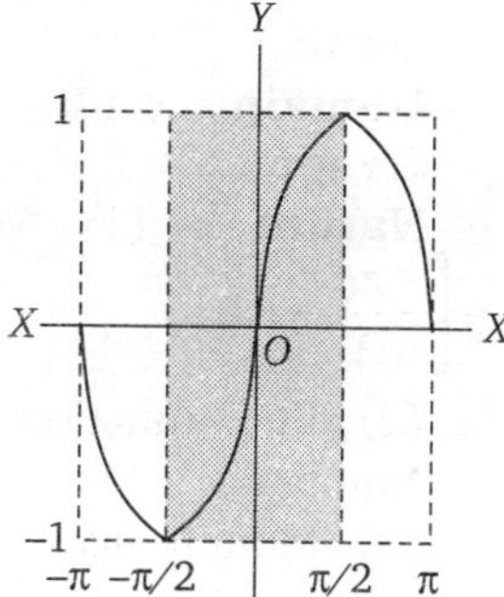

Fig. (38)

(ii) *Cosine Function*

Definition : $y = \cos x$

Domain : $x \in \mathbf{R}$

Range : $y \in [-1, 1]$

Nature : (1) Many to one, one to one in $x \in [0, \pi]$

(2) Onto function for codomain $= [-1, 1]$

(3) Monotonically increasing for all

$$x \in (2n\pi - \pi, 2n\pi)$$

(4) Monotonically decreasing for all

$$x \in (2n\pi, 2n\pi + \pi)$$

(5) Even function

(6) Periodic with period 2π.

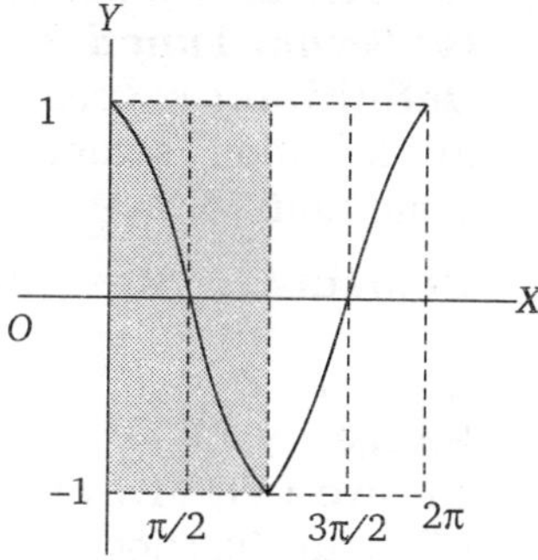

Fig. (39)

(iii) *Tangent Function*

Definition : $y = \tan x$

By definition, for any right angled triangle, tan x represents the ratio of perpendicular to that of its base. Geometrically, tan x means the slope of the curve at any point (x, y) on the curve $y = f(x)$.

Domain : $x \in \mathbf{R} - \left\{(2n+1)\dfrac{\pi}{2}\right\}$, where $n \in \mathbf{Z}$

Range : $y \in \mathbf{R}$

Nature : (1) Many to one, one to one in $x \in \left[-\dfrac{\pi}{2}, \dfrac{\pi}{2}\right]$

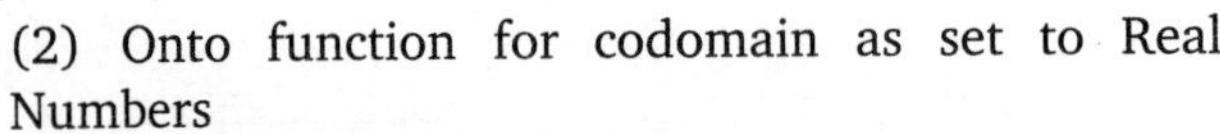

(2) Onto function for codomain as set to Real Numbers

(3) Monotonically increasing for all $x \in \mathbf{R}$

(4) Odd function

(5) Periodic with period π.

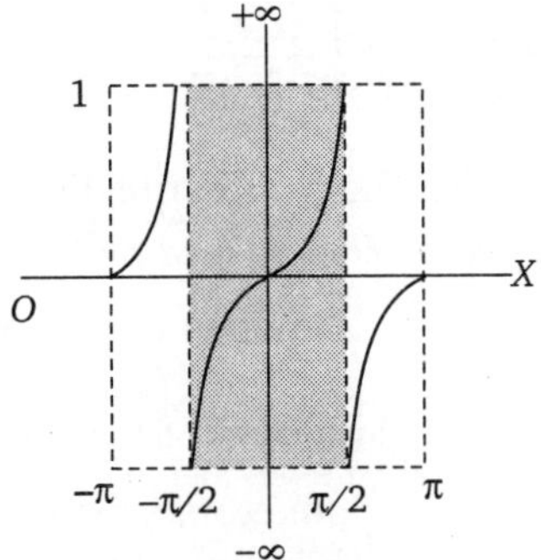

Fig. (40)

(iv) *Cosecant Function*

Definition : $y = \csc x$ or $y = \text{cosec } x$

By definition, cosecant function is the reciprocal of sine function.

Domain : $x \in \mathbf{R} - \{n\pi\}$, where $n \in I$

Range : $y \in (-\infty, -1] \cup [1, \infty)$

Nature : (1) Many to one, one to one in $x \in \left[-\dfrac{\pi}{2}, \dfrac{\pi}{2}\right]$

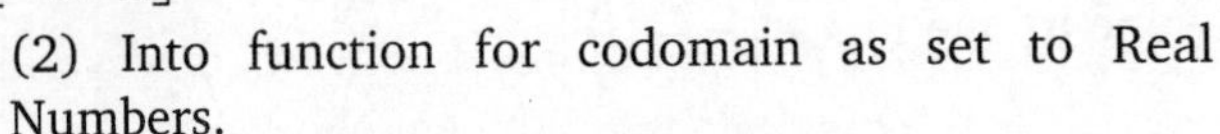

(2) Into function for codomain as set to Real Numbers.

(3) Monotonically decreasing for all

$$x \in \left(2n\pi - \frac{\pi}{2}, 2n\pi + \frac{\pi}{2}\right)$$

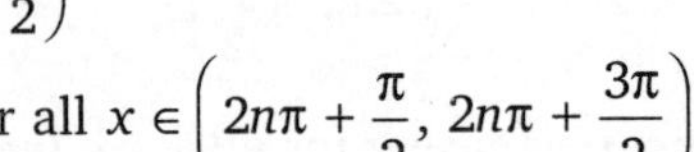

(4) Monotonically increasing for all $x \in \left(2n\pi + \dfrac{\pi}{2}, 2n\pi + \dfrac{3\pi}{2}\right)$

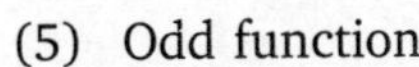

(5) Odd function

(6) Periodic with period 2π.

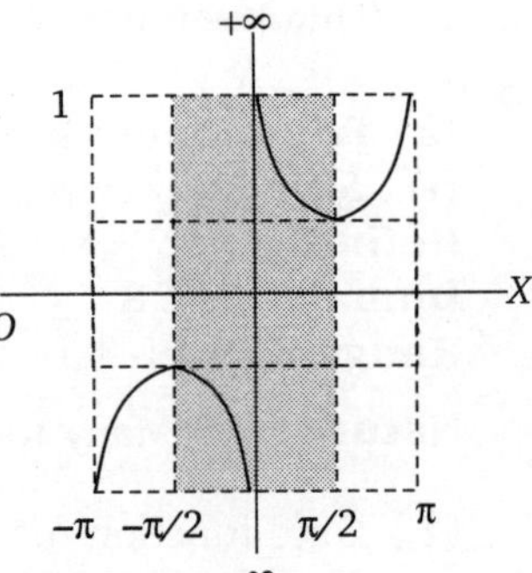

Fig. (41)

(v) *Secant Function*

Definition : $y = \sec x$

By definition, secant function is the reciprocal of cosine function.

Domain : $x \in \mathbf{R} - \left\{(2n+1)\dfrac{\pi}{2}\right\}$, where $n \in \mathbf{Z}$

Range : $y \in (-\infty, -1] \cup [1, \infty)$

Nature : (1) Many to one, one to one in $x \in [0, \pi]$

(2) Into function for codomain as set to Real numbers

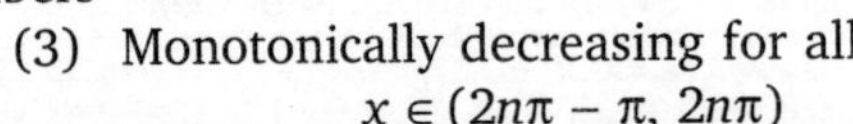

(3) Monotonically decreasing for all

$$x \in (2n\pi - \pi, 2n\pi)$$

(4) Monotonically increasing for all

$$x \in (2n\pi, 2n\pi + \pi)$$

(5) Even function

(6) Periodic with period 2π.

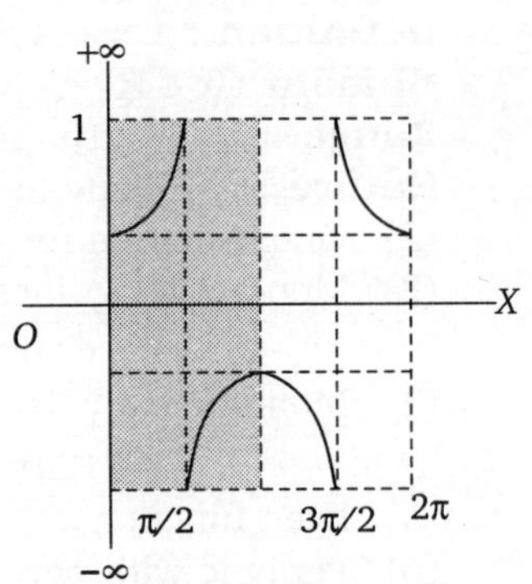

Fig. (42)

(vi) *Cotangent Function*

Definition : $y = \cot x$

By definition, cotangent function is the reciprocal of tangent function.

Domain : $x \in \mathbf{R} - \{n\pi\}$, where $n \in \mathbf{Z}$

Range : $y \in \mathbf{R}$

Nature : (1) Many to one, one to one in $x \in (0, \pi)$

(2) Onto function for codomain as set to Real Numbers

(3) Monotonically decreasing for all $x \in \mathbf{R}$

(4) Odd function.

(5) Periodic with period π.

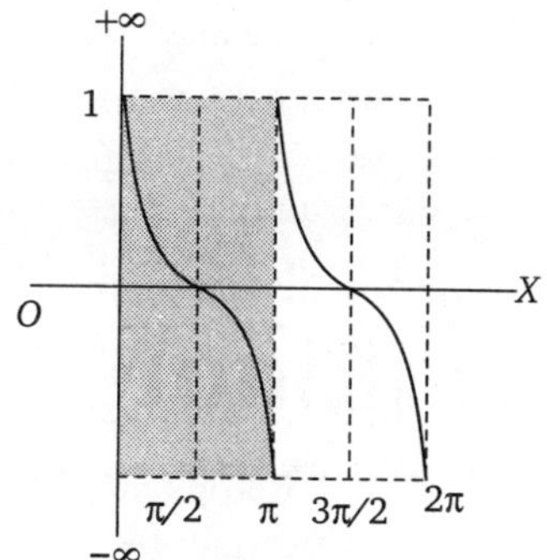

Fig. (43)

1.17 ALGEBRA OF FUNCTIONS

The algebraic operations of addition, subtraction, multiplication and division, yield new functions.

Definitions : Let f and g be two real valued functions with domain D_1 and D_2 respectively.

Let $D = D_1 \cap D_2 \neq \phi$, then

(i) The sum function, denoted by $f + g$, is defined by $(f + g)(x) = f(x) + g(x)$, with domain D.

(ii) The difference function, denoted by $f - g$, is defined by $(f - g)(x) = f(x) - g(x)$, with domain D.

(iii) The product function, denoted by $f.g.$, is defined by $(f.g)(x) = f(x).g(x)$, with domain D.

(iv) The quotient function, denoted by $\frac{f}{g}$, is defined by $\left(\frac{f}{g}\right)(x) = \frac{f(x)}{g(x)}$, with domain D', where $D' = \{x : x \in D, g(x) \neq 0\} \neq \phi$

(v) The reciprocal function denoted by $\frac{1}{f}$ is defined by $\left(\frac{1}{f}\right)(x) = \frac{1}{f(x)}$, with domain D'', where $D'' = \{x : x \in D_1, f(x) \neq 0\}$

(vi) If c is any real number, then scalar multiple of f by c, denoted by cf, is a function $cf(x)$ defined as $(cf)(x) = cf(x)$, with domain D_1.

(vii) If f is a function, then $f.f$ is denoted by f^2, $f^2.f$ is denoted by f^3 and so on. Also $(f)^n(x) = \{f(x)\}^n$, with domain D_1.

1.17.1 Composition of functions

Let $f : A \rightarrow B$ and $g : B \rightarrow C$ be two real valued functions. Then the composite of f and g denoted by $g \circ f$ such that $g \circ f : A \rightarrow C$ defined by $(g \circ f)(x) = g(f(x))$

This is also known as function or resultant of a function.

Similarly, $(f \circ g)(x) = f(g(x))$.

- If $f : X \rightarrow Y$ is a one-one onto mapping, then $f \, o \, f^{-1} = I_y$ and $f^{-1} \, o \, f = I_z$.
- If $f : X \rightarrow Y$ and $g : Y \rightarrow Z$ be two one-one onto mapping, then the mapping of $g \, o \, f$ is also one-one and onto.

- $g \circ f$ may exist while $f \circ g$ may not exist.
- If $g \circ f$ and $f \circ g$ both exist, they may not be equal.
- $g \circ f$ exists if and only if the range of f is a subset of domain of g. Similarly $f \circ g$ exists if range of g is a subset of domain of f.

Theorem 6. *The composition of functions is associative.*

Proof : Let f, g, h be three functions such that $(f \circ g) \circ h$ and $f \circ (g \circ h)$ exists.

Define f, g and h such that $f : A \to B,\ g : B \to C$ and $h : C \to D$

Then, we have, $f : A \to B,\ g : B \to C \quad \Rightarrow \quad f \circ g : A \to C$ and $h : C \to D$

$\Rightarrow \quad (f \circ g) \circ h : A \to D$

Similarly, $f : A \to B,\ g : B \to C,\ h : C \to D$

$\Rightarrow \quad f : A \to B$ and $g \circ h : B \to D \quad \Rightarrow \quad f \circ (g \circ h) : A \to D$

$\Rightarrow \quad (f \circ g) \circ h$ and $f \circ (g \circ h)$ are function from A to D.

Now, to show, $(f \circ g) \circ h = f \circ (g \circ h)$

Let $x \in A$ be arbitrary and $y \in B$, $z \in C$ such that $h(x) = y$ and $g(y) = z$.

Then, $\{(f \circ g) \circ h\}(x) = (f \circ g)\{h(x)\}$

$= (f \circ g)(y) \qquad [\because h(x) = y]$

$= f\{g(h)\} = f(z) \qquad ...(1)$

Again, $\{f \circ (g \circ h)\}(x) = f\{(g \circ h)(x)\}$

$= f[g\{h(x)\}] \qquad [\because h(x) = y]$

$= f\{g(y)\} = f(z) \qquad ...(2)$

From (1) and (2), we conclude that $\{(f \circ g) \circ h\}(x) = \{f \circ (g \circ h)\}(x)$

Since x is arbitrary, therefore $(f \circ g) \circ h = f \circ (g \circ h)$.

Hence, composite of functions is always associative.

- The composition of functions is not commutative, *i.e.*, $f \circ g \neq g \circ f$.

Theorem 7. *The composition of any function with the identity function is the function itself.*

Proof: Let $f : A \to B$ be a function and $I_A : A \to A$ be an identity function.

Since, $I_A : A \to A$ and $f : A \to B$, therefore $f \circ I_A : A \to B$

Let $x \in A$, then $(f \circ I_A)(x) = f(I_A(x)) = f(x)$

(By definition of identity function $I_A(x) = x,\ \forall\, x \in A$)

$\Rightarrow \quad f \circ I_A = f$

Also, $f : A \to B$ and $I_B : B \to B \quad \Rightarrow \quad I_B \circ f : A \to B$.

Let $x \in B$ and let $f(x) = y$, then $y \in B$.

Therefore, $(I_B \circ f)(x) = I_B(f(x)) = I_B(y) = y = f(x)$

$\Rightarrow \quad I_B \circ f = f$

Hence, $f \circ I_A = I_B \circ f = f$.

Theorem 8. *Let $f : A \to B,\ g : B \to A$ be two functions such that $g \circ f = I_A$. Then, f is an injection and g is a surjection.*

Proof : (i) f is an Injection.

Let $x, y \in A$ such that $f(x) = f(y)$, then it implies

$g(f(x)) = g(f(y)) \quad \Rightarrow \quad g \circ f(x) = g \circ f(y)$

$\Rightarrow \quad I_A(x) = I_A(y) \quad (\because \text{ given } g \circ f = I_A) \quad \Rightarrow \quad x = y.$

Since x, y be arbitrary, therefore $f(x) = f(y)$

$\Rightarrow \quad x = y\ \forall x, y \in A \quad \Rightarrow f$ is one-one.

(ii) g is a surjection.

Since $g : B \to A$

Let $x \in A$, then obviously $f(x) \in B \qquad (\because f : A \to B)$

Let $f(x) = y$, then $g(y) = g(f(x)) = g \text{ o } f(x) = I_A(x) = x$
Since x is arbitrary, therefore, for every $x \in A \; \exists \; y = f(x) \in B$, such that $g(y) = x$.
$\Rightarrow g$ is onto.

☛ If $f : A \to B$ and $g : B \to A$ are two functions such that $f \text{ o } g = I_{B'}$ then f is a surjection and g is an injection.

Theorem 9. *Let $f : A \to B$ and $g : B \to A$ be two functions, then*

(i) $g \text{ o } f : A \to C$ is onto $\Rightarrow g : B \to C$ is onto.
(ii) $g \text{ o } f : A \to C$ is one-one $\Rightarrow f : A \to B$ is one-one.
(iii) $g \text{ o } f : A \to C$ is onto and $g : B \to C$ is one-one $\Rightarrow f : A \to B$ is onto.
(iv) $g \text{ o } f : A \to C$ is one-one and $f : A \to B$ is onto $\Rightarrow g : B \to C$ is one-one.

Proof : (i) Let $z \in C$. Now, since $g \circ f : A \to B$ is onto, therefore $\exists \; x \in A$ such that $g \circ f(x) = z \;\Rightarrow g(f(x)) = z$

$\Rightarrow \quad g(y) = z$, where $y = f(x) \in B$

Since z is arbitrary, therefore for all $z \in C$, there exists $y = f(x) \in B$, such that $g(y) = z$.

Hence, $g : B \to C$ is onto.

(ii) Let $x, y \in A$, such that $f(x) = f(y) \Rightarrow g(f(x)) = g(f(y))$

$\Rightarrow \quad g \text{ o } f(x) = g \text{ o } f(y) \quad \Rightarrow \quad x = y \quad (\because g \text{ o } f$ is one-one)

Since x, y is arbitrary, therefore $f(x) = f(y) \Rightarrow x = y \;\forall x, y \in A$.

Hence, f is one-one

(iii) Let $y \in B$.

Then, $g(y) \in C$

Since, $g \text{ o } f : A \to C$ is onto, therefore, for any $g(y) \in C \;\exists\; x \in A$ such that

$g \text{ o } f\,(x) = g\,(y) \quad \Rightarrow \quad f(x) = y. \quad (\because g$ is one-one)

Since y is arbitrary, therefore, for all $y \in B \;\exists\; x \in A$ such that $f(x) = y$.

Hence, $f : A \to B$ is onto.

(iv) Let $y_1, y_2 \in B$, such that $g(y_1) = g(y_2)$

Since $f : A \to B$ is onto and $y_1, y_2 \in B$. So, $\exists\; x_1, x_2 \in A$, such that

$$f(x_1) = y_1 \text{ and } f(x_2) = y_2$$

Now,
$$g(y_1) = g(y_2)$$
$$g(f(x_1)) = g(f(x_2))$$
$$g \text{ o } f(x_1) = g \text{ o } f(x_2)$$

$\Rightarrow \quad x_1 = x_2 \qquad \Rightarrow \quad f(x_1) = f(x_2) \qquad (\because f$ is one-one)

$\Rightarrow \quad y_1 = y_2$

Hence, $g : B \to C$ is one-one.

Theorem 10. *The inverse of bijective function is unique.*

Proof : Let $f : A \to B$ be a bijective function. Let if possible $g : B \to A$ and $h : B \to A$ be two inverses of f. To show $g = h$.

Let $g(y) = x_1$ and $h(y) = x_2$

$g(y) = x_1 \quad \Rightarrow \quad f(x_2) = y \qquad [\because g$ is inverse of $f]$

Also, $h(y) = x_2 \quad \Rightarrow \quad f(x_2) = y$

Therefore, $f(x_1) = f(x_2) \quad \Rightarrow \quad x_1 = x_2 \quad (\because f$ is one-one) $\Rightarrow \quad g(h) = h(y)$

Since y is arbitrary, therefore $g = h \;\forall y \in B$. Hence, $g = h$.

Theorem 11. *If $f : A \to B$ and $g : B \to C$ are two bijective functions, then $g \text{ o } f : A \to C$ is a bijection and $(g \text{ o } f)^{-1} = f^{-1} \text{ o } g^{-1}$.*

Proof : Here, we have $f : A \to B$ is one-one onto and $g : B \to C$ is one-one onto.

Therefore, $g \circ f : A \to C$ is one-one onto $\Rightarrow (g \circ f)^{-1} : C \to A$ exists.
Also, $f : A \to B$ is one-one onto $\Rightarrow f^{-1} : B \to A$ is one-one onto.
and $g : B \to C$ is one-one onto $\Rightarrow g^{-1} : C \to B$ is one-one onto.
Therefore, $f^{-1} \circ g^{-1} : C \to A$.
Now, let $x \in A$, $y \in B$ and $z \in C$, such that $f(x) = y$, $g(y) = z$.
Then, $(g \circ f)(x) = g(f(x)) = g(y) = z \Rightarrow (g \circ f)^{-1}(z) = x$ ...(1)

$$f(x) = y, g(y) = z \Rightarrow f^{-1}(y) = x \text{ and } g^{-1}(z) = y$$

Therefore, $(f^{-1} \circ g^{-1})(z) = f^{-1}(g^{-1}(z)) = f^{-1}(y) = x$ ...(2)
Since $z \in C$ is arbitrary, therefore from (1) and (2), we conclude that

$$(g \circ f)^{-1}(z) = (f^{-1} \circ g^{-1})(z) \; \forall z \in C.$$

Hence, $(g \circ f)^{-1} = (f^{-1} \circ g^{-1})$.

- The inverse of a bijective function is also bijective.
- If $f : A \to B$ is a bijection and $g : B \to A$ is the inverse of f, then $f \circ g = I_B$ and $g \circ f = I_A$ when I_A and I_B are the identity functions on the sets A and B respectively.
- In the above remark, if we take $B = A$, then $f \circ g = g \circ f = I_A$.
- Let $f : A \to B$ and $g : B \to A$ be two functions such that $g \circ f = I_A$ and $f \circ g = I_B$, then f and g are bijective and $g = f^{-1}$.

At a Glance

Term	Meaning
Function	Mapping from one set to another that associate with each member of the starting set exactly one member of the ending set.
Domain	Starting set for a function.
Co-domian	Ending set for a function
Image	Point that result from a mapping
Preimage	Starting point for a mapping
Range	Collection of all images of the domain
Onto (Surjective)	Range is the whole co-domain; every co-domain element has a preimage.
One-to-one (injective)	No two elements in domain map to the same place
Bijection	One to one and onto.
Identity function	Maps each element of a set to itself.
Inverse function	For a bijection, a new function that maps each co-domain element back where it came from.

1.17.2 Difference between Relation and Function

1. If R is a relation from A to B, then domain of R may be a subset of A. But if y is a function from A to B, then domain of f is equal to A.
2. In a relation from A to B, an element of A may be related to more than one element in B. But, in a function from A to B, each element of A must be associated to one and only one element of B. Thus, every function is a relation but, every relation is not necessary a function.

**	Relation (R)	Function (F)
Domain	Domain of R may be subset of A	Domain of f is equal to set A.
Association	An element of A may be related to more than one element of B	Each element of A must be associated to one and only one element of B

ILLUSTRATIVE EXAMPLES

1. *If $f : \mathbf{R} \to \mathbf{R}$ is defined by $f(x) = x^2 - 3x + 2$, find $f(f(x))$.*

Solution. Since $f(x) = x^2 - 3x + 2$.

Therefore,
$$\begin{aligned} f(f(x)) &= f(x^2 - 3x + 2) \\ &= (x^2 - 3x + 2)^2 - 3(x^2 - 3x + 2) + 2 \\ &= x^4 + 9x^2 + 4 - 6x^3 - 12x + 4x^2 - 3x^2 + 9x - 6 + 2 \\ &= x^4 - 6x^3 + 10x^2 - 3x. \end{aligned}$$

2. *Let $A = \{1, 2, 3, 4, 5\}$. Let $f : A \to A$ and $g : A \to A$ be defined by*
$$f(1) = 3, f(2) = 5, f(3) = 3, f(4) = 1, f(5) = 2$$
$$g(1) = 4, g(2) = 1, g(3) = 1, g(4) = 2, g(5) = 3.$$
Find $(f \text{ o } g)$ and $(g \text{ o } f)$.

Solution : Here we have
$$\begin{aligned} (f \text{ o } g)(1) &= f(g(1)) = f(4) = 1 \\ (f \text{ o } g)(2) &= f(g(2)) = f(1) = 3 \\ (f \text{ o } g)(3) &= f(g(3)) = f(1) = 3 \\ (f \text{ o } g)(4) &= f(g(4)) = f(2) = 5 \\ (f \text{ o } g)(5) &= f(g(5)) = f(3) = 3 \\ (g \text{ o } f)(1) &= g(f(1)) = g(3) = 1 \\ (g \text{ o } f)(2) &= g(f(2)) = g(5) = 3 \\ (g \text{ o } f)(3) &= g(f(3)) = g(3) = 1 \\ (g \text{ o } f)(4) &= g(f(4)) = g(1) = 4 \\ (g \text{ o } f)(5) &= g(f(5)) = g(2) = 1. \end{aligned}$$

3. *Let $f, g : \mathbf{R} \to \mathbf{R}$ be two functions defined by $f(x) = \sqrt{x-1}$ and $g(x) = \sqrt{4-x^2}$ $\forall\, x \in \mathbf{R}$. Find*

(i) $f + g$ (ii) $g + f$ (iii) $f - g$ (iv) $g - f$

(v) fg (vi) gf (vii) $\dfrac{f}{g}$ (viii) $\dfrac{g}{f}$

Solution : Since we have $f(x) = \sqrt{x-1}$ and $g(x) = \sqrt{4-x^2}$

Therefore, the domain of $f = [1, \infty[= D_1$ (Say) and, The domain of $g = [-2, 2] = D_2$ (say)
Define $D = D_1 \cap D_2 = [1, \infty[\cap [-2, 2] = [1, 2] \neq \phi$
Then, we have

(i) $(f + g)(x) = f(x) + g(x) = \sqrt{x-1} + \sqrt{4-x^2}$, with domain D

(ii) $(g + f)(x) = g(x) + f(x) = \sqrt{4-x^2} + \sqrt{x-1}$, with domain D

(iii) $(f - g)(x) = f(x) - g(x) = \sqrt{x-1} - \sqrt{4-x^2}$, with domain D

(iv) $(g - f)(x) = g(x) - f(x) = \sqrt{4-x^2} - \sqrt{x-1}$, with domain D

(v) $(fg)(x) = f(x)\, g(x) = \sqrt{x-1} \,.\, \sqrt{4-x^2} = \sqrt{(x-1)(4-x^2)}$, with domain D

(vi) $(gf)(x) = g(x)\, f(x) = \sqrt{4-x^2} \,.\, \sqrt{x-1} = \sqrt{(4-x^2)(x-1)}$, with domain D

(vii) The domain of $\dfrac{f}{g}$ is obtained by deleting those points x at which $g(x) = 0$ from D

Therefore, the domain of $\dfrac{f}{g}$ is $D' = \{x : x \in D, g(x) \neq 0\} = [1, 2[$

$$\therefore \quad \left(\frac{f}{g}\right)(x) = \frac{f(x)}{g(x)} = \frac{\sqrt{x-1}}{\sqrt{4-x^2}} = \sqrt{\frac{x-1}{4-x^2}}$$

(viii) The domain of $\frac{g}{f}$ is given by $D'' = \{x : x \in D, f(x) \neq 0\} =]1, 2]$

$$\text{Therefore, } \left(\frac{g}{f}\right)(x) = \frac{g(x)}{f(x)} = \frac{\sqrt{4-x^2}}{\sqrt{x-1}} = \sqrt{\frac{4-x^2}{x-1}}.$$

4. *If $f(x) = x^2$ and $g(x) = 3x$. Find the value of $(g \circ f)$ for $x = 1, 2, 3$.*

Solution . Here, we have $f(1) = 1^2 = 1$

$$(g \circ f)(1) = g(f(1)) = g(1) = 3 \times 1 = 3$$

$$f(2) = 2^2 = 4$$

Therefore,

$$(g \circ f)(2) = g(f(2)) = g(4) = 3 \times 4 = 12$$

Now, $\quad f(3) = 3^2 = 9$

$$(g \circ f)(3) = g(f(3)) = g(9) = 3 \times 9 = 27$$

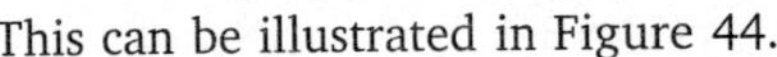

This can be illustrated in Figure 44.

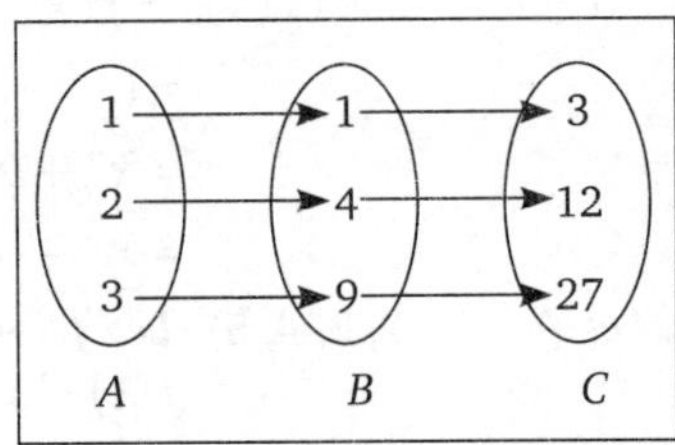

Fig. (44)

5. *Show that $f : \mathbf{R} - \{0\} \to \mathbf{R} - \{0\}$, given by $f(x) = 3/x$ is invertible and it is inverse of itself.*

Solution : (i) f is one-one :

Let $x, y \in \mathbf{R} - \{0\}$ such that $f(x) = f(y)$

Then, $f(x) = f(y) \Rightarrow \frac{3}{x} = \frac{3}{y} \Rightarrow x = y \quad \therefore \; f(x) = f(y) \Rightarrow x = y$

Since x, y be arbitrary, therefore, $f(x) = f(y) \Rightarrow x = y \; \forall \; x, y$

$\Rightarrow \quad f$ is one-one.

(ii) f is onto :

Let y be an arbitrary element of $\mathbf{R} - \{0\}$, then $f(x) = y \Rightarrow \frac{3}{x} = y \Rightarrow x = \frac{3}{y}$

Therefore, for each $y \in \mathbf{R} - \{0\}$, there exist $x = \frac{3}{y} \in \mathbf{R} - \{0\}$, such that

$$f(x) = f\left(\frac{3}{y}\right) = \frac{3}{3/y} = y$$

$\Rightarrow \quad f$ is onto.

Therefore, f bijective and hence invertible.

To find f^{-1}, let $f(x) = y$, then $f(x) = y$

$$\Rightarrow \quad \frac{3}{x} = y \quad \Rightarrow \quad x = \frac{3}{y} \quad \Rightarrow \quad f^{-1}(y) = \frac{3}{y} \quad \Rightarrow \quad f^{-1}(x) = \frac{3}{x} = f(x)$$

Hence, f is inverse of itself.

6. *If the function f: $\mathbf{R} \to \mathbf{R}$ be given by $f(x) = x^2 + 2$ and $g : \mathbf{R} \to \mathbf{R}$ be given by $g(x) = \frac{x}{x-1}$, find $f \circ g$ and $g \circ f$.*

Solution : Here, we observe that the range of f = domain of g and range of g = domain of f. Therefore, $f \circ g$ and $g \circ f$ both exist.

$$\text{Consider, } (f \circ g)(x) = f(g(x)) = f\left(\frac{x}{x-1}\right) = \left(\frac{x}{x-1}\right)^2 + 2 = \frac{x^2}{(x-1)^2} + 2$$

$$\text{and} \quad (g \circ f)(x) = g(f(x)) = g(x^2+2) = \frac{x^2+2}{(x^2+2)-1} = \frac{x^2+2}{x^2+1}.$$

7. *If $f : \mathbf{R} \to \mathbf{R}$ be given by*

$$f(x) = \sin^2 x + \sin^2\left(x + \frac{\pi}{3}\right) + \cos x \,.\, \cos\left(x + \frac{\pi}{3}\right) \forall\, x \in \mathbf{R}$$

and $g : \mathbf{R} \to \mathbf{R}$ be such that $g(5/4) = 1$, then, show that $g \circ f : \mathbf{R} \to \mathbf{R}$ is a constant function.

Solution : Civen that $f(x) = \sin^2 x + \sin^2\left(x + \frac{\pi}{3}\right) + \cos x \cos\left(x + \frac{\pi}{3}\right)$

$$= \frac{1}{2}\left[2\sin^2 x + 2\sin^2\left(x + \frac{\pi}{3}\right) + 2\cos x \cos\left(x + \frac{\pi}{3}\right)\right]$$

$$= \frac{1}{2}\left[1 - \cos 2x - \cos\left(2x - \frac{2\pi}{3}\right) + \cos\left(2x + \frac{\pi}{3}\right) + \cos\frac{\pi}{3}\right]$$

$$= \frac{1}{2}\left[\frac{5}{2} - \cos 2x - \cos\left(2x + \frac{2\pi}{3}\right) + \cos\left(2x + \frac{\pi}{3}\right)\right]$$

$$= \frac{1}{2}\left[\frac{5}{2} - \left\{\cos 2x + \cos\left(2x + \frac{2\pi}{3}\right)\right\} + \cos\left(2x + \frac{\pi}{3}\right)\right]$$

$$= \frac{1}{2}\left[\frac{5}{2} - \cos\left(2x + \frac{\pi}{3}\right)\cos\frac{\pi}{3} + \cos\left(2x + \frac{\pi}{3}\right)\right]$$

$$= \frac{1}{2}\left[\frac{5}{2} - \cos\left(2x + \frac{\pi}{3}\right) + \cos\left(2x + \frac{\pi}{3}\right)\right]$$

$$= \frac{5}{4} \forall\, x \in \mathbf{R}$$

Now, $g \circ f(x) = g(f(x)) = g(5/4) = 1 \;\Rightarrow\; g \circ f(x) = 1 \;\forall\, x \in \mathbf{R}$.

Hence, $g \circ f : \mathbf{R} \to \mathbf{R}$ is a constant function.

8. *If $f : \mathbf{R} \to \mathbf{R}$ be a function given by $f(x) = ax + b \;\forall x \in \mathbf{R}$, find the values of a and b such that $f \circ f = I_{\mathbf{R}}$.*

Solution. Here, we have $f \circ f = I_{\mathbf{R}}$, which implies $F \circ f(x) = I_{\mathbf{R}}(x) \;\forall x \in \mathbf{R}$

$\Rightarrow \quad f(f(x)) = x \;\forall x \in \mathbf{R} \qquad (\because I_{\mathbf{R}}(x) = x)$

$\Rightarrow \quad f(ax + b) = x \;\forall x \in \mathbf{R} \qquad \Rightarrow \quad a(ax + b) + b = x \;\forall x \in \mathbf{R}$

$\Rightarrow \quad (a^2 - 1)\, x + ab + b = 0 \;\forall x \in \mathbf{R} \qquad \Rightarrow \quad a^2 - 1 = 0$ and $ab + b = 0$

$\Rightarrow \quad a = \pm 1$ and $b(a + 1) = 0$

If $a = 1$, then $b = 0$

If $a = -1$, then $b\,(a + 1) = 0 \;\forall b \in \mathbf{R}$.

Therefore, $a = -1$ and b may take any real value.

Hence, either $a = 1$ and $b = 0$, or $a = -1$ and b can take any real value.

9. *Which of the following functions are odd or even or neither*

(i) $f(x) = \tan x + 3 \operatorname{cosec} x + x$ *(ii)* $f(x) = |x| + 1$

(iii) $f(x) = |x - 2|$

Solution : (i) We have $f(x) = \tan x + 3 \operatorname{cosec} x + x$

$\Rightarrow \quad f(-x) = \tan(-x) + 3\operatorname{cosec}(-x) + (-x)$

$= -\tan x - 3 \operatorname{cosec} x - x = -(\tan x + 3 \operatorname{cosec} x + x) = -f(x)$

Therefore, $f(x)$ is an odd function.

(ii) We have $f(x) = |z| + 1 \;\Rightarrow\; f(-x) = |-x| + 1\, |x| + 1 = f(x)$

$\Rightarrow \quad f(x)$ is an even function.

(iii) We have $f(x) = |x - 2| \;\Rightarrow\; f(-x) = |-x - 2| = |-(x + 2)| = |x + 2|$

$\Rightarrow \quad f(-x) \neq f(x)$ or $f(-x) \neq -f(x)$

Therefore, $f(x)$ is neither even, nor odd function.

10. *Let f and g* : $\mathbf{R} \to \mathbf{R}$ *be defined by* (**R** *is the set of real numbers*)

$$f(x) = x + 2, g(x) = \frac{1}{x^2+1}, \text{ compute } f^{-1}(g(x))$$

Solution : Let $f(x) = y$.

Then, we have $f(x) = x + 2 = y$

$\Rightarrow \quad x = y - 2 \qquad \Rightarrow \quad f^{-1}(y) = y - 2$

$\Rightarrow \quad f^{-1}(x) = x - 2 \qquad \Rightarrow \quad f^{-1}(g(x)) = g(x) - 2 = \dfrac{1}{x^2+1} - 2.$

Exercise 1.6

1. If $A = \{a, b, c, d\}$ and f corresponding to the cartesian product
   $\{(a, b), (b, d), (c, a), (d, c)\}$.
   Show that f is one-to-one from A onto A. Find f^{-1}.
2. If A is a non-empty set and $f, g : A \to A$, such that $f \text{ o } g = g \text{ o } f = I_A$, show that f and g are bijections and $g = f^{-1}$.
3. Let $X = \{-2, -1, 0, 1, 2, 3\}$ and $Y = \{0, 1, 2, \ldots, 10\}$ and $f : X \to Y$ be a function such that $f(x) = x^2 \ \forall x \in X$, find $f^{-1}(A)$, where $A = \{0, 1, 2, 4\}$.
4. Find the inverse of the following functions, if exist
   (i) $f(x) = \dfrac{1}{3}x + 4$ (ii) $f(x) = \dfrac{x-1}{x+1}$
   (iii) $f(x) = \sqrt{1-x^2}, 0 \le x \le 1$
5. If $A = \{a, b, c, d\}$ and f corresponds to the cartesian product
   $\{(a, b), (b, d), (c, a), (d, c)\}$. Show that f is one-one from A onto B. Find f^{-1}.
6. If $f : \mathbf{R} \to \mathbf{R}$ is a bijection given by $f(x) = x^3 + 3$, find $f^{-1}(x)$.
7. If $f : \mathbf{R} \to \mathbf{R}$ is defined by $f(x) = 3x - 7$. Show that f is invertible and find f^{-1}.
8. Let $f : A \to B$ be a function, such that
   (i) $A = \{0, -1, -3, 2\}$,
   $B = \{-9, -3, 0, 6\}$ and $f(x) = 3x$.
   (ii) $A = \{1, 3, 5, 7, 9\}$,
   $B = \{0, 1, 9, 25, 49, 81\}$
   and $f(x) = x^2$. Find f^{-1}.
9. Let $f : \mathbf{R} \to \mathbf{R}$ is given by $f(x) = (x+1)^2 - 1, x \ge -1$.
   Show that f is invertible. Also find the set $S = \{x : f(x) = f^{-1}(x)\}$.
10. Find $f^{-1}(3)$, if exists, when $f(x) = x^3 + 4$, where $f : \mathbf{R} \to \mathbf{R}$.
11. If $f = \{(5, 2), (6, 3)\}, g = \{(2, 5), (3, 6)\}$. What is the range of f and g? Find $f \text{ o } g$.
12. If $f(x) = x^2 - 1, g(x) = 3x + 1$, then describe the following functions :
    (i) $g \text{ o } f$ (ii) $f \text{ o } g$
    (iii) $g \text{ o } g$ (iv) $f \text{ o } f$
13. If $f(x) = \dfrac{x-1}{x+1}$, verify that $(f \circ f^{-1})(x) = x$.
14. If $f : \mathbf{R} \to \mathbf{R}$ and $g : \mathbf{R} \to \mathbf{R}$ are defined by $f(x) = x + 2$ and $g(x) = 2x^2 + 5$. Find $f \text{ o } g$ and $g \text{ o } f$
15. If f and g are two real velued functions such that $f(x) = x^2 - 5$ and $g(x) = 2x + 3$, find $f \text{ o } g$.
16. Let $f : \mathbf{R} \to \mathbf{R}$ be decided by $f(x) = \dfrac{x}{x^2+1}$.
    Find $f(f(2))$.
17. If $f(x) = \dfrac{1}{1-x}$ show that $f(f\{f(x)\}) = x$.
18. If $f : \mathbf{R} \to \mathbf{R}$, where $f(x) = x^2 + 2$ and $g : \mathbf{R} \to \mathbf{R}$, where $g(x) = 1 - \dfrac{1}{1-x}$, then find
    (i) $f \text{ o } g$
    (ii) $g \text{ o } f$
19. Let $A = \{x \in \mathbf{R} : 0 \le x \le 1\}$. If $f : A \to A$ is defined by $f(x) = \begin{cases} x, & \text{if } x \in \mathbf{Q} \\ 1 - x, & \text{if } x \notin \mathbf{Q} \end{cases}$
    Then show that $f \text{ o } f(x) = x \ \forall x \in A$.
20. Let $f : \mathbf{Z} \to \mathbf{Z}$ and $g : \mathbf{Z} \to \mathbf{Z}$ be defined by $f(n) = 3n \ \forall n \in \mathbf{Z}$ and $g : \mathbf{Z} \to \mathbf{Z}$ be defined by
    $$g(n) = \begin{cases} n/3, & \text{if } n \text{ is multiple of } 3 \ \forall\, n \in \mathbf{Z} \\ 0, & \text{if } n \text{ is not a multiple of } 3 \ \forall\, n \in \mathbf{Z} \end{cases}$$
    Show that $g \text{ o } f = I_Z$ and $f \text{ o } g \ne I_Z$

ANSWERS

1. $f^{-1} = \{(b, a), (d, b), (a, c), (c, d)\}$ **3** $f^{-1}(A) = \{0, -1, 1, -2, 2\}$,

4.(i) $f^{-1}(x) = 12 - 3x$, (ii) $f^{-1}(x) = \dfrac{1 + x}{1 - x}$, (iii) $f^{-1}(x) = \sqrt{1 - x^2}$

5. $f^{-1} = \{(b, a), (d, b), (a, c), (c, d)\}$ **6.** $f^{-1}(x) = (x - 3)^{1/3}\ \forall x \in \mathbf{R}$

7. $f^{-1}(x) = \dfrac{x + 7}{3}$

8.(i) $f^{-1} = \{(-9, -3), (-3, -1), (0, 0), (6, 2)\}$, (ii) f^{-1} does not exist

9. $S = \{0, -1\}$ **10.** $f^{-1}(3) = -1$ **11.** $f \circ g(2) = 2, f \circ g(3) = 3,$

12.(i) $g \circ f = 3x^2 - 2$, (ii) $9x^2 + 6x$, (iii) $9x + 4$, (iv) $x^4 - 2x^2$

14. $g \circ f = 2x^2 + 8x + 13, f \circ g = 2x^2 + 7$ **15.** $4x^2 + 12x + 4$

16. 10/29 **18.** $f \circ g = \dfrac{x^2}{(x - 1)^2} + 2, g \circ f = \dfrac{x^2 + 2}{x^2 + 1}$

1.18 RECURSIVELY DEFINED FUNCTIONS

In some situations, it is difficult to define an object explicitly. On the other hand, it may be easy to define this object in terms of itself. This process of defining the function is known as recursion, and a function is said to be recursively defined if the function definition refer to itself.

Definition : *A function is said to be recursive if and only if and only if it is obtained from the initial functions using a finite number of operations of compositions, recursion and minimization over regular functions.*

1.18.1 Properties of Recursively Defined Functions

The function definitions of recursively defined functions must have the following properties :

(i) There must be certain arguments for which the function does not refer to itself. These arguments termed as 'base values'.

(ii) Each time the function does refer to itself, the argument of the function must be closer to a base value.

☛ The recursive functions with above two properties are know as 'well-defined'.

1.18.2 Partial and Total Functions

Definition 1 : *"A partial function $f : X \to Y$ is a rule or correspondence which assign to every element of $x \in X$ at most one element $y \in Y$."*

Definition 2 : *"A total function $f : X \to Y$ is a rule or correspondence which assign to every element of $x \in X$ a unique element $y \in Y$."*

For Example : The rule $f : \mathbf{R} \to \mathbf{R}$ defined by $f(x) = +\sqrt{r}, r \in \mathbf{R}$ is a partial function since $f(r)$ is not defined for negative real number, but $g(r) = 2r$ is a total function from **R** to itself.

Definition 3 : *"A function is known as partial recursive if and only if it can be obtained from initial function using finite number of operations of composition, recursion and minimization."*

1.18.3 Primitive Recursive Function

The functions, which can be generated by primitive recursion from a set of initial functions using similar methods which are used in the definition of addition and multiplication, are known as primitive recursive functions.

For primitive recursive expression, the basic are the constant 0, a set of variables $x_1, x_2...$ and a set of function symbols.

Assumptions

(i) 0 is expression.

(ii) Every variable is an expression.

(iii) If f is a function and $t_1, t_2, \ldots t_n$ are expressions, then $f(t_1, t_2, \ldots, t_n)$ is an expression.

1.19 SOME INITIAL FUNCTIONS

(1) Zero Function : *The zero function, denoted by Z is defined by* $Z(x) = 0, \forall x$

For Example : $Z(1) = 0, Z(2) = 0, Z(9) = 0$

(2) Successor Function : *It is defined by* $S(x) = x + 1, \forall x$

For Example : $S(5) = 5+1 = 6, \quad S(7) = 7 + 1 = 8.$

(3) Projection Function : *The projection function is defined as follows :*

$$U_i^n (x_1, x_2, \ldots, x_n) = x_i$$

for n-tuples $(x_1, x_2, \ldots, x_n)$, for $1 \le i \le n, n \in \mathbf{N}$.

For Example : $U_1^3(3, 5, 8) = 3, U_2^3(3, 5, 8) = 5, U_3^3(3, 5, 8) = 8.$

- $U_i^1(x) = x, \forall x \in \mathbf{N}$.
- U_i^1 is simply the identity function.
- U_i^n is also known as generalised identity function.

1.19.1 Composition of Initial Functions

Let f be an n- ary function and $g_1, g_2, \ldots g_n$ are m-ary functions, then $f(g_1, g_2 \cdots, g_n)$ is a m-ary function where

$$f(g_1, g_2, .., g_n)(x_1, x_2, \ldots x_m) = f(g_1(x_1, x_2, \ldots, x_m), \ldots, g_n(x_1, x_2, \ldots, x_m))$$

1.19.2 Primitive Recursive Function

We can define a function f of $m + 1$ variables by recursion if there exists a function g of m variables and a function h of m+ 2 variables as follows :

$$f(x_1, x_2, \ldots, x_m, 0) = g(x_1, x_2, \ldots, x_m)$$
$$f(x_1, x_2, .. x_m, y + 1) = h(x_1, x_2, \ldots x_m, y, f(x_1, x_2, \ldots x_m, y))$$

Here, the value of f at y+1 can be expressed in terms of the value of f at y. Therefore, y is inductive variable. Also, $x_1, x_2, \ldots, x_n$ are parameters which are fixed throughout the definition. Also, both the functions f and g must be known.

Definition : *A function f is called partial recursive if and only if it can be constructed using the initial functions by a finite number of operations of composition and primitive recursion.*

- Every primitive recursive function is a total function.
- It is not always necessary to use only the initial function.

In the construction of a particular recursive function we can use any of the function $f_1, f_2, \ldots, f_n$ which are primitive recursive along with the initial condition to obtain another primitive recursive function provided we restrict to the operation of composition and recursion only.

1.19.3 Some Recursive Functions

(1) Factorial Function

Definition : *The product of the positive integers from 1 to n inclusive is known as 'n factorial'. It is denoted by n ! or* $\underline{|n}$, *i.e.,*

$$n! = n\,(n-1)\,(n-2), \ldots 3.2.1$$

It is also noted that $0! = 1$.

For Example :

$0! = 1,\ 1! = 1,\ 2! = 2,\ 3! = 3 \times 2 \times 1 = 6,\ 4! = 4.3.2.1 = 24,$
$5! = 5 \times 4 \times 3 \times 2 \times 1 = 120 \ldots$ and so on.

In general,

$$n! = n\,(n-1)!$$

☛ If $n = 0$, then $n! = 1$ and if $n > 0$, then $n! = n\,(n-1)!$.

☛ The above definition of $n!$ is recursive since it refers to itself when it uses $(n-1)!$.

(2) Fibonacci Sequence

We can define Fibonacci sequence as follows :

(i) If $n = 0$ or $n = 1$, then $F_n = n$
(ii) If $n > 1$, then $F_n = F_{n-2} + F_{n-1}$

For Example : 0, 1, 1, 2, 3, 5, 8, 13, 21, 34, 55, ...

☛ Fibonacci sequence is another example of recursive definition because we use F_{n-2} and F_{n-1} to define F_n.

(3) Ackermann Function

This is a function with two arguments each of which can be assigned any non-negative integer, i.e., 0, 1, 2, ... The Ackermann function can be defined by the following recurrence relation.

$$A(0, n) = n + 1,\ n = 0, 1, 2, \ldots$$
$$A(m, 0) = A(m-1, 1),\ m = 1, 2, \ldots$$
$$A(m, n) = A(m-1, A(m, n-1)),\ m = 1, 2, \ldots \text{ and } n = 1, 2, \ldots$$

For Example : $A(1, 2) = A(0, A(1, 1))$
$= A(0, A(0, A(1, 0))) = A(0, A(0, A(0, 1)))$
$= A(0, A(0, 2)) = A(0, 3) = 4$

Similarly, we may obtain
$A(0, 0) = 1,\ A(1, 1) = 3,\ A(1, 3) = 5,\ A(2, 2) = 7,\ A(3, 3) = 61$

☛ The Ackermann function is not a primitive recursive function.

☛ The Ackermann function has extraordinary rate of growth. For example,

$$A(4, 4) \sim 2^{2^{2^{65536}}}.$$

1.20 FUNCTIONS FOR COMPUTER SCIENCE

In previous sections, we introduced on an informal basis some functions commonly used in Computer Science applications. In this section we define some others.

1.20.1 Floor and Ceiling Functions

Definition 1 : *Let x be any real number. Then the floor function assign x the largest integer that is less than or equal to x. It is denoted by* $\lfloor x \rfloor$.

Symbolically, for any real number $x \in \mathbf{R}$, $\lfloor x \rfloor = n$

$\Leftrightarrow \quad n \leq x < n + 1.$

For Example : $f(1.5) = \lfloor 1.5 \rfloor = 1$

$f(-3) = \lfloor -3 \rfloor = -3$

Graph of Floor Function : In this figure, the open circles at the edge of each step are used to show that those points are not in the graph.

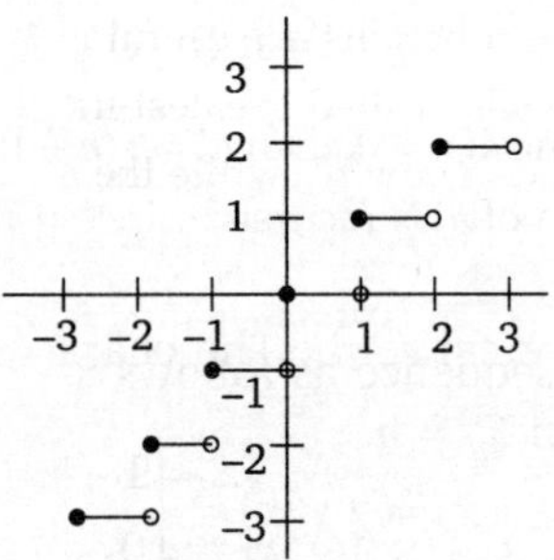

Fig. (45) : $f(x) = \lfloor x \rfloor$

Definition 2 : *Let x be a real number, then ceiling function f(x) assign to x, the smallest integer that is greater than or equal to x. It is denoted by* $\lceil x \rceil$.

Symbolically, for any real number $x \in \mathbf{R}$

$$\lceil x \rceil = n \Leftrightarrow n - 1 < x \leq n.$$

For Example : $\lceil 6.01 \rceil = 7$

$$\left\lceil \frac{-1}{2} \right\rceil = 0$$

Graph of Ceiling Function

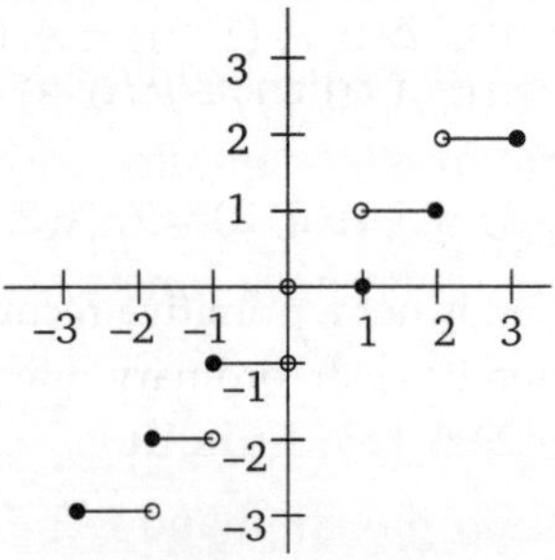

Fig. (46) : $f(x) = \lceil x \rceil$

Useful Properties of Floor and Ceiling Functions

1.	$\lfloor x \rfloor = n \Leftrightarrow n \le x < n+1$
2.	$\lceil x \rceil = n \Leftrightarrow n-1 < x \le n$
3.	$\lfloor x \rfloor = n \Leftrightarrow x-1 < n \le x$
4.	$\lceil x \rceil = n \Leftrightarrow x \le n < x+1$
5.	$x-1 < \lfloor x \rfloor \le x \le \lceil x \rceil < x+1$
6.	$\lfloor -x \rfloor = -\lceil x \rceil$
7.	$\lceil -x \rceil = -\lfloor x \rfloor$
8.	$\lfloor x+n \rfloor = \lfloor x \rfloor + n$
9.	$\lceil x+n \rceil = \lceil x \rceil + n$

- The floor and ceiling functions are useful in storage and data transmissions.
- The floor function is also called greatest integer function.
- The floor of x "rounds x down" while the ceiling of x "rounds x up".

1.20.2 Characteristic Function

Let A be a subset of universal set U. The characteristic function of A is defined as a function from U to $\{0, 1\}$ as follows :

$$f_A(x) = \chi_A(x) = \begin{cases} 1 & \text{if } x \in A \\ 0 & \text{if } x \notin A \end{cases}$$

For Example : Let $U = [1, 2, 3]$ and $A = [1, 2]$
Then, $f_A(1) = 1$, $f_A(2) = 1$, $f_A(3) = 0$
Important Results
If A and B are two sets, then for all x

$$f'_A(x) = 1 - f_A(x)$$
$$f_{A \cap B}(x) = f_A(x)\, f_B(x)$$
$$f_{A \cup B}(x) = f_A(x) + f_B(x) - f_{A \cap B}(x)$$

- For two subsets A and B of a universal set U, then
$f_A(x) = f_B(x) \ \forall\ x \in U$ if and only if $A = B$.

1.20.3 Hash Function

Definition : *A hash function is a function $f : K \to A$ where K is the set of keys and A is the set of physical address. It takes a data item to be stored or to be retrieved and computes the first choice for storage location.*

For Example : The most popular hash function is modulus (mod)
i.e., $f(n) = n \bmod m$.

1.20.4 Sign Function or Non-zero Test Function (sg)

The sign function is a recursive function and is defined by $sg(0) = 0$

$$sg(y+1) = 1$$

- $sg(0) = z(0)$, where $z(0)$ is a zero function.
- $sg(y+1) = S[U_2^2(y, sg(y)]$, where U is a projection function.

1.20.5 Zero test Function, $\overline{sg}$ or Compliment of Sign Function

Zero test function is defined as followed $\overline{sg}$ (0) = 1, $\overline{sg}(y + 1) = 0$

1.20.6 Compare Function

(i) less $(x, y) = sg\ (y - x) = \begin{cases} 1 & \text{if } x < y \\ 0 & \text{if } x \geq y \end{cases}$

(ii) $gr\ (x, y) = sg\ (x - y) = \begin{cases} 1 & \text{if } x > y \\ 0 & \text{if } x \leq y \end{cases}$

(iii) $eq\ (x, y) = sg\ (x - y) = \begin{cases} 1 & \text{if } x = y \\ 0 & \text{if } x \neq y \end{cases}$

1.20.7 Hamming Distance Function

This function gives a measure of difference between two binary words that have the same measure of difference between two binary words that have the same length. It is defined by

$H\ (u, v)$ = The number of positions in which u and v have different bits. Here u and v are binary words of same length.

For Example : H (1010, 1111) = 2 because 1010 and 1111 differ in 2 positions.

1.21 GROWTH OF FUNCTIONS

Definition : *Let f and g be two functions from the set of integers or the set of real numbers to the set of real numbers. Then f of order g written as $f(x)$ is O (g (x)), if there are constants C_1 and C_2 such that $|f\ (x)| \leq C_1\ |g(x)|$, whenever $x > C_2$. This can be read as $f(x)$ is big-oh of $g(x)$.*

1.21.1 Orders of power of a Polynomial Functions

It is known that $1 < x \quad \Rightarrow \quad x < x^2$ and $x^2 < x^3$

Therefore, if $1 < x$, then $1 < x < x^2 < x^3$.

So, we can say that, for any rational numbers r and s, if $x > 1$ and $r < s$, then $x_r < x_s$. Thus, we conclude that

"For any rational numbers r and s, if $r < s$, then x^r is O (x, s)."

ILLUSTRATIVE EXAMPLES

1. *Show that the function $f(x, y) = x + y$ is primitive recursive function. Hence compute the value of $f(2, 4)$.*

Solution : It is given that $f(x, y) = x + y$

If we want to define f by recursion, we define a function g of a single variable and a function h of three variables.

We have
$$\begin{aligned} f\ (x, y + 1) &= x + (y + 1) \\ &= (x + y) + 1 \\ &= f\ (x, y) + 1 \end{aligned}$$

Also, $f\ (x, 0) = x$

Now, $f(x, 0) = x = U_1^1(x)$

$$\begin{aligned} f\ (x, y + 1) &= f\ (x, y) + 1 \\ &= S\ (f\ (x, y)) = S\ (U_3^3\ (x, y, f\ (x, y)) \end{aligned}$$

The function $g(x)$ can be defined by $g(x) = x = f(x, 0)$ and

$$h(x, y, z) = S(U_3^3(x, y, z)) = f(x, y+1)$$

Further, as $g = U_1'$, it is an initial function. h is obtained from U_3^3 and S by using composition and f by recursion by using g and h. Clearly f is primitive recursive function because it is obtained by applying composition and recursion finite number of times to initial function U_1' and U_3' and S. Now, we want to calculate $f(2, 4)$.

Since $f(2, 0) = 2$

So,
$$\begin{aligned} f(2, 4) &= S(f(2, 3)) \\ &= S(S(f(2, 2))) \\ &= S(S(S(f(2, 1)))) \\ &= S(S(S(S(f(2, 0))))) \\ &= S(S(S(S(2)))) \\ &= S(S(S(3))) \\ &= S(S(4)) \\ &= S(5) \\ &= 6. \end{aligned}$$

2. *Show that the proper subtraction is primitive recursive.*

Solution : Define the predecessor function (p) by

$$p(0) = 0$$
$$p(y+1) = y = U_1^2(y, p(y))$$

which is clearly a recursive function and from it, the subtraction function is

$$f(x, 0) = x = U_1'(x)$$
$$f(x, y+1) = p(f(x, y)) = g(x, y, f(x, y))$$

where $g(x, y, z) = p(z) = U_3^3(x, y, z)$.

Hence, proper subtraction is primitive recursive.

3. *Solve the following recursive function using substitution*

$$f(n) = f\left(\frac{n}{2}\right) + 1,\ f(1) = 1,$$

for n to be integer greater than or equal to 1.

Solution : We have $f(n) = f\left(\frac{n}{2}\right) + 1$

$$f(n) = f\left(\frac{n}{4}\right) + 1$$

$$f(n) = f\left(\frac{n}{8}\right) + 1 = f\left(\frac{n}{2^3}\right) + 1$$

In general $f(n) = f\left(\frac{n}{2^i}\right) + 1$

The substitution ends when we get first term of RHS of (1) that is $f\left(\frac{n}{x}\right)$. We can get the first term as $f\left(\frac{n}{x}\right)$ when 2^i becomes equal to n. Let us assume $2^k = n$.

So, after k^{th} substitution $f(n) = f\left(\frac{n}{2^k}\right) + 1$ as $2^k = n$

Hence, $k = \log_n n$.

4. *Let a and b be positive integers and suppose Q is defined recursively as follows*

$$Q(a, b) = \begin{cases} 0 & \text{if } a < b \\ Q(a-b, b) + 1 & \text{if } a \geq b \end{cases}$$

(a) Find (i) Q (2, 5) (ii) Q (12, 5)

(b) What does this function Q do? Also find Q(5861, 7).

Solution : (a)(i) $Q(2, 5) = 0$, since $2 < 5$.

(ii) $Q(12, 5) = Q(7, 5) + 1 = [Q(2, 5) + 1] + 1 = Q(2, 5) + 2 = 0 + 2 = 2$

(b) The function $Q(a, b)$ find the quotient when a is divided by b because each time b is subtracted from a, the value of Q is increased by 1.

Also, $Q(5861,7) = 837$.

5. *Show that $f(x, y) = x^y$ is a primitive recursive function.*

Solution : We have $x^0 = 1$ for $x \neq 0$ and if we put $x^0 = 0$ for $x = 0$.

Further, $x^{y+1} = x^y * x$

Thus, $f(x, y) = x^y$ is defined as $f(x, 0) = 1$

$$f(x, y+1) = x * f(x, y) = U_1^3(x, y, f(x, y)) * U_3^3(x, y, f(x, y))$$

Hence, $f(x, y)$ is a primitive recursive function.

6. *Let n denote a positive integer. Suppose a function L is defined recursively as follows*

$$L(n) = \begin{cases} 0 & \text{if } n = 1 \\ L\left(\left\lfloor L\dfrac{n}{2}\right\rfloor\right) + 1 & \text{if } n > 1. \end{cases}$$

Find L (25) and describe with this function does.

Solution : We can find $L(25)$ recursively as follows :

$$\begin{aligned} L(25) &= L(12) + 1 \\ &= [L(6) + 1] + 1 = L(6) + 2 \\ &= [L(3) + 1] + 2 = L(3) + 3 \\ &= [L(1) + 1] + 3 = L(1) + 4 = 0 + 4 = 4 \end{aligned}$$

We observe that each time n is divided by 2, then the value of L is increased by 1.

Hence, L is the greatest integer such that $2^L \leq n$

Hence, $L(n) = \lfloor \log_2 n \rfloor$.

7. *Using recursion, define the multiplication function * given by $f(x, y) = x * y$.*

Solution : Since multiplication of two natural numbers is simply a repeated addition. Hence, f is primitive recursive.

Here, f is a function of two variables. To define f by recursion, we need a function g of a single variable and a function h of three variables.

Now, $f(x, 0) = 0$

and
$$\begin{aligned} f(x, y+1) &= x * (y+1) \\ &= f(x, y) + x \\ &= S(f(x, y), x) \end{aligned}$$

Also, we can write $f(x, 0) = 0 = z(x)$

and $f(x, y+1) = S(U_3^3(x, y, f(x, y)), U_1^3(x, y, f(x, y)))$

So, by taking $g = z$ and h defined by

$$\begin{aligned} h(x, y, z) &= S[U_3^3(x, y, z), U_1^3(x, y, z)] \\ &= f(x, y+1) \end{aligned}$$

Clearly, we see that f is defined by recursion.

8. *Compute the following A (1, 1), A (1, 2), A (2, 1), A (2, 2).*

Solution : Using the definition of Ackermann function, we have

$$
\begin{aligned}
A(1,1) &= A(0+1, 0+1)\\
&= A(0, A(1,0))\\
&= A(0, A(0,1))\\
&= A(0,2)\\
&= 3\\
A(1,2) &= A(0+1, 1+1)\\
&= A(0, A(1,1))\\
&= A(0,3)\\
&= 4\\
A(2,1) &= A(1+1, 0+1)\\
&= A(1, A(2,0))\\
&= A(1, A(1,1))\\
&= A(1,3)\\
&= A(0+1, 2+1)\\
&= A(0, A(1,2))\\
&= A(0,4)\\
&= 5\\
A(2,2) &= A(1+1, 1+1)\\
&= A(1, A(2,1))\\
&= A(1,5)
\end{aligned}
$$

But,

$$
\begin{aligned}
A(1,5) &= A(0+1, 4+1)\\
&= A(0, A(1,4))\\
&= 1, A(0+1, 3+1)\\
&= 1, A(0, A(1,3))\\
&= 1+1+A(1,3)\\
&= 1+1+1+A(1,2)\\
&= 1+1+1+4\\
&= 7
\end{aligned}
$$

Hence, $A(2, 2) = 7$.

9. *Using the definition of Ackermann function, compute* $A(1, 3)$.

Solution : We have

$$
\begin{aligned}
A(1,3) &= A(0, A(1,2))\\
A(1,2) &= A(0, A(1,1))\\
A(1,1) &= A(0, A(1,0))\\
A(1,0) &= A(0,1)\\
&= 1+1 = 2\\
A(1,1) &= A(0,2)\\
&= 2+1 = 3\\
A(1,2) &= A(0,3)\\
&= 3+1 = 4
\end{aligned}
$$

$\Rightarrow$ $A(1, 2) = 4$

Now, $A(1,3) = A(0,4)$

$= 4 + 1 = 5$

Hence, $A(1, 3) = 5$.

10. *For all real numbers x and all integers m, show that*

$$\lfloor x + m \rfloor = \lfloor x \rfloor + m$$

Solution . Let $n = \lfloor x \rfloor$

Then, by definition of floor function $n \le x < n + 1$

$\Rightarrow$ $n + m \le x + m < n + m + 1.$

Since n and m are integers, so $n + m$ is also an integer.

So, $\lfloor n+m \rfloor = n+m$

But $n = \lfloor x \rfloor$. Hence, by substitution, we have $\lfloor x+m \rfloor = \lfloor x \rfloor + m$.

11. *For any real number x, prove that if x is not an integer, then*

$$\lfloor x \rfloor + \lfloor -x \rfloor = -1.$$

Solution : Let $\lfloor x \rfloor = n$

Then, by definition of floor function $n < x < n+1$

$\Rightarrow \quad -n > -x > -n-1 \qquad \Rightarrow \quad -n-1 < -x < -n$

Now, since $-n-1$ is an integer, so by definition $\lfloor -x \rfloor = -n-1$

Hence, $\qquad \lfloor x \rfloor + \lfloor -x \rfloor = n + (-n-1) = n - n - 1 = -1.$

12. *For any real number x, show that*

$$\lfloor 2x \rfloor = \lfloor x \rfloor + \left\lfloor x + \frac{1}{2} \right\rfloor.$$

Solution : Let us suppose $n \le x < n+1$

If $n + \frac{1}{2} < x$, then $\lfloor 2x \rfloor = 2n+1$

$$\lfloor x \rfloor = n \text{ and } \left\lfloor x + \frac{1}{2} \right\rfloor = n+1$$

Consider $\lfloor 2x \rfloor = 2n+1 = n+n+1 = \lfloor x \rfloor + \left\lfloor x + \frac{1}{2} \right\rfloor.$

13. *Using the definition of order to show that $x^2 + 2x + 1$ is $O(x^2)$.*

Solution . Let us define two functions f and g such that

$$f(x) = x^2 + 2x + 1 \text{ and } g(x) = x^2$$

Now, for all real numbers, $x > 1$

$$|x^2 + 2x + 1| = x^2 + 2x + 1$$
$$\le x^2 + 2x^2 + x^2 \le 4x^2 \le 4|x^2|$$

So, $\qquad |f(x)| \le 4|g(x)|, \forall x > 1$

$\Rightarrow \qquad |f(x)| \le c|g(x)| \; \forall x > k$, where $c = 4, k = 1$

Hence, we conclude that $x^2 + 2x + 1$ is $O(x^2)$.

14. *Use the definition of order to show that $6x^3 - 3x + 4$ is $O(x^3)$.*

Solution : Let us define two functions f and g such that

$$f(x) = 6x^3 - 3x + 4 \text{ and } g(x) = x^3$$

Now, for $x \in R$ and $x > 1$

$$|6x^3 - 3x + 4| \le |6x^3| + |3x| + 4$$
$$\le 6x^3 + 3x + 4 \le 6x^3 + 3x^3 + 4x^3 \le 13x^3 \le 13|x^3|.$$

So, $|f(x)| \le 13|g(x)| \; \forall \; x > 1$

$\Rightarrow \; |f(x)| \le c|g(x)|, \; \forall \; x > k$, where $c = 13, k = 1$

Hence, $6x^3 - 3x + 4$ is $O(x^3)$.

☛ Using the results of above two examples, we conclude the following general result
If $f(x) = a_n x^n + a_{n-1} x^{n-1} + a_{n-2} x^{n-2} + \ldots a_1 x + a_0$

where, $\quad a_0, a_1, \ldots, a_n$ are real numbers, then $f(x)$ is $O(x^n)$.

15. *Consider a recursive function $g : \mathbf{Z}^+ \to \mathbf{Z}$ for all integers $n \geq 1$ such that*

$$g(n) = \begin{cases} 1 & \text{if } n \text{ is } 1 \\ 1 + g(n/2), & \text{if } n \text{ is even} \\ g(3n-1), & \text{if } n \text{ is odd and } n > 1 \end{cases}$$

Is g well defined ?

Solution : By definition of g, we have $g(1) = 1$

$$g(2) = 1 + g(1) = 1 + 1 = 2$$
$$g(3) = g(8) = 1 + g(4) = 1 + (1 + g(2))$$
$$= 1 + (1 + 2) = 4$$

However, $g(5) = g(14) = 1$

$$= g(7) = 1 + g(20)$$
$$= 1 + (1 + g(10))$$
$$= 1 + 1(1 + (1 + g(5)))$$
$$= 3 + g(5)$$

$\Rightarrow$ $3 = 0$, which is not true. Hence, g is not well defined.

Exercise 1.7

1. Define a function
$$f(x, y) = \begin{cases} 1, & \text{if } x = y \\ 0, & \text{if } x \neq y \end{cases}$$
Show that this function is primitive function.
2. Show that the constant function over **N**, *i.e.*, $f(n) = c \;\forall\; n \in \mathbf{N}$, where c is a fixed number is primitive recursive.
3. Show that the following functions are primitive recursive :
(i) $f(n) = 2^n$ (ii) $f(x, y) = |x - y|$
(iii) $f(x) = x^2$
(iv) $f(x) = \begin{cases} 0 & \text{if } x = 0 \\ x - 1 & \text{if } x > 0 \end{cases}$
4. Find the terms t_0, t_1, t_2, t_3, where
(i) $t_n = \lfloor n/2 \rfloor$.
5. Let $f(n) = \lfloor n/2 \rfloor + \lfloor n/3 \rfloor$ for $n \in \mathbf{N}$, show that for $0 \leq n \leq 10$ and for $n = 73$, $f(n) = 0, 0, 1, 2, 3, 3, 5, 5, 6, 7, 8, 60$.
6. For an odd integer n, show that $\lceil n^2/4 \rceil = n^2 + \frac{3}{4}$.
7. Computer $\lfloor x \rfloor$ and $\lceil x \rceil$ for each of the value x
(i) 8 (ii) 6.01
(iii) – 6.2 (iv) 1/2
(v) – 1/2
8. Use the recursive definition of addition and multiplication of two natural numbers, show that $3 + 5 = 8$ and $2 \times 4 = 8$.
9. Define a function $f : \mathbf{Z}^+ \to \mathbf{Z}$ is defined by
$$f(n) = \begin{cases} 1 & \\ f(n/2) & \text{if } n \text{ even} \\ 1 + f(5n - 9), & \text{if } n \text{ is odd and } n > 1 \end{cases}$$
Show that f is not well defined.
10. Show that the functions defined below are of the same order
$f(x) = 3x^4 - 6x^2$ and $g(x) = 5x^4$.
11. Show that the function f and g defined by $f(x) = \log_2 (n^x)$ and $g(n) = \log_e n$, where n is a positive integer, have the same order.
12. Let x and y be two integers and suppose $g(x, y)$ is defined recursively by
$$g(x, y) = \begin{cases} 5 & \text{if } x < y \\ g(x - y, y + 2) + x & \text{if } x \geq y \end{cases}$$
Find $g(2, 7)$, $g(5, 3)$ and $g(15, 2)$.
13. Prove the followings :
(i) $\lceil x^2 \rceil$ is $O(x^2)$
(ii) $\left\lfloor \frac{x\,(x+1)}{2} \right\rfloor$ is $O\,(x^4)$
(iii) $\frac{(x+1)(x+2)}{2}$ is $O(x^2)$
(iv) $1^2 + 2^2 + \ldots + n^2$ is $O(n^3)$
14. Show that the functions f and g defined by $f(n) = 3n^4 - 6n^2$ and $g(n) = 2n^4$, where n is a positive integer, have the same order.

ANSWERS

4.(i) 0, 0, 1, 1 (ii) 0, 1, 2, 3

7.(i) 8,8 (ii) 6, 7 (iii) – 7, – 6 (iv) 0, 1 (v) – 1, 0

12. 5, 10, 42

❐❐❐

Chapter

2

Mathematical Logic

- Sentences
- Statements
- Propositional function Quantifiers
- Connectives
- Type of sentences
- Truth value and truth table
- Tautology
- Contradiction
- Validity of arguments
- Rule of inferences
- Logical identities

2.1 INTRODUCTION

Mathematical logic deals with the topics like statements, negation, connectives, compound statements, conjunction, disjunction, duality, truth table, conditional and bi-conditional statements, valid arguments, tautologies, etc.

In the algebra of sets it has been observed that there are some ceratin primitive concepts associated with undefined terms. The terms true, false, and proposition (statement) are taken here as undefined.

A statement is a declarative sentence which has one and only one of two possible values. These two values are 'true' and 'false' or in other words statements is a declarative sentence which is either true or false but not both. The truth values 'true' and 'false' are denoted by the symbol T and F respectively. Sometimes it is also denoted by 1 (for true) and 0 (for false). The following examples are typical propositions :

(1) 7 is a prime number

(2) When 5 is added to 6 the sum is 7

(3) Living creatures exist on the planet Venus.

In the above examples, (1) is true, (2) is false, while (3) is either true or false but not both. Thus above sentences are statements. Let us consider other examples :

(1) May God bless you ! (a wish)

(2) Please wait here. (a request)

(3) What are you doing ? (an enquiry)

(4) May you live long.

In the above example, we can observe that no truth values can be given to the sentences. That is, they do not declare a definite truth value T or F. Thus they are not statements.

If we assume that the statement is true, then from its content we have that it is false while if the statement is assumed to be false, then from its content we have that it is true.

2.1.1 Statement Letters

It is well known fact that symbols have played a key role in any field, either a field of mathematics or a field or science. Therefore, they have great importance in Mathematics.

Definition : *The symbols, which are taken to represent the statement, are called statement letters. Therefore, we shall use lower case letters p, q, r, ... to represent statement.*

For example : If the statement 'Delhi is the capital of India', is denoted by the letter 'p', then it can be written as

$p =$ Delhi is the capital of India.

From any statement or set of statements, other statements may be formed. The simplest example is that of forming from the statement p the negation of p which is denoted by p'. For any statement p, we define p' or $\neg p$ to be the statement "it is false that p". For example, suppose that p is the statement

$p =$ sleeping is pleasant

The negation of this statement would be the statement.
"it is false that sleeping is pleasant"
or in other words, we can write the negation of p as follows :
"sleeping is not pleasant"
or "sleeping is unpleasant"

2.1.2 Open Statement

A sentence having one or more than one variables becomes a statement after giving some certain value to the variables, is called open statement.

For example : Let us consider $3x + y = 7$.

If we put $x = 2$ and $y = 1$, then above sentence becomes a true statement. Such a sentence is called an open statement.

2.2 PROPOSITIONAL FUNCTIONS QUANTIFIERS

A proposition is a declarative sentence that is either true or false but not both.

Let A be a given set. A propositional function defined on A is an expression $p(x)$, which has the property that $p(a)$ is true or false for each $a \in A$, *i.e.*, $p(x)$ becomes a statement whenever any element $a \in A$ is substituted for the variable x. The set A is called **domain** of $p(x)$ and the set T_p of all the elements of A for which $p(a)$ is True is called the truth set of $p(x)$, *i.e.*,

$$T_p = \{x : x \in A, p(x) \text{ is true}\}$$

For example : Let $p(x) = x + 4 > 1$. Its true set is

$$\{x : x \in \mathbf{N} : x + 4 > 1\} = \mathbf{N}$$

Thus, $p(x)$ is true for every element in **N.**

2.2.1 Universal Quantifiers

Let $p(x)$ be a propositional function defined on a set A. Consider the expression $[\forall\, x \in A, p(x)$ is true$]$. then the symbol $\forall$ (for all) is called the universal quantifier.

2.2.2 Existential Quantifiers

Let $p(x)$ be a propositional function defined on a set A. Consider the expression $[\exists\, x \in A, p(x)$ is true$]$. then the symbol $\exists$ (there exists) is called the existential quantifier.

2.2.3 Multiple Quantifiers

If a proposition have more than one variable then we can quantify it more than once.

For example : $p(x, y) = x^2 - y^2 = (x - y)\ (x + y)$

Multiple universal quantifiers can be arranged in any order without logically changing the meaning of resulting propositions. The same is true for the multiple existential quantifiers.

For example :

$$p(x, y) : x + y = 4 \text{ and } x - y = 2$$
$$= [\exists\, x \in \mathbf{R}, \exists y \in \mathbf{R} : x + y = 4 \text{ and } x - y = 2]$$

ILLUSTRATIVE EXAMPLES

1. *If p (x) is a formula in x, then translate the following :*

(a) $\forall\, x, [p(x)]$ *(b)* $\exists\, x, [p(x)]$ *(c)* $\forall x [\sim p(x)]$ *(d)* $\exists\, x [\sim p(x)]$

Solution. (a) Every x has the property $p(x)$.

(b) There exist x which has property $p(x)$.

(c) No x has property $p(x)$

(d) There is some x which do not have the property $p(x)$.

2. *Translate the following statements, involving quantifiers into formulae :*

(a) All rationals are real *(b) No rationals are real*

(c) Some rationals are real *(d) Some rationals are not real.*

Solution. Let x be an individual variable and $Q(x) \equiv x$ is rational $R(x) \equiv x$ is real.

Then,

(a) $\forall\, x, [Q(x) \rightarrow R(x)]$ (b) $\forall\, x \sim [Q(x) \rightarrow R(x)]$

(c) $\exists\ x, [Q(x) \wedge R(x)]$ (d) $\exists\, x [Q(x) \wedge \sim R(x)]$

2.3 COMPOUND PROPOSITION

A proposition consisting of only a single propositional variable or a single propositional constant is called primary or primitive proposition or simply proposition. They can not be further subdivided. A proposition obtained from the combination of two more proposition by means of logical operators or connectives of two or ore proposition is called compound proposition.

2.3.1 Connectives

Any two statements may be combined in various ways to form new statement. To form new statement, the words, which are used are called connectives. We shall now discuss three most basic and fundamental connectives.These are negation, conjunction and disjunction, which are associated with English words 'Not', 'And', 'or' respectively. Some other connectives will be discussed subsequently. These connectives are shown below :

English Words	Name of Connective	Symbols	Order
Not	Negation	$\sim$ or $\rceil$	1
And	Conjunction	$\wedge$	2
Or	Disjunction	$\vee$	3
One way implication	Conditional	$\Rightarrow$	4
If and only if (iff) (Two way implication)	Bi-conditional	$\Leftrightarrow$	5

2.3.2 Negation

Let p be a symbol for any statement, then negation of p is denoted by $\sim p$ or $\rceil p$ or p'. Let us consider the statement

$$p = I \text{ went to my office yesterday.}$$

Then negation of p is represented by

$$\rceil p = I \text{ did not go to my office yesterday.}$$

2.3.3 Conjunction

Let us consider two statements p and q given by

$$p = \text{Ice is cold, and } q = \text{Blood is green.}$$

These statements may be combined by the connective 'And' to form the new statement is given by

$$p \wedge q = \text{Ice is cold and blood is green.}$$

This new statement is known as the conjunction of p and q. In general, we define the conjunction of p and q for arbitrary statements p and q to be the statement "both p and q" . In wording the word both is often omitted. From this new statement, we observe that the statement is true if both p and q are true, and false, if either one or both of p and q are false.

2.3.4 Disjunction

The two statements p and q *given in above section may also be combined in another way. Which is as follows :*

Either ice is cold or blood is green.

This new statement is referred to as the disjunction of p and q. This disjunction of p and q is denoted by $p \vee q$. The use of "either...or ..." in English is indistinct so in some usages imply "either...or ... or both," but is some other usages imply "either ... or ...", but not both. Let us consider an example :

This creature is either is dog or an animal;

The baby is either a boy or a girl.

The first statement of these is called inclusive disjunction because this allows the possibility that both may happen.The second statement says that both statements can not happen together. Thus this types of disjunction is called exclusive disjunction. The words "or both" are usually omitted, and the word either may be omitted if there is no in distinction.The disjunction of p and q or both are true and false only when both p and q are false.

2.3.5 Conditional

Let us suppose p and q are two statements. Then the statement " $p \Rightarrow q$ " which is read as if "p then q" or "p implies q" is called conditional statements. Here 'p' is called antecedent and 'q' is called consequent. The conditional statement '$p \Rightarrow q$' has four possibilities depends on p and q.

Case I. If p is true, q is false, then "$p \Rightarrow q$" is false.

Case II. If p is true, q is true, then "$p \Rightarrow q$" is true

Case II. If p is false, q is true, then "$p \Rightarrow q$" is true.

Case IV. If p is false, q is false, then "$p \Rightarrow q$" is true.

Alternative Wording for 'Conditional Statement'

- $p \Rightarrow q$ *i.e.*, p implies q
  This can also be expressed in passive voice "q is implied by p"
- "whenever p, we have q". Also q whenever p"
- "p is sufficient for q". Also, "p is sufficient condition for q".
- In order for q is hold, it is enough that we have p".
- "q is necessary for p".
- "p, only if q", *i.e.*, p can happen only if q happens as well.

Meaning of symbols

(i) $p \Rightarrow q$. The arrow symbol $\Rightarrow$ is pronounced "implies"
(ii) $q \Leftarrow p$. The arrow $\Leftarrow$ is pronounced "is implied by".
If the statement "p if and only if q" is true we have the following table :

Condition p	**Condition q**	$\Rightarrow$
True	True	Possible
True	False	Impossible
False	True	Impossible
False	False	Possible

From above table it is clear that for condition p to be true while q is false because $p \Rightarrow q$. Similarly, it is impossible for condition q to be true while p is false because $q \Rightarrow p$. Hence the two conditions p and q must be both true or both false.

2.3.6 Bi-Conditional

Let p and q be two statements, then $p \Leftrightarrow q$ is called bi-conditional statement. This statement can also be written as $(p \Rightarrow q)$ and $(q \Rightarrow p)$ *i.e.* $(p \Rightarrow q) \wedge (q \Rightarrow p)$. This bi-conditional statement is true only if both the statement p and q have same truth values.

Alternative Wording of Biconditional statement (p iff q)

- "p is necessary and sufficient condition for q"
- "p is equivalent to q".
- "$p \Leftrightarrow q$"
- The symbol $\Leftrightarrow$ is an amalgamation of the symbol $\Rightarrow$ and $\Leftarrow$.

☛ The mathematical usage of **and** and **not** corresponds closely with standard English. The use of **or** however does not. In standard English or often suggests a choice of one option or the other but not both. But mathematical or allows the possibility of both. The statement "p or q" means that p is true or q is true or both p and q are true.

2.3.7 Joint Denial

The word "NOR" is a combination of NOT and OR where NOT and OR stands for negation and disjunction. Let p and q be two statements. The "$p \downarrow q$" is called Joint Denial or "NOR" statement and read as "Neither p nor q". This "$p \downarrow q$" can also be written as

$$p \downarrow q \Leftrightarrow \urcorner(p \vee q)$$

The Joint Denial statement $p \downarrow q$ is true only when p and q both are false.

2.3.8 NAND Statement

The word "NAND" is a combination of NOT and AND where NOT and AND stand for negation and conjunction respectively. Then the statement "$p \uparrow q$" is called NAND statement and this can be written as

$$p \uparrow q \Leftrightarrow \rceil (p \wedge q)$$

This statement is false only if both p and q are true.

2.3.9 Types of Conditional Statement

Let p and q be any two statements, then there are some other conditional statements which are related to the conditional $p \Rightarrow q$.

(i) Converse Implication : The statement $q \Rightarrow p$ is called converse implication of the statement of $p \Rightarrow q$.

(ii) Inverse Implication : The statement $\rceil p \Rightarrow \rceil q$ is called inverse implication of the statement $p \Rightarrow q$.

(iii) Contrapositive Implication : The statement $\rceil q \Rightarrow \rceil p$ is called contrapositive of the statement $p \Rightarrow q$.

2.3.10 Use of Brackets

The use of brackets in the statements has an important role. By using brackets in the statements, the meaning or explanation of the statements are completely different. There are some rules related to the brackets.

(i) If the negation (*i.e.*, $\rceil$) is repeated in the statement then there is no need of bracket.

For Example : $\rceil \rceil p$ and $(\rceil (\rceil p))$ both have same meaning.

(ii) If in a statement, the connectives of same type are present, then brackets are applied from the left. For example : $p \vee q \vee r \vee s = [\{(p \vee q) \vee r\} \vee s]$.

(iii) If the different connectives are used in the statement, then we remove the bracket of lower order connective, but we can not remove the bracket of higher order connective.

For example : $p \Rightarrow (q \wedge r) = p \Rightarrow q \wedge r$.

Here order of $\wedge$ is less than the order of $\Rightarrow$. Hence bracket is removed. Let us consider another example

$$p \vee (q \Rightarrow r) \neq p \vee q \Rightarrow r$$

In this example, bracket can not be removed.

2.4 TYPES OF SENTENCES

There are two type of sentences or statements.

(i) Simple Sentence : A statement which has no connectives is called simple sentence or simple statement (or Atomic statement).

For example : He is a boy or she is a girl. Both are simple.

(ii) Compound Sentence : A statement which is formed by two simple statement through the connective is called compound sentence or molecular sentence.

For Example :

(a) If you work hard, then you will get success.

(b) Suresh will play or he will leave ground.

2.4.1 Statement Form

This is a form obtained by simple statements, using finite number of connectives, is called statement form.

For example : Let p, q and r be three simple statements, then the statement $(p \wedge q) \Rightarrow r$ is a statement form. The connective which is used at the end of the statement form is called principal connective.

2.4.2 Principal connective

Thus in the above example, $\Rightarrow$ is the principal connective. The statements on both sides of the principal connective are called **Arguments.** Therefore, in the above example, the statement $(p \wedge q)$ and r are arguments. The statement form is also known as **Truth function.**

2.5 USE OF VENN DIAGRAMS IN CHECKING TRUTH AND FALSITY OF STATEMENTS

In this section, we shall discuss the usage of Venn diagrams to represent truth and falsity of statement of propositions.

ILLUSTRATIVE EXAMPLES

1. *Give the venn diagram for the truth of the following statement "Equivalent triangles are isosceles triangles.)*

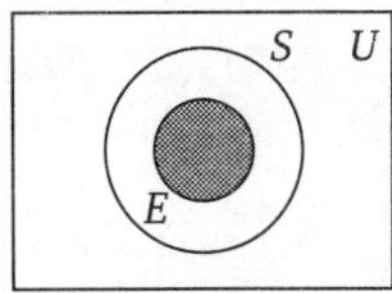

Fig. (1)

Solution. Let U, E and S denotes the set of triangles in a plane, the set of equilateral triangles in a plane and the set of all isosceles triangles in a plane respectively.

Then, $E \subset U$ and $S \subset U$

It is also clear that $E \subset S$. Hence, we have the following Venn diagram to represent the truth of the given statement.

2. *Let U, P and T denotes respectively the set of human beings, the set of policemen and the set of all thieves. Write the truth value of the following statements from the Venn diagram given below :*

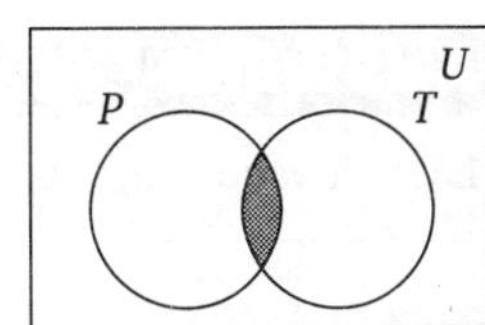

Fig. (2)

(i) *No policeman is a thief.*
(ii) *Thieves are not policeman.*
(iii) *Men who are not policemen are thieves.*
(iv) *Some policemen are thieves.*

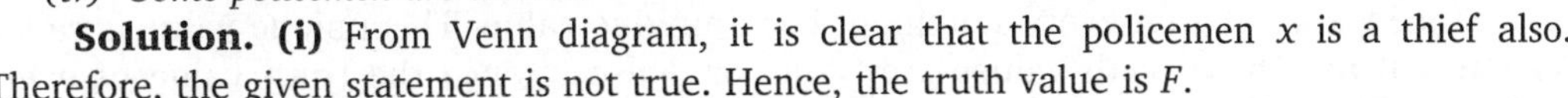

Solution. (i) From Venn diagram, it is clear that the policemen x is a thief also. Therefore, the given statement is not true. Hence, the truth value is F.

(ii) We find that $P \cap T \neq \phi$, therefore, there are some thieves who are also policemen. Hence, the statement is not true and the truth value is F.

(iii) From Venn diagram, it is clear that there are some human beings who are neither policemen nor thieves. Hence, the above statement is not true and the truth value is F.

(iv) Here, it is clear that the policemen x is a thief also. Hence, the given statement is true and the truth value is T.

3. *Use Venn diagrams to check the validity of the following argument*

S_1 : *If a man is a bachelor, he is unhappy.*
S_2 : *If a man is unhappy, he dies young.*
... ...
S : *All bachelors die young.*

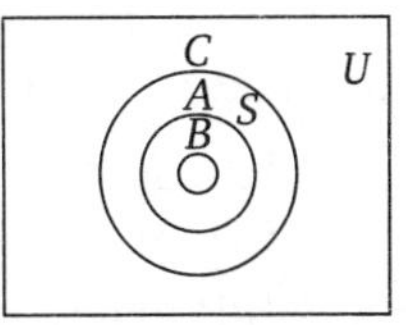

Fig. (3)

Solution. Define the following sets

A = set of all unhappy men
B = set of all bachelors
and C = set of all men who die young.

Then, the truth of the hypothesis S_1 is represented by placing the B inside the set A, *i.e.*, $B \subset A$.

Now, the truth of statement S_2 is represented by placing the set A completely inside the set C, *i.e.*, $A \subset C$.

Now, S_1 and S_2 are true

$\Rightarrow$ $B \subset A$ and $A \subset C$.

$\Rightarrow$ $B \subset C$, *i.e.*, all bachelors die young.

$\therefore$ S is true.

Hence, the given argument is a valid argument.

Exercise 2.1

Use Venn diagram to examine the validity of the following arguments :

1. (i) S_1 : Natural numbers are integers.
S_2 : x is an integer
... ...
S : x is a natural integer.

(ii) S_1 : Natural numbers are integers.
S_2 : x is an integer
... ...
S : x is not a natural integer.

2. S_1 : All squares are rectangles.
S_2 : x is not a rectangle.
... ...
S : x is not a square.

3. S_1 : If it rains, Rijuta will be sick.
S_2 : It did not rain.
... ...
S : Rijuta was not sick.

4. S_1 : If 7 is less than 4, then 7 is not a prime number.
S_2 : 7 is not less than 4.
... ...
S : 7 is a prime number.

5. S_1 : All graduates get a job.
S_2 : Rijuta is a graduate.
... ...
S: Rajuta gets a job.

ANSWERS

1.(i) invalid, (ii) invalid; **2.** valid; **3.** valid; **4.** invalid; **5.** valid.

2.6 TRUTH VALUES AND TRUTH TABLE

Since we know that every statement has a unique value. This unique value is called **Truth value.** These truth values are 'True' or 'False'. Hence the truth value of every statement form is obtained by the truth values of its components. The range of statement form (or Truth function) is $\{T , F\}$. Hence we can say that there will be 2^n truth function having n statements. T and F represent the truth value, True and False respectively. These truth values are also represented by 1 and 0.

To show that the set of statement and the operations of conjunction, disjunction and negation, it is necessary first to define them equal. Let us consider two statement forms 'f' and 'g' having two component statements p and q each. These two statement forms are said to be equal if both have same truth values for each of the four possible ways.

(i) p false and q true.

(ii) p true and q false.

(iii) p and q both are true.

(iv) p and q both are false.

If in anyone of above possible ways, the truth values differ, then f and g are not equal. To understand the meaning and uses of different types of connectives, we analyse the truth values with the help of a table. This table is called truth table.

A truth able displays the relationship between the truth value of the proposition truth tables are especially valuable in the determination of the truth values of propositions from simple proposition.

2.6.1 Truth Table for Negation

Let p be any statement, then $\sim p$ is negation of p. The truth value of negation of p is opposite of truth value of p. Suppose p = Delhi is a capital of India.

Then, $\sim p$ = Delhi is not a capital of India.

The truth table is given below :

p	$\sim P(\urcorner p)$
T	F
F	T

Or

p	$\sim P(\urcorner p)$
1	0
0	1

2.6.2 Truth Table for Conjunction

Let p and q be two simple statements, then $p \wedge q$ is conjunction of p and q. Thus $p \wedge q$ has a true value if both p and q have true values. Let us consider

p = Lucknow is a capital of Uttar Pradesh; q = Delhi is a capital of India.

Then $p \wedge q$ means that Lucknow is a capital of Uttar Pradesh and Delhi is a capital of India. The truth table is given below :

p	q	$p \wedge q$
T	T	T
T	F	F
F	T	F
F	F	F

Or

p	q	$p \wedge q$
1	1	1
1	0	0
0	1	0
0	0	0

2.6.3 Truth Table for Disjunction

Let p and q be any two simple statements, then $p \vee q$ is called disjunction of p and q which is read as p or q. Let us consider

p = Mathematics is very hard subject; q = Commerce is very easy subject.

Then $p \vee q$ can be written as

$p \vee q$ = Mathematics is very hard subject or commerce is very easy subject. The statement $p \vee q$ has the truth value false only if both are false. Thus truth table is given below :

p	q	$p \vee q$
T	T	T
T	F	T
F	T	T
F	F	F

Or

p	q	$p \vee q$
1	1	1
1	0	1
0	1	1
0	0	0

2.6.4 Truth Table for conditional

Let p and q be two simple statements, then $p \Rightarrow q$ is the statement which is conditional. This statement has the truth value false only if antecedent p is true and consequent q is false. The truth table for $p \Rightarrow q$ is given below :

p	q	$p \Rightarrow q$
T	T	T
T	F	F
F	T	T
F	F	T

Or

p	q	$p \Rightarrow q$
1	1	1
1	0	0
0	1	1
0	0	1

2.6.5 Truth Table for Bi-Conditional

Let p and p be any two simple statements. The statement of the type $p \Leftrightarrow q$ is called bi-conditional. This can also be written as $p \Leftrightarrow q = (p \Rightarrow q) \wedge (q \Rightarrow p)$. This statement has the truth value true only if both p and q have same truth values.

Thus the truth table is given below :

p	q	$p \Leftrightarrow q$
T	T	T
T	F	F
F	T	F
F	F	T

Or

p	q	$p \Leftrightarrow q$
1	1	1
1	0	0
0	1	0
0	0	1

2.6.6 Truth table for Joint Denial

Let p and q be two statements. The statement $p \downarrow q$ is called Joint Denial statement, which is read as 'Neither p nor q'. The statement $p \downarrow q$ has truth value 'True' when both p and q are false. Thus the truth table for $p \downarrow q$ is given below :

p	q	$p \downarrow q$
T	T	F
T	F	F
F	T	F
F	F	T

Or

p	q	$p \downarrow q$
1	1	0
1	0	0
0	1	0
0	0	1

2.6.7 Truth Table for NAND Statement

Let p and q be two statements. The statement $p \uparrow q$ is called NAND statement. This statement is false only when both p and q are true. The truth table for NAND statement is given below :

p	q	$p \uparrow q$
T	T	F
T	F	T
F	T	T
F	F	T

Or

p	q	$p \uparrow q$
1	1	0
1	0	1
0	1	1
0	0	1

2.6.8 Table for Symbols

To understand and then to write any statement into symbols there is a table given below :

(i)	If p then q	$p \Rightarrow q$
(ii)	p if q	$q \Rightarrow p$
(iii)	p only if q	$q \Rightarrow p$
(iv)	p unless q	$\rceil q \Rightarrow p$
(v)	p is a sufficient condition for q	$p \Rightarrow q$
(vi)	p is a necessary condition for q	$q \Rightarrow p$
(vii)	A sufficient condition for p is q	$q \Rightarrow p$
(viii)	A necessary condition for p is q	$p \Rightarrow q$
(ix)	In order that p is sufficient that q	$q \Rightarrow p$

(x)	In order that p is necessary that q	$p \Rightarrow q$
(xi)	p if and only if q	$p \Leftrightarrow q$
(xii)	p is a necessary and sufficient condition for q	$p \Leftrightarrow q$

2.6.9 Precedence of logical operators

Rule (1) : Use parentheses to specify the order in which logical operators in a compound proposition are to be applied.

Rule (2) : To reduce the number of parentheses, we specify that the negation operator is applied before all the other logical operators.

Rule (3) : The conjunction operator takes precedence over the disjunction operator (we will continue to use parenthesis) so that the order of disjunction and conjunction operators is clear.

Rule (4) : The conditional and biconditional operator have lower precedence than the conjunction and disconjunction operators $\wedge$ and $\vee$. Consequently, $p \vee q \Rightarrow r$ is the same as $(p \vee q) \Rightarrow r$

Table : Precedence of logical operators

Operator	Precedence
$\rceil$ or $\sim$	1
$\wedge$ $\vee$	2 3
$\Rightarrow$ $\Leftrightarrow$	4 5

☛ We will use parenthesis when the order of the conditional operator and biconditional operator is at issue, although the conditional operator has precedence over the biconditional operator.

ILLUSTRATIVE EXAMPLES

1. *Which of the following expression are statements :*

(i) $1+2+3=1\times 2\times 3$

(ii) $\{2, 3\} \subset \{2, 4, 6\}$

(iii) *May you live long.*

(iv) *7 is a prime number*

(v) $5 \in \{1, 4, 5\}$

(vi) *All roses are white*

(vii) *What is your Name ?*

(viii) *The girls are beautiful.*

(ix) *Go to your home.*

(x) *Blood is red.*

Solution.

(i) Since $1+2+3=1\times 2\times 3$ or $6=6$.
Thus (i) is true. Hence, (i) is a statement.

(ii) $\{2, 3\} \subset \{2, 4, 6\}$.
This expression is false. Hence, it is a statement.

(iii) "May you live long" is not a declarative sentence. Hence, it is not a statement.

(iv) "7 is a prime number" which is declarative sentence having truth value 'True'.
Hence, it is a statement.

(v) $5 \in \{1, 4, 5\}$ is true. Hence, it is a statement.

(vi) "All roses are white". This is a declarative sentence and having truth value 'False'.
Hence, it is a statement.

(vii) "What is your name?" It is not a declarative sentence. Hence, it is not a statement.

(viii) "The girls are beautiful" This sentence is not declarative. Hence, it is not a statement.

(ix) "Go to your home". This is not a declarative sentence. Hence, it is not a statement.

(x) "Blood is red." It is a declarative sentence. Its truth value is True. Hence, it is a statement.

2. *Which of the following sentences are propositions ? What are the truth values of those that are propositions (statement) ?*

(i) Do you speak Hindi ? *(ii) Four is even.*

(iii) Please submit your proposal as soon as possible.

(iv) Do you speak English ? *(v)* $4 - x = 8$.

(vi) Please try to solve the problem.

Solution. We know that a statement is a declarative sentence which is either true or false, but not both. These two values are 'true' and 'false' denoted by symbolical T and F. Thus

(i) not statement (ii) statement (iii) not statement

(iv) not statement (v) not statement (vi) not statement

3. *If* $p \equiv$ *He is a carpenter and* $q =$ *He is making a table.*

Then write down the following statement into symbols :

(i) He is a carpenter and making a table.

(ii) He is a carpenter but is not making a table.

(iii) It is false that he is a carpenter or making a table.

(iv) Neither he is a carpenter nor he is making a table.

(v) He is not a carpenter and he is making a table.

(vi) It is false that he is not a carpenter or is not making a table.

(vii) He is a carpenter or making a table.

Solution. The solution of above compound statements in terms of p and q are given below :

(i) $p \wedge q$ (ii) $p \wedge \neg q$ (iii) $\neg(p \vee q)$ (iv) $\neg p \wedge \neg q$

(v) $\neg p \wedge q$ (vi) $\neg(\neg p \vee \neg q)$ (vii) $p \vee q$

4. *Consider the following :*

p : *This computer is good.*

q : *This computer is cheap.*

Write each of the following statements in symbolic forms :

(i) This computer is good and cheap. *(ii) This computer is not good but cheap.*

(iii) This computer is costly but good. *(iv) This computer is neither good nor cheap.*

(v) This computer is good or cheap.

Solution. (i) $p \wedge q$ (ii) $p' \wedge q$ (iii) $p \wedge q'$

(iv) $p' \vee q'$ (v) $p \vee q$.

5. *Consider the following :*

p : *Question paper is hard.*

p : *I will fail in the examination.*

Then translate the following sentences into symbols :

(i) Question paper is hard then I will fail in the examination.

(ii) If I will not fail in the examination, then question paper is not hard.

(iii) Question paper is not hard if and only if I will fail in the examination.

(iv) If question paper is not hard then I will pass in the examination.

Solution. (i) $p \Rightarrow q$ (ii) $\neg p \Rightarrow \neg q$ (iii) $\neg p \Leftrightarrow q$ (iv) $\neg p \Rightarrow \neg q$

6. *Write the following in symbols :*

(i) Sachin will go out of station or will remain in his house and he will repair his radio.

(ii) The necessary and sufficient condition for an infinite series Σu_n to be convergent is that limit of u_n as n tending to infinity must be zero.

(iii) We shall go to Delhi, but we shall not see the Red Fort.

(iv) Not only the children, but also Mothers and Fathers were killed.

(v) If teams do not arrive or the weather is bad, then there will be no match.

Solution.

(i) Let $p \equiv$ Sachin will go out of station.

$q \equiv$ Sachin will remain in his house.

$r \equiv$ Sachin will repair his radio.

Thus the statement has the symbol $p \vee (q \wedge r)$.

(ii) $p \equiv$ An infinite series Σu_n to be convergent.

$q \equiv$ limit of u_n must be zero as n tending to infinity.

Thus the statement has the symbol $p \Leftrightarrow q$.

(iii) $p \equiv$ We shall go to Delhi.

$q \equiv$ We shall not see the Red Fort.

Thus the statement is $p \wedge \rceil q$.

(iv) $p \equiv$ children were killed.

$q \equiv$ Mothers were killed.

$r \equiv$ Fathers were killed.

Thus the statement is $p \wedge (q \wedge r)$.

(v) $p \equiv$ The teams do not arrive.

$q \equiv$ Whether is bad.

$r \equiv$ there will be no match.

Thus the statement is $(p \vee q) \Rightarrow r$.

7. *If p = Ramesh is a player, q ≡ Mohan is an intelligent boy, then write the following symbols into sentences :*

(i) $p \wedge q$ *(ii)* $\rceil p \wedge \rceil q$ *(iii)* $p \wedge \rceil q$ *(iv)* $\rceil (p \wedge q)$

(v) $\rceil p \Leftrightarrow q$ *(vi)* $p \Rightarrow \rceil q$

Solution. (i) Ramesh is a player and Mohan is an intelligent boy.

(ii) Neither Ramesh is a player nor Mohan is an intelligent boy.

(iii) Ramesh is a player and Mohan is not an intelligent boy.

(iv) It is false that Ramesh is a player and Mohan is an intelligent boy.

(v) Ramesh is not a player if and only if Mohan is an intelligent boy.

(vi) If Ramesh is a player, then Mohan is not an intelligent boy.

8. *If p ≡ Money is evil, q ≡ Wise men are poor, r ≡ beggars are failures. Then translate each of the following statements into symbols :*

(i) Wise men are poor only if money is evil.

(ii) Money is evil unless wise men are poor.

(iii) That beggars are failures is a sufficient condition that money is evil.

(iv) A necessary condition for money to be evil is that beggars are failures.

(v) Money is evil and beggars are failures if wise men are poor.

(vi) Unless beggars are failures, wise men are not poor and money is not evil.

Solution. (i) $p \Rightarrow q$ (ii) $\rceil q \Rightarrow p$

(iii) $r \Rightarrow p$ (iv) $p \Rightarrow r$ (v) $q \Rightarrow (p \wedge r)$ (vi) $\rceil r \Rightarrow \rceil p \wedge \rceil q$.

9. *Write the negation of each of the following statements in terms of symbols :*

(i) It will rain unless the barometer rises.

(ii) I grow fat only if I eat too much.

(iii) A necessary condition that two triangles are equivalent is that they have the same area.

(iv) In order to live well, it is sufficient to be wealthy.

Solution. (i) Let $p \equiv$ It will rain, $q \equiv$ Barometer rises.

Then the statement (i) can be written as $\rceil q \Rightarrow p$.

Thus the negation is $\rceil(\rceil q \Rightarrow p)$.

(ii) Let $p \equiv$ I grow fat, and $q \equiv$ I eat too much.

Then the statement (ii) can be written as $q \Rightarrow p$.

Thus its negation is $\rceil(q \Rightarrow p)$

(iii) Let $p \equiv$ Two triangles are equivalent, and $q \equiv$ They have the same area.

Then the statement (iii) can be written as $p \Rightarrow q$.

Thus its negation is $\rceil(p \Rightarrow q)$

(iv) Let $p \equiv$ Live well, and $q \equiv$ To be wealthy.

Then the statement (iv) can be put as $q \Rightarrow p$.

Thus its negation is $\rceil(q \Rightarrow p)$.

10. *Write the negation of the following :*

(i) If she studies, she will pass in exam.

(ii) If it rains, then they will not go for picnic.

(iii) Every even integer greater than 4 is the sum of two primes.

(iv) Some people have no scooter.

(v) No one wants to buy my house.

Solution. (i) If she will not study, she will not pass in exam.

(ii) If it will not rain, then they will go for picnic.

(iii) Every odd integer greater than 4 is the sum of two primes.

(iv) Some people have scooter.

(v) Everyone wants to buy my house.

11. *Write the negation of the following :*

(i) Anil is not rich and Kanchan is poor.

(ii) A cow is an animal.

(iii) If the determinant of a system of linear equations is zero, then either the system has no solution or it has an infinite number of solution.

Solution.

(i) Anil is rich and Kanchan is not poor.

(ii) A cow is not an animal.

(iii) p : Determinant of a system of linear equation is zero.

q : System has no solution.

r : System has an infinite number of solution.

Its negation is $\rceil p \rightarrow \rceil q \vee \rceil r$.

12. *Write in words the converse, inverse, contrapositive and negation of the implication "If 2 is less then 3, then 1/3 is less than 1/2.*

Solution. Let $p \equiv$ 2 is less than 3. $q \equiv$ 1/3 is less than 1/2.

Then implication is $p \Rightarrow q$:

(i) Converse of $p \Rightarrow q$ is $q \Rightarrow p$. In words $q \Rightarrow p$ means "If 1/3 is less than 1/2, then 2 is less than 3".

(ii) Inverse of $p \Rightarrow q$ is $\rceil p \Rightarrow \rceil q$. Thus in words, $\rceil p \Rightarrow \rceil q$ means "If 2 is not less than 3, then 1/3 is not less than 1/2".

(iii) Contrapositive of $p \Rightarrow q$ is $\rceil q \Rightarrow \rceil p$. Thus in words $\rceil q \Rightarrow \rceil p$ means "If 1/3 is not less than 1/2, then 2 is not less than 3".

(iv) Negation of $p \Rightarrow q$ is $\rceil(p \Rightarrow q)$. Thus in words $\rceil(p \Rightarrow q)$ means "It is false than p implies q".

13. *Let p be a statement "Eight is an even number", and let q be a statement "Candy is sweet". Write in words (i) the implication $p \Rightarrow q$, (ii) its converse, (iii) its inverse, (iv) its contrapositive, (v) its negation.*

Solution. (i) $p \Rightarrow q$ means "If eight is an even number, then candy is sweet."

(ii) Converse of $p \Rightarrow q$ is $q \Rightarrow p$ it means "If candy is sweet, then eight is an even number."

(iii) Inverse of $p \Rightarrow q$ is $\rceil p \Rightarrow \rceil q$. It means "If eight is not an even number, then candy is not sweet".

(iv) Contrapositive of $p \Rightarrow q$ is $\rceil q \Rightarrow \rceil p$. It means "If candy is not sweet, then eight is odd number."

(v) Negation of $p \Rightarrow q$ is $\rceil (p \Rightarrow q)$. In words, we can write "It is false that p implies q" or "Eight is an even number, and candy is not sweet".

☛ $\rceil(p \Rightarrow q) = \rceil(\rceil p \vee q) = p \wedge \rceil q.$

14. *If $p \equiv$ Missiles are costly and $q \equiv$ Grandma chews gum. Write in words, the following statement given in symbol :*

(i) $p \vee \rceil q$ *(ii)* $\rceil p \wedge \rceil q$ *(iii)* $(p \wedge \rceil q) \vee (\rceil p \wedge q)$

Solution. (i) $p \vee \rceil q \equiv$ Either missiles are costly or Grandma does not chew gum.

(ii) $\rceil p \wedge \rceil q \equiv$ Missiles are not costly and Grandma does not chew gum.

(iii) Either missiles are costly and Grandma does not chew gum, or missiles are not costly and Grandma chews gum.

15. *If $p \equiv$ Mathematics is easy and $q \equiv$ Two is less than three. Write in words the following statement given in symbols :*

(i) $\rceil(p \wedge q)$ *(ii)* $\rceil(p \vee q)$ *(iii)* $\rceil p \vee q$ *(iv)* $(p \wedge \rceil q) \vee (\rceil p \wedge q)$

Solution.

(i) It is false that mathematics is easy and two is less than three.

(ii) It is false that either mathematics is easy or two is less than three.

(iii) Either mathematics is not easy or two is not less than three.

(iv) Either mathematics is easy and two is less than three or mathematics is not easy and two is less than three.

16. *If $p \equiv$ It is 10 o' clock, $q \equiv$ the train is late, then state in words the following resultants :*

(i) $q \vee \rceil p$ *(ii)* $\rceil p \wedge q$ *(iii)* $p \wedge \rceil q$ *(iv)* $\rceil(p \wedge q) = \rceil p \vee \rceil q$

(v) $\rceil p \wedge \rceil q$

Solution.

(i) $q \vee \rceil p$: The train is late or it is not 10 o' clock.

(ii) $\rceil p \wedge q$: It is not 10 o' clock and the train is late.

(iii) $p \wedge \rceil q$: It is 10 o' clock and the train is not late.

(iv) $\rceil(p \wedge q) = \rceil p \vee \rceil q$: It is not 10 o' clock, or the train is not late.

(v) $\rceil p \wedge \rceil q$: It is not 10 o' clock and the train is not late.

17. *Consider the following :*

p : You take a course in Discrete Mathematics.

q : You understand logic.

r : You get an A grade on the final exam.

Write in simple sentences the meaning of the following :

(i) $q \Rightarrow r$ *(ii)* $\rceil p \Rightarrow \rceil q$ *(iii)* $(p \wedge q) \Rightarrow \rceil r$ *(iv)* $(p \wedge q) \Rightarrow r$

Solution.

(i) If you understand logics, then you get an A grade in the final exam.

(ii) If you will not take a course in Discrete Mathematics, then you will not understand logic.

(iii) If you take a course in Discrete Mathematics and understand logic, then you may not get an A grade in the final exam.

(iv) If you take a course in Discrete Mathematics and you understand logic, then you get an A grade in the final exam.

18. *Construct a truth table for each of the following functions :*

(i) $(p \wedge q \wedge r) \vee (\neg p \wedge q \wedge \neg r) \vee (\neg p \wedge \neg q \wedge \neg r)$

(ii) $(p \vee q \vee r) \wedge (\neg p \vee q \vee \neg r) \wedge (\neg p \vee \neg q \vee \neg r)$

(iii) $\neg(\neg p \vee (q \wedge r) \wedge \{(p \wedge q) \vee (\neg q \wedge r)\}$

Solution. (i) Truth table

p	q	r	$\neg p$	$\neg q$	$\neg r$	$(p \wedge q \wedge r)$	$(\neg p \wedge q \wedge \neg r)$	$(\neg p \wedge \neg q \wedge \neg r)$	(i)
T	T	T	F	F	F	T	F	F	T
T	T	F	F	F	T	F	F	F	F
T	F	T	F	T	F	F	F	F	F
F	T	T	T	F	F	F	F	F	F
T	F	F	F	T	T	F	F	F	F
F	T	F	T	F	T	F	T	F	T
F	F	T	T	T	F	F	F	F	F
F	F	F	T	T	T	F	F	T	T

(ii) Truth table

p	q	r	$\neg p$	$\neg q$	$\neg r$	$(p \vee q \vee r)$	$(\neg p \vee q \vee \neg r)$	$(\neg p \vee \neg q \vee \neg r)$	(ii)
T	T	T	F	F	F	T	T	F	F
T	T	F	F	F	T	T	T	T	T
T	F	T	F	T	F	T	F	T	F
F	T	T	T	F	F	T	T	T	T
T	F	F	F	T	T	T	T	T	T
F	T	F	T	F	T	T	T	T	T
F	F	T	T	T	F	T	T	T	T
F	F	F	T	T	T	F	T	T	F

(iii) Truth table

p	q	r	$\neg p$	$\neg q$	$p \wedge q$	$(q \wedge r)$	$\neg q \wedge r$	$\neg p \vee (q \wedge r)$	$\{\neg p \vee (q \wedge r)\}$	$\{(p \wedge q) \vee (\neg q \wedge r)\}$	(iii)
T	T	T	F	F	T	T	F	T	F	T	F
T	T	F	F	F	T	F	F	F	T	T	T
T	F	T	F	T	F	F	T	F	T	T	T
F	T	T	T	F	F	T	F	T	F	F	F
T	F	F	F	T	F	F	F	F	T	F	F
F	T	F	T	F	F	F	F	T	F	F	F
F	F	T	T	T	F	F	T	T	F	T	F
F	F	F	T	T	F	F	F	T	F	F	F

19. *Construct the truth table of the following :*

(a) $(p \wedge q) \wedge (q \wedge r) \wedge (r \wedge s)$ *(b)* $\neg(\neg(p \wedge \neg q)$

(c) $(\neg p \wedge (\neg q \wedge s)) \vee (r \wedge s) \wedge (s \vee r)$

Solution. (a)

p	q	r	s	$p \wedge q$	$q \wedge r$	$r \wedge s$	$(p \wedge q) \wedge (q \wedge r) \wedge (r \wedge s)$
T	T	T	T	T	T	T	T
T	T	T	F	T	T	F	F
T	T	F	T	T	F	F	F
T	F	T	T	F	F	T	F
F	T	T	T	F	T	T	F
T	T	F	F	T	F	F	F
T	F	F	T	F	F	F	F
F	T	F	T	F	F	F	F
F	F	T	T	F	F	T	F
T	F	T	F	F	F	F	F
F	T	T	F	F	T	F	F
T	F	F	F	F	F	F	F
F	T	F	F	F	F	F	F
F	F	T	F	F	F	F	F
F	F	F	T	F	F	F	F
F	F	F	F	F	F	F	F

(b)

p	q	$\rceil p$	$\rceil q$	$\rceil p \wedge \rceil q$	$\rceil(\rceil p \wedge \rceil q)$
T	T	F	F	F	T
T	F	F	T	F	T
F	T	T	F	F	T
F	F	T	T	T	F

(c)

p	q	r	s	$\rceil p$	$\rceil q$	$\rceil q \wedge s$	$\rceil p \wedge (\rceil q \wedge s)$ (1)	$r \wedge s$ (2)	$s \vee r$ (3)	(2) $\wedge$ (3) (4)	(1) $\vee$ (4)
T	T	T	T	F	F	F	F	T	T	T	T
T	T	T	F	F	F	F	F	F	T	F	F
T	T	F	T	F	F	F	F	F	T	F	F
T	F	T	T	F	T	T	F	T	T	T	T
F	T	T	T	T	F	F	F	T	T	T	T
T	T	F	F	F	F	F	F	F	F	F	F
T	F	F	T	F	T	T	F	F	T	F	F
T	F	T	F	F	T	F	F	F	T	F	F
F	F	T	T	T	T	T	T	T	T	T	T
T	F	F	F	F	T	F	F	F	F	F	F
F	T	T	F	T	F	F	F	F	F	F	F
T	T	F	T	F	F	F	F	F	T	F	F

F	T	F	F	T	F	F	F	F	F	F	F
F	F	T	F	T	T	F	F	F	T	F	F
F	F	F	T	T	T	T	T	F	T	F	T
F	F	F	F	T	T	F	F	F	F	F	F

20. *Write the negation for the statement* $\forall \in R, x > 3 \Rightarrow x^2 > 9$

Solution. Let $P(x)$ and $Q(x)$ denote '$x > 3$' and $x^2 > 9$.

Then the given statement can be written as

$$\forall \; x\,(P(x) \Rightarrow Q(x))$$

The negation of this statement is

$$\exists\, x\,(P(x) \wedge -Q(x))$$

i.e., there exist a real number x such that $x \leq 3$ and $x^2 \leq 9$.

2.7 TAUTOLOGY

We have already discussed about the compound statements which formed with the help of simple statements using connectives. The truth values of this compound statement depend on the truth values of simple statement substituted for the variables. Thus the truth table of a resulting compound statement gives the summary of all its truth values for all possible choice of values of the variables. Therefore sometimes the truth values of the compound statement may, be "True (*T*)" and sometimes "False (*F*)". But there are some compound statements whose truth values are always *T* or always *F* irrespective of all possible truth values given to the variables.

Definition : *A statement whose truth value is always T (i.e., True) is called a "Tautology" and the statement whose truth value is always False (i.e., F) is called a "Contradiction".*

☛ Negation of a tautology is a contradiction while negation of contradiction is a tautology.

2.7.1 Logical Equivalence

The two compound statements are said to be logically equivalents if both have same truth values for all possible assignments given to the variables. This logically equivalent is also known as tautologically equivalent.

2.8 DUALITY

The two compound statements are said to be dual of each other if either one can be obtained from other by interchanging $\wedge$ and $\vee$ provided both remain valid.

For example : The dual of $(p \vee q) \wedge r$ is $(p \wedge q) \vee r$ and the dual of $\rceil(p \vee q) \wedge \{p \vee \rceil(q \wedge \rceil s)\}$ is $\rceil(p \wedge q) \vee \{p \wedge \rceil(q \vee \rceil s)\}$

2.9 ALGEBRA OF STATEMENTS

Now we shall discuss some tautological laws related to the statements under the algebra of statements. Under this section, we shall also discuss some laws which are tautologies. For simplicity let us take '*t*' for tautology and '*f*' for contradiction.

2.9.1 Commutative Laws

(a) $(p \vee q) \Leftrightarrow (q \vee p)$ (b) $(p \wedge q) \Leftrightarrow (q \wedge p)$

Proof : (a) Truth table for $(p \vee q) \Leftrightarrow (q \vee p)$

p	q	$p \vee q$	$q \vee p$	$(p \vee q) \Leftrightarrow (q \vee p)$
T	T	T	T	T
T	F	T	T	T
F	T	T	T	T
F	F	F	F	T

Thus in the last column all the truth values are T. Hence it is a tautology.
(b) Truth table for $(p \wedge q) \Leftrightarrow (q \wedge p)$

p	q	$p \wedge q$	$q \wedge p$	$(p \wedge q) \Leftrightarrow (q \wedge p)$
T	T	T	T	T
T	F	F	F	T
F	T	F	F	T
F	F	F	F	T

Similarly, $(p \wedge q) \Leftrightarrow (q \wedge p)$ is a tautology.

☛ The notation $\Leftrightarrow$ can be replaced by '$\equiv$' or '='
i.e., $(p \vee q) = (q \vee p)$; $(p \wedge q) = (q \wedge p)$

2.9.2 Associative Laws

(a) $p \vee (q \vee r) \Leftrightarrow (p \vee q) \vee r$ (b) $p \wedge (q \wedge r) \Leftrightarrow (p \wedge q) \wedge r$

Proof : (a) Truth table for $p \vee (q \vee r) \Leftrightarrow (p \vee q) \vee r$

p	q	r	$p \vee q$	$q \vee r$	$p \vee (q \vee r)$	$(p \vee q) \vee r$	$p \vee (q \vee r) \Leftrightarrow (p \vee q) \vee r$
T	T	T	T	T	T	T	T
T	T	F	T	T	T	T	T
T	F	T	T	T	T	T	T
F	T	T	T	T	T	T	T
T	F	F	T	F	T	T	T
F	T	F	T	T	T	T	T
F	F	T	F	T	T	T	T
F	F	F	F	F	F	F	T

Thus $p \vee (q \vee r) \Leftrightarrow (p \vee q) \vee r$ is a tautology.
(b) Truth table for $p \wedge (q \wedge r) \Leftrightarrow (p \wedge q) \wedge r$

p	q	r	$p \wedge q$	$q \wedge r$	$p \wedge (q \wedge r)$	$(p \wedge q) \wedge r$	$p \wedge (q \wedge r) \Leftrightarrow (p \wedge q) \wedge r$
T	T	T	T	T	T	T	T
T	T	F	T	F	F	F	T
T	F	T	F	F	F	F	T
F	T	T	F	T	F	F	T
T	F	F	F	F	F	F	T
F	T	F	F	F	F	F	T
F	F	T	F	F	F	F	T
F	F	F	F	F	F	F	T

Similarly, $p \wedge (q \wedge r) \Leftrightarrow (p \wedge q) \wedge r$ is a tautology.

2.9.3 Distributive Laws

(a) $p \wedge (q \vee r) \Leftrightarrow (p \wedge q) \vee (p \wedge r)$(b) $p \vee (q \wedge r) \Leftrightarrow (p \vee q) \wedge (p \vee r)$.

Proof : (a) Truth table for $p \wedge (q \vee r) \Leftrightarrow (p \wedge q) \vee (p \wedge r)$

p	q	r	$p\wedge q$	$q\vee r$	$p\wedge r$	$p\wedge(q\vee r)$	$(p\wedge q)\vee(p\wedge r)$	$p\wedge(q\vee r)\Leftrightarrow(p\wedge q)\vee(p\wedge r)$
T	T	T	T	T	T	T	T	T
T	T	F	T	T	F	T	T	T
T	F	T	F	T	T	T	T	T
F	T	T	F	T	F	F	F	T
T	F	F	F	F	F	F	F	T
F	T	F	F	T	F	F	F	T
F	F	T	F	T	F	F	F	T
F	F	F	F	F	F	F	F	T

Thus the given expression has the same truth value T. Hence it is a tautology.
(b) Truth table for $p \vee (q \wedge r) \Leftrightarrow (p \vee q) \wedge (p \vee r)$.

p	q	r	$p\vee q$	$q\wedge r$	$p\vee r$	$p\vee(q\wedge r)$	$(p\vee q)\wedge(p\vee r)$	$p\vee(q\wedge r)\Leftrightarrow(p\vee q)\wedge(p\vee r)$
T	T	T	T	T	T	T	T	T
T	T	F	T	F	T	T	T	T
T	F	T	T	F	T	T	T	T
F	T	T	T	T	T	T	T	T
T	F	F	T	F	T	T	T	T
F	T	F	T	F	F	F	F	T
F	F	T	F	F	T	F	F	T
F	F	F	F	F	F	F	F	T

Similarly, $p \vee (q \wedge r) \Leftrightarrow (p \vee q) \wedge (p \vee r)$ is a tautology

2.9.4 Idempotent Laws

(a) $(p \vee p) \Leftrightarrow p$ (b) $(p \wedge p) \Leftrightarrow p$.

Proof : (a) Truth table for $(p \vee p) \Leftrightarrow p$

p	p	$p \vee p$	$(p \vee p) \Leftrightarrow p$
T	T	T	T
F	F	F	T

Here, the last column has same truth value T. Hence it is a tautology.
(b) Truth table for $(p \wedge p) \Leftrightarrow p$

p	p	$p \wedge p$	$(p \wedge p) \Leftrightarrow p$
T	T	T	T
F	F	F	T

Similarly, $(p \wedge p) \Leftrightarrow p$ is a tautology.

2.9.5 Absorption Laws

(a) $p \vee (p \wedge q) \Leftrightarrow p$; (b) $p \wedge (p \vee q) \Leftrightarrow p$

Proof : (a) Truth table for $p \vee (p \wedge q) \Leftrightarrow p$

p	q	$p \wedge q$	$p \vee (p \wedge q)$	$p \vee (p \wedge q) \Leftrightarrow p$
T	T	T	T	T
T	F	F	T	T
F	T	F	F	T
F	F	F	F	T

Since the last column contains all the truth values true. Hence $p \vee (p \wedge (p \wedge q) \Leftrightarrow p$ is a tautology.

(b) Truth table for $p \wedge (p \vee q) \Leftrightarrow p$

p	q	$p \vee q$	$p \wedge (p \vee q)$	$p \wedge (p \vee q) \Leftrightarrow p$
T	T	T	T	T
T	F	T	T	T
F	T	T	F	T
F	F	F	F	T

Similarly, $p \wedge (p \vee q) \Leftrightarrow p$ is a tautology.

2.9.6 De-Morgan's Laws

(a) $\urcorner (p \vee q) \Leftrightarrow (\urcorner p \wedge \urcorner q)$; (b) $\urcorner (p \wedge q) \Leftrightarrow (\urcorner p \vee \urcorner q)$.

Proof : (a) Truth table for $\urcorner (p \vee q) \Leftrightarrow (\urcorner p \wedge \urcorner q)$

p	q	$p \vee q$	$\urcorner p$	$\urcorner q$	$\urcorner (p \vee q)$	$(\urcorner p \wedge \urcorner q)$	$\urcorner (p \vee q) \Leftrightarrow (\urcorner p \wedge \urcorner q)$
T	T	T	F	F	F	F	T
T	F	T	F	T	F	F	T
F	T	T	T	F	F	F	T
F	F	F	T	T	T	T	T

The last column contains all the truth values 'true'. Hence $\urcorner (p \vee q) \Leftrightarrow (\urcorner p \wedge \urcorner q)$ is a tautology.

(b) Truth table for $\urcorner (p \wedge q) \Leftrightarrow (\urcorner p \vee \urcorner q)$

p	q	$\urcorner p$	$\urcorner q$	$(p \wedge q)$	$\urcorner (p \wedge q)$	$(\urcorner p \vee \urcorner q)$	$\urcorner (p \wedge q) \Leftrightarrow (\urcorner p \vee \urcorner q)$
T	T	F	F	T	F	F	T
T	F	F	T	F	T	T	T
F	T	T	F	F	T	T	T
F	F	T	T	F	T	T	T

Similarly, $\urcorner (p \wedge q) \Leftrightarrow (\urcorner p \vee \urcorner q)$ is a tautology.

2.9.7 Detachment Law

$$[(p \Rightarrow q) \wedge p] \Rightarrow q.$$

Proof : Truth table for $[(p \Rightarrow q) \wedge p] \Rightarrow q$

p	q	$p \Rightarrow q$	$(p \Rightarrow q) \wedge p$	$[(p \Rightarrow q) \wedge p] \Rightarrow q$
T	T	T	T	T
T	F	F	F	T
F	T	T	F	T
F	F	T	F	T

The last column contains all the truth values 'True'. Hence it is a tautology.

2.9.8 Chain Laws

$$[(p \Rightarrow q) \wedge (q \Rightarrow r)] \Rightarrow (p \Rightarrow r)$$

Proof : Truth table for chain law

p	q	r	$(p \Rightarrow q)$	$(q \Rightarrow r)$	$(p \Rightarrow q) \wedge (q \Rightarrow r)$	$(p \Rightarrow r)$	$[(p \Rightarrow q) \wedge (q \Rightarrow r)] \Rightarrow (p \Rightarrow r)$
T	T	T	T	T	T	T	T
T	T	F	T	F	F	F	T
T	F	T	F	T	F	T	T
F	T	T	T	T	T	T	T
T	F	F	F	T	F	F	T
F	T	F	T	F	F	T	T
F	F	T	T	T	T	T	T
F	F	F	T	T	T	T	T

Since the last column contains all the truth values 'true'. Hence it is a tautology.

- The chain law can be restated as follows "If p implies q and q implies r, then p implies r".

2.9.9 Identity Laws

If t stands for tautology and f stands for contradiction. Then

(a) $p \wedge t = t \wedge p = p$ (b) $p \vee f = f \vee p = p$.

Proof : Truth table for (a)

p	t	$p \wedge t$	$t \wedge p$
T	T	T	T
F	T	F	F

Truth table for (b)

p	f	$p \vee f$	$f \vee p$
T	F	T	T
F	F	F	F

2.9.10 Complement laws

(a) $p \vee \rceil p = t$ (b) $p \wedge \rceil p = f$

Proof : Truth table for (a)

p	t	$\rceil p$	$p \vee \rceil p$
T	T	F	T
F	T	T	T

Truth table for (b)

p	f	$\rceil p$	$p \wedge \rceil p$
T	F	F	F
F	F	T	F

2.9.11 Functionally Complete set of operations

"A set of operations or connectives is said to be functionally complete if every statement can be expressed entirely in terms of the operations in the set.

Illustrations :

(1) We know that $p \Rightarrow q$ is equal to $\sim pvq$ therefore it is possible to replace each occurrence of $\Rightarrow$ in any statement with an equivalent expression involving $\sim$ and $\vee$.

(2) We know that $p \Leftrightarrow q = (p \Rightarrow q) \wedge (q \Rightarrow p) = (\sim p \vee q) \wedge (\sim q \vee p)$

Therefore the symbol $\Leftrightarrow$ can be replaced by connectives $\vee$, $\wedge$ and $\sim$.

(3) We know that (by Demorgan's law)

$$\sim (p \wedge q) = \sim p \vee \sim q \qquad \text{or} \qquad p \wedge q = \sim (\sim p \vee \sim q)$$

Similarly $p \vee q = \sim (\sim p \wedge \sim q)$

Hence, it is possible to replace $\wedge$ in any statement by connectives $\sim$ and $\vee$. So. any statement can be expressed in an equivalent statement containing $\sim$ and $\wedge$ only, which shows that $\{\wedge, \sim\}$ is functionally complete set of operations. In a similar in an equivalent statement containing $\vee$ and $\sim$ only.

Hence, $\{\vee, \sim\}$ is also functionally complete set.

ILLUSTRATIVE EXAMPLES

1. *Show that $\{\Rightarrow, \sim\}$ is a functionally complete set.*

Solution : From above illustrations we know that $\{\vee, \sim\}$ is a functionally complete set. Therefore, any statement can be expressed in an equivelent statement containing $\vee$ and $\sim$

Since $p \vee q = (\sim p) \Rightarrow q$

$\therefore$ We an replace the connectives $\vee$ in ay statement by $\sim$ and $\Rightarrow$. So, given set $\{\Rightarrow, \sim\}$ is a functionally complete set.

2. *Show that $\{\wedge, \vee\}$ is not a functionally complete set.*

Solution : We know that connectives '$\sim$' can not be expressed entirely in terms of $\vee$ and $\wedge$, therefore, any statement containing '$\sim$' can not be expressed in an equivalent statement containing $\wedge$ and $\vee$ only. Hence, the given set $\{\wedge, \vee\}$ is not a functionally complete set.

2.10 VALIDITY OF ARGUMENTS : METHOD OF PROOFS

The main problem related to the symbolic logic is the investigation of the process of reasoning. We know that in every deductive science, there are no positive declarations having absolute truth. Therefore, we assumed a certain set of statements without proof, and from this set, we obtained some other statements using connectives. We now proceed to investigate those processes which will be accepted as valid in the derivation of a statement, called the conclusion, from other given statements, called premises.

An argument is a process by which a conclusion is obtained from given set of premises. Let $p_1, p_2, \ldots, p_n$ be the all premises and r is the conclusion. Then an argument which yields a conclusion r from premises $p_1, p_2 \ldots p_n$ is valid if and only if the statement $(p_1 \wedge p_2 \wedge \ldots, p_n) \Rightarrow r$ is a tautology. There are three methods to check the validity of a given argument.

Method I : By Truth Table : The first method is to check the validity directly from the truth table, that is for the argument $(p_1 \wedge p_2 \wedge \ldots, \wedge p_n) \Rightarrow r$ the truth table method is used to show that $(p_1 \wedge p_2 \wedge \ldots \wedge p_n) \Rightarrow r$ is a tautology.

Method II : By Simplification Method : In this method we shall have to show that the statement $(p_1 \wedge p_2 \wedge .., \wedge p_n) \Rightarrow r$ can be reduced to 1, using the standard methods of simplification.

Method III : Using Rule of Inference : This method is to reduce the given argument to a series of argument each of which is valid. This method is often the simplest of the three.

Two of the most frequently used valid arguments are the rule of detachment and the law of syllogism. The rule of detachment is given by

$$\begin{array}{c} p \\ p \Rightarrow q \\ \hline q \end{array}$$

Here p and $(p \Rightarrow q)$ are premises and the conclusion is written below a horizontal line, which is q as well. The law of syllogism is given by

$$\begin{array}{c} p \Rightarrow q \\ q \Rightarrow r \\ \hline p \Rightarrow r \end{array}$$

In checking the validity of any argument, we shall also assume that the following rules of substitution are permissible to use.

Rule I : Any valid argument which involves a statement variable will remain valid if every occurrence of a given variable is replaced by a specific statement.

Rule II : Any valid argument will remain valid if any equivalent statement occurrence of a statement is replaced by an equivalent statement.

☛ The validity of any argument does not depend on the truth values 'True' or 'False' of the conclusion. For example, Let us consider two arguments :

1. If ice is warm, then snow is black.
   Ice is warm
   Snow is black
2. 5 is an odd integer.
   If 4 is an even integer, then 5 is an odd integer
   4 is an even integer

Here the first is valid although the conclusion is false, while the second is invalid although the conclusion is true.

2.11 RULE OF DETACHMENT OR MODUS PONENS

The valid argument

$$\begin{array}{c} p \\ p \Rightarrow q \\ \hline q \end{array}$$

is known as rule of detachment. It is also known as modus ponens.

Law of Syllogism. The argument

$$\begin{array}{c} p \Rightarrow q \\ q \Rightarrow r \\ \hline p \Rightarrow r \end{array}$$

is valid argument and known as law of syllogism.

Its implication form is $[(p \Rightarrow q) \wedge (q \Rightarrow r)] \Rightarrow [(p \Rightarrow r)]$

Truth table for law of syllogism

p	q	r	$p \Rightarrow q$	$q \Rightarrow r$	$p \Rightarrow r$	$(p \Rightarrow q) \wedge (q \Rightarrow r)$	$(p \Rightarrow q) \wedge (q \Rightarrow r) \Rightarrow (p \Rightarrow r)$
T	T	T	T	T	T	T	T
T	T	F	T	F	F	F	T
T	F	T	F	T	T	F	T
T	F	F	F	T	F	F	T
F	T	T	T	T	T	T	T
F	T	F	T	F	T	F	T
F	F	T	T	T	T	T	T
F	F	F	T	T	T	T	T

From above table, it is clear that, the law of syllogism is a valid argument.

You must know

- The word *theorem* should not be confused with the word 'theory'. A *theorem* is a specific statement that can be proved. A theory is a broader assembly of ideas and a particular issue.
- The word *theorem* Carries the connotation of importance and generality. For example, Pythogoorean theorem certainly deserves to be called a theorem.
- Some words that are alternatives to *theorem* are given below.

(i) Result : A modest generic word for a theorem. Both important and unimportant theorem can be called results.

(ii) Fact : A very miner theorem. For example '2 +4 = 6' is a fact

(iii) Proposition : A minor theorem. A proposition is more important or more general than a fact but not as prestigious as a theorem.

(iv) Lemma : The lemma are the parts or tools, used to build the more complicated proof of the theorem. So , lemma is a theorem whose main purpose is to help prove another more important theorem.

(v) Corollary : A result with a short proof whose main step is the use of another previously proved theorem.

(vi) Claim : A claim is a theorem whose statement usually appears inside the proof of a theorem. The purpose of a claim is to help to organise key steps in the proof. It is very similar to lemma.

2.12 RULES OF INFERENCE

In the following table, we give some rules of inference. Here, observe that for valid argument, we can use the rules of inference :

Rules of Inference	Implication Form
RI_1 : Addition : $\dfrac{p}{\therefore p \vee q}$	$p \Rightarrow (p \vee q)$
RI_2 : Conjunction : $\dfrac{q}{\therefore p \wedge q}$	$q \Rightarrow p \wedge q$
RI_3 : Simplification : $\dfrac{p \wedge q}{\therefore p}$	$(p \wedge q) \Rightarrow p$

RI_4 : Modus ponens : p $p \Rightarrow q$ $\therefore q$	$(p \wedge (p \Rightarrow q)) \Rightarrow q$
RI_5 : Modus tollens : $\rceil q$ $p \Rightarrow q$ $\therefore \rceil p$	$(\rceil q \wedge (p \Rightarrow q)) \Rightarrow \rceil p$
RI_6 : Disjunctive syllogism : $\rceil p$ $p \vee q$ $\therefore q$	$(\rceil p \wedge (p \vee q)) \Rightarrow q$
RI_7 : Hypothetical syllogism : $p \Rightarrow q$ $q \Rightarrow r$ $\therefore p \Rightarrow r$	$((p \Rightarrow q) \wedge (q \Rightarrow r)) \Rightarrow (p \Rightarrow r)$
RI_8 : Constructive dilemma : $(p \Rightarrow q) \wedge (r \Rightarrow S)$ $p \vee r$ $\therefore q \vee S$	$(p \Rightarrow q) \wedge (r \Rightarrow S) \wedge (p \vee r) \Rightarrow (q \wedge S)$
RI_9 : Destructive dilemma : $(p \Rightarrow q) \wedge (r \Rightarrow S)$ $\rceil q \vee \rceil S$ $\therefore p \vee r$	$(p \Rightarrow q) \wedge (r \Rightarrow S) \wedge (\rceil q \vee \rceil S)$ $\Rightarrow (\rceil p \vee \rceil r)$

ILLUSTRATIVE EXAMPLES

1. *Check the validity of the following argument.*

$$\begin{array}{c} p \\ p \Rightarrow q \\ \underline{q \Rightarrow r} \\ r \end{array}$$

Solution : Method-I : For validity of the above argument, following statement

$$f = [p \wedge (p \Rightarrow q) \wedge (q \Rightarrow r)] \Rightarrow r$$

must be a tautology.

Truth table

p	q	r	$p \Rightarrow q$	$q \Rightarrow r$	$p \wedge (p \Rightarrow q)$	$p \wedge (p \Rightarrow q) \wedge (q \Rightarrow r)$	f
T	T	T	T	T	T	T	T
T	T	F	T	F	T	F	T
T	F	T	F	T	F	F	T
T	F	F	F	T	F	F	T
F	T	T	T	T	F	F	T
F	T	F	T	F	F	F	T
F	F	T	T	T	F	F	T
F	F	F	T	T	F	F	T

From above, it is clear that f is a tautology. Hence given argument is valid.

Method-II : *(Simplification method)*

$f = [p \wedge (p \Rightarrow q) \wedge (q \Rightarrow r)] \Rightarrow r$

$= \sim[(p \wedge (\sim p \vee q) \wedge (\sim q \vee r)] \vee r$ $\quad (\because\ p \Rightarrow q = \sim p \vee q)$

$= \sim p \vee (p \wedge \sim q) \vee (q \wedge \sim r) \vee r$ $\quad (\because\ \sim(p \wedge q) = \sim p \vee v \sim q)$

$= (\sim p \vee p) \wedge (\sim p \vee \sim q) \vee (q \wedge \sim r) \vee r$ $\quad$ (By distribulirity)

$= T \wedge (\sim p \vee \sim q) \vee r \vee (q \wedge \sim r)$ $\quad (\because \sim p \vee p = T)$

$= (\sim p \vee \sim q) \vee \{(v \vee q) \wedge (r \vee \sim r)]$ $\quad (\because p \wedge T = p)$

$= (\sim p \vee \sim q) \vee \{(r \vee q) \wedge T)$ $\quad (\because\ r \vee \sim r = T)$

$= \{\sim p \vee \sim q) \vee (r \vee q)$ $\quad (\because p \wedge T = p)$

$= \{(\sim p \vee \sim q) \vee q\} \vee r$ $\quad (\because p \vee q = q \vee p)$

$= \{\sim p \vee (\sim q \vee q)\} \vee r$ $\quad$ (By associalivity)

$= \{\sim p \vee T\} \vee r$ $\quad (\because\ \sim p \vee p = T)$

$= T$

$\Rightarrow$ Hence, given argument is valid.

Method-III : *(Rule of inference)*

Consider following sequence of argument,

p	a premise
$p \Rightarrow q$	a premise
p	by modus ponens
$q \Rightarrow r$	a premise
r	by modus ponens

Hence, given argument is valid.

2. *Check the validity of the following argument*

(i) $p \Rightarrow q$
$r \Rightarrow \sim q$
―――
$p \Rightarrow \sim r$

(ii) p
$p \wedge q \Rightarrow r \vee s$
$\sim s$
―――
r

Solution : (1) Since the statement $r \Rightarrow \sim q$ is equal to $q \Rightarrow \sim r$ we can replace the premise $r \Rightarrow \sim q$ by $q \Rightarrow \sim r$

Now $P \Rightarrow q$
$q \Rightarrow \sim r$
―――
$p \Rightarrow \sim r$

is valid argument by the law of syllogism. Hence given argument is valid.

(2) We a take as premise for the indirect proof of all given premise except $\sim s$ and the negation of the conclusions $\sim r$. We shall show that the following argument is valid.

p
$p \wedge q \Rightarrow r \vee s$
q
$\sim r$
―――
s

Now,

p	a premise
q	a premise
$p \wedge q$	a conclusion because $p \wedge q \Rightarrow p \wedge q$ is always a tautology.

Further

$p \wedge q$	a valid conclusion
$p \wedge q \Rightarrow r \vee s$	a premise
$r \vee s$	a valid conclusion by modus ponens
$\sim r$	a premise
s	a valid Conclusion ($\because (r \vee s) \Rightarrow \sim r \Rightarrow s$ is a tautology)

3. *Check the validity of the following argument : If Shreesh has completed B.E. or M.B.A., then he is assured of a good job. If Shreesh is assured of a good job, he is happy. Shreesh is not happy. So shreesh has not completed M.B.A.*

Solution. We can name the proposition in the following way :

P denotes "Shreesh has completed B.E. "

Q denotes "Shreesh has completed M.B.A."

R denotes "Shreesh is assured of a good job"

S denotes "Shreesh is happy".

The given premises are :

(i) $(P \vee Q) \Rightarrow R$ (ii) $R \Rightarrow S$ (iii) $\rceil S$

The conclusion is $\rceil Q$

1. $(P \vee Q) \Rightarrow R$ Premise (i)
2. $R \Rightarrow S$ Premise (ii)
3. $(P \vee Q) \Rightarrow S$ Hypothetical syllogism RI_7
4. $\rceil S$ Premise (iii)
5. $\rceil (P \vee Q)$ Modus tollens
6. $\rceil P \wedge \rceil Q$ DeMorgan's law
7. $\rceil Q$ Simplification RI_3

Thus the argument is valid.

4. *Test the validity of the following argument : If milk is black then every crow is white. If every crow is white, then it has four legs. If every crow has four legs, then every buffalo is white and brisk. The milk is black. Therefore the buffalo is white.*

Solution. : We name the propositions in the following way :

P denotes "The milk is black"

Q denotes "Every crow is white"

R denotes "Every crow has four legs"

S denotes "Every buffalo is white"

T denotes "Every buffalo is brisk"

The given premises are :

(i) $p \Rightarrow q$ (ii) $q \Rightarrow r$ (iii) $r \Rightarrow s \wedge t$ (iv) p

The conclusion is S.

Therefore,

1. p Premise (iv)
2. $p \Rightarrow q$ Premise (i)
3. q Modus ponens RI_4
4. $q \Rightarrow r$ Premise (ii)
5. r Modus ponens RI_4
6. $r \Rightarrow s \wedge t$ Premise (iii)
7. $s \wedge t$ Modus ponens RI_4
8. s Simplification RI_3

Hence, the argument is valid.

MISCELLANEOUS ILLUSTRATIVE EXAMPLES

1. *Prove that each of the following is a tautology :*

(a) $p \Rightarrow p$ *(b)* $p \wedge (p \Rightarrow q) \Rightarrow q$

(c) $\rceil p \Rightarrow (p \Rightarrow q)$ *(d)* $[(p \Rightarrow q) \wedge (q \Rightarrow r)] \Rightarrow (p \Rightarrow r)$

(e) $[(p \wedge q) \vee (q \wedge r) \vee (r \wedge p)] \Rightarrow [(p \vee q) \wedge (q \vee r) \wedge (r \vee p)]$

(f) $(p \Rightarrow q) \Rightarrow [(p \vee (q \wedge r) \Leftrightarrow [\{q \wedge (p \vee r)\}]$

Solution. (a) Truth table for $p \Rightarrow p$

p	p	$p \Rightarrow p$
T	T	T
F	F	T

Since the last column contains all the truth values T. Hence $p \Rightarrow p$ is a tautology.

(b) Truth table for $p \wedge (p \Rightarrow q) \Rightarrow q$

p	q	$p \Rightarrow q$	$p \wedge (p \Rightarrow q)$	$p \wedge (p \Rightarrow q) \Rightarrow q$
T	T	T	T	T
T	F	F	F	T
F	T	T	F	T
F	F	T	F	T

Since the last column contains all the truth values T. Hence $p \wedge (p \Rightarrow q) \Rightarrow q$ is a tautology.

(c) Truth table for $\neg p \Rightarrow (p \Rightarrow q)$

p	q	$\neg p$	$p \Rightarrow q$	$\neg p \Rightarrow (p \Rightarrow q)$
T	T	F	T	T
T	F	F	F	T
F	T	T	T	T
F	F	T	T	T

Since the last column of the above table contains all the truth values T (True). Hence given statement is a tautology.

(d) Truth table for $[(p \Rightarrow q) \wedge (q \Rightarrow r)] \Rightarrow (p \Rightarrow r)$

p	q	r	$p \Rightarrow q$	$q \Rightarrow r$	$(p \Rightarrow q) \wedge (q \Rightarrow r)$	$p \Rightarrow r$	$[(p \Rightarrow q) \wedge (q \Rightarrow r)] \Rightarrow (p \Rightarrow r)$
T	T	T	T	T	T	T	T
T	T	F	T	F	F	F	T
T	F	T	F	T	F	T	T
F	T	T	T	T	T	T	T
T	F	F	F	T	F	F	T
F	T	F	T	F	F	T	T
F	F	T	T	T	T	T	T
F	F	F	T	T	T	T	T

Since the last column contains all the truth values T (*i.e.*, True). Hence the given statement is a tautology.

(e) Truth table for $[(p \wedge q) \vee (q \wedge r) \vee (r \wedge p)] \Rightarrow [(p \vee q) \wedge (q \vee r) \wedge (r \vee p)]$

p	q	r	$p \wedge q$	$q \wedge r$	$r \wedge p$	$p \vee q$	$q \vee r$	$r \vee p$	$[(p \wedge q) \vee (q \wedge r) \vee (r \wedge p)]$	$[(p \vee q) \wedge (q \vee r) \wedge (r \vee p)]$	(e)
T	T	T	T	T	T	T	T	T	T	T	T
T	T	F	T	F	F	T	T	T	T	T	T
T	F	T	F	F	T	T	T	T	T	T	T
F	T	T	F	T	F	T	T	T	T	T	T
T	F	F	F	F	F	T	F	T	F	F	T
F	T	F	F	F	F	T	T	F	F	F	T
F	F	T	F	F	F	F	T	T	F	F	T
F	F	F	F	F	F	F	F	F	F	F	T

Since the last column of above, contains all the truth value T (*i. e.* true). Hence the given statement is a tautology.

(f) Truth table for $(p \Rightarrow q) \Rightarrow [(p \vee (q \wedge r) \Leftrightarrow \{ q \wedge (p \vee r)\}]$

p	q	r	$q \wedge r$	$p \vee r$	$p \vee (q \wedge r)$	$q \wedge (p \vee r)$	$p \Rightarrow q$	$\{p \vee (q \wedge r) \Leftrightarrow \{q \wedge (p \vee r)\}$	(f)
T	T	T	T	T	T	T	T	T	T
T	T	F	F	T	T	T	T	T	T
T	F	T	F	T	T	F	F	F	T
F	T	T	T	T	T	T	T	T	T
T	F	F	F	T	T	F	F	F	T
F	T	F	F	F	F	F	T	T	T
F	F	T	F	T	F	F	T	T	T
F	F	F	F	F	F	F	T	T	T

Since the last column of above table contains all the truth values T (*i. e.*, true). Hence the given statement is a tautology.

2. *The necessary and sufficient condition for two statements p and q to be logically equivalent is that $p \Leftrightarrow q$ is a tautology.*

Solution. If the truth value of $p \Leftrightarrow q$ is always T, then $p \Leftrightarrow q$ is a tautology. By the necessary and sufficient condition for the truth value of $p \Leftrightarrow q$ to be T, is that, the truth values of p and q should always be same. Thus p and q are logically equivalent.

3. *Prove that each of the following statement is a tautology :*

(a) $(p \wedge q) \Rightarrow q$ *(b)* $(p \vee \neg p)$ *(c)* $p \vee q \Rightarrow q \vee p$ *(d)* $(\neg(p \wedge \neg p)$

(e) $(p \Rightarrow q) \Leftrightarrow (\neg p \vee q)$ *(f)* $(p \Leftrightarrow q) \Leftrightarrow ((p \Rightarrow q) \wedge (q \Rightarrow p)]$

(g) $(p \Rightarrow q) \Leftrightarrow (\neg q \Rightarrow \neg p)$

Solution. (a) Truth table for $(p \wedge q) \Rightarrow q$

p	q	$p \wedge q$	$(p \wedge q) \Rightarrow q$
T	T	T	T
T	F	F	T
F	T	F	T
F	F	F	T

Since $(p \wedge q) \Rightarrow q$ has its all truth values 'T'. Hence it is a tautology.

(b) Truth table for $p \vee \rceil p$

p	$\rceil p$	$p \vee \rceil p$
T	F	T
F	T	T

Since $p \vee \rceil p$ has always truth value T. Hence it is a tautology.

(c) Truth table for $p \vee q \Rightarrow q \vee p$

p	q	$p \vee q$	$q \vee p$	$p \vee q \Rightarrow q \vee p$
T	T	T	T	T
T	F	T	T	T
F	T	T	T	T
F	F	F	F	T

Since $p \vee q \Rightarrow q \vee p$ has its all truth values T. Hence it is a tautology.

(d) Truth table for $\rceil (p \wedge \rceil p)$

p	$\rceil p$	$p \wedge \rceil p$	$\rceil (p \wedge \rceil p)$
T	F	F	T
F	T	F	T

Since $\rceil (p \wedge \rceil p)$ has its all truth values T. Hence it is a tautology.

(e) Truth table for $(p \Rightarrow q) \Leftrightarrow (\rceil p \vee q)$

p	q	$\rceil p$	$p \Rightarrow q$	$\rceil p \vee q$	$(p \Rightarrow q) \Leftrightarrow (\rceil p \vee q)$
T	T	F	T	T	T
T	F	F	F	F	T
F	T	T	T	T	T
F	F	T	T	T	T

The last column contains all the truth values 'T'. Hence it is a tautology.

(f) Truth table for $(p \Leftrightarrow q) \Leftrightarrow [(p \Rightarrow q) \wedge (q \Rightarrow p)]$

p	q	$p \Rightarrow q$	$q \Rightarrow p$	$p \Leftrightarrow q$	$(p \Rightarrow q) \wedge (q \Rightarrow p)$	$(p \Leftrightarrow q) \Leftrightarrow [(p \Rightarrow q) \wedge (q \Rightarrow p)]$
T	T	T	T	T	T	T
T	F	F	T	F	F	T
F	T	T	F	F	F	T
F	F	T	T	T	T	T

The last column of above table contains all the truth values T. Hence, the given statement is a tautology.

(g)

p	q	$\rceil p$	$\rceil q$	$p \Rightarrow q$	$\rceil q \Rightarrow \rceil p$	$(p \Rightarrow q) \Leftrightarrow (\rceil q \Rightarrow \rceil p)$
T	T	F	F	T	T	T
T	F	F	T	F	F	T
F	T	T	F	T	T	T
F	F	T	T	T	T	T

The last column contains all the truth values. T. Hence, it is a tautology.

4. *Check whether following are tautology or contradiction :*

(a) $[(p \wedge q) \Rightarrow r] \Leftrightarrow [(p \Rightarrow r) \vee (q \Rightarrow r)]$ (b) $(p \Rightarrow q \wedge r) \Rightarrow (\rceil r \Rightarrow \rceil p)$

(c) $(p \Leftrightarrow q \wedge r) \Rightarrow (\rceil r \Leftrightarrow \rceil p)$ (d) $p \wedge (q \vee r) \Leftrightarrow (p \wedge q) \vee (q \wedge r)$

(e) $([p \Leftrightarrow q) \vee (r \Rightarrow p)$ (f) $(p \vee q \vee r) \Leftrightarrow [\{\rceil p \Rightarrow q) \Rightarrow q\} \Rightarrow r]$

(g) $[(p \wedge q) \Rightarrow p] \Rightarrow [q \wedge \rceil q]$ (h) $[\{(p \Rightarrow q) \vee p\} \wedge q] \Rightarrow q$

(i) $p \wedge (q \wedge \rceil p)$

Solution. (a) Truth table for $[(p \wedge q) \Rightarrow r] \Leftrightarrow [(p \Rightarrow r) \vee (q \Rightarrow r)]$

p	q	r	$p \wedge q$	$p \Rightarrow r$	$q \Rightarrow r$	$(p \wedge q) \Rightarrow r$	$(p \Rightarrow r) \vee (q \Rightarrow r)$	(a)
T	T	T	T	T	T	T	T	T
T	T	F	T	F	F	F	F	F
T	F	T	F	T	T	T	T	T
F	T	T	F	T	T	T	T	T
T	F	F	F	F	T	T	T	T
F	T	F	F	T	F	T	T	T
F	F	T	F	T	T	T	T	T
F	F	F	F	T	T	T	T	T

The last column of above table does not contain all the truth values is T. Hence, the given statement is not a tautology.

(b) Truth table for $(p \Rightarrow q \wedge r) \Rightarrow (\rceil r \Rightarrow \rceil p)$

p	q	r	$\rceil p$	$\rceil r$	$q \wedge r$	$p \Rightarrow q \wedge r$	$\rceil r \Rightarrow \rceil p$	$(p \Rightarrow q \wedge r) \Rightarrow (\rceil r \Rightarrow \rceil p)$
T	T	T	F	F	T	T	T	T
T	T	F	F	T	F	F	F	T
T	F	T	F	F	F	F	T	T
F	T	T	T	F	T	T	T	T
T	F	F	F	T	F	F	F	T
F	T	F	T	T	F	T	T	T
F	F	T	T	F	F	T	T	T
F	F	F	T	T	F	T	T	T

The given statement contains all its truth values T. Hence, it is a tautology.

(c) Truth table for $(p \Leftrightarrow q \wedge r) \Rightarrow (\rceil r \Rightarrow \rceil p)$

p	q	r	$\rceil p$	$\rceil r$	$q \wedge r$	$p \Leftrightarrow q \wedge r$	$\rceil r \Rightarrow \rceil p$	$(p \Leftrightarrow q \wedge r) \Rightarrow (\rceil r \Rightarrow \rceil p)$
T	T	T	F	F	T	T	T	T
T	T	F	F	T	F	F	F	T
T	F	T	F	F	F	F	T	T
F	T	T	T	F	T	F	T	T
T	F	F	F	T	F	F	F	T
F	T	F	T	T	F	T	T	T
F	F	T	T	F	F	T	T	F
F	F	F	T	T	F	T	T	T

Similarly, the given statement does not contains always truth value T. Hence, it is not a tautology.

(d) Truth table for $p \wedge (q \vee r) \Leftrightarrow (p \wedge q) \vee (p \wedge r)$

p	q	r	$p \wedge q$	$q \vee r$	$p \wedge r$	$p \wedge (q \vee r)$	$(p \wedge q) \vee (p \wedge r)$	$p \wedge (q \vee r) \Rightarrow (p \wedge q) \vee (p \wedge r)$
T	T	T	T	T	T	T	T	T
T	T	F	T	T	F	T	T	T
T	F	T	F	T	T	T	T	T
F	T	T	F	T	F	F	F	T
T	F	F	F	F	F	F	F	T
F	T	F	F	T	F	F	F	T
F	F	T	F	T	F	F	F	T
F	F	F	F	F	F	F	F	T

The last column contains always truth values T. Hence, the given statement is a tautology.

(e) Truth table for $(p \Rightarrow q) \vee (r \Rightarrow p)$

p	q	r	$p \Rightarrow q$	$r \Rightarrow p$	$(p \Rightarrow q) \vee (r \Rightarrow p)$
T	T	T	T	T	T
T	T	F	T	T	T
T	F	T	F	T	T
F	T	T	T	F	T
T	F	F	F	T	T
F	T	F	T	T	T
F	F	T	T	F	T
F	F	F	T	T	T

Last column of above table contains all its truth values T. Hence, given statement is a tautology.

(f) Truth table for $(p \vee q \vee r) \Leftrightarrow [\{(\rceil p \Rightarrow q) \Rightarrow q\} \Rightarrow r]$

p	q	r	$\rceil p \Rightarrow q$	$(\rceil p \Rightarrow q) \Rightarrow q$	$(((\rceil p \Rightarrow q) \Rightarrow q) \Rightarrow r)$	$p \vee q \vee r$	$(((\rceil p \Rightarrow q) \Rightarrow q) \Rightarrow r) \Rightarrow r$
T	T	T	T	T	T	T	T
T	T	F	T	T	F	T	F
T	F	T	T	F	T	T	T
F	T	T	T	T	T	T	T
T	F	F	T	F	T	T	T
F	T	F	T	T	F	T	F
F	F	T	T	F	T	T	T
F	F	F	T	F	T	F	F

In the above table, the last column contains all the truth values T. Hence, the given statement is not a tautology.

(g) Truth table for $[(p \wedge q) \Rightarrow p] \Rightarrow [q \wedge \rceil q]$

p	q	$\rceil q$	$p \wedge q$	$(p \wedge q) \Rightarrow p$	$q \wedge \rceil q$	$[(p \wedge q) \Rightarrow p] \Rightarrow [q \wedge \rceil q]$
T	T	F	T	T	F	F
T	F	T	F	T	F	F
F	T	F	F	T	F	F
F	F	T	F	T	F	F

The last column contains all the truth values F. Hence, the given statement is contradiction.

(h) Truth table for $[\{(p \Rightarrow q) \vee p\} \wedge q] \Rightarrow q$

p	q	$p \Rightarrow q$	$(p \Rightarrow q) \wedge p$	$[\{(p \Rightarrow q) \vee p\} \wedge q]$	$[\{(p \Rightarrow q) \vee p\} \wedge q] \Rightarrow q$
T	T	T	T	T	T
T	F	F	T	F	T
F	T	T	T	T	T
F	F	T	T	F	T

In above table, last column contains all the truth values T. Hence, given statement is a tautology.

(i) Truth table for $p \wedge (q \wedge \rceil p)$

p	q	$\rceil p$	$q \wedge \rceil p$	$p \wedge (q \wedge \rceil p)$
T	T	F	F	F
T	F	F	F	F
F	T	T	T	F
F	F	T	F	F

The last column contains all the truth value F. Hence, given statement is a contradiction.

5. *Determine whether the following formulas are tautology, contradiction satisfiable :*

(a) $(p \wedge (p \Rightarrow q)) \Rightarrow \rceil q$ (b) $((p \Rightarrow q) \wedge (q \Rightarrow r)) \wedge (p \wedge \rceil r)$

(c) $q \vee (p \wedge \rceil q) \vee (\rceil p \wedge \rceil q)$

Solution. (a) $(p \wedge (p \Rightarrow q)) \Rightarrow \rceil q$

p	q	$\rceil q$	$p \Rightarrow q$	$p \wedge (p \Rightarrow q)$	$(p \wedge (p \Rightarrow q)) \Rightarrow \rceil q$
T	T	F	T	T	T
T	F	T	F	F	T
F	T	F	T	F	F
F	F	T	T	F	T

(b) $((p \Rightarrow q) \wedge (q \Rightarrow r)) \wedge (p \wedge \rceil r)$

p	q	r	$\rceil r$	$p \Rightarrow q$	$q \Rightarrow r$	$(p \Rightarrow q) \wedge (q \Rightarrow r)$ (1)	$p \wedge \rceil r$ (2)	(1) $\wedge$ (2)
T	T	T	F	T	T	T	F	F
T	T	F	T	T	F	F	T	F
T	F	T	F	F	T	F	F	F
F	T	T	F	T	T	T	F	F
T	F	F	T	F	T	F	T	F
F	T	F	T	T	F	F	F	F
F	F	T	F	T	T	T	F	F
F	F	F	T	T	T	T	F	F

(c) $q \vee (p \wedge \rceil q) \vee (\rceil p \wedge \rceil q)$

p	q	$\rceil p$	$\rceil q$	$p \wedge \rceil q$	$q \vee (p \wedge \rceil q)$	$\rceil p \wedge \rceil q$	$q \vee (p \wedge \rceil q) \vee (\rceil p \wedge \rceil q)$
T	T	F	F	F	T	F	T
T	F	F	T	T	T	F	T
F	T	T	F	F	T	F	T
F	F	T	T	F	F	T	T

6. *Prove that* $(p \Rightarrow q) \vee r \equiv (p \vee r) \Rightarrow (q \vee r)$.

Solution. Construct the truth table as follows :

p	q	r	$p \Rightarrow q$	$p \vee r$	$q \vee r$	$(p \Rightarrow q) \vee r$	$(p \vee r) \Rightarrow (q \vee r)$
T	T	T	T	T	T	T	T
T	T	F	T	T	T	T	T
T	F	T	F	T	T	T	T
F	T	T	T	T	T	T	T

T	F	F	F	T	F	F	F
F	T	F	T	F	T	T	T
F	F	T	T	T	T	T	T
F	F	F	T	F	F	T	T

The last two columns have same truth values. Thus corresponding statements of the columns are logically equivalent and

$$(p \Rightarrow q) \vee r \equiv (p \vee r) \Rightarrow (q \vee r).$$

7. *Prove that* $(p \Rightarrow q) \vee (p \Rightarrow r) \equiv p \Rightarrow (q \vee r)$.

Solution. Construct the truth table as follows :

p	q	r	$q \vee r$	$p \Rightarrow q$	$p \Rightarrow r$	$(p \Rightarrow q) \vee (p \Rightarrow r)$	$p \Rightarrow (q \vee r)$
T	T	T	T	T	T	T	T
T	T	F	T	T	F	T	T
T	F	T	T	F	T	T	T
F	T	T	T	T	T	T	T
T	F	F	F	F	F	F	F
F	T	F	T	T	T	T	T
F	F	T	T	T	T	T	T
F	F	F	F	T	T	T	T

The last two columns have same truth values. Hence,

$$(p \Rightarrow q) \vee (p \Rightarrow r) \text{ and } p \Rightarrow (q \vee r)$$

are logically equivalent and $(p \Rightarrow q) \vee (p \Rightarrow r) \equiv p \Rightarrow (q \vee r)$.

8. *Disprove that* $(p \vee q) \vee (p \wedge q) \equiv p$

Solution. Construct the truth table

p	q	$p \vee q$	$p \wedge q$	$(p \vee q) \vee (p \wedge q)$
T	T	T	T	T
T	F	T	F	T
F	T	T	F	T
F	F	F	F	F

The first and the last columns corresponding to the given statement do not contain same truth values. Hence, $(p \vee q) \vee (p \wedge q)$ and p are not logically equivalent.

9. *Prove that :*

(a) $p \Rightarrow q \equiv \neg p \vee q$

(b) $p \Rightarrow (q \Rightarrow r) \equiv (p \wedge q) \Rightarrow r$

(c) $p \Rightarrow (q \wedge r) \equiv (p \Rightarrow q) \wedge (p \Rightarrow r)$

(d) $\neg(p \Leftrightarrow q) \equiv \neg p \Leftrightarrow q \equiv p \Leftrightarrow \neg q$.

Solution. (a) Construct the truth table

p	q	$\neg p$	$p \Rightarrow q$	$\neg p \vee q$
T	T	F	T	T
T	F	F	F	F
F	T	T	T	T
F	F	T	T	T

The last two columns corresponding to the given statements have same truth values. Hence, $(p \Rightarrow q)$ and $\neg p \vee q$ are logically equivalent and $p \Rightarrow q \equiv \neg p \vee q$.

(b) Construct the truth table

p	q	r	$p \wedge q$	$q \Rightarrow r$	$p \Rightarrow (q \Rightarrow r)$	$(p \wedge q) \Rightarrow r$
T	T	T	T	T	T	T
T	T	F	T	F	F	F
T	F	T	F	T	T	T
F	T	T	F	T	T	T
T	F	F	F	T	T	T
F	T	F	F	F	T	T
F	F	T	F	T	T	T
F	F	F	F	T	T	T

The last two columns corresponding to the given statements have same truth values.

$\therefore$ $p \Rightarrow (q \Rightarrow r)$ and $(p \wedge q) \Rightarrow r$ are logically equivalent and $p \Rightarrow (q \Rightarrow r) \equiv (p \wedge q) \Rightarrow r$.

(c) Construct the truth table

p	q	r	$q \wedge r$	$p \Rightarrow q$	$p \Rightarrow r$	$p \Rightarrow (q \wedge r)$	$(p \Rightarrow q) \wedge (p \Rightarrow r)$
T	T	T	T	T	T	T	T
T	T	F	F	T	F	F	F
T	F	T	F	F	T	F	F
F	T	T	T	T	T	T	T

The last two columns have same truth values. Hence, the statements $p \Rightarrow (q \wedge r)$ and $(p \Rightarrow q) \wedge (p \Rightarrow r)$ are logically equivalent and

$$p = (q \wedge r) \equiv (p \Rightarrow q) \wedge (p \Rightarrow r).$$

(d) Construct the truth table

p	q	$\neg p$	$\neg q$	$p \Leftrightarrow q$	$\neg(p \Leftrightarrow q)$	$\neg p \Leftrightarrow q$	$p \Leftrightarrow \neg q$
T	T	F	F	T	F	F	F
T	F	F	T	F	T	T	T
F	T	T	F	F	T	T	T
F	F	T	T	T	F	F	F

The last three columns have same truth values. Hence, $\neg(p \Leftrightarrow q)$, $(\neg p \Leftrightarrow q)$ and $p \Leftrightarrow \neg q$ are logically equivalent and $\neg(p \Leftrightarrow q) \equiv \neg p \Leftrightarrow q \equiv p \Leftrightarrow \neg q$.

10. *Establish the equivalence* $(p \Rightarrow q) \Rightarrow (p \wedge q) = (\rceil p \Rightarrow q) \wedge (q \Rightarrow p)$.

Solution. Construct the truth table

p	q	$\rceil p$	$p \Rightarrow q$	$p \wedge q$	$\rceil p \Rightarrow q$	$q \Rightarrow p$	$(p \Rightarrow q) \Rightarrow (p \wedge q)$	$(\rceil p \Rightarrow q) \wedge (q \Rightarrow p)$
T	T	F	T	T	T	T	T	T
T	F	F	F	F	T	T	T	T
F	T	T	T	F	T	F	F	F
F	F	T	T	F	F	F	F	F

The last two columns have same truth values. Hence, $(p \Rightarrow q) \Rightarrow (p \wedge q)$ and $(\rceil p \Rightarrow q) \wedge (q \Rightarrow p)$ are logically equivalent and

$$(p \Rightarrow q) \Rightarrow (p \wedge q) = (\rceil p \Rightarrow q) \wedge (q \Rightarrow p).$$

11. *Establish the equivalence and write its dual equivalence :*

(a) $(p \wedge q) \vee [\rceil p \vee (\rceil p \vee q)] \equiv \rceil p \vee q$ *(b)* $\rceil p \wedge (\rceil q \wedge r) \vee (q \wedge r) \vee (p \wedge r) \equiv r$

Solution. (a) LHS $= (p \wedge q) \vee [\rceil p \vee (\rceil p \vee q)]$

$\equiv (p \wedge q) \vee [(\rceil p \vee \rceil p) \vee q] = (p \wedge q) \vee [(\rceil p \vee q)]$

$\equiv [p \vee (\rceil p \vee q)] \wedge [q \vee (\rceil p \vee q)]$

$\equiv [p \vee (\rceil p \vee q)] \wedge (\rceil p \vee q)$

$\equiv \rceil p \vee q$ (by absorption) $=$ RHS

Hence, $(p \wedge q) \vee \rceil p \rceil p \vee (p \vee q)] \equiv \rceil p \vee q$.

The dual of this statement is $(p \vee q) \wedge [\rceil p \wedge (p \wedge q)] \equiv \rceil p \wedge q$

(b) LHS $= \rceil p \wedge (\rceil q \wedge r) \vee (q \wedge r) \vee (p \wedge r) \equiv r$

$\equiv [(\rceil p \wedge \rceil q) \wedge r] \vee [\{(q \wedge r) \vee p\} \wedge \{(q \wedge r) \vee r\}]$

$\equiv [(\rceil p \wedge \rceil q) \wedge r] \vee [\{(q \wedge r) \vee p\} \wedge r]$ (By absorption)

$\equiv [(\rceil p \wedge \rceil q) \wedge r] \vee [\{(q \wedge r) \wedge r\} \vee (p \wedge r)]$

$\equiv [(\rceil p \wedge \rceil q) \wedge r] \vee [(q \wedge r) \vee (p \wedge r)]$

$\equiv [(\rceil p \wedge \rceil q) \vee (q \vee p)] \wedge r$

$\equiv [(\rceil p \vee \rceil q) \vee (q \vee p)] \wedge r$

$\equiv r$

Now to get dual of the given statement, interchange $\wedge$ and $\vee$, we get

$$\rceil p \wedge (\rceil q \vee r) \wedge r) q \wedge r \wedge (p \vee r) \equiv r.$$

12. *Show that the following argument is valid*

$$\begin{array}{l} p \\ p \Rightarrow q \\ \underline{q \Rightarrow r} \\ r \end{array}$$

Solution. First Method : Let $f \equiv [p \wedge (p \Rightarrow q) \wedge (q \Rightarrow r)] \Rightarrow r$.

Now, construct the truth table for f

p	q	r	$p \Rightarrow q$	$q \Rightarrow r$	$p \wedge (p \Rightarrow q) \wedge (q \Rightarrow r)$	(f)
T	T	T	T	T	T	T
T	T	F	T	F	F	T
T	F	T	F	T	F	T
F	T	T	T	T	F	T
T	F	F	F	T	F	T
F	T	F	T	F	F	T
F	F	T	T	T	F	T
F	F	F	T	T	F	T

The last column contains all the truth values T. Hence, f is a tautology and hence the given argument is valid.

Second Method :

Let $f \equiv [p \wedge (p \Rightarrow q) \wedge (q \Rightarrow r)] \Rightarrow r$

$\equiv [p \wedge (\urcorner p \vee q) \wedge (\urcorner q \vee r)] \Rightarrow r \equiv \urcorner [p \wedge (\urcorner p \vee q) \wedge (\urcorner q \vee r)] \vee r$

$\equiv [\urcorner p \vee \urcorner(\urcorner p \vee q) \vee \urcorner(\urcorner q \vee)] \vee r \equiv [\urcorner p \vee (p \wedge \urcorner q) \vee (q \wedge \urcorner r)] \vee r$

$\equiv \urcorner p \vee \urcorner q \vee \urcorner r \vee r$

Thus, $f \equiv \urcorner p \vee \urcorner q \vee \urcorner r \vee r = 1$

Since f reduced to 1, this also shows that the argument is valid.

Third Method : Consider the following sequences of arguments

p	a premise
$\underline{p \Rightarrow q}$	a premise
q	by the law of detachment
$\underline{q \Rightarrow r}$	a premise
r	by the law of detachment

Thus are argument is valid.

13. *Check the validity of the argument*

$$\begin{array}{c} p \Rightarrow q \\ \underline{r \Rightarrow \urcorner q} \\ p \Rightarrow \urcorner r \end{array}$$

Solution. Let $f \equiv [(p \Rightarrow q) \wedge (r \Rightarrow \urcorner q)] \Rightarrow (p \Rightarrow \urcorner r)$

Now, we construct truth table for f

p	q	r	$\urcorner q$	$\urcorner r$	$p \Rightarrow q$	$r \Rightarrow \urcorner q$	$(p \Rightarrow q) \wedge (r \Rightarrow \urcorner q)$	$(p \Rightarrow \urcorner r)$	(f)
T	T	T	F	F	T	F	F	F	T
T	T	F	F	T	T	T	T	T	T
T	F	T	T	F	F	T	F	F	T
F	T	T	F	F	T	F	F	T	T
T	F	F	T	T	F	T	F	T	T
F	T	F	F	T	T	T	T	T	T
F	F	T	T	F	T	T	T	T	T
F	F	F	T	T	T	T	T	T	T

Since f column contains only T, the argument is valid.

14. *Test the validity of the argument :*

If two sides of a triangle are equal, then the opposite angles are equal.

Two sides of a triangle are not equal.

∴ The opposite angles are not equal.

Solution. The statement above the horizontal line are two premises. The statement below the horizontal line is the conclusion.

Let p : Two sides of a triangle are equal.

q : The opposite angles of a triangle are equal.

The given argument in symbolic form can be written as

$$\begin{array}{c} p \Rightarrow q \text{ (a premise)} \\ \underline{\urcorner p \text{ (a premise)}} \\ \urcorner q \text{ (conclusion)} \end{array}$$

We shall construct the truth table for statement.

$$[(p \to q) \wedge \rceil p] \to \rceil q$$

p	q	$p \Rightarrow q$	$\rceil p$	$\rceil q$	$(p \Rightarrow q) \wedge \rceil p$	$[(p \Rightarrow q) \wedge \rceil p] \Rightarrow \rceil q$
T	T	T	F	F	F	T
T	F	F	F	T	F	T
F	T	T	T	F	T	F
F	F	T	T	T	T	T

The last column in the table shows that $[(p \Rightarrow q) \wedge \rceil p] \Rightarrow \rceil q$ is not a tautology. Hence the given argument is not valid.

15. *State whether the argument given below is valid or not valid. If it is valid, identify the tautology used :*

I will become famous or I will be the writer.

I will not be a writer.

$\therefore$ *I will become famous.*

Solution. Let p : I will become famous.

q : I will be a writer.

$\therefore$ The given argument in symbolic form can be written as

$p \vee q$ (a premise)

$\sim q$ (a premise)

---

p (conclusion)

The given argument, *i.e.* $(p \vee q) \wedge (\sim q) \to p$ will be valid if the statement $[(p \vee q) \wedge (\sim q)] \to p$ is a tautology. Now we construct the truth table for the above statement

p	q	$p \vee q$	$\sim q$	$(p \vee q) \wedge (\sim q)$	$[(p \vee q) \wedge (\sim q)] \to p$
T	T	T	F	F	T
T	F	T	T	T	T
F	T	T	F	F	T
F	F	F	T	F	T

Since last column contains only Trues, hence the given argument is valid.

Exercise 2.2

1. Which of the following are statements :

(a) Grass is yellow

(b) Is the number 5 a prime ?

(c) Give me a book

(d) $\{x : x^2 = 9\} = \{3, -3\}$

(e) $x^2 + y^2 \geq 0$

(f) White roses are beautiful.

(g) If dogs can bark, then no home guarded by a dog needs to fear intruders.

(h) $\{7x + 5y = 12 : (x, y) \equiv (1, 1), x, y \in \mathbf{N}\}$

2. Let p be the proposition (statement) "x is an even number" and let q be the statement "x is the product of two integers". Translate into symbols each of the following statements

(a) Either x is an even number, or x is a product of two integers.

(b) x is an odd number, and x is a product of two integers.

(c) Either x is an even number and a product of integers, or x is an odd number and is not a product of integers.

(d) x is neither an even number nor a product of integers.

3. Write in reasonable English, the negation of each of the following statements :

(a) Ice is cold, and I am tired.

(b) Either good health is desirable, or I have been misinformed.

(c) Oranges are not suitable for use in vegetable salads.

(d) There is a number which, when added to 6, gives a sum of 13.

4. Define compound statement and explain it by an example.

5. If $p \equiv$ It is about to 6 : 45 p.m.; $q \equiv$ Sangam train is about to depart.

(a) $p \wedge q$ (b) $p \vee q$

(c) $q \vee \rceil p$ (d) $p \wedge \rceil q$

(e) $\rceil(p \wedge q)$ (f) $\rceil(p \vee q)$

(g) $\rceil p \Rightarrow q$ (h) $\rceil q \Rightarrow p$

6. Write the negation of the following :

(a) If the determinant of a system of linear equation is zero, then, either the system has no solution or has an infinite number of solution.

(b) Either today is not a Friday or today is not a Sunday.

7. Use truth tables to verify

(a) The associative law for disjunction

(b) The associative law for conjunction

(c) The distributive law for $\wedge$ and $\vee$, and

(d) The two laws of absorption.

8. Construct the truth tables for the following statements :

(a) $\rceil[\rceil r \vee (p \wedge q)]$

(b) $p \vee (q \wedge r)$

(c) $(p \wedge \rceil q) \vee (\rceil p \wedge q)$

(d) $p \vee q \vee \rceil p$

(e) $(p \wedge q) \vee (\rceil p \wedge \rceil q)$

(f) $\rceil p = \rceil q$

(g) $(p \vee q) \wedge \rceil(p \wedge q)$

9. Prove De Morgan's laws using truth tables.

10. Construct the truth table for the contrapositive of $p \Rightarrow q$.

11. Designate suitable simple statements p and q and translate the following statement into symbols :

(a) If lemons are expensive and sugar is cheap, then sour lemonade is rarely seen.

(b) Sour lemonade is often seen unless sugar is cheap.

(c) A necessary condition for lemons to be cheap is that sugar is expensive.

(d) Sour lemonade is rarely seen only if sugar is cheap.

12. Determine the validity of the following argument : "If wages increase then there will be inflation. The cost of living will not increase, if there is no inflation. Wages will increase, therefore the cost of living will increase.

13. Without using truth table, show that

$$((p \vee q) \wedge \rceil(\rceil p \wedge (\rceil q \vee \rceil r))) \vee (\rceil p \wedge \rceil q) \vee (\rceil p \wedge \rceil r)$$

is a tautology.

14. Determine the validity of the following argument : "If Mary runs for office, she will be elected. If Mary attends the meeting, she will run for office. If mary attends the meeting, she will run for office. Either Mary will attend the meeting or she will go to India, but Mary cannot go to India. Thus Mary will be elected.

15. Prove that the following are tautologies :

(a) $(p \Leftrightarrow q) \wedge (q \Leftrightarrow r) \Rightarrow (p \Leftrightarrow r)$

(b) $\rceil(p \wedge \rceil p)$

(c) $(p \Rightarrow q) \vee r \Leftrightarrow (p \vee r) \Rightarrow (q \vee r)$

(d) $(p \Rightarrow q) \wedge \rceil q$

(e) $[(p \Rightarrow q) \wedge (q \Rightarrow r) \Rightarrow (p \Rightarrow r)$

(f) $p \vee (q \wedge r) \Leftrightarrow (p \vee q) \wedge (p \vee r)$

16. Prove the following equivalences and write its dual.

(a) $\rceil(p \Leftrightarrow q) \equiv (p \wedge \rceil q) \vee (\rceil p \wedge q)$

(b) $p \Rightarrow (q \vee r) \equiv (p \Rightarrow q) \vee (p \Rightarrow r)$

(c) $p \wedge q \equiv q \wedge p$

(d) $p \wedge (q \wedge r) \equiv (p \wedge q) \wedge r$

(e) $(p \wedge \rceil q \vee q) \equiv p \vee q$

(f) $(p \wedge q \Rightarrow r) \equiv (p \Rightarrow r) \vee (q \Rightarrow r)$

17. Define the tautology and duality with example.

18. Prove that conditional operation is neither commutative nor associative.

19. Prove that the following formula is not a tautology.

$$[(p \wedge r) \vee (q \wedge \rceil r)] \Leftrightarrow [(p \wedge r) \vee (\rceil q \wedge \rceil r)].$$

20. Simplify the following :

(a) $(p \vee q) \Leftrightarrow \rceil p$

(b) $p \vee (p \wedge q)$

(c) $\rceil(p \vee q) \vee (\rceil p \wedge q)$

(d) $\rceil(\rceil p \wedge q) \wedge (\rceil p \vee q) \wedge (p \vee q)$.

21. Check the validity of the following :

(a) $p \Rightarrow q$
$r \Rightarrow \rceil q$
$\overline{p \Rightarrow \rceil r}$

(b) p
q
$\rceil p \Rightarrow r$
$q \Rightarrow \rceil r$
$\overline{\rceil r}$

(c) $q \Rightarrow p$
$q \vee s$
$\rceil s$
$\overline{p}$

(d) p
$\rceil p \vee \rceil s) \Rightarrow (\rceil p \wedge \rceil r)$
$\overline{s}$

(e) $r \Rightarrow \rceil q$
$p \Rightarrow q$
$\rceil r \Rightarrow s$
$\overline{p \Rightarrow s}$

(d) p
$\rceil q \vee r$
$\rceil p \Rightarrow q$
$\overline{r}$

22. Write the dual of the following expression :
$(x \wedge 1) \wedge (0 \vee x')$.

ANSWERS

1. (a) Yes, (b) Yes, (c) Yes, (d) Yes, (e) Yes, (f) No, (g) Yes, (h) Yes,

2. (a) $p \vee q$, (b) $\rceil p \wedge q$, (c) $(p \wedge q) \vee (\rceil p \wedge \rceil q)$, (d) $\rceil p \wedge \rceil q$

3. (a) Either ice is not cold or I am not tired, (b) Good health is not desirable and I have not been informed, (c) Oranges are suitable for use in vegetable salads, (d) There is no number which when did not add to 6, does not give a sum of 13

5. (a) It is about 6 : 45 p.m. and Sangam train is about to depart, (b) Either it is about 6 : 45 p.m. or Sangam train is about to depart, (c) Either Sangam Train is about to depart or it is not about to 6 : 45 p.m., (d) It is about to 6 : 45 p.m. and Sangam train is not about to depart, (e) It is false that it is about to 6 : 45 p.m. and Sangam train is about to depart, (f) Neither it is about to 6 : 45 p.m., nor Sangam train is about to depart, (g) If it is not about to 6 : 45 p.m., then Sangam train is about to depart, (h) If Sangam train is not about to depart, then it is about 6 : 45 p.m.

6. (a) It is false that if the determinant of a system of linear equation is zero, then either the system has no solution or has an infinite number of solutions, (b) Today is a Friday and today is a Sunday.

8. (a)

p	q	r	$\rceil r$	$p \wedge q$	$\rceil r \vee (p \wedge q)$	$\rceil[\rceil r \vee (p \wedge q)]$
T	T	T	F	T	T	F
T	T	F	T	T	T	F
T	F	T	F	F	F	T
F	T	T	F	F	F	T
T	F	F	T	F	T	F
F	T	F	T	F	T	F
F	F	T	F	F	F	T
F	F	F	T	F	T	F

(b)

p	q	r	$q \wedge r$	$p \vee (q \wedge r)$
T	T	T	F	T
T	T	F	T	T
T	F	T	F	F
F	T	T	F	F

T	F	F	T	F
F	T	F	T	F
F	F	T	F	F
F	F	F	T	F

(c)

p	q	r	$\rceil p$	$\rceil q$	$p \wedge \rceil q$	$\rceil p \wedge q$	$(p \wedge \rceil q) \vee (\rceil p \wedge q)$
T	T	T	F	F	F	F	F
T	T	F	F	F	F	F	F
T	F	T	F	T	T	F	T
F	T	T	T	F	F	T	T
T	F	F	F	T	T	F	T
F	T	F	T	F	F	T	T
F	F	T	T	T	F	F	F
F	F	F	T	T	F	F	F

(d)

p	q	$\rceil p$	$p \vee q \vee \rceil p$
T	T	F	T
T	F	F	T
F	T	T	T
F	F	T	T

(e)

p	q	$\rceil p$	$\rceil q$	$p \wedge q$	$\rceil p \wedge \rceil q$	$(p \wedge q) \vee (\rceil p \wedge \rceil q)$
T	T	F	F	T	F	T
T	F	F	T	F	F	F
F	T	T	F	F	F	F
F	F	T	T	F	T	T

(f)

p	q	$\rceil p$	$\rceil q$	$\rceil p \Leftrightarrow \rceil q$
T	T	F	F	T
T	F	F	T	F
F	T	T	F	F
F	F	T	T	T

(g)

p	q	$p \vee q$	$p \wedge q$	$\neg p \wedge q$	$(p \vee q) \wedge \neg p \wedge q$
T	T	T	T	F	F
T	F	T	F	T	T
F	T	T	F	T	T
F	F	F	F	T	F

11. If $p \equiv$ lemons are expensive, $q \equiv$ sugar is cheap, $r \equiv$ Lemonade is rarely seen
(a) $(p \wedge q) \Rightarrow r$, (b) $\neg q \Rightarrow \neg r$, (c) $\neg p \Rightarrow \neg q$, (d) $r \Rightarrow q$

16. (a) $\neg(p \Leftrightarrow q) \equiv (p \vee \neg q) \wedge (\neg p \vee q)$, (b) $p \Rightarrow (q \wedge r) \equiv (p \Rightarrow q) \wedge (p \Rightarrow r)$,
(c) $p \vee q \equiv q \vee p$, (d) $p \vee (q \vee r) \equiv (p \vee q) \vee r$
(e) $(p \vee \neg q) \wedge q \equiv p \wedge q$, (f) $(p \vee q \Rightarrow r) \equiv (p \Rightarrow r) \wedge (q \Rightarrow r)$

20. (a) $\neg p \wedge q$, (b) p, (c) $\neg p$, (d) $p \wedge q$

21. (a) valid, (b) valid, (c) valid, (d) valid, (e) valid, (f) invalid.

22. $(x \vee 1) \vee (0 \wedge x')$.

2.13 LOGICAL IDENTITIES

Identity is defined by the logically equivalence between two proposition in such a way that one can be substituted for the other in any proposition in which they occur. The symbol "1" is used for 'tautology' or a true proposition and "0" is used for "contradiction" or a false proposition, if P, Q and R are any propositions then some identities are given below which can be proved by constructing truth tables.

1.	$p \Leftrightarrow (p \wedge p)$	Idempotence of $\wedge$
2.	$p \Leftrightarrow (p \vee p)$	Idempotence of $\vee$
3.	$(p \vee q) \Leftrightarrow (q \vee p)$	Commutativity of $\vee$
4.	$(p \wedge q) \Leftrightarrow (q \wedge p)$	Commutativity of $\wedge$
5.	$(p \vee 1) \Leftrightarrow 1$	
6.	$(p \wedge 1) \Leftrightarrow p$	
7.	$(p \vee 0) \Leftrightarrow p$	
8.	$(p \wedge 0) \Leftrightarrow 0$	
9.	$(p \vee \neg p) \Leftrightarrow 1$	
10.	$(p \wedge \neg p) \Leftrightarrow 0$	
11.	$[(p \vee q) \vee r] \Leftrightarrow [p \vee (q \vee r)]$	Associativity of $\vee$
12.	$[(p \wedge q) \wedge r] \Leftrightarrow [p \wedge (q \wedge r)]$	Associativity of $\wedge$
13.	$\neg(p \vee q) \Rightarrow (\neg p \wedge \neg q)]$	DeMorgan's law
14.	$\neg(p \wedge q) \Rightarrow (\neg p \vee \neg q)]$	DeMorgan's law
15.	$[(p \wedge (q \vee r)] \Leftrightarrow [(p \wedge q) \vee (p \wedge r)]$	Distributivity of $\wedge$ over $\vee$
16.	$[(p \vee (q \wedge r)] \Leftrightarrow [(p \vee q) \wedge (p \vee r)]$	Distributivity of $\vee$ over $\wedge$
17.	$p \Leftrightarrow \neg(\neg p)$	Double negation

18.	$(p \Rightarrow q) \Leftrightarrow (\neg p \vee q)$	Implication
19.	$(p \Leftrightarrow q) \Leftrightarrow [(p \Rightarrow q) \wedge (q \Rightarrow p)]$	Equivalence
20.	$[(p \wedge q) \Rightarrow r] \Leftrightarrow [p \Rightarrow (q \Rightarrow r)]$	Exportation
21.	$[p \Rightarrow q) \wedge (p \Rightarrow \neg q)] \Leftrightarrow \neg p$	Absurdity
22.	$(p \Rightarrow q) \Leftrightarrow (\neg q \Rightarrow \neg p)$	Contrapositive

List of some tautologies which are implications as follows :

1.	$p \Leftrightarrow (p \vee q)$	Addition
2.	$(p \wedge q) \Rightarrow p$	Simplification
3.	$[p \wedge (p \Rightarrow q)] \Rightarrow q$	Modus Ponens
4.	$[(p \Rightarrow q)] \wedge \neg q] \Rightarrow \neg p$	Modus Tollens
5.	$[\neg p \wedge (p \vee q)] \Rightarrow q$	Disjunctive syllogism
6.	$(p \Rightarrow q) \wedge (q \Rightarrow r) \Rightarrow (p \Rightarrow r)$	Hypothetical syllogism
7.	$(p \Rightarrow q) \Rightarrow [(q \Rightarrow p) \Rightarrow (p \Rightarrow r)]$	
8.	$[(p \Rightarrow q) \wedge (r \Rightarrow s)] \Rightarrow [(p \wedge r) \Rightarrow (q \wedge s)]$	
9.	$[(p \Leftrightarrow q) \wedge (q \Leftrightarrow r)] \Rightarrow (p \Leftrightarrow r)$	

2.14 WELL FORMED FORMULAE (WFF)

A well-formed formula (WFF) is not a proposition but when a proposition is substituted in place of a propositional variable, a proposition is obtain. Now a well-formed formula can be defined as

(i) Any propositional variable is a well-formed formula.

(ii) If x is a well-formed formula, then $\neg x$ is also well formed formula.

(iii) If x and y are well formed formulae, then $x \vee y$, $x \wedge y$, $x \Rightarrow y$ and $x \Leftrightarrow y$ are also well-formed formulas.

(iv) Any string of symbols, which is obtained by finitely many application as shown above, is a well defined formula.

If we repalace the proposition variables by propositions in a well-formed formula, we obtain a proposition involving connectives. Ths table, giving the truth values (T or F) of such proposition, which is obtained by replacing the propositional variables by arbitrary propositions is called the truth table of well formed formula. In this truth table, if the well formed formula involve n proposition, then we have 2^n possible combinations of both values of propositions replacing the variables. So, if P and Q are any two propositions, then a well formed formula has four possible combinations of truth value, (*i.e.*, *FF*, *FT*, *TF* and *TT* are possible combinations of well defined formula).

2.14.1 Logical Influence

There are following three rules :

(i) A premise may be introduced at any point in the derivation.

(ii) A formula S may be introduced in a derivation if S is tautologically implied by any one or more of the preceding formula in the derivation.

(iii) If we can derive S from R and a set of premises, then we can derive $R \rightarrow S$ from the set of premises alone.

There are some rules of inference as shown in following table.

2.14.2 Rules of Inference

1.	$p \wedge q \Rightarrow p$	Simplification
2.	$p \wedge q \Rightarrow q$	Simplification
3.	$p \Rightarrow p \vee q$	Addition
4.	$q \Rightarrow p \vee q$	Addition
5.	$\rceil p \Rightarrow p \rightarrow q$	
6.	$q \Rightarrow p \rightarrow q$	
7.	$\rceil(p \rightarrow q) \Rightarrow p$	
8.	$\rceil(p \rightarrow q) \Rightarrow \rceil q$	
9.	$p . q \Rightarrow p \wedge q$	
10.	$\rceil p, p \vee q \Rightarrow q$	Disjunctive syllogism
11.	$p, p \rightarrow q \Rightarrow q$	Modus Ponens
12.	$\rceil q, p \rightarrow q \Rightarrow \rceil p$	Modus Tollens
13.	$p \rightarrow q, q \rightarrow r \Rightarrow p \rightarrow r$	Hypothetical syllogism
14.	$p \vee q, p \rightarrow r, q \rightarrow r \Rightarrow r$	Dilemma

2.14.3 Equivalences

1.	$\rceil\rceil p \Leftrightarrow p$	Double negation
2.	$p \wedge q \Leftrightarrow q \wedge p$	Commutative law
3.	$p \vee q \Leftrightarrow q \vee p$	Commutative law
4.	$(p \wedge q) \wedge r \Leftrightarrow p \wedge (q \wedge r)$	Associative law
5.	$(p \vee q) \vee r \Leftrightarrow p \vee (q \vee r)$	Associative law
6.	$p \vee (q \wedge r) \Leftrightarrow (p \vee q) \wedge (p \vee r)$	Distributive law
7.	$p \wedge (q \vee r) \Leftrightarrow (p \wedge q) \vee (p \wedge r)$	Distributive law
8.	$\rceil(p \wedge q) \Leftrightarrow \rceil p \vee \rceil q$	De' Morgan's law
9.	$\rceil(p \vee q) \Leftrightarrow \rceil p \wedge \rceil q$	De' Morgan's law
10.	$p \vee p \Leftrightarrow p$	
11.	$p \wedge p \Leftrightarrow p$	
12.	$r \vee (p \wedge \rceil p) \Leftrightarrow r$	
13.	$r \wedge (p \vee \rceil p) \Leftrightarrow r$	
14.	$p \rightarrow q \Rightarrow \rceil p \vee q$	
15.	$\rceil(p \rightarrow q) \Leftrightarrow p \wedge \rceil q$	
16.	$p \rightarrow q \Leftrightarrow \rceil q \rightarrow \rceil p$	

17.	$p \to (q \to r) \Leftrightarrow (p \wedge q) \to r$	
18.	$\rceil(p \Leftrightarrow q) \Leftrightarrow p \Leftrightarrow \rceil q$	
19.	$p \Leftrightarrow q \Leftrightarrow (p \to q) \wedge (q \to p)$	
20.	$(p \Leftrightarrow q) \Leftrightarrow (p \wedge q) \vee (\rceil p \wedge \rceil q)$	

ILLUSTRATIVE EXAMPLES

1. *Establish the following tautologies by simplifying the left side to the form of the right side.*

(a) $(\rceil p \wedge q) \vee p \Leftrightarrow 1$ *(b)* $\rceil(\rceil(p \vee q) \Rightarrow \rceil p) \Leftrightarrow 0$

(c) $[(p \Rightarrow \rceil p) \wedge (\rceil p \Rightarrow p)] \Leftrightarrow 0$

Solution. (a) Consider $\rceil(p \wedge q) \vee p \Rightarrow \rceil p \vee \rceil q \vee p \quad \Leftrightarrow (p \vee \rceil p) \vee \rceil q \Leftrightarrow 1 \vee \rceil q \Leftrightarrow 1$

(b) Consider $\rceil(\rceil(p \vee q) \Rightarrow \rceil p \quad \Leftrightarrow \rceil[(p \vee q) \vee \rceil p] \Leftrightarrow \rceil[(p \vee \rceil p) \vee q]$

$\Leftrightarrow \quad \rceil(1 \vee q) \Leftrightarrow \rceil[1] \quad \Leftrightarrow \quad 0$

(c) Consider $[(p \Rightarrow \rceil p) \wedge (\rceil p \Rightarrow p)] \quad \Leftrightarrow [(\rceil p \vee \rceil p) \wedge (p \vee p)$

$\Leftrightarrow \quad \rceil p \wedge p \quad \Leftrightarrow \quad 0$

2. *Prove that* $(p \to q) \wedge (r \to q)$ *and* $(p \vee r) \to q$ *are equivalent formulae.*

Solution. Consider $(p \to q) \wedge (r \to q) \Leftrightarrow (\rceil p \vee q) \wedge (\rceil r \vee q)$

$\Leftrightarrow \quad (\rceil p \wedge \rceil r) \vee q \quad \Leftrightarrow \quad \rceil(p \vee r) \vee q \quad \Leftrightarrow \quad (p \vee r) \to q.$

3. *Find equivalent expression for each of the following expressions, using identities which use only* $\wedge$ *and* $\rceil$ *and are as simple as possible.*

(a) $p \vee q \vee \rceil r$ *(b)* $p \vee [(\rceil q \wedge r) \Rightarrow p]$ *(c)* $p \Rightarrow (q \Rightarrow p)$

Solution. (a) Consider $p \vee q \vee \rceil r \Leftrightarrow \quad \rceil(\rceil p \wedge \rceil q) \vee \rceil r$

$\Leftrightarrow \quad \rceil((\rceil p \wedge \rceil q) \wedge r) \quad \Leftrightarrow \quad \rceil(\rceil p \wedge \rceil q \wedge t)$

(b) Consider $p \vee [(\rceil q \wedge r) \Rightarrow P] \quad \Leftrightarrow \quad \rceil(\rceil p \wedge \rceil[(\rceil q \wedge r) \Rightarrow p)]$

$\Leftrightarrow \quad \rceil[\rceil p \wedge \rceil(\rceil(\rceil q \ \wedge r) \vee p)] \Leftrightarrow \quad \rceil[\rceil p \wedge ((\rceil q \wedge r) \wedge \rceil p)]$

(c) Consider $p \Rightarrow (q \Rightarrow p) \quad \Leftrightarrow \quad p \Rightarrow (\rceil q \vee p)$

$\Leftrightarrow \quad \rceil p \vee (\rceil q \vee p) \quad \Leftrightarrow \quad (\rceil p \vee p) \vee \rceil q$

$\Leftrightarrow \quad 1 \vee \rceil q \quad \Leftrightarrow \quad 1$

4. *Prove that* $((p \vee q) \wedge \rceil(\rceil p \wedge (\rceil q \vee \rceil r))) \vee (\rceil p \wedge \rceil q) \vee (\rceil p \wedge \rceil r)$ *is a tautology.*

Solution. Consider $((p \vee q) \wedge \rceil(\rceil p \wedge (\rceil q \vee \rceil r))) \vee (\rceil p \wedge \rceil q) \vee (\rceil p \wedge \rceil r)$

$\Leftrightarrow \quad ((p \vee q) \wedge \rceil(\rceil p \wedge \rceil(q \wedge r)) \vee \rceil(p \vee q) \vee \rceil(p \vee r)$

$\Leftrightarrow \quad ((p \vee q) \wedge (p \vee (q \wedge r))) \vee \rceil(p \vee q) \vee \rceil(p \vee r)$

$\Leftrightarrow \quad [(p \vee q) \wedge (p \vee q)] \ \vee \rceil[(p \vee q) \wedge (p \vee r)$

$\Leftrightarrow \quad [(p \vee q) \wedge (p \vee r)] \vee \rceil[(p \vee (q \wedge r)$

$\Leftrightarrow \quad [p \vee (q \wedge r)] \vee \rceil[(p \vee (q \wedge r)] \ \Leftrightarrow \quad T.$

5. *Prove that* $\rceil(p \wedge q) \Rightarrow (\rceil p \vee (\rceil p \vee q)) \Rightarrow (\rceil p \vee q)$

Solution. Consider $\rceil(p \wedge q) \Rightarrow (\rceil p \vee (\rceil p \vee q))$

$\Leftrightarrow \quad (p \wedge q) \vee [\rceil p \vee (\rceil p \vee q)] \quad \Leftrightarrow \quad (p \wedge q) \vee (\rceil p \vee q)$

$\Rightarrow \quad (p \wedge q) \vee \rceil p \vee q \quad \Leftrightarrow \quad ((p \wedge q) \vee \rceil p) \vee q$

$\Rightarrow \quad ((p \vee \rceil p) \wedge (q \vee \rceil p) \vee q \quad \Leftrightarrow \quad (1 \wedge (q \vee \rceil p)) \vee r$

$\Rightarrow \quad (q \vee \rceil p) \vee q \quad \Leftrightarrow \quad q \vee \rceil p \Rightarrow \rceil p \vee q.$

6. *Show that* $s \vee r$ *is tautologically implied by*

$$(p \vee q) \wedge (p \to q) \wedge (p \to S).$$

Solution.

1. $p \vee qP$
2. $\rceil p \to q \qquad (\because p \to q \Leftrightarrow (\rceil p \vee q))$
3. $q \to s \qquad p$
4. $\rceil p \to S \quad$ (2), (3), Hypothetical syllogism

5. $\rceil s \to p$ (from (4), using $(p \to q \Leftrightarrow \rceil q \to \rceil p)$

6. $p \to r$ p

7. $\rceil s \to r$ (5), (6) and Hypothetical syllogism

8. $s \vee r$.

7. *Demonstrate that s is a valid inference from the premises* $p \to \rceil q$, $q \vee p$, $\rceil S \to p$ *and* $\rceil r$.

Solution.
1. $q \vee r$ p
2. $\rceil r$ p
3. q (1), (2) disjunctive syllogism
4. $p \to \rceil q$ P
5. $\rceil p$ (3), (4) Contrapositive, Modus Tollens
6. $\rceil s \to p$ p
7. s (5), (6) Modus Tollens.

8. *State whether the following argument is valid or not. If valid, give proof. If not valid, give counter example.*

If a baby is hungry, the baby cries.

If the baby is not mad, then he does not cry.

If a baby is mad, then he has a red face.

Therefore, if a baby is hungry, then he has a red face.

Solution.

H : Baby is hungry

C : Baby cries

M : Baby is mad.

R : Baby has a red face.

Given that

$$\begin{array}{c} H \to C \\ \rceil M \to \rceil C \\ M \to R \\ \hline \therefore \quad H \to R \end{array}$$

Verification.

1. $H \to C$ Premise
2. $\rceil M \to \rceil C$ Premise
3. $C \to M$ (2), Contrapositive
4. $H \to M$ (1), (3), Hypothetical syllogism
5. $M \to R$ Premise
6. $H \to R$ (4), (5) Hypothetical syllogism.

2.15 PREDICATES AND QUANTIFIERS

Assertion which are joined using variables in a "template" that express the property of an object or a relationship between objects are called "Predicates". Predicates may be either variables or constants and are used in control statements in high level languages. Values of individual variables are drawn from a set of values called "Universe of discourse".

To change the predicates into proposition, we should be 'bound' each individual variable of the predicates. There are two ways to change a predicate into proposition.

(i) In this way, we bind the individual variable by assigning a value to it.

Example :

$$P = \text{“}a + b = 6\text{”}, \text{ which is denoted by } P(x, y).$$

If $a = 2$ and $b = 3$, then $P(2, 3)$ is false.

(ii) In this way, we bind the individual variable by "quantification" of the variable. It can be done either by "universal" or "extential".

"The assertion $P(x)$ is true for all values of x".

For all x, $P(x)$ is written as "$\forall_x P(x)$", where $\forall$ is a "universal quantifier". "There exists a value of x for which the assertion $P(x)$ is true". This statement is written as "$\Im_x P(x)$" where $\Im$ is called "external quantifier".

The proposition $\forall_x P(x)$ is equivalent to the conjunction $P(1) \wedge P(2) \wedge P(3) \wedge P(4)$ for the universal consisting of integers 1, 2, 3 and 4 and the proposition $\Im_x P(x)$ is equivalent to the disjunction

$$P(1) \vee P(2) \vee P(3) \vee P(4).$$

The proposition $\Im_x ! P(x)$ is equivalent to the proposition

$$[P(1) \wedge \rceil P(2) \wedge \rceil P(3)] \vee [P(2) \wedge \rceil P(1) \wedge \rceil P(3)] \vee [P(3) \wedge \rceil P(1) \wedge \rceil P(2)]$$

The sequence $\forall_x \forall_y$ can always be replaced by $\forall_y \forall_x$, and the sequence $\Im_x \Im_y$ can always be replaced by $\Im_y \Im_x$, through the order in which individual variables are bound can not always be changed without affecting the meaning of an assertion.

2.16 QUANTIFIERS AND LOGICAL OPERATORS

We assume that I is "universe of discourse" and let

$E(x)$, x is even.
$O(x)$, x is odd.
$P(x)$, x is prime.
$N(x)$, x is non-negative.

Then, we have the followings :

(i) Every integer is even or odd.

$$\forall_x [E(x) \vee O(x)]$$

(ii) The only even prime is two.

$$\forall_x [E(x) \wedge P(x) \Rightarrow x = 2]$$

(iii) Not all primes are odd.

$$\rceil \forall_x [P(x) \Rightarrow O(x)], \Im_x [P(x) \wedge \rceil O(x)].$$

(iv) If an integer is not odd, then it is even.

$$\forall_x [\rceil O(x) \Rightarrow E(x)]$$

The qunatifiers may go anywhere in the transcription of mathematical statements. Now, let $P(x, y, z)$ denote "$xy = z$" for the universe of integers. Informal statements of proposition frequently omit the universal quantification of individual variables.

(i) If $x = 0$, then $xy = x$ for all values of $y \forall_x [x = 0 \Rightarrow \forall_y P(x, y, z)]$

(ii) If $xy = x$ for every y, then $x = 0. \forall_x [\forall_y P(x, y, z) \Rightarrow x = 0]$

Propagation of negation through quantifier sequences is useful in constructing projs and counter examples.

For Example : Consider there exists a z such that $x + z = y$, for every pair of integers x and y.

This is stated as : $\forall_x \forall_y \Im_z [x + z = y]$

This is true for universe of integers I, but not true for the natural numbers N. We establish the falsity for the universe N by showing that its negation is true.

The negation has the form $\rceil \forall_x \forall_y \Im_z [x + z = y]$ which is difficult to interpret.

Now, the equivalent form

$$\Im_x \Im_y \forall_y \rceil [x + z = y] \text{ or } \Im_x \Im_y \forall_z [x + z \neq y]$$

is more tractable and can easily be shown to be true for the non-negative integers by choosing $x > y$.

2.16.1 Logical Relations

1. $\forall_x P(x) \Rightarrow P(C)$, where C is an arbitrary element of universe.
2. $P(C) \Rightarrow \exists_x P(x)$, where C is an arbitrary element of universe.
3. $\forall_x P(x) \Leftrightarrow \neg P(x)$
4. $\forall_x \neg P(x) \Leftrightarrow \neg \exists_x P(x)$
5. $\exists_x \neg P(x) \Leftrightarrow \neg \forall_x P(x)$
6. $[\forall_x P(x) \wedge Q] \Leftrightarrow \forall_x [P(x) \wedge Q]$
7. $[\forall_x P(x) \vee Q] \Leftrightarrow \forall_x [P(x) \vee Q]$
8. $[\forall_x P(x) \wedge \forall_x Q(x)] \Leftrightarrow \forall_x [P(x) \wedge Q(x)]$
9. $[\forall_x P(x) \vee \forall_x Q(x)] \Leftrightarrow \forall_x [P(x) \vee Q(x)]$
10. $[\exists_x P(x) \wedge Q] \Leftrightarrow \exists_x [P(x) \wedge Q]$
11. $[\exists_x P(x) \vee Q] \Leftrightarrow \exists_x [P(x) \vee Q]$
12. $[\exists_x P(x) \wedge Q(x)] \Leftrightarrow [\exists_x P(x) \wedge \exists_x Q(x)]$
13. $[\exists_x P(x) \vee Q(x)] \Leftrightarrow [\exists_x P(x) \wedge Q(x)]$

ILLUSTRATIVE EXAMPLES

1. *Let the universe of discourse be the set of arithmetic assertions with predicates defined as follows :*
$P(x)$ denotes "x is provable"
$T(x)$ denotes "x is true"
$S(x)$ denotes "x is satisfiable".
$D(x, y, z)$ denotes "z is the disjunction $x \vee y$".
Translate the following assertions into English statements :
(a) $\forall_x [P(x) \Rightarrow T(x)]$
(b) $\forall_x [T(x) \vee \neg S(x)]$
(c) $\exists_x [T(x) \wedge \neg P(x)]$
(d) $\forall_x \forall_y \forall_z \{[D(x, y, z) \wedge P(z)] \Rightarrow [P(x) \vee P(y)]\}$
(e) $\forall_x \{T(x) \Rightarrow \forall_y \forall_z [D(x, y, z) \Rightarrow T(z)]\}$

Solution.
(a) Consider $\forall_x [P(x) \Rightarrow T(x)]$. If x is provable, then x is true.
(b) Consider $\forall_x [T(x) \vee \neg S(x)]$
If x is true or it is unsatisfiable.
(c) Consider $\exists_x [T(x) \wedge \neg P(x)]$
There is some x for which x is true and is not provable.
(d) Consider $\forall_x \forall_y \forall_z \{[D(x, y, z) \wedge P(z)] \Rightarrow [P(x) \vee P(y)]\}$
It the assertion $z = x \vee y$ and the assertion $P(z)$ is provable, then either x is provable or y is provable.
(e) Consider $\forall_x \{T(x) \Rightarrow \forall_y \forall_z [D(x, y, z) \Rightarrow T(z)]\}$
If every arithmetic assertion is true, then the assertion $z = x \vee y$ is true.

2. *For each of the following sets of premises, list the relevant conclusions which can be drawn and the rules of inference used in each case.*
(a) All cows are mammals. Some mammals chew their cud.
(b) What's good for the auto industry is good for the country. What's good for the country is good for you. What's good for the auto industry is for you to buy an expensive car.
(c) All even integers are divisible by 2. The integer 4 is even but 3 is not.

Solution. (a) Hence, $C(x)$: x is a cow.

$M(x)$: x is a mammal

$D(x)$: x chew their cud.

$$\forall_x [C(x) \rightarrow M(x)]; \exists_x M(x) \rightarrow D(x)$$

(b) $A(x)$: x is good for the auto industry.

$C(x)$: x is good for the country.

$Y(x)$: x is good for you.

b = "You buying an expensive car" (constant)

$\forall_x [A(x) \rightarrow C(x)]$ ...(1)

$\forall_x [C(x) \rightarrow Y(x)]$ ...(2)

$A(b)$ ...(3)

$A(b) \rightarrow C(b)$ ...(4)

$C(b) \rightarrow Y(b)$ ...(5)

$C(b)$ (3), (4), Modus Ponens ...(6)

$Y(b)$ (5), (6), Modus Ponens ...(7)

$C(b) \wedge Y(b)$ By conjunction

Conclusion : It is good for the country and for you to buy an expensive car.

(c) $E(x)$: x is an even integer.

$D(x)$: x is divisible by 2.

$E(x) \rightarrow D(x)$.

3. *Obtain the assertion which is logically equivalent to $\forall_x P(x)$ but uses only the quantifier $\exists$ and the logical operator $\rceil$. Similarly, express $\exists_x p(x)$ in terms of $\forall$ and $\rceil$. Similarly, express $\exists_x P(x)$ in terms of $\forall$ and $\rceil$.*

Solution. (i) Here, $\forall_x P(x) \Leftrightarrow \rceil(\rceil \forall_x P(x))$

$\Leftrightarrow \rceil(\exists_x \rceil P(x)) \Leftrightarrow \rceil \exists_x \rceil P(x)$

(ii) Here, $\exists_x P(x) \Leftrightarrow \rceil(\rceil \exists_x P(x))$

$\Leftrightarrow \rceil(\forall_x \rceil P(x)) \Leftrightarrow \rceil \forall_x \rceil P(x)$

Normal form : First we shall show how to find the formula from given truth table

P	Q	R	$f(P, Q R)$
T	T	T	T
T	T	F	F
T	F	T	T
F	T	T	F
T	F	F	T
F	T	F	F
F	F	T	T
F	F	F	F

Here, four truth values are true. So, we have

$$f(P, Q, R) = (P \wedge Q \wedge R) \vee (P \wedge \rceil Q \wedge R) \vee (P \wedge \rceil Q \wedge \rceil R) \vee (\rceil P \wedge \rceil Q \wedge R)$$

which is a 'disjunction' of terms, each of which is 'conjunction' of statement variables and their negations. The sum of statement variables and their negation is called 'elementary sum' and the production is called 'elementary production'.

2.17 NORMAL FORMS

2.17.1 Disjunctive Normal Form (DNF)

A formula, which has a sum of elementary products and is equivalent to the given formula, is called Disjunctive Normal Form (DNF) of the given formula.

2.17.2 Conjunctive Normal Form (CNF)

A formula, which has a product of elementary sums and is equivalent to the given formula, is called Conjunctive Normal Form (CNF) of the given formula.

WORKING RULE **(To find Disjunctive Normal Form)**

The procedure to find disjunctive normal form has the following steps :

Step 1 : If the connective $\rightarrow$ and $\rightarrow$ appear in the given formula, then we obtain an equivalent formula in which these connective does not appear.

Example : $\alpha \rightarrow \beta$ is replaced by $(\rceil\alpha \vee \beta)$ and $\alpha \leftrightarrow \beta$ is replaced by $(\alpha \wedge \beta) \vee (\rceil\alpha \wedge \rceil\beta)$ or $(\rceil\alpha \vee \beta) \wedge (\rceil\beta \wedge \alpha)$

Step 2 : If the negation applied to the formula or a part of the formula which is not a statement variable, then we obtain an equivalent formula using De-Morgan's law in which the negation is applied to statement variable only.

Step 3 : We apply the disjunctive law until a sum of elementary products is obtained. This is the 'Disjunctive Normal Form' after applying the idempotent law and suitable re-ordering. Any elementary products which are equivalent to 'F' (False) can be omitted.

- DNF (or CNF) of a given logical expression is not unique.
- DNF s also called 'sum of products' form.
- CNF is also called 'product of sums' form.

ILLUSTRATIVE EXAMPLES

1. *Find the DNF of* $p \Rightarrow (p \Rightarrow q)[\vee \sim(\sim q \vee \sim p)]$

Solution : We have

$$\begin{aligned} p \Rightarrow (p \Rightarrow q) \wedge \sim(\sim q \vee \sim p) & \\ &\equiv \sim p \vee (p \Rightarrow q) \wedge \sim(\sim q \vee \sim p) \\ &\equiv \sim p \vee [(\sim p \vee q) \wedge \sim(\sim q \vee \sim p)] \\ &\equiv \sim p \vee [(\sim p \vee q) \wedge (q \wedge p)] \\ &\equiv \sim p \vee [(\sim p \wedge (q \wedge p) \vee (q \wedge p)] \\ &= \sim p \vee F \vee (p \wedge q) \\ &= \sim p \vee (p \wedge q), \end{aligned}$$

which is the required DNF.

2. *Find the CNF of* $[q \vee (p \wedge q)] \wedge \sim[(p \vee r) \wedge q]$

Solution : We have

$$\begin{aligned} [q \vee (p \wedge q)] \wedge \sim[(p \vee r) \wedge q] &\equiv [q \vee (p \wedge r)] \wedge [\sim(p \vee r) \vee \sim q] \\ &\equiv [q \vee (p \wedge r)] \wedge [(\sim p \wedge \sim r) \sim q] \\ &\equiv (q \vee p) \wedge (q \vee r) \wedge (\sim p \vee \sim q) \wedge (\sim r \wedge \sim q) \end{aligned}$$

which is the required CNF.

3. *Obtain the disjunctive normal form of* $(p \wedge \rceil(q \wedge r)) \vee (p \rightarrow q)$.

Solution. Consider $(p \wedge \rceil(q \wedge r)) \vee (p \rightarrow q)$

$\Leftrightarrow \quad (p \wedge \rceil(q \wedge r)) \vee (\rceil(p \vee q) \qquad \Leftrightarrow \quad (p \wedge (\rceil q \vee r)) \vee (\rceil p \vee q)$

$\Leftrightarrow \quad (p \wedge \rceil q) \vee (p \wedge r) \vee (\rceil p \vee q) \qquad \Leftrightarrow \quad (p \wedge \rceil q \vee \rceil p) \vee (p \wedge T) \vee (p \wedge r)$

$\Leftrightarrow \quad (T \wedge \rceil q) \vee (p \wedge T) \vee (p \wedge r) \qquad \Leftrightarrow \quad \rceil q \vee p \vee (p \wedge r)$

$\Leftrightarrow \quad (p \wedge r) \vee p \vee \rceil q.$

4. *Obtain the disjunctive normal form of* $p \vee (\rceil p \rightarrow (q \vee (q \rightarrow \rceil r)))$.

Solution. Consider $p \vee (\rceil p \rightarrow (q \vee (q \rightarrow \rceil r)))$

$\Leftrightarrow \quad p \vee (\rceil p \rightarrow (q \vee (\rceil q \vee r))) \qquad \Leftrightarrow \quad p \vee (\rceil p \rightarrow [(q \vee \rceil q) \vee (q \vee r)]$

$\Leftrightarrow \quad p \vee (\rceil p \rightarrow [T \vee (q \vee r)] \qquad \Leftrightarrow \quad p \vee (\rceil p \rightarrow (q \vee r))$

$\Leftrightarrow \quad p \vee [\rceil (\rceil p) \vee (q \vee r)] \qquad \Leftrightarrow \quad p \vee [p \vee (q \vee r)]$

$\Leftrightarrow \quad (p \vee p) \vee [p \vee (q \vee r)] \qquad \Leftrightarrow \quad p \vee p \vee q \vee r$

$\Leftrightarrow \quad p \vee q \vee r.$

5. *Obtain the conjunctive normal form of* $\rceil (p \vee q) \leftrightarrow (p \wedge q)$.

Solution. Consider $\rceil (p \vee q) \leftrightarrow (p \wedge q)$.

$\Leftrightarrow \quad [\rceil (p \vee q) \wedge (p \wedge q)] \vee [\rceil (\rceil (p \vee q)) \wedge \rceil (p \wedge q)]$

$\Leftrightarrow \quad [\rceil p \wedge \rceil q) \wedge (p \wedge q)] \vee [(p \vee q) \wedge (\rceil p \vee \rceil q)]$

$\Leftrightarrow \quad [(p \wedge \rceil p \wedge q \wedge \rceil q) \vee (p \vee q) \wedge (\rceil p \vee \rceil q)]$

$\Leftrightarrow \quad (p \vee q) \wedge (\rceil p \vee \rceil q).$

6. *Let the universe be the integers. For each of the following assertions, find a predicate P which makes the implication false.*

(a) $\forall_x \exists_y !P(x, y) \Rightarrow \exists_y !\forall_x P(x, y)$ *(b)* $\exists_y !\forall_x P(x, y) \Rightarrow \forall_x \exists_y !P(x, y)$.

Solution. (a) Consider $\forall_x \exists_y !P(x, y) \Rightarrow \exists_y !\forall_x P(x, y)$, $x + y = 0$

L.H.S. $x + (-1) = 0$

$$x = 1$$
$$x + (-2) = 0$$
$$x = 2$$
$$\text{R.H.S. } \exists_y ! x = 1$$
$$1 + y = 0$$
$$y = -1$$
$$\exists_y \; x = 2,\; 2 + y = 0$$
$$y = -2$$
$$P(x, y) \text{ is } (x + y) = 0$$

(b) Consider $\exists_y !\forall_x P(x, y) \Rightarrow \forall_x \exists_y !P(x, y)$, $xy = 0$

L.H.S. (i) $y = 0$ R.H.S. $x(0) = 0$

$$y = 0 \quad x = 1, 2, 3, \ldots\ldots$$

(ii) $y = 0$

$$y = 0$$
$$P(x, y) \text{ is } xy = 0.$$

7. *Show that* $\exists_x \exists_y P(x, y)$ and $\exists_y \exists_x P(x, y)$ *are equivalent by expanding the expressions into infinite disjunctions.*

Solution. To show $\exists_x \exists_y P(x, y) = \exists_y \exists_x P(x, y)$

$\exists_x \exists_y P(x, y) = [\exists_y P(0, y)] \vee [\exists_y P(1, y)] \vee [\exists_y P(2, y)] \ldots\ldots$

$= [P(0, 0) \vee P(0, 1) \vee P(0, 2)\ldots] \vee [P(0, 1) \vee P(1, 1) \vee P(1, 2)\ldots]$
$\vee [P(2, 0) \vee P(2, 1) \vee P(2, 2) \ldots] \vee \ldots$

$= [P(0, 0) \vee P(1, 0) \vee P(2, 0) \vee \ldots] \vee [P(0, 1) \vee P(1, 1) \vee P(2, 1) \vee \ldots]$
$\vee [P(0, 2) \vee P(1, 2) \vee P(2, 2) \vee \ldots] \vee \ldots$

$= [\exists_x P(x, 0)] \vee [\exists_x P(x, 1)] \vee [\exists_x P(x, 2)] \vee \ldots$

$= \exists_y \exists_x P(x, y).$

Exercise 2.2

1. For each of the following expression, use identities to find equivalent expressions which use only $\vee$ and $\rceil$ and are as simple as possible.
   (a) $(P \wedge Q) \wedge \rceil P$ (b) $[P \Rightarrow (Q \vee \rceil R)] \wedge \rceil P \wedge Q$
2. Show that $S \vee R$ is tautologically implied by $(P \vee Q) \wedge (P \rightarrow R) \wedge (Q \rightarrow S)$.
3. Show that $R \vee S$ follows logically from the premises $C \vee D$, $(C \vee D) \rightarrow \rceil H$, $\rceil H \rightarrow (A \wedge \rceil B)$ and $(A \wedge \rceil B) \rightarrow (R \vee S)$.
4. Show that $R \wedge (P \vee Q)$ is a valid conclusion from the premises $P \vee Q$, $Q \rightarrow R$, $P \rightarrow M$ and $\rceil M$.
5. Find an assertion which is logically equivalent to $\exists_x\, !P(x)$ but which uses only the quantifiers $\forall$ and $\exists$ together with the predicate for equality and logical operator.
6. Find whether the assertion
   $\forall_x [P(x) \Rightarrow Q(x)] \Rightarrow [\forall_x P(x) \Rightarrow \forall_x Q(H)]$ is true or not.
7. Obtain the disjunctive normal form of $P \vee (\rceil P \wedge \rceil Q \wedge R)$.
8. Obtain the disjunctive normal form of $P \rightarrow ((P \rightarrow Q) \wedge \rceil(\rceil Q \vee \rceil P))$.
9. Obtain the disjunctive normal form of $(\rceil P \vee \rceil Q) \rightarrow (\rceil P \wedge R)$.
10. Show that
    $\forall_x \forall_y P(x, y) \rightarrow (\rceil x) \forall_y P(x, y)$
    is logically valid.

ANSWERS

1. $\rceil(\rceil P \vee \rceil Q \vee P)$, $\rceil(P \vee \rceil Q)$
5. $\exists_x\, !P(x) \Leftrightarrow \exists_x [P(x) \wedge \forall_y [P(y) \Rightarrow y = x]$
6. True
7. $(P \vee \rceil Q \vee R)$
8. $(\rceil P \vee Q)$
9. $(P \wedge Q) \vee (\rceil P \wedge R)$

❑❑❑

Chapter

3

Boolean Algebra

- Number system
- Sentences
- Boolean Algebra
- Boolean function
- Canonical form
- Dual canonical form
- Switching circuits
- Karanaugh map
- Logic gates
- Sum of product and product of sums forms
- Normal form
- Half and full substracter

3.1 INTRODUCTION

In algebra, we are generally interested in the study of the properties of the operations and their effects on the object. Certain properties of algebraic system can be viewed as axioms. There are some important concepts related with the algebraic systems, such as isomorphism. We may note that the algebras of sets and logic are the same differing only in their notations. In fact, the propositional logic and set theory can be regarded as instances of a more general theory called Boolean Algebra.

In mathematics, an algebraic structure dealt with the sets, which is associated with one or more rules of combination. These rules are known as binary rules or binary operations. There are four rules of combination for real numbers such as addition, subtraction, multiplication and division. In the algebra of sets, we observed that two rules, intersection and union exist.

Boolean algebra is an algebraic structure which is based on the principles of logics. The idea of this structure was firstly given by British Mathematician George Boole. In case of Boolean algebra, there are mainly three operations (i) And, (ii) Or, and (iii) Not, which are denoted by $\wedge$, $\vee$ and $(\sim)$ respectively.

In the chapter, sometimes, we will use '+', '·', ''' in place of above operations respectively.

3.2 BOOLEAN ALGEBRA

The definition of a Boolean algebra which will be used here is one given by Huntington in 1904.

Definition : A non-empty set B (*i.e.*, a class of elements) together with two binary operations (+) and (·) is a Boolean algebra if and only if the following axioms hold :

P_1 : The operations (+) and (·) are commutative

(i) $a + b = b + a \ \forall \ a, b \in B$

(ii) $a \cdot b = b \cdot a \ \forall \ a, b \in B$

P_2 : The operations (+) and (·) are closed in B

(i) $a + b \in B \ \forall \ a, b \in B$ (ii) $a \cdot b \in B \ \forall \ a, b \in B$

$\boldsymbol{P_3}$: Associative laws hold in B

(i) $a + (b + c) = (a + b) + c \ \forall \ a, b, c \in B$

(ii) $a \cdot (b \cdot c) = (a \cdot b) \cdot c \ \forall \ a, b, c \in B$

$\boldsymbol{P_4}$: There exist in B distinct identity elements 0 and 1 with respect to addition and multiplication respectively

(i) $a + 0 = 0 + a = a \ \forall \ a \in B$ (ii) $a \cdot 1 = 1 \cdot a = a \ \forall \ a \in B$

$\boldsymbol{P_5}$: Each operation is distributive over the other

(i) $a \cdot (b + c) = a \cdot b + a \cdot c \ \forall \ a, b, c \in B$

(ii) $a + (b \cdot c) = (a + b) \cdot (a + c) \ \forall \ a, b, c \in B$

$\boldsymbol{P_6}$: For every a in B there exists an element a' in B such that

$$a + a' = 1 \text{ and } a \cdot a' = 0$$

We observe that the algebra of sets satisfies all these postulates and hence is a Boolean algebra. That is, every Boolean algebra may be interpreted as an algebra of sets for some choice of universal set. Basically, Boolean algebra has two values, *i.e.*, {0, 1} function. Hence, it is denoted by {B, '+', '·', 0, 1, '′'}.

- If we take the binary operation ∪ and ∩ in place of (+) and (·) respectively, then 0 is taken for 0 and I is taken for 1.
- $0' = 1$ and $1' = 0$
- Boolean algebra has at least two elements.
- $a \cdot b$ will be written as *ab.*

In Boolean algebra B, as in the case of the power set algebra $P(S)$, we can define a binary operation of ring sum denoted by $a \oplus b$ and given by

$$a \oplus b = (a \cdot b') + (a' \cdot b) = (a \wedge b') \vee (a' \wedge b)$$

Then an algebraic structure $\{B, \oplus, \wedge\}$ is a ring with identity with $\oplus$ standing for (+) and $\wedge$ standing for (·) and with identity 1 satisfies above axioms from P_1 to P_5 and in which every element is idempotent with respect to ring multiplication is called a Boolean ring. Therefore, we have observed that every Boolean algebra admits a ring structure with respect to the ring sum and the meet operation such that it becomes a Boolean ring.

Theorem 1. (Idempotent laws) *For every element a in a Boolean algebra B,*

$$a + a = a \text{ and } aa = a.$$

Proof. By the existence of identity $a = a + 0$

$= a + aa'$ $(\because \ aa' = 0)$

$= (a + a)(a + a')$ (By distributive law)

$= (a + a)(1)$ $(\because \ a + a' = 1)$

$= a + a$ $(\because \ a \cdot 1 = a)$

Hence, we obtain $a + a = a$.

Similarly, $a = a \cdot 1$ (By the definition of identity)

$= a(a + a')$ $(\because \ a + a' = 1)$

$= aa + aa'$ (By distributive law)

$= aa + 0$ $(\because \ aa' = 0)$

$= aa$ $(\because \ a + 0 = a)$

Hence, $aa = a$.

Theorem 2. (Domination laws) *For every element a in a Boolean algebra B*

$$a + 1 = 1 \text{ and } a \,.\, 0 = 0.$$

Proof. $1 = a + a'$ (By the postulate P_5)

$= a + a'(1)$ $(\because \ a \cdot 1 = a)$

$= (a + a')(a + 1)$ (By distributive law)

$= 1(a + 1)$ $(\because \ a + a' = 1)$

$= a + 1$ $(\because\ 1 \cdot a = a \cdot 1 = a)$

Hence, $a + 1 = 1$

Likewise, $0 = aa'$ (By P_6)

$0 = (aa)\, a'$ (By idempotent law)

$= a0$

Hence, $a0 = 0$.

Theorem 3. (Absorption laws) *For each pair of elements a and b in a Boolean Algebra B*

$$a + ab = a \text{ and } a\,(a + b) = a.$$

Proof. $a = 1a$ $(\because\ a \cdot 1 = 1 \cdot a = a)$

$= (1 + b)\, a$ $(\because\ 1 + a = 1)$

$= 1a + ba$ (By distributive law)

$= a + ba$ $(\because\ 1 \cdot a = a)$

$= a + ab$ $(\because\ ab = ba)$

Thus, $a + ab = a$

Now, $a\,(a + b) = aa + ab$ (By distributive law)

$= a + ab$ $(\because\ aa = a)$

$= a$ (From first result)

Thus, $a\,(a + b) = a$.

Theorem 4. *Complement of any element a in a Boolean algebra B is unique.*

Proof. Let us suppose that x and y are the two complements of a, then

$$\left.\begin{array}{l} a + x = 1,\ ax = 0 \\ a + y = 1,\ ay = 0 \end{array}\right\} \text{(given)}$$

Now, $x = 1x$ $(\because\ 1 \cdot a = a)$

$= (a + y)\, x$ (By given conditions)

$= ax + yx$ (By distributive law)

$= 0 + yx$ (By given condition)

$= yx$ $(\because\ a + 0 = 0 + a = a)$

$= xy$ $(\because$ B is commutative)

$= xy + 0$ $(\because\ a + 0 = a)$

$= xy + ay$ (By given condition)

$= (x + a)\, y$ (By distributive law)

$= 1y$ (By given condition)

$= y$ $(\because\ 1 \cdot a = a)$

and hence, $x = y$.

Therefore, complement of a is unique in B.

Theorem 5. (law of double complement) *For every a in a Boolean algebra,* $(a')' = a$.

Proof. Let a be any element of B, then

$(a')' = 1(a')'$

$= (a + a')\,(a')'$ $(\because\ a + a' = 1)$

$= a\,(a')' + a'\,(a')'$ (By distributive law)

$= a(a')' + 0$ $(\because\ a'\,(a')' = 0)$

$= a(a')' + aa'$ $(\because\ aa' = 0)$

$= aa' + a(a')'$ (By commutative law)

$= a(a' + (a')')$ (By distributive law)

$= a \cdot 1$ $(\because\ a' + (a')' = 1)$

$= a$ $(\because\ a \cdot 1 = a)$

Hence, $(a')' = a$.

Theorem 6. (De Morgan's laws) *For every a and b in a Boolean algebra B :*

(i) $(ab)' = a' + b'$ *(ii)* $(a + b)' = a'\,b'$.

Proof. (i) To prove $a' + b'$ is a complement of ab, we have to prove

$$(ab) + (a' + b') = 1 \quad \text{and } (ab)(a' + b') = 0.$$

Now, $(ab) + (a' + b') = a' + b' + ab$ (By commutative law)

$= (a' + b' + a)(a' + b' + b)$ (By distributive law)

$= (1 + b')(1 + a')$ ($\because$ $a + a' = 1 \cdot b + b' = 1$)

$= (1)(1)$ ($\because$ $1 + a' = 1 \cdot 1 + b' = 1$)

$= 1$

Thus, $(ab) + (a' + b') = 1.$ ...(1)

Next, $(ab)(a' + b') = (ab)a' + (ab)b'$ (By distributive law)

$= a(ba') + a(bb')$ (By associative law)

$= a(a'b) + a(bb')$ (By commutative law)

$= (aa')b + a(bb')$ (By associative law)

$= 0b + a0$ ($\because$ $aa' = 0, bb' = 0$)

$= 0 + 0$ ($\because$ $a0 = 0, 0b = 0$)

$= 0$ ($\because$ $0 + 0 = 0$)

Therefore, $(ab)(a' + b') = 0$ ...(2)

From (1) and (2), we conclude that $a' + b'$ is a complement of (ab).

Hence, $(ab)' = a' + b'$.

(ii) To prove $(a + b)' = a'b'$, we have to prove that

$$(a + b) + a'b' = 1 \quad \text{and} \quad (a + b)a'b' = 0$$

Consider, $(a + b) + a'b' = [(a + b) + a'][(a + b) + b']$ (By distributive law)

$= [a + (b + a')][a + (b + b')]$ (By associative law)

$= [a + (a' + b)][a + (b + b')]$ (By commutative law)

$= [(a + a') + b][a + (b + b')]$ (By associative law)

$= (1 + b)(a + 1)$ ($\because$ $a + a' = 1, b + b' = 1$)

$= (1)(1)$ ($\because$ $a + 1 = 1, 1 + b = 1$)

$= 1.$

Thus, we obtain $(a + b) + a'b' = 1$ ...(1)

next $(a + b)a'b' = a(a'b') + b(a'b')$ (By distributive law)

$= (aa')b' + b(a'b')$ (By associative law)

$= (aa')b' + b(b'a')$ (By commutative law)

$= (aa')b' + (bb')a'$ (By associative law)

$= 0b' + 0a'$ ($\because$ $aa' = 0, bb' = 0$)

$= b'0 + a'0$ (By commutative law)

$= 0 + 0$ ($\because$ $a'0 = 0, b'0 = 0$)

$= 0$ ($\because$ $0 + 0 = 0$)

Thus, we obtain $(a + b)(a'b') = 0.$ ...(2)

From (1) and (2), we conclude that $(a + b)' = a'b'$.

Theorem 7. *Prove that*

(i) $0' = 1$ (ii) $1' = 0.$

Proof. (i) $0' = 0 + 0'$ ($\because$ 0 is an additive identity)

$= 1$

$\therefore$ $0' = 1.$

(ii) $1' = 1 \cdot 1'$ ($\because$ $a = 1 \cdot a$)

$= 0$ ($\because$ $aa' = 0$)

$\therefore$ $1' = 0.$

Hence, the theorem is proved.

Boolean identities : Tabular form

S.No	Identity	Name
1.	$(a')' = a$	law of double complement
2.	$a + a = a$ $a.\, a = a$	Idempotent laws
3.	$a + 0 = a$ $a.1 = a$	Identity laws
4.	$a + 1 = 1$ $a.0 = 0$	Domination laws
5.	$a + b = b + a$ $ab = ba$	Commutative laws
6.	$a + (b + c) = (a + b) + c$ $a(bc) = (ab)c$	Associative laws
7.	$a + bc = (a + b)(a + c)$ $a(b + c) = ab + ac$	Distribution laws
8.	$(ab)' = a' + b'$ $(a + b)' = a'b'$	Demorgan's law
9.	$a + ab = a$ $a(a + b) = a$	Absorption laws
10.	$a + a' = 1$	unit property
11.	$aa' = 0$	zero property

3.2.1 Unique features of Boolean Algebra

There are some unique features of Boolean Algebra different from other branches of algebra. These are :

(1) Boolean algebra has only a finite set of elements namely 1 and 0, (usually called True and False). On the other hand, ordinary algebra deals with a set of an infinite number of elements. Due to this fact, all the variables and constants are allowed to have only two possible values 0 and 1. Therefore, it provides the rules for working in the set {0, 1}.

(2) Boolean algebra does not have operations equivalent to subtraction and division.

(3) Boolean Algebra may also work with unary operation

(4) In Boolean algebra, Cancellation law does not hold

$$a + b = a + c \quad \not\Rightarrow \quad b = c$$

Similarly $$a.\, b = a.\, c \quad \not\Rightarrow \quad b = c$$

3.2.2 Some Fascinating Facts

- The basic operations of Boolean algebra are $\wedge$, $\vee$ and $\rceil$. These operators are also present in many computer languages. Since computer keyboards typically do not have these symbols, the symbol & (for $\wedge$) , 1 (for $\vee$) and ~ (for $\rceil$) are often used instead.
- Boolean algebra also has expressions containing letters and operations. Letters (various) in a Boolean algebra do not stand for numbers. Rather, they stand for the value TRUE and FALSE Therefore, letters in a Boolean algebraic expression can only have two values.
- An 'if-then' statement is logically equivalent to its 'Contra positive'

- An 'if-then' statement is not logically equivalent to its 'converse'.
- The operation '~' is not binary because it works on just one value at a time (*t* is unary)
- Boolean algebra provides the operations and the rules for working with the set {0, 1}
- There are 2^{2^n} different Boolean functions of degree *n*. See the table given below :

Degree	Number
1	4
2	16
3	256
4	65536
5	4294, 967296
6	18446744073709551616

3.3 SUB-BOOLEAN ALGEBRA

Definition : *A non-empty subset B′ of B is said to be a sub-Boolean algebra of a Boolean algebra B if B′ contains 0 and 1 together with two binary operations (+) and (·) and is closed under the above operations and complementation and itself a Boolean algebra.*

☛ Because of De Morgan's laws
(i) $(a \wedge b) = (a' \vee b')'$ or $(a \cdot b) = (a' + b')'$ (ii) $(a \vee b) = (a' \wedge b')'$ or $(a + b) = (a' \cdot b')'$
and because of $1 = a \vee a'$, $0 = a \wedge a'$, it is sufficient to verify a non-empty subset B_1 of B to be a sub-Boolean algebra of a Boolean algebra B either B_1 is closed under {(+) or (∨)} and (′) or B_1 is closed under {(·) or (∧)} and (′). That is
$$a, b' \in B_1 \Rightarrow a \vee b, a' \in B_1 \quad \text{or} \quad a, b' \in B_1 \Rightarrow a \wedge b, a' \in B_1$$

☛ The necessary and sufficient conditions for a non-empty subset B_1 of a Boolean algebra B to be a sub-Boolean algebra is that
(i) $a, b \in B_1 \Rightarrow (a \cdot b) \in B_1$ (ii) $a \in B_1 \Rightarrow a' \in B_1$

3.4 IDEALS OF BOOLEAN ALGEBRA

The study of ideals in a Boolean algebra plays a significant role in the development of the topic.

Definition : A non-empty subset I of a Boolean algebra B is called ideal of B if
(i) $a, b \in I \Rightarrow a + b \in I \ \forall\ a, b$ (ii) $a \in I, s \in B \Rightarrow a \cdot s \in I$
Above properties can also be taken as
(i) $a, b \in I \Rightarrow a \vee b \in I \ \forall\ a, b$ (ii) $a \in I, s \in B \Rightarrow a \wedge s \in I$

☛ Since B is bounded by 0 and 1, therefore {0} is an ideal of B and is called null ideal or zero ideal.

Theorem 1. *Intersection of two ideal of Boolean algebra is again ideal.*

Proof. Let I_1 and I_2 be two ideals of a Boolean algebra B. We shall have to prove that $I_1 \cap I_2$ is an ideal of B. Since both I_1 and I_2 are non-empty subsets of B, then $I_1 \cap I_2$ is non-empty subset of B.

Let $a, b \in I_1 \cap I_2 \Rightarrow a, b \in I_1$ and $a, b \in I_2$
$\Rightarrow a + b \in I_1$ and $a + b \in I_2$ ($\because$ I_1 and I_2 are ideals of B)
$\Rightarrow a + b \in I_1 \cap I_2$
Let $a \in I_1 \cap I_2 \Rightarrow a \in I_1$ and $a \in I_2$
and $s \in B$, then $a \cdot s \in I_1$ and $a \cdot s \in I_2 \Rightarrow a \cdot s \in I_1 \cap I_2$
Thus, if $a, b \in I_1 \cap I_2 \Rightarrow a + b \in I_1 \cap I_2$

and $a \in I_1 \cap I_2, s \in B \quad \Rightarrow \quad a \cdot s \in I_2 \cap I_2$

Hence, $I_1 \cap I_2$ is an ideal of *B*.

Theorem 2. *Union of two ideals is an ideal iff one of them is contained in the other.*

Proof. Let I_1 and I_2 be two ideals of *B* and suppose that one of them is contained in the other, then we have to show that $I_1 \cup I_2$ is an ideal of *B*. Since either $I_1 \subseteq I_2$ or $I_2 \subseteq I_1$.

If $I_1 \cup I_2$ is an ideal of *B*.

If $I_2 \subseteq I_1$, then $I_1 \cup I_2 = I_1$ and I_1 is given an ideal of *B*.

Thus, $I_1 \cup I_2$ is an ideal of *B*.

Conversely, suppose $I_1 \cup I_2$ is an ideal of *B*, then we shall show that either one is contained in the other.

Suppose $I_1 \not\subseteq I_2$ and $I_2 \not\subseteq I_1$

$\Rightarrow$ there exists $a \in I_1$ such that $a \notin I_2$ and there exists $b \in I_2$ such that $b \notin I_1$

$\Rightarrow$ $a, b \in I_1 \cup I_2 \quad \Rightarrow \quad (a + b) \in I_1 \cup I_2$ (Being an ideal of *B*)

$\Rightarrow$ $(a + b) \in I_1$ or $(a + b) \in I_2$

$\Rightarrow$ if $(a + b) \in I_1$ and $b \in I_2 \subseteq B$, then $(a + b)\, b \in I_1$ (I_1 being an ideal of *B*)

$\Rightarrow$ $b \in I_1$ (By absorption law in *B*)

$\Rightarrow$ This gives a contradiction.

Similarly, if taken $a + b \in I_2$, we get a contradiction that $a \in I_2$.

Thus, our assumption that $I_1 \not\subseteq I_2$ and $I_2 \not\subseteq I_1$ are wrong. Hence, one of them is contained in the other.

Theorem 3. *The necessary and sufficient condition for a non-empty subset I of a Boolean algebra to be an ideal of B is that*

(i) $a, b \in I \Rightarrow (a + b) \in I$ *(ii)* $a \in I, x \leq a \Rightarrow x \in I$.

Proof. Condition is necessary : Suppose *I* is an ideal of *B*, then by the definition of an ideal (i) is obtained.

Let $a \in I$, $x \leq a$, then $x = a \cdot x \in I$ (Being an ideal)

Thus, (ii) is also obtained.

Sufficient condition : Suppose (i) and (ii) is satisfied. to show that *I* is an ideal of *B*.

From (i) $a \cdot b \in I \Rightarrow (a + b) \in I$.

Now, we only show that $a \in I$, $s \in B \Rightarrow a \cdot s \in B$.

Since, $a \cdot s \leq a$ and $a \in I$

From (ii), $a \cdot s \in I$.

Hence, *I* is an ideal of *B*.

Definition (Dual Ideals) : A non-empty subset *I* of a Boolean algebra *B* is said to be a dual ideal of *B* if

(i) $a, b \in I \Rightarrow a \cdot b \in I$ (ii) $a \in I, s \in B \Rightarrow (a + s) \in I$.

3.5 BOOLEAN FUNCTION

Before we discuss the study of a Boolean function, we shall define some important words related to the Boolean function.

(i) Constant : *Any single symbol to represent a specified element of Boolean algebra is called a constant.* For example 0 and 1 are the constant of Boolean algebra.

(ii) Variable : *Any literal symbols denoted by a, b, c, x, y, z, ... etc. which are used to represent an arbitrary element of Boolean algebra, known as variables of a Boolean algebra.*

(iii) Monomial : *Any single element of a Boolean algebra is called a monomial* and also those elements which are formed using the operation (·) on two or more than two elements are called monomials.

For example : $x \cdot y$, $x' \cdot y$, $x \cdot y \cdot z$, $x' . y'$, etc.

(iv) Expression or Polynomial : An expression which is built up by a repeated application of Boolean binary operations (+), (·) and (′) a finite number of times to some variables is called Boolean expression.

For example : $x + y + x' \cdot y$ or $x' + y' + x \cdot y$, etc. or

A Boolean expression in n variables $x_1, x_2 \ldots x_n$ in any finite ray using parentheses and symbols 0, 1 and Boolean operations '+', '·' and '′', formed. Thus

(i) 0,1 are Boolean expressions.

(ii) The variables $x_1, x_2, \ldots, x_n$ are Boolean expressions.

(iii) If f_1, f_2 are Boolean expression, then $(f_1 \cdot f_2)$ and $(f_1 + f_2)$ are Boolean expressions.

(iv) Complement of a Boolean expression is a Boolean expression.

Now, we shall denote the Boolean expression formed as above $f(x_1, x_2, \ldots)$ or by f.

(v) Any two Boolean expressions are said to be *equivalent* if one is obtained from other by a finite number of applications of identities of a Boolean algebra. Now, we come across the definition of a Boolean function.

(v) Boolean function : *Any expression which is the combination of a finite set of symbols, that is, a combination of constants and variable by the operations of '+', '·' and '′' is called a Boolean function.* **For example :** $(a' + b')' c + ab' x + 0$ is a Boolean function.

(vi) Minimial Boolean Function : A Boolean function in n variable $x_1, x_2 \ldots x_n$ is said to be minimal if it is the product of n variables provided the rth variable is either taken x_r, or its complement x'_r. **For example :** Let x and y be two variables in Boolean algebra B and the complements of x and y be x' and y'. Then the minimal Boolean functions are given by

$$x \cdot y,\ x' \cdot y,\ x \cdot y',\ x' \cdot y'$$

From above minimal Boolean functions, we conclude that the number of minimal Boolean function in two variables are $2 \times 2 = 2^2 = 4$

Similarly, we consider three variables in a Boolean algebra, then the minimal Boolean function are given by

$$x \cdot y \cdot z,\ x' \cdot y \cdot z,\ x . y' . z,\ x \cdot y \cdot z',\ x' \cdot y' \cdot z,\ x' \cdot y \cdot z',\ x \cdot y',\ z',\ x' \cdot y' \cdot z'$$

Thus, the number of minimal Boolean functions in three variables are

$$2 \times 2 \times 2 = 2^3 = 8$$

3.5.1 Bool's Theorem

Theorem 1. *There are 2^n minimal Boolean functions in n variables.*

Proof. Let $x_1, x_2, x_3, \ldots x_n$ be n variables in a Boolean algebra B and let $x'_1, x'_2, x'_3, \ldots x'_n$ be the complements of the above variables respectively. To form a minimal Boolean function each variable can be selected in two ways, that is either x_r is taken or x'_r. Since there are n variables, thus the number of minimal Boolean function are

$$2 \times 2 \times 2 \times \ldots \times 2 \text{ upto } n \text{ times} = 2^n$$

Hence, the number of minimal Boolean functions are 2^n.

Theorem 2. *No Boolean Algebra can have three distinct elements.*

Proof. Let B be a Boolean Algebra having three elements. Clearly B must have two distinct elements 0 and 1 as identities for + and . respectively. Let a be the third element of Boolean Algebra B.

Now since, B is a Boolean Algebra, therefore $\exists$ on element a' in B such that

$$a + a' = 1 \text{ and } a.a' = 0$$

Now, we have the following three cases :

Case I. If $a' = a$

Then $a + a' = 1 \Rightarrow a + a = 1$ $(\because a' = a)$

$\Rightarrow a = 1$

and $ad' = 0 \Rightarrow a.a = 0$

$\Rightarrow a = 0$

But a is different from 0 and 1. Hence $a' = a$ is not possible

Case II. If $a' = 0$ then $a + a' = 1 \Rightarrow a + 0 = 1 \Rightarrow a = 1$

But a is not equal to 1

$\Rightarrow a' = 0$ is not possible

Case III. If $a' = 1$ then $aa' = 0 \Rightarrow a.1 = 0$

$\Rightarrow a = 0$

But a is not equal to 0

$\Rightarrow a' = 1$ is not possible

Hence, B either has only two elements 0 and 1 or B has four elements because there is an element a in B different from 0 and 1, then B must have another element a' from 0, 1 and a.

Therefore, no Boolean algebra can have exactly three elements.

WORKING RULE To evaluate a Boolean Expression.

Step 1. Perform all Compliment of single terms.

Step 2. Perform all operations within paranthesis.

Step 3. Perform an **and** operation before **or** operation unless paranthesis indicate otherwise.

Step 4. If an expression involve dash (′), perform the operations of the expression first and then compliment the result.

ILLUSTRATIVE EXAMPLES

1. *For every x and y {B, +, ·, 0, 1, ′}, then prove the following :*

(i) $x + (x + y) = x + y$

(ii) $x(xy) = xy$

(iii) $xy = x \Rightarrow xy' = 0$

(iv) $xy' = 0 \Rightarrow x + y = y$

(v) $x + y = y \Rightarrow xy = x$

(vi) $x + x'y = x + y$

(vii) $x' + xy = x' + y$

Solution.

(i) $x + (x + y) = (x + x) + y$ (By associative law)

$= x + y$ $(\because x + x = x)$

Hence, $x + (x + y) = x + y$

(ii) $x(xy) = xy$

L.H.S. $= x(xy)$

$= (xx)y$ (By associative law)

$= xy$ $(\because xx = x)$

= RHS.

Hence, $x(xy) = xy$

(iii) $xy = x \Rightarrow xy' = 0$

Now, $xy' = (xy)y'$ (∵ $xy = x$ given)

$= x(yy')$ (By associative law)

$= x \cdot 0$ (∵ $yy' = 0$)

$= 0$ (∵ $x \cdot 0 = 0$)

Hence, $xy = x \Rightarrow xy' = 0$

(iv) $xy' = 0 \Rightarrow x + y = y$

Since, $xy' = 0$ is given

L.H.S. $= x + y$

$= (x + y) \cdot 1$ (∵ $x \cdot 1 = x$)

$= (y + x)(y + y')$ (By commutative law)

$= y + xy'$ (By distributive law)

$= y + 0$ (∵ $xy' = 0$ given)

$= y =$ RHS

Hence, $xy' = 0 \Rightarrow x + y = y$

(v) $x + y = y \Rightarrow xy = x$

Since, $x + y = y$ is given

L.H.S. $= xy$

$= x \cdot (x + y)$

$= xx + xy$ (By distributive law)

$= x + xy$

$= x$ (By absorption law)

$=$ RHS

Hence, $x + y = y \Rightarrow xy = x$

(vi) $x + x'y = x + y$

L.H.S. $= x + x'y$

$= (x + x')(x + y)$ (By distributive law)

$= 1 \cdot (x + y)$ (∵ $x + x' = 1$)

$= x + y$ (∵ $a \cdot x = x$)

$=$ RHS

Hence, $x + x'y = x + y$

(vii) $x' + xy = x' + y$

LH.S. $= x' + xy$

$= (x' + x)(x' + y)$ (By distributive law)

$= (x + x') \cdot (x' + y)$ (By commutative law)

$= 1 \cdot (x' + y)$ (∵ $x + x' = 1$)

$= x' + y$ (∵ $1 \cdot x = x \cdot 1 = x$)

$=$ RHS

Hence, $x' + xy = x' + y$.

2. *Simplify the following using Boolean algebra :*

(i) $(xy' + z)'$ *(ii)* $(x + y)x'y'$ *(iii)* $xyz + x' + y' + z'$

(iv) $xy + [(x + y')y]'$ *(v)* $(x + x'y) \cdot (x' + xy)$

(vi) $xy + xy' + x'y + x'y'$ *(vii)* $[(x'y')' + x](x + y')'$

Soltuion. (i) $(xy' + z)' = (xy')' \cdot z'$ (By De Morgan's law)

$= [x' + (y')'] \cdot z'$ (By De Morgan's law)

$= (x' + y) \cdot z'$ (∵ $(y')' = y$)

(ii) $(x + y)x'y' = x(x'y') + y(x'y')$ (By distribuive law)

$= (xx')y' + y(x'y')$ (By associative law)

$= (xx')y' + y(y'x')$ (By commutative law)
$= (xx')y' + (yy')x'$ (By associative law)
$= 0 \cdot y' + 0 \cdot x'$ ($\because$ $xx' = 0, yy' = 0$)
$= y' \cdot 0 + x'.0$ (By commutative law)
$= 0 + 0$ ($\because$ $x' \cdot 0 = 0, y'0 = 0$)
$= 0$ ($\because$ 0 is additive identity)

(iii) $xyz + x' + y' + z' = xyz + (xyz)'$ (By De Morgan's law)
$= 1$ ($\because$ $x + x' = 1$)

(iv) $xy + [(x + y')y]' = xy + [(x + y')' + y']$ (By De Morgan's law)
$= xy + [x'(y')' + y']$ (By De Morgan's law)
$= xy + (x'y + y')$ ($\because$ $(y')' = y$)
$= xy + (y' + x'y)$ (By commutative law)
$= xy + (y' + x')(y' + y)$ (By distributive law)
$= xy + (x' + y')(y + y')$ (By commutative law)
$= xy + (x' + y') \cdot 1$ ($\because$ $y + y' = 1$)
$= xy + (x' + y')$ ($\because$ $x \cdot 1 = x$)
$= xy + (xy)'$ (By De Morgan's law)
$= 1$ ($\because$ $x + x' = 1$)

(v) $(x + x'y) \cdot (x' + xy)$
$= [(x + x')(x + y)] \cdot [(x' + x)(x' + y)]$ (By distributive law)
$= [(x + x')(x + y)] \cdot [(x + x')(x' + y)]$ (By commutative law)
$= [1 \cdot (x + y)] \cdot [1\,(x' + y)]$ ($\because$ $x + x' = 1$)
$= (x + y)(x' + y)$ ($\because$ $1 \cdot x = x$)
$= (y + x)(y + x')$ (By commutative law)
$= y + xx'$ (By distributive law)
$= y + 0$ ($\because$ $xx' = 0$)
$= y$ ($\because$ $y + 0 = y$)

(vi) $xy + xy' + x'y + x'y' = x(y + y') + x'(y + y')$ (By distributive law)
$= (x + x') \cdot (y + y')$ (By distributive law)
$= 1 \cdot 1$ ($\because$ $x + x' = 1, y + y' = 1$)
$= 1$ ($\because$ 1 is multiplication identity)

(vii) $[(x'y')' + x](x + y')' = [(x')' + (y')' + x](x'(y')')$ (By De Morgan's law)
$= (x + y + x)(x' \cdot y)$ ($\because$ $(x')' = x, (y')' = y$)
$= (x + x + y)(x'y)$ (By commutative law)
$= (x + y)(x'y)$ ($\because$ $x + x = x$)
$= x(x'y) + y(x'y)$ (By distributive law)
$= (xx')y + (yy)x'$

(By commutative and then associative law)
$= 0 \cdot y + yx'$ ($\because$ $xx' = 0, yy = y$)
$= y \cdot 0 + x'y$ (By commutative law)
$= 0 + x'y$ ($\because$ $y \cdot 0 = 0$)
$= x'y$ ($\because$ $0 + x = x$)

3. *For every x and y in Boolean algebra, prove that* $x + y = 0 \Leftrightarrow x = 0, y = 0$.

Solution. Suppose $x + y = 0$, then we shall have to prove that

$x = 0, y = 0$

Now, $x = x + 0$ ($\because$ $x + 0 = x = 0 + x$)
$= x + yy'$ ($\because$ $yy' = 0$)
$= (x + y) \cdot (x + y')$ (By distributive law)
$= 0 \cdot (x + y')$ ($\because$ $x + y = 0$ given)

$$= (x + y') \cdot 0 \qquad \text{(By commutative law)}$$
$$= 0 \qquad (\because\ x \cdot 0 = 0)$$

and
$$y = y + 0 \qquad (\because\ y = y + 0 = 0 + y)$$
$$= y + xx' \qquad (\because\ xx' = 0)$$
$$= (y + x)(y + x') \qquad \text{(By distributive law)}$$
$$= (x + y)(y + x') \qquad \text{(By commutative law)}$$
$$= 0 \cdot (y + x') \qquad \text{(Given that } x + y = 0)$$
$$= (y + x') \cdot 0 \qquad \text{(By commutative law)}$$
$$= 0 \qquad (\because\ x \cdot 0 = 0)$$

Thus, $x + y = 0 \quad \Rightarrow \quad x = 0, y = 0.$ ...(1)

Conversely, suppose $x = 0$, $y = 0$, then we shall have to show that

$$x + y = 0$$

Since, $x = 0$, $y = 0$, then $x + y = 0 + 0$

$$= 0 \qquad (\because\ 0 \text{ is an additive identity})$$

Thus, $x = 0, y = 0 \quad \Rightarrow \quad x + y = 0$ ...(2)

Hence, from (1) and (2), we get $x + y = 0 \Leftrightarrow x = 0, y = 0$.

4. *Prove the following using Boolean algebra*

(i) $(x + y)' + (x + y')' = x'$ *(ii)* $(x + y)(x' + z) = x'y + xz$

Solution. (i) $(x + y)' + (x + y')' = x'$

$$\text{L.H.S.} = (x + y)' + (x + y')'$$
$$= x'y' + x'(y')' \qquad \text{(By De Morgan's law)}$$
$$= x'y' + x'y \qquad (\because\ (y')' = y)$$
$$= x'(y' + y) \qquad \text{(By distributive law)}$$
$$= x'(y + y') \qquad \text{(By commutative law)}$$
$$= x' \cdot 1 \qquad (\because\ y + y' = 1)$$
$$= x' \qquad (x \cdot 1 = x = 1 \cdot x)$$
$$= \text{RHS}$$

Hence, $(x + y)' + (x + y')' = x'$.

(ii) $(x + y)(x' + z) = x'y + xz$

$$\text{LHS} = (x + y)(x' + z)$$
$$= (x + y)x' + (x + y)z \qquad \text{(By distributive law)}$$
$$= (xx' + yx') + (xz + yz) \qquad \text{(By distributive law)}$$
$$= (0 + yx') + (xz + yz) \qquad (\because\ xx' = 0)$$
$$= yx' + xz + yz \qquad (\because\ 0 + x = x = x + 0)$$
$$= xz + yx' + yz \qquad \text{(By commutative law)}$$
$$= xz + yx' + yz \,.\, 1 \qquad (\because\ x \cdot 1 = x)$$
$$= xz + yx' + yz(x + x') \qquad (\because\ x + x' = 1)$$
$$= xz + yx' + yzx + yzx' \qquad \text{(By distributive law)}$$
$$= xz + yzx + x'y + x'yz \qquad \text{(By commutative law)}$$
$$= xz + yzx + x'y(1 + z) \qquad \text{(By distributive law)}$$
$$= xz + yzx + x'y \cdot 1 \qquad (\because\ 1 + z = 1)$$
$$= xz + xzy + x'y \qquad (\because\ x \cdot 1 = x \text{ and by commutative law})$$
$$= xz(1 + y) + x'y \qquad \text{(By distributive law)}$$
$$= xz \cdot 1 + x'y \qquad (\because\ 1 + y = 1)$$
$$= xz + x'y \qquad (\because\ x \cdot 1 = x)$$
$$= x'y + xz \qquad \text{(By commutative law)}$$
$$= \text{RHS.}$$

5. *Using Boolean algebra prove that*

$xyz + xyz' + xy'z + x'yz = xy + yz + zx$ for all $x, y, z \in B$

Solution. LHS $= xyz + xyz' + xy'z + x'yz$

$= xy(z + z') + xy'z + x'yz$ (By distributive law)

$= xy \cdot 1 + xy'z + x'yz$ (By distributive law)

$= x[(y + y')(y + z)] + x'yz$ (By distributive law)

$= x[1 \cdot (y + z)] + x'yz$ ($\because\ y + y' = 1$)

$= x(y + z) + x'yz$ ($\because\ 1 \cdot x = x$)

$= xy + xz + x'yz$ (By distributive law)

$= xy + (x + x'y)z$ (By distributive law)

$= xy + [(x + x')(x + y)]z$ (By distributive law)

$= xy + [1 \cdot (x + y)]z$ ($\because\ x + x' = 1$)

$= xy + (x + y)z$ ($\because\ 1 \cdot x = x$)

$= xy + xz + yz$ (By distributive law)

$= xy + yz + zx$ (By commutative law)

= RHS.

6. *If $p + x = q + x$ and $p + x' = q + x'$, then show that $p = q$.*

Solution. Given that

$$p + x = q + x$$
$$p + x' = q + x'$$

Then $(p + x)(p + x') = (q + x)(q + x')$

$\Rightarrow\ p + xx' = q + xx'$ (By distributive law)

$\Rightarrow\ p + 0 = q + 0$ ($\because\ xx' = 0$)

$\Rightarrow\ p = q$ ($\because\ x + 0 = x$)

7. *Prove that if $px = qx$ and $px' = qx'$, then $p = q$.*

Solution. Given that $px = qx$ and $px' = qx'$,

Now, $px + px' = qx + qx'$

$\Rightarrow\ p(x + x') = q(x + x')$ (By distributive law)

$\Rightarrow\ p \cdot 1 = q \cdot 1$ ($\because\ x + x' = 1$)

$\Rightarrow\ p = q$ ($\because\ x \cdot 1 = x$)

8. *Prove that in a Boolean algebra, for every triple elements x, y and z*

$$xy + yz + zx = (x + y)(y + z)(z + x)$$

Solution. RHS $= (x + y)(y + z)(z + x)$

$= (x + y)(z + y)(z + x)$ (By commutative law)

$= (x + y)(z + yx)$ (By distributive law)

$= (x + y) \cdot z + (x + y)yx$ (By distributive law)

$= (xz + yz) + x(yx) + y(yx)$ (By distributive law)

$= (zx + yz) + (xx)y + (yy)x$ (By commutative and associative laws)

$= zx + yz + xy + yx$ ($\because\ xx = x,\ yy = y$)

$= yz + zx + xy + xy$ (By commutative laws)

$= yz + zx + xy$ ($\because\ x + x = x$)

$= xy + yz + zx$ (By commutative law)

= LHS

Hence, $xy + yz + zx = (x + y)(y + z)(z + x)$.

9. *For every elements a, b, c in a Boolean algebra prove that if $a + b = a + c$ and $a \cdot b = a \cdot c$, then $b = c$.*

Solution. Given that : $a + b = a + c$ and $ab = ac$

Now, $b = b + ba$ (By absorption law)

$= bb + ba$ ($\because\ bb = b$)

$= b(b + a)$ (By distributive law)

$= b(a + b)$ (By commutative law)

$= b(a + c)$ $(\because\ a + b = a + c)$
$= ba + bc$ (By distributive law)
$= ab + bc$ (By commutative law)
$= ac + bc$ $(\because\ ab = ac)$
$= (a + b)\, c$ (By distributive law)
$= (a + c)\, c$ $(\because\ a + b = a + c)$
$= a \cdot c + c \cdot c$ (By distributive law)
$= ac + c$ $(\because\ c \cdot c = c)$
$= c + ac$ (By commutative law)
$= c$ (By absorption law)

Hence, $b = c$.

10. *Prove that for any a, b and c in a Boolean algebra, the following four expressions are equal :*

(i) $(a + b)(a' + c)(b + c)$ *(ii)* $ac + a'b + bc$
(iii) $(a + b)(a' + c)$ *(iv)* $ac + a'b$

Solution. First we shall prove that (i) = (ii)

LHS $= (a + b)(a' + c)(b + c)$
$= (a + b)(c + a')(c + b)$ (By commutative law)
$= (a + b)(c + a'b)$ (By distributive law)
$= (a + b)\, c + (a + b)\, a'b$ (By distributive law)
$= ac + bc + a(a'b) + b(a'b)$ (By distributive law)
$= ac + bc + (aa')\, b + a'(bb)$ (By commutative and associative laws)
$= ac + bc + 0 \cdot b + a'b$ $(\because\ aa' = 0,\ bb = b)$
$= ac + bc + b \cdot 0 + a'b$ (By commutative law)
$= ac + bc + 0 + a'b$ $(\because\ b \cdot 0 = 0)$
$= ac + bc + a'b$ ($\because$ 0 is an additive identity)
$= ac + a'b + bc$ (By commutative law)
= RHS

Hence, (i) = (ii)

Now, to show that (ii) = (iii)

i.e., $ac + a'b + bc = (a + b)(a' + c)$

RHS $= (a + b)(a' + c)$
$= (a + b)\, a' + (a + b)\, c$ (By distributive law)
$= aa' + ba' + ac + bc$ (By distributive law)
$= 0 + ba' + ac + bc$ $(\because\ aa' = 0)$
$= ba' + ac + bc$ ($\because$ 0 is an additive identity)
$= ac + ba' + bc$ (By commutative law)
$= ac + a'b + bc$ (By commutative law)
= LHS

Hence, (ii) = (iii)

Next, we shall show that (iii) = (iv)

i.e., $(a + b)(a' + c) = ac + a'b$

LHS $= (a + b)(a' + c)$
$= ac + a'b + bc$ ($\because$ (ii) = (iii))
$= ac + a'b + bc \cdot 1$ $(\because\ x \cdot 1 = x)$
$= ac + a'b + bc\,(a + a')$ $(\because\ a + a' = 1)$
$= ac + a'b + (bc)\, a + (bc)\, a'$ (By distributive law)
$= ac + (bc)\, a + a'b + (bc)\, a'$ (By commutative law)
$= ac + b(ca) + a'b + b(ca')$ (By associative law)

$= ac + b(ac) + a'b + b(a'c)$ (By commutative law)
$= ac + b(ac) + a'b + (ba')c$ (By associative law)
$= ac + (ac)b + a'b + (a'b)c$ (By commutative law)
$= ac + a'b$ (By absorption law)
= RHS

Hence, (iii) = (iv).

Consequently, (i) = (ii) = (iii) = (iv).

11. *If p and q are two statements, then express $p \Leftrightarrow q$ in a Boolean algebra.*

Solution. Let $x = p$, $y = q$ and $\rceil p = x'$, $\rceil q = y'$ and $\vee$ stands for +, $\wedge$ stands for '·'

Now, $p \Leftrightarrow q = (p \Rightarrow q) \wedge (q \Rightarrow p)$
$= (\rceil p \vee q) \wedge (\rceil q \vee p)$
$= (x' + y) \cdot (y' + x)$
$= (x' + y) \cdot y' + (x' + y) \cdot x$ (By distributive law)
$= (x'y' + yy') + (x'x + yx)$ (By distributive law)
$= (x'y' + yy') + (xx' + xy)$ (By commutative law)
$= (x'y' + 0) + (0 + xy)$ ($\because$ $xx' = 0$, $yy' = 0$)
$= x'y' + xy$ ($\because$ 0 is an additive identity)
$= xy + x'y'$ (By commutative law)

Hence, $p \Leftrightarrow q = xy + x'y'$.

12. *Rewrite the compound statement $(p \wedge q) \vee [\rceil p\,(p \vee \rceil q) \wedge q]$ in a Boolean algebra and then simplify.*

Solution. Since $\wedge$ stands for '·' and $\vee$ stands for '+', Negation is represented by complement then

$$(p \wedge q) \vee [\rceil(p \vee \rceil q) \wedge q] = xy + [(x + y') \cdot y]'$$

where, $p = x, q = y$

Now, $xy + [(x + y') \cdot y]'$
$= xy + (xy + y'y)'$ (By distributive law)
$= xy + (xy + yy')'$ (By commutative law)
$= xy + (xy + 0)'$ ($\because$ $yy' = 0$)
$= xy + (xy)'$ ($\because$ $a + 0 = a$)
$= 1$ ($\because$ $a + a' = 1$)

3.6 NORMAL BOOLEAN FUNCTION

A Boolean function of n variables is said to be normal if it contains the n variables.
For example : $x + y$, $x' + y'$, $x + x'y$, $x' + xy$ are normal Boolean functions of two variables.

3.7 DISJUNCTIVE NORMAL FORM (OR CANONICAL FORM)

A Boolean function of n variables $x_1, x_2, \ldots, x_n$ for $n > 0$ is said to be disjunctive normal form, if the function is expressed as the sum of the products of the type $f_1(x_1)$, $f_2(x_2) \ldots, f_n(x_n)$, where $f_i(x_i)$ is either taken x_i or x_i' for $i = 1, 2, \ldots, n$ provided no two products are identical. Also, 0 and 1 can be expressed in disjunctive normal form in n variables for any $n \geq 0$.

For example : $f(x, y) = xy + x'y'$
and $f(x, y, z) = xyz + x'yz + xy'z + xyz'$
are disjunctive normal form in two variables and in respectively.

WORKING RULES **(To find DNF)**

Step 1. Simplify the given function to contain minimum number of terms.

Step 2. Multiply each term by as many 1 as the number of missing variables in that term.

Step 3. Putting $x + x' = 1$ for in those terms which do not contain x.

Step 4. Apply distributive law and see that no term is repeated. If so, apply idempotent law

3.7.1 Complete Disjunctive Normal Form

A disjunctive normal form in n variables having 2^n products is called the complete disjunctive normal form in n variables.

For example : $f(x, y) = xy + x'y + xy' + x'y'$

and $f(x, y, z) = xyz + x'yz + xy'z + xyz' + x'y'z + x'yz' + xy'z' + x'y'z'$

are the complete disjunctive normal form, because in two variables there are $2^2 = 4$ products which are in $f(x, y)$ likewise in $f(x, y, z)$.

☛ Every complete disjunctive normal form is identically equal to 1.

Theorem 1. *Every Boolean function having no constants is reduced to disjunctive normal form.*

Proof. Let $f(x_1, x_2, ..., x_n)$ be a Boolean function having no constants in n variables. Suppose function f contains some expressions of the type $(p + q)'$ and $(pq)'$ for some functions p and q. Now, using De Morgan's law those expression yield to $p'q'$ and $p' + q'$, respectively. This process continued until each prime, which appears applies to a single variable x_i. Next applying the distributive law of '$\cdot$' over '+' and f can be reduced to a polynomial. Now, suppose there are some terms which do not contain either x_i or x_i'. For some variable x_i, then those terms may be multiplied by $(x_i + x_i')$ which does not change the function. Continuing this process until we get the missing variable in each term of $f(x_1, x_2, ..., x_n)$, thereafter using $x_i + x_i = x_i$ and $x_i x_i = x_i$ results eliminating the duplicate terms and with this we get the function in disjunctive normal form.

3.8 CONJUNCTIVE NORMAL FORM (OR DUAL CANONICAL FORM)

A Boolean function in n variables $x_1, x_2, ..., x_n$ for $n > 0$ is said to be in conjunctive normal form, if the function is expressed as the product of the type $f_1(x_1), f_2(x_2) ..., f_n(x_n)$, where $f_i(x_i)$ is either taken x_i or x_i' for $i = 1, 2, ... n$ provided no two factors are identical. In addition to it, 0 and 1 are expressed in conjunctive normal form in n variables for $n \geq 0$.

For example :

$f(x, y) = (x + y)(x' + y')$ and $f(x, y, z) = (x + y + z)(x' + y + z)(x + y' + z)$

are in conjunctive normal form in two variables and three respectively.

WORKING RULES **(To find CNF)**

Step 1. Simplify the given functions so that it contains minimum number of terms (use absorption laws, distributive law, unit property and zero property).

Step 2. Add 0 to the simplified function or in each factor if the functions is in factor form.

Step 3. Put this 0 equal to the product of missing variables and its complement.

Step 4. Apply distributive law.

If any variable is missing again in any factor, repeat the above steps and remove the repeated factor with the help of idempotent laws.

3.8.1 Complete Conjunctive Normal Form

A conjunctive normal form in n variables having 2^n factors is called the complete conjunctive normal form in n variables.

For example : $f(x, y) = (x + y)(x' + y)(x + y')(x' + y')$

and $f(x, y, z) = (x + y + z)(x' + y + z)(x + y' + z)(x + y + z')(x + y' + z')$
$(x' + y' + z) \cdot (x' + y + z')(x' + y' + z')$

are in complete conjunctive normal forms.

- Every complete conjunctive normal form is reduced to identically 1.

Theorem 1. *Every Boolean function having no constants is reduced to conjunctive normal form.*

Proof. Let $f(x_1, x_2, ..., x_n)$ be a Boolean function in n variables having no constants. Suppose some terms in $f(x_1, x_2, ..., x_n)$ are of type $(p + q)'$ and $(pq)'$ for some functions p and q, then using De Morgan's law those terms yield to $p'q'$ and $p' + q'$, respectively. This process is continued until each prime, which appears applies to a single variable x_i. Next applying the distributive law of '+' over '.' so that f is reduced to a factored polynomial. Now, suppose there are some factors which do not contain either x_i or x_i'. For some variable x_i, then those factors are added to $x_i x_i'$ without changing the function. Continuing this process until we get the missing variable in each factor of $f(x_1, x_2, ..., x_n)$ and applying the distributive law of '·' over '+' again, thereafter using $x_i + x_i = x_i$ and $x_i x_i = x_i$ for all $i = 1, 2, ..., n$ eliminate the duplicate terms in each factor and with this we get the function in conjunctive normal form.

Theorem 2. *Two Boolean functions are equal if their respective canonical forms are identical.*

Proof. Let us first suppose two canonical forms of a Boolean functions are identical. Since both forms are equal ⇒ they consist of the same terms

⇒ Corresponding Boolean functions are equal.

Conversely, let two Boolean functions be equal, then these functions will have the same value for each possible assignment of 0 and 1, to the variables. Further, each assignment for which the value of the functions is 1, we get a term of the canonical form of the functions

⇒ Normal forms, contains the same terms.

⇒ Boolean functions are identical.

Theorem 3. *If each variable in the complete canonical form in n – variables be arbitrary assigned the value of 0 and 1 then one and only one will have the value 1 and all other terms will have the same value 0.*

Proof. Let f be a function of $x, y, z, ... l$ variables, which are n in numbers. Clearly in the complete additive form if the function there will be a term $x.y.z .... l$ positively.

Then by giving the value 1 to each of the n variables, the value of this term will be 1.1.1 ..... n times = 1

Whereas, in all other terms of the complete additive normal for there will be at least one variable in its compliment form whose value will be 0. Consequently, at least one factor of each term will be zero, reducing the term to its value zero.

Similar argument will be given to any arbitrary alloment of the value 1 or 0 to all the variables. Thus, we conclude that only one term of the complete canonical form will have the value 1 and the rest of all other will have the value 0.

Theorem 4. *Every function can be expressed in additive normal form.*

Proof. If the given function is a polynomial without any bracket then complete the terms missing any variable by multiplying it by 1 and putting the sum of that variable and its complement in place of 1. Then on applying distributive law and simplifying, the given polynomial will be changed to additive normal form. If given function is a polynomial having brackets then bracket can be removed by using absorption law, distributive law and De-Morgan's law.

Therefore, the function will be converted in the form of a polynomial having no brackets and hence it can be converted into additive normal form.

On the hand, if the given function is a monomial, then it can be converted to normal form in all variables by

(a) multiplying it by 1 for each missing variables.

(b) replacing this one by the sum of that variables and its complement.

(c) expanding it by distribution law

Replacing this process by each missing variables and simplifying in the same manner, we get the monomial changed this additive normal form.

Hence, we conclude that every function can be expressed in additive normal form.

ILLUSTRATIVE EXAMPLES

1. *Write the function $f(x, y, z) = (xy' + xz)' + x'$ is disjunctive normal form.*

Solution. $f(x, y, z) = (xy' + xz)' + x'$

$= (xy')' . (xz)' + x'$ (By De Morgan's law)

$= [x' + (y')'] \cdot [x' + z'] + x'$ (By De Morgan's law)

$= (x' + y) \cdot (x' + z') + x'$ $(\because (y')' = y)$

$= (x' + y) \cdot x' + (x' + y) \cdot z' + x'$ (By distributive law)

$= x' \cdot x' + yx' + x'z' + yz' + x'$ (By distributive law)

$= x' + yx' + x'z' + yz' + x'$ $(\because x'x' = x')$

$= x' + x'y + x'z' + yz' + x'$ (By commutative law)

$= x' + x'y + yz' + x'z' + x'$ (By commutative law)

$= x' + x'y + yz' + x'$ (By absorption law)

$= x' + yz' + x'$ (By absorption law)

$= x' + x' + yz'$ (By commutative law)

$= x' + yz'$ $(\because x' + x' = x')$

$= (x'(y + y')(z + z') + yz'(x + x'))$

$= x' \cdot [(y + y')z + (y + y')z'] + yz'x + yz'x'$ (By Distributive law)

$= x' \cdot [yz + y'z + yz' + y'z'] + yz'x + yz'x'$ (By distributive law)

$= x'yz + x'y'z + x'y'z' + yz'x + yz'x'$ (By $x.x = x$)

$= x'yz + xyz' + x'yz' + x'y'z + x'y'z'$ (By commutative law)

Thus, $f(x, y, z) = x'yz + xyz' + x'yz' + x'y'z + x'y'z'$.

2. *Express each of the following in disjunctive normal form :*

(i) $x + x'y$ (ii) $xy' + xz + xy$ (iii) $(u + v + w)(uv + u'w)'$

Solution. (i) $x + x'y = x \cdot 1 + x'y$ $(\because x \cdot 1 = x)$

$= x(y + y') + x'y$ $(\because y + y' = 1)$

$= xy + xy' + x'y$ (By distributive law)

(ii) $xy' + xz + xy = xy' \cdot 1 + xz \cdot 1 + xy \cdot 1$ $(\because x \cdot 1 = x)$

$= xy'(z + z') + xz(y + y') + xy(z + z')$ $(\because x + x' = 1)$

$= xy'z + xy'z' + xzy + xzy' + xyz + xyz'$ (By distributive law)

$= xyz + xyz + xy'z + xy'z + xyz' + xy'z'$ (By commutative law)

$= xyz + xy'z + xyz' + xy'z'$ $(\because x + x = x)$

(iii) $(u + v + w)(uv + u'w)' = (u + v + w)[(uv)' \cdot (u'w)']$ (By De Morgan's law)

$= (u + v + w) \cdot [(u' + v') \cdot (u + w')]$ (By De Morgan's law)
$= (u + v + w)[(u' + v') \cdot u + (u' + v') \cdot w']$ (By distributive law)
$= (u + v + w) \cdot [u'u + v'u + u'w' + v'w')$ (By distributive law)
$= (u + v + w) \cdot (uu' + uv' + u'w' + v'w')$ (By commutative law)
$= (u + v + w) \cdot (0 + uv' + u'w' + v'w')$ ($\because uu' = 0$)
$= (u + v + w) \cdot (uv' + u'w' + v'w')$ ($\because 0 + u = u$)
$= (u + v + w) \cdot uv' + (u + v + w) \cdot u'w' + (u + v + w) \cdot v'w'$ (By distributive law)
$= u(uv') + v(uv') + w(uv') + u(u'w') + v(u'w') + w(u'w')$
$+ u(v'w') + v(v'w') + w(v'w')$ (By distributive law)
$= u(uv') + v(v'u) + (uv')w + u(u'w') + v(u'w')$
$+ w(w'u') + uv'w' + v(v'w') + w(w'v')$ (By commutative law)
$= (uu)v' + (vv')u + uv'w + (uu')w' + vu'w' + (ww')u'$
$+ uv'w' + (vv')w' + (ww')v'$
$= uv' + 0 \cdot u + uv'w + 0 \cdot w' + u'vw' + 0u' + uv'w' + 0w' + 0v'$ $\begin{bmatrix} uu = u', uu' = 0 \\ vv' = 0, ww' = 0 \end{bmatrix}$
$= uv' + u \cdot 0 + uv'w + w' \cdot 0 + u'vw' + u' \cdot 0 + uv'w' + w' \cdot 0 + v' \cdot 0.$
(By commutative law)
$= uv' + 0 + uv'w + 0 + u'vw' + 0 + uv'w' + 0 + 0$ ($\because u \cdot 0 = 0 \ \forall u$)
$= uv' + uv'w + u'vw' + uv'w'$ ($\because$ 0 is an additive identity)
$= uv' \cdot 1 + uv'w + u'vw' + uv'w'$
$= uv'(w + w') + uv'w + u'vw' + uv'w'$ ($\because w + w' = 1$)
$= uv'w + uv'w' + uv'w + u'vw' + uv'w'$ (By distributive law)
$= uv'w + u'vw' + uv'w'$ ($\because x_i + x_i = x_i$)

Hence, $(u + v + w)(uv + u'w)' = uv'w + u'vw' + uv'w'$.

3. *Write each of the following in disjunctive normal form in x, y and z :*
(i) $x + y'$ *(ii)* $x'z + xz'$ *(iii)* $(x + y)(x' + y')$ *(iv)* x

Solution. (i) $x + y' = x \cdot 1 \cdot 1 + y' \cdot 1 \cdot 1$
$= x(y + y')(z + z') + y'(x + x')(z + z')$ ($\because x + x' = 1$)
$= x \cdot [(y + y')z + (y + y')z'] + y'[(x + x')z + (x + x')z']$ (By distributive law)
$= x \cdot [(yz + y'z) + (yz' + y'z')] + y'[(xz + x'z) + xz' + x'z']$
(By distributive law)
$= x \cdot (yz + y'z) + x \cdot (yz' + y'z') + y' \cdot (xz + x'z) + y' \cdot (xz' + x'z')$
(By distributive law)
$= xyz + xy'z + xyz' + xy'z' + y'xz + y'x'z + y'xz' + y'x'z'$
(By distributive law)
$= xyz + xy'z + xyz' + xy'z' + xy'z + x'y'z + xy'z' + x'y'z'$
(By commutative law)
$= xyz + xy'z + xyz' + xy'z' + x'y'z + x'y'z'$ ($\because x_i + x_i = x_i$)

Hence, $x + y' = xyz + xy'z + xyz' + xy'z' + x'y'z + x'y'z'$

(ii) $x'z + xz' = x'z \cdot 1 + xz' \cdot 1$ ($\because x_i \cdot 1 = x_i$)
$= x'z(y + y') + xz'(y + y')$ ($\because y + y' = 1$)
$= x'zy + x'zy' + xz'y + xz'y'$ (By distributive law)
$= x'yz + x'y'z + xyz' + xy'z'$ (By commutative law)

(iii) $(x + y)(x' + y') = (x + y)x' + (x + y)y'$ (By distributive law)
$= (xx' + yx') + (xy' + yy')$ (By distributive law)
$= (0 + yx') + (xy' + 0)$ ($\because xx' = 0, yy' = 0$)
$= yx' + xy'$ ($\because$ 0 is an additive identity)
$= x'y + xy'$ (By commutative law)
$= x'y \cdot 1 + xy' \cdot 1$

$= x'y(z + z') + xy'(z + z')$ $(\because\ z + z' = 1)$

$= (x'yz + x'yz') + xy'z + xy'z'$

$= x'yz + x'yz' + xy'z + xy'z'$

$= x'yz + xy'z + x'yz' + xy'z'$ (By commutative law)

(iv) $x = x \cdot 1 \cdot 1$

$= x(y + y')(z + z')$ $(\because\ y + y' = 1, z + z' = 1)$

$= x \cdot [(y + y')z + (y + y')z']$ (By distributive law)

$= x \cdot [(yz + y'z) + (yz' + y'z')]$ (By distributive law)

$= x(yz + y'z) + x \cdot (yz' + y'z')$ (By distributive law)

$= xyz + xy'z + xyz' + xy'z'$ (By distributive law)

4. *Express the following Boolean functions in disjunctive normal form :*

(i) $f(x, y, z) = [(x + y')(xy'z')]'$

(ii) $f(x, y, z) = xyz + (x + y)(x + z).$

Solution. (i) $f(x, y, z) = [(x + y')(xy'z')']'$

$= (x + y')' + (xy'z')$ (By De Morgan's law)

$= x'(y')' + xy'z'$ (By De Morgan's law and $\because\ (y')' = y$)

$= x'y + xy'z'$

$= x'y \cdot 1 + xy'z'$ $(\because\ y \cdot 1 = y)$

$= x'y(z + z') + xy'z'$ $(\because\ z + z' = 1)$

$= x'yz + x'yz' + xy'z'$ (By distributive law)

Hence, $f(x, y, z) = x'yz + x'yz' + xy'z'$

(ii) $f(x, y, z) = xyz + (x + y)(x + z)$

$= xyz + (x + y)x + (x + y)z$ (By distributive law)

$= xyz + x + xz + yz$ (By absorption and distributive laws)

$= xyz + x \cdot 1 \cdot 1 + xz \cdot 1 + yz \cdot 1$

$= xyz + x(y + y')(z + z') + xz(y + y') + yz(x + x')$

$(\because\ x + x' = 1, y + y' = 1, z + z' = 1)$

$= xyz + x[(y + y')z + (y + y')z'] + xzy + xzy' + yzx + yzx'$

(By distributive law)

$= xyz + x[(yz + y'z) + yz' + y'z'] + xzy + xzy' + yzx + yzx'$

(By distributive law)

$= xyz + xyz + xy'z + xyz' + xy'z' + xzy' + yzx + yzx' + xzy$

(By distributive law)

$= xyz + xyz + xyz + xyz + xy'z + xy'z + xyz' + x'yz + xy'z'$

(By commutative law)

$= xyz + xy'z + xyz' + x'yz + xy'z'$ $(\because\ x_i + x_i = x_i\ \forall i)$

Hence, $f(x, y, z) = xyz + xy'z + xyz' + x'yz + xy'z'$.

5. *Express each of the following in disjunctive normal form in the minimal possible number of variables :*

(i) $(x + y')(y + z')(z + x')(x' + y')$ *(ii)* $(x + y)(x + y')(x' + z)$

Solution. (i) $(x + y')(y + z')(z + x')(x' + y')$

$= (x + y')(y + z')(x' + z)(x' + y')$ (By commutative law)

$= [(x + y')y + (x + y')z'] \cdot [x' + zy']$ (By distributive law)

$= (xy + y'y + xz' + y'z') \cdot (x' + zy')$ (By distributive law)

$= (xy + yy' + xz' + y'z') \cdot (x' + y'z)$ (By commutative law)

$= (xy + 0 + xz' + y'z') \cdot (x' + y'z)$ $(\because\ yy' = 0)$

$= (xy + xz' + y'z') \cdot (x' + y'z)$ $(\because\ 0 + x = x)$

$= (xy + xz' + y'z')x' + (xy + xz' + y'z')y'z$ (By distributive law)

$= (xy)x' + (xz')x' + (y'z')x' + (xy)y'z + (xz')y'z + (y'z')y'z$

(By distributive law)

$= x(yx') + x(z'x') + y'(z'x') + x(yy')z + x(z'y')z + y'(z'y')z$

(By associative law)

$= x(x'y) + x(x'z') + y'(x'z') + x(yy')z + x(y'z')z + y'(y'z')z$

(By commutative law)

$= x(x'y) + x(x'z') + y'(x'z') + x(0)z + x(y'z')z + y'(y'z')z$ $(\because\ yy' = 0)$

$= (xx')y + (xx')z' + (y'x')z' + x(0z) + xy'(z'z) + (y'y')z'z$

(By associative law)

$= (0)y + (0)z' + (x'y')z' + x(0 \cdot z) + xy'(z'z) + (y'y')z'z$

(By commutative law and $xx' = 0$)

$= (0 \cdot y) + (0 \cdot z') + x'y'z' + x(0 \cdot z) + xy'(zz') + y'(zz')$

(By commtative and associative law)

$= (y \cdot 0) + (z' \cdot 0) + x'y'z' + x(z \cdot 0) + xy' \cdot 0 + y' \cdot 0$ (By commutative law)

$= 0 + 0 + x'y'z' + x \cdot 0 + 0 + 0$ $(\because\ x_i \cdot 0 = 0\ \forall i)$

$= x'y'z' + 0$ $(\because\ 0 + x_i = x_i,\ x \cdot 0 = 0)$

$= x'y'z'$ $(\because\ 0 + x = x)$

Hence, $(x + y')(y + z')(z + x')(x' + y') = x'y'z'$ which is the required disjunctive normal form.

(ii) $(x + y)(x + y')(x' + z) = (x + yy') \cdot (x' + z)$ (By distributive law)

$= (x + 0) \cdot (x' + z)$ $(\because\ yy' = 0)$

$= x \cdot (x' + z)$ $(\because\ x + 0 = x)$

$= xx' + xz$ (By distributive law)

$= 0 + xz$ $(\because\ xx' = 0)$

$= xz$ $(\because\ 0 + x = x)$

$= xz \cdot 1$

$= xz\,(y + y')$ $(\because\ y + y' = 1)$

$= (xz)y + (xz)y'$ (By distributive law)

$= x(zy) + x(zy')$ (By associative law)

$= x(yz) + x(y'z)$ (By commutative law)

$= xyz + xy'z$

Hence, $(x + y)(x + y')(x' + z) = xyz + xy'z$ which is the required disjunctive normal form.

6. *Write the complete disjunctive normal form in x, y and z. Determine which terms equal 1 if*
(i) $x = 1$ and $y = z = 0$; (ii) $x = z = 1$ and $y = 0$.

Solution. The complete disjunctive normal form in three variable x, y and z is given by

$f(x, y, z) = xyz = x'yz + xy'z + xyz' + x'y'z + xy'z' + x'yz' + x'y'z'$

(i) Since $x = 1$, then $x' = 1' = 0$ $(\because 1' = 0)$

and $y = z = 0$, then $y' = 0' = 1$, $z' = 0' = 1$

Thus, the term $xy'z'$ equals 1.

(ii) Since $x = 1$, then $x' = 1' = 0$

$z = 1$, then $z' = 1' = 0$

and $y = 0$, then $y' = 0' = 1$.

Thus, the term $xy'z$ equals 1.

7. *Is the statement always true? Justify your answer :*

$$x(y + z') = x(y + w')$$

Solution. If $x(y + z') = x(y + w')$

$\Rightarrow$ $xy + xz' = xy + xw'$ $\Rightarrow$ $xz' = xw'$

$\Rightarrow$ $z' = w'$ $\Rightarrow$ $z = w$.

The statement is true only when $z = w$.

8. *What are Boolean functions? Simplify the given Boolean expression :*

$$(a \cdot b)' \oplus (a \oplus b)'$$

Solution. A Boolean function is an expression which formed with binary variables. The two binary operator OR and AND, the unary operator NOT parenthesis and equal sign.

$(a \cdot b)' \oplus (a \oplus b)'$ $[a \oplus b = ab' + a'b]$

$$= ((a \cdot b)')(a \oplus b)' + (a \cdot b)'((a \oplus b)')'$$
$$= (a \cdot b)(a'b + ab') + (ab)'(a \oplus b)$$
$$= ab(a'b + ab') + (ab)'(a \oplus b)$$
$$= ab(a'b + ab') + (ab)'(a'b + ab')$$
$$= (a'b + ab')1 = a'b + ab'.$$

9. *Obtain the sum of products canonical forms of the Boolean expression :* $x_1x_2' + x_3$.

Solution. We have $x_1x_2' + x_3 = x_1x_2'(x_3 + x_3') + x_3(x_1 + x_1')$

$$= x_1x_2'x_3 + x_1x_2'x_3' + x_1x_3 + x_1'x_3$$
$$= x_1x_2'x_3 + x_1x_2'x_3' + x_1x_3(x_2 + x_2') + x_1'x_3(x_2 + x_2')$$
$$= x_1x_2'x_3 + x_1x_2'x_3' + x_1x_2x_3 + x_1x_2'x_3 + x_1'x_2x_3 + x_1'x_2'x_3$$
$$= x_1x_2'x_3 + x_1x_2'x_3' + x_1x_2x_3 + x_1'x_2'x_3 + x_1'x_2x_3$$

10. *If $f(x, y, z) = xy' + xyz' + x'yz'$, show that :*

$$f(x, y, z) + z' \neq f(x, y, z)$$

Solution. $f(x, y, z) + z' = xy' + xyz' + x'yz' + z'$

$$= xy' + xyz' + z'(x'y + 1)$$
$$= xy' + xyz' + z' \qquad [\because x'y + 1 = 1, z'1 = z']$$
$$= xy' + z'(xy + 1)$$
$$= xy' + z'$$
$$\neq f(x, y, z)$$

Hence, $f(x, y, z) + z' \neq f(x, y, z)$.

11. *Write the following function in conjunctive normal form $f(x, y, z) = (xy' + xz)' + x'$*

Solution. $f(x, y, z) = (xy' + xz)' + x'$

$$= (xy')' \cdot (xz)' + x' \qquad \text{(By De Morgan's law)}$$
$$= (x' + (y')') \cdot (x' + z') + x' \qquad \text{(By De Morgan's law)}$$
$$= (x' + y) \cdot (x' + z') + x' \qquad (\because (y')' = y)$$
$$= (x' + y + x')(x' + z' + x') \qquad \text{(By distributive law)}$$
$$= (x' + x' + y)(x' + x' + z') \qquad \text{(By commutative law)}$$
$$= (x' + y)(x' + z') \qquad (\because x_i' + x_i' = x_i')$$
$$= (x' + y + 0)(x' + z' + 0) \qquad (\because x_i + 0 = x_i)$$
$$= (x' + y + zz')(x' + z' + yy') \qquad (\because zz' = 0, yy' = 0)$$
$$= (x' + y + z)(x' + y + z')(x' + z' + y)(x' + z' + y') \qquad \text{(By distributive law)}$$
$$= (x' + y + z)(x' + y + z')(x' + y + z')(x' + y' + z') \qquad \text{(By commutative law)}$$

Hence, $f(x, y, z) = (x' + y + z)(x' + y + z')(x' + y + z')(x' + y' + z')$

This is the required conjunctive normal form.

12. *Express the conjunctive normal form of the following function :*

$$f(x, y, z) = xyz + x'yz + xy'z' + xyz'.$$

Solution. $f(x, y, z) = xyz + x'yz + xy'z' + xyz'$

$$= [(xyz + x'yz)' + xy'z' + x'yz')']' \qquad (\because (x_i')' = x_i)$$
$$= [(xyz)' \cdot (x'yz)' \cdot (xy'z')' \cdot (x'yz')']' \qquad \text{(By De Morgan's law)}$$

$= (x' + y' + z')(x + y' + z')(x' + y + z)\,[(x + y' + z)]'$

(By De Morgan's law)

$= (x + y + z)(x' + y + z)(x + y' + z')(x' + y + z')$

☛ The complement of any function which is in conjunctive normal form is that function whose factors are exactly those factors of the complete conjunctive normal form which are missing from the given function.

13. *Express each of the following in conjunctive normal form in minimal possible number of variables :*

(i) $x + x'y$ *(ii)* $xy' + xz + xy$

(iii) $(x + y')(y + z')(z + x')(x' + y')$ *(iv)* $xyz + (x + y)(x + z)$

(v) $(x + y)(x + y')(x' + z)$.

Solution. (i) $x + x'y = (x + x')(x + y)$ (By distributive law)

$= 1 \cdot (x + y)$ ($\because x + x' = 1$)

$= x + y$ ($\because 1 \cdot x = x$)

$= (x + y + 0)$ ($\because y + 0 = y$)

$= (x + y + zz')$ ($\because zz' = 0$)

$= (x + y + z)(x + y + z')$ (By distributive law)

Hence, $x + x'y = (x + y + z)(x + y + z')$ which is required conjunctive normal form.

(ii) $xy' + xz + xy = xy' + xy + xz$ (By commutative law)

$= x(y + y') + xz$ (By distributive law)

$= x \cdot 1 + xz$ ($\because y + y' = 1$)

$= x + xz$ ($\because x \cdot 1 = x$)

$= x$ (By absorption law)

$= x + 0$ ($\because x + 0 = x$)

$= x + yy'$ ($\because yy' = 0$)

$= (x + y)(x + y')$ (By distributive law)

$= (x + y + 0)(x + y' + 0)$ ($\because y + 0 = y, y' + 0 = y'$)

$= (x + y + zz')(x + y' + zz')$ ($\because zz' = 0$)

$= (x + y + z).(x + y + z')(x + y' + z)(x + y' + z')$

(By distributive law)

Hence, $xy' + xz + xy = (x + y + z)(x + y + z')(x + y' + z)(x + y' + z')$

This is required conjunctive normal form.

(iii) $(x + y')(y + z')(z + x')(x' + y')$

$= (x + y')(y + z')(x' + z)(x' + y')$ (By commutative law)

$= (x + y' + 0)(y + z' + 0)(x' + z + 0)(x' + y' + 0)$ ($\because x_i + 0 = x_i\ \forall i$)

$= (x + y' + zz')(y + z' + xx')(x' + z + yy')(x' + y' + zz')$ ($\because x_i x_i' = 0\ \forall\ x_i$)

$= (x + y' + z)(x + y' + z')(y + z' + x)(y + z' + x')$

$= (x' + y' + z)(x' + z + y)(x' + y' + z)(x' + y' + z')$ (By distributive law)

$= (x + y' + z)(x + y' + z')(x + y + z')(x' + y + z')$

$= (x' +' + z)(x' + y + z)(x' + y' + z)(x' + y' + z')$ (By commutative law)

$= (x + y' + z)(x + y' + z')(x' + y' + z')(x + y + z')$

$(x' + y + z)(x' + y' + z)(x' + y' + z')$ ($\because x_i \times x_i = x_i\ \forall i$)

This is the required conjunctive normal form.

(iv) $xyz + (x + y)(x + z)$

$= xyz + (x + yz)$ (By distributive law)

$= (x + xyz) + yz$ (By associative law and commutative law)

$= x + yz$ (By absorption law)

$= (x + y)(x + z)$ (By distributive law)

$= (x + y + 0)(x + z + 0)$ ($\because x_i + 0 = x_i\ \forall i$)

$= (x + y + zz')(x + z + yy')$ $(\because\ x_i \cdot x_i = 0\ \forall i)$

$= (x + y + z)(x + y + z')(x + z + y)(x + z + y')$ (By distributive law)

$= (x + y + z)(x + y + z)(x + y + z')(x + y' + z)$ (By commutative law)

$= (x + y + z)(x + y + z')(x + y' + z)$ $(\because\ x_i \cdot x_i = x_i\ \forall i)$

This is the required conjunctive normal form.

(v) $(x + y)(x + y')(x' + z)$

$= (x + y + 0)(x + y' + 0)(x' + z + 0)$ $(\because\ x_i + 0 = x_i\ \forall i)$

$= (x + y + zz')(x + y' + zz')(x' + z + yy')$ $(\because\ x_i x_i' = 0\ \forall\ x_i)$

$= (x + y + z)(x + y + z')(x + y' + z)$
$(x + y' + z')(x' + z + y)(x' + z + y')$ (By distributive law)

$= (x + y + z)(x + y + z')(x + y' + z)$
$(x + y' + z')(x' + y + z)(x' + z + y')$ (By commutative law)

This is the required conjunctive normal form.

14. *Write each of the following in conjunctive normal form in the three variables x, y and z :*

(i) $x + y'$ *(ii)* $x'z + xz'$ *(iii)* $(x + y)(x' + y')$ *(iv)* x

Solution. (i) $x + y' = x + y' + 0$ $(\because\ x_i + 0 = x_i\ \forall i)$

$= x + y' + zz'$ $(\because\ zz' = 0)$

$= (x + y' + z)(x + y' + z')$ (By distributive law)

(ii) $x'z + xz' = x'z + 0 + xz' + 0$ $(\because\ x_i + 0 = x_i\ \forall i)$

$= x'z + xx' + xz' + zz'$ $(\because\ x_i x_i' = 0\ \forall i)$

$= xx' + zx' + xz' + zz'$ (By commutative law)

$= (x + z)x' + (x + z)z'$ (By distributive law)

$= (x + z)(x' + z')$ (By distributive law)

$= (x + z + 0)(x' + z' + 0)$ $(\because\ x_i + 0 = x_i\ \forall i)$

$= (x + z + yy')(x' + z' + yy')$ $(\because\ yy' = 0)$

$= (x + z + y)(x + z + y')(x' + z' + y)(x' + z' + y')$ (By distributive law)

$= (x + y + z)(x + y' + z)(x' + y + z')(x' + y' + z')$ (By commutative law)

(iii) $(x + y)(x' + y') = (x + y + 0)(x' + y' + 0)$ $(\because\ x_i + 0 = x_i\ \forall i)$

$= (x + y + zz')(x' + y' + zz')$ $(\because\ zz' = 0)$

$= (x + y + z)(x + y + z')(x' + y' + z)(x' + y' + z')$ (By distributive law)

(iv) $x = x + 0$ $(\because\ x_i + 0 = x_i\ \forall i)$

$= x + yy'$ $(\because\ yy' = 0)$

$= (x + y)(x + y')$ (By distributive law)

$= (x + y + 0)(x + y' + 0)$ $(\because\ x_i + 0 = x_i\ \forall i)$

$= (x + y + zz')(x + y' + zz')$ $(\because\ zz' = 0)$

$= (x + y + z)(x + y + z')(x + y' + z)(x + y' + z')$ (By distributive law)

15. *Express the following function in conjunctive normal form*

$$f(x, y, z, t) = (x'y + xyz' + xy'z + x'y'z't + t')'$$

Soltuion. $f(x, y, z, t) = (x'y + xyz' + xy'z + x'y'z't + t')'$

$= (x'y)' \cdot (xyz')' \cdot (xy'z)' \cdot (x'y'z't)' \cdot (t')'$ (By De Morgan's law)

$= (x + y')(x' + y' + z)(x' + y + z')(x + y + z + t')t$

(By De Morgan's law and $(x_i')' = x_i\ \forall i$)

$= (x + y' + 0)(x' + y' + z + 0)(x' + y + z' + 0)(x + y + z + t')(t + 0)$

$(\because\ x_i + 0 = x_i\ \forall i)$

$= (x + y' + zz')(x' + y' + z + tt')(x' + y + z' + tt') \cdot (x + y + z + t')(zz' + t)$

$(\because\ x_i x_i' = 0\ \ \forall i$ and $t + 0 = t = t + 0)$

$= (x + y' + z)(x + y' + z')(x' + y' + z + t)(x' + y' + z + t') \cdot (x' + y + z' + t)$
$(x' + y + z' + t')(x + y + z + t')(z + t)(z' + t)$ (By distributive law)

$= (x + y' + z + 0)(x + y' + z' + 0)(x' + y + z + t)(x' + y' + z + t')$
$(x' + y + z' + t)(x' + y + z' + t')(x + y + z + t')\cdot(0 + z + t)(0 + z' + t)$
$(\because\ x_i + 0 = x_i = 0 + x_i\ \forall i)$

$= (x + y' + z + tt')(x + y' + z' + tt')(x' + y + z + t)(x' + y' + z + t')$
$(x' + y + z' + t)(x' + y + z' + t')(x + y + z + t')\cdot(yy' + z + t)(yy' + z' + t)$
$(\because\ x_i x_i' = 0\ \forall i)$

$= (x + y' + z + t)(x + y' + z + t')(x + y' + z' + t)(x + y' + z' + t')$
$(x' + y + z + t)(x' + y' + z + t')(x' + y + z' + t)\cdot(x' + y + z' + t')(x + y + z + t')$
$(y + z + t)(y' + z + t)(y' + z' + t)$ (By distributive law)

$= (x + y' + z + t)(x + y' + z + t')(x + y' + z' + t)(x + y' + z' + t')$
$(x' + y + z + t)(x' + y' + z + t')(x' + y + z' + t)\cdot(x' + y + z' + t')$
$(x + y + z + t')(0 + y + z + t)(0 + y' + z + t)(0 + y' + z' + t)$
$(\because\ 0 + x_i = x_i\ \forall i)$

$= (x + y' + z + t)(x + y' + z + t')(x + y' + z' + t)(x + y' + z' + t')$
$(x' + y + z + t)(x' + y' + z + t')(x' + y + z' + t)\cdot(x' + y + z' + t')(x + y + z + t')$
$(xx' + y + z + t)(xx' + y' + z + t)(xx' + y' + z' + t)$ $(\because\ xx' = 0)$

$= (x + y' + z + t)(x + y' + z + t')(x + y' + z' + t)(x + y' + z' + t')$
$(x' + y + z + t)(x' + y' + z + t')(x' + y + z' + t)\cdot(x' + y + z' + t')$
$(x' + y' + z + t)(x + y + z' + t)(x' + y + z' + t)(x + y' + z' + t)$
$(x' + y' + z' + t)$ $(\because\ x_i x_i = x_i\ \forall i)$.

16. *Change each of the following disjunctive normal form to conjunctive normal form in three variables :*

(i) $uv + u'v + u'v'$ *(ii)* $xyz + xy'z' + x'yz' + x'y'z + x'y'z'$

Solution. (i) $uv + u'v + u'v' = (u + u')v + u'v'$ (By distributive law)
$= 1\cdot v + u'v'$ $(\because\ u + u' = 1)$
$= v + u'v'$ $(\because\ 1\cdot v = v)$
$= (v + u')(v + v')$ (By distributive law)
$= (v + u')\cdot 1$ $(\because\ v + v' = 1)$
$= v + u'$ $(\because\ u\cdot 1 = u)$
$= (v + u' + 0)$ $(\because\ u' + 0 = u')$
$= (v + u' + ww')$ $(\because\ ww' = 0)$
$= (v + u' + w)(v + u' + w')$ (By distributive law)

This is the required conjunctive normal form.

(ii) $xyz + xy'z' + x'yz' + x'y'z + x'y'z'$
$= [(xyz + xy'z' + x'yz' + x'y'z + x'y'z')']'$ $(\because\ (x_i')' = x_i)$
$= [(xyz)'(xy'z')'(x'yz')'(x'y'z')']'$ (By De Morgan's law)
$= [(x' + y' + z')(x' + y + z)(x + y' + z)(x + y + z)]'$ (By De Morgan's law)
$= (x + y' + z')(x' + y + z')(x' + y' + z)(x + y + z)$

This is the required conjunctive normal form.

17. *Convert the following conjunctive normal form into disjunctive normal form :*
$f(x, y, z) = (x + y + z)(x + y + z')(x + y' + z)(x' + y + z')\cdot(x' + y' + z)(x' + y' + z')$

Solution. $f(x, y, z) = (x + y + z)(x + y + z')$
$(x + y' + z)(x' + y + z')\cdot(x' + y' + z)(x' + y' + z')$
$= (x + y + zz')(x + y' + z)(x' + y' + z)(x' + y + z')(x' + y' + z')$
(By commutative and distributive law)
$= (x + y + zz')(xx' + y' + z)(x' + z' + y)(x' + z' + y')$
(By commutative and distributive law)
$= (x + y + zz')(xx' + y' + z)(x' + z' + yy')$ (By distributive law)

$= (x + y + 0)(0 + y' + z)(x' + z' + 0)$ $(\because\ x_i x_i' = 0\ \forall i)$

$= (x + y)(y' + z)(x' + z')$ ($\because$ 0 is an additive identity)

$= (x + y)[(y' + z)x' + (y' + z)z']$ (By distributive law)

$= (x + y)[(y'x' + zx' + y'z' + zz')]$ (By distributive law)

$= (x + y)[(x'y' + x'z + y'z' + 0)]$ (By commutative law and $zz' = 0$)

$= (x + y)(x'y' + x'z + y'z')$ $(\because\ x_i + 0 = x_i)$

$= (x + y)x'y' + (x + y)x'z + (x + y)y'z'$ (By distributive law)

$= x(x'y') + y(x'y') + x(x'z) + y(x'z) + x(y'z') + y(y'z')$ (By distributive law)

$= x(x'y') + y(y'x') + x(x'z) + y(x'z) + x(y'z') + y(y'z')$

(By commutative law)

$= (xx')y' + (yy')x' + (xx')z + (yx')z + x(y'z') + (yy')z'$ (By associative law)

$= 0 \cdot y' + 0 \cdot x' + 0 \cdot z + x'yz + xy'z' + xy'z' + 0 \cdot z'$

(By commutative law and $x_i x_i' = 0\ \forall i$)

$= 0 + 0 + 0 + x'yz + xy'z' + 0$ $(\because\ 0 \cdot x_i = 0\ \forall i)$

$= x'yz + xy'z'$ ($\because$ 0 is an additive identity)

This is the required disjunctive normal form.

18. *Prove that the complete disjunctive normal form of Boolean function is reduced to identically 1.*

Solution. Let us consider a Boolean function in three variables x, y and z which is complete disjunctive normal form.

$f(x, y, z) = xyz + xyz' + xy'z + x'yz + x'y'z + x'yz' + xy'z' + x'y'z'$

$= xy(z + z') + xy'z + x'yz + x'y'z + x'yz' + (x + x')y'z'$ (By distributive law)

$= xy \cdot 1 + xy'z + x'yz + x'y'z + x'yz' + 1 \cdot y'z'$ $(\because\ x + x' = 1 = z + z')$

$= xy + xy'z + x'yz + x'y'z + x'yz' + y'z'$ $(\because\ x_i \cdot 1 = x_i = 1 \cdot x_i)$

$= xy + xy'z + x'y'z + x'yz + x'yz' + y'z'$ (By commutative law)

$= xy + xy'z + x'y'z + x'y(z + z') + y'z'$ (By distributive law)

$= xy + xy'z + x'y'z + x'y \cdot 1 + y'z'$ $(\because\ z + z' = 1)$

$= xy + (x + x')y'z + x'y + y'z'$ (By distributive law and $y \cdot 1 = y$)

$= xy + 1 \cdot y'z + x'y + y'z'$ $(\because\ x + x' = 1)$

$= xy + y'z + x'y + y'z'$ $(\because\ 1 \cdot y = y)$

$= xy + x'y + y'z + y'z'$ (By commutative law)

$= (x + x')y + y'(z + z')$ (By distributive law)

$= 1 \cdot y + y' \cdot 1$ $(\because\ x + x' = 1, z + z' = 1)$

$= y + y'$ $(\because\ 1 \cdot y = 1, y' \cdot 1 = y')$

$= 1$ $(\because\ y + y' = 1)$.

19. *Construct the function f from the following table :*

x	y	z	f	T
1	1	1	0	$x'y'z'$
1	1	0	1	$x'y'z$
1	0	1	1	$x'yz'$
1	0	0	1	$x'yz$
0	1	1	0	$xy'z'$
0	1	0	0	$xy'z$
0	0	1	1	xyz'
0	0	0	0	xyz

where T stands for the term of the function.

Solution. Using Bool's expansion theorem, we have

$$
\begin{aligned}
f &= 0(x'y'z') + 1(x'y'z) + 1(x'yz') + 1(x'yz) + 0(xy'z) + 1(xyz') + 0(xyz)\\
&= x'y'z + x'yz' + x'yz + xyz'\\
&= x'y'z + x'y(z' + z) + xyz'\\
&= x'y'z + x'y\cdot 1 + xyz' \qquad (\because\ z + z' = 1)\\
&= x'y'z + x'y + xyz'\\
&= x'y'z + (x' + xz')y\\
&= x'y'z + y(x' + xz')\\
&= x'y'z + y(x' + x)(x' + z')\\
&= x'y'z + y\cdot 1(x' + z')\\
&= x'y'z + y(x' + z')\\
&= x'y'z + yx' + yz'\\
&= x'y'z + x'y + yz'\\
&= x'(y'z + y) + yz'\\
&= x'(y' + y)(z + y) + yz'\\
&= x'(y' + y)(z + y) + yz'\\
&= x'\cdot 1(z + y) + yz'\\
&= x'(z + y) + yz'\\
&= x'y + x'z + yz'.
\end{aligned}
$$

3.9 REPRESENTATION OF A FINITE BOOLEAN ALGEBRA

An arbitrary Boolean algebra is so much common with an algebra of set, except in presentation of postulates. In addition to it, we can prove that every algebra of sets is a Boolean algebra. Therefore, an arbitrary Boolean algebra can be interpreted as an algebra of sets together with some specially universal set. The theory of representation of Boolean algebra was given by M. H. Stone in the 1930s. Although the laws hold for all Boolean algebra in general. Thus, we shall take the Boolean algebra to be finite only. In 1930s M.H. Stone proved a powerful theorem related to represent the study of Boolean algebra to set theory. He proved that every abstract Boolean algebra is isomorphic to an algebra of sets. That is

Let B be a finite Boolean algebra and there exists a set S, which is finite such that B and the power set $P(S)$ (set of sets) are isomorphic is Boolean algebra.

Let B_1 and B_2 be two Boolean algebras.

- A mapping $f : B_1 \to B_2$ is called a homomorphism if $f(a \vee b) = f(a) \vee f(b)$ or $f(a + b) = f(a) + f(b)$ and $f(a \wedge b) = f(a) \wedge f(b)$ or $f(a \cdot b) = f(a)\, f(b)\ \forall\ a, b \in B_1$
- A one-one and onto mapping which is homomorphism is called **isomorphism.**
- In the power set $P(S)$ (set of sets), which is bounded by ϕ and S, then the singleton subset of S are called **atoms** and **antiatoms.**
- $B = (B, '+', \ldots, 0, 1, ')$ and $P(S) = \{P(S), '\cup', '\cap', \phi, S, '\}$

Theorem 1. ***(Stone Representation Theorem)*** **:** *M.H. Stone proved an important theorem related to represent the Boolean algebra to the algebra of sets.*

Statement. Every finite Boolean algebra is isomorphic to an algebra of sets.

Proof. Let B be a finite Boolean algebra and let S be the set of all atoms of B; then we have to prove that B is isomorphic to $P(S)$ (algebra of sets). Now,

Define a map $f : B \to P(S)$ such that $f(0) = \phi$

$$f(x) = \{a_1, a_2, \ldots, a_m\},\ x = a_1 + a_2 + a_3 + \ldots + a_m\ \forall\ a_i \le x (x \ne 0)$$

Firstly we show that f is **one-to-one**.

Let $x, y \in B$ and suppose $f(x) = f(y)$

$\Rightarrow\quad \{a_1, a_2, \ldots, a_m\} = \{b_1, b_2, \ldots, b_n\},\ a_i \le x,\ b_j \le y$

$\Rightarrow \quad m = n \quad$ and $\; a_i = a_j$

$\Rightarrow \quad x = a_1 + a_2 + \ldots + a_m = b_1 + b_2 + \ldots + b_n = y \quad \Rightarrow \quad x = y$

$\Rightarrow \quad f$ is one-one.

Next, we show that f is onto.

Let $S_1 = \{a_1, a_2, \ldots, a_m\}$ be any arbitrary member of $P(S)$ and $x = a_1 + a_2 + \ldots + a_m$, then

$$f(x) = \{a_1, a_2, \ldots, a_m\}$$

$\Rightarrow \quad f$ is onto.

Finally, we shall show that $f(x + y) = f(x) \cup f(y)$ and $f(x \cdot y) = f(x) \cap f(y)$

Let $f(x, y) = \{\alpha_1, \alpha_2, \ldots, \alpha_r\}$, where $\alpha_i \le x \cdot y$

Since $\alpha_i \le x \cdot y$, then $\alpha_i \le x$, $\alpha_i \le y$ and $x \cdot y = \alpha_1 + \alpha_2 + \ldots + \alpha_r$

If $f(x) \cap f(y) = \{a_1, a_2, \ldots, a_m\} \cap \{b_1, b_2, \ldots, b_n\}$

then, $a_i \le x$, $b_j \le y$, $x = a_1 + a_2 + \ldots + a_m$, $y = b_1 + b_2 + \ldots + b_n$

Suppose, $\{a_1, a_2, \ldots, a_m\} \cap \{b_1, b_2, \ldots, b_n\} = \{\beta_1, \beta_2, \ldots, \beta_k\}$

Since $a_i \le x$, $b_j \le y$, then $\beta_i \le x \cdot y$. Thus, each β_i is equal to some c_j and further since $c_i \le x$ for some a_i and $c_i \le y$ for some b_j.

Thus, each c_i is equal to some d_i, hence $f(x \cdot y) = f(x) \cap f(y)$

Similarly, we can prove that $f(x + y) = f(x) \cup f(y)$

and hence f is an isomorphism.

Theorem 2. *If B is a Boolean algebra and $a \in B$, then B is isomorphic to* $[0, a] \times [a, 1]$.

Proof. Let $f : B \to [0, a] \times [a, 1]$ such that $f(x) = (x \cdot a, x + a)$

where, $\quad x \cdot a \in [0, a]$, $x + a \in [a, 1]$, let $x, y \in B$ and let $x = y$

$\Rightarrow \quad x \cdot a = y \cdot a$ and $x + a = y + a \quad \Rightarrow \quad (x \cdot a, x + a) = (y \cdot a, y + a)$

$\Rightarrow \quad f(x) = f(y)$

Thus, f is well defined.

Now, we have to show that f is one to one and onto and preserves the composition '+' and '·', respectively.

Let us suppose $f(x) = f(y)$

$\Rightarrow \quad (x \cdot a, x + a) = (y \cdot a, y + a) \quad \Rightarrow \quad x \cdot a = y \cdot a$ and $x + a = y + a$

$\Rightarrow \quad x = y \quad \Rightarrow \quad f$ is one to one.

Next, let $(\alpha, \beta) \in [0, a] \times [a, 1]$ be any element then $0 \le \alpha \le a$, $a \le \beta \le 1$, $\alpha, \beta \in B$.

Let us take, $x = \alpha + (\beta \cdot a')$, then

$$\begin{aligned}
f(x) &= f(\alpha + (\beta \,.\, a')) \\
&= ((\alpha + \beta \cdot a') \cdot a, \alpha + (\beta \cdot a') + a) \\
&= (\{\alpha \cdot a + (\beta \cdot a') \cdot a\}, \{\alpha + (\beta + a) \cdot (a' + a)\}) && \text{(By distributive law)} \\
&= (\{\alpha \cdot a + \beta \cdot (a \cdot a')\}, \{\alpha + (\beta + a) \cdot (a' + a)\}) && \text{(By commutative and associative law)} \\
&= (\{\alpha \cdot 0 + \beta \cdot 0\}, \{\alpha + (\beta + a) \cdot 1\}) && (\because\ aa' = 0,\ a + a' = 1) \\
&= (\{\alpha \cdot a + 0\}, \{(\alpha + \beta) + a\}) && (\because\ \beta \cdot 0 = 0,\ a \cdot 1 = 1, \text{ By associative law}) \\
&= (\alpha \cdot a, (\alpha + \beta) + a) \\
&= (\alpha, \beta) && (\because 0 \le \alpha \le a,\ a \le \beta \le 1)
\end{aligned}$$

Thus, f is onto.

Finally, we shall show that $f(x + y) = f(x) + f(y)$ and $f(x \cdot y) = f(x) \cdot f(y)$

Let $x, y \in B$, then

$$\begin{aligned}
f(x, y) &= ((x \cdot y) \cdot a,\ (x \cdot y) + a) && \text{(By definition)} \\
&= ((x \cdot y) \cdot a \cdot a, (x, y) + a) && (\because\ a \cdot a = a) \\
&= ((x \cdot a) \cdot (y \cdot a), (x + a) \cdot (y + a))
\end{aligned}$$

(By commutative and associative laws and distributive law)

Now, $f(x) \cdot f(y) = (x \cdot a, x + a) \cdot (y \cdot a, y + a)) = (x \cdot a)\,(y \cdot a), (x + a)\,(y + a)$

Thus, $f(x \cdot y) = f(x) \cdot f(y)$

and $f(x + y) = ((x + y) \cdot a, (x + y) + a)$ (By definition)

$= (x \cdot a + y \cdot a, (x + y) + a + a)$ (By distributive law and $a + a = a$)

$= (x . a + y . a, (x + a) + (y + a))$

(By commutative law and associative law)

$= (x \cdot a, x + a) + (y \cdot a, y + a) = f(x) + f(y)$

Here, f is an isomorphism.

☛ If B is a Boolean algebra and $a \in B$, then B is isomorphic to $[0, a] \times [0, a]$.

3.10 APPLICATION TO SWITCHING CIRCUITS

In this section, we shall discuss an important application of Boolean algebra, the algebra of circuits involving two state devices. One of such device is a switch or contact. The algebra of circuits has an important role at present in the field of Mathematics as well as in the field of engineering sections. The algebra of circuits associated with the two elements 0 and 1, which are bounds of Boolean algebra. The importance of this subject is to make the circuits and to simplify a complex circuit in a simple circuit, which contains an electronic computer, dial telephone, switching networks and other electronic control devices.

To explain this subject, let us consider the simplest kind of circuits containing only switches. These switches are denoted by a single small letter $a, b, c, x, y, \ldots$ etc. In the 'on' position of the switch, the switch is said to be closed and is denoted by the letter x. While in 'off' position, the switch is said to be open and is denoted by the complement of x, *i. e.*, by x'. Above two positions of switch are shown below in Fig. 1 and Fig. 2 respectively.

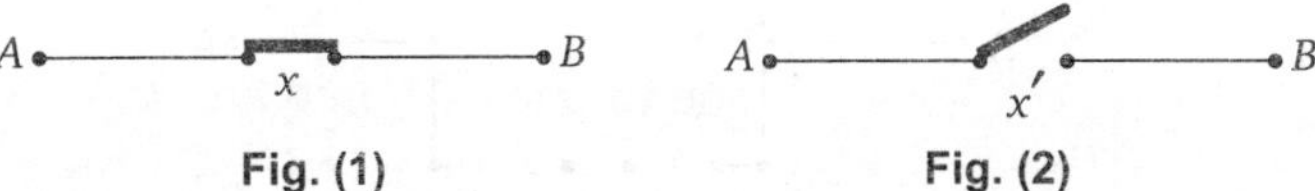

Fig. (1) **Fig. (2)**

For any switch x, there are two possibilities, either x is closed or x is open, then we can say that if x is closed, it has the value 1 and if x is open, it has the value 0. Any switch is represented by the diagram also.

A • ——— • x • ——— • B

Fig. (3)

3.10.1 Combination of Switches in a Circuit

Let us suppose a circuit is consisting of two switches x and y. These switches can be arranged in two different ways. One is in series and the other is in parallel ways.

(i) Series Combination : Suppose two switches x and y are so arranged that the switch x is connected to the switch y directly. This combination is shown below as follows :

A • ——— • x • ——— • y • ——— • B

Fig. (4)

The combined form of above arrangement of switches x and y is denoted by $x \cdot y$. Truth table for the series combination of x and y is given as follows :

x	y	$x \cdot y$
1	1	1
1	0	0
0	1	0
0	0	0

(ii) Parallel Combination : Two switches x and y are so arranged that x is connected to the switch y indirectly shown below as follows :

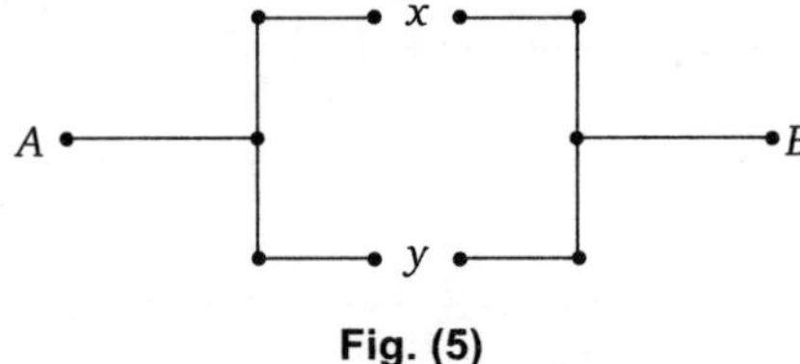

Fig. (5)

The parallel combination of x and y is denoted by $x + y$ and the truth table for this combination is given below :

x	y	$x+y$
1	1	1
1	0	1
0	1	1
0	0	0

(iii) Series Parallel Combination : Suppose three circuits x, y and z are so arranged that x is parallel to the series combination of y and z and shown below as follows :

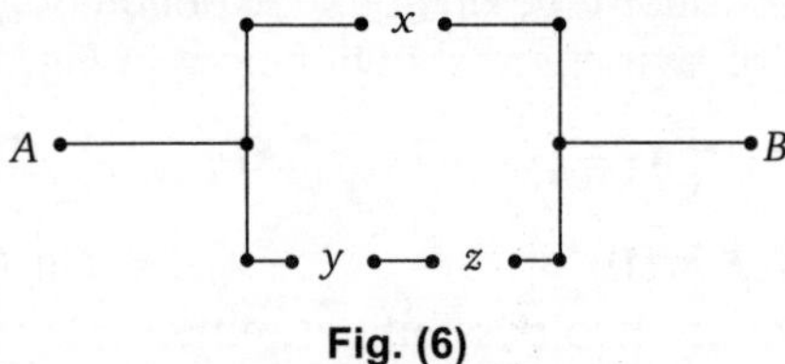

Fig. (6)

This network of x, y and z is represented by the function

$$f(x, y, z) = x + y \cdot z$$

Truth table for this combination is given as follows :

x	y	z	$y \cdot z$	$x + y \cdot z$
1	1	1	1	1
1	1	0	0	1
1	0	1	0	1
0	1	1	1	1
1	0	0	0	1
0	1	0	0	0
0	0	1	0	0
0	0	0	0	0

There are two basic problems related to the application of Boolean algebra to switching circuits, one of them is to simplify any circuit in an algebraic expression and the other is to design any algebraic expression into the circuits.

(iv) Non-Series-Parallel Circuit Combination : It has been observed that every series parallel circuit is expressed to Boolean function and conversely that every

Boolean function is expressed to a series-parallel circuit. Let x, y and z be three switches and be arranged so that forming a non-series-parallel combination and is shown below as follows :

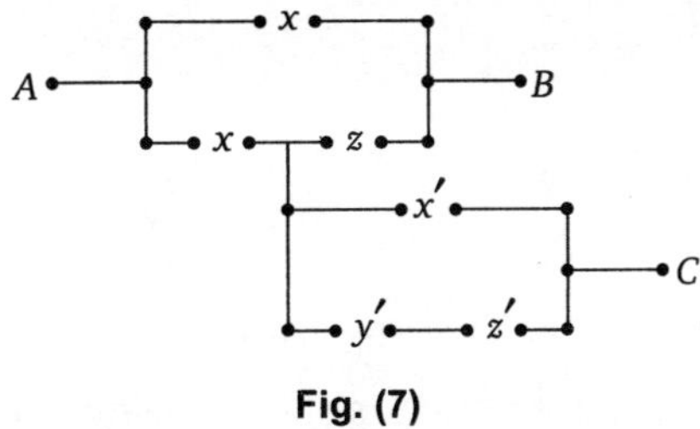

Fig. (7)

The non-series parallel circuit is also called 3-terminal circuit. This 3-terminal circuit is controlling a light by any of the three circuits joining the terminals A, B and C. Therefore, we can find the different functions as follows : From 2-terminal circuit A and B, f_{AB} is the required function. From joining the terminals B and C, we can find f_{BC} and f_{AC} is the function obtained from joining A and C.

In general, we can think of n-terminal circuit, which is formed with switches connected by wires in n terminals and thus we can obtain $\frac{1}{2}n(n-1)$ possible functions of type f_{ij} joining the terminals T_i and T_j for each i and j such that $i \neq j$. The two 2-terminal circuits are equivalent if the function corresponding to the two-terminal circuits are equal. That is, f_{ij} and f_{lm} are the functions joining the terminals T_i and T_j, T_l and T_m respectively. If $f_{ij} = f_{lm}$, then these two 2-terminal circuits are equivalent.

In the theory of an ordinary circuit, there are two transformations; one is wye-to-delta and other is delta-to-wye.

(i) Wye Circuit : As a 3-terminal circuit, we can define a wye circuit as follows, a circuit which is obtained by taking three 2-terminal circuits having one point (terminal) as common is called wye circuit as shown in Fig. (8).

(ii) Delta Circuit : A 3-terminal circuit having any pair of three 2-terminal circuits as common terminals which is shown in Fig. (9).

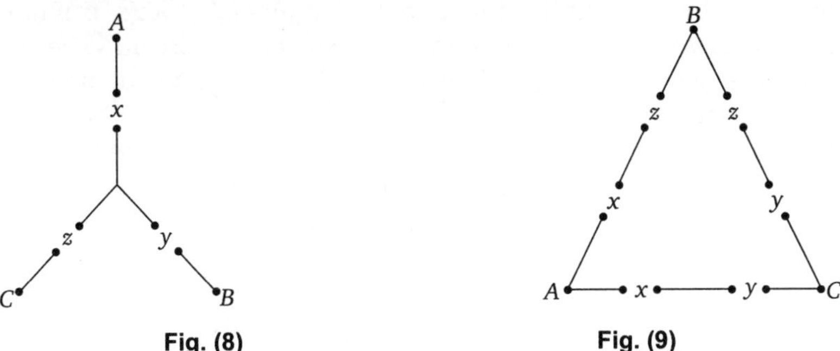

Fig. (8) **Fig. (9)**

The wye-to-delta transformation gives an equivalent circuit in which 2-terminal circuits, formed in each case, which are connected in series for the same pair of switches, as shown in Fig. 10.

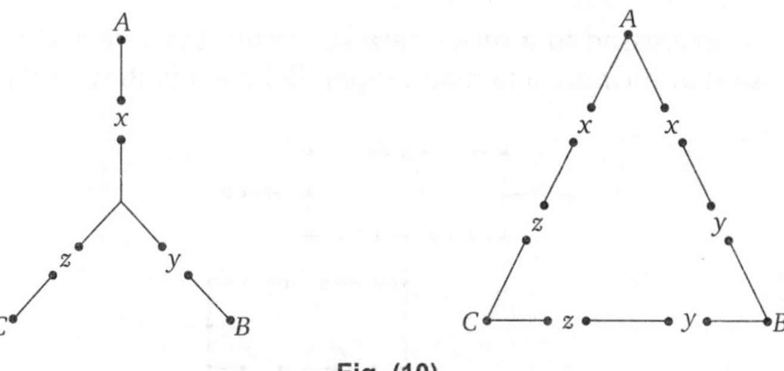

Fig. (10)

(iv) The delta-to-wye transformation is shown below :

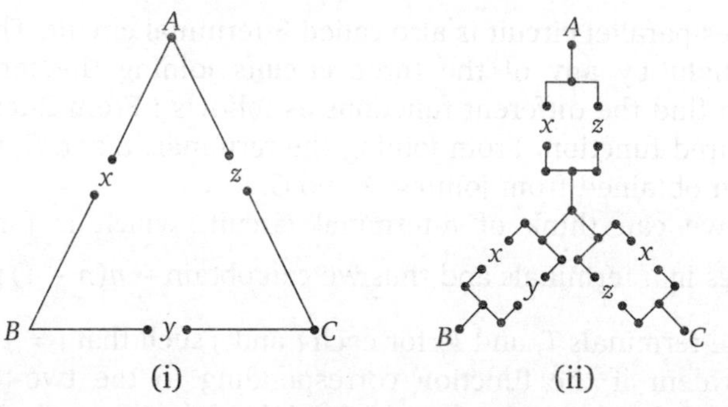

Fig. (11)

This transformation gives an equivalent circuit which is obtained from the distributive law of '+' over '·'. **For example :**

$f_{AB} = x + y \cdot z$ (From first diagram of Fig. 11)

and $f_{AB} = (x + y) \cdot (x + z)$ (From second diagram of Fig. 11)

Thus, $f_{AB} = x + y \cdot z = (x + y) \cdot (x + z)$ (By distributive law)

(v) The Star-to-Mesh Transformation : The generalised form of wye-to-delta transformation is known as the star-to-mesh transformation. In this transformation, star represents an n-terminal circuit having a point common. If this common point is removed, an equivalent circuit is obtained called mesh circuit. The star-to-mesh transformation gives a method for finding the Boolean function for any given circuit, series-parallel and complex circuits.

There are two alternative methods for finding the Boolean function for a circuit, which are easy in simple circuit. The first method to examine the circuit for all possible combinations of closed switches giving a current through the circuit is understood by the broken lines in the circuit given in Fig. 12.

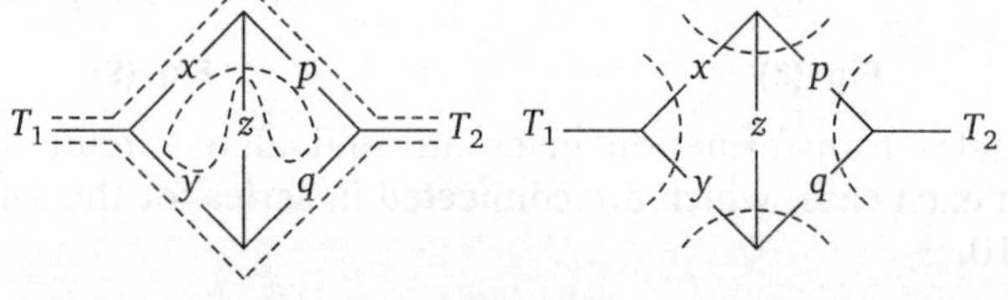

Fig. (12)

These paths for the bridge circuit gives the combination $x \cdot p$, $y \cdot q$, $x \cdot z \cdot q$ and $y \cdot z \cdot p$. Thus, the function $f_{T_1T_2}$ is given by

$$f_{T_1T_2} = xp + yq + xzq + yzp$$

The second alternative method is that, the broken lines through the circuit are drawn in all possible ways such that the given circuit is broken. These combinations are : $x, y; p, q; x, z, p; y, z, q$ as shown in adjoining figure.

Thus, the function $f_{T_1T_2}$ is given by

$$f_{T_1T_2} = (x + y)\cdot(p + q)\cdot(x + z + p)\,(y + z + q)$$

The functions in two alternative methods are equal and have the value 0 if all the switches are open.

ILLUSTRATIVE EXAMPLES

1. *Draw the circuit for the Boolean function governed by the laws of absorption and distributive law for (·) over (+).*

Solution. Absorption laws are

(i) $x\cdot(x+y) = x$ (ii) $x + (x\cdot y) = x$ and distributive law for (·) over (+)

(iii) $x\cdot(y+z) = x\cdot y + x\cdot z$.

(i) Let $f(x, y) = x\cdot(x+y)$

Now, construct a circuit for $f(x, y)$

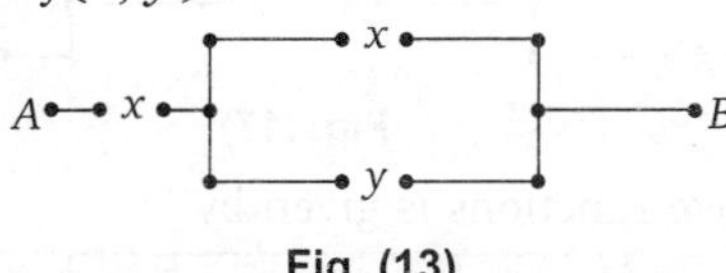

Fig. (13)

and circuit for x

Fig. (14)

Truth table for above circuits

x	y	$x + y$	$x\cdot(x+y)$
1	1	1	1
1	0	1	1
0	1	1	0
0	0	0	0

First and fourth columns have same truth values, thus x and $x(x+y)$ are equal and hence above circuits corresponding these functions are equivalent.

(ii) Let $f(x, y) = x + (x\cdot y)$ construct a circuit for $f(x, y)$

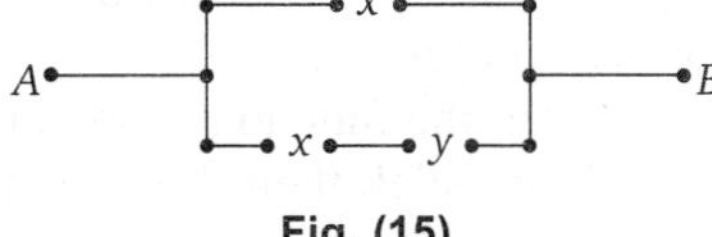

Fig. (15)

and circuit for x

Fig. (16)

Truth tables for above circuits

x	y	$x \cdot y$	$x + x \cdot y$
1	1	1	1
1	0	0	1
0	1	0	0
0	0	0	0

Here, first and last columns have same truth values, so the Boolean functions x and $x + x \cdot y$ are same and hence both circuits corresponding functions are equivalent.

(iii) Distributive law for $(\cdot)$ over $(+)$

$$x \cdot (y + z) = x \cdot y + x \cdot z$$

Now, draw two circuits for $f(x, y, z) = x \cdot (y + z)$

and $f(x, y, z) = x \cdot y + x \cdot z$

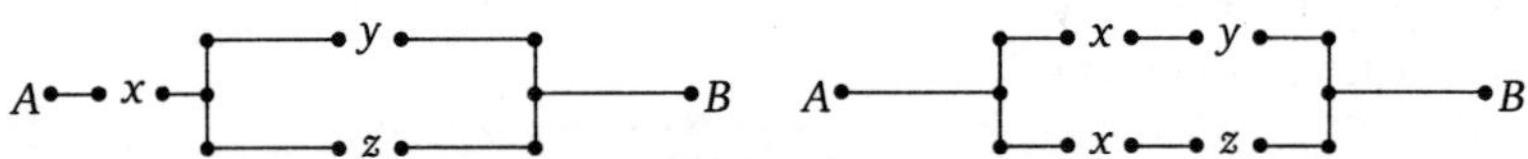

Fig. (17)

The truth table for above functions is given by

x	y	z	$y + z$	$x \cdot (y + z)$	$x \cdot y$	$x \cdot z$	$x \cdot y + x \cdot z$
1	1	1	1	1	1	1	1
1	1	0	1	1	1	0	1
1	0	1	1	1	0	1	1
0	1	1	1	0	0	0	0
1	0	0	0	0	0	0	0
0	1	0	1	0	0	0	0
0	0	1	1	0	0	0	0
0	0	0	0	0	0	0	0

Here fifth and last columns have same truth values, so the functions, $x \cdot (y + z)$ and $x \cdot y + x \cdot z$ are same and hence the circuits corresponding to the above functions are equivalent.

2. *Draw the circuits for each of the following functions, without simplifying the functions*

(i) $abc + ab(dc + ef)$ *(ii)* $a + b\,(c + de) + fg$

(iii) $x[y(z + w) + z(u + v)]$ *(iv)* $(a + b' + c)\,(a + bc') + c'd + d(b' + c)$

Solution. (i) Circuit for $abc + ab(dc + ef)$

If the switches are in series, then the function is obtained by multiplying all the switches variables and if the switches are in parallel, then the switches variables are added. Thus,

(i) 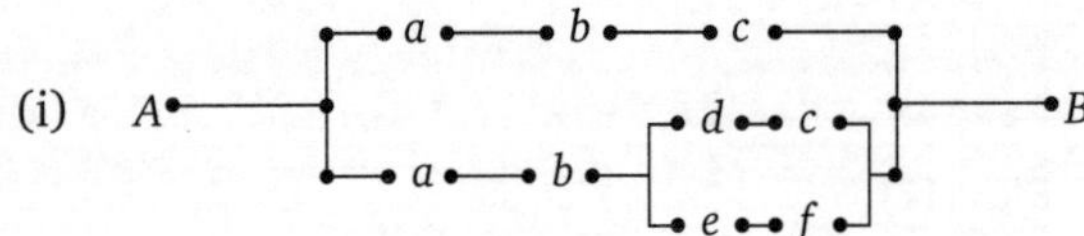

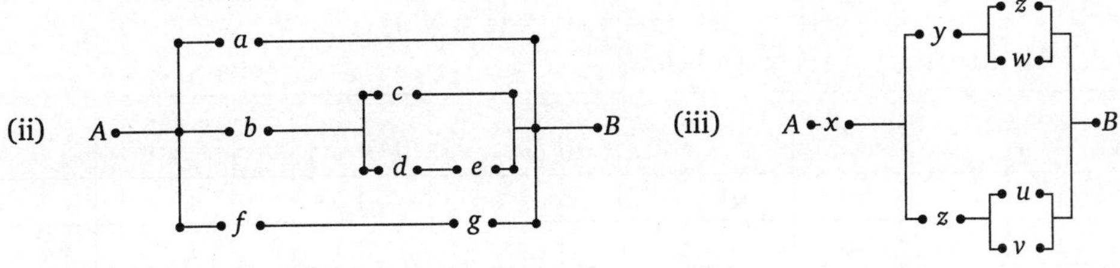

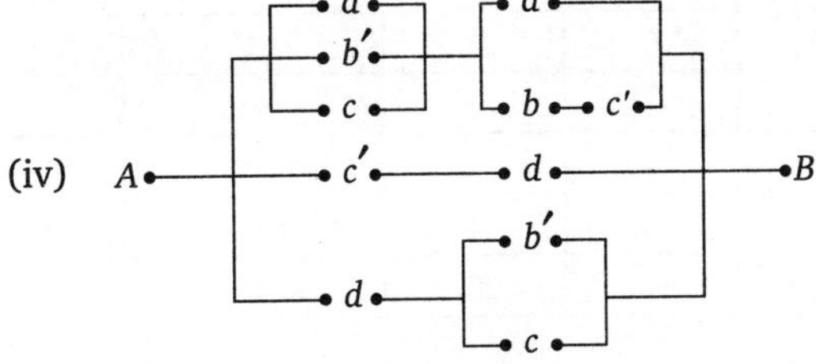

Fig. (18)

3. *Construct a table for closure properties for the following circuits :*

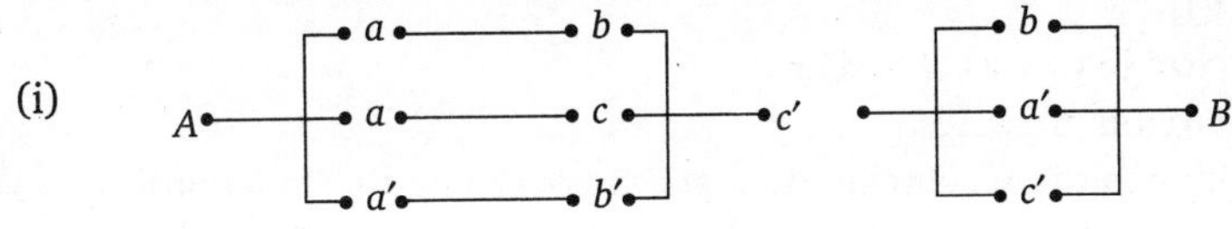

Fig. (19)

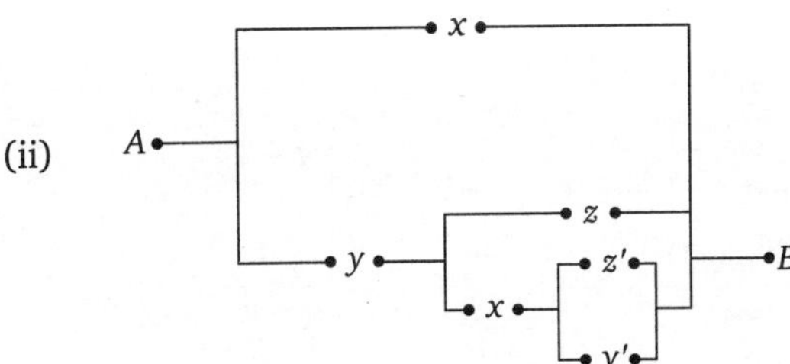

Fig. (20)

Solution. (i) The function corresponding to the first circuit is

$$f(a, b, c) = (ab + ac + a'\,b')\cdot c'\cdot(b + a' + c')$$

Truth Table for $f(a, b, c)$:

a	b	c	a'	b'	c'	ab	ac	$a'b'$	$ab + ac + a'b'$	$b + a' + c'$	f
1	1	1	0	0	0	1	1	0	1	1	1
1	1	0	0	0	1	1	0	0	1	1	1
1	0	1	0	1	0	0	1	0	1	0	0
0	1	1	1	0	0	0	0	0	0	1	0
1	0	0	0	1	1	0	0	0	0	1	0
0	1	0	1	0	1	0	0	0	0	1	0
0	0	1	1	1	0	0	0	1	1	1	1
0	0	0	1	1	1	0	0	1	1	1	1

(ii) The function corresponding to the second circuit is

$$f(x, y, z) = x + y\,[z + x(z' + y')]$$

Truth table for $f(x, y, z)$ is given below :

x	y	z	y'	z'	$z' + y'$	$x(z' + y')$	$z + x(z' + y')$	$y[z + x(z' + y')]$	f
1	1	1	0	0	0	0	1	1	1
1	1	0	0	1	1	1	1	1	1
1	0	1	1	0	1	1	1	0	1
0	1	1	0	0	0	0	1	1	1
1	0	0	1	1	1	1	1	0	1
0	1	0	0	1	1	0	0	0	0
0	0	1	1	0	1	0	1	0	0
0	0	0	1	1	1	0	0	0	0

4. *Draw the circuits for the following polynomials :*

(*i*) $(xy + z)(t + x'y)$ (*ii*) $xyz + xy(sz + tu)$

(*iii*) $x + y(z + st) + uv$ (*iv*) $(x + y)(x' + zy')$

(*v*) $x[y(z + w) + z(u + v)]$

Solution. (i) Circuit for $(xy + z)(t + x'y)$

Step I. There are two circuits in series.

Step II. First circuit contains three switches x, y and z governing the function $(xy + z)$, in which x and y are in series and z is in parallel to the combination of x and y.

Step III. Second circuit contains three switches t, x' and y, in which t is in parallel to the series combination of x' and y.

Hence, the required circuit is given by

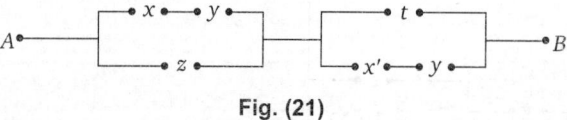

Fig. (21)

(ii) Circuit for $xyz + xy(sz + tu)$

Step I. Here is one circuit in which x, y and z are in series and give the function xyz.

Step II. Now, xyz is in parallel to $xy(sz + tu)$.

Step III. $xy(sz + tu)$ contains two circuits out of these first consists of two switches x and y arranged in series and second circuits, in which the switches s and z arranged in series give function sz, this function is in parallel to the function tu, which are in series too. Hence, the required circuit is given as shown in fig.

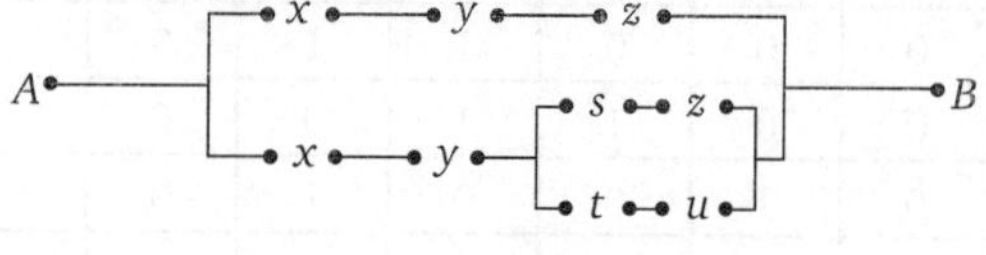

Fig. (22)

(iii) Circuit for $x + y(z + st) + uv$

Step I. Here one circuit exists.

Step II. The circuit consists of three branches, which are in parallel.

Step III. First branch contains one switch x, second branch contains four switches y, z, s and t, in which y is in series with $(z + st)$. In $(z + st)$ z is in parallel with the series combination of s and t. Third branch contains two switches u and v, which are in series.

Hence, the required circuit is shown in fig.

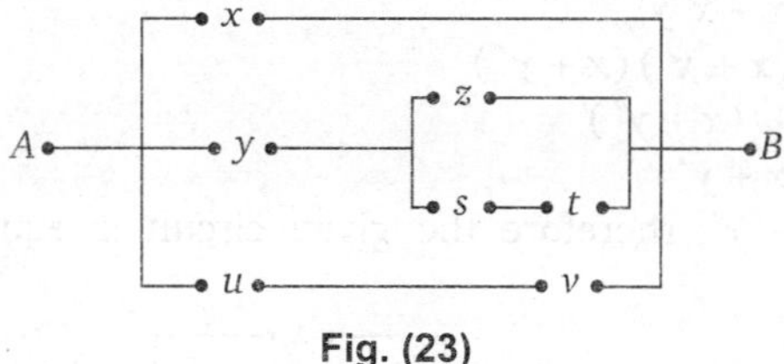

Fig. (23)

(iv) Circuit for $(x + y)(x' + zy')$

Step I. There are two circuits, which are in series.

Step II. First circuit contains two switches x and y arranged in parallel.

Step III. Second circuit contains three swtiches x', z and y' in which x' is arranged in parallel with the series combination of z and y'.

Hence, the required circuit is shown in fig.

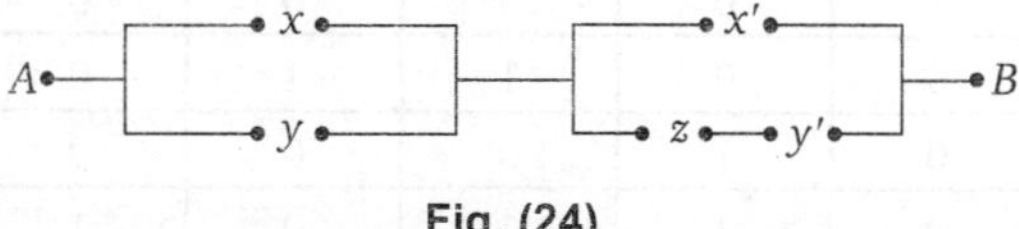

Fig. (24)

(v) Circuit for $x[y(z + w) + z(u + v)]$

Step I. The circuit contains the switch x in series with combination

$$[y(z + w) + z(u + v)]$$

Step II. $[y(z + w) + z(w + v)]$ contains $y(z + w)$ and $z(u + v)$, which are parallel to each other in the circuit.

Step III. $y(z + w)$ contains the switch y in series with parallel combination of z and w.

Step IV. $z(w + v)$ contains the switch z in series with parallel combination of v and u.

Hence, the required circuit is given as shown in following figure :

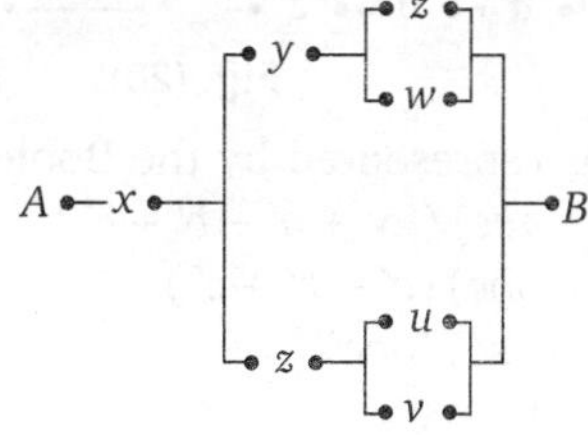

Fig. (25)

5. *Draw a circuit, which is the simplification of the given circuit and verify that both circuits are equivalent.*

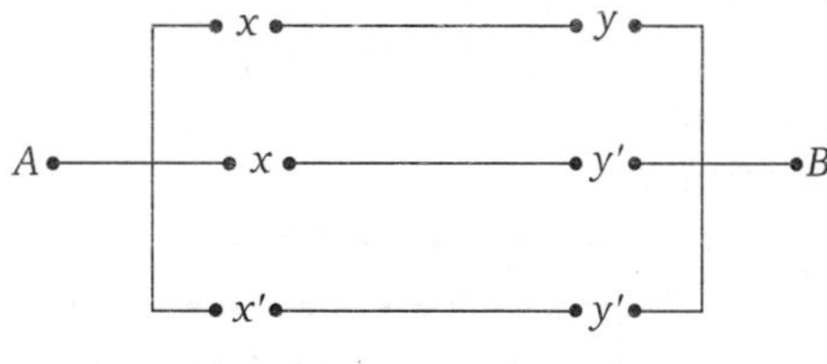

Fig. (26)

Solution. The Boolean function for the given circuit is $f(x, y) = xy + xy' + x'y'$

Now, simplifying this function

$$f(x, y) = xy + xy' + x'y'$$

$$= x(y + y') + x'y' \quad \text{(By distributive law)}$$

$$= x \cdot 1 + x' y' \qquad (\because\ y + y' = 1)$$
$$= x + x' y' \qquad (\because\ x \cdot 1 = x)$$
$$= (x + x')(x + y') \qquad \text{(By distributive law)}$$
$$= 1 \cdot (x + y') \qquad (\because\ x + x' = 1)$$
$$= x + y' \qquad (\because\ 1 \cdot x = x)$$

Thus, $f(x, y) = x + y'$, therefore the given circuit is equivalent to the circuit $x + y'$, which is given by

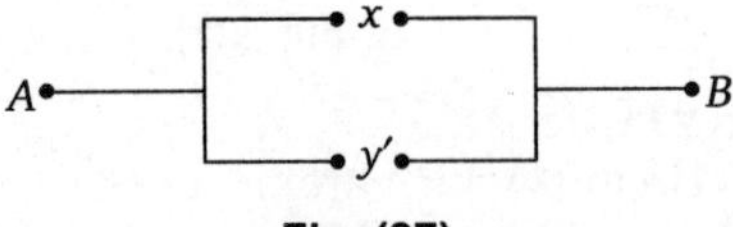

Fig. (27)

Truth table for the verification of $xy + xy' + x'y'$ and $x + y'$:

x	y	xy	x'	y'	xy'	$x'y'$	$xy + xy' + x'y'$	$x + y'$
1	1	1	0	0	0	0	1	1
1	0	0	0	1	1	0	1	1
0	1	0	1	0	0	0	0	0
0	0	0	1	1	0	1	1	1

The functions $xy + xy' + x'y'$ and $x + y'$ have same truth values. Thus, the circuits corresponding the above functions are equivalent.

6. *Simplify the following circuit :*

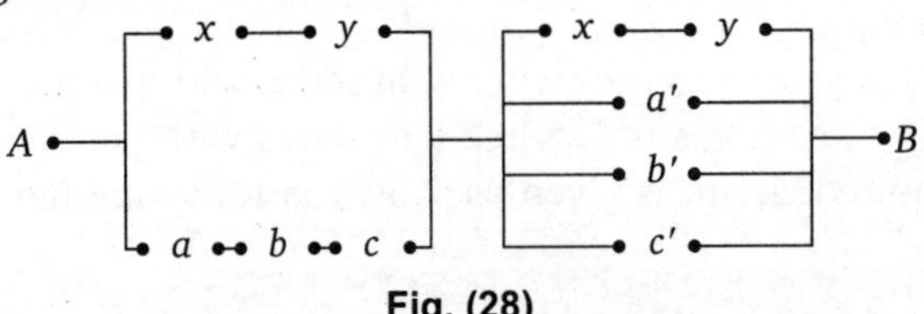

Fig. (28)

Solution. Above circuit is represented by the Boolean function

$$f(x, y, a, b, c) = (xy + abc)(xy + a' + b' + c')$$
$$= xy + (abc)(a' + b' + c') \qquad \text{(By distributive law)}$$
$$= xy + (abc)(abc)' \qquad \text{(By De Morgan's law)}$$
$$= xy + 0 \qquad (\because\ xx' = 0)$$
$$= xy \qquad (\because\ x + 0 = x)$$

Thus, simplified form of $f(x, y, a, b, c) = xy$

$\therefore$ The circuit corresponding to the simplified function xy is given by

A —— x —— y —— B

Fig. (29)

This is the required circuit.

7. *Simplify the following circuit :*

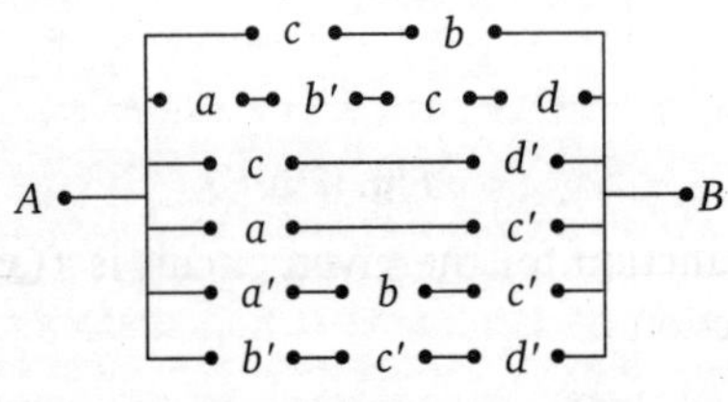

Fig. (30)

Solution. The Boolean function representing the above circuit is given by

$$f(a, b, c, d) = cb + ab' cd + cd' + ac' + a' bc' + b' c' d'$$
$$= c(b + ab' d + d') + c' (a + a' b + b' d')$$
(By commutative and distributive law)
$$= c(b + (ab' + d')(d + d')) + c'\{(a + a')(a + b) + b' d'\}$$
(By distributive law)
$$= c\{b + (ab' + d') \cdot 1\} + c'\{1 \cdot (a + b) + b' d'\} \qquad (\because\ x + x' = 1)$$
$$= c\{b + ab' + d'\} + c'\{a + b + b' d'\} \quad (\because\ x \cdot 1 = x = 1 \cdot x)$$
$$= c\{(b + a)(b + b') + d'\} + c'\{a + (b + b')(b + d')\}$$
(By distributive law)
$$= c(b + a + d') + c'(a + b + d') \quad (\because\ x + x' = 1 \text{ and } x \cdot 1 = x = 1 \cdot x)$$
$$= (c + c')(a + b + d') \quad \text{(By distributive and commutative law)}$$
$$= (a + b + d')\ (c + c' = 1, 1 \cdot x = x)$$

Thus, simplified form of the function is $f(a, b, c, d) = a + b + d'$

Therefore, the simplified circuit is given by

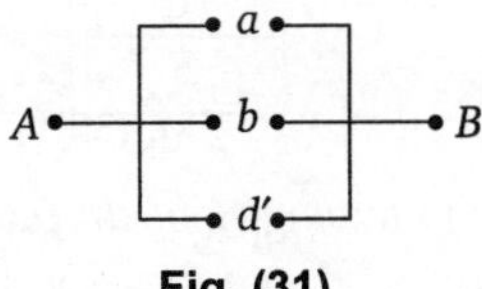

Fig. (31)

8. *Simplify the following bridge circuit :*

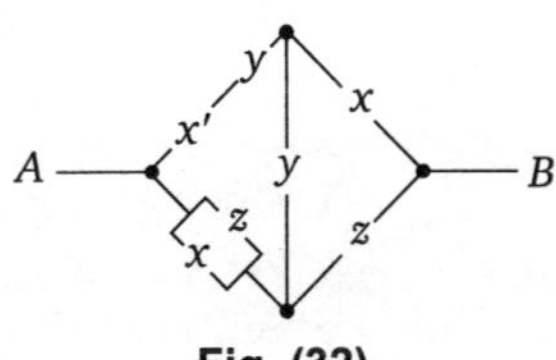

Fig. (32)

Solution. Finding the function f_{AB} using broken lines drawn on given bridge circuit as follows :

Broken lines are drawn in such a way that the circuit may be broken using the combinations x', y, z, x; x, z; $x' \cdot y$, y, z and x, y, z, x. Thus, the function is given by

$$f_{AB} = (x' y + z + x)(x + z)(x' y + y + z)(x + y + z + x)$$
$$= \{(x' y + z + x)(z + x)\}\{(x'y + z + y)(x + z + y)\}$$
(By commutative and associative law)

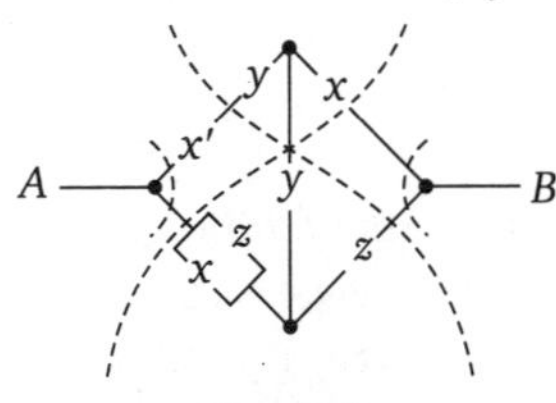

Fig. (33)

$$= (z + x)(z + y) \quad \text{(By absorption law)}$$
$$= z + xy \quad \text{(By distributive law)}$$
$$= xy + z \quad \text{(By commutative law)}$$

Thus, the simplified form of the function f_{AB} is given by $f_{AB} = xy + z$

Hence, the equivalent circuit is

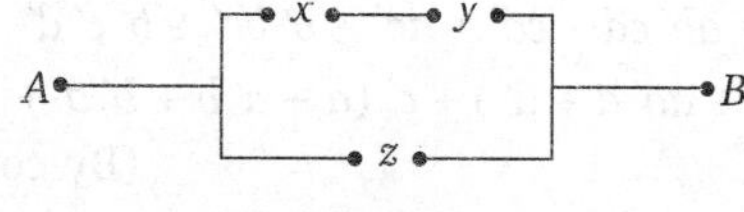

Fig. (34)

9. *Draw a bridge circuit for the following Boolean function :*

$$f = (x'u + x'v's + yu + yv's)(x' + z + w' + v's)(y + z + w' + u)$$

Solution. Since $f = (x'u + x'v's + yu + yu's)(x' + z + w' + v's)(y + z + w' + u)$

$$= (x' + y)(u + v's)(x' + z + w + v's)(y + z + w' + u)$$

Here, from the function, it is observed that broken lines are drawn in such a way that the circuit may be broken. Those broken lines are formed by the combination x', y, u, v', s; x', z, w, v', s; y, z, w', u. Hence, the required bridge circuit is given by

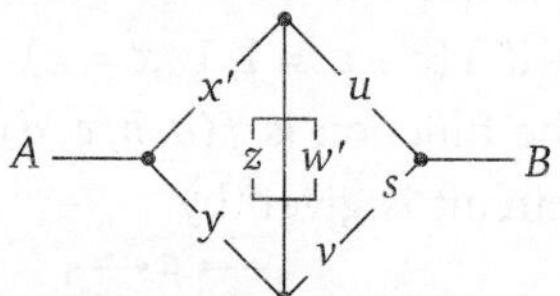

Fig. (35)

10. *Draw a bridge circuit for the following Boolean function :*

$$f = xw' + y'uv + (xz + y')(zw' + uv)$$

Solution. $f = xw' + y'uv + (xz + y')(zw' + uv)$

$= xw' + y'uv + (xz + y') \cdot zw' + (xz + y')uv$ (By distributive law)

$= xw' + y'uv + xz \cdot zw' + y'zw' + xzuv + y'uv$ (By distributive law)

$= xw' + y'uv + y'uv + xzw' + y'zw' + xzuv$

(By associative and commutative laws and $zz = z$)

$= xw' + y'uv + xzw' + y'zw' + xzuv$ ($\because\ x + x = x$)

$= xw' + xzw' + y'uv + y'zw' + xzuv$ (By commutative law)

$= xw' + xw'z + y'uv + y'zw' + xzuv$

(By associative and commutative laws)

$\therefore$ $f = xw' + y'uv + y'zw' + xzuv$ (By absorption law)

Thus, the bridge circuit corresponding $f = xw' + y'uv + y'zw' + xzuv$ is given by

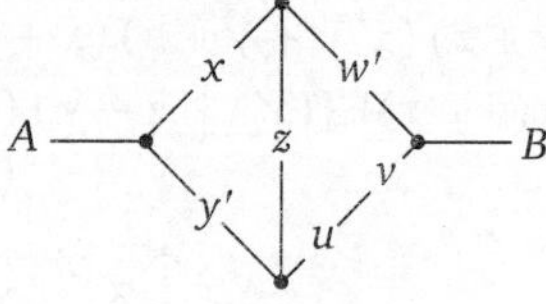

Fig. (36)

11. *Draw a 3-terminal circuit having the following functions :*

$$f = xzw + y'zw$$
$$g = xzw + y'zw + x'y'z.$$

Solution. Express each function as the product of factors

$f = xzw + y'zw$

$= (x + y')zw$ (By distributive law)

and $g = xzw + y'zw + x'y'z$

$= xzw + y'zw + xx'z + x'y'z$ ($\because\ xx' = 0$)

$= (x + y')zw + (x + y')x'z$ (By commutative and distributive law)

$= (x + y')(zw + x'z)$ (By distributive law)

$= (x + y')z\,(w + x')$ (By commutative and distributive law)

Now, draw two 2-terminal circuits for f and g.

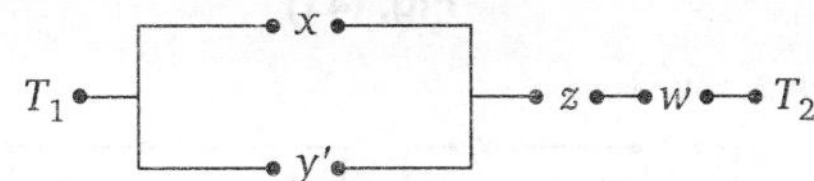

Fig. (37)

$\therefore$ for $f = (x + y')\,zw$ and for $g = (x + y')\,z\,(w + x')$

Fig. (38)

Above two 2-terminals have common factors $(x + y')z$. Thus, 3-terminal circuit is obtained as follows :

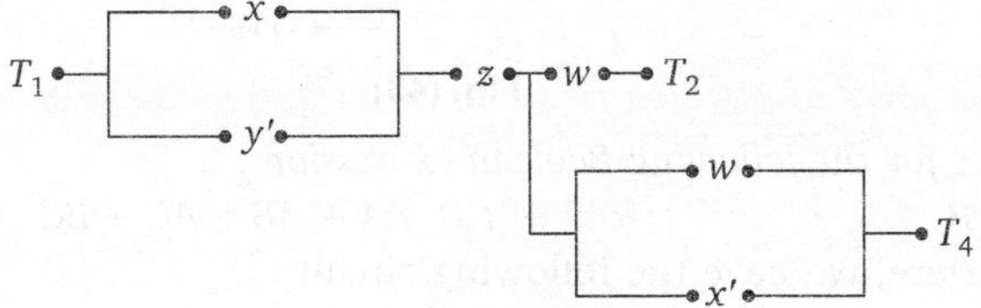

Fig. (39)

12. *Draw a 4-terminal circuit realizing the following three functions :*

$$f = xy'z + (xy' + x'y)\,zw;\; g = xy'zw' + x'yzw';$$
$$h = x'y + (xy' + x'y)(z' + w').$$

Solution. Simplifying the above functions :

$f = xy'z + (xy' + x'y)\,zw$

$= [xy' + (xy' + x'y)w]\,z$ (By commutative and distributive law)

$= [xy'y' + x'yy' + (xy' + x'y)w]\,z$ $(\because yy' = 0, y'y' = y']$

$= [(xy' + x'y)y' + (xy' + x'y)w]\,z$ (By distributive law)

$= (xy' + x'y)(y' + w)\,z$ (By distributive law)

$= (xy' + x'y)\,z\,(y' + w)$ (By commutative law)

and $g = xy'zw' + x'yzw'$

$= (xy' + x'y)\,zw'$ (By distributive law)

and $h = x'y + (xy' + x'y)(z' + w')$

$= x'yy + xy'y + (xy' + x'y)(z' + w')$ $(\because yy = y, y'y = 0)$

$= (x'y + xy')y + (xy' + x'y)(z' + w')$ (By distributive law)

$= (x'y + xy')(y + z' + w')$ (By distributive law)

Now, we draw 2-terminal circuit for f, g and h.

$\therefore$ For $f = (xy' + x'y)\,z\,(y' + w)$

Fig. (40)

For $g = (xy' + x'y)\,zw'$.

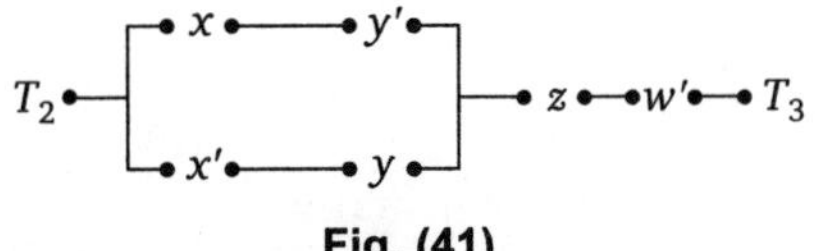

Fig. (41)

For $h = (x'y + xy')(y + z' + w')$

Fig. (42)

Thus, 4-terminal circuit is obtained as follows :

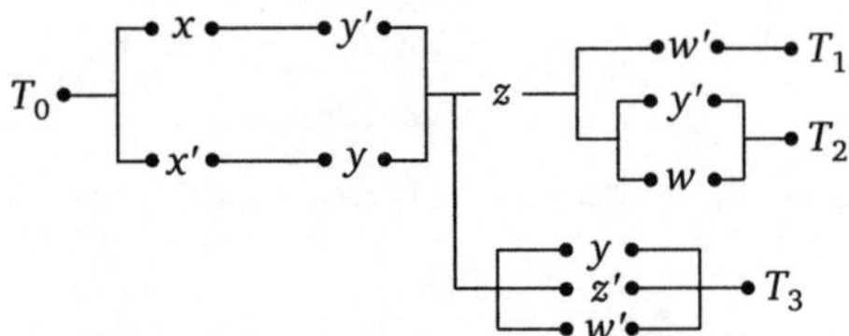

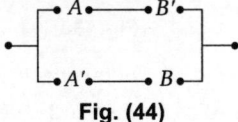

Fig. (43)

13. *Construct circuits for the following Boolean expression :*

(i) $Q = AB' + A'B$ *(ii)* $AA + AB + AC + BC$

Solution : (i) Here, we have the following circuit :

Fig. (44)

(ii) The required circuit is given by

Fig. (45)

3.11 KARNAUGH MAP METHOD FOR SIMPLIFICATION OF BOOLEAN EXPRESSION

The Karnaugh map method is a graphical technique which provides a simple straight forward procedure for simplification of Boolean expression of two, three or four variables, it can also be extended to functions of five, six or more variables. But as the number of variables increases, the excessive number of squares prevents a reasonable selection of adjacent squares and it is difficult to be sure that the best selection has been made. The method was introduced by Maurice Karnaugh in 1953.

A Karnaugh map (*K*-map) is a diagram made up of a number of squares. If the expression contains n variables, the map will have 2^n squares. Each square represents a minterm and 1*s* are written in the corresponding squares for the minterms present in the expression and 0'*s* are written in those squares which correspond to the minterms not present in the expression. Once the map is filled with 0*s* and 1*s*, the canonical sum of products expression for the output can be obtained by ORing together those squares that contain 1.

3.11.1 Two Variables Karnaugh Maps

Since the number of squares are 2, the map will have $2^2 = 4$ squares. The values of one variable, say A, are listed above the top horizontal line and the values of other variable, say B, are listed on the left.

Four possible minterms with two variables A and B

$$AB,\ AB',\ A'B,\ A'B'$$

are represented by the four squares in the map as shown in following figure

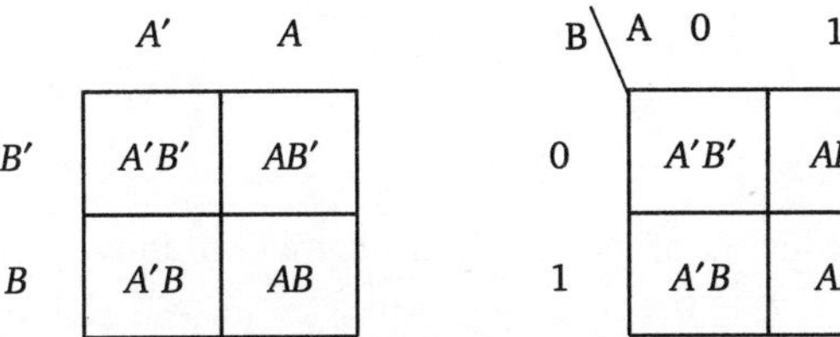

Fig. (46) : Two variables Karnaugh map

- Squares are said to be adjacent if the minterms that they represent differ in exactly one literal.
- The expression can be simplified by properly combining these squares in the K-map which contains 1s. The process for combining 1s is called looping.
- Whenever there are 1s in two adjacent squares in the K-map, the minterms represented by these squares can be looped and it eliminates the variable that appears in complemented and uncomplemented form.

ILLUSTRATIVE EXAMPLES

1. *Use the Karnaugh map method to find a minimal disjunctive normal form (sum-of-product) of the following functions :*

(i) $f(x, y) = xy + xy'$ *(ii)* $f(x, y) = xy + x'y + x'y'$

(iii) $f(x, y) = xy + x'y'$

Solution. (i) We first represent $f(x, y)$ be a Karnaugh map. The Karnaugh map represents $f(x, y) = xy + xy'$ in the following :

	x	x'
y	1	0
y'	1	0

Fig. (47)

We have represented two adjacent squares with 1s in them by a rectangle. This rectangle represents x. Hence,

$$f(x, y) = x$$

(ii) The representation of $f(x, y) = xy + x'y + x'y'$ by Karnaugh map is as follows :

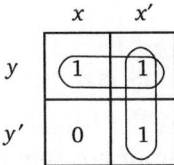

	x	x'
y	1	1
y'	0	1

Fig. (48)

The function $f(x, y)$ contains two pairs of adjacent squares in 1 (indicated by two encircled 1s) which include all the squares of $f(x, y)$ which contains 1. The horizontal pair (encircled) represents y and vertical pair (encircled) represents x'.

Hence, $f(x, y) = y + x' = x' + y$ is the minimal form.

(iii) The Karnaugh map representation of $f(x, y)` = xy + x'y'$ is given below :

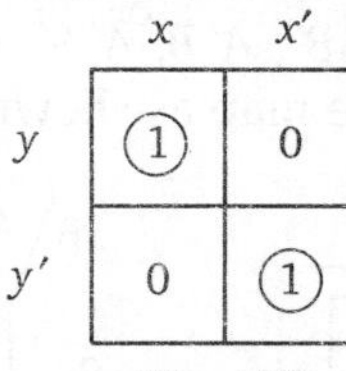

Fig. (49)

Observe the $f(x, y)$ consists of two circles as shown in above figure 49.

Thus, $f(x, y) = xy + x'y'$ is the minimal form.

2. *Find Karnaugh map and simplify the expressions :*

(i) $AB' + A'B'$ (ii) $AB' + A'B$ (iii) $AB' + A'B + A'B'$

Solution. (i) Two adjacent squares $A'B'$ and AB' containing 1 have been grouped together. They have been encircled.

These two terms can be looped that eliminate the A variable since it appears both in complemented and uncomplemented form. This can be verified algebraically as follows :

B \ A	0	1
0	1	1
1	0	0

Fig. (50)

$$AB' + A'B' = (A + A')B'$$

(ii) K-map does not contain any adjacent square of minterms containing 1, hence the expression can not be simplified further.

B \ A	0	1
0	0	1
1	1	0

Fig. (51)

(iii) There are two pairs of 1s and they can be combined as shown in fig.

B \ A	0	1
0	1	1
1	1	0

Fig. (52)

Here, 1 in first column and first row has been enclosed twice, as it is permissible to use the same 1 more than once. Looping of horizontal 1-squares gives the result B' and vertical 1-squares gives A' .

Hence, the given expression reduces to the simplified form as $A' + B'$. This can also be verified algebraically as follows :

$$AB' + A'B + A'B' = AB' + A'(B + B') = AB' + A' \cdot 1 = A' + B'$$

3.11.2 Three Variable Karnaugh Map

A Karnaugh map in three variables is a rectangle divided into eight squares. One of the ways that eight possible minterms are labelled in sequence is shown in following figure 53.

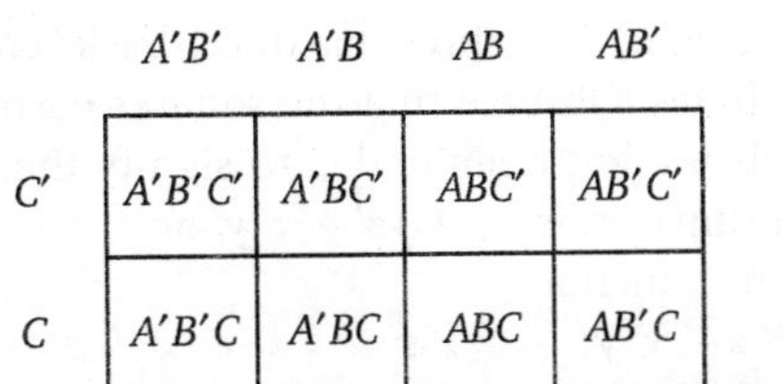

C \ AB	00	01	11	10
0	A'B'C'	A'BC'	ABC'	AB'C'
1	A'B'C	A'BC	ABC	AB'C

Fig. (53) : Karnaugh map for three variables

Given a minterm expansion of a function, it can be plotted on a map by placing 1s in the squares which correspond to minterms present in the expansion and 0s in the remaining squares.

To simplify a sum-of-products expansion in three variables, one has to identify groups of minterms that can be combined. While forming groups of squares containing 1s the follownig considerations must be kept in mind :

(i) The number of squares in a group must be equal to such that 2,4, 8, 16.

(ii) A square containing 1 can be included in as many groups as desired.

(iii) Groups must be largest possible groups, a group of two squares containing 1 should not be made if these squares can be included in group of four squares. A *K*-map that contains a group of four 1s that are adjacent to each other is called a quad. Looping a quad of 1s eliminate the two variables that appear in both complemented and uncomplemented form.

ILLUSTRATIVE EXAMPLES

1. *Find Boolean expression for the given Karnaugh map*

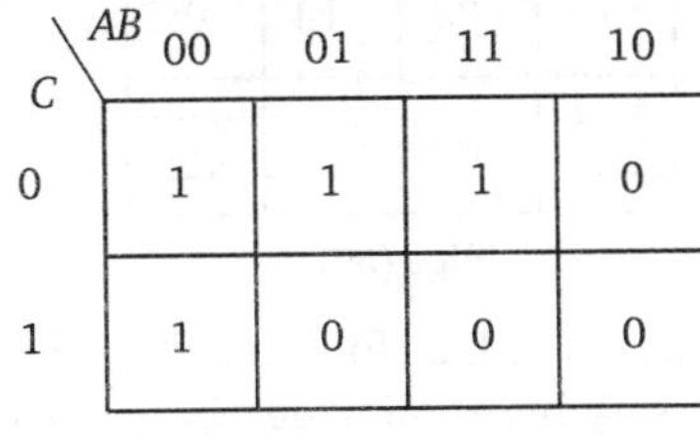

Fig. (54)

Solution. Boolean expression for given Karnaugh map is

$$ABC' + A'B'C + A'BC' + ABC'.$$

2. *Using Karnaugh maps find the minimal form for each of the follownig Boolean functions :*

(i) $f(x, y, z) = xyz + xyz' + x'yz' + x'yz' + x'y'z$

(ii) $f(x, y, z) = xyz + xyz' + xy'z + x'yz$

(iii) $f(x, y, z) = xyz + xyz' + x'yz' + x'y'z' + x'y'z$

Solution. (i) The Karnaugh map corresponding to the given function is given below :

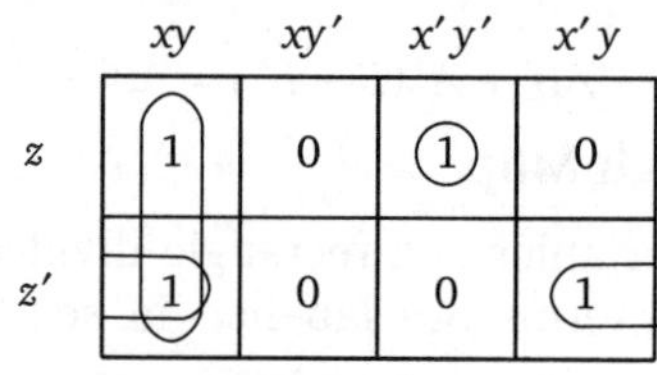

Fig. (55)

From the Karnaugh map, we see that $f(x, y, z)$ has three maximal basic encircled 1*s* containing squares with 1 which are shown by circles. Observe that the squares corresponding to xyz' and $x'yz'$ are adjacent. Thus, the symbols are left open ended to signify that they join in one encircle. The resulting minimal Boolean function is $xy + yz' + x'y'z$

(ii) The Karnaugh map corresponding to the function

$$f = xyz + xyz' + xy'z + x'yz + x'y'z$$

is given below which has five squares with 1*s* in them corresponding to five minterms of f.

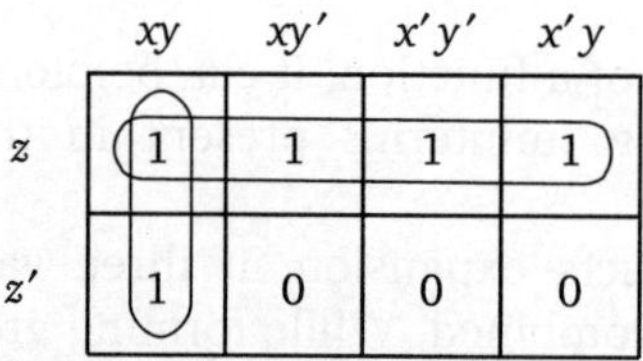

Fig. (56)

From the Karnaugh map, we see that $f(x, y, z)$ has two maximal basis circles containing all the squares with 1, which are shown by circles. One of the maximal basic encircles is the two adjacent squares which represents xy and the other is the 1×4 square which represents z. Both are needed to cover all the squares with 1. So the minimal form of $f(x, y, z)$ is given by

$$f(x, y, z) = xy + z$$

(iii) The Karnaugh map corresponding to the function

$$f(x, y, z) = xyz + xyz' + x'yz' + x'y'z' + x'y'z$$

is given below which has five squares with 1*s* in them corresponding to the five minterms of f.

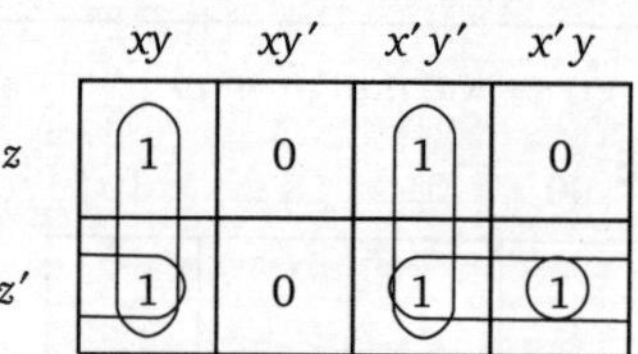

Fig. (57)

As shown by the circles, $f(x, y, z)$ has four maximal basic circles. To cover all the squares with 1*s* in them, it is necessary here to include basic encircles which represent xy and $x'y'$ and only one of two circles which correspond to $x'z'$ and yz'. Thus, $f(x, y, z)$ has two minimal forms :

$$f(x, y, z) = xy + x'y' + x'z' \quad \text{and} \quad f(x, y, z) = xy + x'y' + yz'.$$

3. *Use Karnaugh map to simplify the following :*

$$\chi = A'B'C' + A'B'C + A'BC + A'BC' + AB'C + ABC$$

Soluion. The Karnaugh map of the given function is shown in figure :

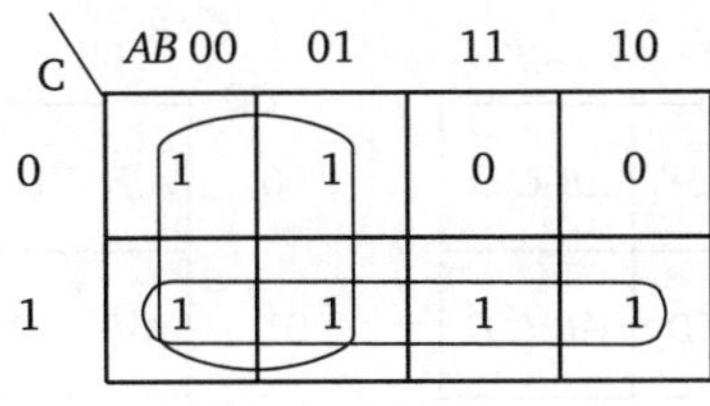

Fig. (58)

Here, two quads have been formed. Here, two 1*s* are a part of both the quads. The quad formed by $A'B'C'$, $A'BC'$, $A'B'C$ and $A'BC$ produces the resultant as A' and the quad formed by $A'B'C$, $A'BC$, ABC and $AB'C$ produces the resultant as C. Hence, the final resultant expressions $X = A' + C$.

4. *Simplify the following Boolean expression by using map method and show circuit diagram for original expression and reduced expression :*

(i) $x'y + xy' + xy$ *(ii)* $x'yz + x'yz' + x'y'z + xy'z$

Solution. (i) Let $F = x'y + xy' + xy$

Using *K*-map, this equation reduced to

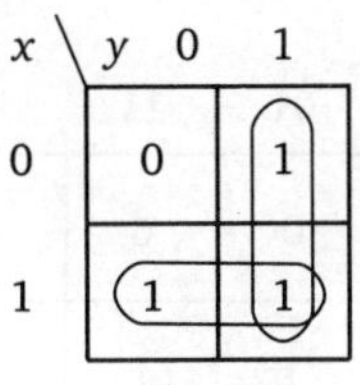

Fig. (59)

This *K*-map contains two pairs of 1*s* that are horizontally and vertically adjacent and each pair is reducing a variable. Thus, the reduced expression is $f = x + y$.

(ii) Let $F = x'yz + x'yz' + x'y'z + xy'z$

Using *K*-map, this equation can be reduced as follows :

x \ yz	00	01	11	10
0	0	1	1	1
1	0	1	0	0

Fig. (60)

This *K*-map contains three pairs of 1*s* that are horizontally and vertically adjacent and each pair is reducing a variable. Thus, the reduced expression is $F = x'z + x'y + y'z$.

3.11.3 Four Variable Karnaugh Map

A Karnaugh map in four variables is a square divided into 16 $(= 2^4)$ squares. The squares represent the 16 minterms in four variables. The definition of adjacent squares must be extended so that not only are left most and right most column adjacent as in 3-variable map, but also the first and last rows are adjacent. The following figure shows Karnaugh map for four variables *A*, *B*, *C* and *D*.

	$A'B'$	$A'B$	AB	AB'
$C'D'$	$A'B'C'D'$	$A'BC'D'$	$ABC'D'$	$AB'C'D'$
$C'D$	$A'B'C'D$	$A'BC'D$	$ABC'D$	$AB'C'D$
CD	$A'B'CD$	$A'BCD$	$ABCD$	$AB'CD$
CD'	$A'B'CD'$	$A'BCD'$	$ABCD'$	$AB'CD'$

$CD \backslash AB$	00	01	11	10
00	$A'B'C'D'$	$A'BC'D'$	$ABC'D'$	$AB'C'D'$
01	$A'B'C'D$	$A'BC'D$	$ABC'D$	$AB'C'D$
11	$A'B'CD$	$A'BCD$	$ABCD$	$AB'CD$
10	$A'B'CD'$	$A'BCD'$	$ABCD'$	$AB'CD'$

Fig. (61) : Karnaugh map for four variables

To simplify a sum-of-product expansion in four variables, one has to verify groups of minterms of squares 2, 4, 8 or 16 containing 1*s* that can be combined. A group of eight 1*s* that are adjacent to one another is called an octet. Looping an octet of 1*s* eliminates three variables that appear in both complemented and uncomplemented form. Some examples of octet are shown in figure :

$CD \backslash AB$	00	01	11	10
00	0	0	0	0
01	1	1	1	1
11	1	1	1	1
10	0	0	0	0

(a) $X = D$

$CD \backslash AB$	00	01	11	10
00	1	1	0	0
01	1	1	0	0
11	1	1	0	0
10	1	1	0	0

(b) $X = A'$

CD \ AB	00	01	11	10
00	1	1	1	1
01	0	0	0	0
11	0	0	0	0
10	1	1	1	1

(c) $X = D'$

CD \ AB	00	01	11	10
00	1	0	0	1
01	1	0	0	1
11	1	0	0	1
10	1	0	0	1

(d) $X = B'$

Fig. (62)

ILLUSTRATIVE EXAMPLES

1. *Use Karnaugh maps to find a minimal form for the following Boolean functions :*

(i) $f(x, y, z, w) = x'yzw + xy'zw' + x'y'zw' + xyz'w' + xy'z'w'$

(ii) $f(x, y, z, w) = xy' + xyz + x'y'z' + x'yzw'$.

Solution. (i) The Karnaugh map representation of the given function is shown below which have five squares with 1s.

	xy	xy'	$x'y'$	$x'y$
zw	0	0	0	1
zw'	0	1	1	0
$z'w'$	1	1	0	0
	0	0	0	0

Fig. (63)

A minimal cover of all 1*s* of the map consists of the three maximal basic circles as shown in the fig. Thus, the minimal form is

$$f(x, y, z, w) = y'zw' + xz'w' + x'yzw$$

(ii) The Karnaugh map representation of the given function is shown below. Observe that there are four squares with 1*s* in them representing xy'. Similarly, there are two squares with 1 representing xyz and so on.

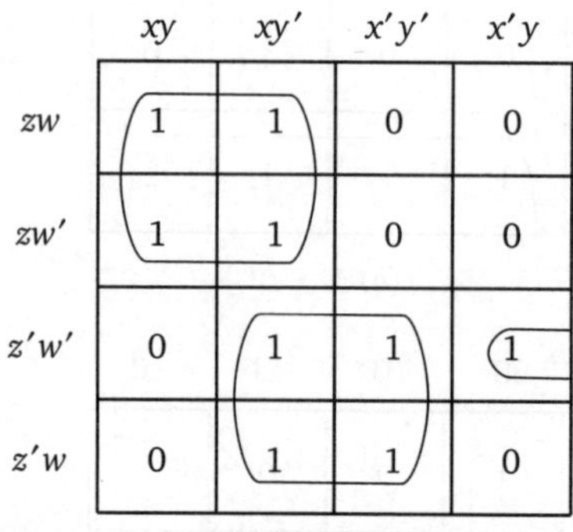

Fig. (64)

The minimum number of maximal basic circles to cover all 1*s* of the map is 3 as shown in fig. Thus, the minimal form is

$$f(x, y, z) = xy + y'z' + yzw'$$

Observe that the upper left 2×2 circles represent xz while the other 2×2 circles represent $y'z'$.

2. *Use Karnaugh map to simplify the following expressions :*

(i) $X = A'B'CD + A'B'CD' + AB'C'D' + AB'CD'$

(ii) $X = A'BC'D' + ABC'D' + A'BCD' + ABCD'$

(iii) $X = A'B'C'D' + AB'C'D' + A'B'CD' + AB'CD'$.

Solution. (i) The Karnaugh map of the given function is shown in following figure. Two pairs of adjacent 1*s* are grouped together. The pair in the first column gives term $A'B'C$ and the pair of top row and bottom row produces the term $AB'D'$. Hence, the simplified result is $X = A'B'C + AB'D'$.

CD \ AB	00	01	11	10
00	0	0	0	1
01	0	0	0	0
11	1	0	0	0
10	1	0	0	1

Fig. (65)

(ii) The Karnaugh map of the given Boolean function is shown in figure. Two 1-squares in the first row and two 1-squares in the last row are adjacent squares and hence they form a quad. The variables B and D' remain unchanged (A and C being complemented and uncomplemented form). The result expression is $X = B \cdot D'$.

CD \ AB	00	01	11	10
00	0	1	1	0
01	0	0	0	0
11	0	0	0	0
10	0	1	1	0

Fig. (66)

(iii) The Karnaugh map of the given Boolean function is shown in figure. Four 1-squares of figure are adjacent squares, the top and bottom rows are considered to be adjacent to each other as are the left most and right most columns. The variables B' and D' remain unchanged (A and C are in complemented and uncomplemented form). The resultant expression is $X = B' \cdot D'$.

CD \ AB	00	01	11	10
00	1	0	0	1
01	0	0	0	0
11	0	0	0	0
10	1	0	0	1

Fig. (67)

3. *Use Karnaugh map to simplify the Boolean expression*

$$a'b'c'd' + ab'c'd' + a'bc'd' + a'b'cd + a'bcd + abcd + ab'cd + a'b'cd' + a'bcd' + abcd' + ab'cd'$$

Solution. The Karnaugh map of the given expression is shown in figure.

cd \ ab	00	01	11	10
00	1	0	0	1
01	0	1	0	0
11	1	1	1	1
10	1	1	1	1

Fig. (68)

The four corner 1-square can be combined to form $b'd'$. The group of eight 1s forming an octet represents C. The pair of 1s which is looped on the map represents the term $a'bd$. Hence, the simplified Boolean expression is $c + b'd' + a'bd$.

3.12 PRIME IMPLICANT

In a sum of product expression, each product term is known or Implicant on a Karnaugh map. Each implicant related to a single 1-square or a group of adjacent 1-square. A prime implicant is an implicant if it can not be combined with another term to eliminate a variable. In following figure

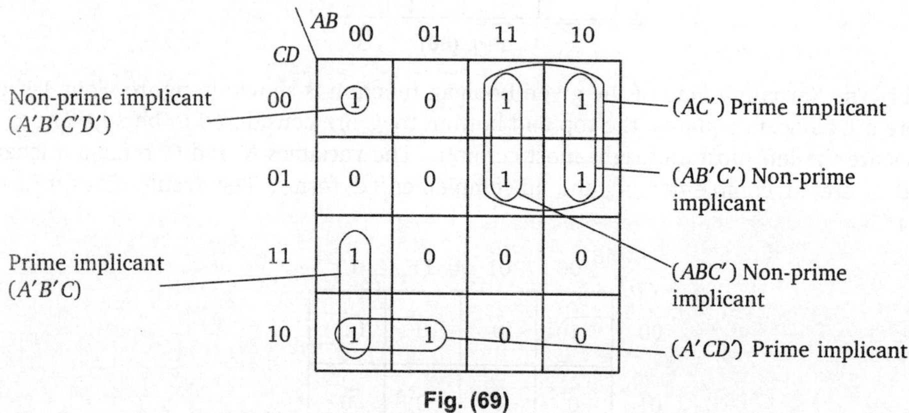

Fig. (69)

$A'B'C$, $A'CD'$ and AC' are prime implicants because they can not be combined with other terms to eliminate a variable. All of the prime implicants of a function can be obtained from a Karnaugh map. A single 1 on a map represents a prime implicant if it is not adjacent to any other 1s. Two adjacent 1s on a map form a prime implicant if they are not contained in a group of four 1s, four adjacent 1s form a prime implication if they are not contained in a group of eight 1s, etc.

If among the minterms subsumming a prime implicant, there is at least one that is covered by this and only by this prime implicant, then the prime implicant is called an essential prime implicant and it must be included in the minimum sum of products. The following figure shows essential and non-essential prime implicants.

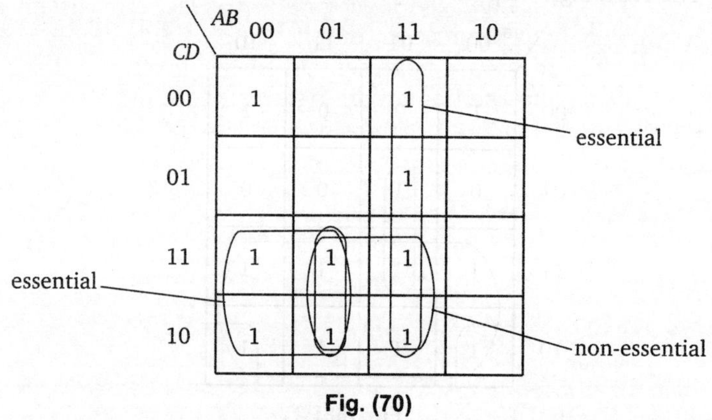

Fig. (70)

Exercise 3.1

1. Describe the Karnaugh map for two and three variables.
2. From the truth table from three variables given, write the corresponding Boolean expression and draw the Karnaugh map.

A	B	C	F
0	0	0	0
0	0	1	0
0	1	0	0
0	1	1	0
1	0	0	0
1	0	1	0
1	1	0	1
1	1	1	1

3. Use the Karnaugh map representation to find the minimal form of each of the following functions :
   (i) $f(x, y) = x'y + xy$
   (ii) $f(x, y, z) = xyz + xy'z + x'yz + x'y'z$
   (iii) $f(x, y, z) = xyz' + xy'z' + x'y'z' + x'y'z + x'yz + x'yz'$
4. For each of the following Boolean expressions, find the minimum sum of products using Karnaugh map :
   (i) $A'B'C' + AB'C' + A'BC' + ABC'$
   (ii) $AB + AB'C + ABC$
   (iii) $AB' + A'B'CD + CD + BC'D + ABCD$
5. Use the Karnaugh map to find a minimal form of each of the following functions :
   (i) $f = xyz'w' + xyz'w + xy'zw + xy'zw' + x'y'zw + x'y'zw' + x'yw'z'$
   (ii) $f = xyzw' + xy'zw' + xy'z'w' + xy'z'w + x'y'zw + x'y'zw' + x'y'z'w' + x'yz'w'$
6. Draw Karnaugh map and simplify the following Boolean expressions :
   (i) $AB + A\overline{B} + \overline{A}\,\overline{B}$
   (ii) $ABC + \overline{A}B\overline{C} + AB\overline{C} + \overline{A}BC$
   (iii) $\overline{A}\overline{B}C + \overline{A}\,\overline{B}\overline{C} + AB\,\overline{C} + \overline{A}B\overline{C}$
   (iv) $A\overline{B}\overline{C} + AB\overline{C} + ABC + \overline{A}\overline{B}\overline{C}$
   (v) $AB\overline{C}\,\overline{D} + A\overline{B}\,\overline{C}\,\overline{D} + AB\overline{C}\,\overline{D} + A\overline{B}\overline{C}D$
   (vi) $\overline{A}\,\overline{B}\,\overline{C}\,\overline{D} + \overline{A}\,B\,\overline{C}\,\overline{D} + \overline{A}\,\overline{B}\,C\,D + \overline{A}\,\overline{B}\,C\,\overline{D} + \overline{A}BCD$
   (vii) $A'B'C' + A'B'C + AB'C + ABC$
   (viii) $BC + AC' + AB' + BCD$

ANSWERS

3.(i) $f = y$ (ii) $f = z$ (iii) $f = z' + x'z$

4.(i) $F = C'$ (ii) $F = AB + AC$ (iii) $F = AB' + BD + CD$

5.(i) $f = y'z + xyz' + yz'w'$ (ii) $f = xzw' + xy'z' + x'y'z + x'z'w$

6.(i) $A + B'$ (ii) B (iii) $BC' + A'B'$ (iv) $AB + B'C'$

(v) AC' (vi) $A'B' + A'CD$ (vii) $A'B' + AC$

3.13 LOGIC GATE

Computer science at the hardware level involves designing devices to produce appropriate outputs from given inputs. A logic gate is an electronic circuit that operates on one or more input signals to produce an output signal. Gates are digital circuits and often called logic circuits. Gates are represented by symbols, the lines entering the symbol from left are inputs, and the line on the right is the output. Placing a small circle on an input or output complements the signals on that line.

3.13.1 OR Gate

An OR gate has two or more inputs but it has only one output. Let x and y be two inputs. The output of OR gate is denoted by $x + y$, where $x + y$ is defined by the following table given below :

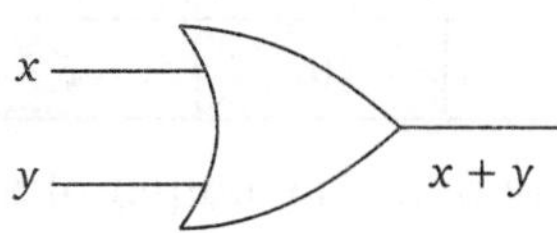

Fig. (71)

Input		Output
x	y	$x + y$
1	1	1
1	0	1
0	1	1
0	0	0

The output of OR gate is 1 if any one of the input is 1, otherwise it is zero. The logical operation of an OR gate can easily be explained with the help of two switches connected in parallel as shown in following figure.

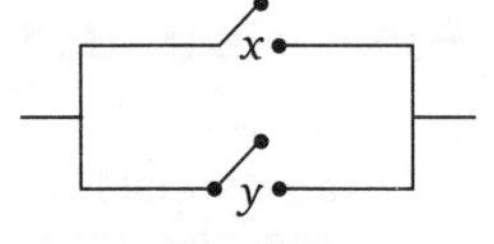

Fig. (72)

3.13.2 AND Gate

It is possible to construct another electronic device called an AND gate which works in a similar way as OR gate except that the output only takes the value 1 when both inputs are 1. The output is equal to the AND product of the logic circuits. The symbol and truth table of the gate is shown in following figure :

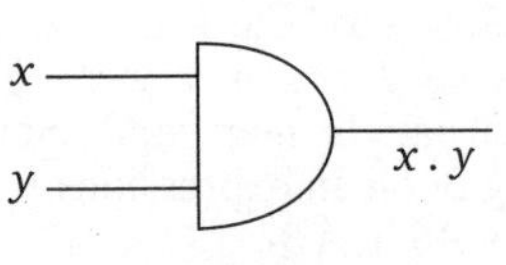

Fig. (73)

Input		Output
x	y	$x \cdot y$
1	1	1
1	0	0
0	1	0
0	0	0

The logical operation of an AND gate can easily be explained with the help of two switches connected in series as shown in following figure.

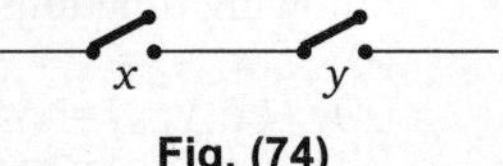

Fig. (74)

3.13.3 NOT Gate

NOT gate changes the input to the complement (′) of it, *i.e.*, the output through the NOT gate is the complement of input. The symbol and truth table of the gate is shown in the following figure :

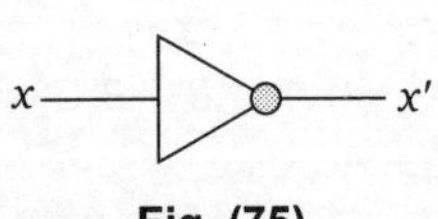

Fig. (75)

Input	Output
x	x'
1	0
0	1

3.13.4 NOR Gate

A NOR gate is equivalent to an OR gate followed by a NOT gate. Following figure represents a NOR gate and its associated truth table :

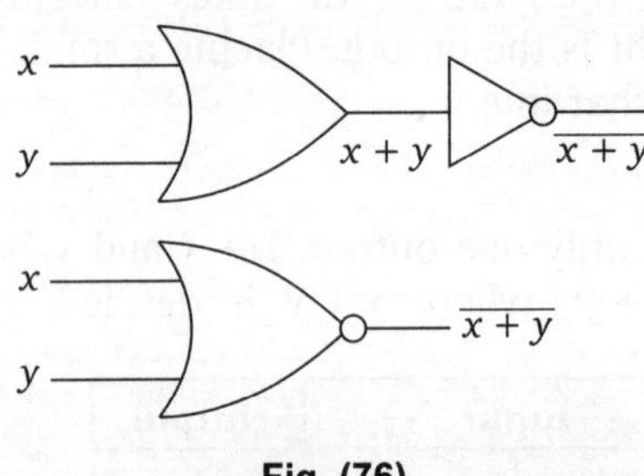

Fig. (76)

Input		Output
x	y	$\overline{x+y}$
1	1	0
1	0	0
0	1	0
0	0	1

A NOR gate also has two or more inputs and one output. The output of a NOR gate is 1 if and only if all the inputs are 0.

3.13.5 NAND Gate

This gate is equivalent to a NOT gate in series with an AND gate. It is represented by the symbol shown in following figure.

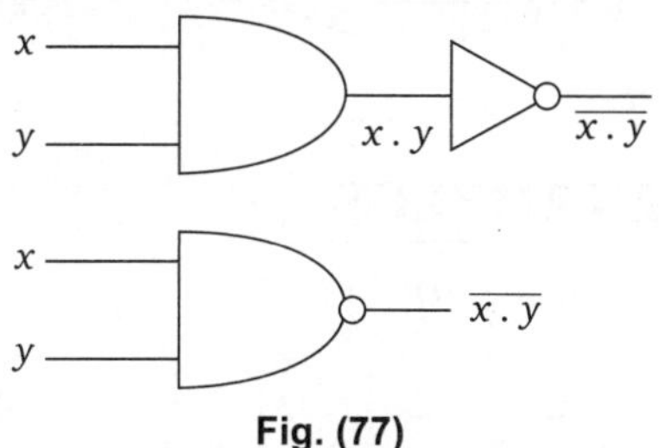

Fig. (77)

Input		Output
x	y	$\overline{x \cdot y}$
1	1	0
1	0	1
0	1	1
0	0	1

3.13.6 Exclusive OR (XOR) Gate

The exclusive-OR gate is different than OR gate because it includes only word that have an odd number of 1*s*. The symbol and truth table of XOR gate are shown in the following figure.

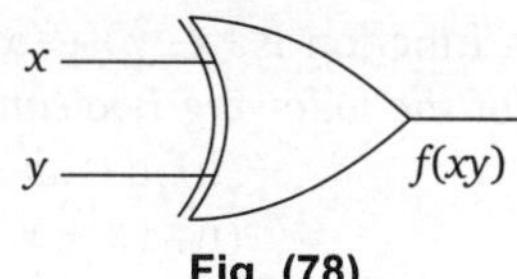

Fig. (78)

This is equivalent to

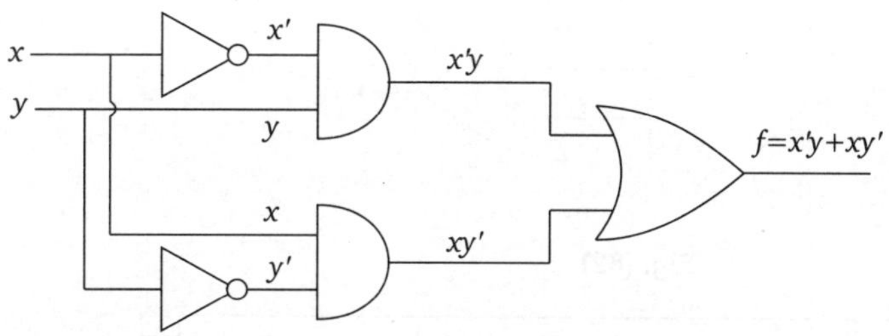

Fig. (79)

x	y	$f = x'y + xy'$
1	1	0
1	0	1
0	1	1
0	0	0

3.13.7 Exclusive NOR (XNOR) Gate

XNOR gate is equivalent to an exclusive OR gate followed by an inverter. The output is 1 only when both inputs are either 0 or 1. The truth table for this gate is shown in the following table

x	y	$f = xy + x'y'$
1	1	1
1	0	0
0	1	0
0	0	1

The logic diagram and logic symbol is shown in figure :

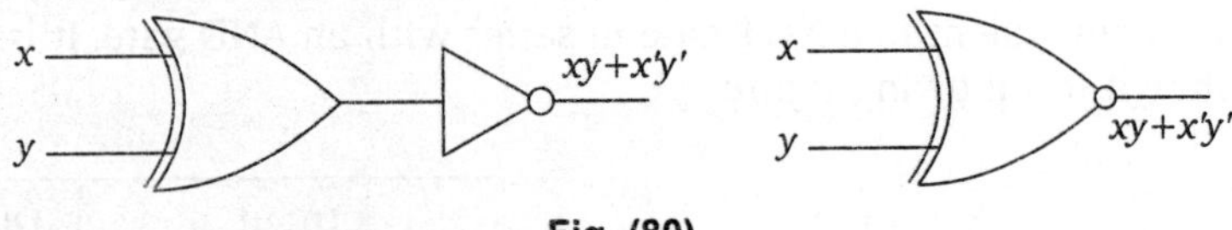

Fig. (80)

ILLUSTRATIVE EXAMPLES

1. *Write the Boolean function corresponding to the following network.*

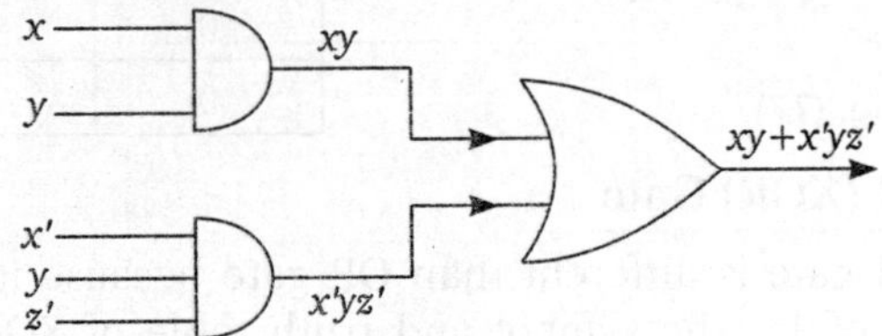

Fig. (81)

Solution. Required Boolean function is $(x \cdot y) + (x' \cdot y \cdot z')$.

2. *Draw the logic circuit for each of the following Boolean expressions :*

(i) $x \cdot y + y \cdot z$ *(ii)* $d \cdot (ab' + a'c)$

(iii) $xy + zx'$ *(iv)* $(x + y) \cdot (x' + y' + z') \cdot (y' \cdot z')$.

Solution.

(i)

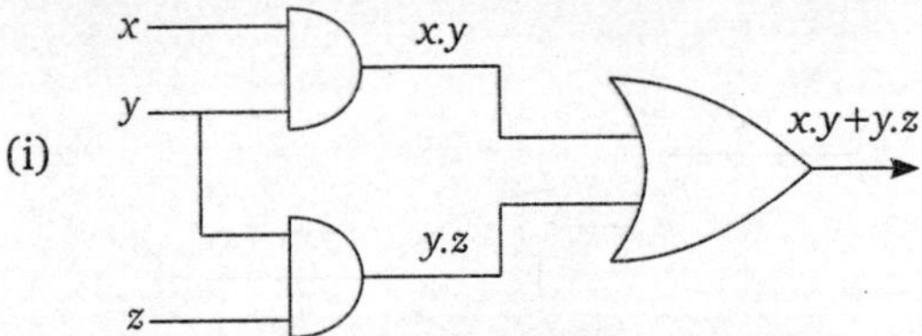

Fig. (82)

(ii)

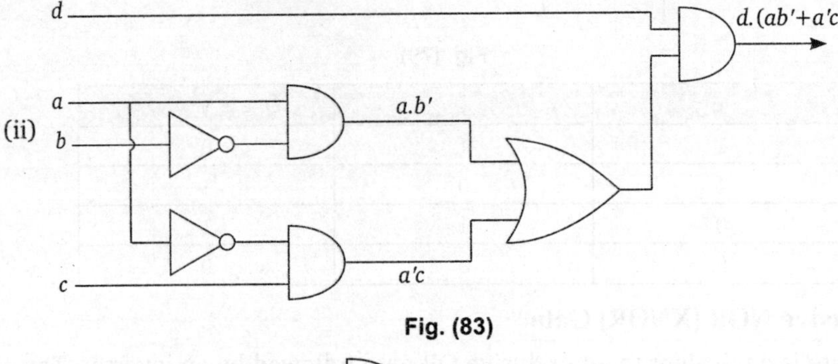

Fig. (83)

(iii)

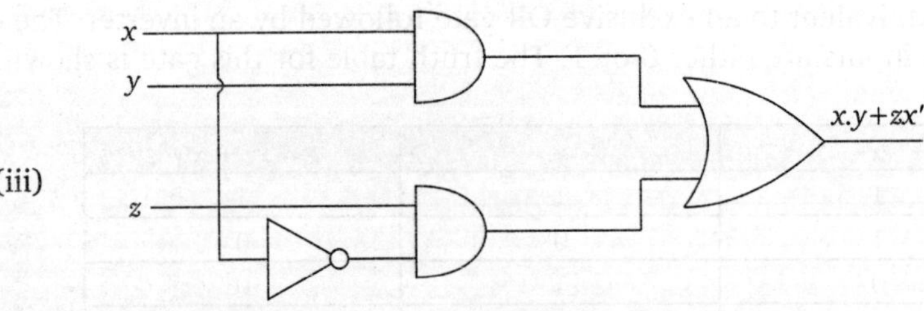

Fig. (84)

(iv)

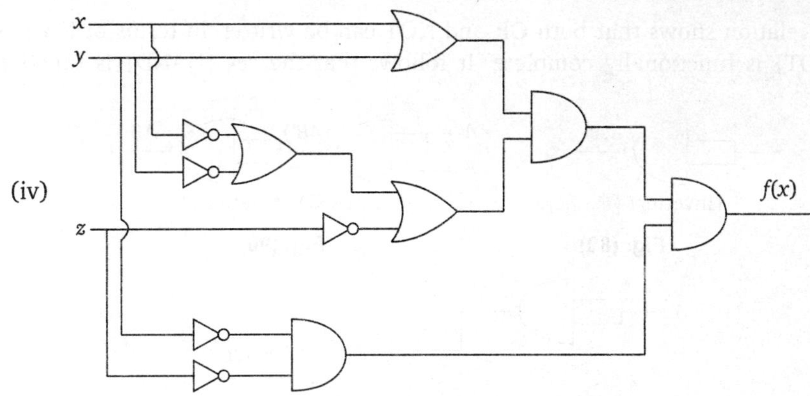

Fig. (85)

3. *Find the networks corresponding to Boolean expressions*

(i) $AB + CD$ (ii) $X'\,Y'\,Z + X'\,YZ + XY'$

Solution. Expression (i) will require two AND gates and one OR gate and (ii) will require three AND gates and OR gate and NOT gate.

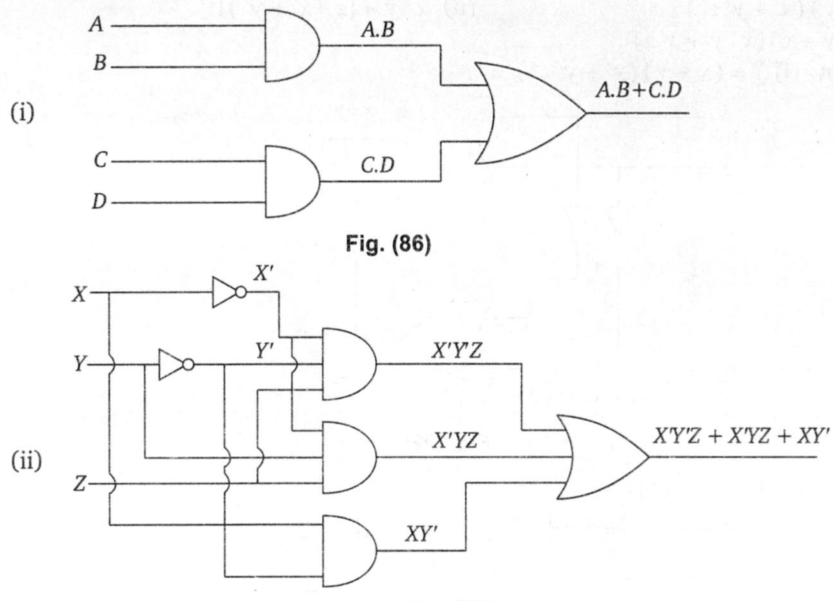

(i) **Fig. (86)**

(ii) **Fig. (87)**

4. *Show that set of gates (AND, NOT) is functionally complete.*

Solution. To show that given set is functionally complete, we need to show that an OR gate can be replaced by a suitable combination of AND gate and NOT gate.

By the use of De Morgan's law, we have $x + y = (x'\,y')'$

This shows that an OR gate can be replaced by one AND gate and three NOT gates as given below :

Fig. (88)

5. *The set {NAND} is a functionally complete set of operations.*

Solution. We observe that A NAND $B = (AB)'$

Therefore, $A' = (AA)' = A$ NAND A

$A + B = ((A')\,(B'))' = A'$ NAND B'

$= (A$ NAND $A)$ NAND $(B$ NAND $B)$.

The above relation shows that both OR and NOT can be written in terms of NAND. But the set {OR, NOT} is functionally complete. It follows that the set {NAND} is functionally complete.

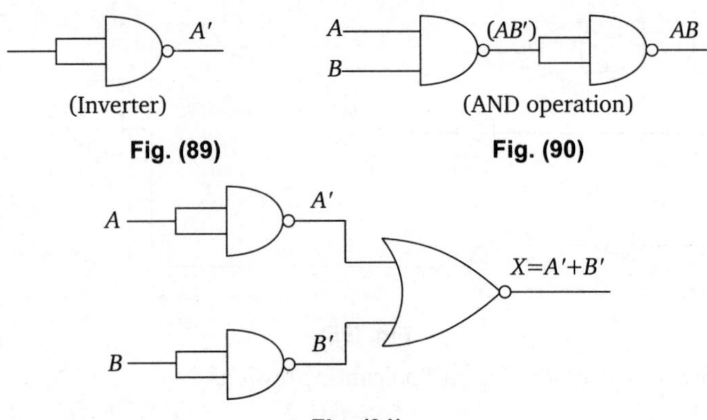

(Inverter) Fig. (89)

(AND operation) Fig. (90)

Fig. (91)

6. *Draw the logic diagram to represent the Boolean expression :*

(i) $(x + y)(x' + y'z')$ *(ii)* $x \cdot y + [z'(x' + y')]$

(iii) $(x + y + z)[x \cdot y + x'z]$

Solution. (i) $f = (x + y)(x' + y' \cdot z')$

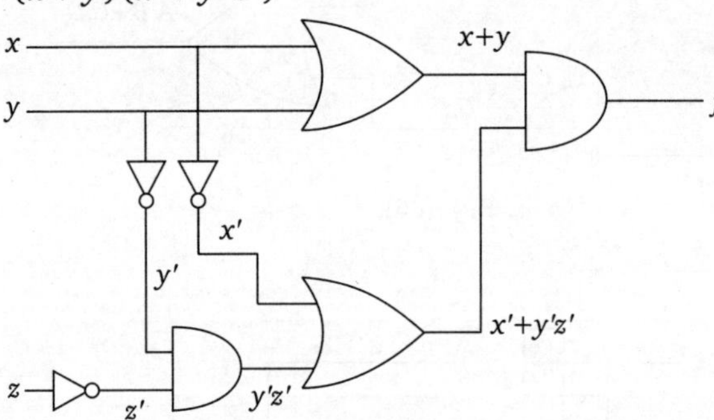

Fig. (92)

(ii) $f = x \cdot y + [z'(x' + y')]$

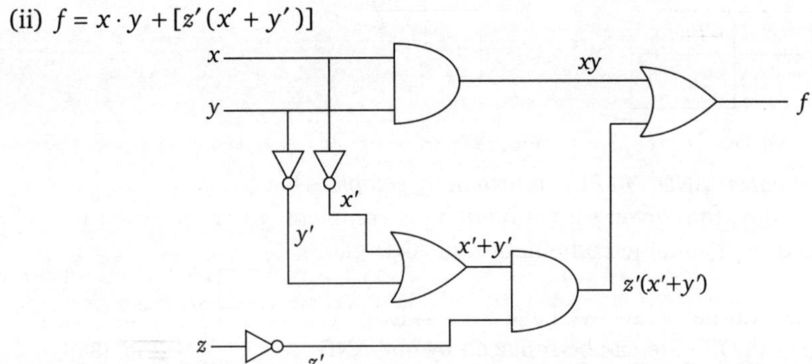

Fig. (93)

(iii) Hence, $f = (x + y + z)[x \cdot y + x'z]$.

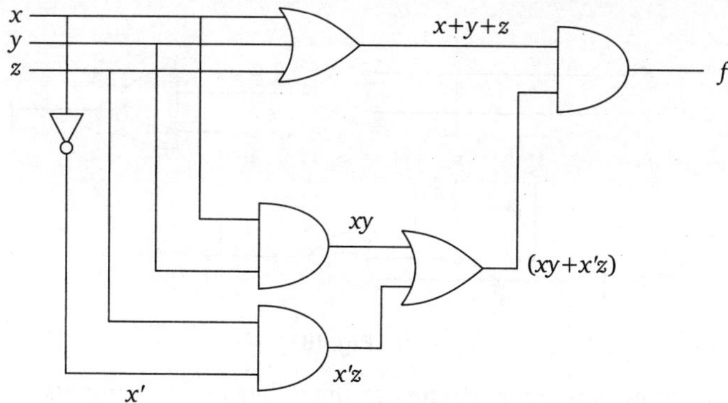

Fig. (94)

7. *Simplify the following Boolean expression and then construct the circuit diagram of reduced Boolean expression :*

(i) $xy + xy' + x'y'$ *(ii)* $xy'z + (z + y)x'$

Solution. (i) $xy + xy' + x'y'$

$$= x(y + y') + x'y' \qquad [\because\ y + y' = 1]$$
$$= x \cdot 1 + x'y'$$
$$= x + x'y'$$

Fig. (95)

(ii) $xy'z + (z + y)x'$

$$= xy'z + x'z + x'y$$
$$= z(xy' + x') + x'y$$
$$= z(x(1 - y) + x') + x'y \qquad [\because\ y + y' = 1]$$
$$= z(x - xy + x') + x'y$$
$$= z(1 - xy) + x'y$$
$$= z - zxy + x'y$$
$$= z - zxy + (1 - x)y$$
$$= z - zxy + y - yx$$
$$= z - xy(z + 1) + y$$
$$= z - xy + y$$
$$= z + y(1 - x) = z + y \cdot x'.$$

Fig. (96)

8. *Draw the logic diagram to represent the Boolean expressions :*

$$(x + y)(x' + y' + z')$$

Solution.

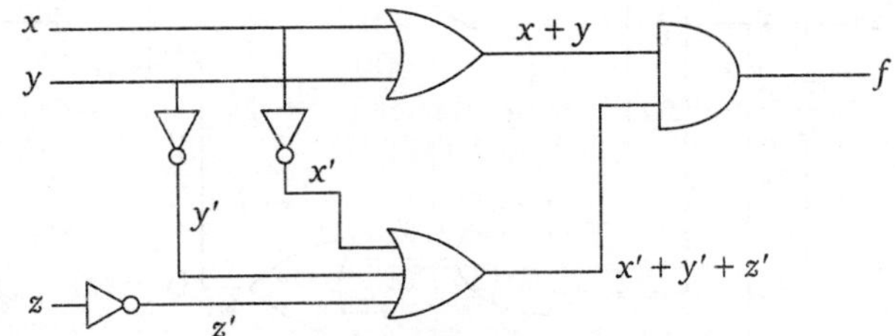

Fig. (97)

9. *Write Boolean expressions corresponding to the following logic circuits :*

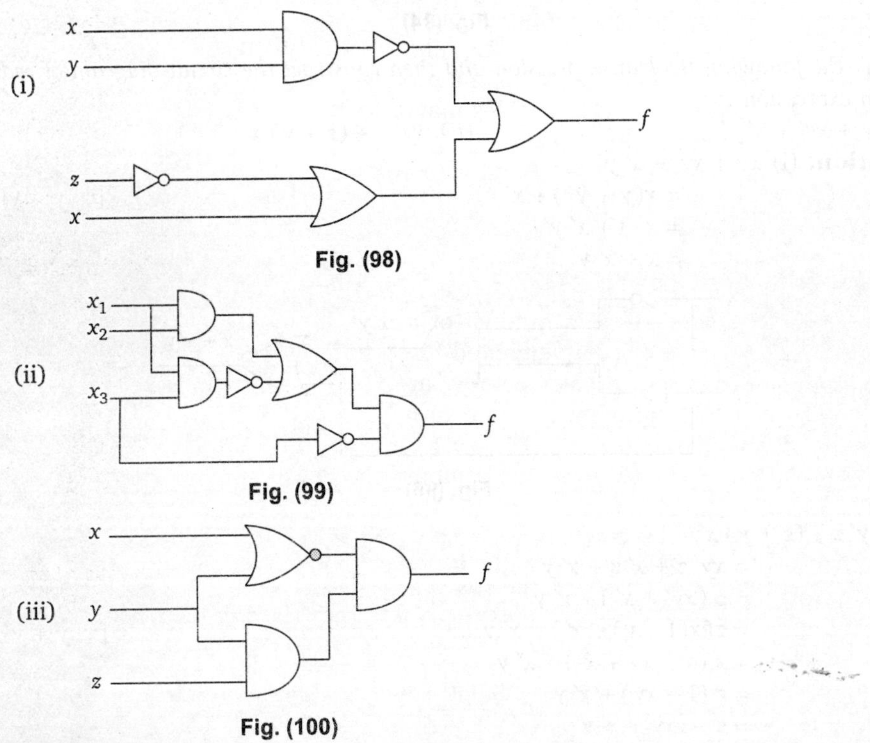

Fig. (100)

Solution. (i) $f = (xy)' + (z' + x)$ (ii) $f = \{x_1x_2 + (x_1x_3)'\} \cdot x_3'$

(iii) $f = (x + y)'\, yz.$

10. *Write a Boolean expression corresponding to the given logic circuit :*

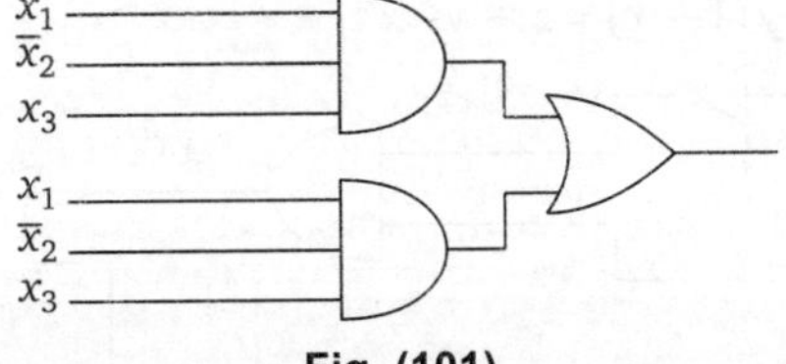

Fig. (101)

Solution. We have $x_1\bar{x}_2x_3 + x_1\bar{x}_2x_3 = x_1x_2x_3.$

11. *Find the output sequence y for an OR gate with input A, B, C (or equivalently) for $Y = A + B + C$ where,*

(i) $A = 100001, B = 100100, C = 110000$

(ii) $A = 11000000, B = 10101010, C = 00000011$

(iii) $A = 00111111, B = 11111100, C = 11000011$

Solution. (i) $Y = 100001 + 100100 + 110000$
$= 100101 + 110000 = 110101$

(ii) $Y = 11000000 + 10101010 + 00000011$
$= 11101010 + 00000011 = 11101011$

(iii) $y = 00111111 + 11111100 + 11000011$
$= 11111111 + 11000011 = 11111111.$

3.14 SUM OF PRODUCTS AND PRODUCT OF SUMS FORM

A Boolean expression E is said to be in a sum of product form if E is sum of two or more products of variables (complemented or uncomplemented), none of which is included in another. Some of the examples of this form are

1. $xz' + x'yz' + xy'z$ **2.** $abc + ac + a'bc'$

A Boolean expression E is said to be in a product of sums if it consists of several sum of terms logically multiplied. The variables may or may not be complemented. Here, are some product of sum expressions

1. $(x + y' + z)(x + z)$ **2.** $(a + b')(c' + a)$

3.15 NORMAL FORM

3.15.1 Minterm

A minterm of n variables is a product of n literals in which each variable appears exactly once in either true or complemented form, but not both. For example, the list of all the minterms of the two variables x and y are $xy, x'y, xy', x'y'$

The list of all of the minterms of three variables x, y and z are

$$xyz, xyz', xy'z, x'yz, xy'z', x'y'z, x'yz', x'y'z'$$

In a similar way, n variables can be combined to form 2^n minterms.

3.15.2 Maxterm

A maxterm of n variables is a sum of n literals in which each variable appears exactly once in either true or complicated form, but not both. For example, all maxterms of two variables x and y

$$x + y, x' + y, x + y', x' + y'$$

All maxterms of three variables x, y and z are

$x' + y' + z', x' + y' + z, x' + y + z', x + y' + z, x + y + z', x' + y + z, x + y + z$

In a similar manner, n variables forming maxterms, with each variable being complemented or uncomplemented, provide 2^n possible combinations.

The canonical sum of products form can be implemented simply by an inter connection of gates.

3.16 BOOLEAN EXPRESSION FROM LOGIC AND SWITCHING NETWORK

The Boolean expression corresponding to any logic and switching network, no matter how complex, may be completely obtained by using the Boolean operations OR, AND and NOT.

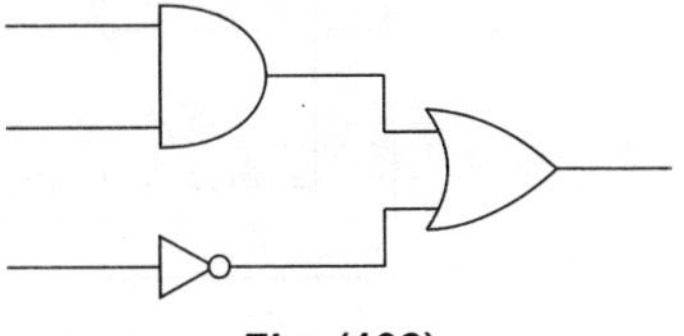

Fig. (102)

This network has three inputs A, B, C and a single output. Utilizing the Boolean expression for each gate, one can describe logic network algebraically. The expression for the AND gate output with inputs A and B is written as $A \cdot B$. The output of NOT gate with input C is written as C'. These two outputs are inputs of the OR gate so that output in the OR sum of the inputs. Thus, we can write the output of the network as

$$f = A \cdot B + C'$$

3.17 ARITHMETIC CIRCUITS

A computer's circuit can respond only to binary numbers. Arithmetic operations, such as addition, subtraction, multiplication, division, etc. are performed in the binary form in a digital computer. Logic circuits of some basic arithmetic operations are discussed in the following.

3.17.1 Half Adder

A half adder is a logic circuit that adds 2 bits. The following table shows the addition of two bits. Column 1 and 2 of table gives the values of two inputs bits, column 3 gives the sum of these two bits and column 4 the carry bit.

$$S = \overline{A}.B + A.\overline{B} = A \oplus B; C = A.B$$

Inputs		Outputs	
A	***B***	***S***	***C***
0	0	0	0
0	1	1	0
1	0	1	0
1	1	0	1

A logic circuit which uses logic gates to implement the half adder is shown in following figure :

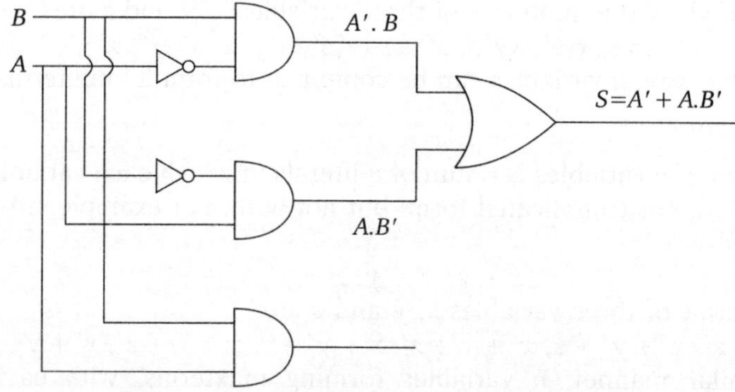

Fig. (104) : A logic circuit realising half adder

It can also be seen from above table that the sum of two binary digits can be represented by the output of an XOR gate, and the carry output can be represented by the output of an AND gate, *i.e.,* if same two inputs are applied to XOR and AND gates, the output of XOR gate will represent the sum and the AND gate will represent the carry.

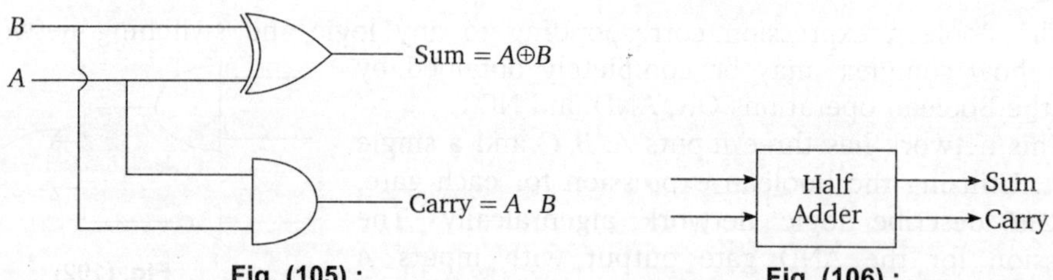

Fig. (105) : Logic circuit of half adder using XOR gate

Fig. (106) : Block diagram of half adder

3.17.2 Full Adder

When adding two binary numbers, we may have a carry from one column to the next. The carry coming out from one column is to be added to the next column. A half adder cannot add 3 bits as it has only 2 input terminals.

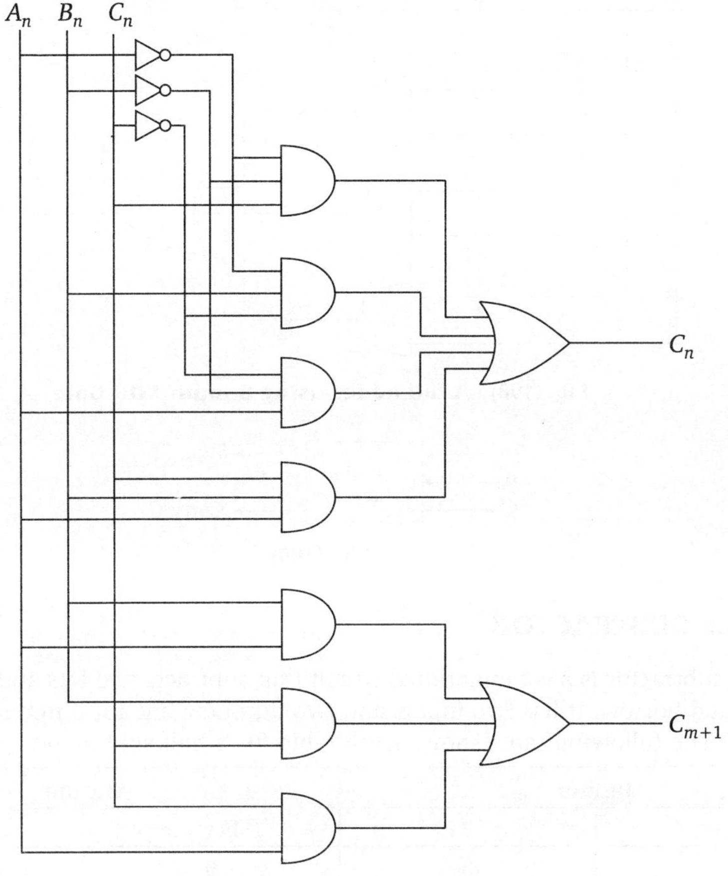

Fig. (107) : Logic circuit for full adder

A full adder is a logic circuit that can add 3 bits at a time.

Again here are two outputs; sum and carry. The following table gives the full adder truth table :

Inputs			**Outputs**	
A_n	B_n	C_n (Carry in)	S_n	C_{n+1} (Carry out)
0	0	0	0	0
0	0	1	1	0
0	1	0	1	0
0	1	1	0	1
1	0	0	1	0
1	0	1	0	1
1	1	0	0	1
1	1	1	1	1

Full adder truth table

The logic circuit for the full adder is shown in figure :

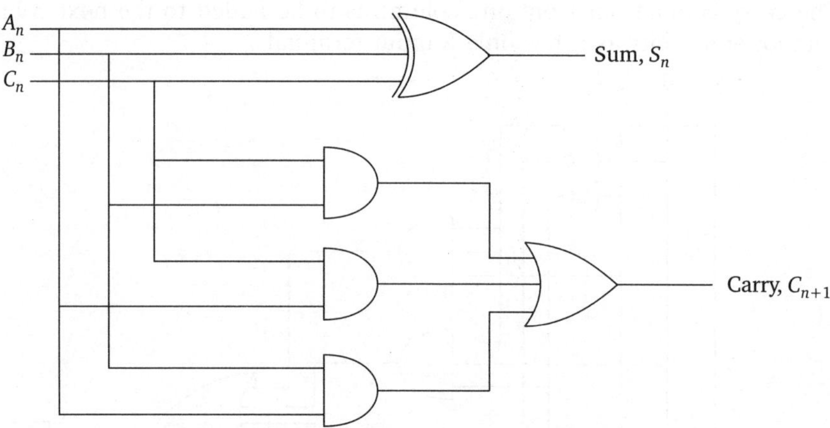

Fig. (108) : A full adder using 3 input *XOR* Gate

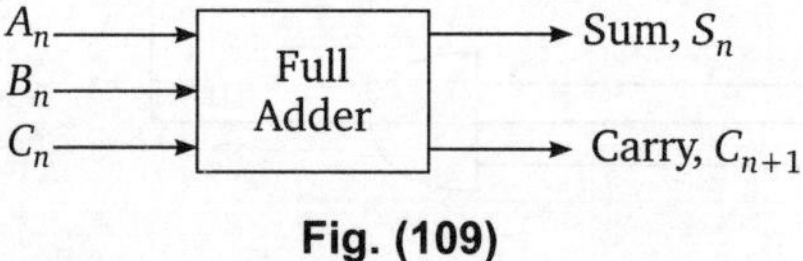

Fig. (109)

3.18 HALF SUBTRACTOR

A half subtractor is a combinational circuit that subtracts two bits and produces their difference and borrow. It has two inputs and two outputs : one for difference and another for borrow. The following table shows truth table for a half subtractor.

Inputs		**Output**	
A	*B*	Difference	Borrow
0	0	0	0
0	1	1	1
1	0	1	0
1	1	0	0

Boolean expression for difference and borrow can be written as

$$\text{Difference} = \overline{A}B + A\overline{B} = A \oplus B \text{ and Borrow} = \overline{A} \cdot B$$

The following figure shows the logic diagram for a half subtractor and the block diagram for a half subtractor.

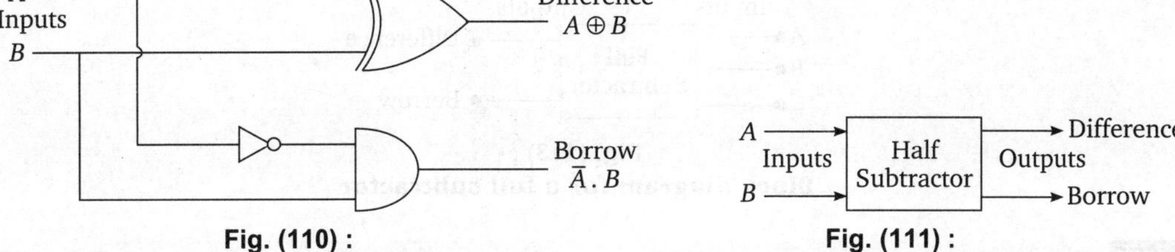

Fig. (110) : Logic circuit for a half subtractor

Fig. (111) : Block diagram for a half subtractor

3.19 FULL SUBTRACTOR

A full subtractor is a combinatiorial circuit that performs a subtraction between two bits, taking into account that a 1 may have been borrowed by a lower significant stage. It has three inputs and two outputs. One for difference and another for borrow. The following table shows the truth table for a full subtractor.

Inputs			**Output**	
A	***B***	***C***	**Difference**	**Borrow**
0	0	0	0	0
0	0	1	1	1
0	1	0	1	1
0	1	1	0	1
1	0	0	1	0
1	0	1	0	0
1	1	0	0	0
1	1	1	1	1

Truth Table for a full subtractor

Boolean expressions difference and borrow can be written as

Difference = $\overline{A}\,\overline{B}C + \overline{A}B\overline{C} + A\overline{B}\,\overline{C} + ABC$

Simplified expression $\Rightarrow$ Difference = $A \oplus B \oplus C$

Borrow = $\overline{A}\,\overline{B}C + \overline{A}B\overline{C} + \overline{A}BC + ABC$

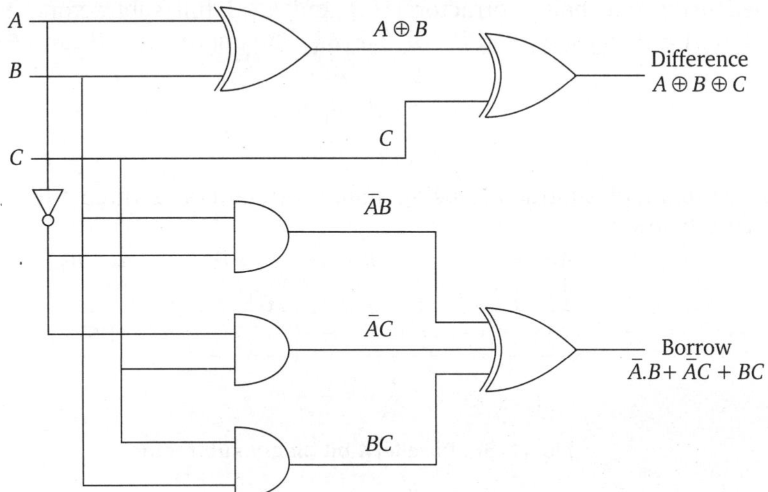

Fig. (112) : Logic circuit for a full subtractor

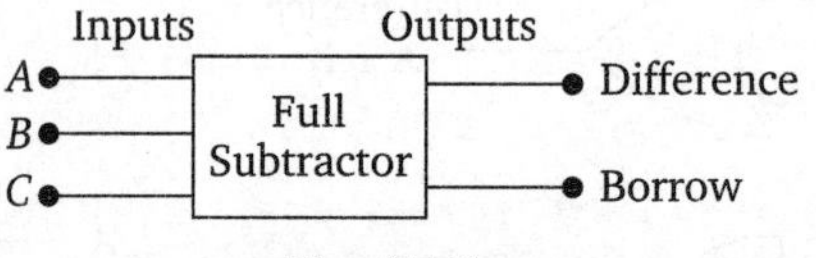

Fig. (113) :
Block diagram for a full subtractor

3.20 BINARY ADDER

A binary adder is a logic circuit that can add two binary numbers. The addition of two binary numbers each of k bits can be accomplished using one half adder (HA) and k-1 full adder (FA). The right most block represents a half adder. Each of the full adders has three inputs and two outputs. The carry output of each adder goes to carry input of the next adder to the left. Consider the example $k = 4$ and two binary numbers to be added $A_3A_2A_1A_0$ and $B_3B_2B_1B_0$.

The answer is :

$$\begin{array}{r} A_3\ A_2\ A_1\ A_0 \\ +\quad B_3\ B_2\ B_1\ B_0 \\ \hline C_4\ S_3\ S_2\ S_1\ S_0 \end{array}$$

The following figure shows a parallel 4 bit binary adder :

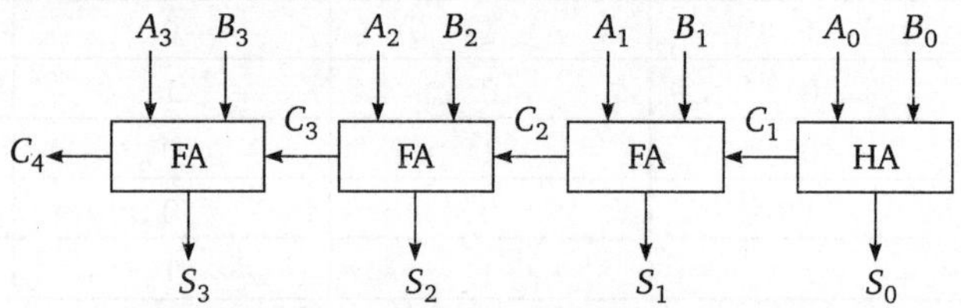

Fig. (114) : Parallel 4 bit binary adder

3.21 BINARY SUBTRACTOR

A subtraction of one k bit binary number B from the other k bit binary number A can be performed using one half subtractor (HS) and $k - 1$ full subtractors (FSs). Let, for example, $k = 4$, $A = A_3A_2A_1A_0$ and $B = B_3B_2B_1B_0$. The answer $A - B$ can be written as

$$\begin{array}{r} A_3\ A_2\ A_1\ A_0 \\ -\quad B_3\ B_2\ B_1\ B_0 \\ \hline D_3\ D_2\ D_1 D_0 \end{array}$$

A four bit parallel subtractor, using half subtractor and three full subtractors is shown in figure below :

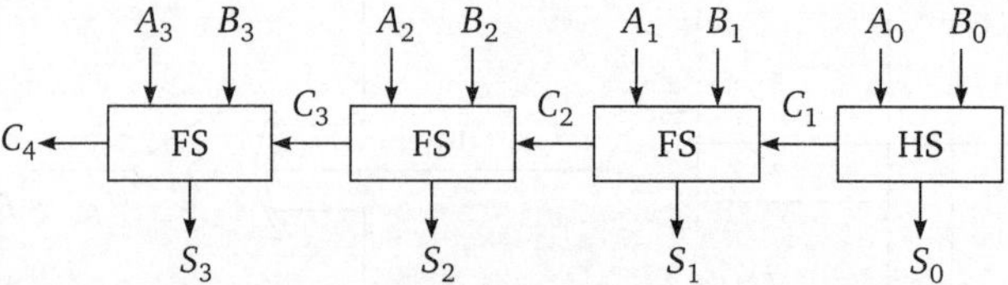

Fig. (115) : Parallel 4 bit binary subtractor

Exercise 3.2

1. In each case, identify the gate whose function is being simulated :

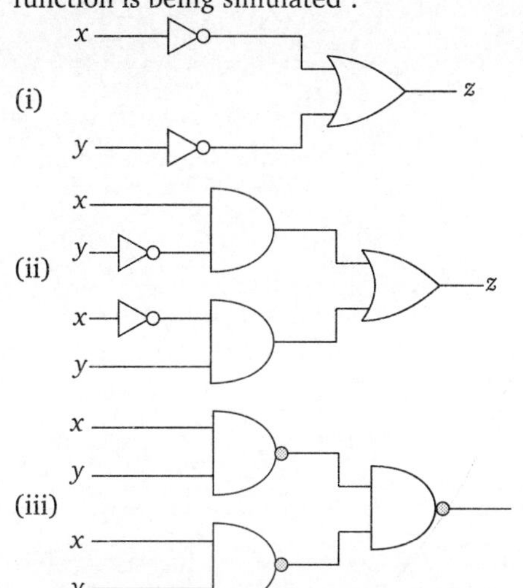

2. Use NAND gate to construct circuits for the following outputs :
(a) x' (ii) $x + y$ (c) xy (d) $x \oplus y$

3. Write Boolean expressions corresponding to the following logic circuits :

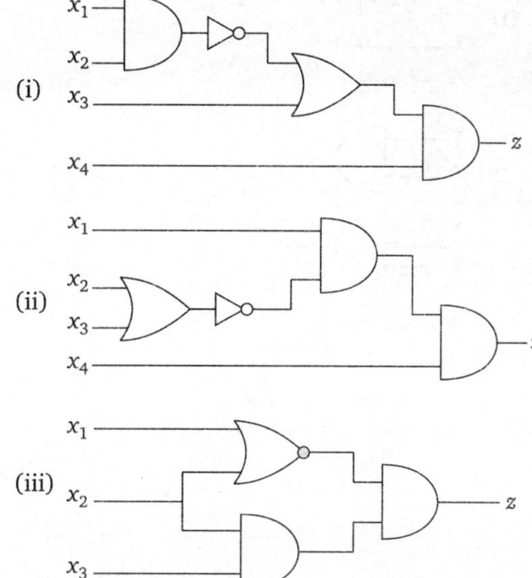

4. Draw logic diagram to represent the Boolean expressions :
(i) $(x_1'x_2')' + x_1x_2$
(ii) $(xy)' + xyz$
(iii) $(x_1 + x_2)\,x_3 + x_4$

ANSWERS

1. (i) $Z = x' + y'$ (ii) $xx' + xy'$ (iii) $\overline{(x.\,y)}\,\overline{(x.\,y)}$

2. (i) x — x'

(ii) x, y — x+y

(iii) x, y — xy

(iv)

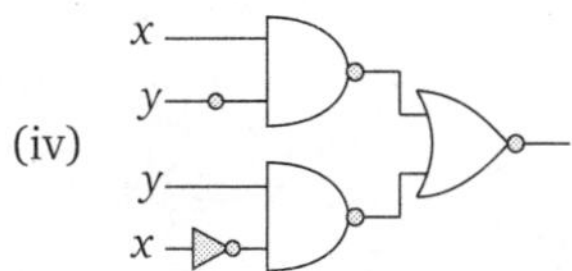

3. (i) $(x_2y_1) + x_3)\,x_4$ (ii) $[x_1\,\overline{(x_2 + x_3)}]\,x_4$
(iii) $\overline{(x_1 + x_2)}\,(x_3x_2)$

4. (i)

(ii)

(iii)

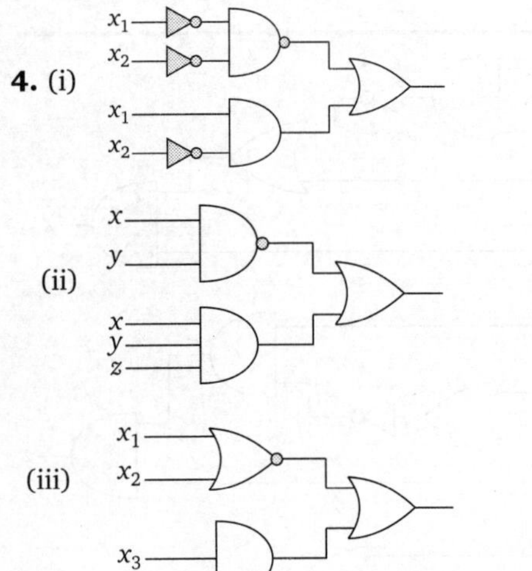

❑❑❑

Chapter 4

Lattice Theory

- Partial order relation
- Well ordered set
- Poset
- Lattice
- Hasse diagram
- Type of lattices
- Homomorphism of lattices
- Modular lattice
- Lattice as algebraic system

4.1 INTRODUCTION

In this chapter, we introduce the concept of lattice as a partially ordered set satisfying certain conditions. Lattice theory have important applications in the theory and design of computers. There are many other areas such as engineering science to which lattice theory is applied.

4.2 PARTIAL ORDER RELATION

Let R be a relation, then R is called partial order relation if

(i) *R is reflexive, i.e.,* $aRa \ \forall \ a \in A$

(ii) *R is antisymmetric, i.e.,* $aRb, bRa \Rightarrow a = b \ \forall \ a, b \in A$

(iii) *R is transitive, i.e.,* $aRb, bRc \Rightarrow aRc \ \forall \ a, b, c \in A$.

For example :

(i) The relation of divisibility (aRb if and only if a / b) is a partial order in $\mathbf{Z}^+$.

(ii) Let $\mathbf{Z}^+$ be the set of integers. The relation $\leq$ (less than or equal to) is a partial order on $\mathbf{Z}^+$. Therefore $(\mathbf{Z}^+, \leq)$ is a poset.

4.2.1 Poset

A set S together with a partial ordering R is called a partially ordered set or poset.

☛ A poset is also called partially ordered set.

4.2.2 Comparability

Two elements a and b in a poset $(S, \leq)$ are called comparable if either $a \leq b$ or $b \leq a$.

4.2.3 Linearly Ordered Set

If every two elements of a poset $(S, \leq)$ are comparable, then set S is called a linearly ordered set and the relation $\leq$ is called a total ordered or linear ordered relation.

☛A linearly ordered set is also called totally ordered set or chain.

4.2.4. Well Ordered Set.

A poset $(S, \leq)$ is called well ordered set if it is totally ordering and every non-empty subset of S has a least element.

- Every well ordered set is totally ordered, because for any subset say $\{a, b\}$ we must have either a or b as its least number. Also, every totally ordered set need not be well ordered
- A finite totally ordered set is also well ordered.

4.2.5 Lower and Upper Bounds

Let $(S, \leq)$ be a poset and let $a, b \in S$. We say that $x \in S$ is a lower bound for a and b provided $x \leq a$ and $x \leq b$. Similarly, we say that $x \in S$ is an upper bound for a and b provided $a \leq x$ and $b \leq x$.

Theorem 1. *In a Boolean algebra* $[B, +, .]$

(i) $a + b = l.u.b\{a, b\}$ (ii) $a.b = g.l.b.\{a, b\}$

Proof. Let $[B, +, .]$ be a Boolean algebra

(i) Since for all $a, b \in B$ we have

$$a + (a + b) = (a + a) + b \quad \text{(By associativity)}$$
$$= a + b \quad (\because a + a = a)$$

Therefore

$$a + (a + b) = a + b \quad \Rightarrow \quad a \leq (a + b) \quad ...(1)$$

Similarly, $b + (a + b) = (b + a) + b$ (By associativity)

$= (a + b) + b$ (By commutativity)

$= a + (b + b)$ (By associativity)

$= a + b$ $(\because b + b = b)$

$\therefore \quad b + (a + b) = a + b \quad \Rightarrow \quad b \leq a + b$ ...(2)

From (1)) and (2), we conclude that $(a + b)$ is the upper bound of a and b. Now let c be any other bound of a and b then

$$a \leq c \text{ and } b \leq c$$

$\Rightarrow \quad a + c = c$ and $b + c = c$

Now $(a + b) + c = a + (b + c)$ (By associativity)

$= a + c$ $(\because b + c = c)$

$= c$ $(\because a + c = c)$

$\therefore \quad a + b \leq c$

$\Rightarrow \quad (a + b)$ is the least upper bound of a and b

$\Rightarrow \quad a + b = l.u.b.\{a, b\}$

(ii) In a similar manner, we have

$a + a.b = a$ and $b + ab = b$ (By absorption law)

$\therefore \quad a.b \leq a$ and $a.b \leq b$

$\Rightarrow \quad a.b$ is the lower bound of a and b.

Let c be some other lower bound of a and b then

$c \leq a, c \leq b \quad \Rightarrow \quad c + a = a.c + b = b$

Now. $c + a.b = (c + a)(c + b)$ (By distributivity)

$= a.b$ $(\because c + a = a \;;\; c + b = b)$

$\Rightarrow \quad c \geq a.b$

$\Rightarrow \quad a.b$ is the greatest lower bound of a and b.

$\Rightarrow \quad a.b = g.l.b\{a, b\}$.

Theorem 2. *A poset has atmost one greatest and one least element.*

Proof. Let x and y be two greatest elements of a poset S.

Since y is greatest $\Rightarrow x \leq y$

similarly, since x is greatest $\Rightarrow y \leq x$. Therefore $x = y$

$\Rightarrow$ If the poset has a greatest element, it only one such element further, since this fact is true for al poset, the duel poset $(s, \geq)$ has atmost one greatest element.

⇒ $(S, \leq)$ also has at most one least element.

☛ Using he above theorem, we can say that "If $(S, \leq)$ is a poset, then a subset R of S has atmost one *l.u.b* and atmost one *g.l.b*.

4.2.6 Greatest Lower Bound and Least Upper Bound

Let $(S, \leq)$ be a poset and let $a, b \in S$. We say that $x \in S$ is a greatest lower bound (Infimum) for a and b provided

(a) x is lower bound for a and b and

(b) If y is a lower bound for a and b, then $y \leq x$.

Similarly, we say that $x \in S$ is a least upper bound for a and b, provided.

(a) x is an upper bound for a and b, and

(b) If y is an upper bound for a and b, then $x \leq y$.

☛ Greatest lower bounds and least upper bounds, if they exist, are unique.

The concept of maximum and minimum are similar to but not the same as minimal and maximal

Term	Meaning
Maximum Maximal	All other elements are below No other element is above
minimum minimal	All other elements are above No other elements is below.

For Example : Consider the poset given in Figure 1 :

(a) Consider elements 8 and 9. Observe that 1, 2 and 5 are upper bounds for 8 and 9 . Since $5 < 1$ and $5 < 2$, we have that 5 is the least upper bound of 8 and 9. On the other hand, 8 and 9 have no lower bounds and consequently no greatest lower bound.

(b) Elements 1 and 2 have lower bounds 5, 8, 9, 10, 11 and 12. The greatest lower bound of 1 and 2 is 5. Element 1 and 2 have no upper bounds and consequently no least upper bound.

(c) Elements 5 and 6 have 2 as the least (and only) upper bound. They have incomparable lower bounds 9 and 11, so they do not have a greatest lower bound.

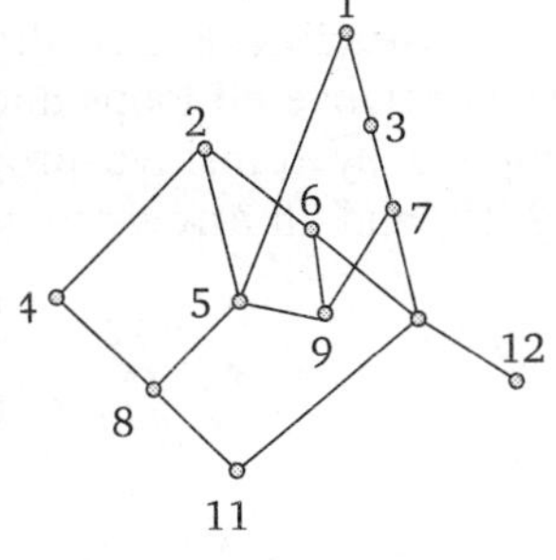

Fig. (1)

(d) Element 9 and 10 have no greatest lower bound and no least upper bound.

(e) Element 4 and 5 have 2 as their least upper bound and 8 as their greatest lower bound.

4.2.7 Meet and Join

Let $(S, \leq)$ be a poset and let $a, b \in S$. If a and b *have a greatest lower bound, it is called the meet of* a and b *and it is denoted by* $a \wedge b$. *If* a and b *have a least upper bound, it is called the join of* a and b *and is denoted by* $a \vee b$.

☛ We use the symbols $\wedge$ and $\vee$ for the meet and join operations because $\wedge$ is abstraction of $\cap$ and $\vee$ is an abstraction of $\cup$.

☛ The symbols $\wedge$ and $\vee$ also stand for the Boolean operations of "and" and "or". To make the similarity of these notation consider the poset P whose ground set is {TRUE, FALSE}. We make the mathematical (also ethical) decision to place truth above false, *i.e.*, we have FALSE < TRUE. In this poset, we have the following Fig. 2. In this poset, we have $T \wedge F = F$ because FALSE is the greatest (and only) lower bound for TRUE and FALSE. Indeed all the following are true :

$T \wedge T = T; T \wedge F = F; F \wedge T = F; F \wedge F = F;$

$T \vee T = T; T \vee F = T; F \vee T = T; F \vee F = F.$

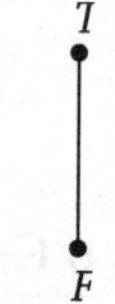

Fig. (2)

- Therefore, the operation $\wedge$ and $\vee$ on $\{T, F\}$ are exactly the same whether we interpret them as **and** and **or** as **meet** and **join.**
- For example, in Fig. 1, the results can be expressed as follows :
  - (i) $8 \wedge 9$ is undefined and $8 \vee 9 = 5$.
  - (ii) $1 \wedge 2 = 5$ and $1 \vee 2 = 5$.
  - (iii) $5 \wedge 6$ is undefined and $5 \vee 6 = 2$.
  - (iv) Bot $9 \wedge 10$ and $9 \vee 10$ are undefined.
  - (v) $4 \wedge 5 = 8$ and $4 \vee 5 = 2$.

4.3 HASSE DIAGRAM

Let S be a partially ordered set and let $a, b \in S$. We say that a is an immediate predecessor of b or b is an immediate successor of a, if $a < b$ but no element of S lies between a and b. It is written as $a << b$.

The Hasse diagram of a finite partially ordered set S is the directed graph whose vertices are the elements of S and there is a directed edge from a to b whenever a << b in S.

- In place of drawing an arrow from a to b, sometimes we place b higher than a and draw line between them. Hasse diagram of a poset S is a picture of S.
- The Hasse diagram of a finite poset S is a directed cycle-free graph.

For Example : Draw the Hasse diagram representing the partial ordering a/b on $S = \{1, 2, 3, 4, 6, 8, 12\}$.

Solution. Firstly, draw the graph for the given partial ordering as given in Figure 3. Then remove all loops and get the graph shown in Figure 4. Finally delete all the edges implied by transitive property which are (1, 4), (1, 6), (1, 8), (1, 12), (2, 8), (2, 12) and (3, 12) and all arrows to get the Hasse diagram as shown in Figure 5.

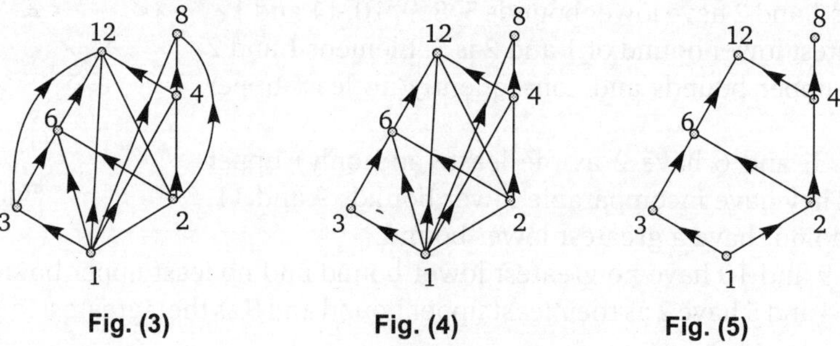

Fig. (3) Fig. (4) Fig. (5)

4.4 LATTICE

A lattice is a poset $(S, \leq)$ in which every pair of elements has a supremum and an infimum. Here we can write $a \wedge b = \inf (\{a, b\})$; $a \vee b = \sup (\{a, b\})$

- All posets are not lattice.
- A lattice can also be defined as follows : "A poset S is said to be lattice if for all elements x and y of S, $x \wedge y$ and $x \vee y$ are defined.

For Example :

(1) Let S be the poset in Figure 6. The $\wedge$ and $\vee$ operation table are given as well.

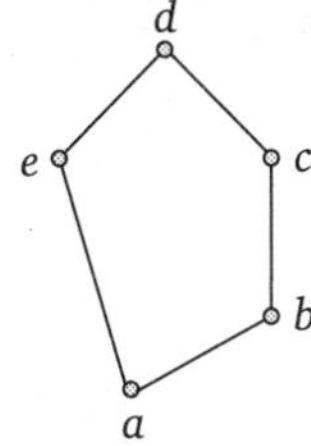

$\wedge$	**a**	**b**	**c**	**d**	**e**
a	a	a	a	a	a
b	a	b	b	b	b
c	a	b	c	c	c
d	a	b	c	d	e
e	a	b	c	d	e

$\vee$	**a**	**b**	**c**	**d**	**e**
a	a	b	c	d	e
b	b	b	c	d	e
c	c	c	c	d	e
d	d	d	d	d	d
e	e	e	e	d	e

Fig. (6)

Since $\wedge$ and $\vee$ are defined for every pair of elements, this poset is a lattice.

(2) Let A be a set and let P is the poset of all subsets of A ordered by containment. In this poset, we have for all $x, y \in P$

$$x \wedge y = x \cap y \text{ and } x \vee y = x \cup y$$

Hence, P is a lattice.

(3) Consider the poset (**Z**, /) (*i.e.*, the set of integers ordered by divisibility). Let $x, y \in \mathbf{Z}$. Then $x \wedge y$ is the greatest common divisor of x and y and $x \vee y$ is their least common multiple. However, (**Z**, /) is not a lattice, because $0 \wedge 0 = \gcd(0, 0)$ is not defined. However, the poset $(\mathbf{Z}^+, /)$ is a lattice. In this case, $\wedge$ and $\vee$ (gcd and lcm) are defined for all pairs of positive integers. Therefore, $(\mathbf{Z}^+, /)$ is a lattice. (This is the example of poset, which is not lattice).

(4) Let $P(S, \leq)$ be a linear order. Notice that for any $x, y \in S$

$$x \wedge y = \begin{cases} x & \text{if } x \leq y \\ y & \text{if } x \geq y \end{cases}$$

We can rewrite this as $x \wedge y = \min\{x, y\}$, where $\min\{x, y\}$ stands for the smaller of x or y.

Similarly, $x \vee y = \max\{x, y\}$ (*i.e.*, the larger of the pair).

4.4.1 Lattices as Algebraic System

A lattice is an algebraic system $(P, \wedge, \vee)$ with two binary operation $\wedge$ and $\vee$ on P, which are both (i) Commutative, (ii) Associative and satisfying (iii) Absorption law, *i.e.*,

$$a \wedge (a \vee b) = a;\ a \vee (a \wedge b) = a$$

☛ The advantage of defining a lattice as an algebraic system is that many concepts which are associated with algebraic system can be used to lattice as well.

4.4.2 Equivalence of two definitions of Lattice

Part 1 : Definition (I). (As posets) $\Rightarrow$ definition (II) (as algebraic system)

Using definition-I, If L is a lattice then for any $a, b \in L$ infimum $\{a. b\} = a \wedge b$ and sup. $\{a. b\} = a \vee b$ exists and unique in L. therefore $\wedge$ and $\vee$ are binary composition on L. Therefore all the three conditions of lattice as algebraic system (i.e., commutative, associative and absorption property) hold for $(L, \wedge, \vee)$

Hence, definition of lattice as poset implies the definition of a lattice as algebra.

Part II. Definition (II) $\Rightarrow$ Definition (I).

Let L be a lattice as algebraic system. Firstly we shall prove that L forms a poset

Define a relation $\leq$ on L such that for all $a, b \in L$

$$a \leq b \quad \Leftrightarrow \quad a \wedge b = a$$

We shall prove that relation $\leq$ is a partial order relation.

(i) $\leq$ is reflexive : For any $a \in L$, we have

$$a \wedge a = a \quad \Rightarrow \quad a \leq a$$
$$\Rightarrow \quad \leq \text{ is reflexive.}$$

(ii) $\leq$ is Antisymmetric : For any $a, b \in L$, suppose that

$a \leq b$ and $b \leq a$

$\Rightarrow$ $a \wedge b = a$ and $b \wedge a = b$

But $a \wedge b = b \wedge a \quad \Rightarrow \quad a = b$

$\Rightarrow$ $\leq$ is antisymetric

(iii) $\leq$ is transitive : Let us suppose that $a \leq b$ and $b \leq c$

$\Rightarrow$ $a \wedge b = a$ and $b \wedge c = b$

$\Rightarrow$ $a \wedge c = (a \wedge b) \wedge c$

$= a \wedge (b \wedge c) = a \wedge b = a$

$\Rightarrow$ $a \leq c$

$\Rightarrow$ $\leq$ is transitive. Hence, $\leq$ is a partial order relation on L.

Now, we show that inf $\{a, b\}$ exist in L

for this we shall prove that

$$\text{inf. } \{a, b\} = a \wedge b$$

We have

$(a \wedge b) \wedge a = a \wedge (b \wedge a)$ (By associativity)

$= a \wedge (a \wedge b)$ (By commutativity)

$= (a \wedge a) \wedge b$ (By associativity)

$= a \wedge b$ (By idempteal law)

$\Rightarrow$ $a \wedge b \leq a$ (By definition of $\leq$)

Similarly $a \wedge b \leq b$

$\Rightarrow$ $a \wedge b$ is a lower bound of $\{a, b\}$

Further, we shall show that $a \wedge b$ is $g.l.b$ of $\{a, b\}$. Let if possible c be any lower bound of $\{a, b\}$. Then $c \leq a$ and $c \leq b$

$\Rightarrow$ $c \wedge a = c$ and $c \wedge b = c$

$\therefore$ $c \wedge (a \wedge b) = (c \wedge a) \wedge b = c \wedge b = c$

$\Rightarrow$ $c \leq a \wedge b$

$\Rightarrow$ $a \wedge b = g.l.b\ \{a, b\} = \text{inf. } \{a, b\}$

Now, we show that $a \leq b \Leftrightarrow a \wedge b = a \Leftrightarrow a \vee b = b$

We have $a \wedge b = c \quad \Rightarrow \quad (a \wedge b) \vee b = a \vee b$

$\Rightarrow$ $b = a \vee b$

Further $a \vee b = b \Rightarrow a \wedge (a \vee b) = a \wedge b \Rightarrow a = a \wedge b$

$a \wedge b = a \Leftrightarrow a \vee b = b$

$\Rightarrow$ $a \leq b \Leftrightarrow a \wedge b = a \Leftrightarrow a \vee b = b$

Hence, by duality we can say that sup. $\{a, b\}$ will be $a \vee b \Rightarrow L$ is a lattice as poset.

4.4.3 Principle of Duality

It states that for any valid statement concerning the properties of lattices we can obtain another valid statement by replacing the relation $\leq$ with $\geq$, the term join operation, with the term meet operation and the meet operation with the term join operation.

Theorem 1. *Let $(L, \leq)$ be a lattice under two binary operations* $\vee$ and $\wedge$. *Let $a, b, c \in L$, then*

(i) $a \vee a = a, a \wedge a = a$ (Idempotent law)

(ii) $a \vee b = b \vee a, a \wedge b = b \wedge a$ (Commutativity)

(iii) $a \vee (b \vee c) = (a \vee b) \vee c, a \wedge (b \wedge c) = (a \wedge b) \wedge c$ (Associativity)

(iv) $a \vee (a \wedge b) = a, a \wedge (a \vee b) = a$ (Absorption law)

Proof : (i) Since we know that $a \leq a \vee a$ ...(1)
Also, if $a \leq a$, we find that $a \vee a \leq a$ ...(2)
Therefore, from (1) and (2), we have $a \vee a = a$.
Similarly, we can show that $a \wedge a = a$.

(ii) Since we know that if $a \wedge b \leq a$ and $a \wedge b \leq b$, then $a \wedge b$ is a lower bound of a and b and if $c \leq a$ and $c \leq b$, then $c \leq a \wedge b$; then $a \wedge b$ is greatest lower bound of a and b. Therefore, we find that a and b are considered symmetric by concept behind l.u.b. and g.l.b. Hence the result.

(iii) Let us suppose $a \vee (b \vee c) = e$ and $(a \vee b) \vee c = f$
Since e is the join of a and $b \vee c$, therefore $a \leq e, b \vee c \leq e$.
Further, we find that $b \leq e, c \leq e$. Also $a \vee b \leq e$.
As from above $a \leq g, b \leq g$. Combining $a \vee b \leq g$ with $c \leq g$, we obtain $(a \vee b) \vee c \leq e$
or $\quad f \leq e$ ...(1)
Similarly, starting with $(a \vee b) \vee c = f$ and proceed as above, we can show that $e \leq f$ ...(2)
From (1) and (2), w conclude that $e = f$
i.e., $\quad (a \vee b) \vee c = a \vee (b \vee c)$
Similarly, we can show that $(a \wedge b) \wedge c = a \wedge (b \wedge c)$

(iv) From the expression $a \vee (a \wedge b) = a$, we find that a is the join of a and $(a \wedge b)$.
Therefore $\quad a \leq a \vee (a \wedge b) =$ l.u.b. $\{a, (a \wedge b)\}$ ...(1)
Now, since $a \leq a$, $a \wedge b \leq a$, therefore $a \vee (a \wedge b) \leq a \vee a$
$$a \vee (a \wedge b) \leq a \qquad [\because a \vee a = a] \qquad ...(2)$$
From (1) and (2), we conclude that $a \vee (a \wedge b) = a$.
Similarly, we can prove that $a \wedge (a \vee b) = a$.

Theorem 2. *Let L be a lattice. For any $a, b \in L$, $a \leq a \vee b$ and $a \wedge b \leq a$.*

Proof : Since we know that the join of a and b is an upper bound of a, therefore $a \leq a \vee b$. Similarly, meet of a and b is the lower bound of a, therefore, we have $a \wedge b \leq a$.

Theorem 3. *Let L be a lattice and let $a, b, c, d \in L$. If $a \leq b$ and $c \leq d$. Then $a \vee c \leq b \vee d$ and $a \wedge c \leq b \wedge d$.*

Proof : Since we know that $b \leq b \vee d$ and $d \leq b \vee d$.
Then, by transitive property $a \leq b \vee d$ and $c \leq b \vee d$
($\because$ It is given that $a \leq b$ and $c \leq d$)
$\Rightarrow \quad b \vee d$ is an upper bound of a and c.
An $a \vee c$ is the l.u.b $\{a, c\}$, we have $a \vee c \leq b \vee d$.
Also, since $a \wedge c \leq a$ and $a \wedge c \leq c$, therefore, again by transitivity
$$a \wedge c \leq b \text{ and } a \wedge c \leq d \qquad (\because a \leq b \text{ and } c \leq d)$$
$\Rightarrow \quad a \wedge c$ is a lower bound of b and d.
But since $b \wedge d$ is the g.l.b. of b and d, therefore, we have $a \wedge c \leq b \wedge d$.

Theorem 4. *Let L be a lattice. Then for any $a, b \in L$, $a \leq b \Leftrightarrow a \wedge b = a \Leftrightarrow a \vee b = b$.*

Proof. Firstly, we shall show that $a \leq b \Leftrightarrow a \wedge b = a$.
Assume that $a \leq b$. Since we have $a \leq a$, therefore $a \leq a \wedge b$ ...(1)
But by definition of $a \wedge b$, we have $a \wedge b \leq a$ ...(2)
From (1) and (2), we conclude that $a \wedge b = a$
which implies $\quad a \leq b \Rightarrow a \wedge b = a$ ...(3)
Conversely, let us assume $a \wedge b = a$ ...(4)
By definition, it is possible only when $a \leq b$.
Combining (3) and (4), we get
$$a \leq b \Leftrightarrow a \wedge b = a$$
Similarly, we can show that $a \leq b \Leftrightarrow a \vee b = b$

Put $a = a \wedge b$, we have $b \vee (a \wedge b) = b \vee a = a \vee b$

But $b \vee (a \wedge b) = b$

Hence, $a \vee b = b$ follows from $a \wedge b = a$. Similarly, we may show that $a \wedge b$ follows from $a \vee b = b$. Hence the result.

Theorem 5. *Let L be a lattice and a, b, c ∈ L. Then* $b \leq c \Leftrightarrow \begin{cases} (i)\ a \wedge b \leq a \wedge c \\ (ii)\ a + b \leq a + c \end{cases}$.

Proof : (i) If $b \leq c$, then from previous theorem $b \leq c \Leftrightarrow b \wedge c = b$.

To show that $a \wedge b \leq a \wedge c$, we would show that $(a \wedge b) \wedge (a \wedge c) = a \wedge b$.

This can be easily shown by the following

$$(a \wedge b) \wedge (a \wedge c) = (a \wedge a) \wedge (b \wedge c) = a \wedge (b \wedge c) = a \wedge b$$

(ii) Do as (i).

Theorem 6. *If the meet operation is distributive over the join operation in a lattice, then the join operation is also distributive over the meet operation.*

Proof : Since the meet operation is distributive, therefore, we have

$a \wedge (b \vee c) = (a \wedge b) \vee (a \wedge c)$.

Consider $(a \vee b) \wedge (a \vee c) = [(a \vee b) \wedge a] \vee [(a \vee b) \wedge c]$

$$= a \vee [(a \vee b) \wedge c]$$
$$= a \vee [(a \wedge c) \vee (b \wedge c)]$$
$$= [a \vee (a \wedge c)] \vee (b \wedge c)$$
$$= a \vee (b \wedge c)$$

$\Rightarrow \quad a \vee (b \wedge c) = (a \vee b) \wedge (a \vee c)$

Hence, the join operation is distributive over the meet operation.

☛ In a similar manner, we may show that "If the join operation is distributive over the meet operation, then the meet operation is also distributive over the join operation".

Theorem 7. *Let (L, ≤) be a lattice and a, b, c ∈ L. Then the following implications hold :*

(i) $a \leq b$ *and* $a \leq c \Rightarrow a \leq b \vee c$ (ii) $a \leq b$ *and* $a \leq c \Rightarrow a \leq b \wedge c$

Proof : (i) We know from the definition of join operation that $b \vee c$ is the supremum of b and c.

Hence, $b \leq b \vee c$.

Thus, $a \leq b$ and $b \leq b \vee c \Rightarrow a \leq b \vee c$ by transitivity.

(ii) Suppose $a \leq b$ and $a \leq c$. $\Rightarrow$ a is a lower bound of $\{b, c\}$.

$\Rightarrow \quad a \leq b \wedge c$, the greatest lower bound of $\{b, c\}$.

☛ Similarly, we can prove the following results.

Let $(L, \leq)$ be a lattice and $(L \geq)$ be its dual. Then for $a, b, c \in L$

(i) $a \geq b$ and $a \geq c \Rightarrow a \geq b \wedge c$

(ii) $a \geq b$ and $a \geq c \Rightarrow a \geq b \vee c$

Theorem 8. *Let (L, ≤) be a lattice. For any a, b, c ∈ L, then following implications hold :*

(i) $a \wedge (b \vee c) \geq (a \wedge b) \vee (a \wedge c)$ (ii) $a \vee (b \wedge c) \leq (a \vee b) \wedge (a \vee c)$

Proof : (i) We know that $a \wedge b \leq a$ and $a \wedge b \leq b \leq b \vee c$

$\Rightarrow \quad a \wedge b$ is a lower bound of $\{a, b \vee c\} \Rightarrow a \wedge b \leq a \wedge (b \vee c)$ ...(1)

because $a \wedge (b \vee c)$ is the greatest lower bound of $\{a, b \vee c\}$

Again, $a \wedge c \leq a$ and $a \wedge c \leq c \leq b \vee c \Rightarrow a \wedge c \leq a \wedge (b \vee c)$ ...(2)

From (1) and (2), $a \wedge (b \vee c)$ is an upper bound of $\{a \wedge b, a \wedge c\}$

$\Rightarrow \quad (a \wedge b) \vee (a \wedge c) \leq a \wedge (b \vee c)$

(ii) This inequality can be proved in a similar manner or using the principle of duality. If we apply the principle of duality, we immediately get

$$a \vee (b \wedge c) \leq (a \vee b) \wedge (a \vee c)$$

by interchanging $\vee$ and $\wedge$ and replacing $\leq$ by $\geq$ in (i).

☛ The inequalities in the above theorem are called **semi-distributive laws.**

Theorem 9 *Let $(L, \leq)$ be a lattice. For any a, b, $c \in L$, the following holds :*

$$a \leq c \Leftrightarrow a \vee (b \wedge c) \leq (a \vee b) \wedge c.$$

This is known as the modular inequality.

Proof : We know from Theorem (4) that

$$a \leq c \Leftrightarrow a \vee c = c$$

substituting c for $a \vee c$ in the RHS of the inequality given in part (ii) of above theorem, we get

$$a \leq c \Rightarrow a \vee (b \wedge c) \leq (a \vee b) \wedge c$$

Theorem 10. *Dual of a lattice is a lattice.*

Proof : Let (L, R) be a given lattice and let $(L, \bar{R})$ be its dual, where $\bar{R}$ is defined as $x \bar{R} y$ iff $y R x$. Then, it can be shown easily that $(L, \bar{R})$ is a poset.

Let $x, y \in L$, then sup $\{x, y\}$ exists in R [$\because$ (L, R) is a lattice]

Let $x \vee y = \sup \{x, y\}$ in (L, R). Then we have $x R (x \vee y)$ and $y R (x \vee y)$

$$(x \vee y) \bar{R} x \text{ and } (x \vee y) \bar{R} y$$

$= x \vee y$ is a lower bound of $\{x, y\}$ in $\{L, \bar{R}\}$

Now, we will show that $x \vee y$ is the greatest lower bound of $\{x, y\}$ in $(L, \bar{R})$.

Let z be any lower bound of (x, y) in $(L, \bar{R})$, then $z \bar{R} x$ and $x \bar{R} y$.

$\Rightarrow$ $x R z$ and $y R z$

$\Rightarrow$ z is an upper bound of $\{x, y\}$ in (L, R).

$\Rightarrow$ $(x \vee y) R z$ as $x \vee y = \sup \{x, y\}$ in (L, R)

$\Rightarrow$ $z \bar{R} (x \vee y)$

$\Rightarrow$ $x \vee y$ is the greatest lower bound of (x, y) in $(L, \bar{R})$.

Similarly, it can be shown that $x \wedge y$ is the least upper bound in $(L, \bar{R})$. Therefore $(L, \bar{R})$ is a lattice.

4.5 TYPES OF LATTICE

4.5.1 Distributive Lattice

The lattice P, denoted by $(P, \wedge, \vee)$ is said to be distributive lattice, if it holds the distributive law, *i. e.*,

$$a \wedge (b \vee c) = (a \wedge b) \vee (a \wedge c) \quad \text{and} \quad a \vee (b \wedge c) = (a \vee b) \wedge (a \vee c), \quad \forall\ a, b, c \in P.$$

☛ All lattices are not distributive.

4.5.2 Complemented Lattice

Let $(L, \wedge, \vee)$ is a lattice and $0, 1 \in L$ such that $0 \leq a \leq 1 \quad \forall\ a \in L$, then

$$a \vee 1 = 1 \qquad a \wedge 1 = a$$

$$a \wedge 0 = 0 \qquad a \vee 0 = a$$

Now, for $a \in L$, there exists $a' \in L$ such that $a \vee a' = 1$.

Thus a' is called complement of a and such a lattice is called complemented lattice.

☛ In a complemented lattice the complement of each element is always unique.

☛ A complement is symmetric in a and a'. This means that a' is complement of a and a is complement of a'. An element $a \in L$ may or may not have a complement. Also, an element belonging to b may have more than one complement.

For Example :

1. The lattice $L = P(S)$, the power set of S, is complemented
2. Complemented lattices are shown in the adjoining Figure 7.

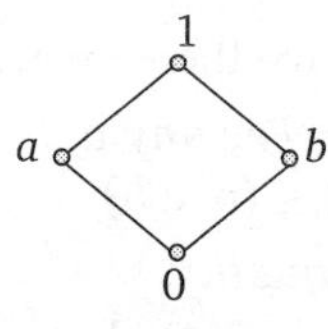

Fig. (7)

Here, complement of a is b.

4.5.3 Bounded Lattice

A lattice is called bounded lattice if it has a least element 0 and a greatest element 1.

For Example :

1. Let $\mathbf{Z}^+$ denotes the set of all positive integers and let $(\mathbf{Z}^+, \leq)$ be a lattice under the partial ordered relation of divisibility. Obviously, this is not a bounded lattice because it does not possess the greatest element, although it has the least element 1.

2. The lattice $(\mathbf{Z}^+, \leq)$ under the partial order relation $\leq$ is not bounded as it has neither a least element nor a greatest element.

3. The lattice $(L, \subseteq)$, where $L = P(S)$, power set of S is bounded. Its least element is ϕ and greatest element is S.

Example 1. Show that every finite lattice L is bounded.

Solution. Let $L = \{a_1, a_2, ..., a_n)$ is a finite lattice then $a_1 \wedge a_2 \wedge .. a_n$ and $a_1 \vee a_2 ... \vee a_n$ are lower and upper bounds of L respectively. Then the lower and upper bounds of L exist. Hence, the finite lattice is bounded.

4.5.4 Complete Lattice

If every non-empty subset of a lattice has a supremum (l.u.b.) and infimum (g.l.b.), then it is said to be a complete lattice.

For Example : If S be the class of all subsets of some universal set and an equivalence relation $\leq$ defined as $X \leq Y \Rightarrow X$ is a subset of Y, then since $X \wedge Y = X \cap Y$ and $X \vee Y = X \cup Y$ is lattice. Further, it follows that every subset has a supremum and infimum. Hence, it is a complete lattice.

Theorem 1. *If $(L, \leq)$ is a lattice with least element 0 and greatest element 1. Then for any $a \in L$*

(i) $a \vee 1 = 1$ *and* $a \wedge 1 = a$ *(ii)* $a \vee 0 = a$ *and* $a \wedge 0 = 0$

Proof : (i) Let a be any element of lattice L. Then we have

$$a \vee 1 \leq 1 \quad ...(1)$$

[$\because$ 1 is the greatest element]

Also $a \vee 1$ is the supremeum of a and 1, we have

$$1 \leq a \vee 1 \quad ...(2)$$

From (1) and (2), we have

$$a \vee 1 = 1$$

Further $a \wedge 1$ is the infimum of a and 1.

$$a \wedge 1 \leq a \quad ...(3)$$

Since $a \leq a$ and $a \leq 1$

$$a \leq a \vee 1 \quad ...(4)$$

(ii) 0 is the least element of L and a is any element of L.

$\therefore$ $0 \leq a$ and $a \leq a \Rightarrow a \vee 0 \leq a$ ...(5)

Also, $a \leq a \vee 0$ ...(6)

From (5) and (6), we have

$$a \vee 0 = a$$

Finally, $a \wedge 0$ is the minimum element of a and 0.

$$a \wedge 0 \leq 0 \quad ...(7)$$

Also, $0 \leq a$ and $0 \leq 0$

$$0 \leq a \wedge 0 \quad ...(8)$$

From (7) and (8), we have

$$a \wedge 0 = 0$$

Theorem 2. *If L is a complete lattice, then it is a bounded lattice.*

Proof : Let L be a complete lattice. Then every non-empty subset of L has least upper bound and greatest lower bound. In particular, L itself has least upper bound and greatest lower bound. Thus, L is a bounded lattice.

Theorem 3. *Every finite lattice is complete.*

Proof : Let $(L, \wedge, \vee)$ be any finite lattice. Let S be any non-empty subset of L. Then S is a finite set. Let $S = \{a_1, a_2, ..., a_n\}$. Then $a_1 \wedge a_2 \wedge ... \wedge a_n$ and $a_1 \vee a_2 \vee ... \vee a_n$ are the supremum and infimum of S in L. Hence, L is complete.

Theorem 4. *Dual of a complete lattice is complete.*

Proof : Let (L, R) be a complete lattice with relation R and let $(L, \bar{R})$ be its dual. Then $(L, \bar{R})$ is a lattice.

Let S be any non-empty subset of $(L, \bar{R})$. Since (L, R) is complete therefore, sup S and inf S exists in (L, R).

Let $a = \inf S$ in (L, R).

Then, $a \, R \, x, \quad \forall \, x \in L$

$x \, \bar{R} \, a, \quad \forall \, x \in L \Rightarrow a$ is an upper bound of S in $(L, \bar{R})$.

Let b be any other upper bound of S in $(L, \bar{R})$.

Then, $x \, \bar{R} \, b, \forall \, x \in S, b \, \bar{R} \, a, \forall \, x \in S$

$\Rightarrow$ $b \, R \, a$ as $a = \inf S$ in (L, R)

$\Rightarrow$ $a \, \bar{R} \, b \Rightarrow a$ is sup of S in $(L, \bar{R})$.

Similarly, we can show that sup S in (L, R) will be inf S in $(L, \bar{R})$. Hence, $(L, \bar{R})$ is complete.

4.6 SUB LATTICE

Let $(P, \wedge, \vee)$ be a lattice and let $Q \subseteq P$. Then the algebra $(Q, \wedge, \vee)$ is said to be a sub lattice of P if and only if Q is closed under both the operation $\wedge$ and $\vee$.

☛ A sub lattice is itself a lattice. Any subset of L which is a lattice need not be a sub lattice. Let us consider any two elements $a, b \in L$ such that $a \leq b$, then the closed interval $[a, b]$ which contains all the elements $x \in L$ such that $a \leq x \leq b$ will be a lattice.

For Example : Let $(L, \leq)$ be a lattice and let $L = \{a, b, c, d, e, f, g, h\}$. The subset of L are $S_1 = \{a, b, d, f\}$, $S_2 = \{c, e, g, h\}$. We find that $(S_1, \leq)$ and $(S_2, \leq)$ are sub lattice of L.

4.6.1 Direct Product

Consider the two lattices $(L, *, \oplus)$ and $(S, \wedge, \vee)$. The algebraic system $(L \times S, +, .)$ in which the binary operation on $L \times S$ is given by

$$(a_1, b_1) . (a_2, b_2) = (a_1 * a_2, b_1 \wedge b_2);$$
$$(a_1, b_1) + (a_2, b_2) = (a_1 \oplus a_2, b_1 \vee b_2)$$

For any $(a_1, b_1), (a_2, b_2) \in L \times S$ is defined as the direct product of the lattices $(L, *, \oplus)$ and $(S, \wedge, \vee)$. Here, the operation + and . on $L \times S$ satisfy the commutativity, associativity and absorption law.

☛ We may write $L \times L$ and $L \times L \times L$ as L^2 and L^3 respectively.

Example 1 : Let $L = \{0, 1\}$ be a chain. Draw the diagram of the product lattice L^2.

Solution. $L^2 = L \times L = \{(0, 0), (0, 1) (1, 0), (1, 1)\}$.

The diagram of the product lattice L^2 is given below :

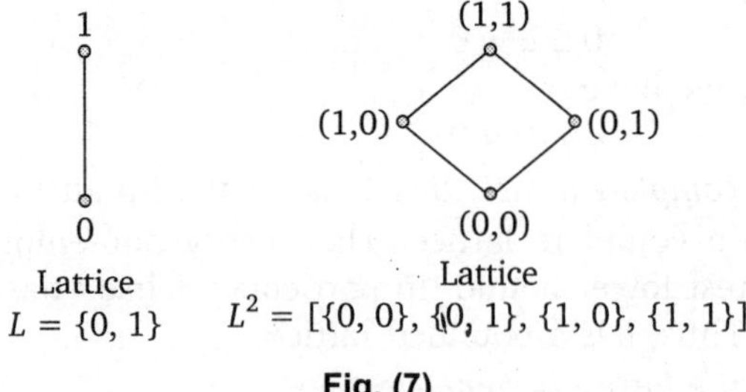

Fig. (7)

4.7 HOMOMORPHISM OF LATTICES

Consider two lattices $(L, *, \oplus)$ and $(S, \wedge, \vee)$. A lattice homomorphism is defined by a mapping $f : L \to S$ such that for any $a_1, a_2 \in L$, we have

$$f(a_1 * a_2) = f(a_1) \wedge f(a_2) \quad \text{and} \quad f(a_1 \oplus a_2) = f(a_1) \vee (a_2).$$

Thus, we see that both the operations of meet and join are preserved.

- If in a lattice only one of these operations are preserved, then it will not be lattice homomorphism.
- A lattice homomorphism is preserved if partial ordering relations $\leq$ and $\geq$ corresponding to the meet and join operations are defined on L and S, respectively, *i.e.*, if $f : L \to S$, then for any element $a_1, a_2 \in L$ such that $a_1 \leq a_2$, we get $f(a_1) \leq f(a_2)$.

4.7.1 The resulting Homomorphism or Isomorphism

A lattice homomorphism $f : L \to S$ is said to be isomorphism if it is bijective (*i.e.*, one-one, onto). In this case, L is said to be isomorphic to S. Here, we have the following observations.

(i) The partial ordering relation $a_1 \leq a_2 \Rightarrow f(a_1) \leq f(a_2)$ will be preserved.

(ii) It is possible to represent two isomorphic lattices by the same diagram by simply replacing the nodes with the images.

(iii) The lattices containing one, two or three elements will be isomorphic to the chain having one, two or three elements, respectively.

4.7.2 Endomorphism and Automorphism

Let $(L, *, \oplus)$ be a lattice. Then a homomorphism $f : L \to L$ is called endomorphism. And if $f : L \to L$ is an isomorphism, then it is said to be automorphism.

Theorem 1. *If the lattices L_1 and L_2 are isomorphic and f denote an isomorphism, then f preserve the ordering relation i.e., for any $a, b \in L_1$*

$$a \leq b \Rightarrow f(a) \leq f(b)$$

Proof. Let $f : L_1 \to L_2$ be an isomorphism

Let $a, b \in L$ such that $a \leq b$

Then $a = a \wedge b$

$\Rightarrow \quad f(a) = f(a \wedge b) = f(a) \wedge f(b)$

$\Rightarrow \quad f(a) \leq f(b)$

$\Rightarrow \quad f$ presences ordering relation.

Theorem 2. *If L_1 and L_2 are two lattices. If $f : L_1 \to L_2$ is an isommorptism of L_1 onto L_2 and a is the least element of L_1 then $f(a)$ is the least element of L_2.*

Proof. Let a be the least element of L_1, then

$$a \leq x \ \forall \ x \in L_1 \Rightarrow f(a) \leq f(x) \quad \forall \ f(a) \in L_2$$

Now since f is onto, any element $y \in L_2$ is of the form $f(x)$ for same $x \in L_1$. Hence, $f(a)$ is the least element of L_1.

Theorem 3. *If $f: L_1 \to L_2$ is an isomorphism and $x, y \in L_1$ such that $x < y$ then there exist $a, b \in L_1$ such that $f(a) = x$, $f(b) = y$ and $a < b$.*

Proof. Since f is onto and $x, y \in M$, therefore then exist a and c in L_1 such that $f(a) = x$ and $f(c) = y$

Now, $f(a \vee c) = f(a) \vee f(c)$

$= x \vee y = y \qquad (\because x < y)$

We know that

$$a \leq a \vee c$$

If $a = a \vee c$ then $f(a) = f(a) \vee f(c) = y$

$\Rightarrow$ $x = y$, which is a contradiction

Hence, $a \neq a \vee c$

$\therefore \qquad a < a \vee c$

Now, we take $b = a \vee c$, then $a < b$ and $f(a) = x$, $f(b) = y$

ILLUSTRATIVE EXAMPLES

1. *Consider the lattice $L = \{1, 2, 3, 6\}$ under divisibility relation and the lattice $(P(s), \subseteq)$ where $S = \{a, b\}$. Then show that lattices L and $P(S)$ are isomorphic.*

Solution : Define a mapping $f: L \to P(S)$ such that

$$f(1) = \phi,\ f(2) = \{a\},\ f(3) = \{b\},\ f(6) = \{a, b\}$$

Then f is one-one and onto

Also, $f(a \wedge b) = f(a) \wedge f(b)$

and $\qquad f(a \vee b) = f(a) \vee f(b)$

$\Rightarrow$ f is an isomorphism and lattice $(L, /)$ is isomorphic to the lattice $(P(S), \subseteq)$

2. *Let $L = \{1, 2, 3, 4, 6, 12\}$. Consider the lattices $(L, /)$ an $(L, \leq)$ then show that lattice $(L, /)$ and $(L, \leq)$ are not isomorphic.*

Solution : Consider any mapping $f: (L, /) \to (L, \leq)$

Then in $(L, /)$

$$3 \wedge 4 = 1 \Rightarrow f(3 \wedge 4) = f(1)$$

But $f(3) \wedge f(4) = f(3)$ or $f(4)$ depending upon whether $f(3)$ is less than or equal to $f(4)$ or $f(4)$ is less than or equal to $f(3)$.

In any case $f(3 \wedge 4) \neq f(3) \wedge f(4)$

$\Rightarrow$ $(L, /)$ and $(L, \leq)$ are not isomorphic.

4.8 MODULAR LATTICE

A lattice M is said to be modular if whenever $a \leq c$, we have

$$a \vee (b \wedge c) = (a \vee b) \wedge c \quad \textbf{(Associative law)}$$

Theorem 1. *Every distributive lattice is modular.*

Proof : Let $(L, \leq)$ be a distributive lattice and $a, b, c \in L$ be such that $a \leq c$. Thus, if $a \leq c$, then $a \vee c$. Now

$$a \vee (b \wedge c) = (a \vee b) \wedge (a \vee c) = (a \vee b) \wedge c$$

Hence, every distributive lattice is modular.

☛ The converse of the above theorem is not true.

Theorem 2. *A sub lattice of a modular lattice is modular.*

Proof : Let S be a sublattice of a modular lattice L. If $a, b, c \in S$ with $a \leq c$, then as $S \subseteq L$, we have $a, b, c \in L$ and therefore

$$a \vee (b \wedge c) = (a \vee b) \wedge c \qquad ...(1)$$

Since S is closed with respect to $\wedge$ and $\vee$, the above result (1) holds in S and hence S is modular.

Theorem 3. *Two lattice L and M are modular if and only if L × M is modular.*

Proof : Let L and M be modular and let $(a_1, b_1), (a_2, b_2), (a_3, b_3) \in L \times M$ be three elements with $(a_1, b_1) \geq (a_3, b_3)$.

Then, $a_1, a_2, a_3 \in L$ and $a_1 \geq a_3$, $b_1, b_2, b_3 \in M$ and $b_1 \geq b_3$ and since L and M are modular, we get

$$a_1 \vee (a_2 \wedge a_3) = (a_1 \vee a_2) \wedge a_3$$
$$b_1 \vee (b_2 \wedge b_3) = (b_1 \vee b_2) \wedge b_3$$

$$\begin{aligned}\Rightarrow \quad (a_1, b_1) \vee [(a_2, b_2) \wedge (a_3, b_3)] &= (a_1, b_1) \vee (a_2 \wedge a_3, b_2 \wedge b_3) \\ &= (a_1 \vee (a_2 \wedge a_3), (b_1 \vee (b_2 \wedge b_3)) \\ &\quad (a_1 \vee a_2) \wedge a_3, (b_1 \vee b_2 \wedge b_3) \\ &= (a_1 \vee a_2, b_1 \vee b_2) \wedge (a_3, b_3) \\ &= ((a_1, b_1) \vee (a_2, b_2)) \wedge (a_3, b_3)\end{aligned}$$

Hence, $L \times M$ is modular.

Conversely, let $L \times M$ be modular. We shall show that both L and M are modular. Let $a_1, a_2, a_3 \in L$ with $a_1 \geq a_3$ and $b_1, b_2, b_3 \in L$ with $b_1 \geq b_3$.

Then, $(a_1, b_1), (a_2, b_2), (a_3, b_3) \in L \times M$ and $(a_1, b_1) \geq (a_3, b_3)$

Since $L \times M$ is modular, we have

$$(a_1, b_1) \vee [(a_2, b_2) \wedge (a_3, b_3)] = [(a_1, b_1) \vee (a_2, b_2)] \wedge (a_3, b_3)$$

or $\quad (a_1, b_1) \vee (a_2 \wedge a_3, b_2 \wedge b_3) = (a_1 \vee a_2, b_1 \vee b_2) \wedge (a_3, b_3)$

or $\quad a_1 \vee (a_2 \wedge a_3), b_1 \vee (b_2 \wedge b_3)) = ((a_1 \vee a_2) \wedge a_3, (b_1 \vee b_2) \wedge b_3)$

$\Rightarrow \quad a_1 \vee (a_2 \wedge a_3) = (a_1 \vee a_2) \wedge a_3$ and $\quad b_1 \vee (b_2 \wedge b_3) = (b_1 \vee b_2) \wedge b_3$

$\Rightarrow \quad L$ and M are modular.

Theorem 4. *The dual of a modular lattice is modular.*

Proof : Let $(L, \leq)$ be a modular lattice. Then, we have to show that its dual $(L, \geq)$ is also modular.

Let $a, b, c \in L$ and $a \geq c$, then we have to prove that

$$a \wedge (b \vee c) = (a \wedge b) \vee c \qquad ...(1)$$

Since $(L, \leq)$ is modular and $c \leq a$, we have

$$c \vee (b \wedge a) = (c \vee b) \wedge a$$

$\Rightarrow \quad c \vee (a \wedge b) = (b \vee c) \wedge a \quad \Rightarrow \quad a \wedge (b \vee c) = (a \wedge b) \vee c$

This is same as (1). Therefore $(L, \geq)$ is modular.

Theorem 5. *The complement of an element of a bounded distributive lattice L is unique.*

Proof : Let $a \in L$. Let if possible a' and a'' be its two complements, then

$a \vee a' = 1 \qquad a \wedge a' = 0$

and $a \vee a'' = 1 \qquad a \wedge a'' = 0$

To show $a' = a''$. Since we know that

$$\begin{aligned} a' &= a' \wedge 1 \\ &= a' \wedge (a \vee a'') \\ &= (a' \wedge a) \vee (a' \wedge a'') && \text{(By distributivity)} \\ &= 0 \vee (a' \wedge a'') && (\because a \wedge a' = 0) \\ &= (a \wedge a'') \vee (a' \wedge a'') && (\because a \wedge a'' = 0) \\ &= (a \vee a') \wedge a'' \\ &= 1 \wedge a'' = a'' \end{aligned}$$

Theorem 6. *Let $(L_1, \leq)$ and $(L_2, \leq)$ be two lattices. Define $L = L_1 \times L_2$. Then $(L, \leq)$ is a lattice.*

Proof : By definition of product space, we can easily verify that $(L_1 \times L_2, \leq)$ will be a poset with partial order $\leq$ defined by $(a_1, a_2) \leq (a_1', a_2')$ provided $a_1 \leq a_1'$ in L_1 and $a_2 \leq a_2'$ in L_2.

Now, to show that if (c_1, d_1) and $(c_2, d_2) \in L$, then $(c_1, d_1) \vee (c_2, d_2)$ and $(c_1, d_1) \wedge (c_2, d_2)$ exist in L. Also, we can easily verify that

$$(c_1, d_1) \vee (c_2, d_2) = (c_1 \vee c_2, d_1 \vee d_2);$$
$$(c_1, d_1) \wedge (c_2, d_2) = (c_1 \wedge c_2, d_1 \wedge d_2)$$

Hence, L is a lattice.

Theorem 7. *The dual of distributive lattice is distributive lattice.*

Proof : Let $(L, \wedge, \vee)$ be a distributive lattice. We will show that its dual $(L, \wedge^*, \vee^*)$ is also distributive where $\wedge^* = \vee$ and $\vee^* = \wedge$. Let $a, b, c \in L$. Then, we show

$$a \wedge^* (b \vee^* c) = (a \wedge^* b) \vee^* (a \wedge^* c)$$

or $$a \vee (b \wedge c) = (a \vee b) \wedge (a \vee c)$$

But the last equality is true because $(L, \wedge, \vee)$ is distributive lattice.

Hence, $(L, \wedge^*, \vee^*)$ is also distributive.

Theorem 8. *Every sublattice of a distributive lattice is distributive.*

Proof : Let S be a sublattice of a distributive lattce L. Let $a, b, c \in S$. Then $a, b\ c \in L$. Thus $a \wedge (b \vee c) = (a \wedge b) \vee (a \wedge c)$ in L.

Since S is closed under $\wedge$ and $\vee$, this equation holds in S. Hence, S is distributive.

Theorem 9. *Dual of a complemented lattice is complemented.*

Proof : Let (L, R) be a complemented lattice with 0 and 1 as least and greatest elements.

Let $(L, \bar{R})$ be the dual of (L, R). Then 1 and 0 are least and greatest elements of $(L, \bar{R})$.

Let $a \in L$ be any element.

Since (L, R) is complemented, therefore there exists $a' \in L$ such that $a \wedge a' = 0$ and $a \vee a' = 1$ in the lattice (L, R).

That is, $0 = \inf\{a, a'\}$ and $1 = \sup\{a, a'\}$ in (L, R).

$\Rightarrow$ $0\,R\,a$, $0\,R\,a'$ and $a\,R\,1$, $a'\,R\,1$ $\Rightarrow$ $a\,\bar{R}\,0$, $a'\,\bar{R}\,0$ and $1\,\bar{R}\,a$, $1\,\bar{R}\,a'$

$\Rightarrow$ 0 is an upper bound of $\{a, a'\}$ in $(L, \bar{R})$ and 1 is a lower bound $\{a, a'\}$ in $(L, \bar{R})$ and 1 is a lower bound of $\{a, a'\}$ in $(L, \bar{R})$.

Let k be any upper bound of $\{a, a'\}$ in $(L, \bar{R})$. Then,

$$a\,\bar{R}\,k \text{ and } a'\,\bar{R}\,k \quad \Rightarrow \quad k\,R\,a \text{ and } k\,R\,a'$$

$\Rightarrow$ $k\,R\,0$ because 0 is infimum of $\{a, a'\}$ in (L, R)

$\Rightarrow$ $0\,\bar{R}\,k$

Thus, 0 is lub of $\{a, a'\}$ in $\{L, \bar{R}\}$. Hence $a \vee a' = 0$ in $(L, \bar{R})$.

Similarly, $a \wedge a' = 1$ in $(L, \bar{R})$.

Thus a' is complement of a in $(L, \bar{R})$.

Hence, $(L, \bar{R})$ is complemented.

Theorem 10. *If $(L, \wedge, \vee)$ is a complemented distributive lattice, then De Morgan's laws*

$$(a \vee b)' = a' \wedge b' \text{ and } (a \wedge b)' = a' \vee b'$$

holds for all $a, b \in L$.

Proof : Observe that L is already bounded by definition of a complemented lattice. Also, complements are unique because of distributivity.

Now, $(a \vee b) \vee (a' \wedge b')$

$$= [(a \vee b) \wedge a'] \wedge [(a \vee b) \vee b'] \qquad \text{(By distributivity)}$$
$$= [(a \vee a') \vee b] \wedge [a \vee (b \vee b')]$$
$$= (1 \vee b) \wedge (a \vee 1) = 1 \wedge 1 = 1$$

and $(a \vee b) \wedge (a \wedge b') = [(a \vee b) \wedge a] \wedge [(a' \wedge b') \wedge b']$
$= [b' \wedge (a' \wedge a)] [a' \wedge (b \wedge b')]$
$= (b' \wedge 0) \vee (a' \wedge 0) = 0 \vee 0 = 0$

Thus $(a' \wedge b')$ behaves as the complement of $(a \vee b)$. By uniqueness of complements, it is the only complement of $a \vee b$.

$$(a \vee b)' = a' \wedge b'$$

Similarly, we can prove the other result, *i.e.*, $(a \wedge b)' = a' \vee b'$.

Theorem 11. *Two bounded lattices* L_1 and L_2 *are complemented iff* $L_1 \times L_2$ *is complemented.*

Proof : Let L_1 and L_2 be two complemented lattices and let 0, 1 and $0'$, $1'$ are least and greatest elements of L_1 and L_2 respectively. Then $(0, 0')$ and $(1, 1')$ will be the least and greatest elements of $L_1 \times L_2$. Let (a_1, a_2) be any element of $L_1 \times L_2$, then $a_1 \in L_1$ and $a_2 \in L_2$.

Since L_1 and L_2 are complemented, there exists $a_1 \in L_1$ and $a_2 \in L_2$ such that $a_1 \wedge a_1' = 0, a \vee a' = 1$ and $a_2 \wedge a_2' = 0'$ and $a_2 \vee a_2' = 1'$. We shall show that (a_1', a_2') is the complement of (a_1, a_2) in $L_1 \times L_2$.

We have $(a_1, a_2) \wedge (a_1', a_2') = (a_1 \wedge a_1', a_2 \wedge a_2')$
$= (0, 0')$.

and $(a_1, a_2) \vee (a_1', a_2') = (a_1 \vee a_1', a_2 \vee a_2')$
$= (1, 1')$

This shows that (a_1', a_2') is the complement of (a_1, a_2) in $L_1 \times L_2$.

Hence $L_1 \times L_2$ is complemented. Conversely, let $L_1 \times L_2$ be complemented, we have to show that L_1 and L_2 are complemented.

Let $a_1 \in L_1$ and $a_2 \in L_2$, thus $(a_1, a_2) \in L_1 \times L_2$.

Since $L_1 \times L_2$ is complemented, there exists $(a_1, a_2) \in L_1 \times L_2$ such that

$(a_1, a_2) \wedge (a_1', a_2') = (0, 0')$ and $(a_1, a_2) \vee (a_1', a_2') = (1, 1')$

$\Rightarrow$ $(a_1 \wedge a_1', a_2 \wedge a_2') = (0, 0')$ and $(a_1 \vee a_1', a_2 \vee a_2') = (1, 1')$

$\Rightarrow$ $a_1 \wedge a_1' = 0, a_2 \wedge a_2' = 0'$ and $a_1 \vee a_1' = 1, a_2 \vee a_2' = 1$

$\Rightarrow$ $a_1 \wedge a_1' = 0, a_1 \vee a_1' = 1$ and $a_2 \wedge a_2' = 0, a_2 \vee a_2' = 1'$

$\Rightarrow$ a_1' and a_2' are complements of a_1 and a_2 respectively. Hence, L_1 and L_2 are complemented.

Theorem 12. *Every chain is a distributive lattice.*

Proof : Let $(L, \leq)$ be a chain and $a, b, c \in L$. Since L is a chain either $a < b$ or $b < a$. If $a < b$, then $a \vee b = b$ and $a \wedge b = a$. Hence, for any two elements $a, b \in L$, $a \wedge b$ and $a \vee b$ exists in L.

Suppose $a < b$.

Case 1 : $b < c$

Now, $a \wedge (b \vee c) = a \wedge c = a$ and $(a \wedge b) \vee (a \wedge c) = a \vee a = a$.

Hence, we have $a \wedge (b \vee c) = (a \wedge b) \vee (a \wedge c)$.

Case 2 : $c < b$

In this case, we have $a < c < b$.

Now, $a \wedge (b \vee c) = a \wedge b = a$ and $(a \wedge b) \vee (a \wedge c) = a \vee a = a$.

Hence, $a \wedge (b \vee c) = (a \wedge b) \vee (a \wedge c)$.

Similarly, if $b \leq a$, then $a \wedge (b \vee c) = (a \wedge b) \vee (a \wedge c)$.

Theorem 13. *In a distributive lattice, if an element has a complement, then this complement is unique.*

Proof : Let if possible an element a has two complements b and c. Then

$$a \vee b = 1, a \wedge b = 0$$
$$a \vee c = 1, a \wedge c = 0$$

We have $b = b \wedge 1$

$$
\begin{aligned}
&= b \wedge (a \vee c) \\
&= (b \wedge a) \vee (b \wedge c) \\
&= 0 \vee (b \wedge c) \\
&= (a \wedge c) \vee (b \wedge c) \\
&= (a \vee b) \wedge c \\
&= 1 \wedge c = c.
\end{aligned}
$$

4.9 COVER OF AND ELEMENT : ATOMS AND IRREDUCIBLE ELEMENTS

Definition (1). *If a and b are two elements of lattice then a is said to be cover b if $b < a$ an there is no element c such that $b < c < a$.*

Definition (2). *Let L be a lattice with a least element 0. An element a in L is called an atom if a covers 0. i.e., $0 < a$ and there is no element c such that $0 < c < a$.*

Definition (3). Let L be a lattice. An element a in L is said to be join irreducible if $a = x \vee y \Rightarrow a = x$ or $a = y$

Also, an element a in L is said to be meet irreducible if

$$a = x \wedge y \Rightarrow a = x \quad \text{or} \quad a = y$$

Illustration :

(1) Consider the hasse diagram. Clearly, in this lattice a and b are atoms. Although, c is not an atom because $0 < a < c$. Also, c is over of a, while 1 is cover of both b and c. Further a, b and c are join irreducible.

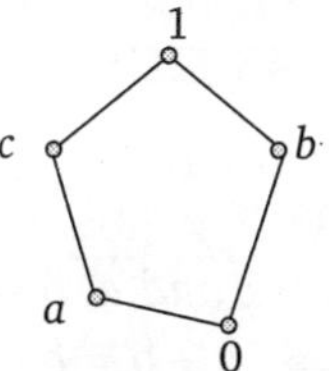

Fig. (8)

(2) Consider the lattice shown in the Hasse diagram. Here, a, b and c are atoms. The elements a, b and c are join irreducible as well as meet irreducible, 0 is join irreducible but not meet irreducible and 1 is meet irreducible but not join irreducible.

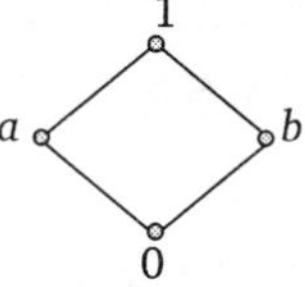

Fig. (9)

(3) Consider the lattice L shown in the given diagram. Clearly
(i) the non-zero elements a, b, d and e are join irreducible. Also c is not join-irreducible
(ii) the elements c, d, e and 1 are meet irreducible
(iii) only a and b are atoms

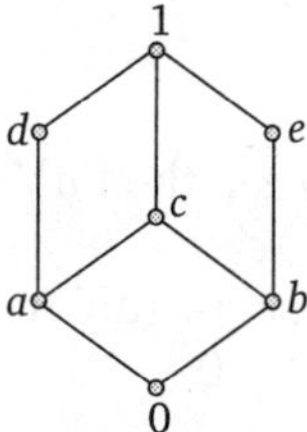

Fig. (10)

Theorem 1. *Let L be a complimented lattice with unique compliment. Then the join-irreducible element of L other than 0 are its atom.*

Proof. Let L be a complimented lattice with unique compliment let us assume that L has more than two elements only (Because of compliment lattice L contains two elements only then these two elements are 0 and 1 and 1 is clearly an atom in this lattice) Let a be any join irreducible element of L which is not 0. We shall show that is a is an atom. Let if possible, a is not an atom, then there exists an element $x \in L$ such that $0 < x < a$

Let a' be the compliment of a.

Now $0 < x < a \Rightarrow \quad x \wedge a' \leq a \wedge a'$

$\Rightarrow \quad 0 \leq x \wedge a' \leq 0 \qquad (\because\ a \wedge a' = 0)$

$\Rightarrow \quad x \wedge a' = 0 \qquad ...(1)$

If $x \leq a'$ then $x \wedge a' = x \Rightarrow 0 = x$ (by (1))

Which is not true $\Rightarrow$ $x \leq a'$ is not possible

$\therefore$ We assume that $x \nleq a'$

If $a' \leq x$ then by (1), $a' = x \wedge a' = 0$

$\Rightarrow a = 1$, which is not join-irreducible in a compliment lattice having more than two elements.

If a' and x are not comparable, then $x \vee a' = 1$

Therefore, $\quad x \wedge a' = 0$ and $x \vee a' = 1$

$\Rightarrow \quad x = a,$ which is again a contradiction

Hence, a is an atom.

4.10 BOOLEAN ALGEBRA AS LATTICE

Let B be a Boolean algebra and $a, b \in B$

Then, we say that $a \leq b$ if $ab' = 0$

Theorem 1. *Let B be a Boolean algebra and a, b ∈ B. Then* $a \leq b$ *if and only if* $a + b = b$

Proof. We know that $ab' = 0 \quad \Leftrightarrow \quad a + b = b$

Therefore, we have

$$a \leq b \text{ if and only if } a + b = b.$$

Theorem 2. *Let B be a Boolean algebra then the relation ≤ defined as* $a \leq b$ *if and only if* $ab' = 0$ *is a partial order on B*

Proof. (i) '≤' is Reflexive

$\therefore \quad aa' = 0 \quad \Rightarrow \quad a \leq a \qquad \forall\ a \in B$

$\Rightarrow \quad \leq$ is reflexive

(ii) ≤ antisymmetric :

Let $a, b \in B$ such that $a \leq b$ and $b \leq a$

$\Rightarrow \quad ab' = 0$ and $ba' = 0$

Now $a = a.1 = a(b + b') = ab + ab'$

$= ab + 0$

$= ab + ba' = ba + ba' = b(a + a')$

$= b.1 = b$

$\Rightarrow \quad \leq$ is antisymmetric

(iii) ≤ is transitive :

Let us suppose that $a \leq b$ and $b \leq c$ then $ab' = 0$ and $bc' = 0$

Now, $\quad ac' = a.1.c' = a(b + b')c'$

$= (ab + ab')c' = a(bc') + (ab')c'$

$= a.0 + 0.c' = 0 + 0 = 0$

$\Rightarrow \quad a \leq c$

Hence $\leq$ is a partial order on B.

Theorem 3. *Let B be a Boolean algebra, then $(B, \leq)$ where $\leq$ is defined as $a \leq b$ if and only if $ab' = 0$ is a lattice. Also, the identities 0 and 1 are the least and the greatest elements of this lattice.*

Proof. We know that $(B, \leq)$ is a partial ordered set therefore, to show that $(B, \leq)$ is a lattice, it is sufficient to prove that for any elements $a, b \in B$, join of a and b is $a + b$ and meet of a and b is ab i.e., we have to show that $\sup\{a, b\} = a \vee b = a + b$ and inf. $\{a.b\} = a \wedge b = ab \;\forall\; a, b \in B$

Since $a.(a+b)' = a(a'b') = (aa')b' = 0.b' = 0$

$\Rightarrow \quad a \leq a + b$

Similarly $b \leq a + b$

$\Rightarrow \quad (a+b)$ is an upper bound of the set $\{a.b\}$

If c be any other upper bound of $\{a.b\}$, then

$$a \leq c \text{ and } b \leq c$$

$\Rightarrow \quad ac' = 0 \quad$ and $\quad bc' = 0$

$\Rightarrow \quad ac' + bc' = 0 \qquad \Rightarrow \quad (a+b)c' = 0$

$\Rightarrow \quad a + b \leq c$

$\Rightarrow \quad a + b$ is the least upper bound of the set $\{a.b\}$ which is the join of a and b denoted by $a \vee b$. Similarly we may show that ab is he infimum of $\{a.b\}$

Therefore inf $\{a.b\} = a \wedge b = ab$

Hence, B is a lattice when + and . are join and meet operations

Further, if $a \in B$, $0.a' = 0$ and hence $0 \leq a$

$\Rightarrow \quad 0$ is the least element of B

Similarly $a.1' = a.0 = 0 \quad \forall\; a \in B \quad \Rightarrow \quad a \leq 1 \quad \forall\, a \in b$

Hence, 1 is the greatest element of B

$\Rightarrow \quad B$ is bounded lattice.

Theorem 4. *Let $(B, +, .)$ be a Boolean algebra, then the lattice $(B, \wedge, \vee)$ when $a \vee b = a + b$ and $a \wedge b = ab$ is bounded, complimented and distributive. Conversely if $(B, \wedge, \vee)$ is bounded, complimented and distributive lattice then $(B, \vee, \wedge)$ is a Boolean algebra where $a + b = a \vee b$, $ab = a \wedge b$ and a' is a compliment of a in $(B, \vee, \wedge)$.*

Proof. Let $(B, +, .)$ be a Boolean algebra, then $(B, \vee, \wedge)$ is a bounded lattice (Th. 3). Since $\vee$ and $\wedge$ are precisely + and . respectively in the definition of boolean algebra which shows that $(B, \vee, \wedge)$ is also distributive and complimented.

Conversely,

Suppose that $(B, \vee, \wedge)$ s bounded, complimented and distributive lattice with 0 and 1 as the least and the greatest elements.

Now for $a, b \in B$, define

$$a + b = a \vee b \qquad \text{and } a.b = a \wedge b$$

Then + and . are commutative with 0 and 1 as their respective identities. Also distributive laws follow from the definitive of a distributive lattice.

Since $(B, \vee, \wedge)$ is a complimented lattice, we can find a compliment of each $a \in B$.

Now, we have $a + a' = 1$ and $aa' = 0$

Hence, $\quad (B, +, .)$ is a Boolean algebra.

Some More Results :

For a Boolean algebra $(B, +, .)$ we have

(i) for every non-zero element b, there exists at least one atom a such that $a \leq b$.

(ii) if a and b are distinct atoms then $ab = 0$

(iii) if b is any non-zero element in B and $a_1, a_2, \ldots\ldots, a_k$ be all atoms such that $a_i \leq b : i = 1, 2, \ldots.. k$ then $b = a_1 + a_2 + \ldots\ldots + a_k$ and this representation is unique.

(iv) The sum of atoms in B equals 1.

ILLUSTRATIVE EXAMPLES

1. *Show that the lattice whose Hasse diagram is given blow is not a Boolean algebra.*

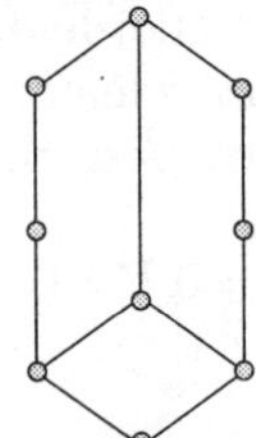

Fig. (11)

Solution : We observe that elements a and e are both compliments of c. But we know that such an element is unique.

Hence, given lattice cannot be a Boolean algebra.

2. *Let* $\mathbf{N} = \{1, 2,, 3...\}$ *be ordered by divisibility. State whether each of the following subsets of* **N** *are linearly ordered :*

(a) $\{24, 2, 6\}$ (b) $\{3, 15, 5\}$ (c) $\{1, 2, 3...\}$ (d) $\{4\}$

Solution. (a) Since 2 divides 6 and 6 divides 24, therefore, the set is linearly ordered.

(b) Since 3 and 5 are not comparable, therefore, the set is not linearly ordered.

(c) Since 2 and 3 are not comparable, the given set N is not linearly ordered.

(d) Every singleton set is linearly ordered.

3. *Consider the Fig. 12. Let L (A) denote the collection of all linearly ordered subsets of A with 2 or more elements. let L (A) be ordered by set inclusion. Draw the Hasse diagram of L (A).*

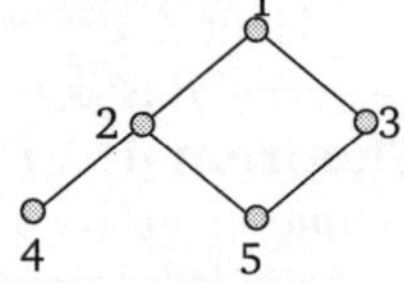

Fig. (12)

Solution. Here the elements of $L(A)$ are as follows :

$\{1, 2, 4\}, \{1, 2, 5\}, \{1, 3, 5\}, \{1, 2\}, \{1, 4\}$

$\{1, 3\}, \{1, 5\}, \{2, 4\}, (2, 5), \{3, 5\}$

The Hasse diagram of $L(A)$ is shown below :

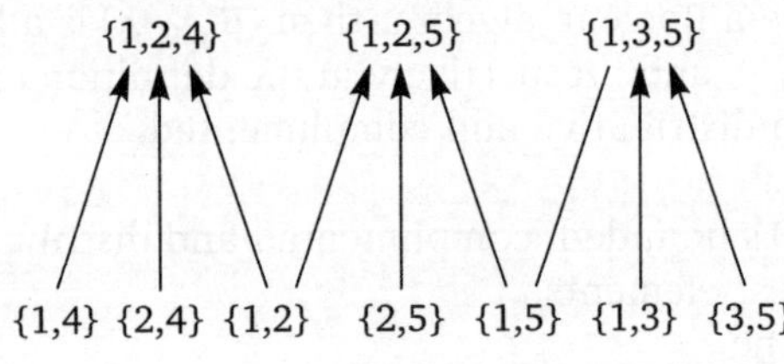

Fig. (13)

4. *(a) Find all minimal and maximal element of A in figure of example 3.*

(b) Does A have a first element or a last element ?

Solution. No element strictly precedes 4 or 5. Therefore 4 and 5 are minimal elements of A. No element strictly succeeds 1, therefore, 1 is a maximal element of A.

(b) A has no first element. Although 4 and 5 are minimal elements of A, neither precedes the other. However, 1 is the last element of A since 1 succeeds every element of A.

5. *Let* $S = \{a, b, c, d, e, f, g\}$ *be ordered in the given figure and let* $X = \{c, d, e\}$.

(a) find the upper and lower bound of X;

(b) find supremum and infimum of X, if they exist.

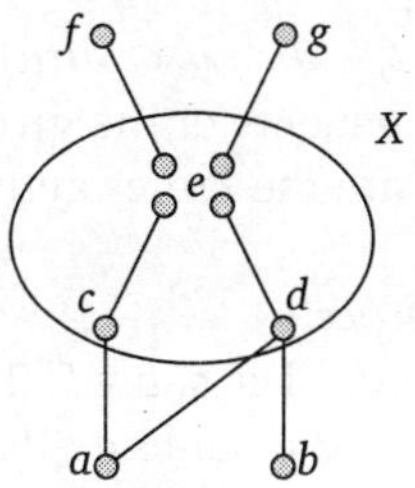

Fig. (14)

Solution. (a) Since the elements e, f and g succeed every element of X, therefore e, f and g are the upper bounds of X. The element a precedes every element of X, therefore, it is the lower bound of X. Also, b is not a lower bound, since b does not precede c. Here, b and c are not comparable.

(b) Since e precedes both f and g, we have $e = \sup (X)$.

Similarly, since a precedes (trivially) every lower bound of X, we have $a = \inf (X)$.

☛ Here, supremum of X belongs to X but infimum of X does not belong to X.

6. *Let S be the ordered set in the adjoining figure. Suppose $A = \{1, 2, 3, 4, 5\}$ is isomorphic to S and $f = \{(a, 1), (b, 3), (c, 5), (d, 2), (e, 4)$ is a similarity mapping from S to A. Draw the Hasse diagram of A.*

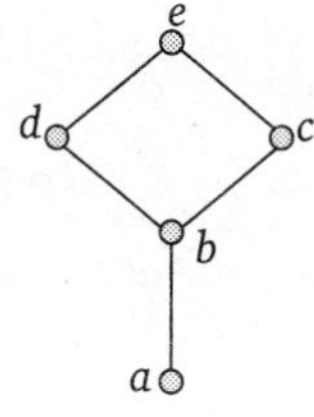

Fig. (15)

Solution. Since the similarity mapping f preserves the order structure of S and therefore f may be seen simply as a relabeling of vertices in the diagram of S. Hence, diagram of A is given as follows :

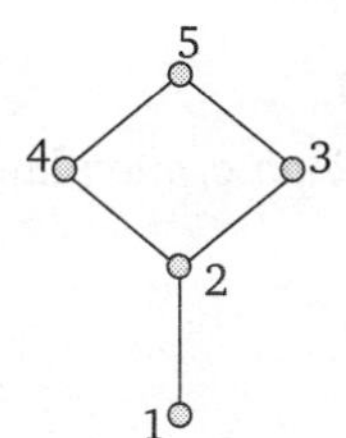

Fig. (16)

6. *The set L of all factors of 12 under divisibility forms a lattice.*

Solution. Here, $L = \{1, 2, 3, 4, 6, 12\}$. Let a, b be any two elements in L, then sup $\{a, b\}$ under divisibility relation is least positive integer such that $a|c$ and $b|c$. Thus, sup $\{a, b\}$ = l.c.m of a and b. Similarly, inf $\{a, b\}$ is the greatest positive integer d such that $d|a$ and $d|b$. Hence, inf $\{a, b\}$ = gcd of a and b. This lattice is represented by the diagram as shown in Figure 17.

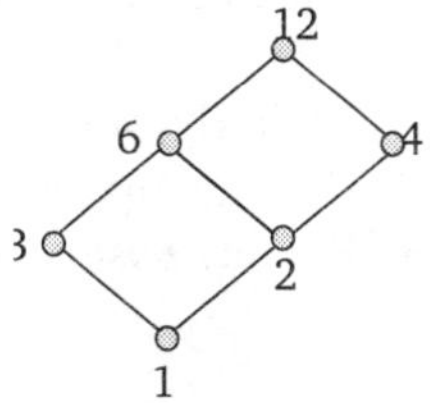

Fig. (17)

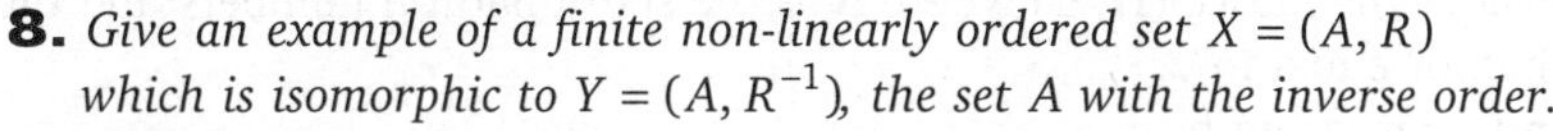

8. *Give an example of a finite non-linearly ordered set $X = (A, R)$ which is isomorphic to $Y = (A, R^{-1})$, the set A with the inverse order.*

Solution. Let R be the partial ordering of $A = \{a, b, c, d, e\}$ shown in Fig. (a). Then Fig. (b) show A with the inverse order R^{-1}. The diagram of R is simply turned upside down to obtain R^{-1}.

a b c d (a) e d e c a (b) b

Fig. (18)

9. *Show that the poset* ({1, 3, 6, 12, 24}, |) *is a lattice.*

Solution. Since every two elements of the given poset have a least upper bound as well as a greatest lower bound, which are the larger and smaller elements respectively, hence, the given poset is a lattice.

10. *Show that every chain is a lattice.*

Solution. Let $(L, \leq)$ be a chain. Let $a, b \in L$. Then, since L is a chain, we have $a \leq b$ or $b \leq a$. Without loss of generality, we assume $a \leq b$. Then $a \vee b = b$ and $a \wedge b = a$. Then both $a \vee b$ and $a \wedge b$ exist in L. Hence, every chain is a lattice.

11. *Show that the posets given below are lattices. Also obtain the Hasse diagram*

(i) $(S_6, /)$ *(ii)* $(S_8, /)$ *(iii)* $(S_{24}, /)$ *(iv)* $(S_{30}, /)$

Solution. (i) Here, we have $S_6 = \{1, 2, 3, 6\}$.

If we take any two elements of S_6, then their lower bound and upper bound will also be in S_6. Therefore, $(S_6, /)$ will be a lattice.

Similarly, we can show that $S_8 = \{1, 2, 4, 8\}$; $S_{24} = \{1, 2, 3, 4, 6, 8, 12, 24\}$ and $S_{30} = \{1, 2, 3, 5, 6, 10, 15, 30\}$ are lattices. The Hasse diagram of these lattices are given below :

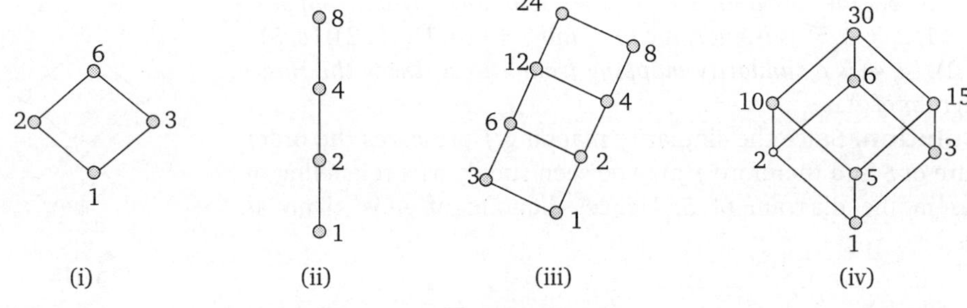

Fig. (19)

12. *Show that the idempotent laws follow from the absorption laws.*

Solution. By absorption laws, we have

$$a \wedge (a \vee b) = a \quad \text{...(1)}$$

and
$$a \vee (a \wedge b) = a \quad \text{...(2)}$$

substituting $a \wedge b$ for b in (1), we get

$$a \wedge (a \vee (a \wedge b)) = a$$

Now, using (2), we get

$$a \wedge a = a$$

Similarly, we can show $a \vee a = a$

Thus idempotent laws follow from the absorption laws.

13. *For any positive integers* $a, b \in N$, *show that*

$$\max\{a, \min(a, b)\} = a$$
$$\min\{a, \max(a, b)\} = a$$

Solution. We consider the lattice $(\mathbf{N}, \leq)$ where $\mathbf{N}$ is the set of natural numbers and the relation $\leq$ is the "less than or equal to".

In this lattice, $\wedge$ and $\vee$ are given by

$$a \vee b = \max\{a, b\}$$
$$a \wedge b = \min\{a, b\}$$

Now, by absorption property

$$a \wedge (a \vee b) = a$$
$$a \vee (a \wedge b) = a$$

Hence, $\min\{a, \max(a, b)\} = a$ and $\max\{a, \min(a, b)\} = a$.

14. *Consider the poset $a = (\{1, 2, 3, 4, 6, 9, 12, 18, 36\}, |)$. Find the greatest lower bound and the least upper bound of the sets $\{6, 18\}$ and $\{4, 6, 9\}$.*

Solution. An integer is a lower bound of $\{6, 18\}$ if 6 and 18 are divisible by this integer. Only such integers are 1 and 6. since $1|6$, 6 is the greatest lower bound of $\{6, 18\}$ is glb $\{6, 18\} = 6$. An integer is an upper bound of $\{6, 18\}$ if and only if it is divisible by 6 and 18 which is 18.

Hence, lub $\{6, 18\} = 18$.

The only lower bound of $\{4, 6, 9\}$ is 1. Hence glb $\{4, 6, 9\} = 1$. The only upper bound of $\{4, 6, 9\}$ is 36. Hence, lub $\{4, 6, 9\} = 36$.

15. *Prove that a finite partial ordered set has*

(i) at most one greatest element *(ii) at most one least element.*

Solution. Assume that a and b are greatest of $(A, \leq)$, since a is the greatest element. We have $b \leq a$. Also b is the greatest element. We have $a \leq b$. Thus $b \leq a$ and $a \leq b$. Since $\leq$ is an antisymmetric for partial ordered set, it follows that $a = b$. Thus, there can not be two different greatest elements of $(A, \leq)$, if it exists. Hence, a finite partial ordered set has at most one greatest element.

16. *Show that in a complemented, distributive lattice, the following are equivalent :*

(i) $a \leq b$ *(ii) $a \wedge b' = 0$* *(iii) $a' \vee b = 1$* *(iv) $b' \leq a'$*

Solution. (i) $a \leq b \Rightarrow a \vee b = b$

$\Rightarrow (a \wedge b) \wedge b' = 0$ $[\because b \wedge b' = 0]$

$\Rightarrow (a \wedge b') \wedge (b \wedge b') = 0$ [By distributivity]

$\Rightarrow a \wedge b' = 0.$

Hence, (i) $\Rightarrow$ (ii).

Again, $a \wedge b' = 0 \Rightarrow (a \wedge b')' = 1 \Rightarrow a' \vee (b')' = 1 \Rightarrow a' \vee b = 1$

Hence, (ii) $\Rightarrow$ (iii).

Now, $a' \wedge b = 1 \Rightarrow (a' \vee b) \wedge b' = b'$ $[\because 1 \wedge b' = b']$

$\Rightarrow (a' \wedge b') \vee (b \wedge b') = b'$ [By distributivity]

$\Rightarrow a' \wedge b' = b'$ $[\because b \wedge b' = 0]$

$\Rightarrow b' \leq a'$

Hence, (iii) $\Rightarrow$ (iv)

Now, $b' \leq a' \Rightarrow a' \wedge b' = b' \Rightarrow (a' \wedge b')' = b$ [By De Morgan's law]

$\Rightarrow a \vee b = b \Rightarrow a \leq b$

$\Rightarrow$ (iv) $\Rightarrow$ (i)

Thus, (i), (ii), (iii), (iv) are equivalent.

17. *Show that the pentagonal lattice is not modular.*

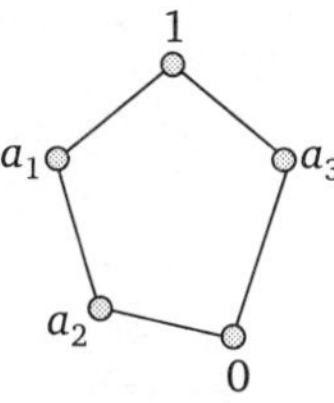

Fig. (20)

Solution. We take $a = a_2$, $b = a_3$, $c = a_1$, then $a < c$, but

$$a \vee (b \wedge c) = a_2 \vee (a_3 \wedge a_1) = a_2 \vee 0 = a_2$$

whereas, $$(a \vee b) \wedge c = (a_2 \vee a_3) \wedge a_1 = 1 \wedge a_1 = a_1$$

and these are unequal.

Hence, pentagonal lattice is not modular.

18. *Let* $L = \{1, 2, 3, 4, 6, 12\}$. *Now consider the lattices* $\{L, |\}$ *and* $(L, \leq)$ *where* $|$ *is "divisibility" relation on L* and $\leq$ *is the usual "less than or equal to" relation on L. The lattice* $\{L, |)$ *and* $(L, \leq)$ *are not isomorphic.*

Solution. Consider any mapping $f\,(L, |) \to (L, \leq)$. Then in $(L, |)$

$$3 \wedge 4 = 1 \quad \Rightarrow \quad f\,(3 \wedge 4) = f\,(1)$$

But $f\,(3) \wedge f\,(4) = f\,(3)$ or $f\,(4)$ depending upon whether $f\,(3)$ is less than or equal to $f\,(4)$ or $f\,(4)$ is less than or equal to $f\,(3)$.

In any case, $f\,(3 \wedge 4) \neq f\,(3) \wedge f\,(4)$.

Thus, $(L, |)$ and $(L, \leq)$ are not isomorphic.

Exercise 4.1

1. Cosider the set **Q** of rational numbers with partial order relation $\leq$, and consider the subset *D* of **Q** defined by
   $D = \{x : x \in \mathbf{Q} \text{ and } Q < x^3 < 15\}$
   (a) Is *D* bounded above or below
   (b) Find sup and inf.
2. Write the dual of each statement
   (a) $(a \wedge b) \vee c = (b \vee c) \wedge (c \vee a)$
   (b) $a \wedge (b \vee c) \geq (a \wedge b) \vee (a \wedge c)$
3. Show that the following weak distributive laws hold for any lattice :
   (a) $a \vee (b \wedge c) \leq (a \vee b) \wedge (a \vee c)$
   (b) $a \wedge (b \vee c) \geq (a \wedge b) \vee (a \wedge c)$
4. Show that lattice *D* of all positive divisors of *n* is a sublattice of the lattice $\mathbf{Z}^+$ under the relation of divisibility.
5. Show that a subset of a linearly ordered poset is a sublattice.
6. Show that a linearly ordered poset is a distributive lattice.
7. Show that the set **Q** of rational numbers is not ordered complete.
8. Show that every finite subset of a lattice has an l.u.b. and g.l.b.
9. Show that in a bounded lattice containing two or more elements, $0 \neq 1$.
10. Let *L* be a lattice, $a, b \in L$ such that $a \leq b$. Is the closed interval $[a, b]$ consisting of all the elements $x \in L$ such that $a \leq x \leq b$ is a sublattice of *L*.
11. Show that the lattice $[\{S_n, \mathbf{R}\} : n = 216]$ is isomorphic to the direct product of lattices $n = 8$ and $n = 27$.
12. Let *L* be a lattice possessing at least two elements. Show that no element of *L* has its own complement.

ANSWERS

1. (a) Yes; **(b)** x, 2;

2. (a) $(a \vee b) \wedge c = (b \wedge c) \vee (c \wedge a)$; **(b)** $(a \vee b) \wedge a = a \vee (b \wedge a)$.

❑❑❑

Chapter 5

Counting Theory

- Counting Principle
- Permutations
- Combination
- Combinatorial identities
- Pigeonhole principle
- Derangements

5.1 INTRODUCTION

Combinatorial analysis deals with permutations, combinations and partitions which determine the number of logical possibilities. Combinatorics is, in essence, the study of arrangements : pairing and grouping, ranking and ordering, selections, elections and allocations. There are following three branches in this subject :

(i) Enumerative Combinatorics

(ii) Existential Combinatorics, and

(iii) Constructive Combinatorics

Enumerative combinatorics is the science of counting which deal with determining the number of possible arrangements of a set of objects under some particular constraints. Existential combinatorics studies problems concerning the existence of arrangements that possess some special property. Constructive combinatorics is the design and study of algorithms for creating arrangements with special properties. Here, we focus our study on enumerative combinatorics.

In this chapter, we present some basic counting techniques that can be used to solve the counting problems.

Now, we begin by describing two basic but useful principles for counting.

5.1.1 First Counting Principle

"If an activity can be constructed in r successive steps and step 1 can be done in n_1 ways; step 2 can then be done in n_2 steps ... and step r can then be done in n_r ways, then the number of possible activities in $n_1, n_2 \ldots n_r$.

- The set of theoretical interpretation of above principle is "If $A \times B$ be the cartesian product of A and B, then $n(A \times B) = n(A) \cdot n(B)$.
- The first counting principle is also known as **Product Rule Principle** or **Fundamental Principle of Multiplication.**
- We may summarize the first counting principle by saying that we multiply together the number of ways of doing each steps when an activity is constructed in successive steps.

5.1.2 Some Examples on First Counting Principle

1. Suppose there are 6 different courses offered in the morning and 8 different courses offered in the afternoon. Then, there will be 6×8 choices for students.
2. If a carpenter has 10 patterns of chairs and 5 patterns of tables, then the ways that he makes a pair of chair and table is 10×5.
3. Let an organization with a membership of 30 choose a president and a vice-president. Then president can be chosen in 30 ways and subsequently, the vice-president in 29 ways. Hence, there are $30 \times 29 = 870$ ways in which the whole choice can be made.
4. Suppose a license plate contains two letters followed by three digits with the first digits not zero, then the ways in which different license plates can be printed, calculated as follows :

   Each letter can be printed in 26 different ways, the first digit in 9 ways and each of the other two digits in 10 ways.

   Hence, 26.26.9.10.10 = 608400 different plates can be printed.

ILLUSTRATIVE EXAMPLES

1. *How many strings of length 2 can be found using the letters ABC if repetitions are allowed?*

Solution. Here, we have three choices for the first letter and three choices for the second letter. Therefore, there are $3 \times 3 = 9$ possible strings, which can be illustrated in the Fig. 1.

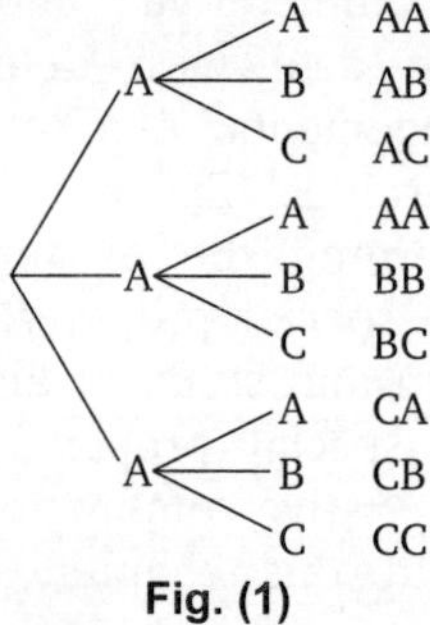

Fig. (1)

2. *In a test paper, there are true false questions. In how many different ways can a student mark the test paper with one answer to each question?*

Solution. Since, each question can be attempted in two ways (either true or false), thus the number of possible ways of answer = 2.2...... 12 times = 2^{12} = 4096.

3. *Show that a set $\{x_1, x_2, x_3 \ldots, x_n\}$ containing n elements has 2^n subsets.*

Solution. Here, a subset can be constructed in n successive steps as follows :

pick or do not pick x_1.

pick or do not pick x_2.

....

pick or do not pick x_n

Each step can be done in two ways. Hence, the number of possible subsets is 2.2............2 (n times) = 2^n.

4. *How many different license plates are available if each plate contains a sequence of two letters followed by four digits (Assume that no sequence of letters are prohibited).*

Solution. Here, we have 26 choices for each of the two letters and 10 that of four each of the four digits. Therefore, the first principle of counting (product rule) these are :

26.26.10.10.10 = 6760000 license places.

5. *How many different eight-bit strings are there? (a bit is either 0 or 1)*

Solution. (a) An eight-bit string can be constructed in eight successive steps.

Select the first bit (zero on one)
Select the second bit (zero or one)
....
Select the eight bit (zero or one)

Since, there are two ways to select each bit by the first counting principle, the total number of eight-bit strings is 2.2 ..... 8 times = 256.

5.1.3 Second Counting Principle

Supose that $X_1, X_2, ..., X_n$ are sets and that the ith set X_i has n_i elements. If $\{X_1, X_2, ..., X_r\}$ is a pairwise disjoint family, then the number of possible elements that can be selected from X_1 or X_2 or .... or X_r is $n_1 + n_2 + ... + n_r$

i.e., the union $\bigcup_{i=1}^{r} X_r$ contains $n_1 + n_2 + ... + n_r$ elements.

- The set theoretical interpretation of above principle is "If A and B are disjoint sets then $n(A \cup B) = n(A) + n(B)$, wehre $n(A)$ denotes the number of elements in A.
- If the events cannot occur simultaneously, then second principle of counting says that the number of ways in which either of two mutually exclusive events can occur is equal to the sum of numbers of ways in which each can occur separately.
- The second counting principle is also known as 'sum rule principle or fundamental principle of addition'.

5.1.4 Some Examples in Second Counting Principle

1. Suppose A is the event choosing a prime number betaeen 10 and 20 and suppose B is the event of choosing an even number between 10 and 20. Then A can occur in 4 ways {11, 13, 17, 19} and B can occur in 4 ways {12, 14, 16, 18}. Then A or B can occur in 4 + 4 = 8 ways, because here even and prime numbers are mutualy exclusive.
2. Suppose there are 10 male teachers and 8 female teachers teaching a class. A student can choose a teacher in 10 + 8 = 18 ways.

ILLUSTRATIVE EXAMPLES

1. *How many eight-bit string being either 101 or 111? (A bit is either 0 or 1).*

Solution. An eight-bit string that begins 101 can be constructed in following five successive steps :

Select the fourth bit (zero or one)
Select the fifth bit (zero or one)
....
Select the eighth bit (zero or one)

Since each of the five bits can be selected in two ways, therefore, by the first counting principle, there are 2.2.2...... 5 times = 2^5 = 32.

Eight bit string that begin with 101. Similarly, there are 32 bit strings that begin with 111.

Now, since there are 32 eight-bit string that begin with 101 and 32 eight-bit string that begin with 111, therefore, there are 32 + 32 = 64 eight-bit string that begin either 101 or 111.

2. *A student can choose a project from one of four lists. The four lists contain 21, 19, 17 and 15 possible projects respectively. How many possible projects are there to choose?*

Solution. The student can choose a project from the first list in 21 ways, from the second list in 19 ways, from the third list in 17 ways and from fourth list in 15 ways. Hence, there are 21 + 19 + 17 + 15 = 72 projects to choose.

3. *In how many ways can we get the sum of 4 or of 8 when two distinguishable dice are rolled? How many ways can we get an even sum*

Solution. The possibilities of getting sum 4 are (1, 3), (2, 2) and (3, 1). Hence, there are 3 ways to get sum 4.

Similarly, we get a sum 8 by the outcomes (2, 6), (3, 5), (4, 4), (5, 3) and (6, 2), *i. e.*, in 5 ways. Therefore, the total number of possibiities in which sum is either 4 or 8 is $3 + 5 = 8$.

Now, to get an even sum, the posibilities are 2, 4, 6, 8, 10 or 12. The possible ways to get the sum 2 is one, to get the sum 4 is three, to get the sum 6 is five, to get the sum 8 is five, to get the sum 10 is three and to get the sum 12 is one. Hence, there are $1 + 3 + 5 + 5 + 3 + 1 = 18$ ways to get an even sum.

4. *In a class there are 18 boys and 6 girls. The teacher wants to select either a boy or a girl to represent the class in a function. How many ways the teacher can make this selection?*

Solution. Here, the teacher is to perform either of the following two jobs :

(i) Select a boy among 8 boys.

or (ii) Select a girl among 6 girls.

The first of these can be performed in 18 ways and the second in 6 ways. By the fundamental principle of addition either of the two jobs can be performed in $18 + 6 = 24$ ways. Hence, the teacher can make the selection of either a boy or a girl in 24 ways.

5. *There are 6 candidates for a classical, 3 for mathematical and 2 for a natural science scholarship.*

(i) In how many ways these scholarship be awarded?

(ii) In how many ways one of these scholarships be awarded?

Solution. Clearly, classical scholarship can be awarded to any one of the six candidates. So, there are 6 ways of awarding the classical. Similarly, mathematical and natural science scholarships can be awarded in 3 and 2 ways respectively. So,

Number of ways of awarding three scholarships

$6 \times 3 \times 2 = 36$ [By fundamental principle of multiplication]

And number of ways of awarding one of the three scholarships $= 6 + 3 + 2 = 11$.

6. *In a monthly test the teacher decides that there will be three questions, one from each of exercise 7, 8 and 9 of the text book. If there are 12 questions in exercise 7, 18 in exercise 8 and 9 in exercise 9. In how many ways can three questions be selected?*

Solution. There are 12 questions in exercise 7. So one question from exercise 7 can be selected in 12 ways. Exercise 8 contains 18 questions. So second question can be selected in 18 ways. There are 9 questions in exercise 9. So, third question can be selected in 9 ways. Hence, there questions can be selected in $12 \times 18 \times 9 = 1944$ ways.

7. *There are 6 multiple choice questions in an examination. How many sequences of answers are possible if the first three have 4 choice and the next three have 5 choice each?*

Solution. Here, we have to perform 6 jobs of answering 6 multiple choice questions. Each one of the first three questions canbe answered in 4 ways and each one of the next three can be answered in 5, different way.

So, the total number of different sequences $= 4 \times 4 \times 4 \times 5 \times 5 \times 5 = 8000$.

8. *For a set of five true/false questions, no student has written all correct answers and no two students have given the same sequence of answers. What is the maximum number of students in the class, for this to be possible.*

Solution. Since a true/false type question can be answered in 2 ways, either by marking it true or false. So, there are 2 ways of answering each of the 5 questions.

$\therefore$ Total number of different sequences of answers $= 2 \times 2 \times 2 \times 2 \times 2 = 2^5 = 32$.

Out of these 32 sequences of answers there is only one sequence of answering all the five questions correctly. But no student has written all the correct answers and different student have given different sequences of answers.

$\therefore$ Maximum number of students in the class = Number of sequences except one sequence in which all answers are correct.

9. *How many numbers are there between 100 and 1000 in which all the digits are distinct?*

Solution. A number between 100 and 1000 has three digits. So , we have to form all possible 3 digit number with distinct digits. We cannot have 0 at the hundred's place. So, the hundred's place can be filled with any of the 9 digits 1, 2, 3... 9. So, there are 9 ways of filling the hundred's place.

Now, 9 digits are left including 0. So, ten's place can be filled with any of the remaining 9 digits in 9 ways. Now, the unit's place can be filled with any of the remaining 8 digits. So, there are 8 ways of filling the unit's place.

Hence, the total number of required numbers $= 9 \times 9 \times 8 = 648$.

10. *How many numbers are there between 100 and 1000 such that 7 is in the unit's place.*

Solution. Every number between 100 and 1000 is a three digit number. So, we have to form 3 digit numbers with 7 at the unit's place by using the digits 0, 1, 2,..., 9. Clearly, repetition of digits is allowed. The hundred's place can be filled with any of the digits from 1 to 9 (zero cannot be there at hundred's place). So hundred's place can be filled in 9 ways.

Now, the ten's place can be filled with any of the digits from 0 to 9. So, ten's place can be filled in 10 ways. Since all the numbers have digit 7 at the unit's place. So, unit's place can be filled in only one way.

Hence, by the fundamental principle of counting the total number of numbers between 100 and 1000 having 7 at the unit's place $= 9 \times 10 \times 1 = 90$.

11. *How many numbers are there between 100 and 1000 such that at least one of their digit is 7 ?*

Solution. Clearly, a number between 100 and 1000 has 3 digits.

$\therefore$ Total number of 3 digit numbers having at least one of their digit is 7

= (total number of three digit numbers)

– (total number of 3 digit numbers in which 7 does not appear at all)

We have to form three digit numbers by using the digits 0, 1, 2, 3,..., 9.

Clearly, hundred's place can be filled in 9 ways and each of the ten's and one's place can be filled in 10 ways. So total numbers of 3 digit numbers $= 90 \times 10 \times 10 = 900$.

Total number of three-digit number in which 7 does not appear at all. Here, we have to form three digit numbers by using the digits 0 to 9 except 7.

So, hundred's place can be filled in 8 ways and each of the ten's and one's place can be filled in 9 ways. So total number of three digit numbers in which 7 does not appear at all $= 8 \times 9 \times 9$.

Hence, total number of 3 digit numbers having at least one of their digit as 7

$$= 9 \times 10 \times 10 - 8 \times 9 \times 9 = 252.$$

12. *How many numbers are there beteen 100 and 1000 which have exactly one of their digit 7 ?*

Solution. A number between 100 and 1000 contains 3 digits. So, we have to form 3 digit number having exactly one of their digit as 7. Such type of number can be divided into following three types.

(I) Those numbers that have 7 in the unit's place but not in any other place.

(II) Those number that have 7 in the ten's place but not in any other place.

(III) Those numbers that have 7 in the hundred's place but not in any other place. Here, required number is the total number of these three types of numbers. We shall now count these three types of numbers separately.

(I) Those three-digit numbers that have 7 in the unit's place but not in any other place. The hundred's place can have anyone of the digits from 0 to 9 except 0 and 7. So, hundred's place can be filled in 8 ways. The ten's place can have anyone of the digits from 0 to 9 except 7. So, the number of ways the ten's place can be filled is 9. The units place has 7. So, it can filled in only one way. Thus, there are $8 \times 9 \times 1 = 72$, numbers of the first kind.

(II) Those three digit numbers that have 7 in the ten's place but not in any other place the number of ways to fill the hundred's place = 8

(By any one of the digits from 1, 2, 3, 4, 5, 6, 8, 9)

The number of ways to fill the ten's place = 1 (By 7 only)

The number of ways to fill the one's place = 9 (By any one of the digits 0, 1, 2, 3, 4, 5, 6, 8, 9).

Thus, there are $8 \times 1 \times 9 = 72$ numbers of the second kind.

(III) Those three digit numbers that have 7 in the hundred's place but not at any other place. The number of ways to fill hundred's place only one and each of the ten's and one's place can be filled in 9 ways. So, there are $1 \times 9 \times 9 = 81$ numbers of the third kind.

Hence, the total number of required type of numbers = 72 + 72 + 81 = 225.

13. *In how many ways can the following prizes be given away to a class of 30 students, first and second in Mathematics, first and second in Physics, first in Chemistry and first in English.*

Solution. Here, we have to give prizes in four subjects and the process of distribution of prizes can be completed by giving prizes in the four subjects.

First and second prizes can be given in Mathematics in (30×29) ways.

First and second prizes can be given in Physics in (30×29) ways.

First prize can be given in Chemistry in 30 ways.

First prize can be given in English in 30 ways.

Hence, the number of ways to give prizes in all the four subjects

$$= (30 \times 29) \times (30 \times 29) \times 30 \times 30 = 68121 \times 10^4$$

14. *How many numbers greater than 1000, but not greater than 4000 can be formed with the digits 0, 1, 2, 3, 4 if : (i) Repetition of digits is allowed? (ii) Repetition of digits is not allowed?*

Solution. (i) Every number between 1000 and 4000 is a four digit number. In thousand's place we can put either 1 or 2 or 3 but not 4. So thousand's place can be filled in 3 ways. Since repetition of digits is allowed, so each of the hundred's place ten's and one's place can be filled in 5 ways. So, total number of numbers between 1000 and 4000, including 1000 and excluding 4000 is $3 \times 5 \times 5 \times 5 = 375$. But we have to find the total number of numbers greater than 1000 but not greater than 4000. Hence, required number of numbers = 375 + 1 (for 4000) – 1 (for 1000) = 375.

(ii) As discussed above thousand's place can be filled in 3 ways. Since repetition of digits is not allowed. So, hundred's place can be filled from the remaining digits in 4 ways. Now, three digits are left, so ten's place can be filled in 3 ways. One's place can be filled in 2 ways.

Hence, required number of numbers = $3 \times 4 \times 3 \times 2 = 72$.

15. *How many three digit odd numbers can be formed by using the digits 1, 2, 3, 4, 5, 6 if*

(i) The repetition of digits is not allowed?

(ii) The repetition of digits is allowed?

Solution. For a number to be odd, we must have 1, 3 or 5 at the unit's place. So, there are 3 ways of filling the unit's place.

(i) Since the repetition of digits is not allowed, the ten's place can be filled with any of remaining 5 digits in 5 ways. Now, four digits are left. So, hundred's place can be filled in 4 ways.

So, required number of numbers = $3 \times 5 \times 4 = 60$

(ii) Since the repetition of digits is allowed, so each of the ten's and hundred's place can be filled in 6 ways. Hence, required number of numbers = $3 \times 6 \times 6 = 108$.

16. *Find the total number of ways in which n-distinct objects can be put into two different boxes so that no box remains empty.*

Solution. Each object can be put either in box B_1 (say) or in box B_2 (say). So there are two choices for each of the n objects. Therefore, the number of choices for n distinct objects is

$$\underbrace{2 \times 2 \times 2 \ldots \times 2}_{n \text{ times}} = 2^n$$

Two of these choice correspond to either the first or the second box being empty. Thus, there are $2^n - 2$ ways in which neither box is empty.

Exercise 5.1

1. In how many ways can an examinee answer a set of ten true/false type questions?
2. A coin is tossed five times and outcomes are recorded. How many possible outcomes are there?
3. There are four places and five post offices. In how many different ways can the parcels be sent by the registered post?
4. A mint prepares metallic calendars specifying months, dates and days in the form of monthly sheets (one plate of each month). How many types of calendars should it prepare to serve for all the possibilities in future years?
5. From Goa to Bombay there are two routes, air and sea. From Bombay to Delhi, there are three routes, air, rail and road. From Goa to Delhi via Bombay, how many kinds of routes are there?
6. A person wants to buy one fountain pen, one ball pen and one pencil from a stationary shop. If there are 10 fountain pen varieties, 12 ball pen varieties and 5 pencil varieties, in how many ways can he select these articles?
7. In a class there are 27 boys and 14 girls. The teacher wants to select 1 boy and 1 girl to represent the class in a fountain. In how many ways can the teacher make this selection?
8. A letter lock consists of three rings each marked with 10 different letters. In how many ways it is possible to make unsuccessful attempt to open the lock?
9. There are 6 multiple choice questions in an examination. How many sequences of answers are possible, if the first three questions have 4 choices each and the next three have 2 each?
10. There are 5 books on Mathematics and 6 books on Physics in a book shop. In how many ways can a student buy (i) a Mathematics book and a Physics book (ii) either a Mathematics books or a Physics book?
11. A number of lock on a suitcase has 3 wheels each labelled with ten digits 0 to 9. If opening of the lock is a particular sequence of three digits with no repeats, how many such sequences will possibles? Also find the number unsuccessful attempts to open the lock.
12. A customer forgets a four-digit code for an Automatic Teller Machine (ATM) in a bank. However remembers that this code consists of digit 3, 5, 6 and 9. Find the largest possible number of trials necessary to obtain the correct code.
13. In how many ways can three jobs I, II and III be assigned to three persons A, B and C if one person is assigned only one job and all are capable of doing each job?
14. How many natural numbers not exceeding 4321 can be formed with digits 1, 2, 3 and 4 if the digits can repeat?
15. How many numbers of six digits can be formed from the digits 0, 1, 3, 5, 7 and 9 when no digit is repeated? How many of them are divisible by 10?
16. If there is six faced die each marked with numbers 1 to 6 on six faces are thrown, find the total number of possible outcomes.
17. A coin is tossed three times and the outcomes are recorded. How many possible outcomes are there? How many possible outcomes if the coin is tossed four times? Five times? n times?
18. Find the number of ways in which 8 distinct toys can be distributed among 5 children.
19. Three dice are rolled. Find the number of possible outcomes in which one die shows 5.

ANSWERS

1. 1024 **2.** 32 **3.** 625 **4.** 14 **5.** 6 **6.** 600 **7.** 378
8. 999 **9.** 512 **10.** (i) 30, (ii) 11 **11.** 720, 719 **12.** 24
13. 6 **14.** 313 **15.** 600, 120 **16.** 216 **17.** 8, 16, 2^n
18. 5^8 **19.** 7^5.

5.2 PERMUTATIONS

"Permutation means arrangement"

Each of the different arrangements which can be made by taking some or all of a number of things is called permutation.

For example : If there are three objects, then the permutations of these objects, taking two at a time, are six.

If we have Δ, O, $*$, then the number of permutations are ΔO, $O*$, $O\Delta$, $*O$, $\Delta *$, $*\Delta$.

So, number of permutations of three different things taken two at a time is 6.

For example : Write down all the permutations of the vowels A, E, I, O, U in English alphabets taking three at a time, starting with A and E only.

Solution. The permutations of vowels A, E, I, O, U taking three at a time and starting with A and E are

AEI, AIE, AEO, AEU, AOE, AUE, AIO, AOI, AIU, AUL, AOU, AUO
EAI, EIA, EAO, EOA, EAU, EUA, EIO, EOI, EIU, EUI, EOU, EUO

Clearly, there are 24 permutations.

(a) Number of permutations of *n* dissimilar things taken *r* at a time

Where n and r are positive integers such that $1 \le r \le n$ denoted by the symbol $P(n\ r)$ or ${}^nP_r$.

$$ {}^nP_r = \frac{n\,!}{(n-r)\,!} = n(n-1)\,(n-2)\ldots(n-r+1) $$

The last factor is $[n-(r-1)] = n-r+1$, where $n\,! = 1.2.3\ldots n$

Note that $n\,! = n\,(n-1)\,! = n\,(n-1)\cdot(n-2)\,!$ etc.

- In the above relation the repetition was not allowed so that if we could fill the first place in n ways then the second could be filled in $(n-1)$, third in $(n-2)\ldots$ and rth in $(n-r+1)$ ways. Hence, the total was $n(n-1)\,(n-2)\ldots(n-r+1)$ ways.
- If repetition is allowed.
- In this case each of the r places can be filled in n ways. Hence, by fundamental theorem all r places can be filled in $n.n.n.n\ldots r$ times $= n^r$ ways.

(b) Number of permutations of *n* dissimilar things taking all at a time

$$ P(n, n) \quad \text{or} \quad {}^nP_n = n(n-1)\,(n-2)\ldots[n-(n-1)] $$
$$ = n(n-1)\,(n-2)\ldots 3.2.1 = n\,! $$

(c) If out of n things, P are exactly alike of one kind q exactly alike of second kind and r exactly alike of third kind and the rest all different than the number of permutations of things taken all at a time $= \dfrac{n\,!}{p\,!\,q\,!\,r\,!}$.

(d) Permutations under certain conditions :

(i) Number of all permutations of n different objects taken r at a time when a particular object is to be always included in each arrangement is $r\,{}^{n-1}P_{r-1}$.

(ii) Number of all permutations of n distinct objects taken r at a time when a particular object never taken in each arrangement is ${}^{n-1}P_r$.

(iii) The number of permutations of n different objects taken r at a time in which two specified objects always occur together is

$$2\,!\,(r-1)^{n-1}P_{r-2}.$$

Condition 1. If there is a difference between clockwise and anticlockwise sequence of objects. Then, if n different things taking r at a time.

The number of circular permutation $= \dfrac{{}^nP_r}{r}$.

Condition 2. When there is no difference between clockwise and anticlockwise sequence of things. In this condition

Taking r things in the n things

The number of circular permutations are $= \dfrac{{}^nP_r}{2r}$.

☛ When $r = n$, it means all the things taken at a time then

(i) In condition I number of circular permutations $= (n-1)\,!$

(ii) In condition II number of circular permutations $= \dfrac{(n-1)\,!}{2}$.

For example : Here, having fixed one thing, the remaining $(n - 1)$ things can be arranged round the table in $(n - 1)\,!$ ways.

☛ If n persons are to be arranged in a row, then the number of arrangements was $n!$ where as for a circular table as shown above the number of arrangements is $(n-1)\,!$.

Particular Case :

Necklace : Number of arrangements of n beads all different to form a necklace or on a circular wire will be $\frac{1}{2}\,(n-1)\,!$

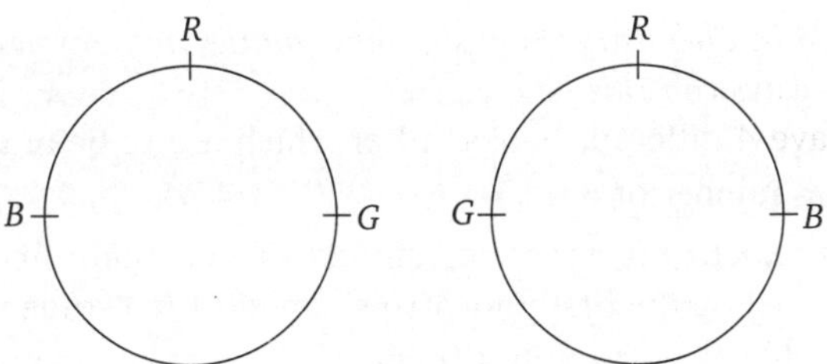

Fig. (2)

R = Ram, G = Ganesh, B = Babu

The above seating arrangement of three persons on the round table are different as shown in figure and that is why we say $(n - 1)\,!$ clockwise and anticlockwise make different arrangements.

In the necklace, we get the same arrangement and that is why we say that total number of arrangement of n beads for forming a necklace is $\frac{1}{2}\,(n-1)\,!$ Here, clockwise or anticlockwise does not change the character of the necklace. It remains the same.

☛ ${}^nP_r = n.{}^{n-1}P_{r-1}$

${}^nP_n = n\,!$

${}^nP_0 = 1.$

5.2.1 Some Special formulae for finding the permutations

Total things	Taking things at a time	Number of permutations
n	r	${}^nP_r = \frac{n!}{(n-r)!}$
n	n	${}^nP_n = n!$
n	p things of same type q things of same type r things of same type and other are diffrent	$\frac{n!}{p!q!r!}$
n	(r) In which each thing can be repeated so many times	n^r
n	(r) While any particular thing always taken	$r.{}^{n-1}P_{r-1}$
n	(r) While any particular thing never taken	${}^{n-1}P_r$
n	(r) In which two specified thing always occur together	$2!(r-1)^{n-r}P_{r-2}$
Circular → n	(r) If there is no difference between clockwise and anticlockwise.	$\frac{{}^nP_r}{r}$
n	(r) If there is a difference between clockwise and anticlockwise.	$\frac{{}^nP_r}{2r}$

5.2.2 Practical Problems on Permutation

1. *Four books, one each in Chemistry, English, Mathematics and Physics are to be arranged on a shelf. In how many ways can this be done?*

Solution. We have 4 different books, all of which are to be aranged on a shelf.

∴ The required number of ways = ${}^4P_4 = 4.3.2.1 = 24.$

2. *Ten students are participating in a race. In how many ways can the first four prizes be won?*

Solution. For awarding the first four prizes, amongst ten students, we need the number of permutations of 10 objects taken 4 at a time.

∴ The required number of ways = ${}^{10}P_4 = 10.9.8.7 = 5040.$

3. *Find how many three letter words can be formed out of the word "Triangle".*

Solution. The word 'TRIANGLE' has 8 different letters. Here, $n = 8$. We want to find 3-letter words.

Therefore, $r = 3$ and the required number of 3 letter words is the number of permutations of 8 objects taken 3 at a time *i.e.*, ${}^8P_3 = \frac{8!}{(8-3)!} = 8.7.6 = 336.$

4. *In how many ways three different rings can be worn in four fingers with at most one in each finger.*

Solution. The total number of ways is same as the number of arrangements of 4 finger, taken 3 at a time. So, required number of ways.

$${}^4P_3 = \frac{4!}{(4-3)!} = \frac{4!}{1!} = 4! = 24.$$

5. *How many three digit numbers are there with no digit repeated (i.e., with distinct digits)?*

Solution. 1st Method : For 3 digit numbers, we have to find the permutations of ten digits 0, 1, 2..., 9 taken 3 at a time, with the condition that 0 is not at the hundred's place (the left-most position)

$\therefore$ The required number of 3 digit numbers $= {}^{10}P_3 - {}^9P_2$

(Here, ${}^{10}P_3$ is the number of all permutations of 10 digits, taken 3 at a time and ${}^9P_2$ is the number of such arrangements with 0 at the hundred's place and therefore, not allowed)

$$= 10.9.8 - 9.8 = 648$$

IInd Method : 3 digit numbers with distinct digits are the following three kinds :

(a) Which do not contain zero : Their number $= {}^9P_3$

(b) Which have zero at ten's place : Their number $= {}^9P_2$

(c) Which have zero at unit's place : Their number $= {}^9P_2$

$\therefore$ Total number of such numbers $= {}^9P_3 + {}^9P_2 + {}^9P_2 = 9.8.7 + 9.8 + 9.8 = 648$

6. *It is required to seat 6 men and 5 women in a row so that the women occupy the even places. How many such arrangements are possible.*

Solution. There should be $6 + 5 = 11$ chairs in a row, out of which 2nd, 4th, 6th, 8th and 10th chairs are even and 5 women are to occupy the same.

$\therefore$ 5 women can be arranged on these 5 chairs in ${}^5P_5 = 5! = 120$ ways.

Again the remaining 6 chairs can be occupied by 6 men in ${}^6P_6 = 6! = 720$ ways.

$\therefore$ Total number of arrangement $= 120 \times 720 = 86400$.

7. *There are six periods in each working day of a school. In how many ways can one arrange 5 subjects such that each subject is allowed at least one period?*

Solution. Five subjects can be arranged in 6 periods in ${}^6P_5$ ways. Now, one period is left and it can be allotted to any one of the five subjects. So, number of ways in which remaining one period can be arranged is 5.

Hence, the total number of arrangements $= {}^6P_5 \times 5 = 3600$.

8. *How many 4 letter words with or without meaning can be formed out of the letters of the word. 'LOGARITHMS' if repetition of letters is not allowed.*

Solution. There are 10 letters in the words 'LOGARITHMS'. So, the number of 4 letter words

= number of arrangement of 10 letters taken 4 at a time $= {}^{10}P_4 = 5040$.

9. *Determine the number of natural numbers smaller than* 10^4, *in the decimal notation in which all the digits are distinct.*

Solution. The required natural numbers consist of 4 digits, 3 digits, 2 digits and one digit.

Total number of 4 digit natural number with distinct digits $= {}^{10}P_4 - {}^9P_3$

Total number of 3 digit natural number with distinct digits $= {}^{10}P_3 - {}^9P_2$

Total number of 2 digit natural number with distinct digits $= {}^{10}P_2 - {}^9P_1$

Total number of one digit natural number = 9

Hence, the required number of natural numbers is equal to

$$({}^{10}P_4 - {}^9P_3) + ({}^{10}P_3 - {}^9P_2) + ({}^{10}P_2 - {}^9P_1) + 9 = 9.9.8.7 + 9.9.8 + 9.9 + 9 = 5274$$

10. *In how many ways 7 pictures can be hung from 5 picture nails on a wall.*

Solution. The number of ways in which 7 pictures can be hung from 5 picture nails on a wall is same as the number of arrangements of 7 things, taking 5 at a time.

Hence, the required number $= {}^7P_5 = \dfrac{7!}{(7-5)!} = \dfrac{7!}{2!} = 2520$.

11. *Find the sum of all the numbers that can be formed with the digits 2, 3, 4, 5 taken all at a time.*

Solution. The total number of numbers formed with the digits 2, 3, 4, 5 taken all at a time = Number of arangement of 4 digits, taken all at a time = ${}^4P_4 = 4! = 24$.

To find the sum of these 24 numbers, we will find the sum of digits at unit's, ten's, hundred's and thousand's places in all these numbers.

Consider the digits in the unit's place in all these numbers. Each of the digits 2, 3, 4, 5 occur in $3!(= 6)$ times in the unit's place.

So, total for the digits in the unit's place in all the numbers $= (2 + 3 + 4 + 5) \times 3! = 84$

Since, each of the digits 2, 3, 4, 5 occur $3!$ times in anyone of the remaining in all numbers, therefore, total desired number $= (2 + 3 + 4 + 5)\ 3! = 84$.

12. *Ten different letters of an alphabet are given words with five letters are formed from these given letters. Determine the number of words which have at least one letter repeated.*

Solution. The number of 5 letters words which can be formed from 10 letter when one or more of its letters is repeated $= 10.10.10.10.10 = 10^5$

The number of 5 letter words which can be formed when none of their letters is repeated

= Number of arrangements of 10 letters by taking 5 at a time.

$= {}^{10}P_5 = 30240$

Hence, the number of 5 letter words which have at least one of their letters repeated

$= 10^5 - 30240 = 69760$.

13. *How many four digits numbers are there with distinct digits?*

Solution. The total number of arangements of ten digits 0, 1, 2, 3, 4, 5, 6, 7, 8, 9 taking 4 at a time is ${}^{10}P_4$. But these arrangements also include those numbers which have 0 at thousand's place. Such numbers are not four digit numbers.

When 0 is fixed at thousand's place, we have to arrange remaining 9 digits by taking 3 at a time. The number of such arangements is ${}^9P_3$.

So, the total number of numbers having 0 at a thousand's lace $= {}^9P_3$

Hence, the total number of four digit numbers

$= {}^{10}P_4 - {}^9P_3 = 5040 - 504 = 4536$.

14. *How many numbers lying between 100 and 1000 can be formed with the digits 1, 2, 3, 4, 5 if the repetition of digits is not allowed?*

Solution. Every number lying between 100 and 1000 is a three digit number. Therefore, we have to find the number of permutations of five digits 1, 2, 3, 4, 5 taken three at a time. Hence, the required number of numbrs $= {}^5P_3 = \frac{5!}{(5-3)!} = \frac{5!}{2!} = 5.4.3 = 60$.

15. *How many different signals can be given using any number of flags from 5 flags of different colours?*

Solution. The signals can be made by using at a time one or two or three or four or five flags.

The total number of signals when r flags are used at a time from 5 flags is equal to the number of arrangements of 5, taking r at a time *i.e.*, ${}^5P_r$. Since r can take values 1, 2, 3, 4, 5. Hence, by the fundamental principle of addition, the total number of signals.

$= {}^5P_1 + {}^5P_2 + {}^5P_3 + {}^5P_4 + {}^5P_5$

$= 5 + 5.4 + 5.4.3 + 5.4.3.2 + 5.4.3.2.1$

$= 5 + 20 + 60 + 120 + 120 = 325$

16. *Three men have 4 coats, 5 waist coats and 6 caps. In how many ways can they wear them?*

Solution. The total number of ways in which three men can wear 4 coats is the number of arrangements of 4 different coats taken 3 at a time. So, three men can wear 4 coats in ${}^4P_3$ ways. Similarly, 5 waist coats and 6 caps can be worn by three men in ${}^5P_3$ and ${}^6P_3$ ways respectively.

Hence, by the fundamental principle of counting, the required number of ways.

$$= {}^4P_3.{}^5P_3.{}^6P_3 = (4!).(5.4.3).(6.5.4) = 172800$$

17. *In how many ways can 6 persons stand in a queue?*

Solution. The number of ways in which 6 different things taken all at a time. Hence, the required numbr of ways = ${}^6P_6 = 6! = 720$

Exercise 5.2

1. Seven athletes are participating in a race. In how many ways can the first three prizes be won?
2. In how many ways can five children stand in a queue?
3. From among the 36 teachers in a school, one principal and one vice principal are to be appointed. In how many ways can this be done?
4. Four letters E, K, S and V are in each, were purchased from a plastic ware house. How many ordered pair letters, to be used as initials, can be formed from them?
5. Four books, one each in Chemistry, Physics, Bilogy and Mathematics are to be arranged in a shelf. In how many ways can this be done.
6. Find the number of different 4 letter words, with or without meanings, that can be formed from the letters of the 'NUMBER'.
7. How many three-digit number are there, with distinct digits, with each digit odd?
8. How many words, with or without meaning, can be formed by using all the letters of the word 'DELHI', using each letter exactly once?
9. How many words, with or withut meaning can be formed by using the letters of the word 'TRIANGLE'.
10. There are two words each of 3 volumes and two words each of 2 volumes, in how many ways can the 10 books be laced on s shelf so that the volumes of the same work re ot separated?
11. There are 6 item in column and 6 items in column B. A student is asked to match each item in column A with an item in column B. How many possible, correct or incorrect, answers are these to this question?
12. How many three digits numbers are there, with no digit repeated?

ANSWERS

1. 10 **2.** 120 **3.** 1260 **4.** 12 **5.** 24 **6.** 360 **7.** 60
8. 120 **9.** 8 ! **10.** 3456 **11.** 720 **12.** 648

5.3 PERMUTATION UNDER CERTAIN CONDITIONS

In this ection we shall discuss permutations where either repetitions of items are allowed or distinction between some of the items are ignored or a particulr item occurs in every arrangement *etc.* Such type of permutations are known as permutation under certain conditions as discussed below.

Type I. Number of al permutations of n different objects taken r at a time, when a particular object is to be always included in each arrangement is $r^{n-1}P_{r-1}$.

Type II. The number of permutations of n distinct objects taken r at a time, when a particular object is never taken in each arrangment, is ${}^{n-1}P_r$.

Type III. The number of permutations of n diffrent objects taken r at a time in which two specified objects always occur together is $2!(r-1)^{n-1}P_{r-2}$.

ILLUSTRATIVE EXAMPLES

1. *In how many ways, can the letters of the word 'PENCIL' be arranged so that.*
(i) N and E are always together *(ii) N is always next to E*
(iii) N and E are never together

Solution. The word PENCIL has 6 different letters which can be arranged in $^6P_6 = 6! = 720$ ways :

(i) When N and E are to be together, we treat these as one letter now the letters or pencil are $P, C, I, L, (N, E)$ *i.e.*, it has 5 letters.
These can be arranged in $^5P_5 = 5! = 120$ ways but the two letters N and E can be aranged themselves among in $^2P_2 = 2$ ways.
$\therefore$ Total number of ways when N and E are always together $= 2.120 = 240$.

(ii) When N is always next to E we keep EN together and consider it as one letter. Now, we have 5 letters to be arranged at 5 places. This can be done in $^5P_5 = 5! = 120$ ways.

(iii) When N and E are never together, the number of ways
= (Total number of ways) – (The number of ways when N and E are together)
$= 720 - 240 = 480$ [Using (i) and (ii)].

2. *In how many ways can 9 examination papers be arranged so that the best and the worst papers are never together?*

Solution. The number of arrangements in which the best and the worst papers never come together can be obtained by subtracting from the total number of arrangements the number of arrangements in which the best and worst come together,

The total number of arrangements of 9 papers $= {}^9P_9 = 9!$

Considering the best and the worst papers as one paper, we have 8 papers which can be arranged $^8P_8 = 8!$ ways. But the best and worst papers can be put together in 2 ! ways.

So, the number of permutation in which the best and the worst papers can be put together (2 !.8!).

Hence, the number of ways in which the best and the worst papers never come together

$$= 9! - 2!.8! = 9.8! - 2.8! = 7.8! = 282240.$$

3. *In how many ways 5 boys and 3 girls can be seated in a row so that no two girls are together?*

Solution. The 5 boys can be seated in a row $^5P_5 = 5!$ ways. In each of these arrangements 6 places are created, shown by the cross-marks, as given below :

$$\times B \times B \times B \times B \times B \times$$

Since, no two girls are to sit toether, so we may arrange 3 girls in 6 places. This can be done in $^6P_3$ ways, *i.e.*, 3 girls can be seated in $^6P_3$ ways.

Hence, the total number of seating arrangements $= {}^5P_5 . {}^6P_3 = 5!6.5.4 = 14400.$

4. *How many even numbers are there with three digits such that if 5 is one of the digits, then 7 is the next digit?*

Solution. We have to determine the total number of even numbers formed by using the given condition. So, at unit's place we can use one of the digits. 0, 2, 4, 6, 8. If 5 is at ten's place then, as per the given condition, 7 should be at unit's place. In such case the number will not be an even number. So 5, cannot be at ten's and one's place. Hence, 5 can be only at hundred's place. Now, two cases arise.

Case I : When 5 is at hundred's place

If 5 is at hundred's place, then 7 will be at ten's place. So unit's lace can be filled in 5 ways by using the digits 0, 2, 4, 6, 8.

So, total number of even numbers = 1.1.5 = 5

Case II : When 5 is not a hundred's place.

Now, hundred's jplace can be filled in 8 ways (0 and 5 cannot be used at hundred's place). In ten's place we can use any one of the ten digits except 5. So, ten's place can be filled in 9 ways. At unit's place we have to use one of the even digits 0, 2, 4, 6, 8. So units place can be filled in 5 ways.

So total number of even numbers = 8.9.5 = 360

Hence, the total number of required even numbers = 360 + 5 = 365.

5. *A code word is to consist of two distinct English alphabets followed by two distinct numbers from 1 to 9. For example CA23 is a code word. How many such code words are these? How many of them end with an even integer?*

Solution. There are 26 English alphabets. So, first two places in the code word can be filled in ${}^{26}P_2$ ways. In last two places we have to use two distinct numbers from 1 to 9. So, last two places can be filled in ${}^{9}P_2$. Hence, by the fundamental principle of counting the total number of code words = ${}^{26}P_2.{}^{9}P_2 = 650.72 = 46800$.

Number of code words ending with an even integer.

In this case the code word can have any of the number 2, 4, 6, 8 at the extreme right position.

So, the extreme right position can be filled in 4 ways. Now, next left position can be filled by two English alphabets in ${}^{26}P_2$ ways.

Hence, total number of code words which end with an even integer

$$= 4.8.{}^{26}P_2 = 4.8.650 = 20800.$$

6. *Five boys and five girls form a line with the boys and girls alternating. Find the number of ways of making the line.*

Solution. 5 boys can be arranged in a line ${}^{5}P_5 = 5!$ ways. Since, the boys and girls are alternating. So, corresponding each of the 5! ways of arrangements of 5 boys we obtain 5 places marked by cross as shown below :

(i) $B_1 \times B_2 \times B_3 \times B_4 \times B_5 \times$ (ii) $\times B_1 \times B_2 \times B_3 \times B_4 \times B_5$

Clearly, 5 girls can be arranged in 5 places marked by cross in $(5!+5!)$ ways.

Hence, the total number of ways of making the line = $5!(5!+5!) = 2(5!)^2$.

7. *How many numbers between 400 and 1000 can be formed with the digits 0, 2, 3, 4, 5, 6 if no digit is repeated in the same number?*

Solution. Numbers between 400 and 1000 consist of three digits with digit at hundred's place greater than or equal to 4. Hundred's place can be filled, by using the digits 4, 5, 6 in 3 ways. Now, ten's and unit's places can be filled by the remaining 5 digits in ${}^{5}P_2$ ways.

Hence, the required number of numbers = $3 \times {}^{5}P_2 = 3 \times \frac{5!}{3!} = 3 \times 20 = 60.$

8. *Find the number of words that can be formed from letters of the word 'LOGARITHM'.*

(i) How many of these being with L?

(ii) How many of these begin with L and end with M?

(iii) How many of these do not begin with L?

(iv) In how many words the vowels are together?

(v) In how many words the vowels are not together.

Solution. The word 'LOGARITHM' has 9 distinct letters.

The number of words formed by these 9 letters = ${}^{9}P_9 = 9!$

(i) For the words which begin with L, we are left with 8 letters and the number of such words = ${}^{8}P_8 = 8!$

(ii) For the words, which begin with L and end with M, we are left with 7 letters and the number of such words $= {}^7P_7 = 7!$.

(iii) For those words which do not begin with L, the first place can be filled with any of the remaining (leaving L) 8 letters. After doing so, we are left with 8 letters to be permuted amongst themselves. This can be done in ${}^8P_8 = 8!$ ways.

∴ Total number of which do not begin with $L = 8.8!$

(iv) There are 3 vowels O, A, I in the word LOGARITHM. When they are to be kept together, we will have 7 different letters $(L, G, R, T, H, M$ and $(O, A, I))$
The number of words formed with them $= {}^7P_7 = 7!$.

But corresponding to each of these words, the vowels in the group (O, A, I) can be arranged in ${}^3P_3 = 3!$ ways.

∴ These required number of words $= 3! \times 7!$.

(v) The number of words in which vowels are not together
= (Total number of words) – (The number of words in which vowels are together)
$= 9! - 3!7! = (9.8 - 6).7! = 66.7!$ [Using (i) and (ii)]

9. *In how many ways can n books on n different subjects be arranged on a book rack so that no two books on particular subject are together?*

Solution. We have n different books. These can be arranged in ${}^nP_n = n!$ ways.

Now, treating the two particular books to be one, we have $n - 1$ books which can be arranged in ${}^{n-1}P_{n-1} = (n-1)!$ ways.

Corresponding to each such arrangement the two books can be arranged amongst themselves in ${}^2P_2 = 2! = 2$ ways.

∴ The two books are together in $2(n-1)!$ ways.

∴ The number of ways in which the two particular books are not together

$$= n! - 2(n-1)! = (n-2)(n-1)!$$

Exercise 5.3

1. How many permutations can be made out of the letters of the word 'EQUATION'? How many of these begin with a vowel?
2. How many words can be formed using the letters of the word 'COURTESY' which begin with C and end with Y?
3. How many words can be formed with the letters of the word 'TRIANGLE' so that vowels always occur in even place?
4. Find the number of words formed from the letters of the word 'HEXAGON' in which (i) Vowels are always together, (ii) Vowels are never together.
5. Find the number of words that can be formed by using the letters of the word 'LAUGHTER'. So that (i) Vowels are never separated, (ii) Consonants are never separated.
6. Find the number of permutations of the letters of the word 'TENDULKAR' in which U is always in the middle.
7. In how many ways can 10 different books be arranged on shelf so that a particular pair of books will be (i) Always together, (ii) Never together.
8. Find the number of permutations of n different things taken r at a time such that two specified things occur together.
9. In how many ways, can 3 books of Mathematics, 4 books of Physics and 2 books of Chemistry be placed on a study table so that the books of the same subjects are always together (it being even that no two books are identical).
10. In how many ways can 5 girls be arranged in a line such that two of them, Rita and Sita, are (i) Always together, (ii) Never together.
11. In how many ways 8 examination papers be arranged so that the best and the worst papers are never together?
12. Find the number of ways in which m gents and n ladies $(m \geq n)$ may be seated in a row. So that no two ladies are together?

ANSWERS

1. 8!, 5.7! **2.** 720 **3.** 2880 **4.** 720, 4320
5.(i) 4320 (ii) 2880 **6.** 8! **7.** $7 \times 9!$, $8\times 9!$ **8.** $2(r-1)\times {}^{n-2}P_{r-2}$
9. 1728 **10.** (i) 48 (ii) 72 **11.** $6\times 7!$ **12.** $m!\times {}^{m+1}P_n$

5.4 PERMUTATION UNDER CERTAIN RESTRICTIONS

In this section we continue our study of permutations and study the following three types of permutations

(a) Permutations in which repetitions are allowed,
(b) Permutations when some of the objects are alike,
(c) Circular permutations.

Type I : The number of permutations of n distinct objects taken r at a time when each may be repeated any number of times, is n^r.

Type II : The number of mutually distinguished permutations of n things, taken all at a time of which p are alike of one kind, q alike of second kind such that $p + q = n$, then $= \dfrac{n!}{p!q!}$.

Type III : Number of circular permutations of n distinct things taken all at a time.

Condition 1. If clockwise and anticlockwise order of arrangements are considered in this condition in n distinct objects taking r at a time, then

Number of circular permutations $= \dfrac{{}^nP_r}{r}$

Condition 2 : When clockwise and anticlockwise arrangements are not considered then in this condition in n distinct objects taking r at a time.

The number of circular permutations are $= \dfrac{{}^nP_r}{2r}$.

- In Type II the number of permutations of n things of which p_1 are alike of one kind; p_2 are alike of second kind, p_3 are alike of third kind ... p_r are alike of rth kind such that $p_1 + p_2 + ... + p_r = n$.

$$\frac{n!}{p_1!p_2!p_3!...p_r!}$$

- In Type II suppose there are r things to be arranged, allowing repetitions. Let further $p_1, p_2, ..., p_r$ be the integers such that the first object occurs exactly p_1 times, the second occurs exactly p_2 times etc. then the total number of permutations of these r objects to the above condition is $\dfrac{(p_1 + p_2 + ... + p_r)}{p_1!p_2!p_3!...p_r!}$

- In the Type III when $r = n$ it means all the things taken at a time then in condition 1 number of circular permutations $= (n-1)!$

n in condition 2 number of permutations $= \dfrac{(n-1)!}{2}$.

ILLUSTRATIVE EXAMPLES

1. *A boy has 5 pockets. In how many ways can he put 4 coins in his pockets?*

Solution. The first coin can be put in any of the 5 pockets and there are 5 ways of pocketing the first coin. Similarly, each one of the other three coins can be pocketed in 5 ways.

$\therefore$ Total number of ways of putting all the 4 coins in his pockets

$$= 5\times 5\times 5\times 5 = 5^4 = 625.$$

2. *In how many ways can 5 letters be posted in 7 letter boxes in a town? If all the letters are not posted in the same letter box, find the corresponding number of ways of posting.*

Solution. Each letter can be posted in 7 ways.

$\therefore$ The number of ways of posting 5 letters $= 7 \times 7 \times 7 \times 7 \times 7 = 7^5 = 16,807$

Now, the number of ways of posting all the letters in the same letter box = 7

$\therefore$ The number of ways of not posting all the letters in the same letter box

$$= 16807 - 7 = 16800.$$

3. *Find the number of 3 digit numbers that can be formed with the digits 1, 3, 5, 7, 9 when (i) a digit may be repeated any number of times, (ii) digits are not repeated.*

Solution. Here, we want to form 3 digit numbers

(i) When repetition of digits is allowed, each of the three places (Unit's, ten's and hundred's) in such number 5 can be filled with any of the five digits (1, 3, 5, 7, 9).

$\therefore$ The required number of 3 digit numbers $= 5 \times 5 \times 5 = 5^3 = 125$

(ii) When repetition of digits is not allowed the required number of 3 digits number

$$= {}^5P_3 = 5.4.3 = 60.$$

4. *How many 4 digit numbers are there, when a digit may be repeated any number of times?*

Solution. We want to form 4 digit numbers, with the help of digits 0, 1, 2, 3,..., 9 when repetition are allowed. Clearly 0 cannot be placed at the thousands place. We can fill this place with any of the digits 1 to 9 *i.e.*, there are 9 ways of filling the thousand's place. Again each of the remaining 3 places (Unit's, ten's, hundred's) can be filled in 10 ways each with 0 to 9 (if repetitions are allowed).

$\therefore$ Total number of 4 digit numbers $= 9 \times 10 \times 10 \times 10 = 9 \times 10^3 = 9000.$

Exercise 5.4

1. In how many ways, can 5 letters be posted in 4 letterboxes.

2. (i) An electric network contains 10 switches such that each switch may have 3 possible connections. How many different switching are there?

(ii) In how many ways can three dice fall if they are thrown simultaneously.

3. In how many ways can 3 prizes be distributed among 4 students, when

(i) no student gets more than one prize.

(ii) a student may get any number of prizes

(iii) no student gets all the prizes.

4. There are 10 varieties of sweets in a restaurant. In how many ways can a dish with 6 pieces of sweets of any variety be arranged?

5. A telegraph post has 5 arms and each arm is capable of 4 distinct positions, including the position of rest. What is the total number of signals that can be made?

6. How many 3 digit numbers can be formed by using the digits 0, 2, 4, 6, 8 while each digit may be repeated any number of times?

7. How many (non zero) numbers less than 10000 can be formed from the digits 0, 1, 2, 4, 6, 8 when a digit may be repeated any number of times?

8. How many different signals can be made from 2 blue, 3 green and 4 scarlet red flags by arranging all of them vertically on a flag staff?

9. There are 6 round stickrs, 2 of them pink and the other 4 orange. It is desired to make a design by posting them in a row. How many such designs are possible?

10. A child has plastic toys bearing the digits 3, 3 and 8. How many three digit numbers can be made.

11. Find the number of arrangements that can be made with the letters of the following words :

(i) APPLE (ii) INDIA

(iii) ENGINEERING

(iv) KURUKSHETRA

(v) INDEPENDENCE

(vi) INTERMEDIATE

12. In how many ways 8 persons be arranged in (i) a line (ii) a circle.

13. Find the number of ways in which 10 different beads can be arranged to form a necklace.

14. In how many ways, can a garland of 15 different flowers be made?same neighbours in any two arrangements?

15. In how many ways, can 8 persons be seated at a round table so that all shall not have the same neighbours in any two arrangements?

ANSWERS

1. 1024 **2.** (i) 3^{10} (ii) 216 **3.** (i) 24 (ii) 64 (iii) 60 **4.** 10^6
5. $4^5 - 1 = 1023$ **6.** 100 **7.** 1295 **8.** 1260 **9.** 15 **10.** 3
11. (i) 60 (ii) 60 (iii) 277200 (iv) 49, 89, 600 (v) 16, 23, 200,
(vi) 1, 99, 58, 400 **12.** (i) 40, 320, (ii) 5040 **13.** 1/29 ! **14.** 7.13 ! **15.** 2520.

5.5 COMBINATIONS

"Combination means Selection"

Definition : *The different groups or selections that can be formed out of a given set of objects by taking some or all of them at a time (without any consideration to the order of their arrangements) are called combinations.*

For Example : List the different combinations formed of three letters *A, B, C* taken two at a time.

Solution. The different combinations formed of three letters *A, B, C* are *AB, AC, BC*.

☛ Any of the following symbols denote the number of combinations of n objects (or things) taken r ($r \le n$) at a time. ${}^nC_r$ or $C(n, r)$ or $\binom{n}{r}$. We shall use ${}^nC_r$ throughout this chapter and the later chapters of this book.

5.5.1 Difference between a Permutation and a Combination

(i) In a combination only selection is made, in a permutation, not only a selection is made but also an arrangement is there in a definite order.

(ii) In a combination, the ordering of the selected objects is immaterial. In a permutation, this ordering is essential.
For example, *AB* and *BA* are same as combinations, but different as permutations.

(iii) Usually (that's, except in trivial cases) the number of permutations exceeds the number of combinations. For instance $C(4, 2) = 6$ and $P(4, 2) = 12$. The trivial case is when $r = 0$ or 1, we have $C(n, 0) = 1 = P(n, 0)$ and $C(n, 1) = n = P(n, 1)$.

(iv) Each combination corresponds to many permutations.
For example, the six permutations 123, 132, 231, 213, 312 and 321 correspond to the same combination 123.

5.5.2 Combination of *n* Different Things Taken *r* at a Time

Theorem 1. *The number of all combinations of n distinct objects taken r at a time is given by*

$${}^nC_r = \frac{n\,!}{(n-r)\,!\,r\,!}$$

Proof. Let x be the number of combinations of n objects taken r at a time. Then

$$x = {}^nC_r \qquad \ldots(1)$$

Since, each combination has r objects, each combination will give rise to $r\,!$ arrangements when arranged amongst themselves.

$\therefore$ x combinations will give rise to $x \times r\,!$ permutations which is the same as the total number of permutations of n objects taken r at a time.

$$\therefore \qquad x\, r\,! = r \;\Rightarrow\; x = \frac{{}^nP_r}{r\,!} \qquad \text{...(2)}$$

By (1) and (2), we get

$$^nC_r = \frac{{}^nP_r}{r\,!} \qquad \text{...(3)}$$

$$\Rightarrow \qquad {}^nC_r = \frac{n\,!}{r\,!(n-r)\,!} \text{ as } {}^nP_r = \frac{n\,!}{(n-r)\,!} \qquad \text{...(4)}$$

We have $^nC_r = \dfrac{n\,!}{(n-r)\,!\,r\,!}$

Putting $r = n$, we obtain $^nC_n = \dfrac{n\,!}{(n-n)\,!\,n\,!} = \dfrac{n\,!}{n\,!\,0\,!} = 1 \quad [\because \; 0\,! = 1]$

Putting $r = 0$, we obtain $^nC_o = \dfrac{n\,!}{(n-0)\,!\,0\,!} = \dfrac{n\,!}{n\,!} = 1$

Thus, $^nC_n = {}^nC_o = 1$

$$^nC_r = \frac{n\,!}{(n-r)\,!\,r\,!} = \frac{1}{r\,!}\left(\frac{n\,!}{(n-r)\,!}\right) = \frac{{}^nP_r}{r\,!}$$

$$^nC_n = 1 = {}^nC_n$$

$$^nC_1 = {}^nP_1 = n$$

Theorem 2. $^nC_r = {}^nC_{n-r} \; (0 \le r \le n)$

Proof. We have $^nC_{n-r} = \dfrac{n\,!}{(n-r)\,![n-(n-r)]\,!}$ [replace r by $n - r$]

$$= \frac{n\,!}{(n-r)\,!\,r\,!} = \frac{n\,!}{r\,!(n-r)\,!} = {}^nC_r$$

We can restate this property as :
If p and q are not negative integers such that $p + q = n$, then $^nC_p = {}^nC_q$
Here, $^nC_p = {}^nC_{n-q} = {}^nC_q$ [By above property]
Conversely, if $^nC_p = {}^nC_q$, then either $p = q$ or $p + q = n$
$\therefore \quad {}^nC_p = {}^nC_q = {}^nC_{n-q} \; [\because \; {}^nC_r = {}^nC_{n-r}]$

Theorem 3. *Prove that* $^nC_r + {}^nC_{r-1} = {}^{n+1}C_r \; (1 \le r \le n)$ *(Pascal's Identity)*

Proof. LHS $= {}^nC_r + {}^nC_{r-1} = \dfrac{n\,!}{r\,!(n-r)\,!} + \dfrac{n\,!}{(r-1)\,![n-(r-1)]\,!}$

$$= n\,!\left[\frac{1}{r\,!(n-r)\,!} + \frac{r}{r(r-1)\,!(n-r+1)\,(n-r)\,!}\right]$$

$$= n\,!\left[\frac{1}{r\,!(n-r)\,!} + \frac{r}{r\,!(n-r)\,!(n-r+1)}\right]$$

$$= \frac{n\,!}{r\,!(n-r)\,!} \cdot \frac{(n-r+1)+r}{(n-r+1)}$$

$$= \frac{n\,!(n+1)}{r\,!(n-r)\,!(n-r+1)}$$

$$= \frac{(n+1)\,!}{r\,!(n-r+1)\,!} = {}^{n+1}C_r.$$

ILLUSTRATIVE EXAMPLES

1. *Evaluate (i)* $^{12}C_4$, *(ii)* $^{15}C_{12}$, *(iii)* $^{49}C_{49}$.

Solution. (i) $^{12}C_4 = \dfrac{12.11.10.9}{4!} = 495$

(ii) $^{15}C_{12} = {}^{15}C_{15-12} = {}^{15}C_3 = \dfrac{15.14.13}{3!} = 455$

(iii) $^{49}C_{49} = \dfrac{49!}{49!(49-49)} = 1.$

2. *If* $^nC_8 = {}^nC_6$, *find* $^nC_2$.

Solution. Given $^nC_8 = {}^nC_6$

Here, $p = 8$, $q = 6$.

We know that, $^nC_p = {}^nC_q \Rightarrow p = q$ or $p + q = n$

Clearly, $p \neq q$

$\therefore \quad p + q = n \quad \Rightarrow \quad 8 + 6 = n \Rightarrow n = 14.$

$$^nC_2 = {}^{14}C_2 = \frac{14.13}{2!} = 91.$$

3. *Find n if* $^{2n}C_3 : {}^nC_2 = 44 : 3$.

Solution. By hypothesis, $\dfrac{^{2n}C_3}{^nC_2} = \dfrac{44}{3}$

$$\Rightarrow \frac{(2n)!}{3!(2n-3)!} \cdot \frac{2!(n-2)!}{n!} = \frac{44}{3} \Rightarrow \frac{2n(2n-1)(2n-2)}{3n(n-1)} = \frac{44}{3}$$

$$\Rightarrow \frac{2(2n-1)2}{3} = \frac{44}{3} \Rightarrow 2n - 1 = 11 \Rightarrow n = 6.$$

4. *Prove that* $^nC_r + 3{}^nC_{r-1} + 3{}^nC_{r-2} + {}^nC_{r-3} = {}^{n+3}C_r$.

Solution. LHS $= {}^nC_r + 3{}^nC_{r-1} + 3{}^nC_{r-2} + {}^nC_{r-3}$

$= ({}^nC_r + {}^nC_{r-1}) + 2({}^nC_{r-1} + {}^nC_{r-2}) + ({}^nC_{r-2} + {}^nC_{r-3})$

$= {}^{n+1}C_r + 2{}^{n+1}C_{r-1} + {}^{n+1}C_{r-2}$ [By $^nC_r + {}^nC_{r-1} = {}^{n+1}C_r$] ...(1)]

$= ({}^{n+1}C_r + {}^{n+1}C_{r-1})\,({}^{n+1}C_{r-1} + {}^{n+1}C_{r-2})$

$= ({}^{n+2}C_r + {}^{n+2}C_{r-1})$ [Using (1)]

$= {}^{n+3}C_r$ [Using (1) again]

= RHS

5. *Find the values of n and r given that* $^nP_r = {}^nP_{r+1}$ *and* $^nC_r = {}^nC_{r-1}$.

Solution. We have $^nP_r = {}^nP_{r+1}$

$$\Rightarrow \frac{n!}{(n-r)!} = \frac{n!}{(n-r-1)!} \Rightarrow \frac{1}{(n-r)(n-r-1)!} = \frac{1}{(n-r-1)!}$$

$\Rightarrow n - r = 1$ ...(1)

Again $^nC_r = {}^nC_{r-1} \Rightarrow \dfrac{n!}{r!(n-r)!} = \dfrac{n!}{(r-1)!(n-r+1)!}$

$$\Rightarrow \frac{1}{r(r-1)!(n-r)!} = \frac{1}{(r-1)!(n-r+1)(n-r)!}$$

$\Rightarrow n - r + 1 = r \Rightarrow n - 2r = 1$

Solving (1) and (2), we get

$$n = 3; r = 2.$$

6. *Prove that the product of m consecutive positive integers is divisible by m !*

Solution. Let the m consecutive positive integers be $n + 1, n + 2, n + 3, \ldots, n + m$. Their product

$$= (n+1)(n+2)(n+3)\ldots(n+m)$$

$$= \frac{n!(n+1)(n+2)(n+3)\ldots(n+m)}{n!}$$

$$= \frac{(n+m)!}{n!} = m!\frac{(n+m)!}{m!n!}$$

$$= m!\frac{(n+m)!}{m![(n+m)-m]!} = m!\,{}^{n+m}C_m$$

which is divisible by $m!$ as ${}^{n+m}C_m$ is a positive integer.

7. *Prove that* $r.{}^nC_r = n\,{}^{n-1}C_{r-1} = (n-r+1)\,{}^nC_{r-1}$ $(1 \le r \le n)$

Solution. First term $= r\,{}^nC_r = r\cdot\dfrac{n!}{r!(n-r)!}$

$$= \frac{n!}{(r-1)!(n-r)!} \qquad [\because r! = r(r-1)!] \qquad \ldots(1)$$

Second term $= n\,{}^{n-1}C_{r-1}$

$$= n\cdot\frac{(n-1)!}{(r-1)![(n-1)-(r-1)]!} = \frac{n!}{(r-1)!(n-r)!} \qquad \ldots(2)$$

Third term $= (n-r+1)\cdot{}^nC_{r-1}$

$$= (n-r+1)\cdot\frac{n!}{(r-1)![n-(r-1)]!}$$

$$= (n-r+1)\cdot\frac{n!}{(r-1)!(n-r+1)\cdot(n-r)!}$$

$$= \frac{n!}{(r-1)!(n-r)!}$$

Equality of these follows from (1), (2) and (3).

8. *Find the value of the expression* ${}^{47}C_4 + \sum_{j=1}^{5} {}^{52-j}C_3$.

Solution. We have, ${}^{47}C_4 + \sum_{j=1}^{5} {}^{52-j}C_3$

$$= {}^{47}C_4 + {}^{51}C_3 + {}^{50}C_3 + {}^{49}C_3 + {}^{48}C_3 + {}^{47}C_3$$

$$= {}^{47}C_3 + {}^{47}C_4 + {}^{48}C_3 + {}^{49}C_3 + {}^{50}C_3 + {}^{51}C_3$$

$$= ({}^{47}C_3 + {}^{47}C_4) + {}^{48}C_3 + {}^{49}C_3 + {}^{50}C_3 + {}^{51}C_3$$

$$= {}^{48}C_4 + {}^{48}C_3 + {}^{49}C_3 + {}^{50}C_3 + {}^{51}C_3 \qquad [\because {}^nC_{r-1} + {}^nC_r = {}^{n+1}C_r]$$

$$= ({}^{48}C_4 + {}^{48}C_3) + {}^{49}C_3 + {}^{50}C_3 + {}^{51}C_3 \qquad [\because {}^nC_{r-1} + {}^nC_r = {}^{n+1}C_r]$$

$$= ({}^{49}C_4 + {}^{49}C_3) + {}^{50}C_3 + {}^{51}C_3 \qquad [\because {}^nC_{r-1} + {}^nC_r = {}^{n+1}C_r]$$

$$= ({}^{50}C_4 + {}^{50}C_3) + {}^{51}C_3 \qquad [\because {}^nC_{r-1} + {}^nC_r = {}^{n+1}C_r]$$

$$= {}^{51}C_4 + {}^{51}C_3 \qquad [\because {}^nC_{r-1} + {}^nC_r = {}^{n+1}C_r]$$

$$= {}^{52}C_4.$$

9. *Let n and r be non-negative integers such that* $1 \le r \le n$ *then show* ${}^nC_r = \frac{n}{r} \cdot {}^{n-1}C_{r-1}$.

Solution. We have ${}^nC_r = \frac{n!}{(n-r)!\,r!} = n \cdot \frac{(n-1)!}{[(n-1)-(r-1)]!\,r\,(r-1)!}$

$$= \frac{n}{r} \cdot \frac{(n-1)!}{[(n-1)-(r-1)]!\,(r-1)!} = \frac{n}{r}\,{}^{n-1}C_{r-1}$$

☛ This property is very useful to find the value of ${}^nC_r$.

For example : ${}^{10}C_3 = \frac{10}{3} \cdot {}^9C_2 = \frac{10}{3} \cdot \frac{9}{2} \cdot {}^8C_1 = \frac{10}{3} \cdot \frac{9}{2} \cdot \frac{8}{1} \cdot 1 = 120$

By using the above property we obtain that :

$${}^nC_r = \frac{n}{r} \cdot \frac{(n-1)}{(r-1)} \cdot \frac{(n-2)}{(r-2)} \cdots \frac{(n-r)}{2} \cdot \frac{n-(r-1)}{1}$$

For example : ${}^9C_4 = \frac{9}{4} \cdot \frac{8}{3} \cdot \frac{7}{2} \cdot \frac{6}{1} = 126.$

10. *If* $1 \le r \le n$, *then* $n\ {}^{n-1}C_{r-1} = (n-r+1) \cdot {}^nC_{r-1}$.

Solution. We have $n \cdot {}^{n-1}C_{r-1} = n \cdot \frac{(n-1)!}{[(n-1)-(r-1)]!\,(r-1)!} = \frac{n!}{(n-r)!\,(r-1)!}$

$$= \frac{(n-r+1) \cdot n!}{(n-r+1)\,(n-r)!\,(r-1)!}$$

$$= (n-r+1)\left[\frac{n!}{(n-r+1)!\,(r-1)!}\right] = (n-r+1) \cdot {}^nC_{r-1}.$$

11. *Show that* ${}^nC_r + 2\,{}^nC_{r-1} + {}^nC_{r-2} = {}^{n+2}C_r$.

Solution. Since we know that ${}^nC_r + {}^nC_{r-1} = {}^{n+1}C_r$

Therefore, ${}^nC_r + 2 \cdot {}^nC_{r-1} + {}^nC_{r-2} = ({}^nC_r + {}^nC_{r-1}) + ({}^nC_{r-1} + {}^nC_{r-2})$

$$= {}^{n+1}C_r + {}^{n+1}C_{r-1} = {}^{n+2}C_r.$$

- ☛ ${}^nC_r = \frac{n!}{(n-r)!\,r!}$
- ☛ ${}^nC_n = {}^nC_o = 1$
- ☛ ${}^nC_r = \frac{{}^nP_r}{r!}$
- ☛ ${}^nC_r = {}^nC_{n-r}$
- ☛ ${}^nC_x = {}^nC_y \Rightarrow\ x = y$ or $x + y = n$
- ☛ ${}^nC_r + {}^nC_{r-1} = {}^{n+1}C_r$ (Pascal's Identity)
- ☛ ${}^nC_r = \frac{n}{r} \cdot {}^{n-1}C_{r-1}$
- ☛ $n \cdot {}^{n-1}C_{r-1} = (n-r+1) \cdot {}^nC_{r-1}$

Exercise 5.5

1. Find the value of each of the following :

(i) ${}^{10}C_8$ (ii) ${}^4C_3$

(iii) ${}^{59}C_{59}$ (iv) ${}^{r+1}C_r$

(v) ${}^{30}C_{27} - {}^{29}C_{27}$

(vi) ${}^{15}C_8 + {}^{15}C_6 - {}^{15}C_6 - {}^{15}C_7$

2. Verify that each of the following is true.

(i) ${}^8C_4 = 2 \cdot {}^7C_4$

(ii) $\frac{{}^9C_4}{{}^8C_3} = \frac{9}{4}$

(iii) ${}^{11}C_5 = 7 \cdot {}^{12}C_5$

(iv) ${}^7C_3 + {}^7C_4 = {}^8C_4$

3. Prove that :

(i) $\sum_{r=1}^{5} {}^5C_r = 31$ (ii) $\sum_{r=3}^{11} {}^9C_{r-2} = 511$

(iii) ${}^2C_1 + {}^3C_1 + {}^4C_1 = {}^3C_1 + {}^4C_2$

(iv) $1 + {}^3C_1 + {}^4C_2 = {}^5C_3$

4. (i) Find r and ${}^rC_8$ given that ${}^{17}C_7 = {}^{17}C_{r-3}$

(ii) Find r and ${}^6C_r$ given that ${}^{13}C_r = {}^{13}C_{2r-5}$

(iii) Find r when ${}^nP_r = 30240$ and ${}^nC_r = 252$

[**Hint :** Use the formula : ${}^nC_r = \frac{{}^nP_r}{r!}$]

(iv) Find r when ${}^nC_{r-1} = {}^nC_{3r}$.

5. Prove that

(i) ${}^{n-1}C_{r-1} + {}^{n-1}C_r = {}^nC_r$

(ii) ${}^nC_r + {}^{n-1}C_{r-1} + {}^{n-1}C_{r-2} = {}^{n+1}C_r$

(iii) ${}^nC_{r-2} + 2\cdot {}^nC_{r-1} + {}^nC_r = {}^{n+2}C_r$

(iv) ${}^nC_m = \sum_{k=0}^{r} {}^rC_k \cdot {}^{n-r}C_{m-k}$

ANSWERS

1.(i) 45 (ii) 364 (iii) 1 (iv) $(r+1)$ (v) 3654 (vi) 0

4.(i) 10;45 (ii) 6; 1 (iii) 5 (iv) $1/4\,(n+1)$

5.6 SOME COMBINATORIAL IDENTITIES

Here, we prove below a few results, already proved with the help of the formula for ${}^nC_r$ with the help of combinatorial arguments.

Theorem 1. *Prove that* ${}^nC_r = {}^nC_{n-r}$ *without using the formula.*

Proof. Every time we select a group of r things out of n given things, we are left with a group of $(n - r)$ things and if we select a group of $(n - r)$ things out of n given things, we are left with a group of r things. Thus, the number of ways of selecting r things out of n things is the same as the number of ways of selection of $(n - r)$ things out of n things. Hence, ${}^nC_r = {}^nC_{n-r}$.

Theorem 2. *Prove by combinatorial arguments that* ${}^{n+1}C_r = {}^nC_r + {}^nC_{r-1}$.

Proof. LHS = ${}^{n+1}C_r$ which is the number of ways of selecting r things out of $(n + 1)$ things. We can make this in the following two ways.

(i) Selection including a particular thing X.

(ii) Selections which exclude this particular thing X.

Every selection in (1) is to include X and therefore, we have to select $(r - 1)$ more things out of the remaining $(n + 1)$ ... $1 = n$ things. This can be done in ${}^nC_{r-1}$ ways.

Again ever selection in (2) is to exclude X and so we have to select r things out of remaining $(n + 1 \ldots 1) = n$ things. This can be done in ${}^nC_r$ ways.

Thus, the total number of selections is ${}^nC_{r-1} + {}^nC_r$.

$\therefore \quad {}^{n+1}C_r \cdot {}^nC_{r-1} + {}^nC_r = {}^nC_r + {}^nC_{r-1}$.

Theorem 3. *Prove that* ${}^nC_o + {}^nC_1 + {}^nC_2 + \ldots + {}^nC_n = 2^n$.

Proof. Let A be any set having n elements. Then ${}^nC_r$ is the number of all the subsets of A having exactly r elements $(0 \le r \le n)$ sum of all the subsets of A

$$= {}^nC_o + {}^nC_1 + {}^nC_2 + \ldots + {}^nC_n \qquad \ldots(1)$$

Now, every subset is uniquely determined by answering the question for each of the n elements, whether that element is or is not in that subset, *i.e.,* for each element there are two ways.

$\therefore$ Total number of such ways $2 \times 2 \times 2 \times \ldots \times 2 = 2^n$ ...(2)

From (1) and (2) we get

$${}^nC_o + {}^nC_1 + {}^nC_2 + \ldots + {}^nC_n = 2^n$$

Corollary : ${}^nC_1 + {}^nC_2 + \ldots + {}^nC_n = 2^n - 1.$

Theorem 4. *Using combinatorial arguments, prove that* $r \cdot {}^nC_r = n \cdot {}^{n-1}C_{r-1}$.

Proof. We know that the number of ways of choosing r objects from the given n objects is ^nC_r. Now, out of each such choice, there are r ways of choosing an object already chosen. Thus, the total number of ways of such a choice = $^nC_r \times r$ ...(1)

Again, we can make this choice in another way.

Let us select particular object X out of the given n objects. This can be done in n ways.

Now, the remaining $(r - 1)$ objects can be selected out of the remaining $(n - 1)$ objects in $^{n-1}C_{r-1}$ ways.

$\therefore$ Total number of ways of such a choice = $n \times {}^{n-1}C_{r-1}$. ...(2)

from (1) and (2) we get,

$$r \times {}^nC_r = n \times {}^{n-1}C_{r-1}.$$

ILLUSTRATIVE EXAMPLES

1. *From a class of 25 students, 4 students are to be chosen for a competition. In how many ways can this be done?*

Solution. We want to choose 4 students out of 25. This can be done in the following ways

$$^{25}C_4 = \frac{25.24.23.22}{4!} = 12650.$$

2. *From 6 boys and 4 girls, 5 students are to be selected for admission to a course. In how many ways, can this be done if there must be exactly 2 girls?*

Solution. We have to select 5 students out of which exactly 2 are to be girls. The number of boys to be selected = 5 – 2 = 3.

Now, we can select, 3 boys from 6 in $^6C_3$ ways and 2 girls from 4 in $^4C_2$ ways.

$\therefore$ Total number of ways = $^6C_3 \times {}^4C_2 = \frac{6.5.4}{3!} \times \frac{4.3}{2!} = 120.$

3. *In a multinational firm, 4 post fall vacant and 30 candidates apply for the post. In how many ways can the selection is made if,*

(i) there is no restriction on selection.

(ii) a particular candidate is always included,

(iii) a particular candidate is always excluded.

Solution. (i) When there is no restriction on selection, we can fill 4 posts by choosing any 4 candidates out of 30 in $^{30}C_4 = \frac{30.29.28.27}{4!} = 27405$ ways

(ii) Since a particular candidate is to be always included, we can select the remaining 3 candidates out of 29.

$\therefore$ The required number of selections = $^{29}C_3 = \frac{29.28.27}{3!} = 3654$

(iii) Since a particular candidate is to be always excluded. Therefore, the choice is to be made for 4 posts from the remaining 29 candidates.

$\therefore$ The required number of selections = $^{29}C_4 = \frac{29.28.27.26}{4!} = 23731.$

4. *A committee of 5 persons is to be formed out of 6 officers and 4 clerks. In how many ways can this be done, when (i) at least 2 clerks are included? (ii) at most 2 clerks are included?*

Solution. (i) When at least 2 clerks are to be included in a committee of 5 persons, we have the following cases :

(a) (2 clerks out of 4) and (3 officers out of 6)
(b) (3 clerks out of 4) and (2 officers out of 6)
(c) (4 clerks out of 4) and (1 officer out of 6)

∴ The number of ways of these selections are

$$^4C_2 \cdot {}^6C_3 = 6.20 = 120$$
$$^4C_3 \cdot {}^6C_2 = 4.15 = 60$$
$$^4C_4 \cdot {}^6C_1 = 1.6 = 6$$

The required number of ways = 120 + 60 + 6 = 186.

(ii) When at most two clerks are to be included in the committee. Such a commitee may consist of 2 clerks and 3 officers

1 clerks and 4 officers

(no clerk) and 5 officers

∴ The number of ways of selection of the committee

$$= {}^4C_2 \cdot {}^6C_3 + {}^4C_1 \cdot {}^6C_4 + {}^4C_o \cdot {}^6C_5$$
$$= 6.20 + 4.15 + 1.6 = 186$$

5. *There are n points in a plane, no three which are collinear which the exception of p points which are collinear. Find the number of (i) Different straight lines and (ii) Different triangles, formed by joining these points.*

Solution. (i) A straight line is obtained by joining pair of points.

∴ The number of straight lines formed by joining n points in pairs = $^nC_2$.

But p be the points that lie in the same straight line.

∴ $^pC_2$ straight lines are lost and instead we get only 1 line on which these points lie.

∴ The required number of straight lines

$$^nC_2 - {}^pC_2 + 1 = \frac{n(n-1)}{2!} - \frac{p(p-1)}{2!} + 1 = \frac{1}{2}[n^2 - n - p^2 + p + 2]$$

(ii) Any three non-collinear points form a triangle.

∴ The number of triangles formed by joining n points taken 3 at a time= $^nC_3$.

Since, p points are collinear, $^pC_3$ triangles are lost.

Thus, the required number of triangles

$$^nC_3 - {}^pC_3 = \frac{1}{6}[n(n-1)(n-2) - p(p-1)(p-2)]$$

6. *(i) How many diagonals are there in a polygon of n sides?*
(ii) A polygon has 44 diagonals. Find the number of its sides?

Solution. (i) A polygon of n-sides has n vertices by joining any two of them, we obtain either a side or a diagonal of the polygon.

∴ The number of straight lines obtained by joining any 2 of its n vertices

$$= {}^nC_2 = \frac{1}{2}n(n-1).$$

These lines include its n sides.

∴ The number of diagonals of the polygon $= \frac{1}{2}n(n-1) - n = \frac{1}{2}n(n-3)$.

(ii) Let the number of sides of the given polygon be n then by part (i),

$$\frac{1}{2}n(n-3) = 44 \quad \text{[Using hypothesis]}$$

$\Rightarrow n^3 - 3n - 88 = 0 \quad \Rightarrow \quad (n-11)(n+8) = 0 \quad \Rightarrow \quad n = 11$ as $n \neq -8$.

∴ This given polygon has 11 sides.

7. *In how many ways 6 red balls and 4 black balls be arranged in row so that no two black balls are together?*

Solution. Let us arrange the red balls (marked R below) and leaving a space between every pair of these red balls $\times R \times R \times R \times R \times R \times R \times$

Now, 4 black balls can be arranged in ${}^7C_4 = {}^7C_3 = \dfrac{7.6.5}{3!} = 35$ ways.

8. *From a class of 12 boys and 10 girls, 10 students are to be chosen for a competition, including at least 4 boys and 4 girls. The two girls who won the prizes last year should be included. In how many ways can this selection be made?*

Solution. For selecting 10 students, last year winner, two girls, are to be always included in addition to at least 4 boys and 4 girls.

$\therefore$ We have to select 8 students (out of 12 boys and 8 girls) choosing at least 4 boys and at least 2 girls. This can be possible in the following case :

(i) 4 boys and 4 girls (ii) 5 boys and 3 girls (iii) 6 boys and 2 girls

$\therefore$ The required number of ways

$$= {}^{12}C_4 \cdot {}^8C_4 + {}^{12}C_5 \cdot {}^8C_2 + {}^{12}C_6 \cdot {}^8C_2$$

$$= 495.70 + 792.56 + 924.28$$

$$= 34650 + 44352 + 25872 = 104874.$$

9. *Out of 21 consonants and 5 vowels, how many different words can be formed each consisting of 2 consonants and 3 vowels?*

Solution. We have to select 2 consonants out of 21 and 3 vowels out of 5.

$\therefore$ The number of ways of making this selection

$$= {}^{21}C_2 \cdot {}^5C_3 = \frac{21.20}{2!} \cdot \frac{5.4.3}{3!} = 210.10 = 2100$$

These are groups containing 2 consonants and 3 vowels.

Each of these 2100 groups contain 5 letters and these can be arranged in $5! = 120$ ways.

$\therefore$ Total number of words formed $= 2100 \times 120 = 252000.$

10. *In any examination, a candidate has to pass in each of the five subjects. In how many ways can he fail?*

Solution. Each subject can be dealt in 2 ways, either the candidate passess in it or fails in it. So, 5 subjects can be dealt in 2^5 ways. But this includes the case in which the candidate passes in all 5 subjects. Excluding this, the number of ways, in which a candidate fails

$$= 2^5 - 1 = 32 - 1 = 31.$$

Exercise 5.6

1. Find the number of ways in which a sub-committee of 5 members be elected from a group of 20 persons.
2. A question paper of mathematics contains 8 questions. Find the number of ways in which an examinee can answer if he is asked to attempt only 6 questions.
3. In an examination paper of mathematics, there are 12 questions. In how many ways can an examinee choose 8 questions in all if first two questions are compulsory ?
4. Find the number of ways in which we can select a committe of 4 persons consisting of exactly 2 women out of the office staff consisting of 7 men and 4 women.
5. A question paper in commerce contains 7 questions in part A and 5 questions in part B. In how many ways, can a candidate attempt the question paper, if he has to select 8 questions in all selecting at least 3 questions from each part ?
6. A cricket club consists of 16 members of which only 6 can bowl. Find the number of ways to choose a cricket eleven so as to include at least 4 bowlers.
7. A man has 12 friends. In how many ways can he invite for a lunch, one or more (*i.e.*, at least one) out of them ?

8. Show that the number of all possible ways, in which a student may attempt one or more questions from eight given questions, each question having an alternative is $3^8 - 1$.

9. There are 7 questions in a paper. In how many ways can a candidate solve two or more questions ?

10. How many different products can be obtained by multiplying two or more of the numbers 3, 5, 7, 9 (without repetition)?

11. In an election, there are 7 candidates of whom 4 are to be elected and a voter can vote for any number of candidates subject to a maximum of four. In how many ways can a voter cast this vote ?

12. In how many ways, can a class of 60 students be divided into 2 sections of 30 students each?

13. In how many ways, can 6 books be equally divided into
(i) 3 groups of 2 each
(ii) 2 groups of 3 each?

14. Find the number of ways in which (i) a selection (ii) an arangement of 4 letters can be made from the letters of the word 'COMBINATION'.

ANSWERS

1. 15, 504 **2.** 28 **3.** 210 **4.** 126 **5.** 420 **6.** 3,312 **5.** 4, 095

9. 120 **10.** 11 **11.** 98 **12.** $\dfrac{60!}{2!(30!)}$

13.(i) 15 (ii) 10 **14.** (i) 136, (ii) 2, 454

5.7 PRINCIPLE OF INCLUSION AND EXCLUSION

(i) For any two events A and B, the inclusion and exclusion formula is

$$P(A \cup B) = P(A) + P(B) - P(A \cap B)$$

where, $P(A)$ denotes the probability of the event A.

Similarly, for n events $(A_1, A_2, ..., A_n)$ the inclusion and exclusion formula is as follows

$$P(A_1 \cup A_2 \ldots \cup A_n) = \sum_{i=1}^{n} P(A_i) - \sum\sum_{i<j} P(A_i \cup A_j)$$
$$+ \sum\sum\sum_{i<j<r} P(A_i \cap A_j \cap A_r) + \ldots \pm P(A_1 \cap {}_2 \ldots \cap A_n)$$

(ii) Let $n(A)$ denotes the number of elements in the set A.

Then for any two sets A and B, the formula is

$$n(A \cup B) = n(A) + n(B) - (n\ A \cap B)$$

for three sets A, B and C, the formula is as follows

$$n(A \cup B \cup C) = n(A) + n(B) + n(C) - n(A \cap B) = n(A \cap C) - n(B \cap C)$$
$$+ n(A \cap B \cap C)$$

ILLUSTRATIVE EXAMPLES

1. *How many numbers from 0 to 999 are divisible by either 5 or 7?*

Solution. Let A denotes the number divisible by 5, B denotes the number divisible by 7. Therefore, the required number which is divisible by 5 and 7 is given by

$$n(A \cup B) = n(A) + n(B) - n(A \cap B)$$
$$= 200 + 142 - 28 \qquad (\because n(A) = 200, n(B) = 142, n\ (A \cap B) = 28)$$
$$= 314.$$

2. *There are 14 books in an almirah, out of which 6 are novels, 7 books are published in 1984 and 3 novels are published in 1984. What is the number of books which are either novels* *or published in 1984 or both? What is the number of non-novels that were not published in 1984 ?*

Solution. Let A denotes the set of books that are novels and B denotes the set of books that are published in 1984. Then, we have

$$n(A) = 6, n(B) = 7, n(A \cap B) = 3$$

Now, we have $n(A \cup B) = n(A) + n(B) - n(A \cap B)$

$$= 6 + 7 - 3 = 10$$

Hence, the number of non-novels that were not published in 1984 is $14 - 10 = 4$.

5.8 THE PIGEONHOLE PRINCIPLE

Theorem 1. *If n pigeons are assigned to m pigeonholes, then at least one pigeonhole contains two or more pigeons* ($m < n$).

Proof. First of all , label the m pigeonholes with the numbers 1 through m and n pigeons with the numbers 1 through n. Now, start with pigeon 1, and assign each pigeon in order to the pigeonhole with the same number. This gives assignments to as many pigeons as possible to the individual pigeonholes. Now, since the number of pigeonholes are less than the number of pigeons ($m < n$), therefore $n - m$ pigeons are left without having assigned a pigeonhole. Hence, there will be at least one pigeonhole which could be assigned to a second pigeon.

☛ The pigeonhole principle can also be stated as "If n pigeonholes are occupied by $n + 1$ or more pigeons, then at least one pigeonhole is occupied by more than one pigeons.

For Example :

1. In a class of 101 students there must be at least two students scoring same marks in a paper having maximum marks 100.
2. In a department of 13 teachers, two of the teachers were born in the same month.
3. In a group of 27 English words, there must be at least two words beginning with the same letter as there are 26 letters in English alphabets.

5.8.1 Generalized Pigeonhole Principle

Theorem 2. *If n pigeons are assigned to m pigeonhole, then one of the pigeonhole must contain at least* $\left\{\left[\frac{n-1}{m}\right]+1\right\}$ *pigeons, where* $\left[\frac{n-1}{m}\right]$ *denotes the largest integer less than or equal to the rational number.*

Proof. Let if possible each pigeonhole does not contain more than $\left[\frac{n-1}{m}\right]$ pigeons, then there will be at most $m \cdot \left[\frac{n-1}{m}\right] + \frac{m\,(n-1)}{m} = n - 1$ pigeons in all, which is a contradiction. Hence, for a given m pigeonholes, one of these must contain at least $\left[\frac{n-1}{m}\right] + 1$ pigeons.

ILLUSTRATIVE EXAMPLES

1. *Show that among any* $(n + 1)$ *positive integers not exceeding* $2n$, *there must be an integer that divides one of the other integers.*

Solution. Write each of the $(n + 1)$ integers $a_1, a_2, \ldots, a_{n+1}$ as a power of 2 times an odd integer, *i.e.*, $a_i = 2b_i$ for $i = 1, 2, \ldots, (n + 1)$, where i is a non-negative integer less than $2n$. Here, we observe that there are only n odd positive integers less than $2n$. Hence, from the pigeonhole principle two of the integers $b_1, b_2, \ldots b_{n+1}$ must be equal. Therefore, there exists integers i and k such that $b_i = b_k$. Let b be the common value of b_i and b_k. Then $a_i = 2b_{j1}$ and $a_k = 2b_{j2}$. If $j_1 < j_2$, then a_i divides a_k and if $j_1 > j_2$ then a_k divides a_i... .

2. *Show that if 30 dictionaries in a library contain a total of 61327 pages, then one of the dictionaries must have at least 2045 pages.*

Solution. Let the pages and dictionaries denote the pigeons and pigeonholes respectively.

We may assign each page of the dictionary, then by the extended pigeonhole principle, one dictionary must contain at least $\left[\dfrac{61327}{30}\right] + 1 = 2045$ pages.

3. *Find the minimum number n of integers to be selected from S = {1, 2, ..., 9} so that*

(i) the sum of two of the n integers is even.

(ii) the difference of two of the n integers is 5.

Solution. (i) We know that the sum of two even integers of two odd integers is even. Consider the subsets {1, 3, 5, 7, 9} and {2, 4, 6, 8} of S as pigeonholes. Hence, by pigeonhole principle $n = 3$.

(ii) Consider the five subsets {1, 6}, {2, 7}, {3, 8}, {4, 9}, {5} of S as pigeonholes. Hence by pigeonhole's principle, $n = 6$ will guarantee that two integers will belong to one of subsets and their difference is 5.

4. *Show that if 20 persons are selected for presenting a cultural programme, then one may select a subset of 3 so that all 3 would be able to present their programmes on the same day of the week.*

Solution. Assign each person to the day of the week on which he would present his programme. The 20 persons (pigeons) in this way are being assigned to 7 pigeonholes (days of the week). Hence, by pigeonhole's principle at least $\left[\dfrac{20-1}{7}\right] + 1 = 3$ of the persons must present their programmes on the same day of the week .

5.9 DERANGEMENTS

At a party, n persons check in their cellular phones. When they leave, the phone returned randomly and unfortunately, no one receives the correct cellular. Let D_n be the number of ways n persons can all receive the wrong phone. Then the sequence $D_1, D_2, \ldots$ satisfy the recurrence relation

$$D_n = (n - 1)(D_{n-1} + D_{n-2})$$

We see that D_n is the number of permutations $m_1, m_2, \ldots, m_n$ of 1, 2, ..., n, where $m_i \neq i$ for $i = 1, 2, \ldots, n$. Such permutations are called derangements. Therefore, "A derangement is a permutation of objectives that leaves no objects in its original position."

Theorem 1. *The number of derangement of a set with n objects*

$$D_n = n!\left[1 - \frac{1}{1!} + \frac{1}{2!} - \frac{1}{3!} + \ldots (-1)^n \frac{1}{n!}\right]$$

elements to a set with n elements.

Proof. Let a permutation have property P_i (if i is fixed). We know that the numbers of derangements is the number of permutations having none of the properties P_i for $i = 1, 2, \ldots, n$

i.e., $$D_n = N(P_1', P_2', P_3', \ldots, P_n')$$

Now, by the principle of inclusion and exclusion, we have

$$D_n = N(P_1' P_2' \ldots P_n') = N - \sum_i N(P_i) + \Sigma N(P_i P_j) \ldots (-1)^n N(P_1 P_2 \ldots P_n) \qquad \ldots(1)$$

where, N denotes the number of permutations of n elements, *i.e.*, $N = n!$

If i is fixed, then $N(P_i) = (n - 1)!$

Similarly, $N(P_iP_j) = (n - 2)$ (Two elements are fixed)

In general $N(P_{i_1} P_{i_2} \ldots P_{i_m}) = (n - m)!$

Since there are $C(n, m)$ ways to select m elements from n elements, we have

$$\Sigma N(P_i) = C(n, 1)\,(n - 1)!$$

$$\Sigma N(P_i\, P_j) = C(n, 2)\,(n - 2)!$$

$$\ldots\ldots\ldots\ldots\ldots\ldots\ldots\ldots\ldots$$

$$\Sigma N(P_{i_1}\, P_{i_2} \ldots P_{i_m}) = C[n, m]\,(n - m)!$$

Putting all these values in (1), we get

$$D_n = n! - C(n, 1)\,(n - 1)! + C(n, 2)\,(n - 2)! + \ldots + (-1)\frac{nn!}{n!\,0!}\,0!$$

$$\Rightarrow \quad D_n = n!\left[1 - \frac{1}{1!} + \frac{1}{2!} + \ldots + (-1)^n \cdot \frac{1}{n!}\right]$$

5. *Find the number of ways of putting 5 letters in the envelope in such a way that no letter is being put in the right envelope.*

Solution. Here, $n = 5$.

Hence, $D_n = n!\left[1 - \frac{1}{1!} + \frac{1}{2!} + \ldots (-1)^n \frac{1}{n!}\right]$

$$= 5!\left[1 - \frac{1}{1!} + \frac{1}{2!} + \ldots (-1)^5 \frac{1}{5!}\right]$$

$$= 5!\left[1 - \frac{1}{1!} + \frac{1}{2!} - \frac{1}{3!} + \frac{1}{4!} - \frac{1}{5!}\right]$$

$$= 5!\left[1 - 1 + \frac{1}{2} - \frac{1}{6} + \frac{1}{24} - \frac{1}{120}\right]$$

$$= 120 - 120 + 60 - 20 + 5 - 1$$

$$= 44$$

Exercise 5.6

1. Show that among any group of five integers (not necessarily consecutive) there are two integers with the same remainder when divided by 4.
2. If there are 13 members in a committee, show that at least 2 of them must have their birthdays in the same month.
3. Show that the number of distributions of n objects into m distinct boxes with the object in each box is a definite order m^n.
4. Consider the inside region of a regular hexagon into six equilateral triangles. Show that if each of the seven points are taken in this region, then two of them must not be farther apart than 1 unit.
5. Show that the minimum number of students required in a class to be sure that at least five will receive the same grades if there are four possible grades A, B, C and D is 5.
6. In a group of six people, each pair of individuals consist of two friends or two enemies. Show that there are either three mutual friends or three mutual enemies in the group.
7. Consider a tournament in which each of n players play against every other player and each player wins at least once. Show that there are at least two players having the same number of wins.
8. Show that any set of seven distinct integers include two integers x and y such that either $x + y$ or $x - y$ is divisible by 10.
9. If we select 13 numbers from the set $\{1, 2, \ldots, 24\}$, then show that there will be two pairs of numbers in which one of them will be a multiple of another.
10. Show that if any five numbers from 1 to 8 are chosen, then two of them will add upto 9.

11. How many solutions are there for the equation $x_1 + x_2 + x_3 + x_4 = 20$?

(i) If all x_i must be non-negative integers

(ii) If all x_i must be non-negative integers and x_4 is at most 10.

12. How many solutions does the equation $x_1 + x_2 + x_3 = 11$ have, where x_1, x_2 and x_3 are non-negative integers?

13. Find the derangements of {1, 2, 3, 4}.

14. Determine the number of different message that can be represented by sequence of 4 dashes and 3 dots.

ANSWERS

12. 78 **13.** 9 **14.** 35

❑❑❑

Chapter

6

Properties of Integers

- Ordering of the Integers
- Mathematical Induction
- Basic representation theorem
- Divisibility theory
- Greatest Common Divisor
- Relatively prime Integers
- Algorithm to find G.C.D. : Investigation of the set of Integers $\{bx + cy\}$
- Greatest common divisor of more than two Integers
- Least common multiple
- Least common multiple of *n* integers
- Fibonacci sequence
- Linear diophantine equations
- Diophantine equation in three or more unknowns
- Diophantine equation of the second degree

6.1 INTRODUCTION

The theory of numbers mainly deals with properties of the natural numbers 1, 2, 3, 4, ... also called the positive integers. These numbers together with the negative integers and zero, form the set of integers. Properties of these numbers have been studied from earliest time. For example, an integer is divisible by 3, if and only if the sum of its digits is divisible by 3 as in the number 852 with sum of digits $8 + 5 + 2 = 15$. The equation $x^2 + y^2 = z^2$ has infinitely many solutions in positive integers, such as $3^2 + 4^2 = 5^2$, whereas $x^3 + y^3 = z^3$ and $x^4 + y^4 = z^4$ have none. There are infinitely many prime numbers, where a prime is a natural number such as 31 that can not be factored into two smaller natural numbers. Thus 33 is not a prime, because $33 = 3. 11$. The fact that the sequence of primes $2, 3, 5, 7, 11, 17, \ldots,$ is endless was known to Euclid. Also known to Euclid was the result that $\sqrt{2}$ is an irrational number.

The theory of numbers is closely tied to the other areas of mathematics, most especially to abstract algebra, but also to linear algebra, combinatorics, analysis, geometry and even topology.

6.2 ORDERING OF THE INTEGERS

The set of positive integers, Z^+, having the following two properties.

(1) The law of Trichotomy : If $a \in Z$, the one and only one of the following is true

(i) $a \in \mathbf{Z}^+$ (ii) $a = 0$ (iii) $-a \in \mathbf{Z}^+$

(2) If $a, b \in \mathbf{Z}^+$, then $a + b \in \mathbf{Z}^+$ and $ab \in \mathbf{Z}^+$

It is clear that $0 \notin \mathbf{Z}^+$ and $1 \in \mathbf{Z}^+$

Definition : *If $a, b \in \mathbf{Z}$ and $a - b \in \mathbf{Z}^+$, then we say that a is greater than b and we write $a > b$.*

If $a < b$ or $a = b$, we write $a \leq b$ and if $a > b$ or $a = b$, we write $a \geq b$.

Clearly, a is positive if and only if $a > 0$ and a is negative if and only if $a < 0$. Also, if $a \in \mathbf{Z}$, then one and only one of the following is true.

$$a \in \mathbf{Z}^+, \; a = 0, \; -a \in \mathbf{Z}^+$$

i.e., $\qquad a > 0, \qquad a = 0, \qquad a < 0$

6.2.1 Well Ordering Principle

Let S be a non-empty subset of Z. If there exists an integer $m \in S$ such that $x \geq m$ for all $x \in S$, then m is said to be the smallest or the least integer in S, *i.e.*, S has a least member. If there exists an integer $n \in S$ such that $x \leq n$ for all $x \in S$, then n is said to be the greatest integer in S.

Statement : *The well ordering principle states that every non-empty subset of the set of positive integers has a least member.*

6.2.2 Archimedian Property

Statement : *For any two positive integers a* and *b, there exists a positive integer n such that na ≥ b.*

Proof : Let if possible there exists no positive integer n such that $na \geq b$. Then, we have for any positive n, $na < b$.

Define $A = \{b - na : n \in \mathbf{N}\}$

Clearly, A is a non-empty subset of positive integers. Then, by well ordering property, we have a smallest number, say $b - n_1 a$ in A.

Now, $b - (n_1 + 1)\,a$ also belongs to A and $b - (n_1 + 1)\,a = b - n_1 a - a < b - n_1 a$.

Therefore, $b - (n_1 + 1)\,a$ is smaller to $b - n_1 a$ in A, which is a contradiction.

Hence, there exists a positive integer n such that $na \geq b$.

6.3 MATHEMATICAL INDUCTION

'Induction' means the method of inferring a general statement from the validity of particular cases. But in mathematics, this kind of inference is not allowed. If a statement is true for a large number of cases, even then we can not say that the general statement is true for all n unless we establish a relation which is always true. For example, consider

$$f(n) = n^2 + n + 41$$

Putting $n = 1, 2, 3, \ldots$ in turn, we obtain 43, 47, 53, ..., 151 which all are prime numbers. On the basis of these results, we assert that the substitution of any positive integer for n in $f(n)$ will always yield a prime number. But this reason is fallacious. In fact $f(n)$ yields a prime number for $n = 1, 2, \ldots, 39$, but for $n = 40$, we have

$$f(40) = 40^2 + 40 + 41$$
$$= 40^2 + 2 \times 40 + 1 = (40 + 1)^2 = 41 \times 41, \text{ which is composite.}$$

This example shows that we can not make general assertion with respect to any n unless results hold for $n = m + 1$, whenver it hold for $n = m$.

Definition : Let $n \in \mathbf{N}$ and $P(n)$ denotes a certain statement or formula or theorem. Then $P(n)$ holds for every natural number n if

(i) it holds for $n = 1$, and

(ii) it holds for $n = m + 1$, wherever it holds for $n = m$.

- In order to prove a certain statement for all natural numbers n, it is essential to establish both the condition (i) and (ii).
- As a matter of fact, condition (i) creates the basis for carrying out induction and condition (ii) gives us right of an unlimited automatic extension of this basis.
- In some problems, the result does not hold, for $n = 1$ or 2, ... but the result is true for a natural number > 1. Let us assume that result is true for a natural number $m > 1$, then

we assume that result is true for $n = k$ (a natural number $> m$). In the next step, we shall show that the result is true for $n = k + 1$ by using the assumption. In this case, this is done, we say that the result is proved for every natural number $\geq m$ by the principle of mathematical induction.

WORKING RULES

Let $P(n)$ be a statement of a theorem for all $n \in \mathbf{N}$. In order to prove any result by method of induction, we proceed as follows :
Step (i) Verify the result is true for $n = 1$
Step (ii) Suppose that result is true for $n = m$
Step (iii) Show that result is true for $n = m + 1$, then $P(n)$ is true for all $n \in \mathbf{N}$.

ILLUSTRATIVE EXAMPLES

1. *By mathematical induction, prove that* $1 + 2 + 3 + \ldots + n = \dfrac{n(n+1)}{2}, \forall\, n \in \mathbf{N}$

Solution. Here, we have $P(n) = 1 + 2 + \ldots + n = \dfrac{n(n+1)}{2}$ ...(1)

For $n = 1, P(1) = \dfrac{1(1+1)}{2} = 1$

Also, for $n = 1$, L.H.S. = 1. Therefore, result is true for $n = 1$.
Let us assume result is true for $n = m$.

i.e., $$1 + 2 + 3 + \ldots + m = \frac{m(m+1)}{2} \qquad ...(2)$$

Adding $(m + 1)$ on both sides of (2), we get

$$1 + 2 + 3 + \ldots + m + (m+1) = \frac{m(m+1)}{2} + (m+1) = (m+1)\left(\frac{m}{2} + 1\right)$$

$$1 + 2 + 3 + \ldots + m + (m+1) = \frac{(m+1)(m+2)}{2} = \frac{(m+1)[(m+1)+1]}{2}$$

Therefore, the result is true for $n = m + 1$.
Hence, by principle of mathematical induction, result (1) is true for all $n \in N$.

2. *Use the principle of mathematical induction to prove that*

$$1 + 4 + 7 + \ldots + (3n - 2) = \frac{n(3n-1)}{2}$$

Solution. We have to show that

$$P(n) = 1 + 4 + 7 + .. + (3n - 2) = \frac{n(3n-1)}{2} \qquad ...(1)$$

Step (1) : For $n = 1$

$$P(1) = 1 = \frac{1(3.1 - 1)}{2} = \frac{1 \times 2}{2} = 1$$

$\therefore$ Result is true for $n = 1$.

Step (2) : Assume that result is true for $n = m$.

i.e., $$P(m) = 1 + 4 + 7 + \ldots + (3m - 2) = \frac{m(3m-1)}{2} \qquad ...(2)$$

Step (3) : Adding $[3(m+1) - 2]$, *i.e.*, $3m + 1$ on both sides of (2), we get

$$1 + 4 + 7 + \ldots + (3m - 2) + (3m + 1) = \frac{m(3m-1)}{2} + 3m + 1$$

$$= \frac{m(3m-1) + 6m + 2}{2} = \frac{3m^2 + 5m + 2}{2}$$

$$= \frac{(m+1)(3m+2)}{2}$$

$$= \frac{(m+1)[3(m+1)-1]}{2}$$

which shows that result is true for $n = m + 1$.

Hence, by the principle of mathematical induction, result (1) is true for all $n \in N$.

3. *Show by induction that* $\dfrac{a^n + b^n}{2} \geq \left(\dfrac{a+b}{2}\right)^n$, $\forall\, n \in \mathbf{N}$, *where a, b are positive real numbers.*

Solution : Let $P(n) : \dfrac{a^n + b^n}{2} \geq \left(\dfrac{a+b}{2}\right)^n, \forall \in \mathbf{N}$

Step (1) : $P(1) = \dfrac{a+b}{2} \geq \left(\dfrac{a+b}{2}\right)$, which is true.

Step (2) : Let $P(n)$ is true for $n = m$.

i.e.,

$$\frac{a^m + b^m}{2} \geq \left(\frac{a+b}{2}\right)^m \qquad ...(1)$$

Step (3) : Multiplying both sides of (1) by $\left(\dfrac{a+b}{2}\right)$, we get

$$\left(\frac{a^m + b^m}{2}\right)\left(\frac{a+b}{2}\right) \geq \left(\frac{a+b}{2}\right)^{m+1}$$

or $\left(\dfrac{a+b}{2}\right)^{m+1} \leq \left(\dfrac{a^m + b^m}{2}\right)\left(\dfrac{a+b}{2}\right) \Rightarrow \left(\dfrac{a+b}{2}\right)^{m+1} \leq \dfrac{a^{m+1} + ab^m + a^m b + b^{m+1}}{4}$

$$\Rightarrow \left(\frac{a+b}{2}\right)^{m+1} \leq \frac{a^{m+1} + b^{m+1}}{4} + \frac{ab^m + a^m b}{4} \qquad ...(2)$$

Now, we have the following two cases

Case (1) : If $a \geq b$

Here, $a^m \geq b^m \Rightarrow a - b \geq 0$ and $a^m - b^m \geq 0$

$\Rightarrow \quad (a-b)(a^m - b^m) \geq 0 \quad \Rightarrow \quad a^{m+1} + b^{m+1} - ab^m - a^m b \geq 0$

$\Rightarrow \quad ab^m + a^m b \leq a^{m+1} + b^{m+1}$

Case (2) : If $a \leq b$

Here, $b^m \geq a^m$

$\Rightarrow \quad b - a \geq 0$ and $b^m - a^m \geq 0 \quad \Rightarrow \quad (b-a)(b^m - a^m) \geq 0$

$\Rightarrow \quad b^{m+1} + a^{m+1} - ab^m - ba^m \geq 0 \quad \Rightarrow \quad ab^m + ba^m \leq a^{m+1} + b^{m+1}$

We observe that in both cases

$$ab^m + ba^m \leq a^{m+1} + b^{m+1}$$

Using (2) and (3), we get

$$\left(\frac{a+b)}{2}\right)^{m+1} \leq \frac{a^{m+1} + b^{m+1}}{4} + \frac{a^{m+1} + b^{m+1}}{4}$$

or $\left(\dfrac{a+b}{2}\right)^{m+1} \leq \dfrac{a^{m+1} + b^{m+1}}{2}$ or $\dfrac{a^{m+1} + b^{m+1}}{2} \geq \left(\dfrac{a+b}{2}\right)^{m+1}$

Therefore, $P(m+1)$ is true.

Hence, by the principle of mathematical induction, $P(n)$ is true for all $n \in \mathbf{N}$.

4. *Let x, y, ∈ **Z** such that $x \neq y$, show that by the method of induction that $(x^n - y^n)$ is divisible by $(x - y)$ for all $n \in \mathbf{N}$.*

Solution. Let $P(n): x^n - y^n$ is divisible by $(x - y)$.

Step (1) : $(x^1 - y^1)$ is divisible by $(x - y)$.

$\Rightarrow$ $P(1)$ is true.

Step (2) : Let $P(m)$ be true such that

$$P(m): x^m - y^m \text{ is divisible by } x - y,$$

i.e., $x^m - y^m = k(x - y)$ for some k.

Step (3) : Consider $x^{m+1} - y^{m+1} = x^{m+1} - x^m y + x^m y - y^{m+1}$

$$= x^m(x - y) + y(x^m - y^m) = x^m(x - y) + ky(x - y)$$

$$= (x - y)(x^m + ky) \text{ which is divisible by } (x - y).$$

Thus, $P(m + 1): (x^{m+1} - y^{m+1})$ is divisible by $(x - y)$ is true.

Hence, by the principle of mathematical induction, $P(n)$ is true for all $n \in \mathbf{N}$.

5. *Use principle of mathematical induction to prove that*

$$\frac{1}{2.3} + \frac{1}{3.4} + \frac{1}{4.5} + \ldots + \frac{1}{(n+1)(n+2)} = \frac{n}{2(n+2)}, \forall n \in \mathbf{N}$$

Solution. Let $P(n): \frac{1}{2.3} + \frac{1}{3.4} + \frac{1}{4.5} + \ldots + \frac{1}{(n+1)(n+2)} = \frac{n}{2(n+2)}, \forall n \in \mathbf{N}$

Step (1) : For $n = 1$

$$P(1) = \frac{1}{2.3} = \frac{1}{2(1+2)} = \frac{1}{6}$$

$\Rightarrow$ $P(1)$ is true.

Step (2) : Let result is true for $n = m$.

i.e., $$P(m): \frac{1}{2.3} + \frac{1}{3.4} + \frac{1}{4.5} + \ldots + \frac{1}{(m+1)(m+2)} = \frac{m}{2(m+2)} \quad \ldots(1)$$

Step (3) : Adding $\frac{1}{(m+2)(m+3)}$ on both sides of (1), we get

$$\frac{1}{2.3} + \frac{1}{3.4} + \frac{1}{4.5} + \ldots + \frac{1}{(m+1)(m+2)} + \frac{1}{(m+2)(m+3)}$$

$$= \frac{m}{2(m+2)} + \frac{1}{(m+2)(m+3)}$$

$$= \frac{1}{m+2}\left[\frac{m}{2} + \frac{1}{m+3}\right] = \frac{1}{m+2}\left[\frac{m^2 + 3m + 2}{2(m+3)}\right]$$

$$= \frac{1}{(m+2)}\left[\frac{(m+1)(m+2)}{2(m+3)}\right]$$

$$\therefore \quad \frac{1}{2.3} + \frac{1}{3.4} + \frac{1}{4.5} + \ldots + \frac{1}{(m+1)(m+2)} + \frac{1}{(m+2)(m+3)} = \frac{m+1}{2(m+3)}$$

$\therefore$ Result is true for $n = m + 1$, *i.e.*, $P(m + 1)$ is true.

Hence, by the principle of mathematical induction, result is true for all $n \in \mathbf{N}$.

6. *Using principle of mathematical induction, show that for all positive integer n*

$$\frac{n^7}{7} + \frac{n^5}{5} + \frac{2}{3}n^3 - \frac{n}{105} \text{ is an integer.}$$

Solution. Let $P(n): \frac{n^7}{7} + \frac{n^5}{5} + \frac{2}{3}n^3 - \frac{n}{105}$ be an integer.

Step (1) : $P(1): \frac{1}{7} + \frac{1}{5} + \frac{1}{3} - \frac{1}{105}$ is an integer.

Since, $\frac{1}{7} + \frac{1}{5} + \frac{2}{3} - \frac{1}{105} = \frac{15 + 21 + 70 - 1}{105} = 1$ is an integer.

$\Rightarrow$ $P(1)$ is true.

Step (2) : Let $P(m)$ be true

i.e., $P(m): \frac{m^7}{7} + \frac{m^5}{5} + \frac{2}{3}m^3 - \frac{m}{105}$ is an integer $= k$ (say) ...(1)

Step (3) : Now, $\frac{(m+1)^7}{7} + \frac{(m+1)^5}{5} + \frac{2(m+1)^3}{3} - \frac{(m+1)}{105}$

$$= \frac{1}{7}(m^7 + 7m^6 + 21m^5 + 35m^4 + 35m^3 + 21m^2 + 7m + 1)$$

$$+ \frac{1}{5}(m^5 + 5m^4 + 10m^3 + 10m^2 + 5m + 1) + \frac{2}{3}(m^3 + 3m^2 + 3m + 1) - \frac{m}{105} - \frac{1}{105}$$

$$= \left(\frac{m^7}{7} + \frac{m^5}{5} + \frac{2}{3}m^3 - \frac{m}{105}\right) + m^6 + 3m^5 + 6m^4 + 7m^3 + 7m^2 + 4m + 1$$

$= k + m^6 + 3m^5 + 6m^4 + 7m^3 + 7m^2 + 4m + 1 =$ an integer [Using (1)]

$\therefore$ $P(m+1): \frac{(m+1)^7}{7} + \frac{(m+1)^5}{5} + \frac{2(m+1)^3}{3} - \frac{(m+1)}{105}$ is an integer, is true.

Hence, by the principle of mathematical induction, result is true for all $n \in \mathbf{N}$.

7. *Using principle of mathematical induction, show that* $(1 + x)^n > 1 + nx$ *for* $n \geq 2$ *and* $x, > -1, x \neq 0.$

Solution. Let $P(n): (1 + x)^n > 1 + nx$ for $n \geq 2$ and $x > -1, x \neq 0$.

Step (1) : For $n = 2$

$P(2): (1 + x)^2 > 1 + 2x$

or $(1 + x^2 + 2x) > 1 + 2x$, which is always true.

Step (2) : Let us assume, result is true for $n = m$.

i.e., $P(m): (1 + x)^m > 1 + mx$ ...(1)

Step (3) : Since, $x > -1 \Rightarrow 1 + x > 0$

Multiplying (1) by $(1 + x)$ on both sides, we get

$\Rightarrow$ $(1 + x)^{m+1} > (1 + mx)(1 + x) \Rightarrow (1 + x)^{m+1} > 1 + x + mx + mx^2$

$\Rightarrow$ $(1 + x)^{m+1} > 1 + (m + 1)x + mx^2 \Rightarrow (1 + x)^{m+1} > 1 + (m + 1)x$ $(\because mx^2 > 0)$

$\therefore$ Result is true for $n = m + 1$.

Hence, by the principle of mathematical induction, result is true for all $n \in \mathbf{N}$.

8. *For* $n > 1$, *show that*

(i) $n! < \left(\frac{n+1}{2}\right)^n$ *(ii)* $\frac{(2n!)}{2^{2n}(n!)^2} \leq \frac{1}{(3n+1)^{1/2}}$

Solution. Let $P(n): n! < \left(\frac{n+1}{2}\right)^n$

Step (1) : For $n = 2$, the inequality is valid since

$2! < \left(\frac{2+1}{2}\right)^2$ *i.e.,* $2 < 9/4$.

Therefore, $P(2)$ is true.

Step (2) : Let result be true for $n = m$.

i.e., $$P(m): m! < \left(\frac{m+1}{2}\right)^m \quad \text{...(1)}$$

Step (3) : Consider $(m+1)! = (m+1)\, m! < (m+1)\left(\frac{m+1}{2}\right)^m$ [Using (1)]

We now prove that

$$(m+1)\left(\frac{m+1}{2}\right)^m < \left(\frac{m+2}{2}\right)^{m+1} \quad \text{...(3)}$$

Inequality (3) can clearly be written as

$$\frac{2^{m+1}}{2^m} < \left(\frac{m+2}{m+1}\right)^{m+1} \quad \text{or} \quad 2 < \left(1 + \frac{1}{m+1}\right)^{m+1}$$

But, by binomial theorem

$$\left(1 + \frac{1}{m+1}\right)^{m+1} = 1 + (m+1)\frac{1}{m+1} + \ldots > 2$$

So, the inequality (3) holds. It now follows from (2) and (3) that

$$(m+1)! < \left(\frac{m+2}{2}\right)^{m+1}$$

Therefore, $P(n)$ is true for $n = m + 1$.

Hence, by the principle of mathematical induction, result is true for all $n \in N$.

(ii) Let $P(n): \dfrac{(2n!)}{2^{2n}(n!)^2} \le \dfrac{1}{(3n+1)^{1/2}} \; (n \ge 1)$

Step (1) : For $n = 1$, both L.H.S. and R.H.S. are equal to 1/2.

Therefore, $P(1)$ is true.

Step (2) : Let us assume the inequality holds for $n = m$.

i.e., $$\frac{(2m!)}{2^{2m}(m!)^2} \le \frac{1}{(3m+1)^{1/2}} \; (m \ge 1) \quad \text{...(1)}$$

Step (3) : Now, we shall prove that inequality holds for $n = m + 1$, for which, we will show that

$$\frac{(2m+2)!}{2^{2m+2}\{(m+1!)\}^2} \le \frac{1}{(3m+4)^{1/2}} \quad \text{...(2)}$$

$$\text{L.H.S} = \frac{(2m+2)(2m+1)(2m)!}{4(m+1)^2 \,.\, 2^{2m} \,.\, (m!)^2} \le \frac{(2m+2)(2m+1)}{4(m+1)^2} \cdot \frac{1}{(3m+1)^{1/2}} \quad \text{[Using (1)]}$$

$$= \frac{2m+1}{2(m+1)} \cdot \frac{1}{(3m+1)^{1/2}} \le \frac{1}{(3m+4)^{1/2}} \quad \text{[Using (2)]}$$

Above will be proved if on squaring, we establish that

$$(3m+4)(2m+1)^2 \le 4\,(m+1)^2(3m+1) \quad \text{...(3)}$$

It should be noted that all the factors are positive

$$3m\{(2m+1)^2 - 4(m+1)^2\} + 4\{(2m+1)^2 - (m+1)^2\} \le 0$$

or $$3m\{-4m-3\} + 4\{3m^2 + 2m\} \le 0 \quad \text{or} \quad -m \le 0$$

Above is true since $m \ge 1$. Therefore, (2) is proved.

Hence, by the principle of mathematical induction, result is true for all $n \in \mathbf{N}$.

9. *Show by induction that the sum of the cubes of three consecutive natural numbers is divisible by 9.*

Solution. Let $P(n)$: Sum of those cubes of three consecutive natural numbers starting from n and divisible by 9.

Step (1) : Here, we have

$P(1)$: Sum of cubes of first three consecutive natural numbers is divisible by 9

$\because \quad 1^3 + 2^3 + 3^3 = 36$, which is divisible by 9.

Therefore, $P(1)$ is true.

Step (2) : Let $P(m)$ be true, *i. e.*, sum of the cubes of three consecutive natural numbers starting with m is divisible by 9.

$\Rightarrow \quad m^3 + (m+1)^3 + (m+2)^3$ is divisible by 9.

$\Rightarrow \quad m^3 + (m+1)^3 + (m+2)^3 = 9k, k \in \mathbf{N}$ ...(1)

Step (3) : Now, we shall prove that $P(m+1)$ is true, *i. e.*,

$(m+1)^3 + (m+2)^3 + (m+3)^3$ is divisible by 9.

Consider $(m+1)^3 + (m+2)^3 + (m+3)^3 = (m+1)^3 + (m+2)^3 + m^3 + 9m^2 + 27m + 27$

$= m^3 + (m+1)^3 + (m+2)^3 + 9(m^2 + 3m + 3)$

$= 9k + 9(m^2 + 3m + 3)$ [Using (1)]

$= 9(k + m^2 + 3m + 3)$

$=$ divisible by 9.

Therefore, $P(m+1)$ is true.

Hence, by the principle of mathematical induction, result is true for all $n \in \mathbf{N}$.

10. *Prove by the method of induction that every even power of every odd number greater than 1 when divided by 8 leaves 1 for a remainder.*

Solution. Step (1) : We first prove that the square of every odd number greater than 1 when divided by 8 leaves 1 for a remainder.

First odd integer greater than 1 is 3 and $3^2 = 9 = 8 \times 1 + 1$

Thus the square of 3, when divided by 8 leaves 1 as a remainder.

Now, assume $\quad (2m+1)^2 = 8k + 1$ ...(1)

where, $k \in \mathbf{N}^+$

Then, we have $(2m+3)^2 - (2m+1)^2 = 8(m+1)$

So that $(2m+3)^2 = (2m+1)^2 = 8(m+1) = 8k + 1 + 8(m+1)$ [Using (1)]

$= 8(k + m + 1) + 1$

Therefore, $(2m+3)^2$ when divided by 8 leaves 1 as a remainder. Then, by mathematical induction, it follows that for all n, $(2n+1)^2$ when divided by 8 leaves 1 as a remainder, *i. e.*, we have proved that

$$(2n+1)^2 = 8k + 1 \quad ...(2)$$

Step (2) : Let us assume that result is true for $m = n$

i. e., $(2n+1)^{2m}$, where $m \in Z^+$, when divided by 8 leaves 1 as remainder.

i. e., assume that

$$(2m+1)^{2m} = 8p + 1 \quad ...(3)$$

where p is a positive integer.

Then $\quad (2n+1)^{2m+2} = (2n+1)^{2n}(2n+1)^2$

$= (8p+1)(8k+1)$ [Using (2) and (3)]

$= 8(8pk + p + k) + 1$

This shows that $(2n+1)^{2m+2}$, when divided by 8 leaves 1 as remainder.

Hence, by the principle of mathematical induction, result is true for all n.

11. *Let $u_1 = 1, u_2 = 1$ and $u_{n+2} = u_{n+1} + u_n$ for $n \geq 1$. Use mathematical induction to show that*

$$u_n = \frac{1}{\sqrt{5}}\left[\left(\frac{1+\sqrt{5}}{2}\right)^n - \left(\frac{1-\sqrt{5}}{2}\right)^n\right] \textit{ for all } n \geq 1$$

Solution. We have to prove that

$$u_n = \frac{1}{\sqrt{5}}\left[\left(\frac{1+\sqrt{5}}{2}\right)^n - \left(\frac{1-\sqrt{5}}{2}\right)^n\right] \text{ for all } n \geq 1.$$

Step (1) : We obviously have

$$u_1 = 1 = \frac{1}{\sqrt{5}}\left[\left(\frac{1+\sqrt{5}}{2}\right) - \left(\frac{1-\sqrt{5}}{2}\right)\right]$$

and
$$u_2 = 1 = \frac{1}{\sqrt{5}}\left[\left(\frac{1+\sqrt{5}}{2}\right)^2 - \left(\frac{1-\sqrt{5}}{2}\right)^2\right]$$

Therefore, result is true for $n = 1$ and $n = 2$.

Step (2) : Let result is true for $n = k$.

i.e.,
$$u_k = \frac{1}{\sqrt{5}}\left[\left(\frac{1+\sqrt{5}}{2}\right)^k - \left(\frac{1-\sqrt{5}}{2}\right)^k\right] \qquad (k = 1, 2, 3, \ldots, m)$$

Now, $\quad u_{m+2} = u_{m+1} + u_m$ for $m \geq 1 \Rightarrow \quad u_{m+1} = u_m + u_{m-1}$ for $m \geq 2$

Therefore, by induction hypothesis on u_k, we have

$$u_{m+1} = u_m + u_{m-1}$$

$$= \frac{1}{\sqrt{5}}\left[\left(\frac{1+\sqrt{5}}{2}\right)^m - \left(\frac{1-\sqrt{5}}{2}\right)^m\right] + \frac{1}{\sqrt{5}}\left[\left(\frac{1+\sqrt{5}}{2}\right)^{m-1} - \left(\frac{1-\sqrt{5}}{2}\right)^{m-1}\right]$$

$$= \frac{1}{\sqrt{5}}\left[\left(\frac{1+\sqrt{5}}{2}\right)^{m-1} - \left\{\frac{1+\sqrt{5}}{2} + 1\right\}\right] - \left(\frac{1-\sqrt{5}}{2}\right)^{m-1}\left\{\frac{1-\sqrt{5}}{2} + 1\right\}$$

$$= \frac{1}{\sqrt{5}}\left[\left(\frac{1+\sqrt{5}}{2}\right)^{m-1}\left(\frac{6+2\sqrt{5}}{4}\right) - \left(\frac{1-\sqrt{5}}{2}\right)^{m-1}\left(\frac{6-2\sqrt{5}}{4}\right)\right]$$

$$= \frac{1}{\sqrt{5}}\left[\left(\frac{1+\sqrt{5}}{2}\right)^{m-1}\left(\frac{1+\sqrt{5}}{2}\right)^2 - \left(\frac{1-\sqrt{5}}{2}\right)^{m-1}\left(\frac{1-\sqrt{5}}{2}\right)^2\right]$$

$$= \frac{1}{\sqrt{5}}\left[\left(\frac{1+\sqrt{5}}{2}\right)^{m+1} - \left(\frac{1-\sqrt{5}}{2}\right)^{m+1}\right]$$

Therefore, result is true for $n = m + 1$.

Hence, by the principle of mathematical induction, result is true for all n.

12. *Show that at any time, the total number of persons on the earth who shake hands an odd number of times is even.*

Solution. To prove this result, we first assign to each hand shake a number in natural order. Then, our assertion is equivalent to the following 'for every n, after a hand shake with number n, the number of people who have made an odd number of hand shaken is even'. This statement depends on n and will be proved by induction. For convenience, we call the people who have made an odd number of hand shakes type A and the rest type B, that is of type B are those people who had made an even number of hand shake.

After the hand shake with number 1, we have two people of type A, an even number. After the mth hand shake, let the number of people of type A be even and let the hand shake number $(m + 1)$ take place. Now, there are following three cases, when the hand shake number $(m + 1)$ will occur between

(i) Two people of type A

(ii) Two people of type B

(iii) A person of type A and a person of type B

In case (i) two persons of type A and one handshake to their odd number of handshake and becomes of type B; In case (ii), two persons of type B becomes of type A and in case (iii) a person of type A becomes of type B and C person of type B is changed into type A. Thus the number of people of type A, either decreases by two or increases by two or remains unchanged. In any cases, the number remain even and proof is complete.

13. *Use mathematical induction to show that* $2 + 4 + 6 + \ldots + 2n = n^2 + n$

Solution. Here, we have n terms in L.H.S.

and R.H.S. $= n^2 + n$

Step (1) : For $n = 1$

$$2.1 = 1 + 1 \Rightarrow 2 = 2$$

$\Rightarrow$ Result is true for $n = 1$.

Step (2) : Let result be true for $n = k$, *i.e.*,

$$2 + 4 + 6 + \ldots + 2k = k^2 + k$$

Step (3) : To show result is true for $n = k + 1$, *i.e.*

$$2 + 4 + 6 + \ldots + 2(k + 1) = (k + 1)^2 + (k + 1) \quad (1)$$

L.H.S $= 2 + 4 + 6 + \ldots + 2k + 2(k + 1)$

$= k^2 + k + 2k + 2$ [Using Step (ii)]

$= k^2 + 3k + 2$

Now, R.H.S. $= (k + 1)^2 + (k + 1)$

$= k^2 + 1 + 2k + k + 1$

$= k^2 + 3k + 7$

$\Rightarrow$ L.H.S. = R.H.S.

Therefore, (1) is satisfied.

$\Rightarrow$ Result is true for $n = k + 1$.

Hence, by the principle of mathematical induction, result is true for all n.

6.3.1 Principles of Finite Induction

First principle : Let A be a subset of postive integers N with the properties

(i) $1 \in A$ and

(ii) whenever $m \in A \Rightarrow m + 1 \in A$, then, $A = N$.

Second Principle : Let A be a subset of positive integers N with the properties

(i) $1 \in A$ and

(ii) If m is a positive integer such that $1, 2, \ldots, m \in A \Rightarrow (m + 1) \in A$, then, $A = N$

☛ The well ordering principle is equivalent to the principle of finite induction.

ILLUSTRATIVE EXAMPLES

1. *Show that* ${}^nC_r + {}^nC_{r-1} = {}^{n+1}C_r$, $1 \le r \le n$ *(Pascal's rule)*

Solution. We know that $\frac{1}{r} + \frac{1}{n - r + 1} = \frac{n - r + 1 + r}{r(n - r + 1)} = \frac{n + 1}{r(n - r + 1)}$

Multiplying both sides by $\dfrac{n!}{(r-1)!(n-r)!}$ we get

$$\frac{n!}{r(r-1)!(n-r)!}+\frac{n!}{(r-1)!(n-r+1)(n-r)!}=\frac{(n+1)n!}{(r-1)!r(n-r+1)(n-r)!}$$

which implies $$\frac{n!}{r!(n-r)!}+\frac{n!}{(r-1)!(n-(r-1))}=\frac{(n+1)!}{r!(n+1-r)!}$$

Hence, ${}^nC_r + {}^nC_{r-1} = {}^{n+1}C_r,$

2. *Show that each binomial coefficient* ${}^nC_r, \forall\ n \geq 1$ *and for all r satisfying* $0 \leq r \leq n$ *is an integer.*

Solution. For $n = 1$, we have $r = 0$ or 1 and ${}^nC_r = {}^1C_0 = 1$; ${}^nC_r = {}^1C_1 = 1$

Clearly, both are integers.

Now, suppose that the result is true for $n - 1 \geq 1$ and for all r satisfying $0 \leq r \leq n - 1$, *i.e.*, ${}^{n-1}C_r$ is an integer.

Now, we shall show the result for n.

Let $r \leq n$ be an integer.

If $r = n$, then, ${}^nC_r = {}^nC_n = 1$

If $r < n$, then $r \leq n - 1$ and ${}^nC_r = {}^{n-1}C_{r-1} + {}^{n-1}C_r.$

But according to our assumption, $n - 1C_r = {}^{n-1}C_{r-1}$ are integers. Thus, ${}^nC_r$ is an integer.

Hence, by the principle of mathematical induction, the coefficients ${}^nC_r$ are integers.

3. *Show that product of any r consecutive integers is divisible by r!*

Solution. Let $P = (n-r)(n-r+1)\ldots(n-1)$ be the product of r consecutive integers. Then

$$P = \frac{(n-r)(n-r+1)\ldots(n-1)(n-r-1)\ldots 2.1}{(n-r-1)\ldots 2.1}$$

$$= \frac{1.2..(n-r-1)(n-r)\ldots(n-1)}{2.1..(n-1-r)}$$

$$= \frac{(n-1)!}{(n-1-r)!} = \frac{(n-1)!r!}{r!(n-1-r)!} = {}^{n-1}C_r.\, r!$$

Now, since ${}^{n-1}C_r$, is an integer, then clearly P is divisible by $r!$.

6.3.2 Some Interesting Facts About Numbers

(1) Every natural number upto 1000 can be expressed as a sum of four squares of natural numbers.

(2) No n^{th} power is a sum of fewer than n^{th} powers.

(3) Leonard Euler conjecture asserts that no n^{th} power is a sum of fewer than n, n^{th} powers. In particular, for $n = 3$, this would assert that no cube is the sum of two smaller cubes.

(4) The Goldback conjecture asserts that every even integer greater than 2 is the sum of two primes. For example, $4 = 2 + 2$, $6 = 3 + 3$, $20 = 7 + 13$, $50 = 3 + 47$, $100 = 29 + 71$

(5) Every prime p is a divisor of $(p-1)! + 1$.

(6) Every prime number of the from $(4n + 1)$ is a sum of two squares.

6.4 BASIC REPRESENTATION THEOREM

Theorem 1. *For a fixed integer k, every positive integer n has a unique representation of the form* $n = a_0k^r + a_1k^{r-1} + \ldots + a_r,\ a_0 \neq 0,\ 0 \leq a_i < n, i = 0, 1, 2\ldots, r$

Proof : Existence of such representation

Here, we have the following cases

Case (1) : $n < k$ then n is the unique representation of n.

Case (2) : $n = k$, then $n = 1.\ k + 0$. This is again a unique representation.

Case (3) : $n > k$, then there exists unique integers q and r such that $n = qk + r$, $0 \le r < k$.

If $q < k$, we take $q = a_0$ and $r = a$. Then $n = a_0\ k + a_1$, which is a unique representation of n.

If $q \ge k$, then there exist integers q_1 and r_1 such that $q = q_1\ k + r_1$, $0 \le r_1 < k$. If $q_1 < k$, then

$$n = (q_1 k + r_1)\ k + r$$
$$= q_1 k^2 + r_1 k + r$$

Taking $q_1 = a_0$, $r_1 = a_1$ and $r = a_2$.

Then, $\quad n = a_0\ k^2 + a_1\ k + a_2$ is the required representation.

Further, if $q_1 \ge k$, we continue the process and this process will terminate after a finite number of steps and we get

$$n = a_0 k^r + a_1 k^{r-1} + \ldots + a_r$$

6.4.1 Uniqueness of Such Representation

Let, if possible, n has two distinct representations

$$n = a_0 k^{r_1} + a_1 k^{r_1 - 1} + \ldots + a_{r-1},\ a_0 \ne 0,\ 0 \le a_i < k,\ 0 \le i \le r_1 - 1 \qquad \ldots(1)$$

and $\quad n = b_0 k^{r_2} + b_1 k^{r_2 - 1} + .. + b_{r_2 - 1},\ b_0 \ne 0,\ 0 \le b_i < k,\ 0 \le i \le r_2 - 1 \qquad \ldots(2)$

Since r_1 and r_2 are integers then by law of trichotomy, we have either

$$r_1 > r_2 \quad \text{or} \quad r_1 = r_2 \quad \text{or} \quad r_1 < r_2$$

If $r_1 > r_2$, then $\quad k^{r_1} = (k-1) \sum_{i=0}^{r_1 - 1} k^i + 1$

Thus, $\quad n = k^n - \left[(k-1) \sum_{i=0}^{r_1} n^i + 1 \right] + b_0\ k^{r_2} + \ldots + b_{r_2}$

$$= C_0\ k^{r_1} + C_1\ k^{r_1 - 1} + \ldots + C_{r_1},\ 0 \le C_i < k,\ C_0 \ne 0,\ 0 \le i \le r_1 - 1 \qquad \ldots(3)$$

Therefore, we may assume two different representations of n with same number of terms as follows

$$x = a_0\ k^{r_1} + a_1\ k^{r_1 - 1} + \ldots + a_{r_1},\ 0 \le a_i \le k,\ a_0 \ne 0,\ 0 \le i \le r_1 - 1 \qquad \ldots(4)$$

and $\quad x = b_0\ k^{r_1} + b_1\ k^{r_1 - 1} + \ldots + b_{r_1},\ 0 \le b_i < k,\ b_0 \ne 0,\ 0 \le i \le r_1 - 1 \qquad \ldots(5)$

Let i be the largest integer such that $a_i \ne b_i$, then $a_{i+1} = b_{i+1}, \ldots a_{r_1} = b_{r_1}$

Then, from (4) and (5) we conclude that

$$(a_0 - b_0)\ k^{r_1} + (a_1 - b_1)\ k^{r_1 - 1} + \ldots + (a_i - b_i)\ k^i = 0$$

Since, $k > 1$ and $i < r_1$, we have

$$a_i - b_i = -k\,[(a_0 - b_0)\ k^{r_1 - i - 1} + \ldots + (a_{i-1} - b_{i-1})]$$

This shows that $k \mid (a_i - b_i)$, which is not true because a_i and b_i both are less than k. This contradiction shows that representation must be unique.

ILLUSTRATIVE EXAMPLES

1. *Write 506 in decimal system.*

Solution. We have $506 = 50 \times 10 + 6$

Also, $50 = 5 \times 10$

Thus, we can write $506 = 5 \times 10 \times 10 + 6 = 5 \times 10^2 + 0 \times 10 + 6$

Hence $(506) = (506)_{10}$.

2. *Write 25 with base 2.*

Solution. We have the following division scheme

		Remainder
2	25	
2	12	1
2	6	0
2	3	0
2	1	1
	0	1

Hence, $(25)_{10} = (11001)_2$

3. *Write 347 with base 8.*

Solution. We have the following division scheme

		Remainder
8	347	
8	43	3
8	5	3
	0	5

Hence $(347)_{10} = (533)_8$.

4. *Show that if $a_r k^r + a_{r-1} k^{r-1} + \ldots + a_0$ is a representation of n to the base k, then* $0 < n \le k^{r+1} - 1$.

Solution. Here, we have
$$0 < n = a_r k^r + a_{r-1} k^{r-1} + \ldots + a_0$$
$$\le 1.\, k^r + 1.\, k^{r-1} + \ldots + 1$$
$$= k^r + k^{r-1} + \ldots + 1 = \frac{k^{r+1} - 1}{k - 1}.$$
$$\le \frac{k^{r+1} - 1}{2 - 1} \qquad [\because k \ge 2]$$
$$= k^{r+1} - 1$$

Exercise 6.1

Prove the following by the principle of mathematical induction : (Ques. 1 to 9)

1. $n(n+1)(2n+1)$ is divisible by 6, $\forall\, n \in \mathbf{N}$.

2. $1 + 4 + 7 + \ldots + (3n - 2) = \frac{1}{2} n(3n - 1)$, $\forall\, n \in \mathbf{N}$.

3. $\frac{1}{1.2} + \frac{1}{2.3} + \frac{1}{3.4} + \ldots + \frac{1}{n(n+1)} = \frac{n}{n+1}$, $\forall\, n \in \mathbf{N}$.

4. $(2^{3n} - 1)$ is divisible by 7, $\forall\, n \in \mathbf{N}$.

5. $(10^{2n-1} + 1)$ is divisible by 11.

6. $n < 2^n, \forall\, n \in \mathbf{N}$

7. For $n \in \mathbf{N}$, $10^n + 3(4)^{n+2} + 5$ is divisible by 9.

8. $1^3 + 2^3 + 3^3 + \ldots + n^3 = \left(\frac{n(n+1)}{2}\right)^2$

9. (i) $7 + 77 + 777 + \ldots + \underset{n \text{ digit}}{777\ldots7} = \frac{7}{81}(10^{n+1} - 9n - 10)$

(ii) $\frac{1}{3.7} + \frac{1}{7.11} + \frac{1}{11.15} + \ldots + \frac{1}{(4n-1)(4n+3)} = \frac{n}{3(4n+3)}$

(iii) $1.6 + 2.9 + 3.12 + \ldots + n(3n + 3) = n(n+1)(n+2)$

10. Show that $7^{2n} + (2^{3n-3})\, 3^{n-1}$ is divisible by 25, $n \in \mathbf{N}$.

11. (i) Show that $5^{2n+2} - 24n - 25$ is divisible by 576.

(ii) Show that $27^n + 3.5^n - 5$ is divisible by 24 for all $n \geq 0$

12. If p is a natural number, then show that $p^{n+1} + (p+1)^{2n-1}$ is divisible by $p^2 + p + 1$ for every positive integer n.

13. Prove that binomial theorem
$$(x+a)^n = x^n + {}^nC_1\, x^{n-1}\, a + {}^nC_2\, x^{n-2}\, a^2 + \ldots + {}^nC_r\, x^{n-r}\, a^r + \ldots + a^n$$

14. If $x^3 = x + 1$, then show that $x^{3n} = a_n x + b_n - c_n x^{-1}$, where, $a_{n+1} = a_n + b_n$; $b_{n+1} = a_n + b_n + c_n$, $c_{n+1} = a_n + b_n$

15. Use the principle of mathematical induction to show that

(i) $5^{2n} - 1$ is divisible by 24.

(ii) $4^n - 3n - 1$ is divisible by 9.

(iii) $10^{2n-1} + 1$ is divisible by 11

(iv) $a^{2n-1} - 1$ is divisible by $a - 1$

(v) $n^2 - n + 41$ is prime

16. Show that

(i) $1.4.7 + 2.5.8 + 3.6.9 + \ldots + n(n+3)(n+6) = \frac{\pi}{4}(n+1)(n+6)(n+7)$

(ii) $1 + 3 + 5 + \ldots + (2n-1) = n^2$

17. If n straight lines in a plane are such that no two of them are parallel and no three of them are concurrent, show that they intersect each other in $\frac{n(n-1)}{2}$ points.

18. Let $P(n)$ be the statement "The arithmetic mean of n and $(n+2)$ is the same as their geometric mean". Show that $P(1)$ is not true. Also, show that if $P(n)$ is true, then $P(n+1)$ is also true. How does this contradict the principle of induction ?

19. Show that $(x + a_1)(x + a_2)(x + a_3) \ldots$
$$\ldots (x + a_n) = x^n + P_1\, x^{n-1} + P_2 x^{n-2} + \ldots + P_{n-1} x + P_n$$
where, $P_1 = \Sigma a_i$, $P_2 = \Sigma a_i a_j$, $P_3 = \Sigma a_i a_j a_k$, $1 \leq i \leq n$, $1 \leq i \leq j \leq n$, $1 \leq i \leq j \leq k \leq n$, $P_n = a_1\, a_2 \ldots a_n$

20. Let $P(n)$ be the statement : $2^n \geq 3n$. If $P(r)$ is true, show that $P(r+1)$ is true. Do you conclude that $P(n)$ is true for all $n \in \mathbf{N}$.

21. Show that

(i) $\frac{n^7}{7} + \frac{n^5}{5} + \frac{n^3}{3} + \frac{n^2}{2} - \frac{37}{210}n$, is a positive integer, $\forall\, n \in \mathbf{N}$

(ii) $\frac{n^{11}}{11} + \frac{n^5}{5} + \frac{n^3}{3} + \frac{62}{165}$ is positive integer, $\forall\, n \in \mathbf{N}$

(iii) $\frac{n^5}{5} + \frac{n^3}{3} + \frac{7n}{15}$, is a natural number.

22. Show by the method of induction that for all $n \in \mathbf{N}$, 3^{2n}, when divided by 8, the remainder is always 1.

23. Show that the product of any r consecutive natural numbers is always divisible by $r!$

24. Show that $\sum_{k=0}^{n} k^2\, {}^nC_k = n(n+1)\, 2^{n-2}$ for $n \geq 1$.

25. (i) For what natural number n, the inequality $2^n > 2n + 1$ is valid ?

(ii) For what natural number n, the inequality $2^n > n^2$ is valid ?

26. For all integers $n \geq 2$, prove that
$$\sum_{r=1}^{n-1} i\,(i+1) = n\,(n-1)(n+1)/\,3.$$

27. Prove that if $a_r\, k^r + a_{r-1}\, k^{r-1} + \ldots + a_0$ is a representation of n to the base k, then $0 < n \leq k^{r+1} - 1$

28. Show that $\frac{(2n)!}{n!\,(n+1)!}$ is an integer.

ANSWERS

25. (i) for $n \geq 3$; (ii) for $n = 1$ and for all natural numbers $x \geq 5$.

6.5 DIVISIBILITY THEORY

An algorithm is a step by step process, complete in a finite number of steps, for solving a given problem. By the division algorithm, we mean that process with which the student became familiar in arithmetic. Divisors, multiples and prime and composite numbers are concepts that have been known and studied at least since the time of Euclid, about 350 BC. In this chapter, we shall discuss divisibility theory of integers.

6.5.1 Division Algorithm

Definition : *An integer b is divisible by an integer a, not zero, if there is an integer x such that b = ax and we write a | b. In case, b is not divisible by a, we write a ∤ b.*

Theorem 1. *(a) For any given integer a (the dividend) and any given non-zero integer b (the divisor) there exist integers q (the quotient) and r (the remainder) such that*

$$a = qb + r \quad ...(1)$$

and

$$0 \le r < |b| \quad ...(2)$$

(b) q and *r are unique.*

Proof : Since a lies between two consecutive integers of the sequence

$$\ldots -2|b|, -|b|, 0, |b|, 2|b|, \ldots$$

we may assume $q|b| \le a < (q+1)|b|$

Then, $a - q|b| \ge 0$, $a - q|b| < |b|$. Let $a - q|b| = r$. Then, $0 \le r < |b|$ and therefore, we have

$$a = qb + r, \text{ when } b > 0.$$

and

$$a = (-q)\, b + r \text{ when } b < 0.$$

Hence, the existence of q and r is proved. Now, we shall prove the uniqueness, as follows :

Let there exists another representation given by $a = q_1 b + r_1, 0 \le r_1 < |b|$, then

$$(q - q_1)\, b = r_1 - r, \; 0 \le |r_1 - r| < |b|$$

i.e., $|q - q_1||b| < |b|$ thus $|q - q_1| < 1$. Since q, q_1 are both integers, $q = q_1$ and consequently $r = r_1$ which implies that q and r are unique.

- The integer r is called the least non-negative remainder or briefly the remainder of a divided by b. If $r = 0$, then $a = qb$ and hence a is a multiple of b.
- If $a | b$ and $0 < a < b$, then a is called a proper divisor of b.
- It is understood that we never use 0 as the left member of the pair of integers in $a | b$. On the other hand, not only may 0 occur as the right member of the pair, but also in such instances we always have divisibility. Thus, $a | 0$ for every integer a not zero.
- The notation $a^k \| b$ is sometimes used to indicate that $a^k | b$ but $a^{k+1} \nmid b$

Theorem 2. *(1) a | b implies a | bc for any integer c.*

(2) a | b and b | c imply a | c. (Transitive property)

(3) a | b and *a | c imply a | (bx + cy) for any integers x* and *y. (Linearity property)*

(4) a | b and *b | a imply a = ± b.*

(5) a | b , a > 0, b > 0 imply a ≤ b.

(6) if m ≠ 0, a | b implied by ma | mb. (Multiplication property)

(7) 1 | a (1 divides every integer)

(8) a | 0. (Every integer divides 0)

Proof : The proof of these results follow at once from the definition of divisibility. To give a sample proof, consider (3). Since $a | b$ and $a | c$ are given, this implies that there are integers r and s such that $b = ar$ and $c = as$. Therefore, $bx + cy$ can be written as $a(rx + sy)$ which proves that a is a divisor of $bx + cy$.

- The property (3) can be extended to any finite set as follows : $a | b_1, a | b_2, \ldots, a | b_n$ imply $a \mid \sum_{j=1}^{n} b_j x_j$ for any integers x_j.

Theorem 3. *If c = ax + by and d|a but d∤c, then d ∤ b.*

Proof : Here, we have $d | a$ implies there exists an integer q_1 such that $a = dq_1$

Therefore, $c = ax + by = dq_1 x + by$

Let if possible $d | b$, then there exists an integer q_2 such that $b = dq_2$.

$\therefore \quad c = dq_1 x + dq_2 y = d(q_1 x + q_2 y) \quad \Rightarrow \quad d | c$

Hence, contrapositively $d \nmid c \Rightarrow d \nmid b$.

Theorem 4. *For any two integers a* and $b > 0$ *there exists integers* q_1 and r_1 *such that*

$$a = bq_1 + cr_1,\ 0 \le r_1 < b/2,\ c = +1 \text{ or } -1.$$

Proof. By division algorithm, we have

$$a = bq + r,\ 0 \le r < b$$

Now there are following cases.

Case 1 : If $r < b/2$. If we take $q = q_1$, $r = r_1$, $c = 1$, then (1) gives

$$a = bq_1 + cr_1,\ 0 \le r_1 < b/2$$

Case 2 : If $r > b/2$. Then $0 < b - r < b/2$. If we take $q = q_1 + 1$, $r = b - r_1$ and $c = -1$, then (1) gives

$$a = b(q_1 + 1) - (b - r_1) = bq_1 + cr_1,\ 0 \le r_1 < b/2$$

Case 3 : If $r = b/2$. If $q = q_1$, $r = r_1$ and $c = 1$, then from (1), we have

$$a = bq_1 + cr_1,\ r_1 = b/2$$

Further, if we replace q_1 by $q + 1$, r_1 by $b - r$ and c by -1, we get

$$a = b(q + 1) - (b - r) = bq_1 + cr_1,\ r_1 = b/2$$

Deductions : (1) Every integer having one of the following form

(a) $3q$ or $(3q \pm 1)$ (Taking $b = 3$ in the above theorem)

(b) $4q$, $(4q \pm 1)$ or $(4q \pm 2)$ (Taking $b = 4$ in the above theorem)

(c) $5q$, $(5q \pm 1)$ or $(5q \pm 2)$ (Taking $b = 5$ in the above theorem)

(2) Every odd integer having one of the following form

(a) $2q + 1$ (b) $2q - 1$ (c) $4q \pm 1$ (d) $\pm(4q + 1)$

Theorem 5. *Every square number is of the form* $9k$ *or* $9k + 1$ *where k is an integer.*

Proof : Since any integer can be written in the form $3q$ or $3q \pm 1$,

$$(3q)^2 = 9q^2 = 9k \quad \text{and} \quad (3q \pm 1)^2 = 3(3q^2 \pm 2q) + 1 = 3k + 1.$$

Deduction : If n is a positive odd integer and $n = ab$, then

$$n = ab = \left(\frac{a+b}{2}\right)^2 - \left(\frac{a-b}{2}\right)^2$$

Since a and b are both odd, then $\frac{a+b}{2}$ and $\frac{a-b}{2}$ are integers. Therefore, we can say that if a positive odd integer a can be decomposed into a product of two divisors, then a can be written as the difference of two square numbers.

Theorem 6. *The squqre of an odd integer is of the form* $8q + 1$.

Proof : Let k be any odd integer. Then, we have $k = 4q_1 + 1$

or $\quad k = -(4q_1 + 1)$ for integer q_1.

Then, $\quad k^2 = [\pm(4q_1 + 1)]^2 = 16q_1^2 + 8q_1 + 1 = 8(2q_1^2 + q_1) + 1$

$= 8q + 1$, where $q = q_1^2 + q_1$, again an integer.

Hence, square of an odd integer is of the form $8q + 1$.

Theorem 7. *The product of any three consecutive integers is a multiple of 3.*

Proof : Since any integer can be written in the form $3n$ or $3n \pm 1$, the difference of two integers is of the same form is a multiple of 3 and therefore, not less than 3. But the difference of any two of three consecutive integers is less than 3, so that the three consecutive integers are respectively of the above three forms, among which one is of the form $3n$, *i.e.*, a multiple of 3.

Deduction : From above theorem, it follows that the product of three consecutive integers is a multiple of 3.2. A generalization of this property gives the following result :

"The product of any three consecutive integers is divisible by 3 !".

Theorem 8. *Let p be a positive integer greater than 1. Then every positive integer a can be written uniquely in the form.*

$$a = C_n\ p^n + \ldots + C_1 p + C_0 \qquad \ldots(1)$$

where, $n \geq 0$, C_i *is an integer* $0 < C_i < p$, $C_n \neq 0$. *p is called the base of a, which is denoted by* $(C_n\ C_{n-1} \ldots C_1\ C_0)_p$.

Proof : We shall prove this theorem by induction on a.

When $a = 1$, we have $n = 0$ and $C_0 = 1$. Then (1) is true for $a = 1$. Now assume that theorem is true for any integer less than a. Since $p > 1$, $a > 0$, therefore, a must lie between two certain consecutive numbers of the following sequence.

$$p^0, p^1, p^2, \ldots p^n, \ldots$$

i.e., there exists a unique integer n, such that

$$p^n \leq a < p^{n+1}$$

Then, by division algorithm, we have

$$a = C_n p^n + r,\ 0 \leq r < p^n$$

Clearly, $p > C_n > 0$, if $r = 0$, then

$$a = C_n p^n + 0.\, p + \ldots + 0.\, p + 0$$

If $r \neq 0$, then by induction hypothesis

$$r = b_t p^t + \ldots + b_1 p + b_0,\ t < n, \text{ where } 0 \leq b_i < p$$

Therefore, $a = C_n p^n + b_t p^t + \ldots + b_1 p + b_0$ and (1) is true.

Uniqueness : To prove uniqueness, let us assume that there is another representation

$$a = d_m p^m + \ldots + d_1 p + d_0 \qquad \ldots(2)$$

with $m \geq 0, 0 \leq d_i < p$. If C_i and d_i are not equal, by subtracting (1) from (2), we get

$$0 = e_s p^s + \ldots + e_1 p + e_0$$

where, s is the largest value of i for which $C_i \neq d_i$, so that $e_s \neq 0$. If $s = 0$, then $C_1 = C_0 = 0$, which is a contradiction.

If $s > 0$, we have $\quad |e_i| = |C_i - d_i| < p - 1, i = 0, 1, \ldots, s - 1$

and $\quad e_s p^s = -(e_{s-1} p^{s-1} + \ldots + e_0)$

Therefore $p^s < |e_s p^s| = |e_{s-1} p^{s-1} + \ldots + e_0| < (p-1)(p^{s-1} + \ldots + p + 1) = p^s - 1$

which is again a contradiction. Hence, we conclude that C_i and d_i are all equal, *i.e.*,

$$n = m, C_i = d_i, i = 0, 1, 2, \ldots, n.$$

Hence, the representation is unique.

Deduction : If we take $p = 2$, then every positive integer may be represented as the sum of distinct powers of 2, *i.e.*,

$$a = C_n\ 2^n + \ldots .\ C_1.2 + C_0$$

where, each C_i is either 0 or 1.

For example : We consider $a = 2107$. If $p = 10$, then

$$2107 = 2\,(10)^3 + (10)^2 + 7 = (2107)_{10}$$

If $p = 12$, since $2107 = 175 \times 12 + 7$, $175 = 14 \times 12 + 7$, $14 = 12 + 2$ then $(2107) = (1277)_{12}$

Similarly, if $p = 2$, then $(2107) = (100000011111)_2$

ILLUSTRATIVE EXAMPLES

1. *If* $(a - s) | (ab + st)$, *then show that* $(a - s) | (at + bs)$.

Solution. Since $(ab + st) - (at + bs) = (a - s)(b - t)$ and the hypothesis is that $ab + st$ is a multiple of $a - s$. Thus, $at + bs$ is a multiple of $a - s$.

2. *Show that, one of every three consecutive integers is divisible by 3.*

Solution. Let $n, n + 1, n + 2$ be any three consecutive integers. Then, we know that n is of the form $3q$, $(3q + 1)$ or $(3q - 1)$.

If $n = 3q$, then clearly it is divisible by 3. If $n = 3q + 1$, then

$$n + 2 = 3q + 1 + 2 = 3q + 3 = 3(q + 1)$$

which is again divisible by 3.

Finally, if $n = 3q - 1$, then $n + 1 = 3q - 1 + 1 = 3q$, which is also divisible by 3.

Hence, one of every three consecutive integers is divisible by 3.

3. *Show that if n is an even number, then $3^n + 1$ is divisible by 2; if n is odd number, then $3^n + 1$ is divisible by 2^2, if n is any number, whether even or odd, then $3^n + 1$ is not divisible by 2^m with $m \geq 3$.*

Solution. We know that the square of an odd number minus 1 is a multiple of 8.

Now, when $n = 2m$, we have

$$3^n = 3^{2m} = (3^m)^2 = 8a + 1$$

and therefore $3^n + 1 = 2(4a + 1)$

when $n = 2m + 1$, we have

$$3^n + 1 = 3^{2m+1} + 1 = 3(8a + 1) + 1 = 4(6a + 1)$$

Since $4a + 1$ and $6a + 1$ are odd, the statement is true.

4. *If a and b are any two odd integers, then one of the two numbers* $\frac{a+b}{2}$ and $\frac{a-b}{2}$ *is odd and the other is even.*

Solution. Let us assume that $a = 2k_1 + 1$ and $b = 2k_2 + 1$

where, k_1 and k_2 are any two integers.

Then $$\frac{a+b}{2} = \frac{2k_1 + 1 + 2k_2 + 1}{2} = (k_1 + k_2) + 1 \qquad ...(1)$$

also, $$\frac{a-b}{2} = \frac{(2k_1 + 1) - (2k_2 + 1)}{2} = (k_1 + k_2) \qquad ...(2)$$

From (1) and (2), we conclude that if k_1 and k_2 both are even (or both odd), then $\frac{a+b}{2}$ is an even integer and $\frac{a-b}{2}$ is an odd integer.

Hence, if a and b are any two integers, then one of the two numbers $\frac{a+b}{2}$ and $\frac{a-b}{2}$ is odd and the other is even.

5. *Show that if a is any positive integer, then $a^2 + a + 1$ is not a square number.*

Solution. Since, we have $a^2 < a^2 + a + 1 < a^2 + 2a + 1 = (a + 1)^2$

The next square number greater than a^2 is $(a + 1)^2$. Hence, $a^2 + a + 1$ is not a square number.

☛ If an integer a is a square of some other integer, then a is called a square number.

6. *If a is an odd integer, then show that* $\frac{a^4 + 4a^2 + 11}{16}$ *is an integer.*

Solution. Let us suppose $a = 2n + 1$

Then, consider $$\frac{a^4 + 4a^2 + 11}{16} = \frac{(2n + 1)^4 + 4(2n + 1)^2 + 11}{16}$$

$$= \frac{16n^4 + 32n^3 + 40n^2 + 24n + 16}{16}$$

$$= n^4 + 2n^3 + 1 + \frac{n(5n+3)}{2}$$

Clearly, $\frac{n(5n+3)}{2}$ is an integer, if n is even integer. If n is an odd integer, then $5n + 3$ is an even integer and therefore, $\frac{n(5n+3)}{2}$ is again an even integer. Hence, the given quantity is an integer.

7. *Show that if $1 < a_1 < a_2 < \ldots < a_{n-1} < a_n$, then there exist i and j with $i < j$ such that $a_i | a_j$.*

Solution. Let $a_i = 2^n b_i$, $n_i \geq 0$, b_i is odd. Since, among 1, 2, ..., 2n, there are only n distinct odd numbers $b_1, b_2, \ldots, b_{n+1}$ are not all distinct. In other words, among them there are some equal odd numbers. Let $b_i = b_j$. Then, $a_i | a_j$.

8. *Show that the number of the form $\frac{a(a^2+2)}{3}$ is an integer where a is an integer greater than or equal to 1.*

Solution. Since, we know that every integer a is of the form $3n$, $3n + 1$ or $3n + 2$.

If $a = 3n$, then $\frac{a(a^2+2)}{3} = \frac{3n[(3n)^2+2]}{3} = n(9n^2+1)$, which is an integer.

If $a = 3n + 1$, then $\frac{a(a^2+2)}{3} = \frac{(3n+1)[(3n+1)^2+2]}{3} = (3n+1)(3n^2+2n+1)$, which is again an integer.

If $a = 3n + 2$, then $\frac{a(a^2+2)}{3} = \frac{(3n+2)[(3n+2)^2+2]}{3} = (3n+2)(3n^2+4n+2)$, which is also an integer. Hence, a number of the form $\frac{a(a^2+2)}{3}$ is an integer, where $a \geq 1$ is an integer.

Exercise 6.2

1. If integer b divides a positive integer a then show that b is not necessarily greater than a.
2. Show that we can choose two integers from any given three integers such that their sum and difference are even numbers.
3. If a is an odd integer, then show that $\frac{a^4 + (a+2)^3 + (a+4)^2 + 1}{12}$ is an integer.
4. Show that if $d_1, \ldots, d_k$ are all positive divisors of n, then $(d_1 \ldots d_k)^2 = n^k$.
5. Show that $n^2 - n$ is divisible by 2 for every integer n; that $n^2 - n$ is divisible by 6; that $n^5 - n$ is divisible by 30.
6. If $a > 0$, $n \geq 2$, then show that a^n can be expressed as the sum of a consecutive positive integers.
7. Show that if x and y are odd, then $x^2 + y^2$ is even but not divisible by 4.
8. Show that $a^{n+1} - (a-1)n - a$ can be divided by $(a-1)^2$.
9. Show that for any positive integer n, there exists at least n consecutive integers such that each of them has a divisor which is a square number.
10. Show that there are no positive integers $a, b, n > 1$ such that $(a^n - b^n) | (a^n + b^n)$.
11. Find 9 integers such that they form an arithmetic progression and the sum of the squares of each of them is a square integer.
12. Show that $(n-1)^2 | (n^k - 1)$ if and only if $(n-1) | k$.

6.6 GREATEST COMMON DIVISOR

If c is a divisor of a and a divisor of b simultaneously, *i.e.*, $c|a$ and $c|b$, then c is called a common divisor of a and b. For example, 1 is a common divisor of a and b. Since any non-zero integer has only a finite number of divisors, then a and b (both not zero) also have a finite number of common divisors, the largest integers among which is called the greatest common divisor of a and b and is written as (a, b).

Definition : *Let a and b any two given integers (both not zero), then the greatest common divisor of a and b denoted by (a, b) is the positive integer d such that*

(i) $d|a$ *and* $d|b$ *(ii) if* $c|a$ *and* $c|b$ *then* $c|d$.

The greatest common divisor of any two integers, both not zero can be found by following the Euclidean algorithm.

6.6.1 Euclid's Algorithm

Let a and b be any two positive integers, then we obtain an integer $k \geq 1$ such that

$$\left.\begin{aligned} a &= q_1 b + r_1 \;;\; 0 \leq r_1 < b \\ b &= q_2 r_1 + r_2 \;;\; 0 \leq r_2 < r_1 \\ &\dots \dots \dots \\ r_{k-2} &= q_k\, r_{k-1} + r_k : 0 < r_k < r_{k-1} \\ r_{k-1} &= q_{k+1}\, r_k \end{aligned}\right\}$$

From the first equation, we have $(a, b) = (b, r)$ and therefore,

$$(a, b) = (b, r) = (r_1, r_2) = \dots = (r_{k-1}, r_k) = r_k$$

Hence, r_k is the required g.c.d. (a, b) that is to say we can find the greatest common divisor by using Euclidean algorithm.

6.6.2 Absolutely Least Remainder Algorithm or Minimal Algorithm

Let a and b be two positive integers such that $a > b$. Then, there exist integers Q_1 and R_1 such that

$$a = bQ_1 + e_1\, R_1,\; 0 < R_1 < \frac{b}{2} \quad \dots(1)$$

Again, there exists integers Q_2 and R_2 such that

$$b = R_1 Q_2 + e_2 R_2,\; 0 < R_2 \leq \frac{R_1}{2} \quad \dots(2)$$

Continuing this process, we get

$$R_1 = R_2\, Q_3 + e_3 R_3,\; 0 < R_3 \leq \frac{R_2}{2} \quad \dots(3)$$

$$\vdots$$

$$R_{n-3} = R_{n-2}\, Q_{n-1} + e_{n-1}\, R_{n-1},\; 0 < R_{n-1} \leq \frac{R_{n-2}}{2} - (n-1)$$

$$R_{n-2} = R_{n-1}\, Q_n + e_n\, R_n,\; R_n = 0 \quad \dots(4)$$

where $e_1, e_2, \dots e_n$ all are either +1 or −1.

Finally, since $a > b > R_1 > R_2 \dots > R_n$ form a decreasing sequence of non-negative integers. Therefore, $R_n = 0$ for some integer n. The g.c.d. of a and b will be R_{n-1} as in Euclid's algorithm.

Theorem 1. *Let a and b be positive integers such that* $a > b$ *and let* $r_k = 0$ *in Euclid's algorithm. Then,* r_{k-1} *is the g.c.d. of a and b.*

Proof. From third equation of (1), we have

$$r_{k-2} = r_{k-1}\, q_k \text{ which implies } r_{k-1} | r_{k-2}$$

Further, we have

$$\begin{aligned} r_{k-3} &= r_{k-2}\, q_{k-1} + r_{k-1} \\ &= r_{k-1}\, q_k\, q_{k-1} + r_{k-1} \\ &= r_{k-1}\, [q_k\, q_{k-1} + 1] \end{aligned}$$

which implies

$$r_{k-1} | r_{k-3}$$

Continuing this process, finally, we get

$$r_{k-1} | a \text{ and } r_{k-1} | b$$

Now let c divides a and b. Since, $a = bq_1 + r_1$, then c divides b and r_1. Also, $b = r_1 q_2 + r_2$, which implies c divides r_1 and r_2. Continuing the process, we get c divides r_{k-1}.

Hence g.c.d. $(a, b) = r_{k-1}$.

Theorem 2. *Any common divisor of a and b is a divisor of their greatest common divisor (a, b).*

Proof : If $c|a$, $c|b$, then from (1), we have $c | r_1$. Also since $c | b$ and $c | r_1$, then $c | r_2$. Continuing this process, at least we obtain $c | r_k$. Hence, $c | (a, b)$.

Theorem 3. $(a, b)\, c = (ac, bc), c > 0$

Proof : By multiplying each equation of (1) by c, the integers a, b and r_1 become ac, bc and $r_i\, c$ respectively and hence, $(ac, bc) = (a, b)\, c$.

Theorem 4. *If* $(a, b) = 1$, *then* $(ac, b) = (c, b)$.

Proof : Since $(ac, b) | ac$ and $(ac, b) | bc$, we have

$$(ac, b) | (ac, bc) = (a, b)| \ |c| = |c|$$

But, $(ac, b) | b$, therefore $(ac, b) | (c, b)$

Further, since $(c, b) | ac$ and $(c, b) | b$, then $(c, b) | (ac, b)$. Hence, $(ac, b) = (c, b)$.

Deductions : From above theorem, we can easily obtain the following useful properties :

(1) If $b | ac$ and $(a, b) = 1$, then $b | c$ and therefore $(ac, b) = b$ which together with theorem (4) gives $(c, b) = b$. Therefore $b | c$. On the other hand, if $(a, b) = 1$, then $(a, b^n) = 1$.

(2) If $a | c, b | c$ and $(a, b) = 1$, then $ab | c$. In fact, from $a | c$, it follows that $c = ac_1$ so that $b | ac_1$. Consequently $b | c_1$ and $ab | ac_1$ that is $ab | c$.

(3) If $(a, c) = 1$ and $(b, c) = 1$, then $(ab, c) = 1$. For, by theorem (13), we have

$$(ab, c) = (b, c) = 1.$$

(4) If $(a, b) = 1$, then $(ab, a + b) = 1$. For, from $(a, a + b) = 1$, $(b, a + b) = 1$, it follows that $(ab, a + b) = 1$.

Theorem 5. *If a and b are any two integers not both zero, then (a, b) uniquely exists.*

Proof : We know that the (a, b) is not affected by the sign of a and b. Therefore, we assume that both a and b are positive and $a \geq b$. By division algorithm, we have

$$a = bq_1 + r_1, 0 \leq r_1 < b \qquad ...(1)$$

Now, there are following cases :

If $r_1 = 0$, then $b | a$ and $(a, b) = b$. Therefore, (a, b) exists.

If $r_1 \neq 0$, then by division algorithm, we have $b = r_1 q_2 + r_2$, $0 \leq r_2 < r_1$ ...(2)

If $r_2 = 0$ then $r_1 | b$ and so from (1)

$$a = (r_1\, q_2)\, q_1 + r_1$$
$$= r_1 (q_2\, q_1 + 1)$$

which implies $r_1 | a$

Further, let $s | a$, $s | b \Rightarrow s | a - bq_1 \Rightarrow s | r_1$ [Using (1)]

$\therefore$ $(a, b) = r_1$, which shows the existence of (a, b) at this stage.

If $r_2 \neq 0$, we again apply the same process.

After n steps, we get zero remainder.

Thus, we get a sequence of integers r_i such that

$$0 \leq r_n < r_{n-1} \ldots < r_2 < r_1 < b$$
$$r_{n-2} = r_{n-1}\, q_n + r_n,\ n \geq 3$$

and

$$r_{n-1} = q_{n+1}\, r_n$$

Therefore, $r_n \mid r_{n-1}, r_n \mid r_{n-2} \dots r_n \mid b$ and $r_n \mid a$.

Further, if s a is a common factor of a and b, then $s \mid a$ and $s \mid b$, which implies

$$s \mid a - bq$$
$$\Rightarrow \quad s \mid r_1$$
$$\Rightarrow \quad s \mid r_2$$
$$\dots$$
$$\Rightarrow \quad s \mid r_n$$

Hence, $(a, b) = r_n$.

Now, we shall prove the uniqueness of (a, b).

Let, if possible, d_1 and d_2 are two g.c.d.'s of a and b.Then, by definition

$$d_1 \geq d_2 \text{ and } d_2 \geq d_1$$

i.e.,
$$d_1 = d_2$$

Hence, (a, b) is unique.

Theorem 8. *If g is the greatest common divisor of* a and b, *then there exists integers* x and y *such that* $g = (a, b) = ax + by$.

Proof : Consider the linear combination $ax_0 + by_0$, where x_0 and y_0 range over all integers. This set of integers $\{ax_0 + by_0\}$ includes positive and negative values, and also 0 by the choice $x_0 = y_0 = 0$. Choose x and y such that $ax + by$ is the least positive integer l in the set, so $l = ax + by$.

Next, we prove that $l \mid a$ and $l \mid b$, let if possible $l \nmid a$. $l \nmid a$ implies there exist integers q and r such that

$$a = lq + r$$

with $0 < r < l$. Therefore, we have $r = a - lq = a - q\,(ax + by)$

$$= a\,(1 - qx) + b\,(-qy)$$

and thus r is in the set $\{ax_0 + by_0\}$. This contradicts the fact that l is the least positive integer in the set $\{ax_0 + by_0\}$.

Now, since g is the greatest common divisor of a and b, we may write

$$a = gA,\ b = gB \text{ and } l = ax + by = g\,(Ax + By).$$

Thus, $g \mid l$ which gives $g \leq l$. Now $g < l$ is impossible, since g is the greatest common divisor, therefore, $g = l = ax + by$.

Deductions : (1) The greatest common divisor g of a and b can be characterized in the following two ways.

(a) It is the least positive value of $ax + by$ where x and y range over all integers.

(b) It is the positive common divisor of a and b that is divisible by every common divisor.

(2) If an integer d is expressible in the form $d = ax + by$, then d is not necessarily the g.c.d. of (a, b). However, it does not follow from such an equation that (a, b) is a divisor of d. In particular, if $ax + by = 1$ for some integers x and y, then $(a, b) = 1$.

Theorem 7. *For any positive integer m*

$$(ma, mb) = m\,(a, b)$$

Proof : By theorem-15, we have (ma, mb) = least positive value of max + mby

$$= m. \quad [\text{least positive value of } ax + by]$$
$$= m\,(a, b)$$

Deductions : (1) If $d \mid a$ and $d \mid b$ and $d > 0$, then $\left(\dfrac{a}{d}, \dfrac{b}{d}\right) = \dfrac{1}{d}\,(a, b)$

(2) If $(a, b) = g$, then $\left(\frac{a}{g}, \frac{b}{g}\right) = 1$

Proof : If $g = (a, b)$, then, we have $g \mid a$ and $g \mid b$

Therefore, $\frac{a}{g}$ and $\frac{b}{g}$ both are integers.

Also, $g = (a, b)$

$$= \left(g.\frac{a}{g}, g\frac{b}{g}\right) = g\left(\frac{a}{g}, \frac{b}{g}\right) \quad \Rightarrow \quad \left(\frac{a}{g}, \frac{b}{g}\right) = 1$$

Theorem 8. *Let $a > 1$, and m, n be positive integers. Then* $(a^m - 1, a^n - 1) = a^{(m,n)} - 1$

Proof : When $m = n$, then result is obvious.

Now, suppose that $m > n$, $m = qn + r$, then

$$a^m - 1 = (a^n - 1)\, a^{m-n} + a^{m-n} - 1$$
$$= (a^n - 1)\, a^{m-n} + (a^n - 1)a^{m-2n} + a^{m-2n} - 1$$
$$= (a^n - 1)\, a^{m-n} + a^{m-2n} + \ldots + a^{m-qn}) + a^r - 1$$

Hence, $(a^m - 1, a^n - 1) = (a^n - 1, a^r - 1) = (a^r - 1, a^{r_1} - 1) = \ldots$

$$= (a^d - 1, a^0 - 1) = a^d - 1$$

where, $d = (m, n)$

6.7 RELATIVELY PRIME INTEGERS

If the greatest common divisors of a and b is 1, then a and b are said to be relatively prime. Also, $a_1, a_2, \ldots, a_n$ are said to be relatively prime in pairs. If $(a_i, a_j) = 1$ for all $i = 1, 2, \ldots, n$ and $j = 1, 2, \ldots, n$ with $i \neq j$.

- The fact that $(a, b) = 1$ is sometimes expressed by saying that a and b are co-prime or by saying that a is prime to b.
- If $(a_i, a_j) > 1$, whenever $i \neq j$, the numbers $a_1, \ldots, a_n$ are said to be relatively prime in pairs. If $(a_1, .., a_n)$ are relatively prime in pairs, then $(a_1, \ldots, a_n) = 1$.

Theorem 1. *The integers a and b are relatively prime if and only if there exists integers x and y such that* $ax + by = 1$.

Proof : Let us first suppose there exists integers x and y such that $ax + by = 1$. To show a and b are relatively prime.

Let d be the common divisor of a and b, say $a = p.\, d$ and $b = qd$.

Then, $(px + qy)\, d = 1$, shows that d must be a unit.

$\therefore \quad g = (a, b) = \pm 1$, where g is the g.c.d. of (a, b).

Hence, a and b are relatively prime.

Conversely if $(a, b) = 1$, then Euclid algorithm guarantees the existence of integers x and y such that $ax + by = 1$.

Deductions : (1) If $(a, b) = g$ and $a = Ag$, $b = Bg$, then $(A, B) = 1$.

Proof : By the Euclid algorithm, integers x and y exist, therefore $g = ax + by$

Then, from $g = (Ax + By)\, g$, we get $AX + By = 1$.

Then, by above theorem, it follows that $(A, B) = 1$.

(2) If $(a, b) = 1$ and $(a, c) = 1$, then $(a, bc) = 1$.

Proof : Using above theorem, x_1 and y_1 exist so that $ax_1 + by_1 = 1$ and x_2 and y_2 exist so that $ax_2 + cy_2 = 1$.

Then, $1 = 1.\, 1 = (ax_1 + by_1)(ax_2 + cy_2)$

$$= a(x_1\, ax_2 + x_1\, cy_2 + by_1\, x_2) + bc\,(y_1\, y_2).$$

Therefore, integers $x_3 = x_1\, ax_2 + x_1\, cy_2 + by_1\, y_2$ and $y_3 = y_1 y_2$ exist so that

$$ax_3 + bc\, y_3 = 1.$$

Hence, again by previous theorem 18, we have

$$(a, bc) = 1$$

Theorem 2. *If a and b are relatively prime and if a divides bc, then a must divide c.*

Proof : Since a and b are relatively prime, therefore, by definition, we have $(a, b) = 1$.

Therefore, there exists integers x and y such that $1 = ax + by$. Hence,

$$c = c\,(ax + by) = acx + bcy$$

but a divides bc, therefore $bc = aP$, which shows that

$$c = a\,(cx + Py)$$

Hence, a divides c.

Theorem 3. *(Euclid's Lemma)* : *If $a|bc$ and $(a, b) = 1$, then $a\,|\,c$.*

Proof : We have $(a, b) = 1$

Therefore, there exists integers x and y such that

$$ax + by = 1$$

$\Rightarrow \quad c\,(ax + by) = c\,.\,1 \qquad \Rightarrow \quad c = cax + cby$

Now, $\quad a|ac$ and $a|bc \qquad \Rightarrow \quad a\,|\,acx + bcy$

$\Rightarrow \quad a|c$

Deduction : (1) If a and b are integers, p is a prime such that $p|ab$ and $p\nmid a$, then $p|b$.

(2) If $p\,|\,a_1\,a_2 \ldots a_n$, then there exists some i such that $p|a_i$.

☛ If a and b are not relatively prime, then Euclid's Lemma does not holds good.

Theorem 4. *If a and b are any two integers, not both zero. Then a positive integer $g = (a, b)$ if and only if*

(i) $g|a$ and $g|b$ (ii) whenever $c\,|a$ and $c\,|b$, then $c\,|g$.

Proof : Let, $g = (a, b)$. Then clearly $g|a$ and $g|b$. Therefore, condition (i) is satisfied.

Now, since $g = (a, b)$, therefore, there exist integers x and y, such that $g = ax + by$. Now, $c\,|a$ and $a\,|b$ implies $c\,|ax + by$. Thus, condition (ii) is satisfied.

Conversely, let g be any positive integer satisfying conditions (i) and (ii). To show $g = (a, b)$. Using equation (ii), we have, if c is a common divisor of a and b, then $c\,|g$, therefore $g \geq c$ and hence, g is the greatest common divisor of a and b, *i.e.*, $(a, b) = g$

Theorem 5. *For any integer x, $(a, b) = (b, a) = (a, -b) = (a, b + ax)$.*

Proof : Denote (a, b) by d and $(ab + ax)$ by g. Here, it is clear that $(a, -b) = d$.

Now, we know that there exist integers x_0 and y_0 such that

$$d = ax_0 + by_0$$

Therefore, we can write $d = a\,(x_0 - xy_0) + (b + ax)\,y_0$ which show that the greatest common divisors of a and $b + ax$ is a divisor of d, *i.e.*, $g \mid d$. Now, we shall prove that $d \mid g$. Since $d|\,a$ and $d \mid b$, we have $d \mid (b + ax)$. Since, we know that every common divisor of a and $b + ax$ is a divisor of their g.c.d., *i.e.*, a divisor of g. Thus $d\,|g$. Hence, we conclude that $d = \pm\, g$. However, d and g are both positive, by definition so $d = g$.

ILLUSTRATIVE EXAMPLES

1. *Find the greatest common divisor of 525 and 231.*

Solution. We have $525 = 2 \times 231 + 63$

$$231 = 3 \times 63 + 42$$
$$63 = 1 \times 42 + 21$$
$$42 = 2 \times 21$$

Hence, $(525, 231) = 21$

2. *Find g.c.d. of 396 and 671.*

Solution. We can write $671 = 2 \times 396 - 121$

$$396 = 3 \times 121 + 33$$
$$= 121 = 4 \times 33 - 11$$
$$33 = 3 \times 11 + 0$$

Hence, $(396, 671) = 11$

6.8 ALGORITHM TO FIND G.C.D. : INVESTIGATION OF THE SET OF INTEGERS $\{bx + cy\}$

The investigation of the set of integers $\{bx + cy\}$ to find a smallest positive element is not practical for large values of b and c. If b and c are small, value of g, x_0 and y_0 such that $g = bx_0 + cy_0$ can be found by inspection.

For example, if $b = 10$ and $c = 6$, it is clear that $g = 2$ and one pair of values for x_0, y_0 is 2 and –3. But if b and c are large, we can not use inspection method. Theorem 22 can be used to calculate g effectively and also to get values of x_0 and y_0. We now discuss an example to show how Theorem 22 can be used to calculate the greatest common divisor.

Consider the case $b = 963$, $c = 657$. If divide c into b, we get a quotient $q = 1$ and remainder $r = 306$. Therefore, $b = cq + r$ or $r = b - cq$, in particular $306 = 963 - 1\,(657)$.

Now, $(b, c) = (b - cq, c)$ by replacing a and x by c and $-q$ in theorem 22, therefore, we have

$$(963, 657) = (963 - 1\,(657), 657) = (306, 657)$$

The integer 963 has been replaced by the smallest integer 306 and we repeat the procedure. Therefore, we divide 306 into 657 to get a quotient 2 and a remainder 45 and

$$(306, 657) = (306, 657 - 2(306)) = (306, 45)$$

Next 45 is divided into 306 with quotient 6 and remainder 36. Then 36 is divided into 45 with quotient 1 and remainder 9. Thus, we conclude that

$$(963, 657) = (306, 657) = (306, 45) = (36, 45) = (36, 9)$$
$$\Rightarrow \quad (963, 657) = 9$$

and we can express 9 as a linear combination of 963 and 657 by sequentially writing each remainder as a linear combination of the two original numbers.

$$306 = 963 - 657$$
$$45 = 657 - 2\,(306) = 657 - 2\,(963 - 657) = 3\,(657) - 2\,(963)$$
$$36 = 306 - 6\,(45) = (963 - 657) - 6\,(3 \times 657 - 2 \times 963)$$
$$= 13\,(963) - (19\,(657)$$
$$9 = 45 - 36 = 3\,(657) - 2\,(963) - [13\,(963) - 19\,(657)]$$
$$= 22\,(657) - 15\,(963)$$

Thus, we can find $g = 9$, $x = -15$, $y = 22$.

These values of x_0 and y_0 are not unique $-15 + 657k$ and $22 - 963k$ will do where k is any integer.

- To find the greatest common divisor of b and c of any two integers b and c, we now generalize what is done in the special case above. The process will also give integers x_0 and y_0 satisfying the equation $bx_0 + cy_0 = (b, c)$.
- The case $c = 0$ is special because $(b, 0) = |b|$. for $c \neq 0$, we have $(b, c) = (b, -c)$ and hence, we presume that c is positive.

ILLUSTRATIVE EXAMPLES

1. *Find the greatest common divisor of 42823 and 6409.*

Solution. Here, we apply the Euclidean algorithm. We divide c into b, where $b = 42823$ and $c = 6409$. When 6409 is divided into 42823, we get 6.6816976, therefore quotient is 6. To get the reminder, we multiply 6 by 6409 to get 38454, and we subtract this from 42823 to get the remainder 4369. Thus, $q_i = 6$ and $r_1 = 4369$.

Continuing the same process, if we divide 4369 into 6409, we get a quotient $q_2 = 1$ and remainder $r_2 = 2040$. Further, dividing 2040 into 4369 gives $q_3 = 2$ and $r_3 = 289$. Again, dividing 289 into 2040 give $q_4 = 7$ and $r_4 = 17$.

Since 17 is a divisor of 289, the solution is that the g.c.d. is 17.

The above process can be put in tabular form as follows :

$$\begin{aligned} 42823 &= 6\,(6409) + 4369 && = (42823, 6409) \\ 6409 &= 1\,(4369) + 2040 && = (6409, 4369) \\ 4369 &= 2\,(2040) + 289 && = (4369, 2040) \\ 2040 &= 7\,(289) + 17 && = (2040, 289) \\ 289 &= 17\,(17) && = (289, 17) = 17 \end{aligned}$$

2. *Find g.c.d. of* 256 and 1166 *and express g.c.d. as linear combination of* 256 and 1166.

Solution. Here we have $1166 = 256\,(4) + 142$

$$\begin{aligned} 256 &= 142\,(1) + 114 \\ 142 &= 114\,(1) + 28 \\ 114 &= 28\,(4) + 2 \\ 28 &= 2\,(14) + 0 \end{aligned}$$

Thus, $(256, 1166) = 2$

Also,
$$\begin{aligned} 2 &= 114 - 28\,(4) \\ &= 114 - 4\,(142 - 114\,(1)) \\ &= 114 - 4\,(142) + 4\,(114) \\ &= 5\,(114) - 4\,(142) \\ &= 5\,(256 - 142(1)) - 4\,(142) \\ &= 5\,(256) - 5\,(142) - 4\,(142) \\ &= 5\,(256) - 9\,(142) \\ &= 5\,(256) - 9\,(1166) + 36\,(256) \\ &= 41\,(256) - 9\,(1166) \\ &= 256x + 1166y \end{aligned}$$

where, $x = 41$ and $y = -9$. Hence g.c.d. 2 has been expressed as linear combination of 256 and 1166.

3. *Find integer* x and y *to satisfy* $42823x + 6409y = 17$.

Solution. We find integers x_i and y_i such that

$$42823x_i + 6409y_i = r_i$$

Here, it is natural to consider $i = 1, 2, \ldots$, but to initiate the process, we also consider $i = 0$ and $i = -1$. We put $r_{-1} = 42823$ and write

$$42823\,(1) + 6409\,(0) = 42823$$

Similarly, we put r_0 6409 and write

$$42823\,(0) + 6409\,(1) = 6409$$

We multiply the second of these equation by $q_1 = 6$ and subtract the result from the first equation to obtain

$$42823\,(1) + 6409\,(-6) = 4369$$

Multiply this equation by $q_2 = 1$ and subtract it from the preceding equation, we find

$$42823\,(-1) + 6409\,(7) = 2040$$

We multiply this by $q_3 = 2$ and subtract the result from the preceding equation to find that

$$42823\,(3) + 6409\,(-20) = 289$$

Next, multiply this by $q_4 = 7$ and subtract the result from the preceding equation to find that

$$42823.\,(-22) + 6409.\,(147) = 17$$

On dividing 17 into 289, we find that $q_5 = 17$ and that $289 = 17 \times 17$. Therefore, r_4 is the last positive remainder so that $g = 17$ and we may take $x = -22$, $y = 147$.

4. *Find g.c.d. of 28 and 49. Express it as linear combination of these numbers.*

Solution. We have $49 = 28\,(1) + 21$

$$28 = 21\,(1) + 7$$
$$21 = 7\,(3) + 0$$

which implies that $r_2 = 7$ is the required greatest common divior. Also,

$$\begin{aligned} 7 &= 28 - 21.\,(1) \\ &= 28 - (49 - 28\,(11)) \\ &= 28.\,(2) - 49\,(1) \\ &= 28x + 49y \end{aligned}$$

when $x = 1, y = -1$.

5. *Find* $g = (b, c)$ *where* $b = 5033464705$ *and* $c = 3137640337$ *and determine* x and y *such that* $bx + cy = g$.

Solution. Proceeding same as above, we get the following calculation scheme :

	5033464705	1	0
1	3137640337	0	1
1	1895824368	1	–1
1	1241815699	–1	2
1	654008399	2	– 3
1	587807570	–3	5
8	66200829	5	– 8
1	58200938	– 43	69
7	7999891	48	– 77
3	2201701	– 379	608
1	1394788	1185	– 1901
1	806913	– 1564	2509
1	587875	2749	– 4410
2	219038	– 4313	6919
1	149799	11375	– 18248
2	69239	– 15688	25167
6	11321	42751	– 68582
8	1313	– 272194	436659
1	817	2220303	– 3561854
1	496	–2492497	3998513
1	321	4712800	– 7560367
1	175	–7205297	11558880
1	146	11918097	–19119247
5	29	– 19123394	30678127
2	91	107535067	–172509882

Hence, $g = 1$ and we may take $x = 107535067$, $y = -172509882$.

6.9 GREATEST COMMON DIVISOR OF MORE THAN TWO INTEGERS

In the previous section, we introduced the concept of the greatest common divisor of two integers. Now, in the same way, we can also define greatest common divisor of more than two integers, not all zeroes. The g.c.d. of integers $a_1, a_2, \ldots a_n$ not all zero, is the largest integer which is divisor of each of these integers, it exists uniquely and is denoted by $(a_1, a_2, \ldots a_n)$.

If $(a_1, a_2, \ldots a_n) = 1$, then we say that the integers $a_1, a_2, \ldots a_n$ are mutually relatively prime. If each pair of integers a_i and a_j from the set is relatively prime, then these integers are called pairwise relatively prime.

Clearly, if integers are pairwise relatively prime, then they must be mutually relatively prime. The converse is not true. For example :

Since, (16, 10, 15) = 1, therefore, 6, 10 and 15 are mutually relatively prime, but any two of these integers are not relatively prime. Hence, they are not pairwise relatively prime.

WORKING RULES

Step 1. Two find the greatest common divisor (a, b, c) of three integers a, b and c, we shall first find the *g. c. d.* (a, b) of a and b and then find the g.c.d. $((a, b), c)$ of (a, b) and c; the result is the required (a, b, c).

☛ Similarly, the infinite set of integers $a_1, a_2, \ldots a_n \ldots$, also has the greatest common divisor $(a_1, a_2, \ldots, a_n \ldots)$ which can be obtained by using the same procedure.

ILLUSTRATIVE EXAMPLES

1. *Find the greatest common divisor of 136, 221, 391.*

Solution. We have
$$\begin{aligned}(136, 221, 391) &= (136, (221, 391))\\ &= (136, 17)\\ &= 17\end{aligned}$$

We can also compute it as follows :
$$\begin{aligned}(136, 221, 391) &= (136, 221 - 136, 391 - 2(136))\\ &= (138, 85, 119) = (51, 85, 34)\\ &= (17, 17, 34) = 17\end{aligned}$$

Moreover, since $5(136) - 3(221) = 17$
$$(-22).17 + 1(391) = 17$$

We have $\qquad 136.(-110) + 221(66) + 391(1) = 17$

2. *Let $a = qc + r$, $b = q_1c + r_1$, show that $(a, b, c) = (r, r_{1,} c)$.*

Solution. We have
$$\begin{aligned}(a, b, c) &= ((a, c), b) = ((c, r), b) = (r, (b, c))\\ &= (r, (c, r_1)) = (r, r_{1,} c)\end{aligned}$$

6.10 LEAST COMMON MULTIPLE

Let $m \neq 0$ be a multiple of a and b. Then m is called a common multiple of a and b. Clearly ab is a common multiple of a and b. Among the common multiple of a and b there is no greatest integer, but there is a unique positive least integer, which is called the least common multiple of a and b and denoted by $[a, b]$.

Definition : *If a and b are two non-zero integers, then a positive integer m is called their least common multiple if*

(i) $a \mid m$ and $b \mid m$ and

(ii) there exists a positive integer n such that if $a|n$ and $b|n$, then $m \le n$, equivalently $m|n$.

Theorem 1. *A common multiple of a and b is a multiple of the least common multiple* $[a, b]$.

Proof : Let k be a common multiple of a and b. Dividing k by $[a, b] = m$, we get

$$k = qm + r, 0 \le r < m$$

Since $a | k$ and $a | m$, so $a | r$. Similarly, we can show that $b | r$. If $r \ne 0$, then r is a common multiple of a and b. This contradicts the assumption that m is the least common multiple. Therefore, $r = 0$ and $k = qm$, *i.e.*, k is a multiple of m.

Theorem 2. *If* $a > 0$, $b > 0$ *be two integers, then* $[a, b]\ (a, b) = a.b$

Proof : Let $[a, b] = m$ and $(a, b) = d$. Since, $a | m$ and $b | m$, we have $ab | ma$, $ab | mb$ and hence $ab | (ma, mb)$, that is $ab | md$.

Also, since $a | ab/d$ and $b | ab/d$, that is ab/d is a common multiple of a and b. By previous theorem, $m | ab/d$, therefore md/ab and $ab = md$.

Verification : For example, $[6,9] = 18$, $(6, 9) = 3$, here $18 . 3 = 6 . 9$

Deduction : If $(a, b) = 1$, then, we have $[a, b] = ab$ so that ab is the least common multiple of a and b. Conversely, if the least common multiple of a and b is ab, then $(a, b) = 1$. Hence, a necessary and sufficient condition for $[a, b] = ab$ is $(a, b) = 1$.

Theorem 3. *If* $k > 0$ *is a common multiple of a and b, then* $\left(\frac{k}{a}, \frac{k}{b}\right) = \frac{k}{[a, b]}$.

Proof : Since, we have $\left(\frac{k}{a}, \frac{k}{b}\right) | ab | = (kb, ka) = k\,(a, b)$

$$= k.\frac{a.b}{[a, b]} \qquad \left(\because (a, b) = \frac{ab}{[a, b]}\right)$$

Hence, $\left(\frac{k}{a}, \frac{k}{b}\right) = \frac{k}{[a, b]}$

Deductions : (1) If $k = [a, b]$, then $\left(\frac{k}{a}, \frac{k}{b}\right) = 1$. Conversely, if $\left(\frac{k}{a}, \frac{k}{b}\right) = 1$, then $k = [a, b]$.

(2) If $k > 0$, then a necessary and sufficient condition for $k = [a, b]$ is $\left(\frac{k}{a}, \frac{k}{b}\right) = 1$.

6.10.1 Formula for Computing the Least Common Multiple

We have $[a, b] = \frac{ab}{(a, b)}$

Therefore, we first find (a, b) and then use the formula to get the required $[a, b]$.

Theorem 1. *Let a and b two positive integers, then*

$$(a + b)\ [a, b] = b\ [a, a + b].$$

Proof : We have $b\,[a, a + b] = \frac{ba\,(a + b)}{(a, a + b)} = \frac{(a + b)\,ab}{(a, b)} = (a + b)\,[a, b]$

Theorem 2. *The least common multiple of two non-zero integers is unique.*

Proof : Let a and b be any two non-zero integers. Let, if possible, there are two least common multipliers m_1 and m_2 of a and b. By definition a and b divide both m_1 and m_2.

Again, by definition of *l.c.m.*, m_1 divides m_2 and m_2 divides m_1.

Hence, $m_1 = m_2$ *i.e.*, least common multiplier is unique.

6.11 LEAST COMMON MULTIPLE OF *n* INTEGERS

The least common multiple of $[a, b, c]$ of three integers a, b and c can be found by using the following formula

$$[a, b, c] = [[a, b], c] \qquad \ldots(1)$$

i.e., we first find the least multiple $[a, b]$ of a and b, and then the least common multiple $[[a, b], c]$ of $[a, b]$ and c, which is the required $[a, b, c]$.

Generalization : The above formula (1) can be generalized to the case of n integers $a_1, a_2, \ldots, a_n$ as follows

If $[a_1, a_2] = m_2, [m_2, a_3] = m_3, \ldots, [m_{n-1}, a_n] = m_n$

Then, $[a_1, a_2, \ldots, a_n] = m_n$

If $a_1, a_2, \ldots, a_n$ are mutually relatively prime, then

$$[a_1, a_2, \ldots, a_n] = a_1 a_2 \ldots a_n$$

Theorem 1. *Let a, b, c be three positive integers. Then* $[a, b, c] = \dfrac{abc}{(ab, bc, ca)}$

Proof : We know that

$$[a, b, c] = [[a, b], c] = \frac{[a, b].c}{([a, b], c)} = \frac{abc}{(a, b)([a, b], c)} = \frac{abc}{(ab, (a, b)\, c)} = \frac{abc}{(ab, bc, ca)}$$

☛ The above formula is also true for any number of integers.

ILLUSTRATIVE EXAMPLES

1. *Find the g.c.d and l.c.m of 119 and 272.*

Solution. Here, we have $272 = 119\,(2) + 34$

$$119 = 34\,(3) + 17$$

$$34 = 17\,(2) + 0$$

which implies that 17 is the g.c.d of 119 and 272.

Also, $[119, 272] = \dfrac{119 \times 272}{(119, 272)}$ $\left[\because [a, b] = \dfrac{a.b}{(a, b)}\right]$

$$= \frac{119 \times 272}{17} = 1904$$

2. *Find the l.c.m. of 136, 221 and 391.*

Solution. We have $[136, 221, 391] = [[136, 221], 391]$

$$= \left[\frac{136 \times 221}{17}, 391\right] = [1768, 391] \qquad (\because \text{g.c.d.of 136 and 221 is 17})$$

$$= \frac{1768 \times 391}{17} = 40664$$

3. *Find the l.c.m. of 8 , 12, 15, 20 and 25.*

Solution. Using $[a, b] = \dfrac{ab}{(a, b)}$

We have $[8, 12] = \dfrac{8 \times 12}{(8, 12)} = \dfrac{8 \times 12}{4} = 24$

$$[24, 15] = \frac{24 \times 15}{(24, 15)} = \frac{24 \times 15}{3} = 120$$

$$[120, 20] = \frac{120 \times 20}{(120, 20)} = \frac{120 \times 20}{20} = 120$$

$$[120, 25] = \frac{120 \times 25}{(120, 25)} = \frac{120 \times 25}{5} = 600$$

Hence, l.c.m. of 8, 12 15, 20 and 25 is given by $[8, 12, 15, 20, 25] = 600$.

6.12 FIBONACCI SEQUENCE

The sequence $a_1, a_2, a_3, \ldots$ in which $a_1 = 1$, $a_2 = 1$ and $a_n = a_{n-1} + a_{n-2}$ for every $n > 2$ is called a Fibonacci sequence.

Th first few Fibonacci numbers are

$$1, 1, 2, 3, 5, 8, 13, 21, 34, 55, 89, 144, \ldots$$

It is surprising to discover that the Fibonacci numbers can be extracted form Pascal's triangle by adding the numbers along the north-east diagonals as below :

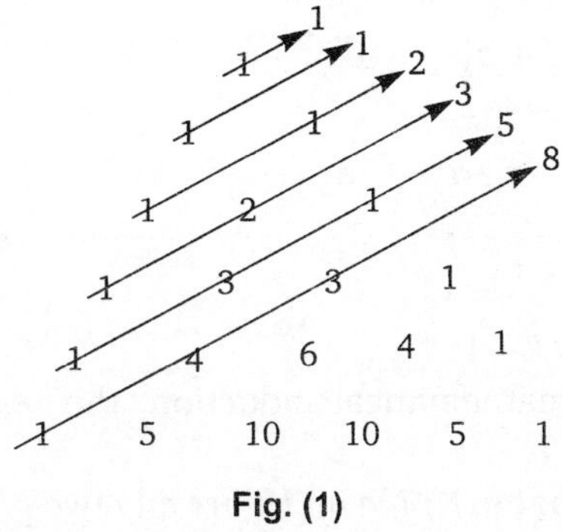

Fig. (1)

Interestingly enough, Fibonacci numbers appear in nature, music, geography and geometry. They can be found in the spiral arrangement of seeds in sunflowers, the scale patterns of pine cones, the number of petals in flowers and the arrangement of leaves on trees.

Fibonacci numbers can be obtained explicitly using Binet's formula

$$F_n = \frac{\alpha^n - \beta^n}{\alpha - \beta}, \text{ where } \alpha = \frac{1+\sqrt{5}}{2}, \beta = \frac{1-\sqrt{5}}{2}$$

which are root of quadratic equation $x^2 = x + 1$.

6.12.1 Lucas Number

The Lucas number L_n are determined by the relation $L_1 = 1, L_2 = 3$ and $L_n = L_{n-1} + L_{n-2}$ for $n > 2$

We have $L_n = \left(\frac{1+\sqrt{5}}{2}\right)^n + \left(\frac{1-\sqrt{5}}{2}\right)^n$

Lucas numbers can also be obtained by Binet's formula

$$L_n = \alpha^n + \beta^n, \text{ where } \alpha = \frac{1+\sqrt{5}}{2}, \ \beta = \frac{1-\sqrt{5}}{2}$$

☛ The Fibonacci number a_n and Lucas number L_n satisfy the same recurrence relation but with different initial conditions.

Theorem 1. *If a_n is the nth form of the Fibonacci sequence and $\alpha = \frac{1+\sqrt{5}}{2}$, then $a_n > \alpha^{n-1}$ for all $n > 1$.*

Proof : We can easily verify that $\alpha^2 = \left(\frac{1+\sqrt{5}}{2}\right)^2 = \frac{1+\sqrt{5}}{2} + 1 = \alpha + 1$

Thus, for any $k > 1$, we have

$$\alpha^k = \alpha^{k-2}.\alpha^2 = \alpha^{k-2}[\alpha+1] = \alpha^{k-1} + \alpha^{k-2}$$

Now, $a_2 = 2$ and $\alpha = \dfrac{1+\sqrt{5}}{2}$ $\quad\therefore\quad a_2 > \alpha = \alpha^{2-1}$

i.e., result is true for $n = 2$

Further, from the definition of Fibonacci sequence, we have

$$a_3 = a_2 + a_1 > \alpha + 1 = \alpha^2 = \alpha^{3-1}$$

i.e., result is true for $n = 3$.

Now, suppose that result is true for $n = 2, 3, \ldots\ldots, k$, then

$$a_k > \alpha^{k-1} \text{ and } a_{k-1} > \alpha^{k-2}$$

Therefore,
$$\begin{aligned} a_{k+1} &= a_k + a_{k-1} > \alpha^{\kappa-1} + \alpha^{\kappa-2} \\ &= \alpha^{\kappa-2}[\alpha+1] \\ &= \alpha^{k-2}.\alpha^2 \\ &= \alpha^{\kappa} \\ &= \alpha^{k+1} - 1 \end{aligned}$$

i.e., result is true for $n = k+1$.

Hence, by principle of mathematical induction, the result is true for all values of $n > 1$.

Theorem 2. ***(Lame's Theorem)*** **:** *If a and b are any two positive integers with a > b and n is the number of divisions in Euclid's algorithm, then $n \le 5p$, where p is the number of digits in b.*

Proof : Suppose that a and b be two positive integers and n be the number of divisions in Euclid's algorithm. Then, we get a positive number $r_{n-1}, r_{n-2}, \ldots\ldots, r_2, r_1, b, a$ in increasing form. Comparing this sequence with Fibonacci sequence, we have

$$r_{n-1} > 1 = a_1 \qquad \ldots(1)$$

Also,
$$r_{n-2} = r_{n-1}\,.\,q_1 \text{ and } q_2 \ge 2$$

$\therefore$
$$r_{n-2} \ge a_2 \qquad \ldots(2)$$

Again for $k > 2$, we have

$$r_{n-k} = r_{n-(k-1)}\,q_{n-(k-2)} + r_{n-(k-2)} \ge r_{n-(k-1)} + r_{n-(k-2)} \qquad \ldots(3)$$

Now, putting $k = 3, 4, \ldots\ldots, n-1$ in (3), we get

$$\begin{aligned} r_{n-3} &\ge r_{n-2} + r_{n-1} \ge a_2 + a_1 = a_3 \\ r_{n-4} &\ge r_{n-3} + r_{n-2} \ge a_3 + a_2 = a_4 \\ &\vdots \\ r_1 &\ge r_3 + r_4 \ge a_{n-2} \\ r_1 &\ge r_2 + r_3 \ge a_{n-1} \end{aligned}$$

Also, $\quad b = r_1\,q_2 + r_2 \ge r_1 + r_2 \ge a_{n-1} + a_{n-2} = a_n > \alpha^{n-1}$, where $\alpha = \dfrac{1+\sqrt{5}}{2}$

Thus, $\log_{10} b > (n-1)\log_{10}\alpha = (n-1)\log_{10}\left(\dfrac{1+\sqrt{5}}{2}\right) = \dfrac{1}{5}$ $\qquad \ldots(4)$

Since, p is the number of digits in b, therefore,

$$p = \log_{10} b \qquad \ldots(5)$$

Using (4) and (5), we conclude that

$$p > (n-1)/5 \;\Rightarrow\; n < 5p + 1 \;\Rightarrow\; n \le 5p$$

Deduction : $n \le \dfrac{1}{\log_{10}\alpha}, \alpha = \dfrac{1+\sqrt{5}}{2}$

Theorem 3. *If a and b are any two positive integers with $a > b$ and n is the number of divisions in minimal algorithm for a and b, then $(n-1) < \frac{10}{3}p$, where p is the number of digits in b.*

Proof : Let a and b be two positive integers such that $a > b$. Then we have

$$R_1 \le \frac{b}{2}$$

$$R_2 \le \frac{R_1}{2} \le \frac{b}{2^2}$$

$$R_3 \le \frac{R_2}{2} \le \frac{b}{2_3}$$

$$\vdots \qquad \vdots$$

$$R_{n-1} \le \frac{R_{n-2}}{2} \le \frac{b}{2^{n-1}}$$

also, $R_{n-1} \ge 1$

$\therefore$ $b \ge 2^{n-1}.R_{n-1}$

$\ge 2^{n-1}$

Taking log of both the sides, we get

$$\log_{10} b \ge (n-1)\log_{10} 2$$

$$\ge (n-1).\frac{3}{10} \qquad \left(\because \log_{10} 2 > \frac{3}{10}\right)$$

Also, $p > \log_{10} b$

which conclude that

$$p > (n-1)\frac{3}{10}$$

i.e., $$(n-1) < \frac{10}{3p}$$

Theorem 4. *(Kronecker's Theorem)* : *The number of divisions in minimal algorithm for two positive integers a and b is not greater than the number of divisions in Euclid's algorithm for the same integers, i.e., $M(a,b) \le E(a,b)$, where $M(a,b)$ is the number of division in minimal algorithm and $E(a,b)$ is the number of divisions in Euclid's algorithm.*

Proof : To prove this theorem, we shall use the principle of mathematical induction.

Let, $a = 2, b = 1$. Then, we have

$$2 = 2.1 + 0$$

$\Rightarrow$ $E(a,b) = 1$ and $M(a,b) = 1$ $\Rightarrow$ $M[a,b] \le E(a,b)$

$\Rightarrow$ Result is true for $a = 2$.

Now, for $a = 3, b = 1$, we have

$$3 = 3.1 + 0$$

$\Rightarrow$ $E(a,b) = 1, M(a,b) = 1$ $\Rightarrow$ $M(a,b) \le E(a,b)$

For $a = 3$, $b = 2$, we have

$$3 = 2.1 + 1, \quad 2 = 1.2 + 0$$

$\Rightarrow$ $E(a,b) = 2, M(a,b) = 1$ $\Rightarrow$ $M(a,b) \le E(a,b)$

$\Rightarrow$ Result is true for $a = 3$.

Now, suppose that theorem is true for $a = 2, 3, \ldots k-1$. To show that result is true for $a = k$, let b be a positive integer less than k. Then, we have

$$k = bQ + eR,$$

where $e = \pm 1$ and $0 \le R \le b/2$

Thus $\quad M(k,b) = 1 + M(b,R)$ ...(1)

$k = bq + r, \; 0 \le r < b \; \Rightarrow \; (k,b) = 1 = E(b,r)$ ...(2)

Now, there are following three possibilities :

(i) $0 = R = r$: In this case, we have

$$M(k,b) = 1 \text{ and } E(k,b) = 1$$

$\Rightarrow$ Theorem holds for $a = k$.

(ii) $0 < R = r$: In this case, we have

$$E(k,b) = 1 = 1 + E(b, r)$$

but, $b < k$ and $R < b$, then by our assumption

$$E(b, R) \ge M(b, R)$$

From (2), (3) and (1), we have

$$E(k, b) = 1 + E(b, R)$$
$$\ge 1 + M(b, R) = M(k, B)$$

$\Rightarrow$ Theorem holds for $a = k$.

(iii) $0 < R < r$: Let $r = b - R$

Put this value in (2), we get

$$E(k,b) = 1 + E(b, b-R) \quad ...(4)$$

Further, $b < k$ and $b - R < b$, therefore,by induction

$$E(b, b-k) \ge M(b, b-k)$$
$$E(k,b) = 1 + E(b, b-R) \quad ...(5)$$

From (4) and (5), we have

$$E(k,b) = 1 + E(b, b-R)$$
$$\ge 1 + M(b, b-R)$$
$$\ge 1 + M(b,R) = M(k,b)$$

$\Rightarrow$ Theorem holds for $a = k$.

Hence, by principle of mathematical induction, we have, the result is true for $a > 1$.

Exercise 6.3

1. Show that if $ac \mid bc$, then $a \mid b$.
2. Let n be an odd integer greater than 1, show that $n^3 - n$ is a multiple of 24.
3. Prove that if a and b are positive integers satisfying $(a, b) = [a, b]$, then $a = b$.
4. Show that common divisor of integers $a_1, a_2, \ldots, a_n, \ldots$ is a divisor of their greatest common divisor.
5. Let $n \ge 2$ and k be any positive integers. Prove that $(n-1)^2 \mid (n^k - 1)$ if and only if $(n-1) \mid k$.
6. If $m > 0$, $n > 0$ and m is an odd integer, show that $(2^m - 1)(2^n + 1) = 1$.
7. If $(a, b) = 1$, show that $(a - b, a + b) = 1$ or 2.
8. If a and b are non-zero integers, then there exist four integers h, k, r, s such that $hs - kr = 1$, $ak + bs = 0$.
9. Show that if $(a, b) = 1$, then $(a + b, a^2 - ab + b^2) = 1$ or 3.
10. Show that $M(a, b) = M(a \pm b, b)$
11. If $\frac{a}{3} < b < \frac{2}{5}a$, then show that $M(a, -a-b) = 1 + M[b, b-(3b-a)]$.
12. If $(a, b) = d$, then d is the least positive integer among all the integers of the form $ax + by$, *i.e.*, $d = ax + by$, where x and y are any integers.
13. For Fibonacci sequence, show that $a_1 + a_2 + a_3 + \ldots. + a_n = a_{n+2} - 1$.
14. For Fibonacci sequence, show that $a_{n+1}\, a_{n-1} - a_n^{\,2} = (-1)^n$.
15. For Fibonacci sequence, show that $a_{m+n} = a_{m-1}a_n + a_m a_{n+1}$ for any positive integer m and n and hence prove that $a_m \mid a_n$ if $m \mid n$.

16. If the sum of two reduced fractions is an integer, say $\left(\frac{a}{b}\right)+\left(\frac{c}{d}\right)=n$, then show that $|b|=|d|$.

17. Show that for every $n \geq 1$, there exists uniquely determined $a > 0$, $b > 0$ such that $n = a^2b$, where b is a square free integer.

ANSWERS

2. As $n^3 - n = (n-1)\,n\,(n+1)$ is the product of three consecutive integers, it is a multiple of 3. If n is an odd integer, $n-1$ and $n+1$ both must be even and one of them is a multiple of 4. Hence, $(n-1)(n+1)$ is a multiple of 8. Thus, $n^3 - n$ is the multiple of 3. $8 = 24$.

4. Let b be the common divisor and d is the greatest common divisor. As $[b, d] = c$, then $d \leq c$. Again, as $b \mid a_i$, $d \mid a_i$, then $c \mid a_i$ and hence c is the common divisor of $a_1, \ldots, a_m$, but d is the greatest common divisor, hence, $c \leq d$. Hence, $c = d$ and $b \mid d$.

5. Use $n^k = ((n-1)+1)^k$.

6. Let $(2^m - 1, 2^m + 1) = d$. Then $2^m = kd + 1, 2^n = ld - 1, kl > 0$. Therefore, $2^{mn} = (kd+1)^n = td + 1, 2^{mn} = (ld-1)^m = sd - 1$, where $(s-t)d = 2$. Thus $d \mid 2$, *i.e.*, $d = 1$ or 2, but as $2^m - 1, 2^n + 1$ are both odd integers, hence, $d = 1$.

7. Let $(a - ba + b) = d$. Then, $a - b = kd, a + b = ld$. Therefore, $2a = (k+l)\,d, 2b\,(k-l)\,d$. Hence, $2 = (2a, 2b) = d\,(k + l, k - l)$ or $d/2$.

8. Let a and b be two positive integers. Since $[a, b](a, b) = ab$ letting $[a, b] = ak$, $s = -a$, which is divisible by (a, b), we get $ak + bs = 0$. Also, since $l = ab / a(a, b) = b / (a, b)$, then $(k, s) = 1$ and hence $hs + (-k)r = 1$.

15. Let $n = mq$ and induct on q.

6.13 LINEAR DIOPHANTINE EQUATIONS

An equation which has two or more than two unknowns is called an indeterminate or Diophantine equation, after the name of Greek mathematician Diophantine. Generally, a system of equations is called indeterminate or Diophantine, if the number of equations is less than that of the unknowns. For such type of equation, we only look for the solutions in a restricted class of numbers such as positive integers, negative integers, or integers. In this chapter, we shall discuss the simplest Diophantine equations; when do they have solutions? When do we describe them explicitly.

Definition : *An equation of the form*

$$ax + by = c \qquad \ldots(1)$$

with $a \neq 0, b \neq 0$ and c integers, is called a linear diophantine equation in two unknowns x and y.

6.13.1 The Equation $ax + by = c$

Here, we want to find all pairs of integers x, y which satisfy the equation

$$ax + by = c \qquad \ldots(1)$$

in which a, b and c are given integers with $a \neq 0, b \neq 0$.

If $a = b = c = 0$, then every pair (x, y) of integers is a solution of (1), whereas if $a = b = 0$ and $c \neq 0$, then (1) has no solution. The following is a fundamentai theorem which lets us know when an indeterminate equation has solutions and when it does not.

Theorem 1. *The linear diophantine equation*

$$ax + by = c \quad ...(1)$$

a, b, c being integers, has integer solutions if and only if $d|c$ *where* $d = g.c.d$ *of a and* $d = (a, b)$. *Moreover if* $x = x_0, y = y_0$ *is a particular solution, then any solution can be written as*

$$x = x_0 + \frac{b}{d}.t, y = y_0 - \frac{a}{d}.t, \text{ where } t \text{ is any integer.}$$

Proof : Let us first suppose $d|c$, then we have $c = rd$, where r is any integer.
Now, since $(a, b) = d$, then by definition, there exist integers x_1 and y_1 such that

$$ax_1 + by_1 = d \quad ...(1)$$

Multiplying both sides of (1) by $\frac{c}{d}$, we have

$$\frac{c}{d}.ax_1 + \frac{c}{d}.by_1 = d.\frac{c}{d} = c$$

$$\Rightarrow \quad c = a\left(\frac{c}{d}x_1\right) + b\left(\frac{c}{d}y_1\right) = ax + by$$

$\Rightarrow \quad \left(\frac{c}{d}x_1\right)$ and $\left(\frac{c}{d}y_1\right)$ satisfy the equation (1).

Thus, linear diophantine equation has a solution.
Conversely, let us suppose that the equation $ax + by = c$ has a solution, say (x_0, y_0).

Then, $ax_0 + by_0 = c$
But, $ax_0 + by_0$ must be a multiple of d, *i.e.*, $ax_0 + by_0 = rd$, where r is any integer.
Therefore, $\quad c = rd \quad \Rightarrow \quad d|c$
Further, if $x = x_0, y = y_0$ is a solution of (1), then

$$ax_0 + by_0 = c$$

Subtracting (1) from this equation, we get

$$a(x_0 - x) + b(y_0 - y) = 0, i.e., a(x_0 - x) = b(y - y_0)$$

$$\Rightarrow \quad a(x_0 - x_1) = b(y_1 - y_0) \quad ...(2)$$

for $(x, y) = (x_1, y_1)$
Now, since $(a, b) = d$
there exist integers r_1 and r_2 such that $a = r_1 d, b = r_2 d$.
Putting these values in (2), we get

$$r_1 d[x_1 - x_0] = -r_2 d(y_1 - y_0) \quad \Rightarrow \quad r_1(x_1 - x_0) = -r_2(y_1 - y_0)$$

$$\Rightarrow \quad \frac{x_1 - x_0}{r_2} = -\frac{y_1 - y_0}{r_1} = t \text{ (for some integer)} \quad ...(3)$$

Therefore, by division algorithm, we can write $y_1 = y_0 - tr_1$, for some integer t

or
$$y_1 = y_0 - \frac{a}{d}t$$

Now, from (3), we have

$$r_1(x_1 - x_0) = +r_2 r_1 t \quad \Rightarrow \quad x_1 - x_0 = +r_2 t$$

$$\Rightarrow \quad x_1 = x_0 + r_2 t = x_0 + \frac{b}{d}t$$

Hence, $x_1 = x_0 + \frac{b}{d}t$ and $y_1 = y_0 - \frac{a}{d}t$ is the general solution of (1).

Deductions : (1) If $(a, b) = 1$, the solution of (1) can be written as $x = x_0 + bt$, $y = y_0 - at$ where, $x = x_0, y = y_0$ is a solution of (1).

(2) If (x_0, y_0) is one solution of $ax + by = c$, $(a, b) = d$, then $x_1 = x_0 + \frac{b}{d}.t$, $y_1 = y_0 - \frac{a}{d}.t$ is the general solution of (1).

(3) If (x_0, y_0) is one solution of $ax + by = c, (a, b) = 1$, then $x_1 = x_0 + \frac{b}{d}.t$, $y_1 = y_0 - \frac{a}{d}.t$ is the general solution of (1).

(4) If (x_0, y_0) is one solution of $ax + by = c, (a, b) = 1$, then $x_1 = x_0 + bt, y_1 = y_0 -$ at is the general solution of (1).

ILLUSTRATIVE EXAMPLES

1. *Determine if the linear diphantine equation*
(i) $12x + 18y = 30$, *(ii)* $2x + 3y = 4$ *and (iii)* $6x + 8y = 25$ *are solvable.*

Solution. Comparing the given equations with $ax + by = c$, we have

(i) $a = 12, b = 18, c = 30$ and $(12, 18) = 6$ and $6 / 30$, so the linear diophantine equation $12x + 18y = 30$ has a solution.

(ii) $a = 2, b = 3, c = 4$ and $(2, 3) = 1$ and $1 / 4$ so the linear diophantine equation $2x + 3y = 4$ also have a solution.

(iii) $a = 6, b = 8, c = 25$ and $(6, 8) = 2$, but $2 \nmid 25$, so the linear diophantine equation $6x + 8y = 25$ is not solvable.

2. (Mahavira Puzzle) : *Twenty three weary travelers entered the outskirts of a lush green and beautiful forest. They found 63 equal heaps of plantains (fruit) and seven single fruits. They divided them equally. Find the number of fruits in each heap.*

Solution. Let x denote the number of plantains in a heap and y the number of plantains received by a traveler. Then according to the given problem, we get the diophantine equation $63x + 7 = 23y$

The linear equation in Mahavira's puzzle is $63x - 23y = -7$, the g.c.d. of 63 and 23 is 1, *i.e.*, $(63, 23) = 1$, so by deductions 1 of the diophantine equation has the solution of type $x = x_0 + bt$ and $y = y_0 - at$

To find a particular solution x_0, y_0, first we express the g.c.d. 1 as a linear combination of 63 and 23. To do this, we apply euclidean algorithm.

$$63 = 2.23 + 17$$
$$23 = 1.17 + 6$$
$$17 = 2.6 + 5$$
$$6 = 1.5 + 1$$
$$5 = 5.1 + 0$$

Now, use the first four equations in reverse order

$$\begin{aligned} 1 &= 6 - 1.5 \\ &= 6 - 1(17 - 2.6) \\ &= 3.6 - 1.17 \\ &= 3(23 - 1.17) - 1.17 \\ &= 3.23 - 4.17 \\ &= 3.23 - 4(63 - 2.23) \\ &= -4.63 + 11.23 \end{aligned}$$

Multiple both side by -7 to get -7×1 for R.H.S. of Diophantine equations

$$\begin{aligned} -7 &= (-7)(-4)63 + (-7).11.23 \\ &= 63.28 - 23.77 \text{ which shows that} \end{aligned}$$

$$x_0 = 28 \text{ and } y_0 = 77$$

from $x = x_0 + bt = 28 - 23t$ and $y = y_0 - at = 77 - 63t$ are general solution of given diophantine equation. t is an arbitrary integer.

3. *Find the general solution of* $70x + 112y = 168$.

Solution. Here, the given equation is

$$70x + 112y = 168 \quad \ldots(1)$$

Firstly, we shall find the gcd of 70 and 112 in the following manner,

$$112 = 70\,(1) + 42$$
$$70 = 42\,(1) + 28$$
$$42 = 28\,(1) + 14$$
$$28 = 14(2) + 0$$
$$\Rightarrow \quad (70, 112) = 14$$

Since, $14 \mid 168$, therefore, equation (1) has a solution.

Dividing (1) by 14, we get

$$5x + 8y = 12$$

We can easily see that $x = -4$ and $y = 4$ satisfy the above equation. Its general solution is given by

$$x_1 = x_0 - \frac{112}{14}t = -4 - 8t \quad \text{and} \quad y_1 = y_0 - \frac{70}{14}t = 4 + 5t$$

4. *Solve the diophantine equation* $525x + 231y = 42$ *...(1)*

Solution. We can easily find that

$$(525, 231) = 21$$

Therefore, dividing (1) by 21, we get

$$25x + 11y = 2$$

Again, as $(25, 11) = 1$, by the Euclid's algorithm we have

$$25\,(4) + 11\,(-9) = 1$$

Hence, $x = 2 \,.\, 4 = 8$, $y = 2(-9) = -18$ is a solution of the given equation.

Therefore, the required general solution is given by

$$x = 8 + 11t\,,\; y = -18 - 25t$$

Clearly, there are no positive integer solutions.

6.13.2 Euler Method for Solving Linear Diophantine Equations

There are so many ways of obtaining a particular solution. When the coefficient of (1) are not large, it can sometimes be found by inspection. Besides this, we use the process of successively diminishing the coefficients.

1. *Find the positive integer solution of* $7x + 19y = 213$.

Solution. Dividing the given equation by the smaller coefficient 7, we get

$$x = \frac{213 - 19y}{7} = 30 - 2y + \frac{3 - 5y}{7}$$

Since, x is an integer, y is also an integer. Therefore

$$\frac{3 - 5y}{7} = u$$

is also, an integer. Now, we have

$$5y + 7u = 3$$

Dividing it by 5, we have

$$y = \frac{3 - 7u}{5}, = -u + \frac{3 - 2u}{5} \quad \text{or} \quad 2u + 5v = 3$$

Clearly, $u = -1$, $v = 1$ is a solution, Hence, $x = 25, y = 2$. Thus, the general solution of the given equation is

$$x = 25 + 19t,\ y = 2 - 7t$$

Since, we require the solution to be positive, *i.e.*, $25 + 19t > 0, 2 - 7t > 0$

We require $-\frac{25}{19} < t < \frac{2}{7}$

Thus, $t = 0$ or $t = -1$. Hence, the required positive integer solutions are

$$x = 25,\ y = 2,\ x = 6,\ y = 9$$

Theorem 1. *If* $ax + by = c, (a, b) = 1$ ...(1)
b is numerically smaller of the two coefficients a and b and a_1 *and* c_1 *are the minimal remainders of a and c respectively with respect to* $|b|$. *Then, (1) can be written in the form*

$$a_1 x + |b|\, x_1 = c_1 \text{ where } |a_1| \le \frac{|b|}{2} \text{ and } |c_1| \le \frac{|b|}{2}.$$

Proof : As per given, a_1 and c_1 are minimal remainders of a and c with respect to $|b|$, we have

$$a = |b| q_1 + a_1,\ 0 < |a_1| \le \frac{|b|}{2}$$

$$c = |b| q_2 + c_1,\ 0 < |c_1| \le \frac{|b|}{2}$$

Therefore, (1) reduces to $(|b| q_1 + a_1) x + by = |b| q_2 + c_1$

or $$a_1 x + |b| \left(q_1 x + \frac{b}{|b|} y - q_2 \right) = c_1$$

Putting $x_1 = q_1 x + \frac{b}{|b|} y - q_2$, the above equation reduces to $a_1 x + |b|\, x_1 = c_1$.

ILLUSTRATIVE EXAMPLES

1. *Find the general solution of* $21x + 13y = 1791$ *(By Euler method).*

Solution. Here, we have $21x + 13y = 1791 \Rightarrow y = \frac{1791 - 21x}{13}$ [∵ 13 < 21]

$$= \frac{138 \times 13 - 3 - 26x + 5x}{13}$$

$$= 138 - 2x + \frac{5x - 3}{13} \qquad \text{...(1)}$$

Putting $y_1 = \frac{5x - 3}{13}$, we get

$$13y_1 - 5x = -3 \Rightarrow x = \frac{3 + 13y_1}{5} \qquad \text{...(2)}$$

$$= \frac{5 - 2 + 15y_1 - 2y_1}{5} = 1 + 3y_1 - \frac{2 + 2y_1}{5}$$

Now, putting $x_1 = \frac{2 + 2y_1}{5} \Rightarrow 5x_1 = 2 + 2y_1$, *i.e.*, $y_1 = \frac{5x_1 - 2}{2}$

Putting $x_1 = 0$, we get $y_1 = -1$. Therefore, from (3), we have

$$x = \frac{3 + 13(-1)}{5} = \frac{3 - 13}{5} = -\frac{10}{5} = -2$$

Again from (2), we get

$$y = \frac{1791 - 21(-2)}{13} = \frac{1791 + 42}{13} = \frac{1833}{13} = 141$$

Hence, the general solution of (1) is given by

$$x = -2t,\ y = 141 + 13t.$$

2. *Find the possible solution of* $11x + 5y = 79$. ...(1)

Solution. Clearly, we have $(11, 5) = 1$.

Now, $11 = 5 . 2 + 1$

$79 = 5. 16 - 1$

Then, (1) can be written as

$$(5 . 2 + 1)\, x + 5y = 5 \times 16 - 1 \quad \Rightarrow \quad 5[2x + y - 16] + x = -1$$

$\Rightarrow \quad 5u + x = -1,$ where $u = 2x + y - 16$

Putting $u = 0$, we get $x = -1$, then from (1), we get $y = 18$.

$\Rightarrow \quad x = -1, y = 18$ is one solution.

The general solution is given by

$$x = -1 + 5t, \; y = 18 - 11t$$

Since, we require the solution to be positive, therefore, we have to find the value of t for which x and y are positive.

Putting $t = 1$, we get

$$x = -1 + 5 \times 1 = 4$$
$$y = 18 - 11 \times 1 = 7$$

Further, for $t = 2$ and > 2, y will be positive.

Hence, the only positive solution is $x = 4$ and $y = 7$.

3. *Find the general solution of* $311x - 112y = 73$.

Solution. Here, the given equation is

$$311x - 112y = 73 \qquad ...(1)$$

Also, $(311, 112) = 1$

Here, 112 is numerically smaller.

Consider 311 and 73. Then

$$311 = 112\,(3) - 25$$
$$73 = 112\,(1) - 39$$

Then, (1) reduces to

$$(112 \times 3 - 25)\, x - 112y = 112(1) - 39)$$

$$\Rightarrow \quad -25x + 112(3x - y - 1) = -39 \quad \Rightarrow \quad -25x + 112u = -39 \qquad ...(2)$$

where, $u = 3x - y - 1$

Equation (2) is satisfied by $u = 3$, $x = 15$.

Thus, we have $-25x + 112u = -39, u = 3, x = 15$ ...(3)

In this equation neither of the two coefficient is unity. Therefore (2) can be treated in the same way as (1).

Consider 112 and 39.

$$112 = 25\,(4) + 12$$
$$39 = 25\,(2) - 11$$

Then, (2) can be written as

$$-25x + (25 \times 4 + 12)\,u = -(25\,(2) - 11)$$

$$\Rightarrow \quad 25(-x + 4u + 2) + 12u = 11 \quad \Rightarrow \quad 25v + 12u = 11 \qquad ...(4)$$

where, $v = -x + 4u + 2$

Equation (4) is satisfied by $v = -1$ and $u = 3$.

Therefore, we have $25v + 12u = 11$

$$v = -1$$
$$u = 3 \qquad ...(5)$$

Again, neither of the two coefficient is unity. Therefore, consider 25 and 11 such that

$$25 = 12 \times 2 + 1$$
$$11 = 12 \times 1 - 1$$

Then, (4) can be written as

$$(12 \times 2 + 1)v + 12u = 12(1) - 1$$

$\Rightarrow \quad 12[2v + 4 - 1] + v = -1 \quad \Rightarrow \quad 12\omega + v = -1$ ...(6)

where, $\omega = 2v + u - 1$

Here, one coefficient is unity. Putting $\omega = 0$, we get

$$v = -1$$
$$u = 3$$
$$x = 15$$
$$y = 41$$

$\Rightarrow \quad x = 15$ and $y = 41$ is one solution.

Here, the general solution is given by

$$x = 15 + 112t,\ y = 41 + 311t, \text{t is integer.}$$

4. *Find the solution of linear Diophantine equation* $172x + 20y = 1000$.

Solution. Applying the Euclidean algorithm to find the g.c.d (172, 20), we find that

$$172 = 8.20 + 12$$
$$20 = 1.12 + 8$$
$$12 = 1.8 + 4$$
$$8 = 2.4$$

so g.c.d. (172, 20) = 4.

Also, 4 / 1000, so the solution of given problem exist.

To obtain the integer 4 as a linear combination of 172 and 20, we proceed backward through the previous calculation as follows :

$$\begin{aligned} 4 &= 12 - 8 \\ &= 12 - (20 - 12) \\ &= 2.12 - 20 \\ &= 2(172 - 8.20) - 20 \\ - &= 2.172 + (-17).20 \end{aligned}$$

Multiply by 250 on both sides, we get

$$1000 = 250[2.172 + (-17).20] = 500.172 + (-4250).20$$

so that $x = 500$ and $y = -4250$ provide one solution of the given Diophantine equation. All other solutions are given by

$$x = 500 + \left(\frac{20}{4}\right)t = 500 + 5t$$

$$y = -4250 - \left(\frac{172}{4}\right)t = -4250 - 43t, \text{ for some integer t.}$$

To find the positive integers solution, we must choose the value of t such that it satisfy the inequalities

$$500 + 5t > 0; -43t - 4250 > 0$$

Solving these we get

$$-98\frac{36}{43} > t > -100$$

so integer value of $t = -99$, thus, only positive solution is

$$x = 500 + 5(-99) = 5$$
$$y = -4250 - 43(-99) = 7$$

6.14 DIOPHANTINE EQUATION IN THREE OR MORE UNKNOWNS

We can solve the Diophantine equation in three or more unknowns in a similar manner.

The whole process can be understood by the following example.

Example 1. Solve $50x + 45y + 36z = 10$.

Solution. The given equation can be decomposed into two equations given by

$$50x + 45y = 5t,\ 5t + 36z = 10$$

As $50t + 45(-t) = 5t$, $5(-70) + 36(10) = 10$, the solutions of the above two equations are respectively.

$$\begin{cases} x = t + 9k_1, & t = -70 + 36k_2 \\ y = -t - 10k_1 & z = 10 - 5k_2 \end{cases}$$

On eliminating t, we get the required solutions

$$x = -70 + 9k_1 + 36k_2$$
$$y = 70 - 10k_1 - 36k_2$$
$$z = 10$$

or

$$x = 2 + 9k_1 + 36k_2$$
$$y = -2 - 10k_1 - 36k_2$$
$$z = -5k_2,$$ where k_1 and k_2 are any integers.

Aliter. The above equation can be solved by the process of successively eliminating the coefficients.

Since 36 is the smallest coefficient, therefore we can write the given equation as

$$36(x + y + z) + 14x + 9y = 10$$

Let $x + y + z = k_1$, then

$$14x + 9y + 36k_1 = 10$$

i.e.,
$$9(x + y + 4k_1) + 5x = 10$$

Again, let $x + y + 4k_1 = 5k_2$ we get

$$5x + 45k_2 = 10, \text{ i.e., } x + 9k_2 = 2$$

Hence, the required solution is given by

$$x = 2 - 9k_2$$
$$y = -2 - 4k_1 + 14k_2$$
$$z = 5k_1 - 5k_2$$

where k_1 and k_2 are any integers.

☛ Two sets of expansion, obtained above are obviously equivalent.

Theorem 1. *The linear Diophantine equation* $a_1 x_1 + \ldots + a_n x_n = c, a_i, c$ *being all integers has integer solution if and only if* $d = (a_1, \ldots, a_n) | c$.

Proof : The necessary part of the theorem is obvious. Now, we prove the sufficient part.

Let $b_1, b_2, \ldots, b_n$ be such that

$$a_1 b_1 + \ldots + a_n b_n = d$$

Putting $c = dc_1$, we have

$$a_1(b_1 c_1) + \ldots + a_n(b_n c_1) = dc_1 = c$$

$\Rightarrow \quad x_1 = b_1 c_1, \ldots, x_n = b_n c_1$ is the solution of the given equation.

Theorem 2. *Let a and b be two relatively positive primes, then the necessary and sufficient condition that the equation*

$$ax + by = n,\ 0 \le n < ab \qquad \ldots(1)$$

has no non-negative integer solution $n = ab - ka - la$, *k, l being integers.*

Proof : Let $n = ab - ka - la$, $0 \le n < ab$, $k \ge 1, l > 1$. If $x \ge 0$, $y \ge 0$ is the integer solution of (1) *i.e.*, $ax + by = ab - ka - lb$, then

$$a(x + k) + b(y + l) = ab$$

Now, $b | (x + k), a | (y + l)$

$\therefore \quad x + k \ge b, y + l \ge a \quad$ or $\quad a(x + k) + b(y + l) > 2ab,$

which is a contradiction. Hence, the sufficient condition holds. Now, we shall prove the necessary condition.

Consider the following three properties :

(1) If (1) has non-negative integer solution, the solution is unique.

(2) Find the number of n such that (1) has non-negative integer solution. If (1) has solution, the solution is unique, therefore, the solution (x, y) are in one-one correspondence with n. Therefore, the number of n is equal to the number of integer points which are in the region as shown in Figure 2. Clearly, this number is $\frac{(a+1)(b+1)-2}{2}$.

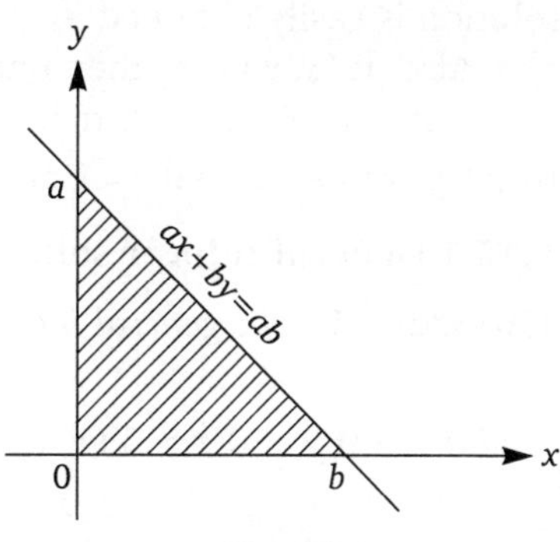

Fig. (2)

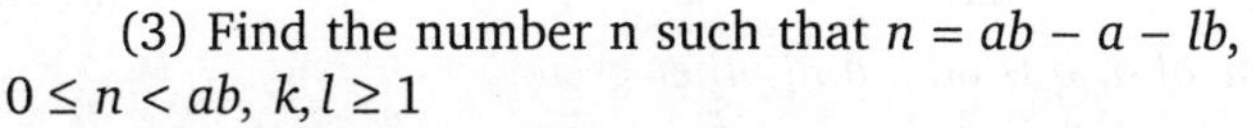

(3) Find the number n such that $n = ab - a - lb$, $0 \le n < ab,\ k, l \ge 1$

This number is not greater than the number of equations of the form (1) which have no non-negative integer solution. By the same reason, the number of n equals that for the integer points (k, l) in the region as shown in Figure 3.

Clearly, it is $\frac{(a-1)(b-1)}{2}$.

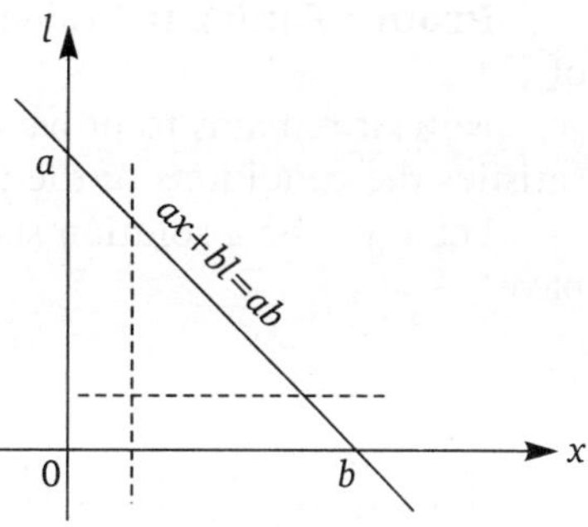

Fig. (3)

Hence, the total number of n such that (1) has non-negative integer solution, whether the positive or not is

$$\frac{(a+1)(b+1)-2}{2} + \frac{(a-1)(b-1)}{2} = ab$$

It is just the total number of integers in the interval $[0, ab[, i.e.,$ (1) has no non-negative integer solution only when n is of the form as in 3.

Hence, the necessary condition holds.

An Important Result

Let a, b and c be integers with not both a and b equal to 0 and let $g = g.c.d(a, b)$. If $g \nmid c$, then the equation

$$ax + by = c \qquad ...(1)$$

has no solution in integers. If $g \mid c$, then this equation has infinitely many solutions. If the pair (x_1, y_1) is one integral solution, then all others are of the form $x = x_1 + kb \mid g$, $y = y - ka \mid g$, where k is any integer.

6.15 DIOPHANTINE EQUATION OF THE SECOND DEGREE

The Diophantine equation of the second degree

$$x^2 + y^2 = z^2 \qquad ...(1)$$

is called the Shang-gao or Pythagores equation.

If one of x, y, z is zero, say $x = 0$, then $y = \pm z$.

Therefore, we may assume that x, y, z are all positive. Also, we may assume that $(x, y) = 1$, for if $(x, y) = d$, then $d | z$ and

$$\left(\frac{x}{d}\right)^2 + \left(\frac{y}{d}\right)^2 = \left(\frac{z}{d}\right)^2$$

where $\left(\frac{x}{d}, \frac{y}{d}\right) = 1$. Hence, if we can find the solution of (1), with $(x, y) = 1$, its general solution is easily obtained.

Also, if $(x, y) = 1$, then one of x and y must be even and other must be odd.

If x, y are both odd, then $x^2 \equiv y^2 \equiv 1 \pmod 4$ and $z^2 = x^2 + y^2 \equiv 2$ (mod4) which is not possible, since $z^2 = 1$ or $0 \pmod 4$. Hence, we may assume y to be even.

6.15.1 General Integer solution of Pythagores Equation

Theorem 1. *The general integer solution of Pythagores equation*

$$x^2 + y^2 = z^2 \qquad \text{...(1)}$$

with $x > 0, y > 0, z > 0, (x, y) = 1$ and y is even, is given by

$$x = a^2 - b^2,\ y = 2ab,\ z = a^2 + b^2 \qquad \text{...(2)}$$

where, $a > b > 0$, $(a, b) = 1$ and one of a, b is odd, and other even.

Proof : Putting the values of (2) in (1), we can easily verify that (2) is the solution of (1).

Now, it remains to prove that every solution that exist from suitably chosen a and b satisfies the conditions of the theorem.

Let x, y, z be a solution satisfying the given conditions. Since y is even, from (1), we have

$$\frac{z+x}{2} \cdot \frac{z-x}{2} = \left(\frac{y}{2}\right)^2 \qquad \text{...(3)}$$

But as $\left(\frac{z+x}{2}, \frac{z-x}{2}\right) \Big| \left(\frac{z+x}{2} + \frac{z-x}{2}\right) = z$

$$\left(\frac{z+x}{2}, \frac{z-x}{2}\right) \Big| \left(\frac{z+x}{2} - \frac{z-x}{2}\right) = x$$

and $(x, y) = 1$

$$\left(\frac{z+x}{2}, \frac{z-x}{2}\right) = 1$$

Let us assume $\frac{z+x}{2} = a^2$, $\frac{z-x}{2} = b^2$

Clearly, $a > b$, $(a, b) = 1$. Thus $x = a^2 - b^2$, $z = a^2 + b^2$, $y = 2ab$

Since z is odd, one of a, b is even and other is odd. Therefore, a, b satisfy the condition given in Theorem.

- In general, if $(x, y) = d$, then $(x, y, z) = d$. Thus, the general solution of (1) is given by $x = \pm d(a^2 - b^2), y = \pm 2abd$, $z = \pm d(a^2 + b^2)$.
- All rational points on the unit circle can be written as $\left(\pm\frac{a^2 - b^2}{a^2 + b^2}, \pm\frac{2ab}{a^2 + b^2}\right)$ or $\left(\pm\frac{2ab}{a^2 + b^2}, \pm\frac{a^2 - b^2}{a^2 + b^2}\right)$, where $(a, b) = 1$.

6.15.2 General Integer Solution of the Equation $x^2 + y^2 + z^2 = w^2, (x, y, z, w) = 1$

Theorem 1. *The general integer solution of the equation*

$$x^2 + y^2 + z^2 = w^2, (x, y, z, w) = 1$$

is given by

$$x = a^2 - b^2 + c^2 - d^2, y = 2ab + 2cd$$
$$z = 2ad - 2bc, w = a^2 + b^2 + c^2 + d^2$$

Proof : Without loss of any generality, we may assume that y, z are both even and x, w both are odd. Let $\left(\frac{y}{2}, \frac{z}{2}\right) = r$, $(x, w) = s$ and $y = 2ry_1$, $z = 2rz_1$, $x = sx_1$, $w = sw_1$. As x_1, w_1 are both odd, the given equation reduces to

$$r^2(y_1^2 + z_1^2) = s^2 \frac{w_1 + x_1}{2} \cdot \frac{w_1 - x_1}{2}$$

Since, $(r, s) = 1, \left(\frac{w_1 + x_1}{2}, \frac{w_1 - x_1}{2}\right) = (w_1 - x_1) = 1$, we have

$$r = r_1 . r_2 . (r_1, r_2) = 1$$
$$r_1^2 | \frac{w_1 + x_1}{2}, r_2^2 | \frac{w_1 - x_1}{2}$$

Now, $$y_1^2 + z_1^2 = s^2 \frac{w_1 + x_1}{2r_1^2} \cdot \frac{w_1 - x_1}{2r_2^2}$$

We know that odd prime factors of $y_1^2 + z_1^2$ are one of the form $4k + 1$. Further, prime factor $4k + 1$ can be expressed as the sum of two square numbers. Thus, there are integers a_1', b_1', c_1', d_1' such that

$$\frac{w_1 + x_1}{2r_1^2} = a_1'^2 + b_1'^2 = c_1'^2 + d_1'^2 \quad \Rightarrow \quad y_1^2 + z_1^2 = (a_1^2 + b_1^2)(c_1^2 + d_1^2)$$

Now, $y_1^2 + z_1^2 = (y_1 + \sqrt{-1}\, z_1)(y_1 - \sqrt{-1}\, z_1)$

and $y_1 + \sqrt{-1}. z_1 = (a_1 + \sqrt{-1}. c_1)(b_1 + \sqrt{-1}. d_1)$

$y_1 - \sqrt{-1}. z_1 = (a_1 - \sqrt{-1}. c_1)(b_1 - \sqrt{-1}. d_1)$

Then, we get

$$y_1 = a_1 b_1 - c_1 d_1, \ z_1 = a_1 d_1 + b_1 c_1$$
$$x_1 = \frac{1}{s}\{r_1^2(a_1^2 + c_1^2) - r_2^2(b_1^2 + d_1^2)\}$$
$$w_1 = \frac{1}{s}\{r_1^2(a_1^2 + c_1^2) + r_2^2(b_1^2 + d_1^2)\}$$

Hence, we have

$$x = a^2 - b^2 + c^2 - d^2, y = 2ab + 2cd$$
$$z = 2ad + 2bc, w = a^2 + b^2 + c^2 + d^2$$

where, $a = r_1 a_1$, $b = r_2 b_1, c = r_1 c_1, d = r_2 d_1$ are all non-negative integers.

Theorem 2. *If x, y, z is a solution of $x^2 + y^2 = z^2$ and $(x, y, z) = 1$, then of x, y, there is a multiple of 3 and a multiple of 4. Also, one of x, y, z is a multiple of 5. Hence, xyz is a multiple of 60.*

Proof : Let us suppose that both x, y are not multiples of 3. Further, since an integer not a multiple of 3 is of the form $3n \pm 1$, with square

$$(3n \pm 1)^2 = 3(3n^2 \pm 2n) + 1 = 3k + 1$$

therefore, $x^2 + y^2$ is of the form $3k - 1$, which is not a square. It is impossible since z^2 is a square. Therefore, one of the x, y is a multiple of 3.

Further, suppose that x, y both are not multiple of 4. But we know that an integer not a multiple of 4 is of the form $4n \pm 1$ or $4n \pm 2$ with squares

$$(4n \pm 1)^2 = 8(2n^2 \pm n) + 1 = 8k + 1$$

$$(4n \pm 2)^2 = 8k + 4$$

If x, y are both of the form $4n \pm 1$ or one is of the form $4n \pm 1$ and the other $4n \pm 2$, then $x^2 + y^2$ is of the form $8k + 2$ or $8k + 5$, which are not squares. Hence, one of x, y must be a multiple of 4.

Again, if none of x, y, z is a multiple of 5, then they must have the forms $5n \pm 1, 5n \pm 2$.

But as $(5n \pm 1)^2 = 5k + 1, (5n \pm 2)^2 = 5k - 1$

$$x^2 + y^2 \text{ is of the form } 5k - 2, 5k + 2 \text{ or } 5k.$$

In the former, $x^2 + y^2$ is not a square, in the latter, z is a multiple of 5, these contradict the assumption and therefore, one of x, y, z must be a multiple of 5.

6.15.3 Solution of General diophantine Equation of Second Degree

Here, we will discuss how to solve the general Diophantine equation of the second degree, by illustrating the method as in following example.

ILLUSTRATIVE EXAMPLES

1. *Find the integer solution of $x^2 + xy + y^2 - 3x + 4y - 6 = 0$.*

Solution. We can treat the given equation as a quadratic equation in x. Then, we have

$$x = \frac{1}{2}\left\{-(y-3) \pm \sqrt{5y^2 - 22y + 33}\right\}$$

Since, x is an integer, $5y^2 - 22y + 33$ must be a square. We can easily see that when $y = 1$, it is a square number 16. Thus, $x = \frac{1}{2}(2 \pm 4) = 3, -1$

Therefore, the required solutions are given by

$$x = 3, x = -1$$
$$y = 1, y = 1$$

Similarly, when $y = 2$, it is again a square number 9, and when $y = 4$, it is a square number 25. Thus, we also obtain the required solution as

$$x = 2, y = 2, \quad x = -1, \quad y = 2, \quad x = 2, \quad y = 4 \text{ and } x = -3, y = 4$$

6.15.4 Equation $x^n + y^n = z^n$

The equation $x^n + y^n = z^n$ has no solution with non-zero integers x, y and z if $n > 2$. This is the well known Fermat's last or Fermat's great theorem.We know that any integer greater than 2 is divisible by 4 or by an odd prime. Thus, if we can prove that when $n = 4$ and when n is any prime, the equation has no solution, then Fermat's last theorem is true.

We give the proof for $n = 4$ below.

Theorem 1. *The Diophantine equation $x^4 + y^4 = z^2$ has no solution with non-zero positive integers x, y, z.*

Proof : Let if possible $(x^2)^2 + (y^2)^2 = z^2$ has positive integer solution. Suppose $(x, y) = 1, y$ is even and that z is the smallest of the z in all solutions. We can easily prove that $x^2 = a^2 - b^2$, $y^2 = 2ab, z = a^2 + b^2$, where, $(a, b) = 1$ and exactly one of a, b is even. If a is even, then b is odd.

Therefore, $1 = x^2 = a^2 - b^2 \equiv -1 \pmod 4$, which is not possible

Therefore, a must be odd and b even.

Again, we obtain $x = p^2 - q^2, b = 2pq, a = p^2 + q^2$, where $(p, q) = 1, p > q > 0$ and both p and q are not odd.

For $y^2 = 2ab$, we have $y^2 = 2pq(p^2 + q^2)$

Since, any two of p, q and $p^2 + q^2$ are relatively prime and hence each must be a square $p = r^2, q = s^2, p^2 + q^2 = t^2$ we then have $r^4 + s^4 = t^2$.

Now, $z = a^2 = b^2 > a = t^2 > t$ *i.e.*, z is greater than t.

which is a contradiction. Therefore, there is no non-zero solution.

☛ Similarly, we can prove that for any positive integer n, the Diophantine equation $x^n + y^n = z^n$ has no positive solution larger than n.

ILLUSTRATIVE EXAMPLES

1. *Show that the positive integer solution of* $x^{-1} + y^{-1} = z^{-1}$, $(x, y, z) = 1$ *must have the form*

$$x = a(a+b),\ y = b(a+b),\ z = ab, \text{ where } a, b > 0,\ (a, b) = 1$$

Solution. If (x, y, z) is a solution of the given equation $(x, y) = c$, i.e., $x = ca$, $y = cb$, $(a, b) = 1$. Then $z^{-1} = x^{-1} + y^{-1} = \frac{a+b}{cab}$, thus $z = \frac{c\,ab}{a+b}$

Since, $(a, b) = 1$ is given, therefore, $(ab, a+b) = 1$. Thus, $(a+b)|c$

Setting $c = c'(a+b)$, then $z = c'ab$

Now, since $(x, y, z) = 1$ we have

$$(ca, cb, c'ab) + (c(a, b), c'ab) = (c'(a+b), c'(ab)) = c' = 1$$

i.e., $$c = (a+b), c' = 1.$$

2. *Show that the integer solution of* $x^2 + 2y^2 = z^2, (x, y) = 1$, *can be expressed as*

$$x = \pm(a^2 - 2b^2),\ y = 2ab,\ z = a^2 + 2b^2$$

Solution. From the given equation, we can easily find $2y^2 = z^2 - x^2 = (z - x)(z + x)$

Since, $(x, z) = 1$, then $(z + x, z - x) = 1$ or 2.

Now, since x, y can only be both odd, hence $(z + x, z - x) = 2$.

Therefore, $$y^2 = \frac{z+x}{2}(z - x) \qquad \text{...(1)}$$

But $z + x, z - x$ can not both be multiple of 4. If $z + x$ is not so, then $\frac{z+x}{2}$ is odd, so that $\left(\frac{z+x}{2}, z - x\right) = 1$. Therefore, odd $\frac{z+x}{2}$ and even $z - x$ are both square numbers. Suppose $\frac{z+x}{2} = a^2, z - x = (2b)^2$.

Then, $x = a^2 - 2b^2, y = 2ab, z = a^2 + 2b^2$

If $z - x$ is not a multiple of 4, then from (1)

Letting $z + x = (2b)^2, \frac{z-x}{2} = a^2$, we get

$$x = -(a^2 - 2b^2)$$
$$y = 2ab$$
$$z = a^2 + 2b^2$$

3. *Show that the positive solution of the equation* $x^{-2} + y^{-2} = z^{-2}$, $(x, y, z) = 1$ *is given by* $x = a^4 - b^4$, $y = 2ab(a^2 + b^2), z = 2ab(a^2 - b^2)$ *where,* $a > b > 0$, $(a, b) = 1$. *a, b can not be both even or both odd.*

Solution. As per given, we have $(x, y, z) = 1$, so that $(x^2, y^2, z^2) = 1$.

Then, proceeding same as example (1), we get

$$x^2 = r(r + s),\ y^2 = s(r + s),\ z^2 = rs$$

where, $r, s > 0,\ (r, s) = 1$

Now, from $z^2 = rs$, we know that r, s are both square numbers and from $x^2 = r(r+s)$, we find that $r + s$ is also a square number . Putting

$$r = r_1^2, z = s_1^2, r + s = t_1^2$$

we have $$r_1^2 + s_1^2 = t_1^2,\ r, s > 0,\ (r, s) = 1$$

Then, using Theorem –1 of Pythagores equation, we get

$$r_1 = a^2 - b^2,\ s_1 = 2ab, t_1^2 = a^2 + b^2$$

where $a > b > 0, (a, b) = 1$, one of a, b is odd and the other is even.

Therefore, $$x = r_1 t_1 = a^4 - b^4$$
$$y = s_1 t_1 = 2ab\,(a^2 + b^2)$$
$$z = r_1 s_1 = 2ab\,(a^2 - b^2)$$

Exercise 6.4

1. Find the general solution of $170x - 455y = 625$.
2. Find the general solution of $39x - 56y = 11$.
3. Find all positive solution of $5x + 14y = 620$.
4. Solve the following equations :
   (a) $2072x + 1813y = 2849$
   (b) $8x - 18y + 10z = 16$.
   (c) $4x + 10y + 14z + 6t = 20$.
5. Find the sum of all positive integers each of which has 2 digits and has remainder 4 when divided by 4.
6. Find the value of a if $x^2 + y^2 + axy$ has positive integer solution.
7. Show that the equation $213x + 441y = 10002$ has solution in integers but none where both x and y are positive integers.
8. Show that if $ax + by = n$ has any solution in integers we may assume the problem reduced to the form $Ax + By = N$, where $(A, B) = 1$ and $B > 0$.
9. Show that the equation $x^2 + y^2 + z^2 = x^2y^2$ has no integer solution. except $x = y = z = 0$.
10. If n is any positive integer, show that $x^2 + y^2 = z^2$ always has positive integer solution.

ANSWERS

1. $x_1 = 1 + 91t, y_1 = -1 + 34t$.
2. $x_1 = 29 + 56t, y_1 = 20 + 34t$
3. $x_1 = 110 - 4t, y_1 = 5 + 5t, t = 1, 2, \ldots, 7$.
4. (a) $x_1 = -3 + 7t, y_1 = 5 - 8t$ (b) $x_1 = 4 - 9t_1 - 10t_2, y_1 = 2 - 4t_1 - 5t_2, z_1 = 2$.
   (c) $x_1 = -18 - 54c + 21b + 5a, y_1 = 6 + 18c - 7b - 2a, z_1 = 1 + 3c - b, t_1 = 3c$
5. Let $4k + 1$ be a number containing 2 digits. Since $10 \le 4k + 1 < 100$ or $2\frac{1}{4} < k < 24\frac{3}{4}$, there are 22 integer values between 3 and 24 that k can take. Hence, we get 22 integers which all contain 2 digits and form an arithmetic progression whose sum is 1210.
6. Let $(x, y) = d,\ x = dx', y = dy'$. Then $x'^2 + y'^2 = ax'y'$.
   Hence, $x' \mid y'^2, y' \mid x'^2$ or $x' = 1, y' = 1 \Rightarrow a = 2$.
7. Let a, b, c be any system of solutions of the Phythagores equation, then $(ac^{n-1})^2 + (bc^{n-1})^2 = (c^2)^n$.

□□□

Chapter 7

Discrete Numeric Functions and Generating Functions

- Discrete numeric function
- Accumulated sum
- Convolution of numeric function
- Asymptotic behaviour
- Big OH
- Big-omega and big-theta notations
- Generating function
- Combinatorial Problem

7.1 INTRODUCTION

We know that a function is a binary relation that assigns to each element in the domain a unique value which is an element in the range. In this chapter, we shall discuss the discrete numeric functions and generating functions.

7.2 DISCRETE NUMERIC FUNCTION

Definition : *These are defined to be the functions whose domain is the set of natural numbers, and whose range is the set of real numbers.*

Notations : We shall use bold face lower case letters to denote numeric functions. For a numeric function a, we use $a_0, a_1,, a_2, \dots a_r \dots$, to denote the values of the function at 0, 1, 2, .., r,.. . Thus a_r is the value of the numeric function at r.

We may also specify a numeric function by exhaustively listing its values $(a_0, a_1, a_2, \dots)$ that is by writing $a = (a_0, a_1, a_2, \dots, a_r, \dots .)$.

Generally, we need to use a representation that is not infinitely long. Thus, $a_r = 3r^2 + 2$ for every $r \geq 0$

$$b_r = \begin{cases} 2r & 0 \leq r \leq 11 \\ 3^r - 1 & r \geq 11 \end{cases}$$

$$c_r = \begin{cases} -4 & r = 3, 5, 7 \\ 0 & \text{otherwise} \end{cases}$$

$$d_r = \begin{cases} 2 + r & 0 \leq r \leq 5 \\ 2 - r & r > 5, r \text{ is odd} \\ 2/r & r > 5, r \text{ is even} \end{cases}$$

$$e_r = \begin{cases} 1 + r & 0 \leq r \leq 4 \\ 1 - r & r \geq 5, r \text{ is odd} \\ 1/4 & r > 4, r \text{ is even} \end{cases}$$

$$f_r = \begin{cases} 3r & 0 \leq r \leq 11 \\ 3^r & r \geq 12 \end{cases}$$

are numeric functions.

7.2.1 Application of Discrete Numetic Functions

Numeric functions find their great applications in many day-to-day physical problems, such as, in banking or in accountancy, air flight manipulations, electrical power supply, financial operation etc.

For example :

1. Let a businessman deposit Rs. 100 in his saving account in State Bank of India which promises to pay 7% interest on this amount per year. The amount in the account at the end of each year can be represented by a numeric function a having its values as $(100, 107, 114.49, \ldots)$ in successive years. This can be briefly written as

$$a = a_r = 100\,(1.7)^r, r \geq 0$$

2. Let $a = a_r$ denote the difference between the output voltage and a reference voltage, after r hours of an electric power supply. If $+a_r$ denotes the voltage above its reference voltage, after r hours of operation, and $-a_r$ denote the voltage below the reference voltage (after r hours of operation), then $|a|$ = deviation of the output voltage from the reference voltage.

ILLUSTRATIVE EXAMPLES

1. *Let a_r denotes the altitude of an aircraft in thousands of meters at the rth minute. Let the aircraft takes off after spending 10 minutes on the ground, climbs up at a uniform speed to an altitude of 10 thousand meters in 10 minutes, starts to descend uniformly 2 hours of flying time and lands 10 minutes later. Then sketch the numeric function.*

Solution. Since a_r denotes the altitude of an aircraft in thousand of meters at r^{th} minute. We are given that in first 10 minutes it is on the ground and climbs 10000 meters in 10 minutes.

Thus, numeric function is :

$$a_r = \begin{cases} 0, & 0 \leq r \leq 10 \\ r - 10, & 11 \leq r \leq 20 \\ 10, & 21 \leq r \leq 140 \\ 150 - r, & 141 \leq r \leq 150 \\ 0, & r \geq 150 \end{cases}$$

This is the required numeric function of the altitude of the aircraft.

2. *A ball is dropped to the floor from a height of 40m. Suppose that the ball always rebounds to each half of the height from which it falls :*

(i) Let a_r denotes the height it reaches in the rth rebound. Sketch the numeric function a.

(ii) Let b_r denotes the loss in height during the r^{th}rebound. Express b_r in terms of a_r. Also sketch the numeric function b.

(iii) A second ball is dropped from a height of 15m to the same floor at the same time as the first ball reaches the highest point of its second rebound. Let c_r denote the second ball reaches to its rth rebound. Express c_r in terms of a_r.

Solution. Initial height = 40 meter

(i) The height reached by the ball in first rebound, $a_1 = \frac{1}{2}(40)\,m$

The height reached by the ball in the second rebound

$$a_2 = \frac{1}{2}(a_1) = \frac{1}{2} \times \frac{1}{2} \times 40 = \frac{1}{2^2}(40)$$

Similarly, the height reached by the ball in the rth rebound

$$a_r = \frac{1}{2^r}(40)$$

Thus the required numeric function : $a_r = \frac{40}{2^r}$.

(ii) The loss in height during the first rebound.

$$b_1 = 40 - a_1 = 40 - \frac{1}{2}(40) = 40\left(1 - \frac{1}{2}\right) m$$

The loss in height during the second rebound

$$b_2 = 40\left(a - \frac{1}{2^2}\right) m$$

Similarly, the loss in height during the rth rebound

$$b_r = 40\left(1 - \frac{1}{2^r}\right) m = 40 - \frac{40}{2^r} = 40 - a_r$$

(iii) The height reached by the first ball after the second rebound

$$a_2 = \frac{40}{2^2} = 10\, m$$

Therefore, numeric function corresponding to the first ball after second rebound :

$$a_r = \frac{10}{2^r}$$

Now, initial height of the second ball = 15 m

So, the numeric function corresponding to the second ball.

$$c_r = \frac{15}{2^r}$$

Now, $$\frac{c_r}{a_r} = \frac{15/2^r}{10/2^r} = \frac{3}{2}$$

$\therefore \quad c_r = \frac{3}{2}a^r$, which is the required relation between a_r and c_r.

3. *A person deposits Rs. 100 in a savings account at an interest rate 5% per year compounded annually. Obtain the amount at the end of r years. Sketch the numeric function.*

Solution : The amount after one year

$$a_1 = 100\left(1 + \frac{5}{100}\right) = \text{Rs. } 100\ (1.05) \qquad \left[\text{Use } A = P\left(1 + \frac{r}{100}\right)^n\right]$$

Also, the amount after two years

$$a_2 = 100\left(a + \frac{5}{100}\right)^2 = \text{Rs. } 100\ (1.05)^2$$

Similarly, the amount after r years

$$a_r = 100\left(1 + \frac{5}{100}\right)^r = \text{Rs. } 100\ (1.05)^r$$

Hence, the required numeric function : $a_r = 100(1.05)^r$

4. *In a process control system, a monitoring device measures the temperature inside a chemical reaction chamber once every 10 sec. Let a_r denote the rth reading in degree in centigrade. Determine an expression for a_r if it is known that the temperature rises from 100° to 125°C at a constant rate in the first 100 seconds and stays at 125°C afterwards.*

Solutions. The uniform rate of rising the temperature $= \frac{125 - 100}{100} = 0.025.$

The temperature at the first reading,

$$a_1 = (100 + 0.25 \times 1)°$$

and the temperature at the second reading,

$$a_2 = (100 + 0.25 \times 2)°$$

Similarly, the temperature at the rth reading, $a_r = (100 + 0.25 \times r)°$

Hence, the required numeric function, $a_r = 100 + 0.25r$.

7.3 MANIPULATION OF NUMERIC FUNCTIONS

7.3.1 Sum and Product of Two Numeric Functions

The sum of two numeric functions a_r and b_r is a numeric function c_r whose value at r is equal to the sum of the values of the given numeric functions at r.

The product of two numeric functions a_r and b_r is a numeric function d_r whose value at r is equal to the product of the values of the given numeric function at r.

For example :

Consider the two numeric functions a and b where

$$a_r = \begin{cases} 0 \quad , & 0 \le r \le 2 \\ 2^{-r} + 5 \quad , & r \ge 3 \end{cases} \text{ and } b_r = \begin{cases} 3 - 2^r, & 0 \le r \le 1 \\ r + 2, & r \ge 2 \end{cases}$$

Let $c = a + b$, then

$$c_r = a_r + b_r = \begin{cases} 3 - 2^r, & 0 \le r \le 1 \\ 4, & r = 2 \\ 2^{-r} + r + 7, & r \ge 3 \end{cases}$$

Let $d = a - b$, then

$$d_r = a_r b_r = \begin{cases} 0, & 0 \le r \le 2 \\ r.2^{-r} + 2^{-r+1} + 5r + 10, & r \ge 3 \end{cases}$$

Now, if a denotes the monthly income of a husband and b denotes the monthly income of his wife, then $a + b$ will be their joint monthly income.

Let a denote the balance in a savings account in each month and b denote the monthly interest rate, which fluctuates from month to month. Then, ab will be the interest earned in each month.

ILLUSTRATIVE EXAMPLES

1. *Let **a** and **b** be two numeric functions given by*

$$a_r = \begin{cases} 0, & 0 \le r \le 4 \\ 2^{-r} + 3, & r \ge 5 \end{cases}; \; b_r = \begin{cases} 1 - 2^r, & 0 \le r \le 2 \\ 2^r + 2, & r \ge 3 \end{cases}$$

*Find **a** + **b** and **ab**.*

Solution : Let **a + b = c** and **a.b = d**

Then, $c_r = a_r + b_r$

$$= \begin{cases} 1 - 2^r + 0 = 1 - 2^r, & 0 \le r \le 2 \\ 2^r + 2 + 0 = 2^r + 2, & 3 \le r \le 4 \\ 2^r + 2 + 2^{-r} + 3 = 2^r + 2^{-r} + 5, & r \ge 5 \end{cases}$$

$$d_r = a_r . b_r = \begin{cases} (1 - 2^r).0 = 0, & 0 \le r \le 2 \\ (2^r + 2).0 = 0, & 3 \le r \le 4 \\ (2^r + 2)(2^{-r} + 3), & r \ge 5 \end{cases}$$

$$= \begin{cases} 0, & 0 \le r \le 2 \\ 0, & 3 \le r \le 4 \\ 1 + 2.2^{-r} + 3.2^{r} + 6, & r \ge 5 \end{cases}$$

$$= \begin{cases} 0, & 0 \le r \le 4 \\ 7 + 2.2^{-r} + 3.2^{r}, & r \ge 5 \end{cases}$$

2. *Let **a** be a numeric function such that a_r is equal to remainder when the integer r is divide by 7. Let **b** be a numeric function such that it is equal to 0 if the integer r is divisible by 3 and is equal to 1 otherwise.*

(i) Let $c_r = a_r + b_r$. For what values of r is $c_r = 0$? For what values of r, $c_r = 7$?

(ii) Let $d_r = a_r . b_r$. For what values of r is $d_r = 0$? For what values of r, $d_r = 1$?

Solution : Here, we have

$$a_r = \begin{cases} 0 & r = 7i \\ 1 & r = 7i + 1 \\ 2 & r = 7i + 2 \\ 3 & r = 7i + 3 \quad i = 0, 1, 2, \ldots \\ 4 & r = 7i + 4 \\ 5 & r = 7i + 5 \\ 6 & r = 7i + 6 \end{cases}$$

and $$b_r = \begin{cases} 0 & r = 3i \\ 1 & r \text{ otherwise} \end{cases} \quad i = 0, 1, 2, \ldots$$

(a) Now, $c_r = a_r + b_r$

$c_r = 0$, where both a_r and $b_r = 0$

$\Rightarrow$ $r = 7i$ and $r = 3i$ $\Rightarrow$ $r = 21i$

$c_r = 7$ when $a_r = 6$ and $b_r = 1$

When $r = 7i + 6$ is not a multiple of 3.

(b) $d_r = a_r . b_r$

$= 0$, when either $r = 7i$ or $r = 3i$

and $d_r = 1$, when $r = 7i + 1$ but not a multiple of 3.

7.3.2 Modulus of a Numeric Function

Let **a** be a numeric function. Then modulus of **a** denoted by $|\mathbf{a}|$ is a numeric function defined as $|\mathbf{a}| = \begin{cases} a_r, & \text{if } a_r \text{ is non – negative} \\ -a_r, & \text{if } a_r \text{ is negative} \end{cases}$

For example : Let **a** be a numeric function defined as $a_r = (-1)^r \left(\frac{8}{6}\right), r \ge 0$ obtain $|\mathbf{a}|$.

Solution : Let $b = |\mathbf{a}|$, then $b_r = \frac{8}{6}, r \ge 0$

7.3.3 Definition of $S^i a$ and $S^{-i} a$ for a Numeric Function

Let a be a numeric function and i be a positive integer. Then $S^i a$ is a numeric function b such that its value at r is 0 for $r = 0, 1, 2, \ldots, i - 1$ and is a_{r-1} for $r \ge i$

$S^{-i} a$ is also a numeric function **c** such that its value at r is a_{r+1} for $r \ge 0$.

We can define $S^i a$ and $S^{-i} a$ as follows.

$$S^i a = \begin{cases} 0, & 0 \le r \le i-1 \\ a_{r-1}, & r \ge i \end{cases};$$

$$S^{-i} a = \{a_{r+1} \quad r \ge 0$$

For example : Let $a_r = \begin{cases} 1, & 0 \le r \le 10 \\ 2, & r \ge 11 \end{cases}$

Let b be $S^5 a$, then $b_r = \begin{cases} 0, & 0 \le r \le 4 \\ 1, & 5 \le r \le 15 \\ 2, & r \ge 16 \end{cases}$

Also, as another example, let **a** describe the altitude of an aircraft as given in the foregoing. Then $S^{17} a$ describes the altitude when take off is defined by 17 minutes.

Now, let $a_r = \begin{cases} 1 & 0 \le r \le 10 \\ 2 & r \ge 11 \end{cases}$

Let b be $S^{-7} a$, then $b_r = \begin{cases} 1 & 0 \le r \le 3 \\ 2 & r \ge 4 \end{cases}$

Also, as another example, let **a** be a numeric function where a_r is the number of applicants whose height is r inches above the minimum height required to join the Army. It follows that the numeric function $S^{-2} a$ describes the number of applications in the height groups if two inches have been added to the minimum height requirement.

ILLUSTRATIVE EXAMPLES

1. *Let* **a** *be a numeric function defined as*

$$a_r = \begin{cases} 1, & 0 \le r \le 20 \\ 2, & r \ge 21 \end{cases}, \text{ find } S^{11} a \text{ and } S^{-11} a.$$

Solution : We have $b = S^{11}_a = \begin{cases} 0 & 0 \le r \le 10 \\ a_{r-11} & r \ge 11 \end{cases}$ (By definition of $S^i a$)

Now, $a_{r-11} = \begin{cases} 1 & 0 \le r-11 \le 20 \\ 2 & r-11 \ge 21 \end{cases}$

$$= \begin{cases} 1 & 11 \le r \le 31 \\ 2 & r \ge 32 \end{cases}$$

Therefore, $S^{11} a = \begin{cases} 0 & 0 \le r \le 10 \\ 1 & 11 \le r \le 31 \\ 2 & r \ge 32 \end{cases}$

$S^{-11} a = \{a_{r+11} \; r \ge 0\}$ (By definition of $S^{-i} a$)

Now, $a_{r+11} = \begin{cases} 1 & 0 \le r+11 \le 20 \\ 2 & r+11 \ge 21 \end{cases}$

$\Rightarrow$ $a_{r+11} = \begin{cases} 1 & -11 \le r \le 9 \\ 2 & r \ge 10 \end{cases}$

$\Rightarrow$ $a_{r+11} = \begin{cases} 1 & 0 \le r \le 9 \\ 2 & r \ge 10 \end{cases}$, since $r \ge 0$

Hence, $c = S^{-11} a = \begin{cases} 1 & 0 \le r \le 9 \\ 2 & r \ge 10 \end{cases}$

7.4 THE ACCUMULATED SUM

The accumulated sum of a numeric function **a** *is a numeric function whose value at r is equal to* $\sum_{i=0}^{r} a_i$.

For example :

1. Let **a** describe the monthly earnings of an employee. Let **b** be the accumulated sum of **a.** Then **b** gives his accumulated earnings by month.

2. Let $a_r = 100\,(1.07)^r,\ r \geq 0$

Let b be the accumulated sum of **a.** Then

$$b_r = \sum_{i=0}^{r} a_i = \sum_{i=0}^{r} 100(1.07)^i = \frac{100000}{7}[(1.07)^{r+i} - 1],\ r \geq 0$$

Note that if we deposit \$100 each year in a savings account of an annual compounded interest rate of 7 percent, then b_r, is the total amount we have in the account after r years.

ILLUSTRATIVE EXAMPLES

1. *Let* **a** *be a numeric function given by* $a_r = 100(1.05)^r,\ r \geq 0$. *Find accumulated sum of* **a**.

Solution : Let b be the accumulated sum of a. Then we have

$$b_r = \sum_{i=0}^{r} a_i = \sum_{i=0}^{r} 100\,(1.05)^r$$

$$= 100\,[(1.05)^0 + (1.05)^1 + \ldots + (1.05)^r]$$

$$= 100\,[1 + 1.05 + \ldots + (1.05)^r]$$

$$= 100\left[\frac{(1.05)^{r+1} - 1}{(1.05) - 1}\right] r \geq 0 \quad \left[\because \text{ Sum of G. P.} = \frac{a\,(r^n - 1)}{r - 1}\right]$$

2. *Suppose that Ravi deposits Rs. 100 each year in a saving account at an annual compound interest rate of 8 percent. What is the total amount in Ravi's saving account after r years?*

Solution : The numeric function which describes the amount in the saving account at the end of the rth year of Rs. 100 was deposited into the account at the beginning is given by **a**, we have

$$a_r = 100\,(1.08)^r,\ r \geq 0$$

It is clear that total amount in Ravi's saving account after r years is given by the accumulated sum of the numeric function **a**. Let **b** denote the accumulated sum of **a**. Then

$$b_r = \sum_{i=0}^{r} a_i = \sum_{i=0}^{r} 100\,(1.08)^i = \frac{10000}{8}[(1.08)^{r+1} - 1],\ r \geq 0$$

gives the total amount in Ravi's saving account after r years.

7.5 THE FORWARD DIFFERENCE AND THE BACKWARD DIFFERENCE OF THE NUMERIC FUNCTIONS

The forward difference of a numeric function **a** is a numeric function denoted by Δa, whose value at r is given by

$$\Delta a_r = a_{r+1} - a_r,\ r \geq 0$$

Similarly, the backward difference of a numeric function **a** is a numeric function denoted by ∇a, whose value at r is given by

$$\nabla\, a_r = a_r - a_{r-1},\ r \geq 1$$

and whose value is equal to a_0 at 0.

☛ If the numeric function **a** describes the total business income of a company in each month, then Δa will describe the increase of income from the rth month to the $(r+1)^{th}$ month and ∇a will describe the increase of income for the rth month over the $(r-1)^{th}$ month.

ILLUSTRATIVE EXAMPLES

1. *Let **a** be a numeric function defined as*

$$a_r = \begin{cases} 0, & 0 \le r \le 2 \\ 2^{-r} + 5, & r \ge 3 \end{cases}$$

Calculate Δa and ∇a.

Solution : Let b denote the forward difference of **a**, *i.e.*, Δa, then

$$b_r = \begin{cases} 0, & 0 \le r \le 1 \\ 41/8 & r = 2 \\ -2^{-(r+1)} & r \ge 3 \end{cases}$$

Let c denote the backward difference of a, *i.e.*, ∇a, then

$$c_r = \begin{cases} 0, & 0 \le r \le 2 \\ 41/8 & r = 3 \\ -2^{-r} & r \ge 4 \end{cases}$$

2. *Let **a** be a numeric function such that*

$$a_r = \begin{cases} 0, & 0 \le r \le 3 \\ 2^r + 3 & r \ge 4 \end{cases}$$

Calculate Δa and ∇a.

Solution : Let $b = \Delta a$ be a forward difference of a. Then, we know that

$$b_r = \Delta a_r = a_{r+1} - a_r \; r \ge 0$$

So,
$$a_{r+1} = \begin{cases} 0 & 0 \le r+1 \le 3 \Rightarrow -1 \le r \le 2 \\ 2^{r+1} + 3 & r+1 \ge 4 \Rightarrow r \ge 3 \end{cases}$$

$$\Rightarrow \quad a_{r+1} = \begin{cases} 0 & 0 \le r \le 2 \\ 2.\,2^r + 3 & r \ge 3 \end{cases}$$

and
$$a_{r-1} = \begin{cases} 0 & 0 \le r-1 \le 3 \Rightarrow 1 \le r \le 4 \\ 2^{r-1} + 3 & r-1 \ge 4 \Rightarrow r \ge 5 \end{cases}$$

$$\Rightarrow \quad a_{r-1} = \begin{cases} 0 & 1 \le r \le 4 \\ \dfrac{2^r}{2} + 3 & r \ge 5 \end{cases}$$

$$b_r = a_{r+1} - a_r = \begin{cases} 0 - 0 = 0 & 0 \le r \le 2 \\ 2.2^3 + 3 - 0 = 19 & r = 3 \\ 2.2^r + 3 - 2^r - 3 = 2^r & r \ge 4 \end{cases}$$

$$c_r = a_r - a_{r-1} = \begin{cases} 0 - 0 = 0 & 0 \le r \le 3 \\ 2^4 + 3 - 0 = 19 & r = 4 \\ 2^r + 3 - \dfrac{2^r}{2} - 3 = \dfrac{2^r}{2} = 2^{r-1} & r \ge 5 \end{cases}$$

☛ $S^{-1}(\nabla a) = \Delta a.$

7.6 CONVOLUTION OF TWO NUMERIC FUNCTIONS

If **a** *and* **b** *are two numeric functions, then their convolution denoted by* **a*** **b** *is a numeric function* **c** *defined by*

$$\mathbf{c} = \mathbf{a} * \mathbf{b} = a_0 b_r + a_1 b_{r-1} + \ldots + a_r b_0 = \sum_{i=0}^{r} a_i\, b_{r-i}$$

ILLUSTRATIVE EXAMPLES

1. *Let* **a** *and* **b** *be the two numeric functions such that*

$$a_r = 3^r \qquad r \geq 0$$
$$b_r = 2^r \qquad r \geq 0$$

Obtain **a*b** .

Solution : Let c be the convolution of **a** and **b**. Then

$$c_r = \sum_{i=0}^{r} 3^i 2^{r-i},\ r \geq 0$$

7.7 ASYMPTOTIC BEHAVIOUR OF NUMERIC FUNCTIONS

Asymptotic behaviour of a numeric function means how the value of the function varies for large value of r.

For Example : $a_r = 6, r \geq 0$

Here, the value of the numeric function remains constant for increasing r. On the other hand, for

$$b_r = 6r^2,\ r \geq 0$$

The value of the function increases for increasing r and is proportional to r^2, while for

$$c_r = 6 \log r,\ r \geq 0$$

The value of the function increases for increasing r and is proportional to $\log r$. Finally for

$$d_r = \frac{6}{r},\ r \geq 0$$

The value of the function decreases for increasing r is proportional to $1/r$ and approaches 0 as a limit.

7.7.1 Asymptotic Dominance

In many occasions, we are interested in comparing the asymptotic behaviour of two numeric function S. Therefore, we introduce the notation of asymptotically dominance. *Let* **a** *and* **b** *be two numeric functions. We say that* **a** *asymptotically dominates* **b**, *or* **b** *is asymptotically dominated by* **a**, *if there exists positive constants k and m such that*

$$|b_r| \leq m a_r,\ \textit{for } r \geq k$$

For example :

1. Let **a** and **b** be two numeric functions such that

$$a_r = r + 1, \quad r \geq 0$$
$$b_r = \frac{1}{r} + 7,\ r \geq 0$$

Then **a** asymptotically dominates **b** because for $k = 7$ and $m = 1$

$$|b_r| \leq a_r \text{ for } r \geq 7.$$

On the other hand, **b** does not dominate **a** asymptotically, since for any choice of k and m, there exists r_0 such that $r_0 \geq k$ and $|a_{r_0}| > m b_{r_0}$.

2. Let **a** and **b** be two numeric functions such that

$$a_r = \frac{1}{100} r^2 - 1000$$

$$b_r = \begin{cases} -1{,}000{,}000 & 0 \le r \le 10 \\ -100{,}000 r & r > 10 \end{cases}$$

Because for $k = 10^6 + 1$ and $m = 10$

$$|b_r| \le 10a, \text{ for } r \ge 10^6 + 1$$

We can conclude that **a** asymptotically dominates **b**.

☛ A numeric function **a** asymptotically dominates the numeric function **b** means **a** grows faster than **b**. It means for sufficiently large r, the absolute value of **b**, does not exceed a fixed proportion of a_r.

For Example. Suppose we deposit Rs. 1 in two separate bank accounts. After r years, our total deposit in account A becomes $1 + 0.2r$ rupees and our total deposit in account B becomes $(1 + 0.03)^{4r}$ rupees. Let

$$a_r = 1 + 0.2r \qquad r \ge 0$$

$$b_r = (1 + 0.03)^{4r} \qquad r \ge 0$$

Here, it is clear that **b** asymptotically dominates **a** since, for $k = 9$ and $m = 1$

$$|1 + 0.2r| \le (1 + 0.03)^{4r} \text{ for } r \ge 9$$

That is, the growth rate of our money in account B is higher than that in account A. Although in the first few years the total amount in B is less than that in account A, the total in account B exceeds that in account A in long run.

ILLUSTRATIVE EXAMPLES

1. *Let **a** and **b** be two numeric functions such that*

$$a_r = 2r + 3, r \ge 0 \text{ and } b_r = \frac{3}{r} + 5, r \ge 0$$

*Show that **a** dominates **b** asymptotically.*

Solution : Let us take $m = \frac{2}{3}$ and $n = 5$, then we have

$$|b_r| \le \frac{2}{3} a_r \text{ for } r \ge 5.$$

Hence, the numeric function **a** asymptotically dominates **b**.

2. *Let **a** and **b** be two numeric functions such that*

$$a_r = \frac{r^2}{10} - 10 \text{ and } b_r = 100r.$$

*Then show that **a** asymptotically dominates **b**.*

Solution : For $k = 1$ and $n = 10^3 + 1 = 1001$, we have

$$|b_r| = 1001 \times 100 \le \frac{(1001)^2}{10} - 10 = a_r, \text{ whenever } r \ge 1001$$

Hence, **a** asymptotically dominates **b**.

3. *Let **a** and **b** be two numeric functions such that*

$$a_r = r^2 + r, r \ge 0 \text{ and } b_r = r^2, r \ge 0$$

*Show that **a** asymptotically dominates **b** and **b** also asymptotically dominates **a**.*

Solution : We first show that **a** asymptotically dominates **b**.

For $n = 1$ and $k = 1$, we have

$$|b_r| = r^2 \le r^2 + r = a_r, \text{ for } r \ge 1$$

Hence, **a** asymptotically dominates **b**.

We now show **b** asymptotically dominates **a**. For $n = 1$ and $k = 2$. We have

$$|r^2 + r| \leq r^2 + r^2 = 2r^2 \text{ for } r \geq 1.$$

Thus, $|a_r| \leq 2b$, when ever $r \geq 1$. Hence, **b** asymptotically dominates **a**.

4. *Let* **a** *and* **b** *be two numeric functions such that*

$$a_r = 3r^2 + r + 1,\ r \geq 0 \text{ and } b_r = 3r^2 - r^{1/2} - 7,\ r \geq 0$$

Show that **a** *asymptotically dominates* **b** *and* **b** *asymptotically dominates* **a**.

Solution : There exists $m = 3/2$ and $n = 1$ such that

$$|b_r| \leq ma_r, \text{ for } r \geq n$$

This shows **a** asymptotically dominates **b**.

Also, there exists $m = 2/3$ and $n = 4$ such that $|a_r| \leq mb_r$, for $r \geq n$

This shows that **b** asymptotically dominates **a.**

5. *Give two numeric functions* **a** *and* **b** *which are not equal such that* **a** *asymptotically dominates* **b** *and* **b** *also asymptotically dominates* **a**.

Solution : Consider the following numeric functions **a** and **b**, where

$$a_r = r^2 + r + 1,\ r \geq 0$$

$$b_r = \frac{r^2}{2} - r - 2,\ r \geq 0$$

For $k = 1$ and $n = 1$, we have

$$|b_r| \leq ka_r, \text{ whenever } r \geq 1.$$

Thus **a** asymptotically dominates **b**.

Again for $k = 4$ and $n = 7$, we have

$$|a_r| \leq kb_r, \text{ whenever } r \geq 7$$

because $r^2 + r + 1 \leq 4(r^2/2 - r - 2)$ whenever $r \geq \sqrt{5r+9}$ and we know $r \geq \sqrt{5r+9}$, when $r \geq 7$.

Thus **b** also asymptotically dominates **a.**

6. *Let* **a**, **b** *and* **c** *be three numeric functions such that* $a_r = 2r - 1,\ r \geq 0;\ b_r = \frac{3}{r} + 3,\ r \geq 1;$ $c_r = r \log r,\ r \geq 1$

Discuss the asymptotic dominance of these numeric functions.

Solution : (i) For $m = \frac{3}{2}$ and $n = 4$, we get

$$|b_r| \leq ma_r,\ \forall\ r \geq n$$

This shows that **b** is asymptotically dominated by **a.**

(ii) **a** is asymptotically dominated by **c** because for $m = 2$ and $n = 3$ such that $|a_r| \leq mc_r$, $\forall r \geq n$. This shows that **a** is asymptotically dominated by **c**.

Similarly, we can show that **b** is asymptotically dominated by **c**.

- For any numeric function **a**, $|a|$ asymptotically dominates a.
- If **b** is asymptotically dominated by **a**, then for any constant α, αb is also asymptotically dominated by **a**.
- If **b** is asymptotically dominated by **a**, then for any integer i, $S^i b$ is asymptotically dominated by $S^i a$
- A numeric function **a** asymptotically dominates the numeric function **b** means **a** grows faster then **b**. It means for sufficiently large r, the absolute value of **b**, does not exceed a fixed proportion of a_r.
- If both **b** and **c** are asymptotically dominated by **a**, then for any constants α and β, $\alpha b + \beta c$ is also asymptotically dominated by **a**.
- If **c** is asymptotically dominated by **b** and **b** is asymptotically dominated by **a,** then **c** is asymptotically dominated by **a**.

- It is possible that **a** asymptotically dominates **b** and **b** also asymptotically dominates **a**, for example, let $a_r = r^2 + r + 1, r \geq 0; b_r = 0.05r^2 - r^{1/3} - 9, r \geq 0$
- It is possible that neither **a** asymptotically dominates **b**, nor **b** asymptotically dominates **a**. For example, let
$$a_r = \begin{cases} 1 & r \text{ is even} \\ 0 & r \text{ is odd} \end{cases}; \; b_r = \begin{cases} 0 & r \text{ is even} \\ 1 & r \text{ is odd} \end{cases}$$
- It is possible that both **a** and **b** asymptotically dominates **c**, while **a** does not asymptotically dominates **b**, nor does **b** asymptotically dominates **a**. For example, let
$$a_r = \begin{cases} 1 & r = 3i \text{ or } 3i+1 \\ 0 & r = 3i+2 \end{cases}; \; b_r = \begin{cases} 1 & r = 3i \text{ or } 3i+2 \\ 0 & r = 3i+1 \end{cases}; \; c_r = \begin{cases} 1 & r = 3i \\ 0 & \text{otherwise} \end{cases}$$

7.8 ORDER A OR BIG-OH OF A

For a given numeric function **a**, $O(a)$ denotes the set of all numeric functions that are asymptotically dominated by **a**. $O(a)$ is called 'order **a**' or 'big-oh of **a**'. If the numeric function **b** is asymptotically dominated by **a** then we say that $b \in O(a)$ and it is read as **b** is of order **a**.

For Example

(1) Let $\mathbf{a} = r^2$

$\mathbf{b} = 10r^2 + 25$

$\mathbf{c} = 5 - r$

$\mathbf{d} = \frac{1}{2}r^2 \log r - r^2$

We note that **b** is $O(a)$, **c** is $O(a)$,but **d** not $O(a)$. Also, **a** is $O(d)$, as **b** and **c**.

(2) If $\mathbf{a} = 2r^2$ and $\mathbf{b} = 3r^2 + 1$, then clearly $\mathbf{b} \in O(a)$ or **b** is $O(a)$ and $\mathbf{a} \in O(b)$ or **a** is $O(b)$.

(3) If $\mathbf{a} = r^2$ and $\mathbf{b} = r^2 \log r - 2r^2$, then $b \notin O(a)$ or b is not $O(a)$.

- When numeric function **a** can be expressed in a simple closed form such as $\mathbf{a} = r^2$, we shall also write $O(a)$ as $O(r^2)$. Thus , when we say that a numeric function **b** is $O(r \log r)$,we mean that **b** is asymptotically dominated by the numeric function $\mathbf{a} = r \log r$. We want to remind that **b** is $O(a)$ only means that **b** does not grow faster than **a**; indeed **b** could grow much slower than **a**. Thus a statement such as "**b** grows too fast because **b** is $O(2^r)$" is not meaningful because we only know that **b** does not grow faster than 2^r. It is possible that **b** grows much slower than 2^r.
- For any numeric function **a, a** is $O(|\mathbf{a}|)$.
- If **b** is $O(a)$, then for any constant α, $\alpha.\mathbf{b}$ is also $O(a)$.
- If **b** is $O(a)$, then for any integer i, $S^i b$ is $O(S^i b)$.
- If both **b** and **c** are $O(a)$, then for any constants α and β, $\alpha\mathbf{b} + \beta\mathbf{c}$ is also $O(a)$.
- If **c** is $O(b)$ and **b** is $O(a)$, then **c** is $O(a)$.
- It is possible that **a** is $O(b)$ and **b** is also $O(a)$.
- It is possible that **a** is not $O(b)$ and **b** is also $O(a)$.
- It is possible that **c** is both $O(a)$ and $O(b)$, while **a** is not $O(b)$ and **b** is not $O(a)$.

ILLUSTRATIVE EXAMPLES

1. *Find the numeric function* **c** *made up of the letters* $\{x, y, \alpha, \beta, r\}$ *with the first portion made of English letters and the second portion made up of Greek letters.*

Solution : Let **a** be the numeric function made up of English letters and **b** be the numeric function made up of Greek letters.

Then, $a_r = 2^r,\ 0 \le r \le 2$ and $b_r = 3^r,\ 0 \le r \le 3$

Now, the required numeric function c will be given by

$$c_r = \sum_{i=0}^{r} 2^r 3^{r-i}$$

2. *Let **a** be a numeric function given by*

$$a_r = \begin{cases} 2, & 0 \le r \le 3 \\ 2^{-r} + 3, & r \ge 4 \end{cases}$$

(a) determine S^2a and $S^{-2}a$, *(b) determine Δa and ∇a.*

Solution : (a) We have $S^2a = \begin{cases} 0 & 0 \le r \le 1 \\ 2 & 2 \le r \le 5 \\ 2^{-(r+2)} + 3 & r \ge 6 \end{cases}$

$$S^{-2}a = \begin{cases} 2 & 0 \le r \le 1 \\ 2^{-r-2} & r \ge 2 \end{cases}$$

(b) We have $a_r = \begin{cases} 2 & 0 \le r \le 3 \\ 2^{-r} + 3 & r \ge 4 \end{cases}$

Therefore

$$a_{r+1} = \begin{cases} 2 & 0 \le r+1 \le 3 \Rightarrow -1 \le r \le 2 \\ 2^{-r-1} + 3 & r+1 \ge 4 \Rightarrow r \ge 3 \end{cases}$$

$$\Rightarrow \quad a_{r+1} = \begin{cases} 2 & 0 \le r \le 2 \\ 2^{-r-1} + 3 & r \ge 3 \end{cases}$$

$$\Rightarrow \quad a_{r-1} = \begin{cases} 2 & 0 \le r-1 \le 3 \Rightarrow r \le 4 \\ 2^{-r+1} + 3 & r-1 \ge 4 \Rightarrow r \ge 5 \end{cases}$$

Now, $\Delta a = a_{r+1} - a_r = \begin{cases} 2-2=0 & 0 \le r \le 2 \\ 2^{-4} + 3 - 2 = \dfrac{17}{16} & r = 3 \\ 2^{-r-1} + 3 - 2^{-r} - 3 = \dfrac{2^{-r}}{2} - 2^{-r} = -\dfrac{1}{2}.2^{-r} = -2^{-r-1}, & r \ge 4 \end{cases}$

and $\nabla a = a_r - a_{r-1} = \begin{cases} 0 & 0 \le r \le 3 \\ 2^{-4} + 3 - 2 = \dfrac{17}{16} & r = 4 \\ 2^{-r} + 3 - 2^{-r+1} - 3.2^{-r} - 2.2^{-r} = -2^{-r} & r \ge 5 \end{cases}$

3. *Let **a** be a numeric function such that $a_r = r^3 - 2r^2 + 3r + 2$. Find Δa, $\Delta^2 a$, $\Delta^3 a$ and $\Delta^4 a$.*

Solution : We have $a_r = r^3 - 2r^2 = 3r + 2$

$$a_r = a_0 r^{(3)} + a_1 r^{(2)} + a_2 r^{(1)} + a_3 \qquad \text{(In factorial notation)}$$

Now finding the values of a_0, a_1, a_2 and a_3 by the method of detached coefficients or synthetic division as follows :

1	1 0	– 2 1	3 – 1	2 $= a_3$
2	1 0	–1 2	2 $= a_1$	
3	1 0	1 $= a_1$		
	1 $= a_0$			

Now, we have

$$a_r = r^{(3)} + r^{(2)} + 2r^{(1)} + 2$$

($\because$ In case of factorial notation, Δ worked as differential operator)

$$\Delta a_r = 3r^{(2)} + 2r^{(1)} + 2$$

$$\Delta^2 a_r = 6r^{(1)} + 2$$

$$\Delta^3 a_r = 6$$

$$\Delta^4 a_r = 0.$$

4. *Let* $\mathbf{c} = \mathbf{ab}$. *Show that* $\Delta c_r = a_{r+1}(\Delta b_r) + b_r(\Delta a_r)$.

Solution : Let $\mathbf{c} = \mathbf{ab}$.

$$\begin{aligned}\Delta_{c_r} &= c_{r+1} - c_r \\ &= a_{r+1}b_{r+1} - a_r b_r \\ &= a_{r+1}b_{r+1} - a_{r+1}b_r + a_{r+1}b_r - a_r b_r \\ &= a_{r+1}[b_{r+1} - b_r] + b_r[a_{r+1} - a_r] \\ &= a_{r+1}(\Delta b_r) + b_r(\Delta a_r).\end{aligned}$$

5. *Let* $\mathbf{a}$ *and* $\mathbf{b}$ *be two numeric functions. The quotient of* $\mathbf{a}$ *and* $\mathbf{b}$ *denoted by* $\mathbf{a/b}$ *is a numeric function whose value at r is equal to* $\frac{a_r}{b_r}$. *Let* $\mathbf{d} = \frac{\mathbf{a}}{\mathbf{b}}$. *Show that*

$$\Delta d_r = \frac{b_r\,(\Delta a_r) - a_r\,(\Delta b_r)}{b_r\, b_{r+1}}$$

Solution : We have $\Delta d_r = d_{r+1} - d_r$

$$\begin{aligned}&= \frac{a_{r+1}}{b_{r+1}} - \frac{a_r}{b_r} \\ &= \frac{a_{r+1} - a_r b_{r+1}}{b_{r+1}b_r} \\ &= \frac{a_{r+1}b_r - b_r a_r + b_r a_r - a_r b_{r+1}}{b_{r+1}b_r}\end{aligned}$$

(On adding & subtracting $b_r a_r$ in numerator)

$$\begin{aligned}&= \frac{b_r\,(a_{r+1} - a_r) - a_r\,(b_{r+1} - b_r)}{b_{r+1}b_r} \\ &= \frac{b_r\,(\Delta a_r) - a_r\,(\Delta b_r)}{b_r.\, b_{r+1}}\end{aligned}$$

6. *Determine* $a * b$ *where* $\mathbf{a}$ *and* $\mathbf{b}$ *are two numeric functions such that*

$$a_r = \begin{cases} 1; & 0 \le r \le 2 \\ 0; & r \ge 3 \end{cases};$$

$$b_r = \begin{cases} 1; & 0 \le r \le 2 \\ 0; & r \ge 3 \end{cases}$$

Solution : Let $c = a * b = \sum_{i=0}^{r} a_i b_{r-i}$

Then, $c = a_0 b_r + a_1 b_{r-1} + a_2 b_{r-2} + \ldots + a_r b_0$

$$c_r = \begin{cases} a_0b_0 = 1 & r=0 \\ a_0b_1 + a_1b_0 = 2 & r=0 \\ a_0b_2 + a_1b_1 + a_2b_0 = 3 & r=2 \\ 2 & r=3 \\ 1 & r=4 \\ 0 & r \geq 5 \end{cases}$$

$[\because\ c_0 = a_0b_0 = 1.1 = 1]$
$[\because\ c_1 = a_0b_1 + a_1b_0 = 1.1 + 1.1 = 2]$

7. *Find a * b where **a** and **b** are two numeric functions such that*

$$a_r = \begin{cases} 0 & 0 \leq r \leq 2 \\ 2^r & r \geq 3 \end{cases}; \ b_r = \{2^r \quad r \geq 0$$

Solution : Let $c = a * b = \sum_{i=0}^{r} a_i b_{r-i}$

Then, $c = a_0b_r + a_1b_{r-1} + a_2b_{r-2} + \ldots + a_rb_0$

$$c_r = \begin{cases} a_0b_0 = 0 & r=0 \\ a_1b_0 + b_1a_0 = 0 & r=1 \\ a_0b_2 + a_1b_1 + a_2b_0 = 0 & r=2 \\ a_0b_3 + a_1b_2 + a_2b_1 + a_3b_0 = 0+0+0+2^r.2^r = 2^{2r} & r=3 \end{cases}$$

Further, for $r = 4$,

$$a_0b_4 + a_1b_3 + a_2b_2 + a_3b_1 + a_4b_0 = 0+0+0+2^r.2^r + 2^r.2^r = 2.2^{2r}$$

At $r = 5$,

$$a_0b_5 + a_1b_4 + a_2b_3 + a_3b_2 + a_4b_1 + a_5b_0$$
$$= 0+0+0+2^r.2^r + 2^r.2^r + 2^r.2^r = 3.2^{2r}$$

Hence, $c_r = \begin{cases} 0 & 0 \leq r \leq 2 \\ (r-2)2^r & r \geq 3 \end{cases}$

8. *Every particle inside a nuclear reactor splits into two particles in each second. Suppose one particle is injected into the reactor every second beginning at t = 0. How many particles are there in the reactor at r^th second?*

Solution : The number of particle at $t = 0$ is given by $a_0 = 1$.

Also, the number of particle at $t = 1$.

$$a_1 = 2 \times 1 + 1 = 3$$

and the number of particles at $t = 2$

$$a_2 = 3 \times 2 + 1 = 7$$

Similarly at $t = 3$

$$a_3 = 7 \times 2 + 1 = 15$$

Let S is the sum of particles in r seconds, then we have

$$S = 1 + 3 + 7 + 15 + \ldots + T_r$$
$$S = 1 + 3 + 7 + \ldots + T_{r-1} + T_r$$
$$0 = 1 + 2 + 4 + 8 + \ldots + T_r$$
$$T_r = 1 + 2 + 4 + 8 + \ldots \qquad \text{(Which is a G.P. with common ratio 2)}$$
$$= 2^{r+1} - 1. \text{ Hence, } a_r = 2^{r+1} - 1$$

$\because$ By using $T_r = ar^{n-1}$.

9. *Let the numeric function $\alpha = \alpha_0 + \alpha_1 r + \alpha_2 r^2 + \ldots + \alpha_n r^n$. Show that α is $O(r^n)$.*

Solution : We have $|\alpha| = |\alpha_0 + \alpha_1 r + \alpha_2 r^2 + \ldots + \alpha_n r^n|$

$$\leq |\alpha_0| + |\alpha_1||r| + |\alpha_2| r^2 + \ldots + |\alpha_n| r^n \qquad \text{(By triangle inequality)}$$
$$\leq [|\alpha_0| + |\alpha_1| + |\alpha_2| + \ldots + |a_n| r^n$$

Now we take $m = |\alpha_0| + |\alpha_1| + \ldots + |a_n|$, then

$$|\alpha| = |\alpha_0 + \alpha_1 r + \ldots + \alpha_n r^n| \le mr^n \text{ for } r \ge 1.$$

Hence, the numeric function α is asymptotically dominated by r^n and therefore α is $O(r^n)$.

10. *Let* **a, b** *and* **c** *be numeric functions such that* **a*b= c**. *Given that*

$$a_r = \begin{cases} 1 & r = 0 \\ 2 & r = 1 \\ 0 & r \ge 2 \end{cases} \text{ and } c_r = \begin{cases} 1 & r = 0 \\ 0 & r \ge 1 \end{cases}$$

Find b_r.

Solution : We have

$$c_r = a_r * b_r = \sum_{i=0}^{r} a_i b_{r-i} = a_0 b_r + a_1 b_{r-1} + a_2 b_{r-2} + \ldots a_r b_0$$

$\Rightarrow \quad c_0 = a_0 b_0 \qquad$ *i.e.,* $1 = 1 \times b_0 \quad \Rightarrow \quad b_0 = 1 = (-2)^0$

Now, $c_1 = a_0 b_1 + a_1 b_0 \qquad \Rightarrow \quad 0 = 1b_1 + 2b_0$

$\Rightarrow \quad 0 = b_1 + 2 \qquad \Rightarrow \quad b_1 = -2 = (-2)^1$

Also, $c_2 = a_0 b_2 + a_1 b_1 + a_2 b_0 \qquad \Rightarrow \quad 0 = 1.\, b_2 + 2(-2) + 0.1$

$\Rightarrow \quad b_2 = 2^2 = (-2)^2$

Hence, by symmetry, we have $b_r = (-2)^r$

11. *Let a person deposit Rs. 100 in his savings account at the rate of 5% interest in the beginning, Rs. 110 at the end of first year, Rs. 120 at the end of second year, Rs. 100 (1 + 0.1r) at the end of* r^{th} *year and so on. Find the total amount in the account at the end of* r^{th} *year.*

Solution : Let the numeric function **a** describes the yearly deposit.

i.e., $a_r = 100(1 + 0.1r),\ r \ge 0$

Let the numeric function **b** describes the amount in the savings account at the end of rth year, if 1 rupee was deposited into the account at the beginning, then

$$b_r = (1.05)^r,\ r \ge 0$$

So, the total amount in the account at the end of the rth year

$$c_r = \sum_{i=0}^{r} a_i b_{r-i}$$

It means if Rs. a_i was deposited in the savings account at the end of the ith year, it will becomes Rs. $a_i b_{r-i}$ after the $(r - i)$ years.

12. *Let* **a** *and* **b** *be two numeric functions given by*

$$a_r = r + O\left(\frac{1}{r}\right);\ b_r = \sqrt{r} + O\left(\frac{1}{\sqrt{r}}\right)$$

Show that $c_r = a_r b_r = r^{3/2} + O(\sqrt{2})$

Solution : We have $c_r = a_r b_r = \left[r + O\left(\frac{1}{r}\right)\right]\left[\sqrt{r} + O\left(\frac{1}{\sqrt{r}}\right)\right]$

$$= r^{3/2} + rO\left(\frac{1}{\sqrt{r}}\right) + \sqrt{r}O\left(\frac{1}{r}\right) + O\left(\frac{1}{r}\right)O\left(\frac{1}{\sqrt{r}}\right)$$

$$= r^{3/2} + O(\sqrt{r}) + O\left(\frac{1}{\sqrt{r}}\right) + O\left(\frac{1}{r^{3/2}}\right) = r^{3/2} + O(\sqrt{r})$$

[Taking highest power of r]

13. *We note that* $O(1) \subset O(\log r) \subset O(r) \subset O(r^r) \subset O(\alpha^r) \subset O(r!)$*. We ask the reader to confirm the results.*

Solution : Let A and B be two sets of numeric functions. We use the following notations to denote sets of numeric functions.

$$A + B = \{a + b : a \in A,\ b \in B\}$$
$$\alpha A = \{\alpha\, a : a \in A\}$$
$$A.B = \{ab : a \in A,\ b \in B\}$$

We state the following results and leave the proofs to the reader.

(1) If **b** is $O(a)$, then $O(b)$ is a subset of $O(a)$. Consequently, if **b** is $O(a)$ and **a** is $O(b)$, then sets $O(a)$ and $O(b)$ are equal.

(2) For any **a**, $O(a) + O(a) = O(a)$.

(3) If $\mathbf{b} \in O(a)$, then $O(a) + O(b) = O(a)$.

(4) For any constant α, $\alpha\, O(a) = O(\alpha a) = O(a)$.

(5) For any **a** and **b**, $O(a)\,O(b) = O(ab)$.

An Important Example

Let $\mathbf{a} = \frac{1}{3}r^3 + \frac{1}{2}r^2 + r$

Clearly, **a** is $O(r^3)$. Also, **a** is in the set of the numeric functions.

$$\left\{\frac{1}{3}r^3\right\} + O(r^2)$$

as well as the set of numeric functions

$$\left\{\frac{1}{3}r^3 + \frac{1}{2}r^2\right\} + O(r)$$

In the literature, we frequently write

$$\mathbf{a} = O(r^3)$$

(i) $\mathbf{a} = \frac{1}{3}r^3 + O(r^2)$ (ii) $\mathbf{a} = \frac{1}{3}r^3 + \frac{1}{2}r^2 + O(r)$ (iii) $\mathbf{a} \in O(r^3)$

$$\mathbf{a} \in \left\{\frac{1}{3}r^3\right\} + O(r^2)$$

$$\mathbf{a} \in \left\{\frac{1}{3}r^3 + \frac{1}{2}r^2\right\} + O(r)$$

Note that (i), (ii) and (iii) provide different information on the numeric function a. Because

$$\left\{\frac{1}{3}r^3 + \frac{1}{2}r^2\right\} + O(r) \subset \left\{\frac{1}{3}r^3\right\} + O(r^2) \subset O(r^3),$$

(iii) is the strongest and (i) is the weakest of the three relations concerning the behaviour of numeric function a. In other words, (iii) implies (ii) which in turn, implies (i), but not conversely.

7.9 BIG OMEGA AND BIG THETA NOTATION

7.9.1 Big-Omega

For a given numeric function ***a****,* $\Omega(a)$ *[big omega] is the set of all numeric functions* ***b*** *for which there exists two positive constants n and m such that*

If $|b_r| \geq m\, a_r$ *for* $r \geq n$

☛ If **b** is $\Omega(a)$, then **b** grows at least as fast as a.

For example : Let **a, b, c** and **d** be four numeric functions such that

$$a_r = 2r + 5, r \geq 0; b_r = r - 2, r \geq 0$$
$$c_r = r \log r, r \geq 0; d_r = r^3/2, r \geq 0$$

Both **b** and **c** are in $\Omega(a)$. However, neither **b** or **c** is in $\Omega(d)$ in the literature instead of saying **b** is in $\Theta(a)$, we often say **b** is $\Omega(a)$.

7.9.2 Big-Theta

For a given numeric function **a**, *$\Theta(a)$ [big theta] is the set of all numeric function* **b** *for which there exists positive constants m_1, m_2 and n such that*

$$m_1 a_r \leq |b_r| \leq m_2 a_r \text{ for } r \geq n$$

- If **b** is $\Theta(a)$, then it grows at the same rate as **a**.
  On the other hand the time complexity of the problem of selecting the largest of r numbers, and the time complexity of the problem of selecting the largest and smallest of r numbers are both $\Theta(r)$.

Exercise 7.1

1. Let **a** be a numeric function such that
$$a_r = \begin{cases} 2, & 0 \leq r \leq 3 \\ 2^{-r} + 5, & r \geq 4 \end{cases}$$
Find $S^2 a$ and $S^{-2} a$.

2. Determine the convolution $a * b$ where $a_r = 1$, for all $r \geq 0$ and $b_r = 2^r$ for all $r \geq 0$.

3. Let **a** and **b** be two numeric functions such that $a_r = 5^r, r \geq 0$ and $b_r = 3^r, r \geq 0$. Find the convolution of **a** and **b**.

4. Find $a * b$ where $a_r = \begin{cases} 1, & 0 \leq r \leq 2 \\ 0, & r \geq 3 \end{cases}$,
$$b_r = \begin{cases} 1, & 0 \leq r \leq 2 \\ 0, & r \geq 3 \end{cases}$$

5. Let $\mathbf{a} = 3^r$ and $\mathbf{b} = 2^r$ show that **a** dominates **b** asymptotically. Does **b** dominate **a** asymptotically. Show further that $a * b$ dominates both **a** and **b** asymptotically.

6. Show that every bounded numeric functions is $O(1)$.

7. Let **a** and **b** be numeric function such that
$$a_r = \begin{cases} r^2, & r \text{ is even} \\ r, & r \text{ is odd} \end{cases} \quad \text{and} \quad b_r = r^2$$
Show that **b** asymptotically dominates **a** but **a** does not asymptotically dominates **b**.

8. Let **a** be a numeric function, where $a_r = r^3 - 3r^2 + 3r + 2$. Determine Δa, $\Delta^2 a$, $\Delta^3 a$.

7.10 GENERATING FUNCTION

We introduce now an alternative way to represent numeric functions. For a numeric function $(a_0, a_1, a_2, \ldots, a_r, \ldots)$, we define an infinite series

$$a_0 + a_1 z + a_2 z^2 + \ldots + a_r z^r + \ldots$$

which is called the **generating function** of the numeric function a.

Definition : *Let $a_0, a_1, a_2, \ldots a_n, \ldots$ be sequence of real numbers. A series in powers of S, such as*

$$A(S) = a_0 + a_1 S + a_2 S^2 + \ldots = \sum_{i=0}^{\infty} a_i S^i$$

is called a generating function of the sequence $\{a_n\}$.

7.3.1 Application of Generating Function

Generating functions are important for solving counting problem. The generating functions arise in two different manners, one as counting device and second from the investigation of recurrence solution. Let us concentrate first on generating function of various series.

It is an easy job to obtain the generating functions of a given numeric function.

For example, if **a** be a numeric function given by

$$\mathbf{a} = (9^0, 9^1, 9^2, \ldots, 9^r, \ldots)$$

then its corresponding generating function $A(z)$ is

$$A(z) = 9^0 + 9z + 9^2z^2 + 9^3z^3 + \ldots + 9^rz^r + \ldots$$

or in brief, this can be written as

$$A(z) = \frac{1}{1-9z}$$

Clearly, given a numeric function, we can easily obtain its generating function and conversely.

ILLUSTRATIVE EXAMPLES

1. *Find the generating function of following series :*

$$1, -1, 1, -1, 1, -1, \ldots$$

Solution : Let the given series be directed by $\{a_n\}$, *i.e.*,

$$a_0 = 1,\ a_1 = -1,\ a_2 = 1,\ a_3 = -1, \ldots$$

Hence, the generating function will be

$$G(x) = \sum_{n=0}^{\infty} a_n x^n = a_0 + a_1x + a_2x^2 + \ldots$$

$$= 1 - x + x^2 - x^3 + x^4 - \ldots$$

or $$= \frac{1}{1+x}$$ (Being the sum of infinite G.P.)

Hence, $\frac{1}{1+x}$ generates the real number series $1, -1, 1, -1, 1, -1, \ldots$

2. *Find the generating function which generates 1, 0, 0, 1, 0, 0, 1, 0, 0,...*

Solution : Let $a_0 = 1,\ a_1 = 0,\ a_2 = 0,\ a_3 = 1, \ldots$

Therefore, the generating function will be

$$G(x) = \sum_{n=0}^{\infty} a_n x^n = a_0 + a_1x + a_2x^2 + a_3x^3 + \ldots$$

$$= 1 + 0 + 0 + x^3 + 0 + 0 + x^5 + \ldots$$

$$= 1 + x^3 + x^5 + \ldots$$ [Infinite G.P.]

$$= \frac{1}{1-x^3}$$

Hence, $\frac{1}{1-x^3}$ generates the given real series.

3. *What is the generating function for the sequence 1, 1, 1, 1, 1, 1, 1?*

Solution : The generating function of 1, 1, 1, 1, 1, 1, 1 is given by

$$G(x) = \sum_{n=0}^{\infty} a_n x^n = 1 + x + x^2 + x^3 + x^4 + x^5 + x^6$$

$$= \frac{x^7 - 1}{x - 1} \text{ or } \frac{1 - x^7}{1 - x}$$

4. *Find the generating function of the numeric function $a_r = 2^r,\ r \geq 0$.*

Solution : The generating function of the given numeric function is given by

$$G(x)=\sum_{n=0}^{\infty} a_n x^n = 2^0 + 2^1 x + 2^2 x^2 + \ldots + 2^r x^r + \ldots$$

$$= \frac{1}{1-2x}.$$

5. *Find the generating function of the numeric function $a_r = 1$, $r \geq 0$.*

Solution : The required generating function

$$G(x)=\sum_{n=0}^{\infty} a_n x^n = 1 + x + x^2 + \ldots + x^r + \ldots$$

$$= \frac{1}{1-x} \text{ for } |x| < 1.$$

6. *Find the generating function for $f_n = 3^n$, $n \geq 0$.*

Solution : The required generating function

$$G(x) = 1 + 3x + (3x)^2 + (3x)^3 + \ldots$$

$$G(x) - 1 = 3x + (3x)^2 + (3x)^3 + \ldots$$

$$\frac{G(x)-1}{3x} = 1 + 3x + (3x)^2 + \ldots = G(x)$$

$$G(x) - 1 = 3x.G(x)$$

$$G(x)[1-3x] = 1$$

$$G(x) = \frac{1}{1-3x}$$

Therefore, the required generating function is $\frac{1}{1-3x}$.

7.11 COMBINATORIAL PROBLEM (AS APPLICATION OF GENERATING FUNCTION)

Let the value of the numeric function **a** be a_r ; and $a_r = c(n, r)$.

Therefore, the generating function $A(z)$ of **a** can be written as

$$A(z) = c(n, 0) + c(n, 1)z + c(n, 2)z^2 + \ldots + c(n, r)z^r + \ldots + c(n, n)z^n \quad \ldots(1)$$

If we consider n objects as the sum of k and $n - k$ objects; and r objects as $r = i + r - i$, then we can write

$$c(n, r) = \sum_{i=0}^{r} c(k, i)\, c(n-k, r-i)$$

Therefore, it is possible to write $a = E.F.$ in which

$$E(z) = c(k, 0) + c(k, 1)z + c(k, 2)z^2 + \ldots + c(k, k)z^k$$

and $$F(z) = c(n-k, 0) + c(n-k, 1)z + c(n-k, 2)z^2 + \ldots + c(n-k, n-k)z^{n-k}$$

Making the repetition, we have

$$A(z) = (1 + z)^n \quad \ldots(2)$$

As the generating function of the numeric function

$$c(1, 0), c(1, 1), 0, 0, \ldots, \text{ is } (1 + z)$$

Now, we have the following observations :

(1) Putting $z = 1$ in (2), we may find [equation (1)]

$$c(n, 0) + c(n, 1) + \ldots + c(n, r) + \ldots + c(n, n) = 2^n \quad \ldots(3)$$

This shows that the number of ways of selecting one, two...or n objects from n objects is 2^n.

(2) Putting $z = -1$ in equation (2) and making use of equation (1), we have

$$c(n,0) - c(n,1) + c(n,2) + \ldots + (-1)^r c(n,r) + \ldots (-1)^n c(n,n) = 0$$

that is $c(n,0) + c(n,2) + c(n,4) + \ldots$

$$= c(n,1) + c(n,3) + c(n,5) + \ldots \qquad \ldots(4)$$

This shows that the number of ways of selecting an even number of objects from n objects is equal to the number of ways of selecting an odd number of objects.

ILLUSTRATIVE EXAMPLES

1. *Find a generating function for a_r, the number of ways to select r balls from a pile of three green, three white and three blue and three red balls.*

Solution : The generating function will be a multiplication of four factors corresponding to each colour green, white, blue and red. Since there are three balls of each colour, each factor will be $(1 + x + x^2 + x^3)$. Hence, the required generating function is $(1 + x + x^2 + x^3)^4$.

2. *Let a_r = the number of ways of selecting r objects from n objects with unlimited repetition. Prove that*

$$(1 + z + z^2 + \ldots + z^n) = c(n + r - 1, r)$$

Solution : Since each object is selected as many times as we wish, therefore a_r will be equal to the coefficient of z^r in $(1 + z + z^2 + z^3 \ldots)^n$.

Now, $$(1 + z + z^2 + z^3 + \ldots)^n = \left(\frac{1}{1-z}\right)^n = (1-z)^{-n}$$

Hence, we have

$$a_r = (-1)^r \frac{(-n)(-n-1)\ldots(-n-r+1)}{r!} = \frac{(n+r-1)!}{r!(n-1)!} = c(n+r-1, r).$$

☛ The above proof follows due to the fact that each of the factors of $(1 + z + z^2 + z^3 + \ldots)$ contribute to the number of times one of the object is selected.

3. *Find the generating function of the numeric function $a_r = c(m, r), r \geq 0$*

Solution : The generating function of the given numeric function is

$$G(x) = c(m,0) + c(m,1)x + c(m,2)x^2 + \ldots + c(m,m)x^m$$
$$= (1+x)^m$$

4. *Using generating function, find the number of ways of selecting six objects from three types of objects with repetition of up to four objects of each type.*

Solution : The generating function of a_r, the number of ways to select r objects from three types of objects with repetition of up to four objects of each type is

$$G(x) = (1 + x + x^2 + x^3 + x^4)^3$$

Now, we are required to find a_6, the coefficient of x^6 in $G(x)$. We can write $G(x)$ as

$$G(x) = \left(\frac{1-x^5}{1-x}\right)^3 = (1-x^5)^3(1-x)^{-3}$$
$$= (1 - 3x^5 + 3x^{10} - x^{15})(1-x)^{-3}.$$

Coefficient of x^6 in $G(x)$ = Coefficient of x^6 in $(1-x)^{-3} - 3 \times$ Coefficient of x in $(1-x)^{-3}$

$$= \frac{3.4\ldots8}{6!} - 3 = 25.$$

7.12 EXPONENTIAL GENERATING FUNCTIONS

Let $\{a_n\}$ be a sequence of real numbers. Then the exponential generating function is defined as :

$$A = \sum_{n=0}^{\infty} \frac{a^n}{n!} x^n$$

- The generating function without exponential is termed as ordinary generating function.

7.12.1 Properties of Generating Functions

(i) Let $\{a_n\}$ and $\{b_n\}$ be two sequences with generating functions $A(S)$ and $B(S)$ respectively. Then let $c_n = a_n + b_n$, *i.e.*, $\{c_n\}$ be a sequence. The generating function of $\{c_n\}$ is given by

$$C(S) = A(S) + B(S)$$

For example : Let $a_n = 2^n$, $b_n = 3^n$ and $c_n = a_n + b_n$

Then, $c(S) = A(S) + B(S) = \frac{1}{1-2S} + \frac{1}{1-3S} = \frac{2-5S}{1-5S+6S^2}$

(ii) Convolution : If the $\{c_n\}$ is defined as $c_n = a_n * b_n$, then $\{c_n\}$ is called convolution of $\{a_n\}$ and $\{b_n\}$.

where $$c_r = a_0 b_r + a_1 b_{r-1} + \ldots + a_{r-1} b_2 + a_r b_0$$

c_r is the coefficient of S^r in

$$(a_0 + a_1 S + a_2 S^2 + ..)(b_0 + b_1 S + b_2 S^2 + \ldots)$$

Now, if $A(S)$ and $B(S)$ are the respective generating functions then

$$C(S) = A(S).B(S)$$

- Convolution is an associative and commutative operation and distributed with respect to addition.
- If **b** is $O(a)$, then for any constant α, $\alpha.\mathbf{b}$ is also $O(a)$.
- If **b** is $O(a)$, then for any integer i, $S^i b$ is $O(S^i b)$.
- (i) If the numeric function $\mathbf{c} = \mathbf{a} + \mathbf{b}$, then $C(x) = A(x) + B(x)$, where $A(x)$, $B(x)$ and $C(x)$ are the generating functions of the numeric functions **a,b** and **c** respectively.
- (ii) If the numeric function $\mathbf{b} = \alpha\mathbf{a}$, α is a constant, then $B(x) = \alpha A(x)$.
- (iii) If for a numeric function **a**, **b** is a numeric function given by $b_r = \alpha^r a_r$, then $B(x) = A(\alpha x)$
- (iv) If $A(x)$ is the generating function of the numeric function a, then $x^i A(x)$ is the generating function of $S^i a$ and $x^{-i}[A(x) - a_0 - a_1 x - \ldots a_{i-1} x^{i-1}]$ is the generating function of $S^{-1} a$.
- (v) If $\mathbf{b} = \Delta a$, then $B(x) = \frac{1}{x}[A(x) - a_0] - A(x)$.
- (vi) If $\mathbf{b} = \nabla a$, then $B(x) = A(x) - xA(x)$.
- (vii) If $\mathbf{c} = \mathbf{a} * \mathbf{b}$, then $C(x) = A(x).B(x)$.

7.12.2 Some Important Sequences with Their Generating Functions

	Sequence	Generating Function
(i)	$a_n = \frac{1}{n!}$	$A(S) = e^s$
(ii)	$a_n = {}^mC_r$ for fixed m	$A(S) = (1 + S)^n$
(iii)	$a_n = 1$ for all n	$A(S) = (1 - S)^{-1}$
(iv)	${}^nC_k p^k q^{n-k}$, $k = 0, 1, 2, \ldots, n$ $p + q = 1$, $p, q > 0$	$A(S) = (p + qS)^n$
(v)	$q^k p \quad k = 0, 1, 2, \ldots, \infty$ $p + q = 1$, $p, q > 0$	$A(S) = \frac{P}{1 - qS}$
(vi)	$[1, 2, 2^2, 2^3, \ldots, 2^n, \ldots]$	$A(S) = \frac{1}{1 - 2S}$
(vii)	$\left[1, \frac{2}{1!}, \frac{4}{2!}, \frac{8}{3!}, \frac{16}{4!}, \ldots\right]$	e^{2x}

ILLUSTRATIVE EXAMPLES

1. *Find the generating functions of the following :*

(i) 2, 4, 8, 16, 32,... *(ii) 2, –2, 2, –2,...*

(iii) 1, 0, 1, 0, ... *(iv) 0, 1, –2, 4, –8,...*

Solution :

(i) The required generating function is given by

$$G(x) = \sum_{n=0}^{\infty} a_n x^n = 2 + 4x + 8x^2 + 16x^3 + \ldots$$

$$= 2[1 + 2x + (2x)^2 + (2x)^3 + \ldots]$$

(Infinite G.P. with common ratio $2x$, therefore use $S = \frac{a}{1 - r}$)

$$= \frac{2}{1 - 2x}, \; |2x| < 1$$

(ii) The required generating function is given by

$$G(x) = 2 - 2x + 2x^2 - 2x^3 + \ldots$$

$$= 2(1 - x + x^2 - x^3 + \ldots)$$

$$= \frac{2}{1 + x}, \; |-x| < 1 \qquad \left(\because S = \frac{a}{1 - r}\right)$$

(iii) The required generating function is given by

$$G(x) = 1 - 0x + 1x^2 - 0x^3 + 1x^4 + \ldots$$

$$= 1 + x^2 + x^4 + \ldots$$

$$= \frac{1}{1 - x^2}, \; |x^2| < 1.$$

(iv) The required generating function is given by

$$G(x) = 0 + x - 2x^2 + 4x^3 - 8x^4 + \ldots$$

$$= x(1 - 2x + (2x)^2 - (2x)^3 + \ldots)$$

$$= \frac{x}{1+2x}, |2x| < 1.$$

2. *Find the generating functions of the following numeric functions :*

(i) $a_r = 2r + 3, r = 0, 1, 2$ *(ii)* $a_r = {}^{r+4}C_r, r = 0, 1, 2$

(iii) $a_r = \frac{1}{(r+1)!}, r = 0, 1, 2$ *(iv)* $a_r = {}^8C_r, r = 0, 1, 2$

Solution : (i) The required generating function is given by

$$G(x) = \sum_{r=0}^{\infty} a_r x^r = \sum_{r=0}^{\infty} (2r+3)\, x^r$$

$$G(x) = 3 + 5x + 7x^2 + 9x^3 + \ldots + (2r+3)x^r$$

$$\Rightarrow \quad xG(x) = 3x + 5x^2 + 7x^3 + \ldots$$

Therefore,

$$G(x)\,[(1-x)] = 3 + 2x + 2x^2 + 2x^3 + \ldots$$

i.e.,

$$G(x)\,(1-x) = 3 + \frac{2x}{(1-x)}$$

$$\Rightarrow \quad G(x) = \frac{3}{(1-x)} + \frac{2x}{(1-x)^2} = \frac{3-3x+2x}{(1-x)^2} = \frac{3-x}{(1-x)^2}$$

(ii) $a_r = {}^{r+4}C_r$

$$G(x) = \sum_{r=0}^{\infty} a_r x^r = {}^4C_r + {}^5C_1 x + {}^6C_2 x^2 + {}^7C_3 x^3 + \ldots$$

$$= 1 + 5x + 15x^2 + \ldots$$

$$= (1-x)^{-5}, \text{ where } |x| < 1.$$

(iii) Here, we have $a_r = \frac{1}{(r+1)!}$

$$\therefore \quad \text{Generation of function } G(x) = \sum_{r=0}^{\infty} a_r x^r = 1 + \frac{1}{2!}x + \frac{1}{3!}x^2 + \frac{1}{4!}x^3 + \ldots$$

$$= \frac{1}{x}\left[x + \frac{x^2}{2!} + \frac{x^3}{3!} + \ldots\right]$$

$$= \frac{1}{x}\left[1 + x + \frac{x^2}{2!} + \frac{x^3}{3!} + \ldots - 1\right]$$

$$= \frac{1}{x}[e^x - 1] \quad \text{(By definition of exponential function)}$$

$$= \left(\frac{e^x - 1}{x}\right)$$

(iv) The required generating function is given by

$$G(x) = \sum_{r=0}^{\infty} a_r x^r$$

$$a_r = {}^8C_r$$

$$G(x) = {}^8C_0 + {}^8C_1 x + {}^8C_2 x^2 + \ldots + {}^8C_8 x^8$$

$$= (1+x)^8 \quad \text{(By definition of binomial expansion)}$$

3. *Find the generating functions of the numeric function* $a_r = 2^r + 3^r$, $r \geq 0$.

Solution : The generating function is given by

$$G(x) = \sum_{r=0}^{\infty} a_r x^r = \sum_{r=0}^{\infty} (2^r + 3^r)x^r$$

$$= \sum_{r=0}^{\infty} 2^r x^r + \sum_{r=0}^{\infty} 3^r x^r \qquad [\because G(x) = G_1(x) + G_2(x)]$$

Here, $$G_1(x) = \sum_{r=0}^{\infty} 2^r x^r = 1 + 2x + 2^2x^2 + \ldots = \frac{1}{1-2x}$$

Similarly, $G_2(x) = \dfrac{1}{1-3x}$

$$G(x) = \frac{1}{1-2x} + \frac{1}{1-3x} \qquad [\because G(x) = G_1(x) + G_2(x)]$$

4. *Find the generating function for the infinite sequence 1, α, α^2, α^3,...where α is a fixed constant.*

Solution : The generating function for this sequence is

$$G(x) = 1 + \alpha x + \alpha^2x^2 + \alpha^3x^3 + \ldots$$

$$G(x) - 1 = \alpha x + \alpha^2x^2 + \alpha^3x^3 + \ldots$$

$$\frac{G(x)-1}{\alpha x} = 1 + \alpha x + \alpha^2x^2 + \alpha^3x^3 + \ldots = G(x)$$

$$G(x) - 1 = \alpha x\, G(x)$$

$$G(x)\,(1 - \alpha x) = 1$$

$$G(x) = \frac{1}{1-\alpha x}$$

Therefore, the generating function is $\dfrac{1}{1-\alpha x}$.

5. *Determine the generating function of the numeric function*

$$a_r = \begin{cases} 2^r, & r \text{ is even} \\ -2^r, & r \text{ is odd} \end{cases}$$

Solution : The required generating function is given by

$$G(x) = \sum_{r=0}^{\infty} a_r x^r$$

$$= a_0 + a_1x + a_2x^2 + \ldots$$

$$= 1 + (-2)x + (2)^2x^2 + (-2)^3x^3 + \ldots$$

$$= 1 + 2^2x^2 + 2^4x^4 + \ldots + (-2)x + (-2)^3x^3 + \ldots$$

(On separating the positive and negative terms)

$$= \frac{1}{1-4x^2} - \frac{2x}{1-4x^2}$$

$$= \frac{1-2x}{(1-4x^2)} = \frac{1}{(1+2x)}$$

6. *Determine the numeric function corresponding to each of the following generating function*

(i) $G(x) = \dfrac{1}{5-6x+x^2}$ *(ii)* $G(x) = \dfrac{(1+x^2)}{(1-x)^4}$

(iii) $G(x) = \dfrac{2}{1-4x^2}$ *(iv)* $G(x) = \dfrac{x^4}{1-2x}$

Solution : (i) We have $G(x) = \dfrac{1}{5-6x+x^2} = \dfrac{1}{(5-x)(1-x)}$

(By resolving into partial fraction)

$$= \frac{1}{4}\left[\frac{1}{1-x} - \frac{1}{5-x}\right] = \frac{1}{4}\left[\frac{1}{1-x}\right] - \frac{1}{4}\left[\frac{1}{5-x}\right]$$

$$= \frac{1}{4}\left[\frac{1}{1-x}\right] - \frac{1}{20}\left[\frac{1}{1-\frac{x}{5}}\right]$$

$$= \frac{1}{4}(1-x)^{-1} - \frac{1}{20}\left(1-\frac{x}{5}\right)^{-1}$$

$\therefore$ General term $= \dfrac{1}{4}x^r - \dfrac{1}{20}\dfrac{x^r}{5^r}$

$$a_r = \frac{1}{4} - \frac{1}{20}\frac{1}{5^r} \quad \text{(for } x = 1)$$

$$= \frac{1}{4}\left[1 - \frac{1}{5}.\frac{1}{5^r}\right] = \frac{1}{4}\left[1 - \frac{1}{5^{r+1}}\right]$$

Aliter :

Consider $5 - 6x + x^2$

$$= 5 - 5x - x + x^2$$

$$= 5(1-x) - x(1-x)$$

$$= (5-x)(1-x)$$

$$\therefore \quad T_{r+1} = \frac{(-1)(-2)(-3)\ldots(-1-r+1)(-x)^r}{r!}$$

$$T_{r+1} = \frac{1.2\ldots r}{r!}x^r = x^r$$

$$T_{r+1} = \frac{(-1)\ldots(-r)}{r!}\left(-\frac{x}{5}\right)^r = \frac{r!\,x^r}{r!\,5^r} = \frac{1}{5^r}x^r$$

(ii) $G(x) = \dfrac{(1+x^2)}{(1-x)^4} = (1+x^2)[1-x]^{-4}$

General term of $(1-x)^{-4} = T_{r+1}$

$$= \frac{(-4)(-5)(-6)\ldots(-4-r+1)(-x)^r}{r!}$$

$$= \frac{4.5.6\ldots(r+3)x^r}{r!}$$

$$= \frac{(r+1)(r+2)(r+3)\,x^r}{3!}$$

$$= \frac{(r+1)(r+2)(r+3)}{6}x^r$$

So, $a_r = \dfrac{(r+1)(r+2)(r+3)}{6} + \dfrac{(r-1)\,r\,(r+1)}{6}$

$$= \frac{(r+1)}{6}[(r+2)(r+3) + r(r-1)]$$

$$= \frac{r+1}{6}[r^2 + 5r + 6 + r^2 - r]$$

$$= \frac{r+1}{6}[2r^2 + 4r + 6]$$

$$= \frac{(r+1)}{3}(r^2 + 2r + 3)$$

(iii) $$G(x) = \frac{2}{1-4x^2} = \frac{2}{(1-2x)(1+2x)}$$

$$G(x) = \frac{1}{1-2x} + \frac{1}{1+2x}$$

General term of $(1-2x)^{-1} = T_{r+1} = \frac{n(n-1)\ldots(-1-r+1)(-2r)^r}{r!} = 2^r x^r$

and general term of $(1+2x)^{-1} = (-2)^r x^r$

Therefore, $$a_r = \begin{cases} 0 & r \text{ is odd} \\ 2^{r+1} & r \text{ is even} \end{cases}$$

(iv) $$G(x) = \frac{x^4}{1-2x} = x^4[1-2x]^{-1}$$

General term of $(1-2x)^{-1} = 2^r x^r$

So, general term corresponding to the function $x^4(1-2x)^{-1} = 2^{r-4}$

or $$a_r = \begin{cases} 0 & 0 \le r \le 4 \\ 2^{r-4} & r \ge 4 \end{cases}$$

7. *Let $c = a * b$, where $a_r \ge 2^r$, $r \ge 0$ and $b_r \ge 3^r$, $r \ge 0$. Find the generating function of c.*

Solution : We have $C(r) = \sum_{r=0}^{\infty} A(r). \sum_{r=0}^{\infty} B(r)$

$$C(x) = A(x).B(x)$$

$$A(x) = 1 + 2x + 2^2x^2 + 2^3x^3 + \ldots = \frac{1}{1-2x}$$

$$B(x) = \frac{1}{1-3x}$$

Hence, $C(x) = \frac{1}{(1-2x)} \cdot \frac{1}{(1-3x)}$ [Using $C(x) = A(x).B(x)$]

8. *Let a_r denote the number of ways a sum of r can be obtained when two indistinguishable dice are rolled. Find the generating function.*

Solution : We have $$a_r = \begin{cases} 0 & 0 \le r \le 1 \\ r-1 & 2 \le r \le 7 \\ 5 & r = 8 \\ 4 & r = 9 \\ 3 & r = 10 \\ 2 & r = 11 \\ 1 & r = 12 \end{cases}$$

$$G(x) = \sum_{r=0}^{12} a_r x^r = a_0 + a_1 x + \ldots + a_{12}x^{12}$$

$$= x^2 + 2x^3 + 3x^4 + 4x^5 + 5x^6 + 6x^7 + 5x^8 + 4x^9 + 3x^{10} + 2x^{11} + x^{12}$$

$$= x^2[1 + 2x + 3x^2 + 4x^3 + 5x^4 + 6x^5 + 5x^6 + 4x^7 + 3x^8 + 2x^9 + x^{10}]$$

9. *Using generating function, find the number of ways to select 10 balls from a large pile of red, white and blue balls iff*

(i) the selection has at least two balls of each colour.

(ii) the selection has at most two red balls.

(iii) the selection has an even number of blue balls.

Solution :

(a) This selection problem is equivalent to the number of integer solution of the equation

$$n_1 + n_2 + n_3 = 10\, n_i \geq 2$$

Here, n_1 represents the number of red balls, n_2 the number of white balls and n_3 the number of blue balls.

The generating function is

$$G(x) = (x^2 + x^3 + \ldots)^3 = x^6 (1 + x + x^2 + \ldots)^3 = x^6 (1 - x)^{-3}$$

Coefficient of x^{10} in $G(x)$ = Coefficient of x^4 in $(1 - x)^{-3} = {}^6C_4 = 15$

(b) The generating function is given by

$$G(x) = (1 + x + x^2)(1 + x + x^2 + \ldots)^2 = (1 + x + x^2)(1 - x)^2$$

Coefficient of $x^{10} = {}^{11}C_{10} + {}^{10}C_9 + {}^9C_8 = 30$

(c) The generating function is given by

$$G(x) = (1 + x + x^2 + \ldots)^2 (1 + x^2 + x^4 + x^6 + \ldots)$$

Corresponds to having 0, 2, 4, 6,... (even) number of blue balls in the selection.

We can write $G(x) = (1 + x^2 + x^4 + \ldots)(1 - x)^{-2}$

$\therefore$ Coefficient of x^{10} in $G(x)$ = Sum of coefficients of $x^{10}, x^8, x^6, x^4, x^2$ and 1 in $(1 - x)^{-2}$

$$= {}^{11}C_{10} + {}^9C_8 + {}^7C_6 + {}^5C_4 + {}^3C_2 + 1 = 36.$$

10. *Evaluate the sum* $1^2 + 2^2 + 3^2 + \ldots + r^2$

Solution : We have $\dfrac{1}{1 - z} = 1 + z + z^2 + z^3 + \ldots + z^r + \ldots$ ...(1)

Differentiating both sides of (1) w.r.t. z, we get

$$\frac{1}{(1 - z)^2} = 1 + 2z + 3z^2 + 4z^3 + \ldots + rz^r + \ldots \quad \ldots(2)$$

Differentiating (2) w.r.t. z of the multiplying by z, we get

$$\frac{d}{dz} \frac{z}{(1 - z)^2} = 1^2 + 2^2 z + 3^2 z^2 + 4^2 z^3 + \ldots + r^2 z^{r-1} + \ldots$$

and that $z \dfrac{d}{dz} \dfrac{z}{(1 - z)^2} = 0^2 + 1^2 z + 2^2 z^2 + 3^2 z^3 + 4^2 z^4 + \ldots + r^2 z^r + \ldots$

Thus, $z\,(d/dz)[z/(1 - z)^2]$, which is equal to $[z(1 + z)/(1 - z)^3]$ is the generating function of the numeric function $(0^2, 1^2, 2^2, 3^2, \ldots r^2, \ldots)$ and therefore $z\,(1 + z)/(1 - z)^4$ is the generating function of the numeric function

$$(0^2, 0^2 + 1, 0^2 + 1^2 + 2^2, 0^2 + 1^2 + 2^2 + 3^2, \ldots, 0^2 + 1^2 + 2^2 + 3^2 + \ldots + r^2, \ldots).$$

According to the binomial theorem, the coefficient of z^r in $1/(1 - z)^4$ is

$$\frac{(-4)(-4 - 1)(-4 - 2)\ldots(-4 - r + 1)(-1)^r}{r!} = \frac{4 \times 5 \times 6 \times \ldots \times (r + 3)}{r!} = \frac{(r + 1)(r + 2)(r + 3)}{1.2.3}$$

Therefore, the coefficient of z^r in the expansion of $[z(1 + z)/(1 - z)^4]$ is

$$\frac{r(r+1)(r+2)}{1.2.3} + \frac{(r-1)r(r+1)}{1.2.3.} = \frac{r(r+1)(2r+1)}{6}$$

that is $1^2 + 2^2 + 3^2 + \ldots + r^2 = \dfrac{r(r+1)(2r+1)}{6}$

Thus, $1^2 + 2^2 + \ldots + r^2 = \dfrac{r(r+1)(2r+1)}{6}$

11. *Show that the generating function for a, such that $a_r = {}^{2r}C_r$ is $(1-4x)^{-1/2}$.*

Solution : The generating function is given by

$$G(x) = \sum a_r x^r = 1 + {}^2C_1 x + {}^4C_2 x^2 + {}^6C_3 x^3 + \ldots$$
$$= 1 + 2x + 6x^2 + 20x^3 + \ldots$$
$$= (1-4x)^{-1/2}.$$

12. *Suppose the sequence of $< a_n >$, where $0 \le n \le \infty$ has generating functions $f(x)$.*

(i) *What sequence is generated by the function $g(x) = (1+x)f(x)$?*

(ii) *And what sequence is generated by the function $h(x) = \dfrac{f(x)}{1+x}$?*

Solution : (i) $f(x) = a_0 + a_1 x + a_2 x^2 + \ldots + a_r x^r + \ldots$

$$g(x) = (1+x)\, f(x) = (1+x)(a_0 + a_1 x + a_2 x^2 + \ldots + a_r x^r + \ldots)$$
$$= a_0 + (a_0 + a_1)x + (a_1 + a_2)x^2 + \ldots + (a_{r-1} + a_r)x^r + \ldots$$

Thus, sequence corresponding to $g(x)$, say $< b_r >$ is

$$b_r = \begin{cases} a_0, & r = 0 \\ a_{r-1} + a_r & r \ge 1 \end{cases}$$

(ii) $h(x) = \dfrac{f(x)}{1+x} = (a_0 + a_1 x + a_2 x^2 + \ldots + a_r x^r + \ldots)(1+x)^{-1}$

$$= (a_0 + a_1 x + a_2 x^2 + \ldots + a_r x^r + \ldots)(1 - x + x^2 \ldots + (-1)^r x^r + \ldots)$$
$$= a_0 + (a_1 - a_0)x + (a_2 - a_1 + a_0)x^2 + \ldots +$$
$$(a_r + a_{r-1}(-1) + a_{r-2}(-1)^2 + \ldots + a_1(-1)^{r-1} + a_0(-1)^r)x^r$$

Hence, if $< C_n >$ is sequence corresponding to the generating function $h(x)$, then

$$C_r = a_r + a_{r-1}(-1) + a_{r-2}(-1)^2 + \ldots + a_1(-1)^{r-1} + a_0(-1)^r,\ r \ge 0.$$

13. *Use generating function to show that $\sum_{k=0}^{n} c(n,k)^2 = c(2n, n)$, where n is a positive integer.*

Solution : Since ${}^{2n}C_n$ is the coefficient of x^n in the expansion of $(1+x)^{2n}$

Also, $(1+x)^{2n} = [(1+x)^n]^2 = [c(n,0) + c(n,1)x + c(n,2)x^2 + \ldots + c(n,n)x^n]^2$

The coefficient of x^n above the expansion is

$$c(n,0)\,c(n,n) + c(n,1)\,c(n,n-1) + \ldots + c(n,n)\,c(n,0) = c(n,0)^2 + c(n,1)^2 + \ldots + c(n,n)^2$$
$$= \sum_{k=0}^{n} c(n,k)^2 \qquad [\because\ c(n, n-k) = c(n,k)]$$

Therefore, $c(2n, n)$ and $\sum_{k=0}^{n} c(n,k)^2$ both are coefficients of x^n in the expansion of $(1+x)^{2n}$.

Hence, $\sum_{k=0}^{n} c(n,k)^2 = c(2n, n)$.

14. *If $G(x) = a_0 + a_1x + a_2x^2 + \ldots + a_rx^r + \ldots$ be the generating function of numeric function $a = (a_0, a_1, a_2, \ldots, a_r, \ldots)$, then show that $\frac{1}{(1-x)}G(x)$ is the generating function of the numeric function **b** which is accumulated sum of **a**.*

Solution : We know that $\frac{1}{1-x} = 1 + x + x^2 + \ldots + x^r + \ldots$ ...(1)

Hence, $\frac{1}{1-x}$ is the generating function for the numeric function (1, 1, 1, 1,...).

Also, we are given that $G(x) = a_0 + a_1x + a_2x^2 + \ldots + a_rx^r + \ldots$ ...(2)

is the generating function for **a**. Multiplying (1) and (2), we get

$$\frac{1}{1-x}G(x) = a_0 + (a_0 + a_1)x + (a_0 + a_1 + a_2)x^2 + \ldots + (a_0 + a_1 + \ldots + a_r)x^r + \ldots$$

$\Rightarrow$ $\frac{1}{(1-x)}G(x)$ is the generating function for the numeric function **b** where $b_r = \sum_{i=1}^{r} a_i$

Thus, **b** is accumulated sum of **a** and $\frac{1}{(1-x)}G(x)$ is the generating function of **b**.

15. *Obtain the generating function of the numeric function $a_r = 3^{r+2}$, $r \geq 0$.*

Solution : We have

$$G(x) = \sum_{r=0}^{\infty} a_rx^r = 3^2 + 3^3x + 3^4x^2 + \ldots = 3^2[1 + 3x + 3^2x^2 + \ldots] = \frac{a}{1-3x}.$$

16. *Find the number of solutions of $n_1 + n_2 + n_3 = 17$, where n_1, n_2, n_3 are non-negative integers such that $2 \leq n_1 \leq 5$, $3 \leq n_2 \leq 6$, $4 \leq n_3 \leq 7$.*

Solution : The number of solutions with the given constraints is the coefficient of x^{17} in the expansion of

$$(x^2 + x^3 + x^4 + x^5)(x^3 + x^4 + x^5 + x^6)(x^4 + x^5 + x^6 + x^7)$$

or $x^9[1 + x + x^2 + x^3]^3$ or $x^9[1 + x]^3[1 + x^2]^3$

or $x^9[1 + 3x + 3x + x^3][1 + 3x^2 + 3x^4 + x^6]$

Obviously, the coefficient of x^{17} is 3. Hence, there are three solutions.

17. *Determine the number of ways in which $2t + 1$ marbles can be distributed among three distinct boxes so that no box will contain more than to marbles. We claim that the coefficient of z^{2t+1} in $A(z) = (1 + z + z^2 + \ldots + z^t)^3$.*

Solution : The coefficient of z^{2t+1} in $A(z)$ is the number of ways to make up the term z^{2t+1} from the three factors $1 + z + z^2 \ldots + z^t$. The contribution from each factor $1 + z + z^2 + \ldots + z^t$ can be $1, z, z^2 \ldots$ or $1, z, z^2 \ldots$ or z^t corresponding to having none, one, two,... or t marbles in a box. Since

$$(1 + z + z^2 + \ldots + z^t)^3 = \left(\frac{1 - z^{t+1}}{1 - z}\right)^3$$

$$= (1 - 3z^{t+1} + 3z^{2t+2} - z^{3t+3})(1 - z)^{-3} \quad \ldots(1)$$

The coefficient of z^{2t+1} in (1) is the coefficient of z^{2t+1} in $(1 - z)^{-3}$ minus three times the coefficient of z^t in $(1 - z)^{-3}$ thus it is

$$c(3 + 2t + 1 - 1, 2t + 1) - 3c(3 + t - 1, t)$$

which is simplified as $c(2t + 3, 2t + 1) - 3c(t + 2, t)$

18. *Find the generating function of the sequence y_0, y_1, y_2 defined as follows :*

$$y_n + 2y_{n-1} - 15y_{n-2} = 0 \text{ for } n \geq 2$$

with $y_0 = 0$, $y_1 = 0$.

Solution : We have $y_n + 2y_{n-1} - 15y_{n-2} = 0$.

The generating function is $G(x) = y_0 + y_1x + y_2x^2 + \ldots = \sum_{n=2}^{\infty} y_nx^n$

Multiplying (1) by x^n and sum over all $n \geq 0$, we get

$$\sum_{n=2}^{\infty} y_nx^n + 2\sum_{n=2}^{\infty} y_{n-1}x^n - 15\sum_{n=2}^{\infty} y_{n-2}x^n = 0$$

$$\Rightarrow \quad [G(x) - y_0 - y_1(x^2)] + 2[xG(x) - y_0] - 15x^2G(x) = 0.$$

Since $y_0 = 0$, $y_1 = 1$, we get

$$[G(x) - x] + 2[xG(x) - 15x^2G(x)] = 0$$

$$G(x)[1 + 2x - 15x^2] = x$$

$$G(x) = \frac{x}{1 + 2x - 15x^2} = \frac{1}{(1+3x)(1+5x)}$$

Therefore, the generating function $G(x) = \dfrac{x}{(1+3x)(1+5x)}$

Exercise 7.2

1. Let $\mathbf{c} = \mathbf{ab}$, show that $\Delta c_r = a_{r+1}\Delta b_r + b_r\Delta a_r$.
2. Let $\mathbf{a}$ and $\mathbf{b}$ be two numeric functions such that $a_r = r + 1$ and $b_r = \alpha^r$ for all $r \geq 0$. Determine Δab.
3. Let a_r denote the number of ways of sum of r can be obtained when two indistinguishable dice are rolled. Determine $G(x)$.
4. Find the generating function of following sequences :
   (i) 1, 0, 1, 0, 1, 0,...
   (ii) 1, –1, 1, –1, 1, –1, 1, –1,...
   (iii) 1, 2, 3, 4, 5, ...
   (iv) $1, S^1, S^2, S^3, S^4$
   (v) 0, 2, 0, 4, 0, 6, 0, 8, . . .
5. Find a_n if $A(S) = \dfrac{2 + 3S - 6S^2}{1 - 2S}$.
6. If $A(S) = \dfrac{2}{1 - 4S^2}$, then find a_n.
7. How many solutions to an equation $a + b = 4$ are there if a must be the element of set $A = \{0, 1, 2, 3, 4\}$ and b must be an element of $B = \{2, 4, 6\}$?
8. In how many ways can $3r$ balls are selected from $2r$ red balls, $2r$ blue balls and $2r$ white balls.
9. Find the generating function for a_r, the number of ways to selecting r objects from n objects with limited repetitions. Hence find a_r.
10. Using the generating function, find the number of distributions of 12 identical objects into
    (a) Five different boxes with at most four objects in each box.
    (b) Seven different boxes with at least one object in each box.
    (c) Three different boxes with at most five objects in the first box.
11. Let $G(x) = (1+x)^{2n} + n(1+x)^{2n-1} + \ldots + x^n(1+x)^n$, find a_r.
12. Find a simple expression for the generating function of each of the following discrete numeric functions
    (i) 1, – 2, 3, – 4, 5, – 6
    (ii) 1, 2/3, 3/9, 27,...$(r + 1)3^r$, . . .
    (iii) 1, 1, 2, 2, 3, 3, 4, 4, . . .
    (iv) $0 \times 1, 1 \times 2, 2 \times 3, 3 \times 4, \ldots$
    (v) $0 \times S^0, 1 \times S^1, 2 \times S^2, \ldots, r \times S^r, \ldots$
13. Determine the discrete numeric function corresponding to each of the following generating functions :
    (i) $A(z) = \dfrac{1}{1 - z^3}$
    (ii) $A(z) = (1 + z)^n + (1 - z)^n$

(iii) $A(z) = \dfrac{(1+z)^2}{(1-z)^4}$

(iv) $A(z) = \dfrac{1}{5-6z+z^2}$

(v) $A(z) = \dfrac{z^2}{5-6z+z^2}$

(vi) $A(z) = \dfrac{7z^2}{(1-2z)(1+3z)}$

(vii) $A(z) = \dfrac{1+z^2}{4-4z-z^2}$

(viii) $A(z) = \dfrac{1}{(1-z)(1-z^2)(1-z^3)}$

14. Find the generating function of the Fibonacci sequence $\{f_n\}$ defined by

$f_n = f_{n-1} + f_{n-2}$, $f_0 = 0$, $f_1 = 1$.

15. Using generating function, evaluate the sum

$3.2.1 + 4.3.2 + 5.4.3 + \ldots + r(r+1)(r-1)$

16. Show that f is $O(g)$ but g is not $O(f)$, where $f(n) = 2^n$ and $g(n) = 3^n$.

17. Find the number of solutions of the equation $x_1 + x_2 + x_3 = 17$, where x_1, x_2 and x_3 are non-negative integers such that
$2 \le x_1 \le 5$, $3 \le x_2 \le 6$, $4 \le x_3 \le 7$.

18. Find the generating function for a_r = the number of non-negative integral solutions
$l_1 + l_2 + l_3 + l_4 + l_5 = r$
where, $0 \le l_1 \le 3$, $0 \le l_2 \le 3$, $2 \le l_3 \le 6$,
$2 \le l_4 \le 6$, l_5
is odd and $1 \le l_5 \le 9$.

❑❑❑

Chapter 8

Recurrence Relations

- Recurrence relation
- Order and degree
- Solution of a recurrence relation
- Homogeneous solutions
- Total and particular solution
- Generating function method
- Simultaneous recurrence relations

8.1 INTRODUCTION

We give the name recursion to the technique of defining a function, a set, or an algorithm in terms of itself, where it is generally understood that the definition will be in terms of previous values of similar objects. Describing an object in a simple definition form is usually called the basis for recursion.

In this chapter, we shall apply, what we have just learned first to the specification of discrete numeric functions, and then to the specifications of algorithms.

8.2 RECURRENCE RELATIONS

Consider the following rabbit problem. Each month, beginning when female of a pair of rabbits is two months old, gives birth to a pair of rabbits (one male, one female). Two months later, she (new pair) will also begin producing a new pair in each month and so on. The problem is to find the number of rabbits at the end of each year.

The above problem coined by famous mathematician Fibonacci in the year 1202.

The solution to the above problem can be had from following relation:

$$F(n) = F(n-1) + F(n-2) \qquad \ldots(1)$$

where $F(n)$ denotes the number of pair after n months have passed. Also, $F(n)$ is known as Fibonacci number which has following series

$$1, 1, 2, 3, 5, 8, 13, \ldots$$

Then what is the troublesome part of the problem. Well, the problem arises in finding out the value of $F(500)$. To get $F(500)$, we need two values *i.e.*, $F(499)$ and $F(498)$, but then how to get these two values. Well do you think it simple to get for $F(499)$, we need further two values, *i.e.*, $F(498)$ and $F(497)$ and like this we have to find all values upto $F(1)$, then we can find $F(500)$.

Actually, the relation (1) helps us to make it simple. Such relation is one of the example of recurrence relation.

8.2.1 Towers of Hanoi

Towers of Hanoi is a popular puzzle. There are three pegs mounted on a board together with disks of different sizes. Initially these disks are placed on the first peg in order of size, with the largest disk at bottom and the smallest disk at top. The task is to move the disks from first peg to third peg using middle peg. The rules of puzzle are :

1. only one disk can be moved at a time.
2. No disk can be placed on the top of a smaller disk.

Let H_n denote the number of moves needed to solve puzzle with n disks. Let us define H_n recursively. Clearly $H_1 = 1, n \geq 1$. Consider top $(n-1)$ disks. We can move top $(n-1)$ disks, following the rules to middle peg using H_{n-1} moves using third as an auxiliary. That leaves the largest disk at first peg. It takes one move to transfer it to third peg.

Now, the $n-1$ disks at middle peg can be moved to third peg using the first peg as auxiliary peg in H_{n-1} moves.

Hence, total number of moves H_n can be recursively defined as

$$H_n = 1, \text{ if } n = 1$$

$$H_n = 2H_{n-1} + 1, \text{ otherwise.}$$

Definition : *A linear recurrence relation for the sequence $\{a_n\}$ is an equation that express a_n in terms of one or more of the previous terms of the sequence and may be function of n.*

$$a_n = c_1 a_{n-1} + c_2 a_{n-2} + \ldots + f(n) \qquad \ldots(2)$$

In expression (1), $F(n)$ and in expression (2), a_n are same thing. In expression (2), $f(n)$ is a function of n.

WORKING RULES

The recursive definitions of a relation a_n with domain B consists of these steps : for $k \geq 1$.

Step 1. Basic : A few initial values of the relation $a_0, a_1, \ldots, a_{k-1}$ are specified. An equation that specifies such initial values is called as initial condition.

Step 2. Recursive : A formula to compute a_n from the k preceding relational values $a_{n-1}, a_{n-2}, \ldots, a_{n-k}$ is made. Such a formula is called recursive formula or recurrence relation. Hence, the recursive function definition has one or more initial conditions and a recurrence relation.

8.3 ORDER OF THE RECURRENCE RELATION

The order of the recurrence relation or difference equation is defined to be the difference between the highest and lowest subscripts of $f(x)$ or a_r or y_k.

For example :

The equation $| 3a_r + 2 | a_{r-1} = 0$ is a first order recurrence relation.

8.4 DEGREE OF THE RECURRENCE RELATION OR DIFFERENCE EQUATION

The degree of recurrence relation or difference equation is defined to be the highest power of $f(x)$ or a_r or y_k.

For example :

1. The recurrence relation $y_{k+3} + 2y_{k+2} + 4y_{k+1} + 2y_k = k(x)$ has the degree 1 because the highest power of y_k is 1 and its order is 3.

2. The recurrence relation $a_r^4 + 2a_{r-1}^2 + 6a_{r-2}^2 + 4a_{r-3} = 0$ has the degree 4, as the highest power of a_r is 4.

8.5 SOLUTION TO A RECURRENCE RELATION

Solving a recurrence relation means finding an explicit expression for a_r as a function of n only. This explicit expression should not involve any other term of the series, while we are solving recurrence relation, the most important role is played by boundary conditions. But what we mean by boundary conditions? You must remember the Fibonacci series discussed above. The problem (Fibonacci) needs at least two values, *i.e.*, $F(1)$ and $F(2)$, both of them are 1. Hence, the recurrence relation generated for Fibonacci series of two boundary conditions.

A recurrence relation is also called a difference equation, and those two terms will be used interchangeably. For a numeric function $(a_0, a_1, a_2, \ldots, a_r, \ldots)$ are equation relating a_r, for any r, to one or more of the a_i's $i < r$ is called a recurrence relation.

- In many discrete computation problems, it is sometimes easier to obtain a specification of a numeric function in terms of a recurrence relation than to obtain a general expression for the value of the numeric function at r or a closed-form expression for its generating function. It is clear that according to the recurrence relation, we can carry out a step by step computation to determine a_r from $a_{r-1}, a_{r-2}, \ldots$ to determine a_{r+1} from $a_r, a_{r-1}, \ldots$ and so on, provided that the value of the function at one or more points is given so that the computation can be initiated. These given values of the function are called the boundary conditions.

ILLUSTRATIVE EXAMPLES

1. *Solve the following recurrence relation for given boundary conditions.*

$$a_n = a_{n-1} + 2a_{n-2},\ n \geq 2,\ a_0 = 2,\ a_1 = 7.$$

Solution : The given recurrence relation is

$$a_n = a_{n-1} + 2a_{n-2} \qquad \ldots(1)$$

Multiplying both sides by S^n and summing up to n from 2 to ∞ (because the given recurrence relation is valid for $n \geq 2$)

$$\sum_{n=2}^{\infty} a_n S^n = \sum_{n=2}^{\infty} a_{n-1} S^n + 2 \sum_{n=2}^{\infty} a_n - 2S^n \qquad \ldots(2)$$

Define
$$A(S) = \sum_{n=2}^{\infty} a_n S^n \qquad \ldots(3)$$

Compare left hand side term of (2) with (3). Also, compare the first and second terms at right hand side of (2) and (3).

Writing (2) as following

$$\sum_{n=2}^{\infty} a_n S^n = S \sum_{n=2}^{\infty} a_{n-1} S^{n-1} + 2S^2 \sum_{n=2}^{\infty} a_{n-2} - 2S^{n-2} \qquad \ldots(4)$$

Applying (4)

$$A(S) - a_0 - a_1 S = S(A(s) - a_0) + 2S^2 A(S) \qquad \ldots(5)$$

Simplifying (5), we have

$$A(S) = \frac{a_0 + a_1 S - a_0 S}{1 - S - 2S^2}$$

but $a_0 = 2$ and $a_1 = 7$ (given)

Hence, $A(S) = \dfrac{2 + 5S}{1 - S - 2S^2} = \dfrac{2 + 5S}{(1 - 2S)(1 + S)}, A(S) = \dfrac{3}{1 - 2S} - \dfrac{1}{1 + S} \qquad \ldots(6)$

Equation (6) is required generating function for the recurrence relation (1). Hence comparing the coefficient of S^n on both sides $a_n = 3 \cdot 2^n - (-1)^n$ which is the required solution.

2. *The recurrence relation for Fibonacci numbers is given by*

$$f_n = f_{n-1} + f_{n-2} \quad \text{and } f_1 = f_2 = 1$$

Solve the above recurrence relation.

Solution : The characteristic equation corresponding to the given relation is $x^2 - x - 1 = 0$ and its solutions are

$$\alpha = \frac{1+\sqrt{5}}{2} \text{ and } \beta = \frac{1-\sqrt{5}}{2}, \text{ we can verify } \alpha + \beta = 1 \text{ and } \alpha\beta = -1.$$

Its general solution is

$$f_n = A\alpha^n + B\beta^n$$

To find A and B, we have

$$f_n = A\alpha^n + B\beta^n$$
$$f_1 = A\alpha + B\beta = 1$$
$$f_2 = A\alpha^2 + B\beta^2 = 1$$

Solving these two equations, we get

$$A = \frac{a}{1+\alpha} = \frac{(1+\sqrt{5})/2}{(5+\sqrt{5})/2} = \frac{1+\sqrt{5}}{5+\sqrt{5}} = \frac{(1+\sqrt{5})(5-\sqrt{5})}{(5+\sqrt{5})(5-\sqrt{5})}$$
$$= \frac{5+5\sqrt{5}-\sqrt{5}-5}{25-5} = \frac{1}{\sqrt{5}}$$

and similarly, we get

$$B = -\frac{1}{\sqrt{5}}$$

Thus,
$$f_n = \frac{1}{\sqrt{5}}\left[\left(\frac{1+\sqrt{5}}{2}\right)^n - \left(\frac{1-\sqrt{5}}{2}\right)^n\right]$$

which is De Moivre's formula for the Fibonacci sequence.

8.6 LINEAR RECURRENCE RELATIONS WITH CONSTANT COEFFICIENTS

A recurrence relation of the form

$$c_0 a_r + c_1 a_{r-1} + c_2 a_{r-2} + \ldots + c_k a_{r-k} = f(r) \quad \ldots(1)$$

where c_i's are constants, is called a linear recurrence relation with constant coefficients. The recurrence relation in (1) is known as k^{th} order recurrence relation, provided that both c_0 and c_k are non-zero.

For Example.

$$a_r + 3a_{r-1} = 2^r$$

is a first-order linear recurrence relation with constant coefficient. Also, both

$$3a_r - 5a_{r-1} + 2a_{r-2} = r^2 + 5 \quad \ldots(2)$$

and
$$a_r + 7a_{r-2} = 0$$

The second order linear recurrence relation with constant coefficients. Consider the recurrence relation in (2). Suppose we are given that $a_3 = 3$ and $a_4 = 6$. We can compute a_5 as

$$a_5 = -\frac{1}{3}[-5 \times 6 + 2 \times 3 - (5^2 + 5)] = 18$$

We can compute a_6 as

$$a_6 = -\frac{1}{3}[-5 \times 18 + 2 \times 6 - (6^2 + 5)] = \frac{119}{3}$$

and so on.

Also, we can compute

$$a_2 = -\frac{1}{2}[3 \times 6 - 5 \times 3 - (4^2 + 5)] = 9$$

$$a_1 = -\frac{1}{2}[3 \times 3 - 5 \times 9 - (3^2 + 5)] = 25$$

$$a_0 = -\frac{1}{2}[3 \times 9 - 5 \times 25 - (2^2 + 5)] = \frac{107}{2}$$

and so on. We conclude that (2), together with the value $a_3 = 3$ and $a_4 = 6$ completely specifies the discrete numeric function a.

In general, for a kth order linear recurrence relation with constant coefficients as shown in (1), if k consecutive values of the numeric function a, $a_{m-k}, a_{m-k+1}, \ldots a_{m-1}$ are known for some m, the value of a_m can be calculated according to (1), namely

$$a_m = -\frac{1}{c_0}[c_1 a_{m-1} + c_2 a_{m-2} + \ldots + c_k a_{m-k} - f(m)]$$

Also, the value of a_{m+1} can be calculated as

$$a_{m+1} = -\frac{1}{c_0}[c_1 a_m + c_2 a_{m-1} + .. + c_k a_{m-k+1} - f(m+1)]$$

and the values of $a_{m+2}, a_{m+3}, \ldots$ can be computed in a similar manner. Also the value of a_{m-k-1} can be calculated as

$$a_{m-k-1} = -\frac{1}{c_k}[c_0 a_{m-1} + c_1 a_m + \ldots + c_{k-1} a_{m-k} - f(m-1)]$$

and the value of a_{m-k-2} can be calculated as

$$a_{m-k-2} = -\frac{1}{c_k}[c_0 a_{m-2} + c_1 a_{m-2} + \ldots + c_{k-1} a_{m-k-1} - f(m-2)]$$

- For a kth order linear recurrence relation with constant coefficients, fewer than k values of the numeric function will not be sufficient to determine the numeric function uniquely. For example, let

$$a_r + a_{r-1} + a_{r-2} = 4 \qquad \ldots(1)$$

- If we are given that $a_0 = 2$, we can find many numeric functions that will satisfy the recurrence relation as well as the given boundary condition. Thus

  2, 0, 2, 2, 0, 2, 2, 0, 2, 2, 0, ...
  2, 2, 0, 2, 2, 0, 2, 2, 0, 2, 2, ...
  2, 5, –3, 2, 5, –3, 2, 5, –3, 2, ...

  are all possibilities. Yet more than k value of the numeric function might make it impossible for the existence of a numeric function that satisfies the recurrence relation and the given boundary conditions. For example, for the recurrence relation in (1), if we are given that

$$a_0 = 2, a_1 = 2, a_2 = 2$$

  then obviously a_0, a_1 and a_2 do not satisfy the recurrence relation. Consequently, no a can satisfy (1) and the boundary conditions.

8.7 HOMOGENEOUS SOLUTIONS

We know that the solution of a linear difference equation with constant coefficients is the sum of two parts, the homogeneous solution, which satisfies the difference equation when the RHS of the equation is set to 0 and the particular solution, which satisfies the difference equation with $f(r)$ on the RHS. In other words, the discrete numeric function that is the solution of the difference equation is the sum of two discrete numeric functions, one is the homogeneous solution and the other is the particular solution. Let $(a^{(h)}, a_1^{(h)}, \ldots, a_r^{(h)}, \ldots)$ denote the homogeneous solution and $(a_0^b, a_1^b, \ldots, a_r^b, \ldots)$ denote the particular solution to the difference equation. Since

$$c_0 a_r^{(h)} + c_1 a_{r-1}^{(h)} + \ldots c_k a_{r-k}^h = 0 \qquad \ldots(1)$$

and

$$c_0 a_r^b + c_1 a_{r-1}^b + \ldots + c_k a_{r-k}^b = f(r)$$

We have $c_0\,(a_r^{(h)} + a_r^{(p)}) + c_1\,(a_{r-1}^h + a_{r-1}^p) + \ldots + c_k\,(a_{r-k}^h + a_{r-k}^p) = f(r)$

- The total solution, $a = a^h + a^p$ satisfies the difference equation.
- A homogeneous solution of a linear difference equation with constant coefficients is of the form $A\alpha_1^r$, where α_1 is called a characteristic root and A is a constant determined by the boundary conditions. Substituting $A\alpha^r$ for a_r in the difference equation with the RHS of the equation set to 0, we obtain
$$c_0 A\alpha^r + c_1 A\alpha^{r-1} + c_2 A\alpha^{r-2} + \ldots + c_k A\alpha^{r-k} = 0$$
which is called the characteristic equation of the difference equation.
- If α_1 is one of the roots of the characteristic equation. $A\alpha_1^r$ is a homogeneous solution to the difference equation.

WORKING RULES (for solving homogeneous equation)

Step 1. Write the characteristic equation of the given relation.

Step 2. Find all the characteristic roots of the characteristic equation. Let it be $\alpha_1, \alpha_2, \ldots, \alpha_n$.

Step 3. If all roots are distinct, *i.e.*, $\alpha_1 \neq \alpha_2 \neq \ldots \neq \alpha_n$. Then the general solution is $A_1\alpha_1^k + A_2\alpha_2^k \ldots + A_n\alpha_n^k = 0$, where $A_1, A_2, \ldots, A_n$ all are constants and have to be determined by initial conditions.

Step 4. If there is a multiple root, *i.e.*, $\alpha_1 = \alpha_2 \neq \alpha_3 \neq \alpha_4 \ldots \neq \alpha_n$, then solution is $(A_1 + A_2 k)\alpha_1^k + A_3\alpha_3^k + \ldots + A_n\alpha_n^k + \ldots$ (Note that here α_1 is a multiple root).

Step 5. Find the values of the constant $A_1, A_2, \ldots, A_n$ by using the given initial conditions.

Step 6. To get the required solution substitute the values of $A_1, A_2, \ldots, A_n$ in the general solution.

ILLUSTRATIVE EXAMPLES

1. *Solve the recurrence relation* $t_n = 4(t_{n-1} - t_{n-2})$ *subject to initial condition* $t_n = 1$ *for* $n = 0$ *and* $n = 1$.

Solution : The characteristics polynomial for the recurrence relation

$$x^2 - 4x + 4 = 0$$

$\therefore \quad x^2 - 2x - 2x + 4 = 0 \qquad \therefore \quad x(x-2) - 2(x-2) = 0$

$\therefore \quad (x-2)(x-2) = 0$

$x = 2$ with multiplicity 2.

Thus, we obtain the general solution as

$$t_n = n\ 2^n$$

The initial condition give

$$c_1 + 0c_2 = 1 \quad \ldots(1)$$
$$2c_1 + 2c_2 = 1 \quad \ldots(2)$$

Solving these two equations, we get

$$c_1 = 1 \text{ and } c_2 = -1/2$$

Therefore, the solution is $t_n = 2n - n\ 2^{n-1}$

2. *Solve the following recurrence relation :*

$$a_n = 1 \qquad n = 2$$
$$a_n = 2a_{n-1} + 1 \qquad n > 2$$

Solution : The recurrence relation is $a_n = 2a_{n-1} + 1$
substituting for a_{n-1}, we get

$$a_n = 2(2a_{n-2} + 1) + 1 = 2^2\ a_{n-2} + 2 + 1$$

substituting for a_{n-2}, we get

$$= 2^3\ a_{n-3} + 2^2 + 2 + 1$$

After i^{th} substitution, we get

$$= 2^i\ a_{n-i} + 2^{i-1} + 2^{i-2} + \ldots + 2 + 1$$

When $i = n - 1$, we get

$$= 2^{n-1} a_1 + 2^{n-2} + \ldots + 2^2 + 2 + 1$$
$$= 2^{n-1} + 2^{n-2} + \ldots + 2^3 + 2^2 + 2 + 1$$

Hence, $a_n = 2^n - 1$.

3. *Predict a solution for the following recurrence relation :*

$$a_n = 0, \qquad \textit{for } n = 0$$
$$a_n = a_{n-1} + 4n, \qquad \textit{for } n \geq 1.$$

Solution : The recurrence relation is $a_n = a_{n-1} + 4n$ substituting for a_{n-1}, we get

$$a_n = [a_{n-2} + 4(n-1)] + 4n$$
$$= a_{n-2} + 4(n-1) + 4n$$
$$= a_{n-3} + 4(n-2) + 4(n-1) + 4n$$

After i^{th} substitutions, we get

$$a_n = a_{n-i} + 4(n-i+1) + 4(n-i+2) + \ldots + 4(n-1) + 4n$$

For $i = n$, we get

$$t_n = t_0 + 4(1) + 4(2) + \ldots + 4(n-2) + 4(n-1) + 4n$$
$$= 0 + 4(1 + 2 + \ldots + n) = \frac{4n(n+1)}{2}$$

Hence, $a_n = 2n(n+1)$, $n \geq 0$ is a solution of given recurrence relation.

4. *Solve the following recurrence relation :*

$$a_n = 5a_{n-1} - 8a_{n-2} + 4a_{n-3}, \textit{for } n \geq 3$$

subject to the initial conditions $a_n = n$, if $n = 0, 1$ and 2.

Solution : First, let us rewrite the recurrence relation

$$a_n - 5a_{n-1} + 8a_{n-2} - 4a_{n-3} = 0$$

The characteristics polynomial is

$$x^3 - 5x^2 + 8x - 4 = 0$$
$$(x-1)(x-2)^2 = 0$$

The roots are, therefore $x = 1$ and $x = 2$ with multiplicity 2.
The general solution is given by

$$a_n = C_1(1)^n + C_2(2)^n + C_3 n(2)^2$$

That is $a_n = C_1\, 1^n + C_2\, 2^n + C_3\, n \,.\, 2^n.$

By initial conditions, we get

$$C_1 = C_2 = 0 \text{ for } n = 0$$
$$C_1 + 2C_2 + 2C_3 = 1 \text{ for } n = 1$$
$$C_1 + 4C_2 + 8C_3 = 2 \text{ for } n = 2.$$

Solving these three equations, we get $C_1 = -2, C_2 = 2$ and $C_3 = -1/2.$

Therefore, the solution is $a_n = 2^{n+1} - n2^{n-1} - 2$

5. *Solve the following recurrence relation* $a_n = 7a_{n-1} - 10a_{n-2}$ *subject to the initial conditions*

$$a_n = 5 \quad \textit{for } n = 0$$
$$a_n = 16 \quad \textit{for } n = 1.$$

Solution : Let us rewrite the recurrence $a_n - 7a_{n-1} + 10a_{n-2} = 0$

The characteristics polynomial is $x^2 - 7x + 10 = 0$

Solving this polynomial, we get root as $r_1 = 2$ and $r_2 = 5.$

Thus the general solution is $a_n = C_1 2^n + C_2 5^n$

Using initial conditions, we get

$$C_1 + C_2 = 5, \text{ for } n = 0$$

and $$2C_1 + 5C_2 = 16, \text{ for } n = 1$$

Solving these two equations, we get

$$C_1 = 3, C_2 = 2.$$

Therefore, the solution is $a_n = 3(2)^n + 2(5)^n$

6. *Find solution of recurrence relation* $a_n = -3a_{n-1} - 3a_{n-2} - a_{n-3}$ *with initial conditions*

$$a_0 = 1, \quad \textit{for } n = 0$$
$$a_1 = -2, \quad \textit{for } n = 1$$
$$a_3 = -1, \quad \textit{for } n = 2$$

Solution : Let us rewrite the equation as

$$a_n + 3a_{n-1} + 3a_{n-2} + a_{n-3} = 0.$$

The characteristics polynomial is

$$x^3 + 3x^2 + 3x + 1 = 0$$
$$(x+1)^3 = 0$$

Hence, the root is $r = -1$ with multiplicity of three.

Therefore, the general solution is

$$a_n = C_1 (-1)^n + C_2 (-1)^n \,.\, n + C_3(-1)^2 n^2.$$

We get the following equation by initial conditions

$$C_1 = 1$$
$$-C_1 - C_2 - C_3 = -2$$
$$C_1 + 2C_2 + 4C_3 = -1$$

Solving these three equations, we get

$$C_1 = 1, C_2 = 3 \text{ and } C_3 = -2$$

Hence, the required solution of recurrence is

$$a_n = (1 + 3n - 2n^2)(-1)^n.$$

8.8 TOTAL AND PARTICULAR SOLUTIONS

Definition (1) : *The particular solution of the given linear difference equation is the solution which satisfies the difference equation with* $f(r)$ *on the RHS. This is denoted as* $a^{(p)} = (a_0^{(p)}, a_1^{(p)}, \ldots, a_r^{(p)}, \ldots)$

Definition (2) : *The total solution of the given difference equation is the sum of the homogeneous solution and the particular solution, i.e.,*

$$a = a^{(h)} + a^{(p)}$$

For homogeneous solution, we put $a_r = \alpha^r$ in the RHS of the given linear difference equation with constant coefficient and thus get an equation of the form

$$c_0\alpha^k + c_1\alpha^{k-1} + \ldots + c_k = 0$$

This equation is called the characteristic equation of the recurrence relation. The solution of this equation are called the characteristic roots of the recurrence relation.

Case 1. If $\alpha_1, \alpha_2, .., \alpha_k$ are distinct real roots of the characteristic equation then the homogeneous solution of the linear difference equation is given by

$$a_r^{(h)} = A_1\alpha_1^r + A_2\alpha_2^r + \ldots + A_k\alpha_k^r$$

where $A_1, A_2, \ldots, A_k$ are constants which are to be determined with the help of boundary conditions.

Case 2. If $\alpha_1, \alpha_2, \ldots, \alpha_k$ are distinct roots with multiplicities $m_1, m_2, \ldots, m_k$ respectively. $m_i \geq 1, i = 1, 2, \ldots, k$ and $m_1 + m_2 + \ldots + m_k = k$, then the homogeneous solution of the linear difference equation is given by

$$a_r^{(h)} = (A_{1,0} + A_{1,1}\, r + \ldots + A_{1,m}\, r^{m_1 - 1})\alpha_1^r + (A_{2,0} + A_{2,1}\, r + \ldots + A_{2,m_2}\, r^{m_2 - 1})\alpha_2^r$$
$$+ \ldots + (A_{k,0} + A_{k,1}\, r + \ldots + A_{k,m_k}\, r^{m_k - 1})\alpha_k^r$$

8.8.1 Method of Finding Particular Solution

There is no general method for determining the particular solution of a difference equation. In simple cases, the particular solution can be obtained by the method of inspection.

WORKING RULES

Step 1. If $f(r) = r^n$, then the particular solution is taken in the form

$$P_1r^n + P_2r^{n-1} + \ldots + P_{n+1}$$

where $P_1, P_2, \ldots, P_{n+1}$ are constants which are determined by comparing the coefficients of different power of r.

Step 2. If $f(r) = n^r$, then the general form of the particular solution is assumed to be P^rn. This form is applicable, when n is not a characteristic root.

Step 3. If n is a characteristic root of multiplicity $(m - 1)$, then the general form of the particular solution is assumed to be $r^{m-1}\, pn^r$.

Step 4. If $f(r)$ is the combination of the above two terms, then the corresponding general terms of the particular solution is assumed to be of the form of combination of the above forms.

☛ A characteristic equation of kth degree has k characteristic roots. Suppose the roots of the characteristic equation are distinct. In this case, it is easy to verify that

$$a_r^{(h)} = A_1\alpha^r + A_2\alpha_2^r + \ldots + A_k\alpha_k^r$$

is also a homogeneous solution to the difference equation, where $\alpha_1, \alpha_2, \ldots \alpha_k$ with distinct characteristic roots and $A_1, A_2, \ldots, A_k$ are constants which are to be determined by the boundary conditions.

8.8.2 Form of Particular Solutions for Given RHS of the Recurrence Relation

	Type of R.H.S. of recurrence Relation	Form of P.I.
1.	C, a constant	A, α constant
2.	$C\alpha^n$, and if α is not a characteristic root	$A \cdot \alpha^n$
3.	$C\alpha^n$, and if α is a root of multiplicity m	$An^m \cdot \alpha^n$
4.	$C_0 + C_1k$, a linear function	$A_0 + A_1k$, a linear function
5.	$C_0 + C_1k + C_2k^2 + \ldots + C_mk^m$, an mth degree polynomial	$A_0 + A_1k + A_2k^2 + \ldots + A_mk^m$
6.	Cn^sa^n and if α is not a root of characteristic polynomial	$(A_0 + A_1n + \ldots + A_sn^s)\alpha^n$
7.	(a) $Cn^s\alpha^n$ if α is a root of multiplicity m (b) Cn^s and if 1 is not a root (c) Cn^s and if 1 is a root of multiplicity m	$n^m(A_0 + A_1n + \ldots + A_sn^s)\alpha^n$ $(A_0 + A_1n + \ldots + A_sn^s)$ $n^m(A_0 + A_1n + \ldots + A_sn^s)$

8.9 NON-HOMOGENEOUS LINEAR DIFFERENCE EQUATION

We have two methods to find the particular solution of non-homogeneous difference equations. These are as follows :

(i) Undetermined coefficients method

(ii) E and Δ operator method

8.9.1 Undetermined Coefficients Method

This method is used to find the particular solution of non-homogeneous linear difference equations, whose right hand side term $R(n)$ consists of terms of special forms. In this method, first we assume the general form of the particular solution according to the form of $R(n)$ consisting a number of unknown constant coefficients, which have to be determined.

8.9.2 E and Δ Operator Method

Consider the recurrence relation

$$a_0\,E^2y_n + a_1Ey_n + a_2y_2 = \phi(n)$$

$$(a_0E^2 + a_1E + a_2)\,y_n = \phi(n)$$

$$f(E)\cdot y_n = \phi(n)$$

$$\Rightarrow \qquad \text{P. I.} = \frac{1}{f(E)}\cdot\phi(n)$$

Case (1) : When $\phi(n) = a^n$

(a) $\text{P. I.} = \frac{1}{f(E)}\,a^n$. Put $E = a$

$$= \frac{a^n}{f(a)} \text{ if } f(a) \neq 0$$

(b) If $f(a) = 0$

(i) $\quad (E - a)\,y_n = a^n$

$$\text{P. I.} = \frac{1}{E-a}\,a^n + na^{n-1}$$

(ii) $\quad (E - a)^2\,y_n = a^n$

$$\text{P. I.} = \frac{1}{(E-a)^2} a^n = \frac{n(n-1)}{1!} a^{n-2}$$

(iii) $(E-a)^m y_n = a^n$

$$\text{P. I.} = \frac{1}{(E-a)^n} a^n = \frac{n(n-1)\ldots(n-\overline{m-1})}{m!} a^{n-m}$$

Case (2) : When $\phi(n) = e^{ak} = \beta^k$, where $\beta = e^a$

$$\text{P. I.} = \frac{1}{\phi(E)} e^{ak} = \frac{1}{\phi(E)} \beta^k = \frac{1}{\phi(\beta)} \beta^k$$

Case (3) : When $\phi(n)$ = constant $\frac{1}{\phi(E)} = A \cdot \frac{1}{\phi(1)}$

Case (4) : When $\phi(n) = n^P$

Particular solution $= \frac{1}{f(E)} \cdot n^P = \frac{1}{f(1+\Delta)} n^P = f(1+\Delta)^{-1} n^P$.

Case (5) : When $\phi(n) = a^n \psi(n)$, where $\psi(n)$ is a polynomial.

Then, Particular solution $= \frac{1}{f(E)} a^n \psi(n) = a^n \frac{1}{f(aE)} \psi(n)$

ILLUSTRATIVE EXAMPLES

1. *Find the particular solution of the difference equation*

$$a_{r+2} - 2a_{r+1} + a_r = 3r + 5.$$

Solution : $a_{r+2} - 2a_{r+1} + a_r = 3r + 5$ ...(1)

The homogeneous solution of the recurrence relation is $a_r = C_1 + C_2 r$

Corresponding to the term $3r + 5$, we assume the general form of the solution as $A_1 r + A_2$, but due to occurrence of these terms in homogeneous solution of (1), we multiply this by suitable power of r so that none of the term will occur in the homogeneous solution of (1). Thus multiply by r^2.

Hence, the general form of the solution becomes

$$= A_1 r^3 + A_2 r^2 + \ldots \quad \ldots(2)$$

Putting this solution in LHS of equation (1), we get

$$= A_1 (r+2)^3 + A_2 (r+2)^2 - 2A_1 (r+1)^3 - 2A_2 (r+1)^2 + A_1 r^3 + A_2 r^2$$

$$= A_1 (r^3 + 8 + 6r^2 + 12r) + A_2 (r^2 + 4 + 4r) - 2A_1 (r^3 + 1 + 3r^2 + 3r)$$
$$- 2A_2 (r^2 + 1 + 2r) + A_1 r^3 + A_2 r^2$$

$$= (12A_1 + 4A_2 - 6A_1 + 4A_2) r + (8A_1 + 4A_2 - 2A_1 - 2A_2) \quad \ldots(3)$$
$$= (6A_1) r + (6A_1 + 2A_2).$$

Equating equation (3) with LHS of equation (1), we get

$$6A_1 = 3 \quad (\because A_1 = 1/2)$$
$$6A_1 + 2A_2 = 5 \quad (\because A_1 = 1)$$

Therefore, the particular solution is $\frac{1}{2} r^3 + r^2$

2. *Find the particular solution of the difference equation* $a_{r+2} - 5a_{r+1} + 6a_r = 5^r$.

Solution : We have $a_{r+2} - 5a_{r+1} + 6a_r = 5^r$ ...(1)

The general form of the solution $= A \cdot 5^r$.

Find the value of A, put this solution on LHS of the equation (1), then this becomes

$$= A \cdot 5^{r+2} - 5A \cdot 5^{r+1} + 6 \cdot A \cdot 5^r$$

$$= 25\,A.5^r - 25A.5^r + 6A.5^r$$

$$= 6A.5^r.$$

Equating equation (1) to RHS of equation (1), we get

$$A = 1/6.$$

Therefore, the particular solution of the recurrence equation is $= \frac{1}{6}.\,5^r$

3. *Solve the recurrence relation* $a_{r+2} - 5a_{r+1} + 6a_r = 5^r$.

Solution : We have $a_{r+2} - 5a_{r+1} + 6a_r = 5^r$ ...(1)

$$(E^2 - 5E + 6)\,a_r = 5^r$$

The characteristic equation is

$$x^2 - 5x + 6 = 0$$

$$x^2 - 3x - 2x + 6 = 0$$

$$x(x-3) - 2(x-3) = 0$$

$$(x-3)(x-2) = 0$$

Homogeneous solution is $C_1 2^r + C_2 3^r$.

$$\text{P.I.} = \frac{1}{E^2 - 5E + 6}.\,5^r = \frac{1}{25 - 25 + 6}\,5^r = \frac{1}{6}\,5^r \qquad (\because E = 5)$$

P.I. = Particular solution is $\frac{1}{6}\,5^r$.

4. *Solve the recurrence relation* $u_{n+2} - 7u_{n+1} + 10u_n = 12e^{3n} + 4^n$.

Solution : We have $u_{n+2} - 7u_{n+1} + 10u_n = 12e^{3n} + 4^n$

$$(E^2 - 7E + 10)\,u_n = 12\,e^{3n} + 4^n$$

The characteristic equation is $x^2 - 7x + 10 = 0$

$\Rightarrow \quad x^2 - 5x - 2x + 10 = 0 \qquad \Rightarrow \quad x(x-5) - 2(x-5) = 0$

$\Rightarrow \quad (x-5)\,(x-2) = 0 \qquad \Rightarrow \quad x = 2, x = 5.$

The homogeneous solution is $C_1(2)^n + C_2(5)^n$

Particular solution is

$$\text{P.I.} = \frac{1}{E^2 - 7E + 10}\,(12e^{3n} + 4^n)$$

$$= 12.\frac{1}{E^2 - 7E + 10}\,e^{3n} + \frac{1}{E^2 - 7E + 10}\,4^n$$

$$= 12.\frac{e^{3n}}{e^6 - 7e^3 + 10} - \frac{4^n}{2}$$

$$\text{P.I.} = \frac{12\,e^{3n}}{e^6 - 7e^3 + 10} - \frac{4^n}{2}.$$

5. *Solve the recurrence relation* $y_{n+2} - 4y_{n+1} + y_n = 3$

Solution : We have $y_{n+2} - 4y_{n+1} + y_n = 3$

$$(E^2 - 4E + 1)\,y_n = 3.$$

$$\text{P.I.} = \frac{1}{(E^2 - 4E + 1)}.\,3 = 3.\frac{1}{(E^2 - 4E + 1)}.\,1^n = 3.\frac{1}{(1 - 4 + 1)}.\,1 = -\frac{3}{2}$$

Particular solution $= -\frac{3}{2}$

6. *Find the particular solution of the following recurrence relation $y_{n+1} - y_n = n^2$.*

Solution : We have $y_{n+1} - y_n = n^2$

$$(E - 1)\, y_n = n^2$$

Particular solution $= \dfrac{1}{(E - 1)} \cdot n^2 = \dfrac{1}{(1 + \Delta - 1)} n^2 = \dfrac{1}{\Delta} n^2$

$$= \frac{1}{\Delta}[n\,(n-1) + n]$$

$$= \frac{1}{\Delta}[n^{(2)} + n^{(1)}] = \frac{n^{(3)}}{3} + \frac{n^{(2)}}{2}$$

$$= \frac{1}{3} n\,(n-1)\,(n-2) + \frac{n(n-1)}{2}$$

$$= \frac{n\,(n-1)}{6}[2\,(n-2) + 3]$$

$$= \frac{n\,(n-1)(2n-1)}{6}$$

Particular solution $= \dfrac{n(n-1)(2n-1)}{6}$

7. *Find the particular solution of the following recurrence relation $y_{n+2} - 2y_{n+1} + y_n = 2^n \cdot n^2$*

Solution : We have $y_{n+2} - 2y_{n+1} + y_n = 2^n \cdot n^2$ ...(1)

$$(E^2 - 2E + 1) = 2^n \cdot n^2$$

Particular solution $= \dfrac{1}{(E^2 - 2E + 1)} 2^n\, n^2 = \dfrac{1}{(E-1)^2} 2^n \cdot n^2 = 2^n \cdot \dfrac{1}{(2E - 1)} n^2\, (E \to 2E)$

$$= 2^n \cdot \frac{1}{(2 + 2\Delta - 1)^2} n^2 = 2^n \frac{1}{(2\Delta + 1)^2} n^2$$

$$= 2^n \cdot \frac{1}{(1 + 2\Delta)^2} n^2 = 2^n\,(1 + 2\Delta)^{-2}\,[n\,(n-1) + n]$$

$$= 2^n \left[1 - 4\Delta + \frac{(-2)(-3)}{2} 4\Delta^2 + \ldots\right] \{[n]^2 + [n]\}$$

$$= 2^n\,[[n]^2 - [n] = 4 \times 2[n] - 4 + 12 \times 2]$$

$$= 2^n\,[n(n-1) + n - 8n - 4 + 24] = 2^n\,[n^2 - 8n + 20]$$

Particular solution $= 2^n\,[n^2 - 8n + 20]$

8. *Solve the recurrence relation $a_r + 4a_{r-1} + 4a_{r-2} = r^2 - 3r + 5$.*

Solution : We have $a_r + 4a_{r-1} + 4a_{r-2} = r^2 - 3r + 5$

$$(E^2 + 4E + 4)\, a_r = r^2 - 3r + 5$$

The characteristic equation is $x^2 + 4x + 4 = 0$

$\Rightarrow \quad (x + 2)^2 = 0 \qquad \Rightarrow \quad x = -2, -2.$

Homogeneous solution is $a_{r(n)} = (C_1 + C_2 r) \cdot (-2)^r$

$$(E^2 + 4E + 4)\, a_r = r^2 - 3r + 5$$

Particular solution $= \dfrac{1}{(E^2 + 4E + 4)}[(r)^2 - 3(r) + 5]$

$$= \frac{1}{(E+2)^2}(r^2 - 3r + 5)$$

$$= \frac{1}{(3+\Delta)^2}[[r]^2 - 2[r] + 5]$$

$$= \frac{1}{9}\frac{1}{\left(1+\frac{\Delta}{3}\right)^2}[[r]^2 - 2[r] + 5]$$

$$= \frac{1}{9}\left(1+\frac{\Delta}{3}\right)^{-2}([r]^2 - 2[r] + 5)$$

$$= \frac{1}{9}\left(1-\frac{2\Delta}{3}+\frac{3\Delta^2}{9}\right)([r]^2 - 2[r] + 5)$$

$$= \frac{1}{9}\left[[r]^2 - \frac{10}{3}[r] + 7\right]$$

$$= \frac{1}{9}\left(r^2 - \frac{13}{3}r + 7\right)$$

$$a_{r(P)} = \frac{1}{9}\left(r^2 - \frac{13}{3}r + 7\right)$$

Therefore, the solution is $a_r = a_{r(n)} = a_{r(P)}$

$$a_r = (C_1 + C_2 r)(-2)^r + \frac{1}{9}\left(r^2 - \frac{13}{3}r + 7\right)$$

8.10 SOLUTION BY THE METHOD OF GENERATING FUNCTION

Sometimes, if the generating function is determined, an expression for the value of the numeric function can easily be obtained. Therefore, instead of solving a difference equation by an expression for the value of a numeric function, we can also determine the generating function of the given numeric function from the difference equation.

8.10.1 Method of Generating Function

Consider the recurrence relation $a_k + C_1 s_{k-1} + \ldots + C_n s_{k-n} = 0$ where $C_1, C_2, \ldots, C_n$ are constant with $C_n \neq 0$ and $k \geq n$.

Let $S(z) = \sum_{k=0}^{\infty} a_k z^k$

WORKING RULES

Step 1. Multiply each term of the recurrence relation by z^n, summing from n to ∞ and then replace all infinite sums by equivalent expression and transform the recurrence relation into an algebraic equation $S(z) = \dfrac{P(z)}{Q(z)}$

where, $P(z) = a_0 + (a_1 + C_1 a_0) z + (a_2 + C_1 a_1 + C_2 a_0) z^2 + \ldots + (a_{n-1} C_1 a_{n-2} + \ldots + C_{n-1} a_0) z^{n-1}$

and $Q(z) = 1 + C_1 z + C_2 z^2 + \ldots + C_n z^n$

Step 2. After getting the values of $P(z)$ and $Q(z)$, transform $S(z)$ back to obtain the coefficient of a_n in the following two ways :

(a) Factorize $Q(z)$ and then use partial fractions to get $S(z)$ as a sum of familiar series and hence that a_k is the sum of the coefficients of known series.

(b) If the initial and boundary conditions are given, then we can find as many coefficient of $S(z)$ as we desire, even if we can not factorize $Q(z)$ by long division of $Q(z)$ into $P(z)$.

8.10.2 Closed Form Expressions of the Generating Functions for Some Sequences

	Sequence	Generating Function
1.	1	$1/1-z$
2.	$(-1)^n$	$1/1+z$
3.	a^n	$\frac{1}{1-az}$
4.	$(-a)^n$	$\frac{1}{1+az}$
5.	$C(k, n)$	$(1+z)^k$
6.	$C(k-1+n, n)\, a^n$	$\frac{1}{(1-z)^k}$
7.	n	$\frac{z}{(1-z)^2}$
8.	$n+1$	$\frac{1}{(1-z)^2}$
9.	n^2	$\frac{z(1+z)}{(1-z)^3}$
10.	n^3	$\frac{z(1+4z+z^2)}{(1-z)^4}$
11.	na^n	$\frac{az}{(1-az)^2}$
12.	$(n+1)\, a^n$	$\frac{1}{(1-az)^2}$
13.	ba^n	$\frac{b}{(1-az)}$
14.	n^2a^n	$\frac{az(1+az)}{(1-az)^3}$
15.	$n(n+1)$	$\frac{2z}{(1-z)^3}$
16.	$\frac{1}{n\,!}$	e^z
17.	bna^n	$\frac{abz}{(1-az)^2}$

ILLUSTRATIVE EXAMPLES

1. *Use generating function to solve the recurrence relation $a_n = a_{n-1} + n$ with $a_0 = 1$.*

Solution : Multiply both sides of the given recurrence relation by z^n and summing over $n = 1$ to ∞, we get

$$\sum_{n=1}^{\infty} a_n z^n = \sum_{n=1}^{\infty} a_{n-1} z^n + \sum_{n=1}^{\infty} n z^n$$

$$\Rightarrow \quad A(z) - a_0 = zA(z) + \frac{z}{(1-z)^2}$$

where, $\quad A(z) = a_0 + a_1 z + a_2 z^2 + \ldots + a_r z^r + \ldots$

$$\Rightarrow \quad (1-z)\,A(z) = 1 + \frac{z}{(1-z)^2} \qquad (\because a_0 = 1)$$

$$\Rightarrow \quad A(z) = \frac{1}{(1-z)} + \frac{z}{(1-z)^3} \qquad \Rightarrow \quad A(z) = (1-z)^{-1} + z(1-z)^{-3}$$

$$\Rightarrow \quad a_n = 1 + {}^{n+1}C_{n-1}$$

Because coefficient of z^n is $(1-z)^{-1}$ to 1 and in $z(1-z)^{-3}$ is ${}^{n+1}C_{n-1}$. But ${}^{n+1}C_{n-1} = {}^{n+1}C_2$. Therefore, we can write $a_n = 1 + {}^{n+1}C_2$.

2. *Consider the recurrence relation $a_r = 3a_{r-1} + 2$, $r \geq 1$ with boundary condition $a = 1$.*

Solution : Multiply the given equation by z^r, we get

$$a_r z^r = 3a_{r-1} z^r + 2z^r, \; r \geq 1$$

On summing the result, we get

$$\sum_{r=1}^{\infty} a_r z^r = 3 \sum_{r=1}^{\infty} a_{r-1} z^r + 2 \sum_{r=1}^{\infty} z^r \qquad \ldots(1)$$

Now, if $A(z) = \sum_{r=1}^{\infty} a_r z^r$, then

$$\sum_{r=1}^{\infty} a_r z^r = A(z) - a_0 \quad \text{and} \quad \sum_{r=1}^{\infty} a_{r-1} z^r = z \sum_{r=1}^{\infty} a_{r-1} z^{r-1} = z\,A(z)$$

Also, $\sum_{r=1}^{\infty} z^r = \dfrac{z}{1-z}$ (Being the sum of infinite G.P.)

Put all these values in (1), we get

$$A(z) - a_0 = 3zA(z) + \frac{2z}{1-z} \quad i.e., \; (1-3z)\,A(z) = \frac{2z}{1-z} + 1$$

$$\Rightarrow \quad A(z) = \frac{1+z}{(1-3z)(1-z)} = \frac{z}{(1-3z)} - \frac{1}{(1-z)}$$

Hence, we have $a_r = 2 . 3^r - 1$, $r \geq 0$

3. *Solve the recurrence relation $a_r^2 - 2a_{r-1}^2 = 1$ with $a_0 = 2$*

Solution : Let $b^r = a_r^2$. Then given relation becomes $b_r - 2b_{r-1} = 1$ with $b_0 = 4$.

Now, multiplying both sides by z^r and summing from $r = 1$ to ∞, we obtain

$$\sum_{r=1}^{\infty} b_r z^r - 2 \sum_{r=1}^{\infty} b_{r-1} z^r = \sum_{r=1}^{\infty} z^r$$

$$\Rightarrow \quad [B(z) - b_0] - 2z\,B(z) = \frac{z}{1-z} \qquad \Rightarrow \quad (1-2z)\,B(z) = \frac{z}{1-z} + 4 = \frac{4-3z}{1-z}$$

$$\Rightarrow \quad B(z) = \frac{4-3z}{(1-2z)(1-z)} = 5 \cdot \frac{1}{1-2z} - \frac{1}{1-z}$$

$$\Rightarrow \quad b_r = 5 . 2^r - 1 \qquad \therefore \quad a_r = \sqrt{5 . 2^r - 1}$$

4. *Use generating function to solve the recurrence relation* $a_r = a_{r-1} + a_{r-2}$ *with* $a_1 = 2$ *and* $a_2 = 3$.

Solution : Observe that a_0 is not given. Therefore the recurrence relation is not valid for $r = 2$ and is valid only for $r \geq 3$. From the given relation, we have for $r = 2$

$$a_2 = a_1 + a_0$$

If we choose $r_0 = 1$, then the given recurrence relation is valid for $r \geq 2$. In this case, we multiply both sides of the given relation by z^r and summing from $r = 2$ to $r = \infty$, we obtain

$$\sum_{r=2}^{\infty} a_r z^r = \sum_{r=2}^{\infty} a_{r-1} z^r + \sum_{r=2}^{\infty} a_{r-2}\, z^r$$

$$\Rightarrow \quad [A(z) - a_0 - a_1(z)] = z[A(z) - a_0] + z^2 A(z)$$

$$\Rightarrow \quad A(z) = \frac{1+z}{1-z-z^2} \qquad (\because a_0 = 1, a_1 = 2)$$

$$= \frac{1+z}{(1-\alpha\, z)(1-\beta\, z)}$$

where, $\alpha = \frac{1}{2}(1+\sqrt{5})$ and $\beta = \frac{1}{2}(1-\sqrt{5})$

$$\Rightarrow \quad A(z) = \left(\frac{1+\beta}{\alpha-\beta}\right)\left(\frac{1}{1-\alpha z}\right) + \left(\frac{1+\beta}{\beta-\alpha}\right)\left(\frac{1}{1-\beta\, z}\right) \text{ (By partial fraction)}$$

$$\Rightarrow \quad A(z) = \left(\frac{5+3\sqrt{5}}{10}\right)(1-\alpha\, z)^{-1} - \left(\frac{-5+3\sqrt{5}}{10}\right)(1-\beta z)^{-1}$$

$$\Rightarrow \quad a_r = \left(\frac{5+3\sqrt{5}}{10}\right)\alpha^r - \left(\frac{-5+3\sqrt{5}}{10}\right)\beta^r, r \geq 0$$

$$\therefore \quad a_r = \left(\frac{5+3\sqrt{5}}{10}\right)\cdot\left(\frac{1+\sqrt{5}}{2}\right)^r - \left(\frac{-5+3\sqrt{5}}{10}\right)\cdot\left(\frac{1-\sqrt{5}}{2}\right)^r, r \geq 0$$

8.11 GENERAL PROCEDURE FOR DETERMINING THE GENERATING FUNCTION

Consider the difference equation $c_0 a_r + c_1 a_{r-1} + c_2 a_{r-2} + \ldots + c_k a_{r-k} = f(r)$ which is valid for $r \geq s$, where $s \geq k$.

Multiplying both sides of this equation with z^r and summing the result from $r = s$ to $r = \infty$, we get

$$\sum_{r=s}^{\infty} [c_0 a_r + c_1 a_{r-1} + c_2 a_{r-2} + \ldots + c_k a_{r-1}]\, z^r = \sum_{r=s}^{\infty} f(r)\, z^r$$

Now, since $\sum_{r=s}^{\infty} c_0 a_r z^r = c_0\, [(A(z) - a_0 - a_1 z - a_2 z^2 \ldots - a_{s-1} z^{s-1}]$

$$\sum_{r=s}^{\infty} c_1 a_{r-1} z^r = c_1 z\, [A(z) - a_0 - a_1 z - a_2 z^2 - \ldots - a_{s-2} z^{s-2}]$$

... ...

$$\sum_{r=s}^{\infty} c_k a_{r-k} z^r = c_k z^k\, [A(z) - a_0 - a_1 z - a_2 z^2 - \ldots - a_{s-k-1} z^{s-k-1}]$$

We have $A(z) = \dfrac{1}{c_0 + c_1 z + \ldots + c_k z^k}$

$$\left[\sum_{r=s}^{\infty} f(r)\, z^r + c_0(a_0 + a_1 z + a_2 z^2 + \ldots + a_{s-1} z^{s-1}\right]$$

$$+c_1 z\,[a_0 + a_1 z + a_2 z^2 + \ldots + a_{s-2} z^{s-2} + \ldots] + \ldots$$

$$+ c_k z^k\,[a_0 + a_1 z + a_2 z^2 + \ldots + a_{s-k-1} z^{z-k-1}]$$

8.12 SIMULTANEOUS RECURRENCE RELATIONS

Now, we consider a set of simultaneous recurrence relations and solve it using generating functions.

ILLUSTRATIVE EXAMPLES

1. *Use generating function to solve the following simultaneous recurrence relations :*

$$a_r = 3a_{r-1} + 2b_{r-1}$$

$$b_r = a_{r-1} + b_{r-1}, \text{ with } a_0 = 1 \text{ and } b_0 = 0.$$

Solution : Multiplying both equations by z^r and summing from $r = 1$ to ∞, we get

$$\sum_{r=1}^{\infty} a_r . z^r = 3 \sum_{r=1}^{\infty} a_{r-1} z^r + 2 \sum_{r=1}^{\infty} b_{r-1} z^r$$

and $$\sum_{r=1}^{\infty} b_r z^r = \sum_{r=1}^{\infty} a_{r-1} z^r + \sum_{r=1}^{\infty} b_{r-1} z^r$$

$$A(z) - 1 = 3z\, A(z) + 2zB(z) \qquad \ldots(1)$$

and $$B(z) = zA(z) + zB(z) \qquad \ldots(2)$$

Solving (1) and (2), we obtain

$$A(z) = \frac{1-z}{1-4z+z^2} = \frac{3+\sqrt{3}}{6[1-(2+\sqrt{3})\,z]} + \frac{3-\sqrt{3}}{6[1-(2-\sqrt{3})\,z]}$$

and $$B(z) = \frac{z}{1-4z+z^2} = \frac{\sqrt{3}}{6[1-(2+\sqrt{3})\,z]} - \frac{\sqrt{3}}{6[1-(2-\sqrt{3})\,z]}$$

$$a_r = \frac{3+\sqrt{3}}{6}(2+\sqrt{3})^r + \frac{3-\sqrt{3}}{6}(2-\sqrt{3})^r$$

and $$b_r = \frac{\sqrt{3}}{6}(2+\sqrt{3})^r - \frac{\sqrt{3}}{6}(2-\sqrt{3})^r$$

MISCELLANEOUS ILLUSTRATIVE EXAMPLES

1. *Compute first four terms of the sequence defined by each of the following recurrence relations :*

(a) $a_n = a_{n-1}^2$ *for* $n \geq 2$

$a_n = 2$ *for* $n = 0$

(b) $a_n = a_{n-1} - a_{n-3}$ *for* $n \geq 3$

$a_n = 1$ *for* $n = 0$

$a_n = 2$ *for* $n = 1$

$a_n = 0$ *for* $n = 2$

Solution : (a) First four sequences are

$$a_1 = 2$$
$$a_2 = 4$$
$$a_3 = 16$$
$$a_4 = 256$$

(b) We compute first few terms as

$$a_0 = 1$$
$$a_1 = 2$$
$$a_2 = 0$$
$$a_3 = 1$$
$$a_4 = 3$$

2. *Solve the recurrence relation* $a_r = a_{r-1} + 2a_{r-2}$ *with* $a_0 = 2$ *and* $a_1 = 10.$

Solution : Putting $a_n = \alpha^n$ in the given recurrence relation, we have

$$\alpha^r - \alpha^{r-1} - 2\alpha^{r-2} = 0 \qquad \text{or} \qquad \alpha^2 - \alpha - 2 = 0.$$

This is the characteristic equation. Its roots are 2 and –1. Therefore the solution of the given relation is given by

$$a_r = A_1 2^r + A_2(-1)^r$$

where A_1 and A_2 are constants and have to be determined by using given initial conditions. Using the conditions $a_0 = 2$ and $a_1 = 5$, we have

$$2 = A_1 + A_2 \quad \text{and} \quad 10 = 2A_1 - A_2$$

Solving these two equations, we get $A_1 = 4$ and $A_2 = -2$. Hence, the required solution will be

$$a_r = 2^2 2^r - 2(-1)^r = 2^{r+2} - 2(-1)^r$$

3. *Solve the recurrence relation* $9y_{k+2} - 6y_{k+1} + y_k = 0$

Solution : We have $9y_{k+2} - 6y_{k+1} + y_k = 0.$

The characteristic equation is $9x^2 - 6x + 1 = 0$

$$9x^2 - 3x - 3x + 1 = 0$$
$$3x(3x - 1) - 1(3x - 1) = 0$$
$$(3x - 1)(3x - 1) = 0$$
$$x = 1/3, 1/3.$$

Therefore, the homogeneous solution is $y_k = (C_1 + C_2 k)\left(\frac{1}{3}\right)^k$.

4. *Solve the recurrence relation* $a_r - 6a_{r-1} + 8a_{r-2} = 0$

Solution : The characteristic equation $x^2 - 4x - 2x + 8 = 0$

$$x(x - 4) - 2(x - 4) = 0$$
$$(x - 4)(x - 2) = 0$$
$$x = 2, x = 4$$

Therefore, the homogeneous solution of the recurrence relation is

$$a_r = C_1 2^r + C_2 4^r,$$

where C_1 and C_2 are constants.

5. *Solve the difference equation* $a_r = a_{r-1} + a_{r-2}$ *with* $a_0 = 0$ *and* $a_1 = 1.$

Solution : The corresponding characteristic equation is

$$\alpha^2 - \alpha - 1 = 0 \quad \text{(Proceed same as in Example 4)}$$

whose roots will be $\alpha_1 = \dfrac{1+\sqrt{5}}{2}$ and $\alpha_2 = \dfrac{1-\sqrt{5}}{2}$

Therefore the required solution is given by

$$a_r = A_1\left(\frac{1+\sqrt{5}}{2}\right)^r + A_2\left(\frac{1-\sqrt{5}}{2}\right)^r$$

where A_1 and A_2 are constants. Using the initial conditions, we have

$$0 = A_1 + A_2$$

$$1 = A_1\left(\frac{1+\sqrt{5}}{2}\right) + A_2\left(\frac{1-\sqrt{5}}{2}\right)$$

Solving these equations, we get

$$A_1 = \frac{1}{\sqrt{5}} \text{ and } A_2 = -\frac{1}{\sqrt{5}}$$

Hence, the required solution is given by

$$a_r = \frac{1}{\sqrt{5}}\left(\frac{1+\sqrt{5}}{2}\right)^r + \frac{1}{\sqrt{5}}\left(\frac{1-\sqrt{5}}{2}\right)^r$$

6. *Solve the recurrence relation* $a_{r+4} + 2a_{r+3} + 3a_{r+2} + 2a_{r+1} + a_r = 0.$

Solution : We have $a_{r+4} + 2a_{r+3} + 3a_{r+2} + 2a_{r+1} + a_r = 0$

The characteristic equation is $x^4 + 2x^3 + 3x^2 + 2x + 1 = 0$

$$\Rightarrow \quad (x^2+x+1)(x^2+x+1) = 0 \Rightarrow \quad x = \frac{-1 \pm i\sqrt{3}}{2}, \frac{-1 \pm i\sqrt{3}}{2}$$

Therefore, the homogeneous solution is given by

$$a_r = (C_1 + C_2 r)\left(\frac{-1+i\sqrt{3}}{2}\right)^r + (C_3 + C_4 r)\left(\frac{-1-i\sqrt{3}}{2}\right)^r$$

7. *Solve the recurrence relation* $a_r = a_{r-1} + 6a_{r-2}$ *with* $a_0 = 3, a_1 = 6.$

Solution : The corresponding characteristic equation is

$$\alpha^2 - \alpha - 6 = 0$$

The roots of this equation are given by $\alpha_1 = 3, \alpha_2 = -2.$

Therefore, the required solution is given by

$$a_r = A_1 3^r + A_2(-2)^r$$

The constants A_1 and A_2 are determined with the help of given initial conditions. We have

$$3 = A_1 + A_2$$
$$6 = 3A_1 - 2A_2$$

Solving these equations, we get $A_1 = \frac{12}{5}, A_2 = \frac{3}{5}$. Hence, the solution is given by

$$a_r = \frac{12}{5}3^r + \frac{3}{5}(-2)^r$$

8. *Solve the recurrence relation* $y_k - y_{k-1} - y_{k-2} = 0.$

Solution : We have $y_k - y_{k-1} - y_{k-2} = 0.$

The characteristic equation is $x^2 - x - 1 = 0$

$$\Rightarrow \qquad x = \frac{1 \pm \sqrt{1+4}}{2} = \frac{1 \pm \sqrt{5}}{2}$$

Therefore, the homogeneous solution of the equation is

$$y_k = C_1\left[\frac{1+\sqrt{5}}{2}\right]^k + C_2\left[\frac{1-\sqrt{5}}{2}\right]^k.$$

9. *Solve the difference equation* $a_r + 6a_{r-1} + 12a_{r-2} + 8a_{r-3} = 0.$

Solution : The corresponding characteristic equation is

$$\alpha^3 + 6\alpha^2 + 12\alpha + 8 = 0 \quad \text{or} \quad (\alpha+2)^3 = 0$$

This gives that $\alpha = -2$ is a triple characteristic root. Therefore, the required solution is given by

$$a_r = (A_1 r^2 + A_2 r + A_3)(-2)^r.$$

10. *Show that the solution of the difference equation* $y_{n+2} + 2y_{n+1} + 4y_n = 0$ *is*

$$y_n = 2^n A \cos \frac{2n\pi}{3} + 2^n B \sin \frac{2n\pi}{3}$$

where A and B are arbitrary constants.

Solution : The given recurrence relation is $y_{n+2} + 2y_{n+1} + 4y_n = 0$

The characteristic equation is $x^2 + 2x + 4 = 0$

$$\Rightarrow \quad x = \frac{-2 \pm \sqrt{4-16}}{2} = -1 \pm i\sqrt{3}$$

$\Rightarrow \quad y_n = r^n [A \cos n\theta + B \sin n\theta]$ where $r = \sqrt{(-1)^2 + (\sqrt{3})^2} = 2$

and $\theta = \tan^{-1}\left(\dfrac{\sqrt{3}}{-1}\right) = \dfrac{2\pi}{3}$

Hence $y_n = 2^n \left[A \cos \dfrac{2n\pi}{3} + B \dfrac{2n\pi}{3} \right]$

11. *Solve the recurrence relation* $9a_r - 6a_{r-1} + a_{r-2} = 0$ *satisfying the conditions* $a_0 = 0$ *and* $a_1 = 2$.

Solution : We have $9a_r - 6a_{r-1} + a_{r-2} = 0$

The characteristic equation is $9x^2 - 6x + 1 = 0 \qquad \Rightarrow \quad (3x-1)^2 = 0$

$$x = 1/3, 1/3.$$

Therefore, the homogeneous solution is given by

$$a_r = (C_1 + C_2 r)\left(\frac{1}{3}\right)^r$$

Putting $r = 0$ and $r = 1$, in equation (1), we get

$$a_0 = C_1 = 0$$

$$a_1 = (C_1 + C_2)\frac{1}{3} = 2$$

$$C_1 + C_2 = 6$$

$$C_2 = 6.$$

Hence, $a_r = 6r\left(\dfrac{1}{3}\right)^r$.

12. *Solve the difference equation* $a_r + 5a_{r-1} + 6a_{r-2} = 4r^2 - 6r + 1$.

Solution : This is a non-homogeneous linear difference equation. Its corresponding characteristic equation is given by $\alpha^2 + 5\alpha + 6 = 0$

Its characteristic roots are $\alpha = -2, -3$. Therefore, the homogeneous solution is

$$a_r^{(h)} = A_1(-2)^r + A_2(-3)^r \qquad \ldots(1)$$

For particular solution, we take $a_r = P_1 r^2 + P_2 r + P_3$

Using this value of a_r in the given difference equation, we get

$P_1 r^2 + P_2 r + P_3 + 3[P_1(r-1)^2 + P_2(r-1) + P_3] + 6[P_1(r-2)^2 + P_2(r-2) + P_1] = 4r^2 - 6r + 1$

or $\quad 12P_1 r^2 + (P_2 - 10P_1 + 5P_2 - 24P_1 + 6P_2)\, r + P_3 + 5P_1 - 5P_2$

$$+ 5P_3 + 24P_1 - 12P_2 + 6P_3 = 4r^2 - r + 1$$

or $\quad 12P_1 r^2 + (12P_2 - 34P_1)r + 12P_3 - 17P_2 + 29P_1 = 4r^2 - r + 1$

This gives $12P_1 = 4$, $12P_2 - 34P_1 = -6$, $12P_3 - 17P_2 + 24P_1 = 1$

Solving these equations we get

$$P_1 = 1/3, P_2 = 8, P_3 = 4$$

$$\therefore \qquad a_r^{(p)} = 1/3 \,.\, r^2 + 8.r + 4 \qquad \ldots(2)$$

From (1) and (2), we have

$$a_r = a_r^{(h)} + a_r^{(p)} = A_1(-2)^r + A_2(-3)^r + 1 / 3r^2 + 8r + 4$$

13. *Find the solution of the recurrence relation $a_r = 3a_{r-1} + 2r$.*

Solution : This is a non-homogeneous linear difference equation. So, we have to obtain its homogeneous solution as well as particular solution. Its corresponding characteristic equation is given by $\alpha - 3 = 0$

This gives $\alpha = 3$ as the characteristic root. Therefore homogeneous solution is

$$a_r^{(h)} = A_1 . 3^r \quad ...(1)$$

For particular solution, we take $a_r = P_1 r + P_2$ where P_1 and P_2 are constants. Using this value of a_r in the given difference equation, we have

$$P_1 r + P_2 - 3(P_1(r-1) + P_2) = 2r \quad \text{or} \quad -2P_1 r + (3P_1 - 2P_2) = 2r$$

Comparing the coefficient of r and constant terms, we get

$$-2P_1 = 2 \quad \text{and} \quad 3P_1 - 2P_2 = 0$$

This gives $P_1 = -1$ and $P_2 = -3/2$. Therefore

$$a_r^{(p)} = -r - \frac{3}{2} \quad ...(2)$$

From (1) and (2), we conclude that $a_r = a_r^{(h)} + a_r^{(p)} = A_1 3^r - r - \dfrac{3}{2}$

14. *Solve the difference equation $a_r = 6a_{r-1} - 11a_{r-2} + 6a_{r-3}$ with the initial conditions $a_0 = 1$, $a_1 = 2$ and $a_3 = 6$.*

Solution : The characteristic equation of the given difference equation is

$$\alpha^3 - 6\alpha^2 + 11\alpha - 6 = 0.$$

Its characteristic roots are 1, 2, 3. Therefore the required solution is given by

$$a_r = A_1 1^r + A_2 2^r + A_3 3^r = A_1 + A_2 2^r + A_3 3^r$$

To find the values of constants A_1, A_2 and A_3, we use the initial conditions which give

$$1 = A_1 + A_2 + A_3$$
$$2 = A_1 + 2A_2 + 3A_3$$
$$3 = A_1 + 4A_2 + 9A_3$$

Solving the given equation, we get $A_1 = 1$, $A_2 = -1$ and $A_3 = 1$. Hence the solution of the given difference equation is $a_r = 1 - 2^r + 3^r$.

15. *Solve the difference equation $a_r - 5a_{r-1} + 6a_{r-2} = 7^r$.*

Solution : This is a linear non-homogeneous difference equation. Its corresponding characteristic equation is $\alpha^2 - 5\alpha + 6 = 0$.

Its characteristic roots are $\alpha = 2, 3$. Therefore, the homogeneous solution is

$$a_r^{(h)} = A_1 2^r + A_2 3^r \quad ...(1)$$

For particular solution, we take

$$a_r = P.7^r, \text{ where } P \text{ is a constant.}$$

Using this value of a_r in the given difference equation, we get

$$P(7^r - 5P7^{r-1} - 1) + 6P7^{r-2} = 7^r$$

or $49P - 35P + 6P = 49$ or $20P = 49$ or $P = 49 / 20$

Therefore, $a_r^{(p)} = (49 / 20)\, 7^r$ ...(2)

From (1) and (2), we have $a_r = a_r^{(h)} + a_r^{(p)} = A_1 2^r + A_2 3^3 + (49 / 20)\, 7^r$.

16. *Solve the difference equation $a_r - 6a_{r-1} + 9a_{r-2} = r.3^r$.*

Solution : The characteristic equation is given by $\alpha^2 - 6\alpha + 9 = 0$, $(\alpha - 3)^2 = 0$

This gives that $\alpha = 3, 3$ are the roots of the characteristic equation.

Therefore, the homogeneous solution is given by $a_r^{(h)} = (A_1 r + A_2)\, 3^r$ ...(1)

For particular solution, we take $a_r = r^2(P_1 r + P_2)\, 3^r$

because the characteristic root 3 is of multiplicity 2, using this value of a_r in the given difference equation, we get

$$(P_1 r^3 + P_2 r^2)3^3 - 6(P_1(r-1)^3 + P_2(r-1)^2)$$

$$3^{r-1} + 9[P_1(r-2)^3 + P_2(r-2)^2]\, 3^{r-2} = r3^r$$

$$\Rightarrow \quad 9[P_1 r^3 + P_2 r^2] - 18[P_1(r^3 - 3r^2 + 3r - 1) + P_2(r^2 - 2r + 1)]$$
$$+ 9[P_1(r^3 - 6r^2 + 12r - 8) + P_2(r^2 - 4r + 4)] = 9r$$

$$\Rightarrow \quad 6P_1 r - 6P_1 + 2P_2 = r$$

This gives

$$6P_1 = 1$$
$$-6P_1 + 2P_2 = 0$$

or $$P_1 = \frac{1}{6},\ P_2 = \frac{1}{2}$$

Therefore, $$a_r^{(p)} = r^2\left[\frac{1}{6}r + \frac{1}{2}\right] 3^r \qquad ...(2)$$

Hence, from (1) and (2), we conclude that

$$a_r = a_r^{(h)} + a_r^{(p)} = (A_1 r + A_2)\, 3^r + r^2\left(\frac{1}{6}r + \frac{1}{2}\right) 3^r.$$

17. *Find the particular solution of the difference equation* $a_r + a_{r-1} = 5r\, 2^r$.

Solution : Let the general form of the particular solution is

$$a_r = (rP_1 + P_2)\, 2^r.$$

Substituting this value of a_r in the given difference equation, we get

$$(rP_1 + P_2)\, 2^r + ((r-1)\, P_1 + P_2)\, 2^{r-1} = 5r\, 2^r \quad \text{or} \quad 3rP_1 + (3P_2 - P_1) = 10^r$$

Comparing both sides, we get

$$3P_1 = 10 \quad \text{and} \quad 2P_2 - P_1 = 0$$

This gives $$P_1 = \frac{10}{3} \text{ and } P_2 = \frac{10}{9}.$$

Therefore, $$a_r^{(p)} = \frac{10}{9}(3r - 1)\, 2^r$$

18. *Find the general solution of* $a_r - 4a_{r-1} + 4a_{r-2} = 2^r + r.2^r$.

Solution : The given recurrence relation is $a_r - 4a_{r-1} + 4a_{r-2} = 2^r + r.2^r$ ...(1)

whose homogeneous equation is $a_r - 4a_{r-1} + 4a_{r-2} = 0$

The characteristic equation is $\alpha^2 - 4\alpha + 4 = 0 \Rightarrow \alpha = 2, 2$

$\therefore$ The homogeneous solution is $a_r^{(h)} = (B_1 + B_2 r)\, 2^r$

To find the particular solution, let us assume $a_r^{(p)} = r^2(A_1 r + A_2)\, 2^r$

Put this value in (1), we get

$$r^2(A_1 r + A_2)\, 2^r - 4(r-1)^2\{A_1(r-1) + A_2\}\, 2^{r-1}$$
$$+ 4(r-1)^2\{A_1(r-2) + A_2\}\, 2^{r-2} = (r+1)\, 2^r$$

After simplification, we get

$$6A_1 r\, 2^r = r\, 2^r \quad \text{and} \quad (-6A_1 + 2A_2)\, 2^r = 2^r$$

which gives $$A_1 = \frac{1}{6} \text{ and } A_2 = 1$$

$\therefore$ The particular solution is $a_r^{(p)} = r^2\left(\frac{r}{6}+1\right)2^r$

Hence, the complete solution is $a_r = (B_1 + B_2 r)\, 2^r + r^2\left(\frac{r}{6}+1\right)2^r$

19. *Solve* $a_r - 6a_{r-1} + 8a_{r-2} = r\, 4^r$.

Solution : The associated homogeneous relation of the given equation is

$$a_r - 6a_{r-1} + 8a_{r-2} = 0 \quad \ldots(1)$$

The characteristic equation is $\alpha^2 - 6\alpha + 8 = 0 \Rightarrow \alpha = 2, 4 \;\therefore\; \alpha_r^{(h)} = B_1 2^r + B_2 4^r$

Now to find particular solution.

Let us assume $a_r^{(p)} = r(A_0 + A_1 r)\, 4^r$ as R.H.S. involves an exponential function, *i. e.*, $r\, 4^r$.

Put $a_r^{(p)}$ in (1), we get

$$r(A_0 + A_1 r)\, 4^r - 6(r-1)(A_0 + A_1(r-1))\, 4^{r-1} + 8\,(r-2)(A_0 + A_1(r-2)\, 4^{r-2}) = r\, 4^r$$

$$\Rightarrow \quad 4^{r-2}[16r(A_0 + A_1 r) - 24(r-1)(A_0 + A_1)(r-1)]$$

$$+ 8(r-2)(A_0 + A_1(r-2)) = r\, 4^{r-2}\,.\,4^2$$

$$\Rightarrow \quad 16r(A_0 + A_1 r)\, 4^r - 24(r-1)(A_0 + A_1(r-1)) + 8\,(r-2)(A_0 + A_1(r-2)\,) = 16r$$

Examples Based on Generating Functions

20. *Solve the recurrence relation* $a_r = 2a_{r-1} + 3, r \geq 1$ *with the initial condition* $a_0 = 1$.

Solution : Let $G(x)$ be the generating function for the numeric function a_r such that

$$G(x) = \sum_{r=0}^{\infty} a_r\, x^r \quad \ldots(1)$$

$$\Rightarrow \quad x\,G(x) = x \sum_{r=0}^{\infty} a_r\, x^r = \sum_{r=0}^{\infty} a_r x^{r+1} = \sum_{r=1}^{\infty} a_{r-1}\, x^r \quad \ldots(2)$$

Multiplying both sides of given recurrence relation by x^r and taking summation, we have

$$\sum_{r=1}^{\infty} a_r x^r = 2\sum_{r=1}^{\infty} a_{r-1} x^r + 3\sum_{r=1}^{\infty} x^r$$

$$G(x) - a_0 = 2xG(x) + 3\,[x + x^2 + x^3 + \ldots] \quad \text{[Using (1) and (2)]}$$

$$G(x) - 1 = 2xG(x) + \frac{3x}{1-x} \quad [\because a_0 = 1]$$

$$G(x)\,[1 - 2x] = 1 + \frac{3x}{1-x}$$

$$G(x) = \frac{1}{1-2x} + \frac{3x}{(1-x)(1-2x)}$$

$$G(x) = \frac{1}{1-2x} + \frac{3}{1-2x} - \frac{3}{1-x} \quad \text{(By partial fraction)}$$

$$G(x) = \frac{4}{1-2x} - \frac{3}{1-x}$$

$$G(x) = 4\,[(1-2x)^{-1}] - 3(1-x)^{-1}$$

Now, general term of $(1-2x)^{-1} = 2^r\, x^r$

and general term of $(1-x)^{-1} = x^r$

Therefore, $\quad G(x) = 4\sum_{r=0}^{\infty} 2^r\, x^r - 3\sum_{r=0}^{\infty} x^r$

Hence, $a_r = 4.2^r - 3$.

21. *Solve the difference equation* $a_r - 5a_{r-1} + 6a_{r-2} = 2^r + r,\ r \geq 2$ *with the initial condition* $a_0 = 1,\ a_1 = 1.$

Solution : The given recurrence relation is $a_r - 5a_{r-1} + 6a_{r-2} = 2^r + r, r \geq 2$

Let $\quad G(x) = \sum_{r=0}^{\infty} a_r x^r = a_0 + a_1 x + \sum_{r=2}^{\infty} a_r\, x^r$

$\Rightarrow \quad xG(x) = \sum_{r=0}^{\infty} a_r x^{r+1} = \sum_{r=1}^{\infty} a_{r-1}\, x^r$ and $\quad x^2G(x) = \sum_{r=2}^{\infty} a_{r-2} x^r$

Multiplying the given recurrence relation by x^r and summing the results from $r = 2$ to ∞

($\because$ relation is valid for $r \geq 2$).

$$\sum_{r=2}^{\infty} a_r x^r - 5 \sum_{r=2}^{\infty} a_{r-1} x^r + 6 \sum_{r=2}^{\infty} a_{r-2} = \sum_{r=2}^{\infty} 2^r x^r + \sum_{r=2}^{\infty} r.\, x^r$$

$$[G(x) - a_0 - a_1(x)] - 5x\,[G(x) - a_0] + 6\,[x^2G(x)] = \sum_{r=2}^{\infty} 2^r\, x^r + \sum_{r=2}^{\infty} r.\, x^r$$

$$[G(x) - 1 - r] - 5x\,[G(x) - 1] + 6x^2G(x) = \sum_{r=2}^{\infty} 2^r x^r + \sum_{r=2}^{\infty} r\; x^r \qquad (\because a_0 = 1 = a_1)$$

$$G(x)[1 - 5x + 6x^2] - 1 - x + 5x = \frac{4x^2}{1-2x} + \frac{2x^2 - x^3}{(1-x)^2}$$

$$\sum_{r=2}^{\infty} a_r x^r = 2x^2 + 3x^3 + 4x^4 + 5x^5 + \ldots$$

$$G(x) = 2x^2 + 3x^3 + 4x^4 + 5x^5 + \ldots$$

$$\Rightarrow \quad G(x)\,.\, x = 2x^3 + 3x^4 + 4x^5 + \ldots$$

Therefore, $G(x)[1 - x] = 2x^2 + x^3 + x^4 + x^5 + \ldots$

$$\Rightarrow \quad G(x)[1 - x] = 2x^2 + \frac{x^3}{1-x} \quad \text{(Being the sum of infinite G.P.)}$$

$$\Rightarrow \quad G(x) = \frac{2x^2}{1-x} + \frac{x^3}{1-x^2}$$

$$\frac{2x^2(1-x) + x^3}{(1-x)^2} = \frac{2x^2 - 2x^3 + x^3}{(1-x)^2} = \frac{2x^2 - x^3}{(1-x)^2} = \frac{x^2\,(2-x)}{(1-x)^2}$$

$$G(x)\,[(1-3x)(1-2x)] - 1 + 4x = \frac{4x^2}{1-2x} + \frac{2x^2 - x^3}{(1-x)^2}$$

$$G(x)[(1-3x)(1-2x)] = 1 - 4x + \frac{4x^2}{1-2x} + \frac{2x^2 - x^3}{(1-x)^2}$$

$$G(x) = \frac{1-4x}{(1-3x)(1-2x)} + \frac{4x^2}{(1-2x)^2(1-3x)} + \frac{2x^2 - x^3}{(1-x)^2(1-3x)(1-2x)}$$

$$= \frac{(1-4x)(1-2x)(1-x)^2 + 4x^2(1-x)^2 + (2x^2 - x^3)(1-2x)}{(1-3x)(1-2x)^2(1-x)^2}$$

$$= \frac{(1-4x)(1-2x)(1-x)^2 + 4x^2(1-x)^2 + (2x^2 - x^3)(1-2x)}{(1-3x)(1-2x^2)(1-x)^2}$$

$$= \frac{1 - 8x + 27x^2 - 35x^3 + 14x^4}{(1-x)^2(1-2x)^2(1-3x)}$$

$$= \frac{5/4}{1-x} + \frac{1/2}{(1-x)^2} - \frac{3}{1-2x} - \frac{2}{(1-2x)^2} + \frac{17/4}{(1-3x)}$$

(By resolving into partial fraction)

Therefore, we conclude that $a_r = \frac{5}{4} + \frac{1}{2}(r+1) - 3.2^r - 2(r+1).2^r + \frac{17}{4} 3r^r$.

22. *Solve the difference equation $a_{r+2} - 3a_{r+1} + 2a_r = 0$, with the initial condition $a_0 = 2$, $a_1 = 3$.*

Solution : Here, the given equation is $a_{r+2} - 3a_{r+1} + 2a_r = 0$

Let $G(x) = \sum_{r=0}^{\infty} a^r x^r \Rightarrow [G(x) - a_0 - a_1(x)] - 3x[G(x) - a_0] + 2x^2 G(x) = 0$

$\Rightarrow \quad G(x)[(1-x)(1-2x)] = 2 - 3x$

$$G(x) = \frac{2-3x}{(1-x)(1-2x)} = \frac{1}{(1-x)} + \frac{1}{(1-2x)}$$

$$G(x) = \sum_{r=0}^{\infty} a^r + \sum_{r=0}^{\infty} 2^r x^r$$

$a_r = 1 + 2^r$.

Exercise 8.1

1. Solve the following recurrence relations :

(i) $a_n = a_{n-1} - a_{n-2};\ n \geq 2,\ a_0 = 3,\ a_1 = 5$.

(ii) $a_n = a_{n-1} - n;\ n \geq 1,\ a_0 = 0$.

(iii) $a_n = 3a_{n-1} + 2;\ n \geq 1,\ a_0 = 1$

(iv) $a_n = a_{n-1} + a_{n-2};\ n \geq 2,\ a_0 = 1,\ a_1 = 3$

(v) $a_n = -6a_{n-1} - 12a_{n-2} - 8a_{n-3};$ $n \geq 3,\ a_0 = 1,\ a_1 = -2,\ a_2 = 8$.

(vi) $a_n + 2a_{n-1} = n + 3$ if $a_0 = 0,\ n \geq 1$.

(vii) $a_n = 3a_{n-2} - 2a_{n-3};$ $n \geq 3,\ a_0 = 1,\ a_1 = a_2 = 0$

2. Find the first five terms of the sequence defined by $a_r = a_{r-1}^2,\ a_1 = 2$

3. For the discrete function $a_r = 2r + 3$, find the recurrence relation.

4. Find the solution to the recurrence relation $a_r + 3a_{r-1} + 3a_{r-2} + a_{r-3} = 0$, with the initial conditions $a_0 = 0,\ a_1 = -2$ and $a_2 = -1$.

5. Find all solutions of recurrence relation $a_r = 5a_{r-1} - 6a_{r-2} + 7^r$.

6. Solve the following recurrence relations :

(i) $a_r - 7a_{r-1} + 16a_{r-2} - 12a_{r-3} = 0$ for $n \geq 3$ given that $a_0 = 1,\ a_1 = 4,\ a_2 = 8$.

(ii) $a_r - 4a_{r-1} + 4a_{r-2} = 0$ given that $a_0 = 1$ and $a_1 = 6$

(iii) $a_r - 7a_{r-1} + 10a_{r-2} = 3^r$ given that $a_0 = 0$ and $a_1 = 1$.

(iv) $a_r + 6a_{r-1} + 9a_{r-2} = 3$ given that $a_0 = 0$ and $a_1 = 1$.

(v) $a_r + a_{r-1} + a_{r-2} = 0$ ·ven that $a_0 = 0$ and $a_1 = 2$.

(vi) $a_r - a_{r-1} - a_{r-2} = 0$ given that a and $a_1 = 1$.

(vii) $a_r - 2a_{r-1} + 2a_{r-2} - a_{r-3} = 0$ given that $a_0 = 2,\ a_1 = 1$ and $a_2 = 1$.

7. Determine the particular solution for the difference equations :

(i) $a_r - 3a_{r-1} + 2a_{r-2} = 2^r$

(ii) $a_r - 4a_{r-1} + 4a_{r-2} = 2^r$

(iii) $a_r - 2a_{r-1} = f(r)$, where $f(r) = 7r$

(iv) $a_r - 2a_{r-1} = 7r^2$

(v) $a_r - a_{r-1} = 7r$

(vi) $a_r - a_{r-1} = 7r^2$

8. Solve the recurrence relation $a_r + 3a_{r-1} + 2a_{r-2} = f(r)$ given

$$f(r) = \begin{cases} 1, & r = 5 \\ 0, & \text{otherwise} \end{cases}$$

with boundary conditions $a_0 = a_1 = 0$.

❑❑❑

Chapter 9

Graph Theory

9.1 INTRODUCTION

Graph theory is a branch of mathematics which deal the problems with the help of graphs. It has applications in data structures, operation research, computer science, electronics and mechanical engineering and other branches of science. It is an applied branch of mathematics. A graph can be used to represent almost any physical situation involving discrete objects and a relationship among them.

9.2 GRAPHS AND THEIR REPRESENTATION

A graph can be thought of as a drawing or diagram consisting of a collection of vertices (dots or points) together with edges (lines) joining certain pair of these vertices.

Definition : *A graph G consists of following two things :*

(i) A set $V = V(G)$, whose elements are called vertices, points or nodes of G.

(ii) A set $E = E(G)$ of unordered pairs of distinct pairs called edges of G.

Such a graph is denoted by $G(V, E)$.

- Vertices are also sometimes called points, nodes or dots.
- If e is an edge with end vertices u and v, then e is said to join u and v.
- The definition of a graph allows the possibility of the edge having identical end vertices, *i. e.*, it is possible to have a vertex u joined to itself by an edge.

9.2.1 Some Examples on Graphs

(1) Let $G = (V, E)$, where $V = \{a, b, c, d, e\}$,

$e = \{e_1, e_2, \ldots e_8\}$ and the ends of the edges are given by

$e_1 \leftrightarrow (a, b)$, $e_2 \leftrightarrow (b, c)$, $e_3 \leftrightarrow (c, c)$,

$e_4 \leftrightarrow (c, d)$, $e_5 \leftrightarrow (b, d)$, $e_6 \leftrightarrow (d, e)$,

$e_7 \leftrightarrow (b, e)$, $e_8 \leftrightarrow (e, b)$

Then, graph is given by

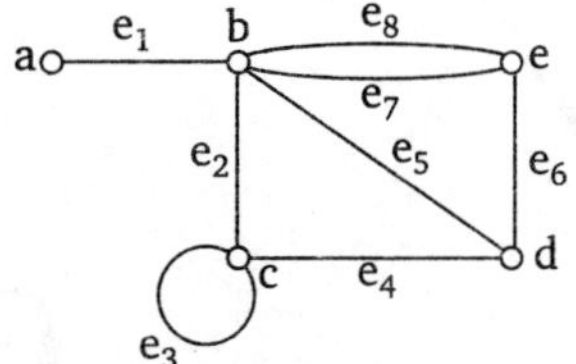

Fig. (1)

(2) The following figure is a graph with five vertices and seven edges :

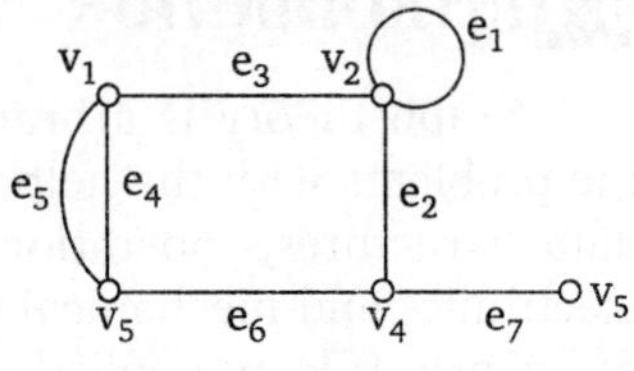

Fig. (2)

9.2.2 Formal definition of a Graph

The definition of graph given above works quite well if we have the visual representation of the graph before us to show which edges connect which vertices. Without the picture, however, we need a concise way to convey this information. In this way, we define the graph formally as follows :

Definition : *A graph G is an ordered triplet* (V, E, g), *where*

V = *a non-empty set of vertices*

E = *a set of edges*

g = *a function associating with each edge a an ordered pair* (x, y) *of vertices called the end points of a.*

For Example : Consider the following graph.

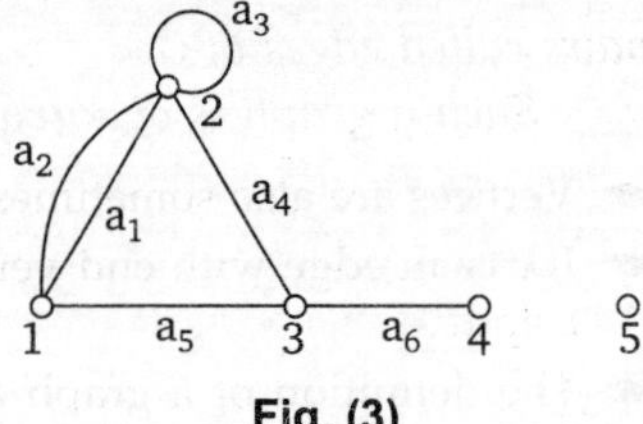

Fig. (3)

In this graph the function g associating edges with end points perform the following mappings

$$g(a_1) = 1 - 2$$
$$g(a_2) = 1 - 2$$
$$g(a_3) = 2 - 2$$
$$g(a_4) = 2 - 3$$
$$g(a_5) = 1 - 3$$

and $$g(a_6) = 3 - 4$$

9.3 MULTIGRAPH

Consider the adjoining graph.

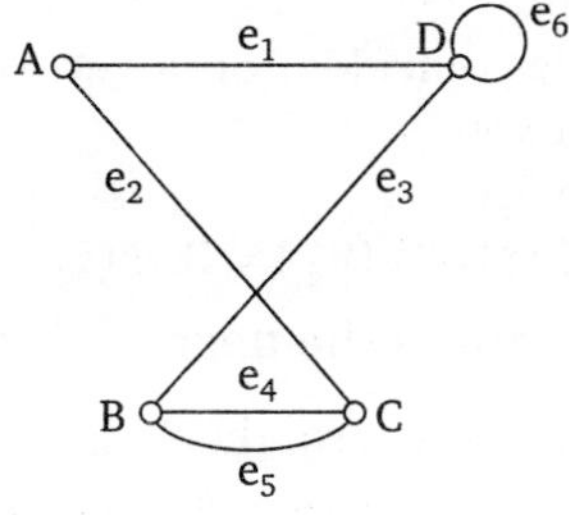

Fig. (4)

In this graph, the edges e_4 and e_5 are known as multiple edges since they connect the same end points and the edge e_6 is called a **loop** since its end points are the same vertex. Such type of diagram is known as multigraph.

9.4 GRAPH TERMINOLOGY

9.4.1 Parallel-edges

More than one edge associated with a given pair of vertices, *i.e.*, if two (or more) edges of a graph *G* have the same end points, then these edges are known as parallel edges.

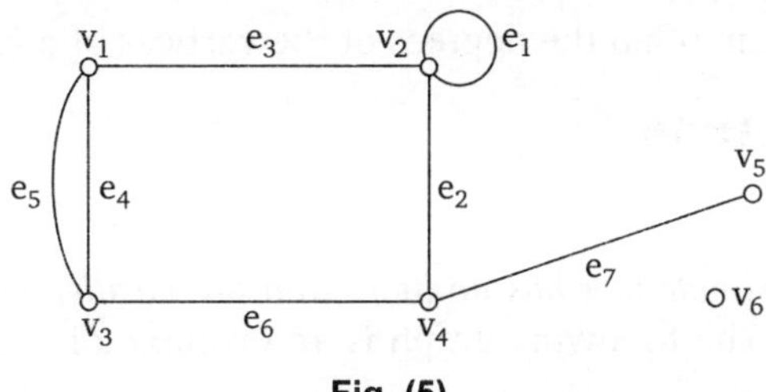

Fig. (5)

For Example : In Figure 5, e_4 and e_5 are parallel edges.

9.4.2 Self-loop

An edge having the same vertex as both its end vertices and edges to be associated with a vertex pair $(v_i\ v_j)$.

For Example : In figure 5, e_1 is self loop.

9.4.3 Isolated Vertex

A vertex of graph *G*, which is not the end of any edge is known as isolated vertex.

For Example : In figure 5, v_6 is an isolated vertex.

9.4.4 Adjacent Vertex

Two vertices which are joined by an edge are called adjacent or neighbours. The set of all such neighbours of vertex v is called neighbourhood set of v and it is denoted by $N(v)$.

For Example : In figure 5, v_3 and v_4 are adjacent.

9.4.5 Incidence and Degree

When a vertex v_i is an end vertex of some edge e_j, v_i and e_j are said to be incident with each other. In Figure 5, e_2, e_6 and e_7 are incident with vertex v_4. Two non-parallel edges are said to be adjacent if they are incident on a common vertex.

For Example : In figure 5, e_2 and e_4 are adjacent.

Further, the number of edges incident on a vertex v_i with self-loop counted twice is called the degree $d(v_i)$ of vertex v_i.

For Example : In figure 5,

$$d(v_1) = d(v_3) = d(v_4) = 3, d(v_2) = 4, d(v_5) = 1.$$

9.4.6 Important Conclusions about the degree of the graph

(1) If the degree of a vertex v is zero, then v is called an isolated vertex of graph G.

(2) If the degree of a vertex v is one, then the vertex v is called a pendent vertex or a leaf and the edge incident with the pendent vertex of G is called a pendent edge.

(3) Let $v_1, v_2, \ldots v_n$ be the vertices of a graph G with $\deg(v_i) \leq deg(v_{i+1})$ for all i $(1 \leq i \leq n)$. Then the sequence $\deg(v_1), \deg(v_2), \ldots, \deg(v_n)$ is called a degree sequence of a graph G and two graphs may have a same degree sequence with different graphical representation.

(4) A vertex of a graph is called odd or even depending on whether its degree is even or odd.

(5) We can define

$\delta(G) = \min\{\deg(v_1) : v_1 \in V\}$

= minimum of all the degrees of the vertices of a graph G and

$\Delta G = \max\{\deg(v_1) : v_1 \in V\}$

= maximum of all the degrees of the vertices of a graph G

9.5 TYPES OF GRAPHS

9.5.1 Simple Graph

A graph is called simple if it has no loops and no parallel edges.

For Example : The following graph is an example of simple graph.

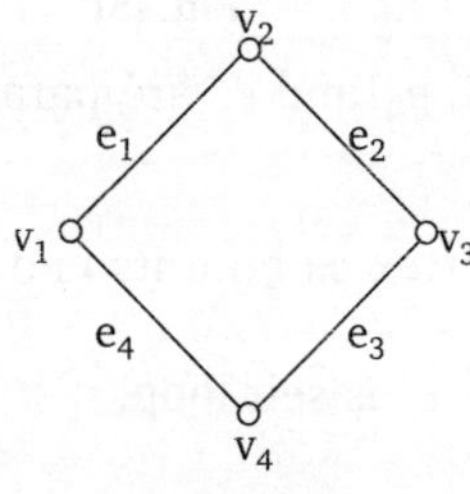

Fig. (6)

9.5.2 Finite and Infinite Graph

A graph with a finite number of vertices as well as a finite number of edges is called a finite graph. A graph which is not finite is known as infinite graph.

For Example :

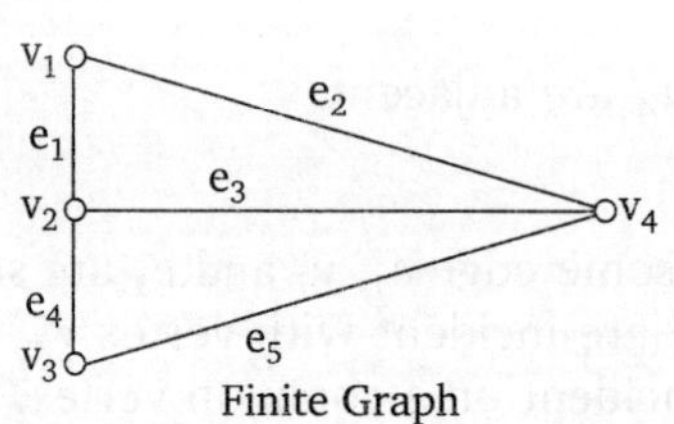

Finite Graph

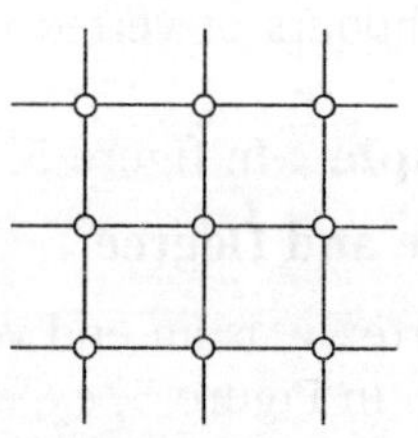

Portion of Infinite Graph

Fig. (7)

9.5.3 Trivial Graph or Null Graph

A finite graph with one vertex and no edges, *i.e.*, a single point is called the trivial graph.

9.5.4 Pseudo Graph

A graph having loops but no multiple edges is known as pseudo graph.

9.5.5 Labelled Graph

A graph G is called a labelled graph if its edges are labelled with name or data. We can write labels in place of an ordered pair in the edge set.

For Example : $G = [\{1, 2, 3, 4, 5\}, \{e_1, e_2, e_3, e_4, e_5\}]$

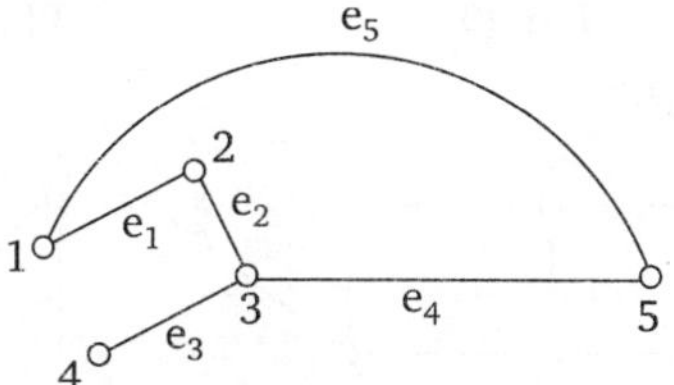

Fig. (8)

9.5.6 *k*-Regular Graph

If for some positive integer k, deg $(v) = k$ for every vertex v of the graph G, then G is called k-regular graph.

Facts about the *k*-regular Graph

(1) A 3-regular graph is known as cubic graph.

(2) $\delta(G) = \Delta(G) = k$, *i.e.*, all the vertices of G have the same degree k.

(3) A regular graph of degree zero has no edge.

(4) In a regular graph of degree 1, every component has exactly one line.

(5) If G is a 2-regular graph, then every component has a cycle.

(6) Every cubic graph has an even number of vertices.

9.5.7 Complete Graph

A complete graph is a simple graph in which each pair of distinct vertices is joined by an edge. Thus, a graph with n vertices is said to be complete if it has as many edges as possible provided there are no loops and no parallel edges.

If the complete graph has vertices $v_1, v_2, \ldots, v_n$, then the edge set E can be written as

$$E = \{(v_i, v_j) : v_i \neq v_j \,; i, j = 1, 2, \ldots, n\}$$

9.5.8 Bipartite Graph

Let G be a graph. If the vertex set V of G partitioned into two non empty subsets X and Y (*i.e.*, $X \cup Y = V$ and $X \cap Y = \phi$) in such a way that each edge of G has one end in X and one end in Y, then G is called bipartite. The partition $V = X \cup Y$ is known as bipartition of G.

9.5.9 Complete Bipartite Graph

It is a simple bipartite graph G with bipartition $V = X \cup Y$ in which every vertex in X is joined to every vertex of Y. If X has m vertices and Y has n vertices, such a graph is denotd by $k_{m,n}$. A complete bipartite graph on n vertices is denoted by k_n.

- Since each of the m vertices in the partition set X of $k_{m,n}$ is adjacent to each of the n vertices in the partition set Y, $k_{m,n}$ has mn edges.
- The graph $k_{1,n}$ is called a star graph.
  For Example : Complete Bipartite Graph:

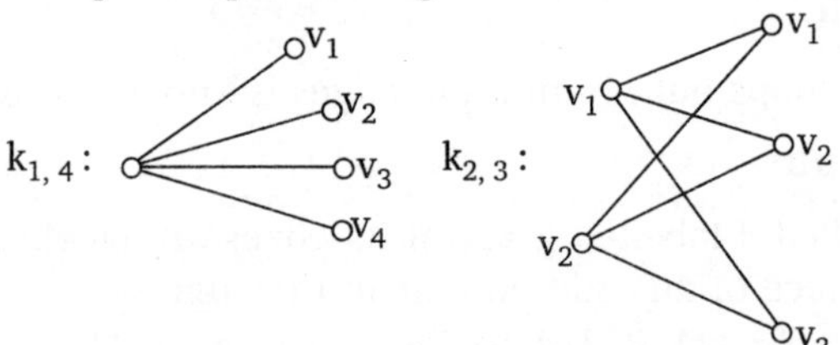

Fig. (9) Fig. (10)

Some Bipartite Graph

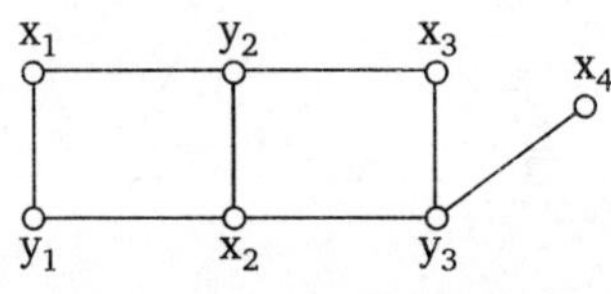

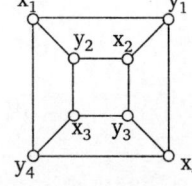

Fig. (11) Fig. (12)

9.5.10 Cycle Graph

For $n \geq 3$, the cycle C_n consists of n vertices $v_1, v_2, \ldots, v_n$ and edges $\{v_1, v_2\}, \{v_2, v_3\}, \ldots, \{v_{n-1}, v_n\}$ and $\{v_n, v_1\}$.

For Example : The cycles C_3, C_4, C_5 and C_6 are given as follows :

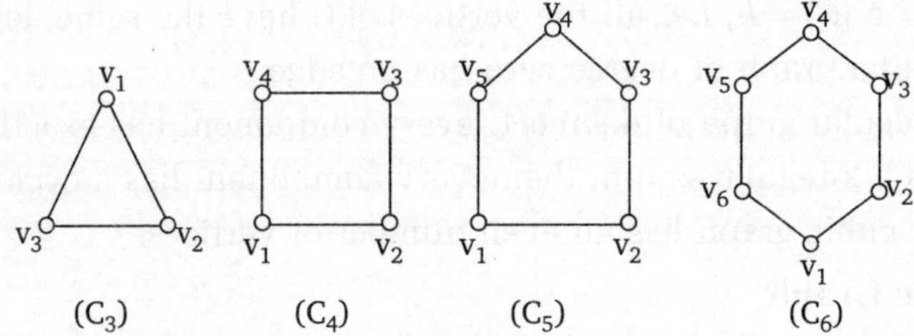

Fig. (13)

9.5.11 Wheel Graph

We can obtain wheel W_n by adding an extra vertex to the cycle C_n, for $n \geq 3$. We connect this vertex to each of the n vertices in C_n by new edges.

For Example : The wheels W_3, W_4, W_5 and W_6 are given as follows :

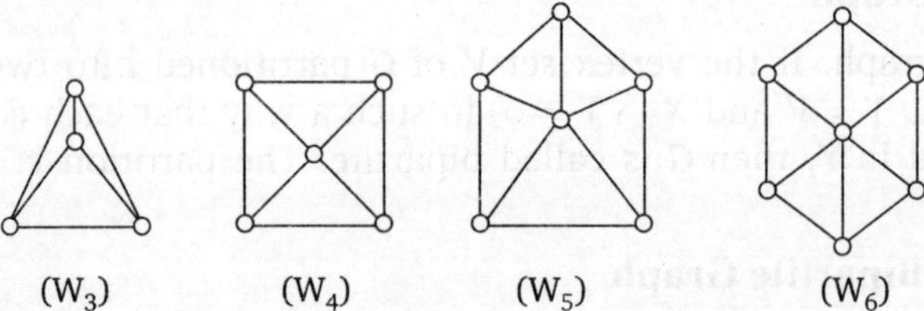

Fig. (14)

9.5.12 Path Graph

We can obtain a path graph of order n by deleting an edge from a cycle graph C_n. It is denoted by P_n.

For Example :

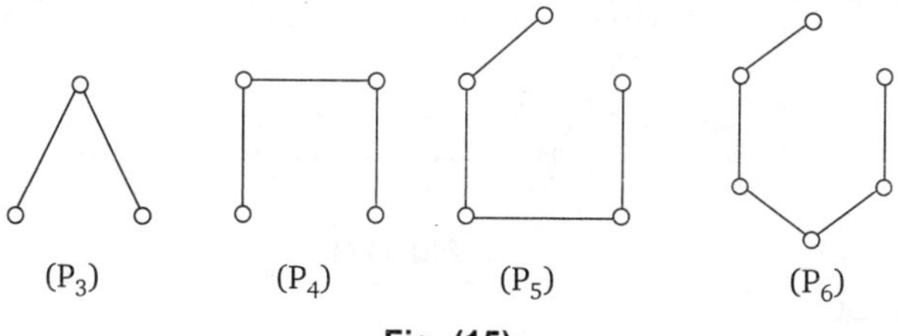

Fig. (15)

9.5.13 N-Cube Graph

It is denoted by Q_n. Therefore, Q_n is the graph that has vertices representing the 2^n bit string of length n. Here, two vertices are adjacent if and only if the bit strings that they represent differ in exactly one bit position. The graph Q_n has 2^n vertices and $n.2^{n-1}$ edges.

For Example :

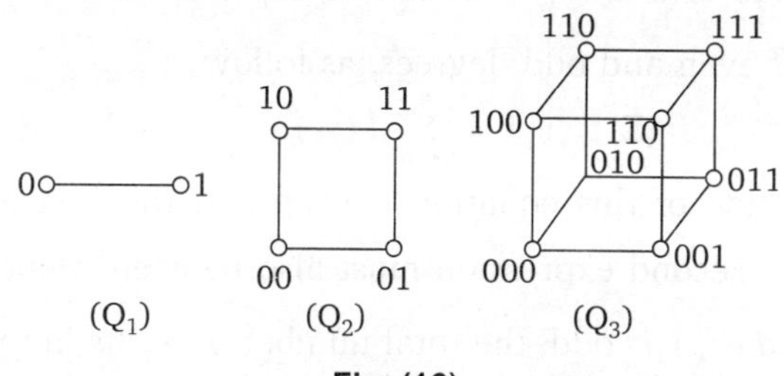

Fig. (16)

9.5.14 Quotient Graph

Let $G = (V, E)$ be a graph and R is an equivalence relation on the set V. Then to construct the quotient graph G^R, we proceed as follows :

The vertices of G^R a: e the equivalence classes of V produced by R. If $[V]$ and $[W]$ are the equivalence classes of vertices v of graph G, then there is an edge G^R from $[V]$ to $[W]$ if and only if some vertex iı $[V]$ is connected to some vertex in $[W]$ in G. Thus we get G^R by merging all the vertices iı. each equivalence class into a single vertex and combining any edges that are superimpo ;ed by such a process.

Theorem 1. (First Th eorem of Graph Theory) : *For any graph G, the sum of the degree of all vertices is twice the number of edges in G, i.e.,*

$$\sum_{i=1}^{n} \deg(v_i) = 2e$$

Proof : Let $G = (V, E)$ be a graph with e edges and n vertices $v_1, v_2, \ldots, v_n$. Then $\sum_{i=1}^{n} \deg(v_i) = \sum_{i=1}^{n}$ (Number of edges adjacent to v_i in G). In a graph, each edge is adjacent to exactly two vertices of G and while taking the sum in right hand side of the above equation, each edge count twice and every edge of G contributes to the degree of a vertex of G. Hence, the right hand side of above equation is equal to $2e$, where e is the number of edges of graph G.

☛ The above theorem is also known as **hand shaking lemma.**

For Example : Consider the following graph

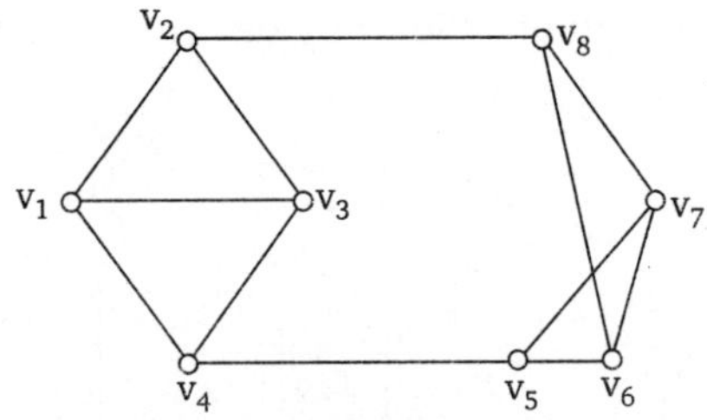

Fig. (17)

We observe that

$d(v_1) = 3, d(v_2) = 3, d(v_3) = 3, d(v_4) = 3, d(v_5) = 3,$
$d(v_6) = 3, d(v_7) = 3, d(v_8) = 3$

$$\sum_{i=1}^{8} d(v_i) = d(v_1) + d(v_2) + d(v_3) + d(v_4) + d(v_5) + d(v_6) + d(v_7) + d(v_8)$$

$$= 3 + 3 + 3 + 3 + 3 + 3 + 3 + 3 = 24 = 2 \times 12 = 2 \times \text{number of edges}$$

Theorem 2. *The number of vertices of odd degree in a graph is always even.*

Proof : If we consider the vertices with odd and even degree separately, the quantity in the left side of $\sum_{i=1}^{n} d(v_i) = 2e$ can be expressed as the sum of two sums, each taken over vertices of even and odd degrees, as follows :

$$\sum_{i=1}^{n} d(v_i) = \sum_{even}^{n} d(v_i) + \sum_{odd}^{n} d(v_i)$$

Since left hand side of this equation is even and the first expression on the right hand side is even, the second expression must also be even. Hence $\sum_{odd}^{n} d(v_i)$ = an even number, in this each $d(v_k)$ is odd, the total number of terms in the sum must be even to make the sum an even number. This proves the theorem.

☛ It is not true in general that a graph must have an odd number of even vertices.

ILLUSTRATIVE EXAMPLES

1. *Draw the diagram of the following graph G (V, E) where*

(i) $V = \{A, B, C, D\}$, $E = \{AB, AC, BC, BD, CD\}$

(ii) $V = \{A, B, C, D, E\}$, $E = \{AB, AC, BC, DE\}$

Solution. We have the following diagrams :

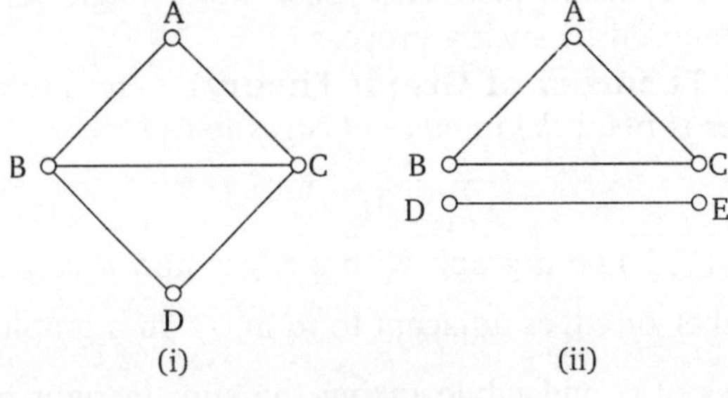

Fig. (18)

2. *Describe the graph G in the adjoining diagram.*

Solution. We observe that

The set of vertices, $V = \{a, b, c, d, e\}$ and set of edges,

$$E = \{ab, ac, ad, bc, be, cd, ce, bd\}$$

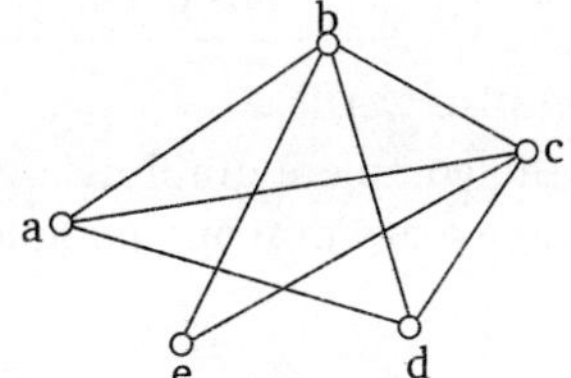

Fig. (19)

Also, deg $(a) = 3$, deg $(b) = 4$, deg $(c) = 4$, deg $(d) = 3$, deg $(e) = 2$,

Clearly, deg (a) + deg (b) + deg (c) + deg (d) + deg $(e) = 16 = 2 \times$ number of edges

3. *Write down the degree of each vertices of the following graph.*

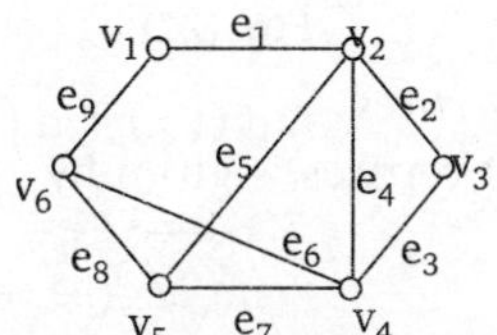

Fig. (20)

Solution. deg $(v_1) = 2$, deg $(v_2) = 4$, deg $(v_3) = 2$, deg $(v_4) = 4$, deg $(v_5) = 3$, deg $(v_6) = 3$.

4. ***(Utility Problem) :*** *Nine members of a new club meet each day for lunch at a round table. They decide to sit so that every member has different neighbours at each lunch. How many days can this arrangement last ?*

Solution. The given situation can be represented by a graph. Each member can be represented by a vertex and an edge joining two vertices represent the relationship of sitting next to each other.

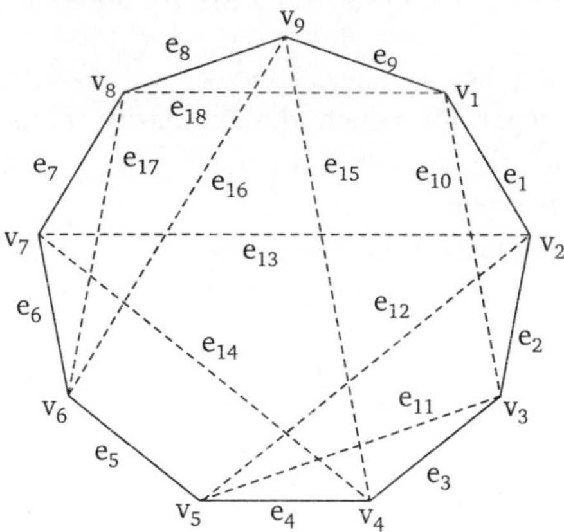

Fig. (21)

In this figure, two sitting arrangement have been shown. These are

$v_1e_1v_2e_2v_3e_3v_4e_4v_5e_5v_6e_6v_7e_7v_8ve_8v_9e_9v_1$ (Solid line)

and $v_1e_{10}v_3e_{11}v_5e_{12}v_2e_{13}v_7e_{14}v_4e_{15}v_9e_{16}v_6e_{17}v_8e_{18}v_1$ (Dotted line)

Further, there are only two more arrangements possible by graph theoretic consideration. They are

$v_1v_5v_7v_3v_9v_2v_8v_4v_6v_1$ and $v_1v_7v_9v_5v_8v_3v_6v_2v_4v_1$

☛ In general, it can be shown that for n people, the number of such possible arrangement is $\frac{n-1}{2}$, if n is odd and $\frac{n-2}{2}$, if n is even.

5. *Draw complete graphs on at most six vertices.*

Solution. It is known that the complete graph is a graph in which each pair of distinct vertices is joined by an edge. Here, we are drawing complete graphs of one, two, three, four, five and six vertices.

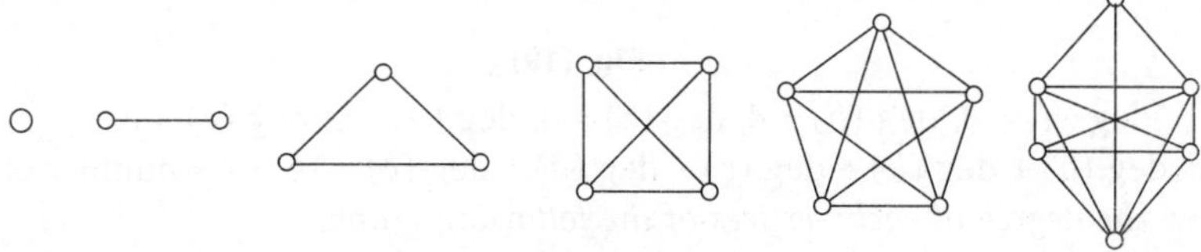

Fig. (22)

6. *Construct a cubic graph on 10 vertices.*

Solution. Cubic graph on 10 vertices is given by

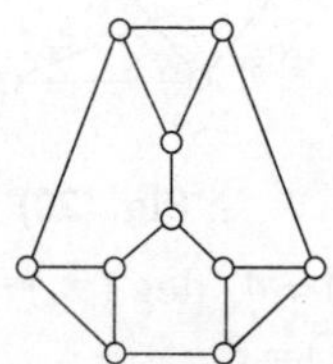

Fig. (23)

7. *Suppose G is a non-directed graph with 12 edges. If G has 6 vertices each of degree 3 and the rest have degree less than 3, what is the minimum number of vertices G can have.*

Solution. It is given that the number of edges in a graph $=12$.

$\Rightarrow \quad \Sigma \deg(v_1) = 2 \times 12 = 24$

Now, we have 6 vertices of degree 3. Let x denote the number of vertices each of whose degree is less than 3

So, $\Sigma \deg(v_1) \leq 6 \times 3 + 3x \quad \Rightarrow \quad 24 \leq 18 + 3x \quad \Rightarrow \quad 3x > 6, i.e., x > 2$

The least positive integer for which the regularity $x > 2$ holds is $x = 3$. Hence, the minimum number of vertices G can have is $3 + 6 = 9$.

8. *Find the size of k-regular graph.*

Solution. Let G be a regular graph. By defintion of regularity of G, we have

$$\deg(v_1) = k, \text{ for all } v_i \in V(G)$$

But it is known that

$$2q = \sum_{i=1}^{n} \deg(v_i) \Rightarrow 2q = \sum_{i=1}^{n} k = nk \Rightarrow q = \frac{nk}{2}.$$

9. *If $G = (V, E)$ in a (p, q) graph, show that $\delta \leq \frac{2q}{p} \leq \Delta$.*

Solution. Let $V = \{v_1, v_2, \ldots, v_p\}$, then we have

$$\Rightarrow \quad p\delta \leq \sum_{i=1}^{p} \deg(v_1) \leq p\Delta \Rightarrow \quad p\delta \leq 2g_1 \leq p\Delta \quad \Rightarrow \quad \delta \leq \frac{2g_1}{p} \leq \Delta.$$

10. *Show that a simple complete graph with 8n vertices has n (n – 1) edges.*

Solution. We can select any vertex from n vertices in ${}^{n}C_1$ ways. Let us take vector v_1 since each vertex is connected to every other vertices. Here, v_2 is connected to vertices $v_1, v_2, \ldots v_n$ by $(n-1)$ ways.

Since there are n vertices, there will be $n(n-1)$ edges. As the graph is taken to be simple without any loop, hence $n(n-1)$ is the maximum number of edges. If there are n loops, then there will be $n + (n-1).\, n = n^2$ edges considering each loop as a single edge.

11. *If R is the equivalence relation defined by the partition [{a, f}, {e, b, d}, {c}]. Find the quotient graph G^R.*

Solution. The required quotient graph G^R is given below :

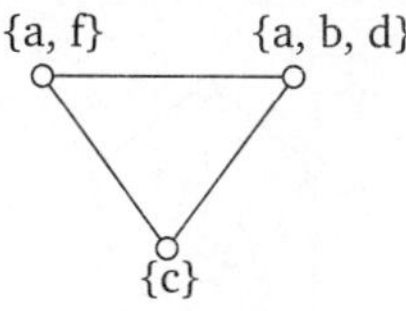

Fig. (24)

12. *Let R = {(1, 1), (2, 2), (3, 3), (4, 4), (5, 5), (6, 6), (7, 7), (8, 8), (9, 9), (10, 10), (11, 11), (12, 12), (13, 13), (14, 14), (15, 15), (16, 16), (1, 10), (10, 1), (3, 12), (12, 3), (5, 14), (14, 5), (2, 11), (11, 2), (4, 13), (13, 4), (6, 15), (15, 6), (7, 16), (16, 7), (8, 9), (9, 8)}. Draw the quotient graph G^R.*

Solution.

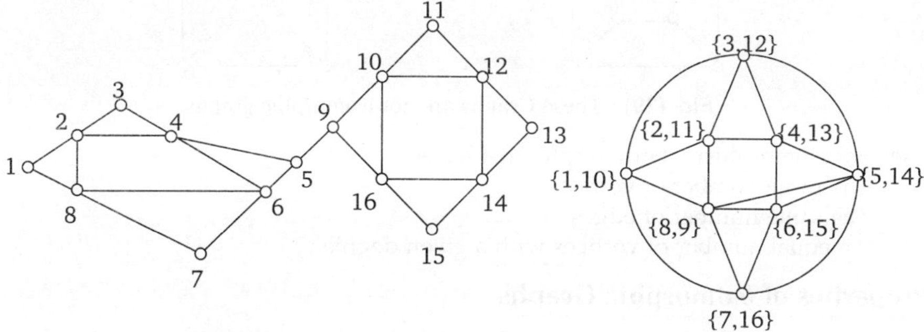

Fig. (25) Fig. (26)

13. *Consider the adjacent graph :*

(a) Describe formally the graph G in the diagram, that is, find the set V (G) of vertices of G and the set E (G) of edges of G.

(b) Find the degree of each vertex.

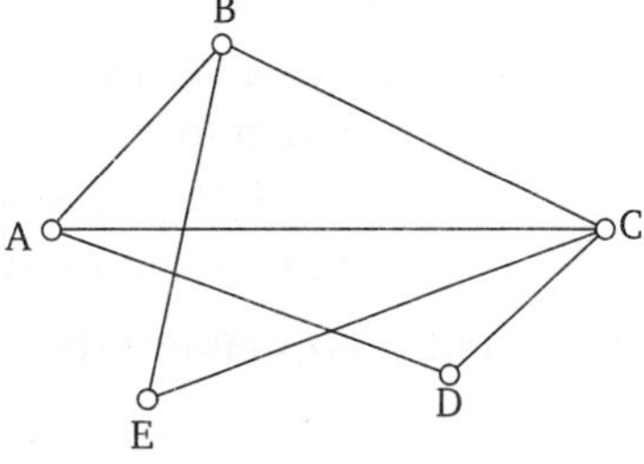

Fig. (27)

Solution.

(a) There are five vertices :

Therefore, $V(G) = \{A, B, C, D, E\}$

There are seven pairs $\{x, y\}$ of vertices where the vertex i connected with vertex j.

Hence,

$$E(G) = [\{A, B\}, \{A, C\}, \{A, D\}, \{B, C\}, \{B, E\}, \{C, D\}, \{C, E\}].$$

(b) The degree of a vertex is equal to the number of edges to which it belong, *e.g.*, $deg(A) = 3$, since A belongs to three edges $\{A, B\}, \{A, C\}, \{A, D\}$.

Similarly, deg $(B) = 3$, deg $(C) = 4$, deg $(D) = 2$, deg $(E) = 2$.

9.6 GRAPH ISOMORPHISM

Definition : *Two graphs* $G_1 = (V_1, E_1)$ and $G_2 = (V_2, E_2)$ *are said to be isomorphic to each other if they have one-to-one correspondence between their edges and their vertices such that their incidence relationship must be preserved.*

For Example :

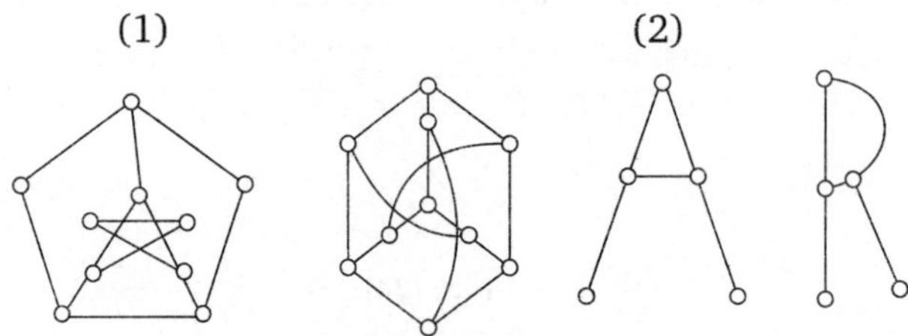

Fig. (28) : Isomorphic Graphs

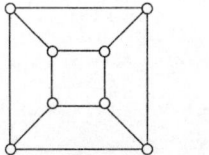

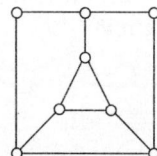

Fig. (29) : These Graphs are not isomorphic graphs

- For isomorphism, both graph must have
  The same number of vertices
  The same number of edges
  An equal number of vertices with a given degree.

9.6.1 Properties of Isomorphic Graphs

If G_1 is isomorphic to G_2, *i.e.*, $G_1 \cong G_2$, then there exists a function $\phi : V(G_1) \to V(G_2)$ with following properties :

(i) $\phi(x) = \phi(y) \Rightarrow x = y, \forall\, x, y \in V(G_1)$ *i.e.*, ϕ is one-one.

(ii) For every vertex $y \in V(G_2)$, there exists a vertex $x \in V(G_1)$ such that $\phi(x) = y$ *i.e.*, ϕ is onto.

(iii) $(x, y) \in E(G_1)$ if and only if $(\phi(x), \phi(y)) \in E(G_2)$ *i.e.*, ϕ is one-one.
i.e., structure preserving property is satisfied.

9.6.2 Self Complementary Graph

A graph $G = (V, E)$ is said to be self-complementary if it is isomorphic to its complements.

9.7 HOMEOMORPHIC GRAPH

Given any graph G, we can obtain a new graph by dividing an edge of G with additonal vertices. Two graphs are said to be homeomorphic if they can be obtained from the same graph or isomorphic graphs by this method.

Definition : *Two graphs* G_1 and G_2 *are said to be homeomorphic graphs if* G_2 *can be obtained from* G_1 *by a sequence of subdivision of the edge of* G.

For Example :

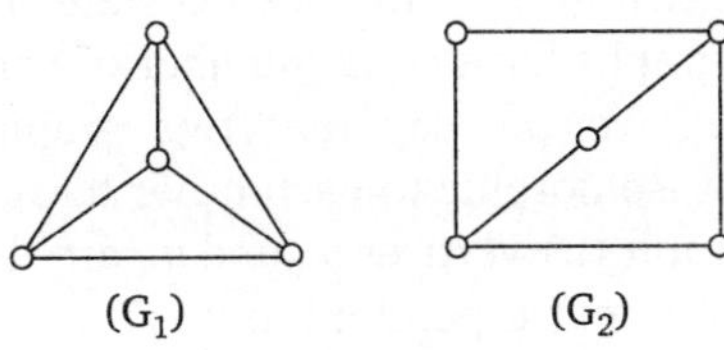

(G_1) (G_2)

Fig. (30)

Here, G_1 and G_2 are homeomorphic because G_1 can be obtained from G_2 by introducing vertices of degree 2 on edges (v_1, v_3) and (v_2, v_4).

9.8 AUTOMORPHISM

Definition : *Let G be a graph. Then an isomorphism of a graph G onto itself is called an automorphism of G.*

ILLUSTRATIVE EXAMPLES

1. *Show that following two graphs are isomorphic.*

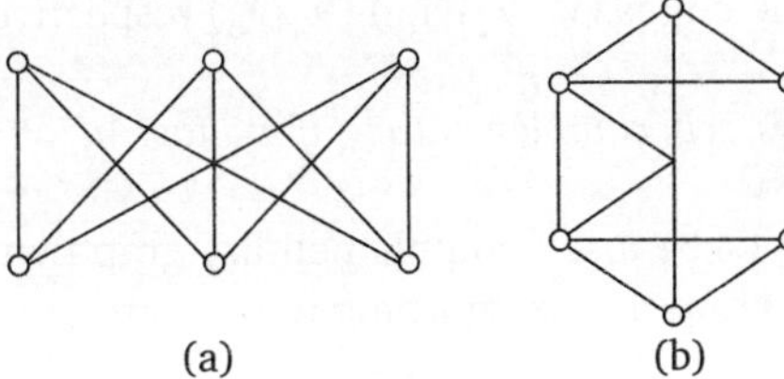

(a) (b)

Fig. (31)

Solution. We observe that

(i) Both have same number of vertices, *i. e.*, six.

(ii) Both have same number of edges, *i. e.*, nine.

(iii) All vertices of both the graphs are of three degree.

Hence, both graphs (a) and (b) are isomorphic to each other.

2. *Check this pair of graphs is isomorphic or not ?*

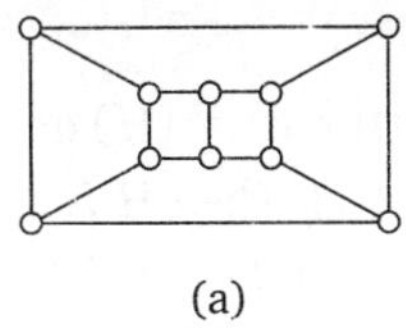

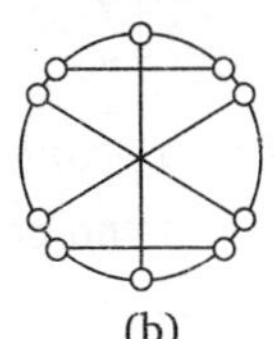

(a) (b)

Fig. (33)

Solution. We know that for isomorphism, following conditions must be satisfied :

(i) Both graphs have same number of vertices, *i. e.*, 10

(ii) Both graphs have same number of edges *i. e.*, 15

(iii) All vertices of both the graphs are of same degree, *i. e.*, 3.

Hence, given graphs (a) and (b) are isomorphic.

3. *Show that the following graphs are not isomorphic to each other.*

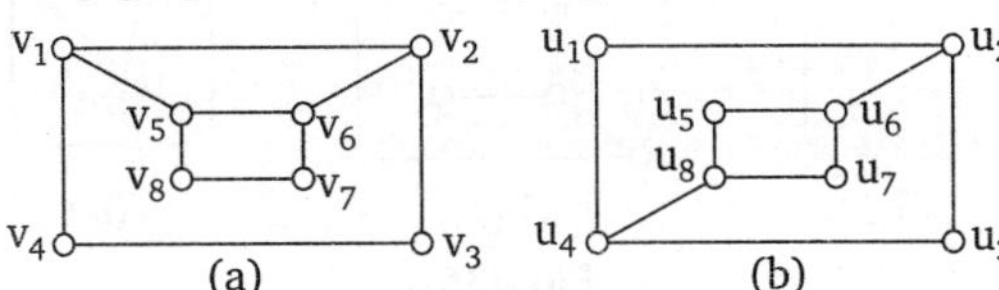

(a) (b)

Fig. (34)

Solution. In the above two graphs, four vertices are of degree three and four vertices are of degree two. Also both graphs have equal number of vertices (*i.e.*, 8) and equal number of edges (*i.e.*, 10). Therefore, we can say that two graphs satisfy all the conditions of isomorphism, but they are not isomorphic to each other because in graph (a), vertices v_5 and v_6 are of degree 3 while in graph (b) vertices u_6 and u_8 are degree 3. Hence, they can not be isomorphic to each other until v_5 corresponding to u_8.

4. *Show that the graphs given below are homeomorphic.*

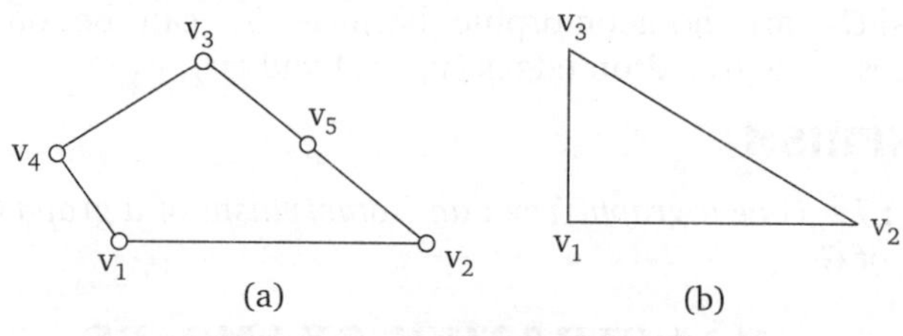

Fig. (35)

Solution. The graph (a) can be obtained from the graph (b) by adding vertices v_4 and v_5 of degree 2 on the edges (v_1, v_2) and (v_2, v_3) respectively. Hence, above two graphs are homeomorphic.

5. *Show that a graph G is self complementary if it has 4n or 4n + 1 vertices, where n is a non-negative integer.*

Solution. Let $G = (V, E)$ be a self complementary graph with m points. It is given that G is self complementary, therefore G is isomorphic to $\bar{G}$.

$$\Rightarrow \quad |E(G)| = |E(\bar{G})| \qquad \Rightarrow \quad E(G) + |E(\bar{G})| = \frac{m(m-1)}{2}$$

$$\Rightarrow \quad 2|E(G)| = \frac{m(m-1)}{2} \qquad \Rightarrow \quad E(\bar{G}) = \frac{m(m-1)}{4}$$

$\Rightarrow$ Now, $\frac{m(m-1)}{4}$ is an integer and one of m or $(m-1)$ is odd.

Therefore, m or $(m-1)$ is a multiple of G.

Hence, n is of the form $4n$ or $4n + 1$.

9.9 SUB-GRAPHS

Definition (1) *Let H be a graph with vertex set V (H) and edge set E (H). Also, let G be a graph with vertex set V (G) and edge set E (G). Then H is said to be a subgraph of G if*

$$V(H) \subseteq V(G) \text{ and } E(H) \subseteq E(G)$$

In other words, a graph H is said to be subgraph of G if all the vertices and all the edges of H are in G and each edge of H has the same end vertices in H as in G.

Definition (2) *A subgraph H of a graph G is any graph obtained from G by a deletion of edges and/or vertices from G. Note that when a vertex v is deleted, all edges terminating at v must also be deleted.*

For Example : Consider the following graphs

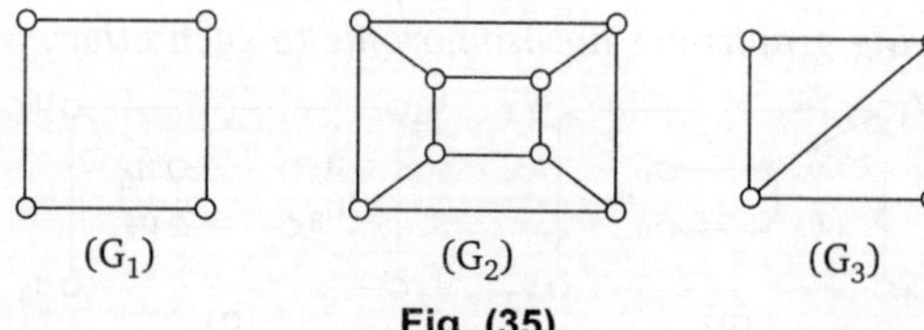

Fig. (35)

Clearly, G_1 is a subgraph of both G_2 and G_3, is not a subgraph of G_2.

- The concept of subgraph is similar to the concept of subset in set theory. As a subset, it is contained in the set, similarly a subgraph is thought of as being contained in another graph.
- If H is a subgraph of G, then G is called a **supergraph** of H.
- The simplest types of subgraphs of a graph G are those obtained by the deletion of a vertex or an edge.

9.9.1 Proper Subgraph

If H is a subgraph of G, then we write $H \subseteq G$. When $H \subseteq G$ but $H \neq G$, *i.e.*, $V(H) \neq V(G)$ or $E(G) \neq E(G)$, then H is called a proper subgraph of G.

The subgraph, other than proper is known as improper subgraph.

9.9.2 Some Important Conclusions

(i) Every graph is its own subgraph.
(ii) A single vertex in G is a subgraph of G.
(iii) A single edge in G together with its end vertices, is a subgraph of G.
(iv) A subgraph of a subgraph of G is also a subgraph of G.

9.10 TYPES OF SUBGRAPHS

9.10.1 Disjoint Subgraphs

(a) Vertex Disjoint Subgraph : *Let $G = (V, E)$ be a graph, then two subgraphs H_1 and H_2 of G are called vertex disjoint if H_1 and H_2 have no vertex in common (Also, they do not have any edge in common).*

Mathematically, for vertex disjoint subgraphs, we have

$$V(H_1) \cap V(H_2) = \phi \text{ and } E(H_1) \cap E(H_2) = \phi$$

For example :

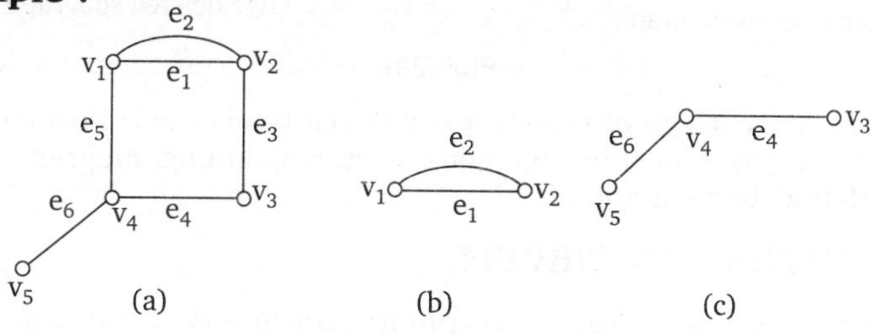

Fig. (36)

Clearly (b) and (c) are vertex disjoint subgraphs of graph (a).

(b) Edge Disjoint Subgraph : *Let $G = (V, E)$ be a graph. Two subgraphs H_1 and H_2 of G are said to be edge disjoint if they do not have any edge in common (they may have vertices in common).*

Mathematically, If $E(H_1) \cap E(H_2) = \phi$ then H_1 and H_2 are edge disjoint subgraphs.

For Example : Consider the following graphs

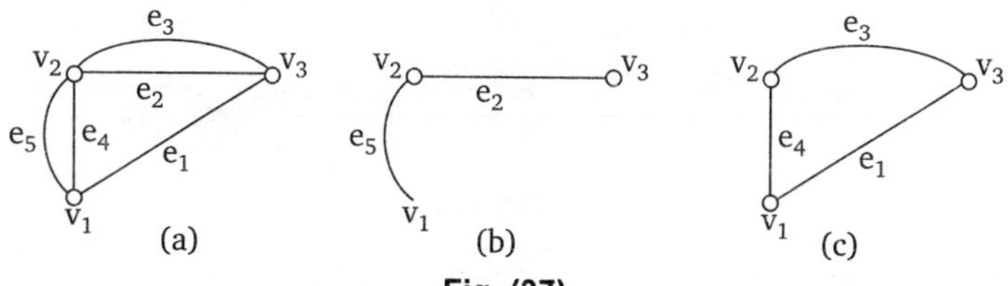

Fig. (37)

Clearly (b) and (c) are edge disjoint subgraphs of graph (a).

9.10.2 Induced Subgraph

Let $G = (V, E)$ be a graph. Also, let U be a subset of V. Then the subgraph $G\ (U)$ of G induced by U is defined to be the graph having vertex set U and edge set containing those edges of G that have both ends in U. Similarly, if F is a non-empty subset of the edge set E of G, then the subgraph $G\ (F)$ of G induced by F is the graph whose vertex set is the set of ends of edges in F and whose edge set is F.

9.10.3 Spanning Subgraph

We know that if $V\ (H) \subset V\ (G)$, then H is an improper subgraph of G and if $V\ (H) = V\ (G)$. Then H is said to be spanning subgraph of G.

Definition : *A subgraph of G is said to be spanning subgraph if it contains all vertices of G.*

9.10.4 Vertex deleted Subgraph and edge Deleted Subgraphs

Let $G\ (V, E)$ be a graph. If from G, we delete a subset U and V and all the edges which have a vertex in U as an end point, then $(G - U)$ is called a vertex deleted subgraph and similarly, if a subset F of E is deleted from G, then $(G - F)$ denotes the subgraph of G with vertex set V and edge set $E - F$, then $(G - F)$ is called edge deleted subgraph.

For example :

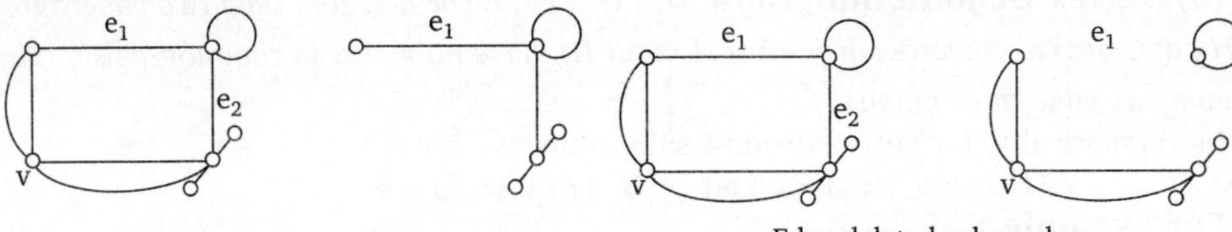

Fig. (38)

- By deleting from a graph G all loops and in each collection of parallel edges all edges but one in the collection, we get a simple spanning subgraph of G, called the **underlying simple graph** of G.

9.11 WALKS, PATHS AND CIRCUIT

Definition : *A walk in a graph G is a finite sequence* $W = v_0 e_1\ v_1 e_2\ v_2 \dots v_{k-1} e_k\ v_k$ *whose terms are alternately vertices and edges such that for* $1 \le i \le k$, *the edge* e_i *has ends* v_{i-1} and v_i.

Each edge e_i is immediately preceded and succeeded by the two vertices with which it is incident.

The vertices $v_1, \dots v_{k-1}$ in the above walk W, are called its internal vertices. The integer k, the number of edges in the walk, is called length of W.

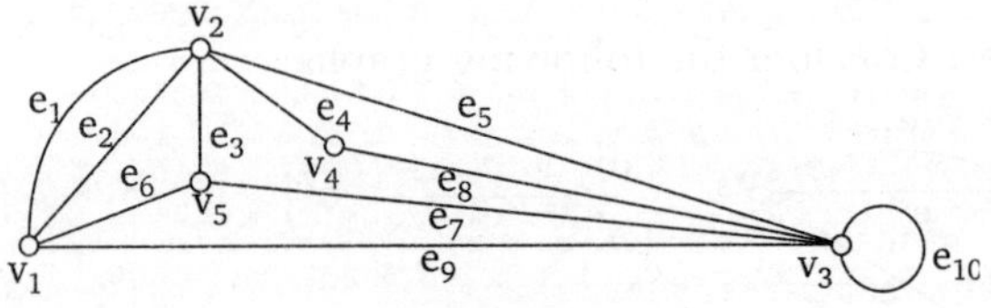

Fig. (39)

In the figure, $W_1 = v_1e_1\ v_2e_5\ v_3e_{10}\ v_3e_5\ v_2e_3\ v_5$ and $W_2 = v_1e_1\ v_2e_1\ v_1e_1\ v_2$ are both walks of length 5 and 3 respectively from v_1 to v_5 and from v_1 to v_2 respectively.

Given two vertices u and v of a graph G, a $u - v$ walk is called closed or open depending on whether $u = v$ or $u \neq v$.

Definition : *If the vertices $v_0, v_1, \dots v_k$ of the walk $W = v_0e_1\ v_1e_2 \dots e_k\ v_k$ are distinct then W is called path.*

$W = v_2\ e_4\ v_4\ e_8\ v_3\ e_7\ v_5\ v_6\ v_1$ is a path in a graph G.

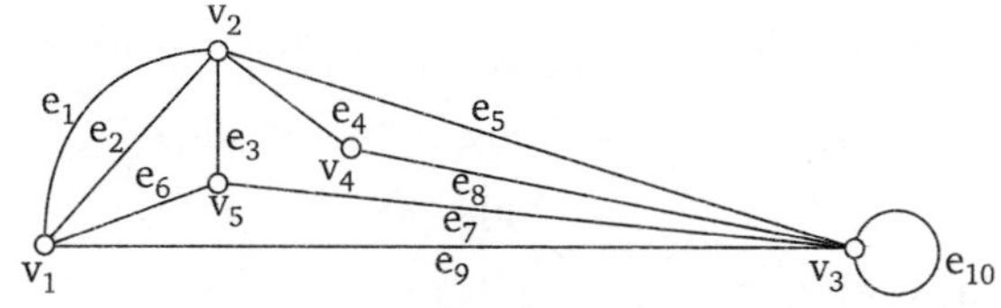

Fig. (40)

- A path of n vertices has length $n-1$.

Definition : *A closed walk in which no vertex appears more than once, except the initial and the final vertex, is called circuit.*

- A circuit is closed and non-intersecting walk.

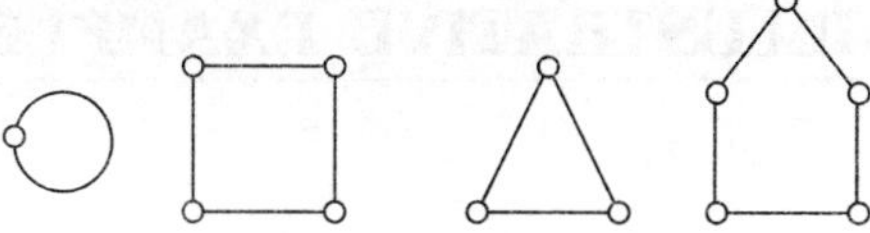

Fig. (41) : Four Different Circuit

- A trivial walk is one containing no edge.
- If the edges $e_1, e_2, \dots e_k$ of the walk $W = v_0\ e_1\ v_1\ e_2\ v_2 \dots e_k\ v_k$ are distinct then W is called a trail.
- Circuit is also known as cycle, elementary cycle, circular path and polygon.
- Circuit has at least one edge.
- In circuit, every vertex have degree two.

9.11.1 Some Important Facts

(1) If a path consists of a single edge, then its length is one.

(2) In a path, the terminal vertices are of degree 1 while each intermediate vertex is of degree 2.

(3) A self loop cannot be included in a path while a self loop can appear in a walk.

(4) If the path or circuit is a subgraph of some graph, then the degree of vertices are counted with respect to the edge included in the path or circuit only.

(5) A circuit (or cycle) is also called an elementary cycle or a circular path or a polygon.

(6) Every self loop is a circuit but the converse is not true.

(7) A graph without cycles is called an acyclic graph.

(8) A path does not intersect itself.

9.11.2 Flowchart of Walk, Path and Circuits

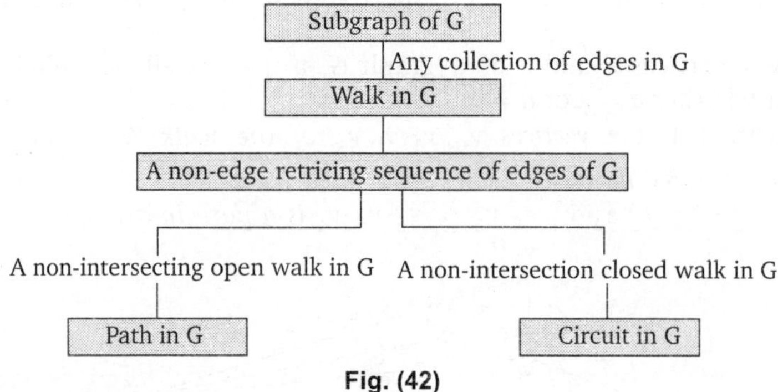

Fig. (42)

9.11.3 More Definitions

Definition 1. *The distance between any two vertices u, v in a graph to be denoted by d (u, v) is the length of the shortest path between u* and *v.*

Definition 2. *The diameter of a connected path is the maximum distance between any two vertices in a graph.*

ILLUSTRATIVE EXAMPLES

1. *Find all walks of the given graph*

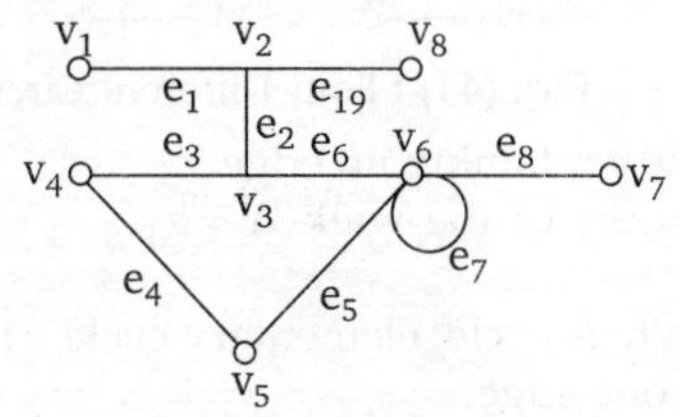

Fig. (43)

Solution. We observe that :

(i) w_1 is an open walk from v_1 to v_4 of length 7, *i. e.*, $l(w) = 7$

$w_1 : v_1\, e_1\, v_2\, e_9\, v_8\, e_9\, v_2\, e_6\, v_3\, e_6\, v_6\, e_6\, v_3\, e_3\, v_4$

w_1 is not a trial because edge e_6 is repeated and not a path.

(ii) w_2 is an open walk from v_5 to v_7 of length 3

$w_2 : v_5\, e_5\, v_6\, e_7\, v_6\, e_8\, v_7$

w_2 is also a trail but not a path because the vertex v_6 is repeated.

(iii) w_3 is an open walk from v_1 to v_7 of length 4

$w_3 : v_1\, e_1\, v_2\, e_2\, v_3\, e_6\, v_6\, e_8\, v_7$

w_3 is a trial as well as a path because no edge and no vertex is repeated.

(iv) v_4 is a closed walk of length 5.

$w_4 : v_4\, e_3\, v_3\, e_6\, v_6\, e_7\, e_5\, v_5\, e_4\, v_4$

w_4 is a closed trail but not closed path because v_6 is repeated two times.

(v) w_5 is a closed walk of length 4.

$w_5 : v_3\, e_3\, v_4\, e_4\, v_5\, e_5\, v_6\, e_6\, v_3$

w_5 is a closed trail as well as closed path.

(vi) $w_6 : v_1\, e_1\, v_2\, e_8\, v_7$

w_6 is not a walk because v_2 and v_7 are not the end vertices of the edge e_8 that is cited between them.

Since w_6 is neither a trail not a path, therefore we can not define the length.

(vii) $w_7 : v_1\ e_1\ v_2\ e_9\ v_8\ e_9$

w_7 is not a walk as sequence of w_7 is ended at an edge e_9 of G

(viii) $w_8 : v_1\ e_1\ v_2\ v_8\ e_9\ v_2$

w_8 is a not a walk as no edge occur between the vertex v_2 and v_8

2. *Find all walks of the given graph*

(i) *Find all paths from the vertex A to the vertex F.*

(ii) *All trails from A to F.*

(iii) *The distance between A and F*

(iv) *Diameter of the graph.*

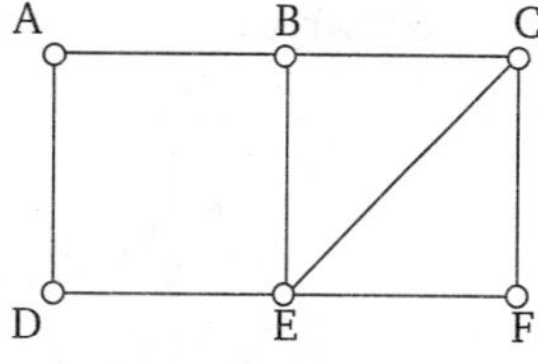

Fig. (44)

Solution.

(i) Paths from A to F are

$P_1 = \{AB, BE, EF\}$, $P_2 = \{AB, BC, CF\}$,

$P_3 = \{AD, DE, EF\}$, $P_4 = \{AB, BE, EC, CF\}$

$P_5 = \{AD, DE, EB, BC, CF\}$, $P_6 = \{AD, DE, EC, CF\}$,

$P_7 = \{AB, BC, CE, EF\}$

(iii) All the above paths and two more are

$T_8 = \{AD, DE, EB, BC, CE, EF\}$, $T_9 = \{AD, DE, EC, CB, BE, EF\}$

(iii) The distance between A and F is 3, $d(A, F) = \{AB, BE, EF\}$.

(iv) $d(A, B) = 1$, $d(B, C) = 1$, $d(A, C) = 2$, $d(C, F) = 1$,

$d(A, D) = 1$, $d(B, E) = 1$, $d(D, E) = 1$, $d(A, E) = 2$,

$d(B, F) = 2$, $d(D, F) = 2$, $d(A, F) = 3$, $d(C, D) = 2$

$d(E, F) = 1$, $d(C, E) = 1$, $d(B, D) = 2$

3. *Draw all the circuits of the following graph.*

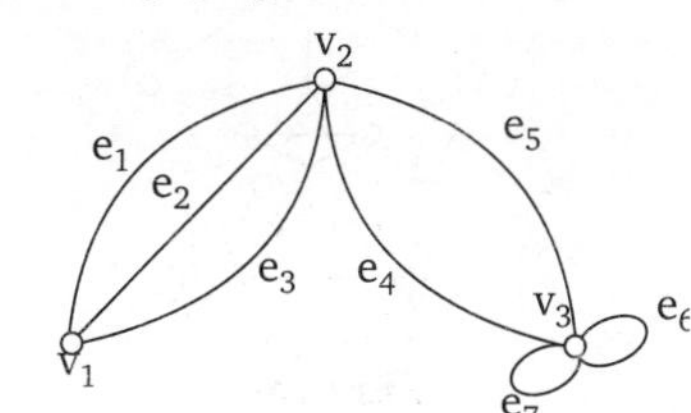

Solution.

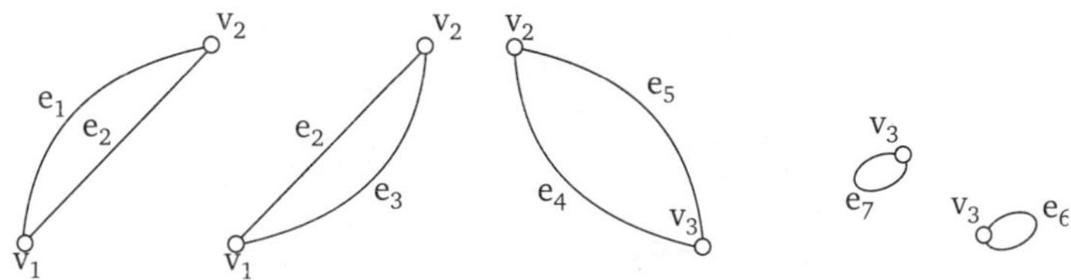

Fig. (45)

(i) $v_1\ e_1\ v_2\ e_2\ v_1$ (ii) $v_1\ e_2\ v_2\ e_3\ v_1$ (iii) $v_2\ e_5\ v_3\ e_4\ v_2$ (iv) $v_3\ e_7\ v_3$

(v) $v_3\ e_6\ v_3$

4. *List all path from v_1 to v_8.*

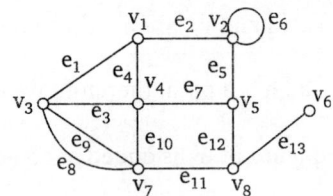

Fig. (46)

Solution.

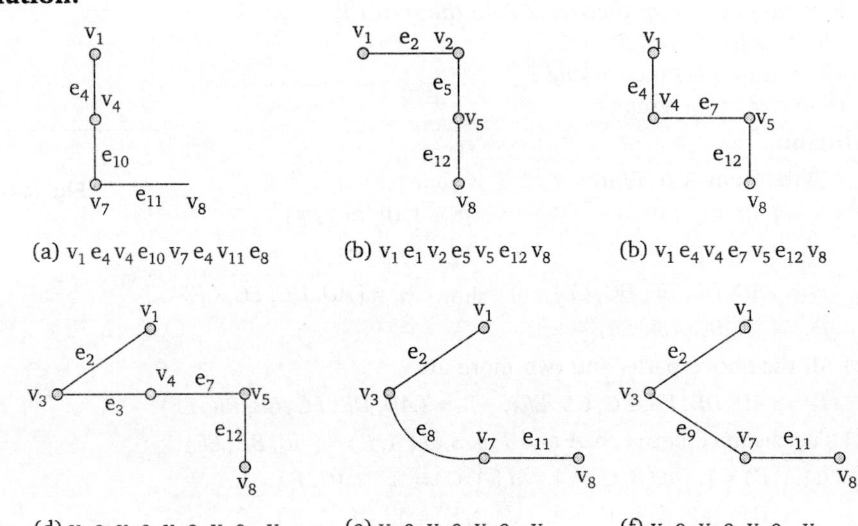

(a) $v_1\, e_4\, v_4\, e_{10}\, v_7\, e_4\, v_{11}\, e_8$ (b) $v_1\, e_1\, v_2\, e_5\, v_5\, e_{12}\, v_8$ (b) $v_1\, e_4\, v_4\, e_7\, v_5\, e_{12}\, v_8$

(d) $v_1\, e_2\, v_3\, e_3\, v_4\, e_7\, v_5\, e_{12}\, v_8$ (e) $v_1\, e_2\, v_3\, e_8\, v_7\, e_{11}\, v_8$ (f) $v_1\, e_2\, v_3\, e_9\, v_7\, e_{11}\, v_8$

5. *Draw a circuit from the graph which is of length nine.*

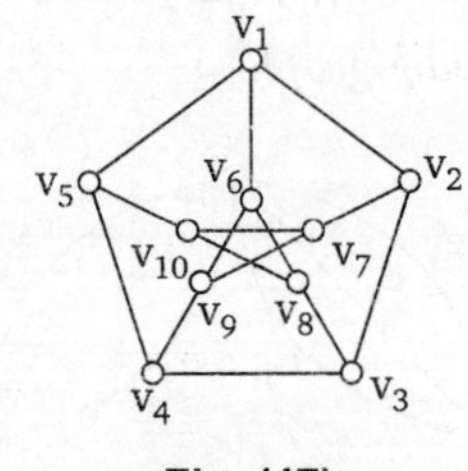

Fig. (47)

Solution.

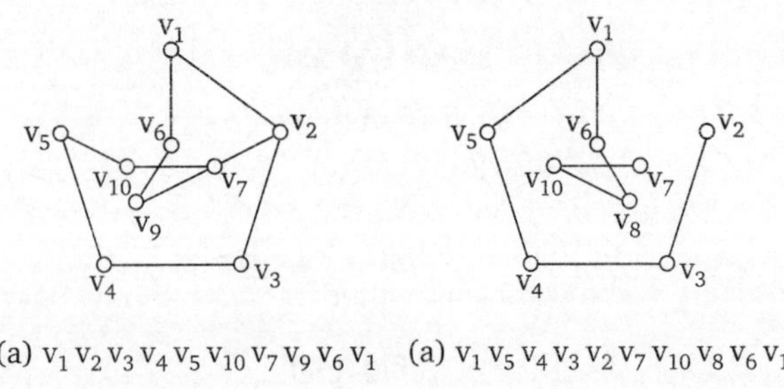

(a) $v_1\, v_2\, v_3\, v_4\, v_5\, v_{10}\, v_7\, v_9\, v_6\, v_1$ (a) $v_1\, v_5\, v_4\, v_3\, v_2\, v_7\, v_{10}\, v_8\, v_6\, v_1$

Fig. (48)

These are the only two cycles of length nine of the given graph.

Exercise 9.1

1. Define the following with a graph :
   (a) graph, (b) vertex, (c) self loop, (d) parallel edge,
   (e) walk, (e) path, (f) circuit
2. Let G be a graph with n vertices and exactly $n - 1$ edges. Prove that G has either a vertex of degree 1 or an isolated vertex.
3. Let G be a graph :

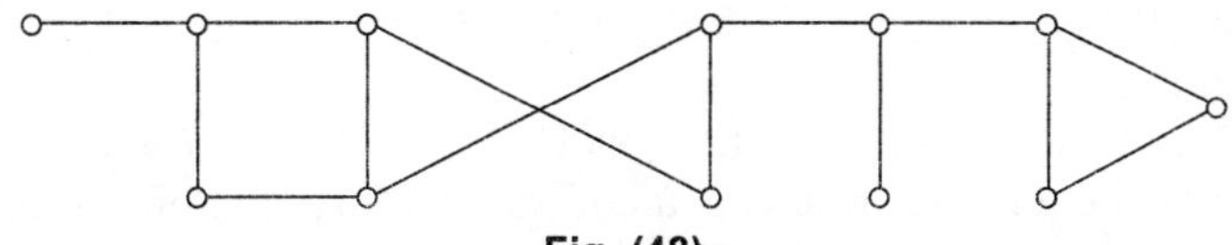

Fig. (48)

   (a) Find a closed walk of length 6. Is your walk a trail ?
   (b) Find an open walk of length 12. Is your walk a path ?
   (c) Find a closed trail of length 6. Is your trail a cycle ?
   (d) What is the length of the longest cycle in G ?
   (e) What is the length of a longest path in G? How many paths in G are there of this length ?
4. In each of collection of three graphs shown in figure (a), there is exactly one isomorphic pair. Find each pair and justify that your answer is correct. Find the odd one out.

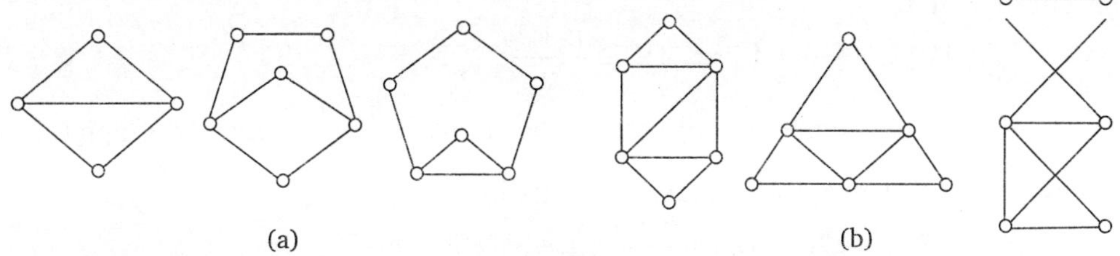

Fig. (49)

5. Find which of these are isomorphic pairs ?

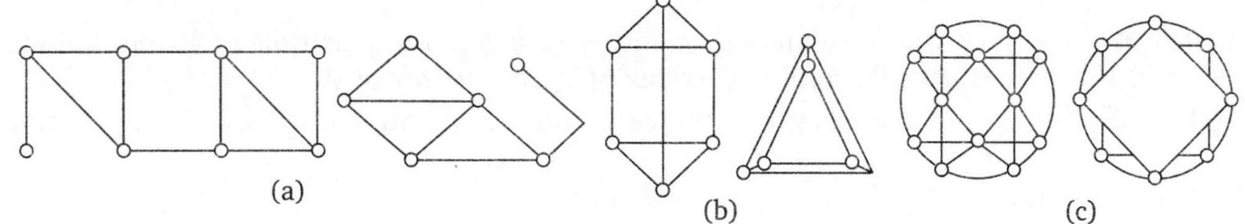

Fig. (50)

6. Prove that any two simple connected graphs with n vertices all of degree two are isomorphics.

9.12 CONNECTED, DISCONNECTED GRAPHS AND COMPONENTS

A graph G is said to be connected if there is at least one path between every pair of vertices in G. Otherwise G is disconnected. A null graph of more than one vertex is disconnected.

It is easy to see that a disconnected graph consists of two or more connected graphs. Each of these connected subgraphs is called a component. This graph consists of six components.

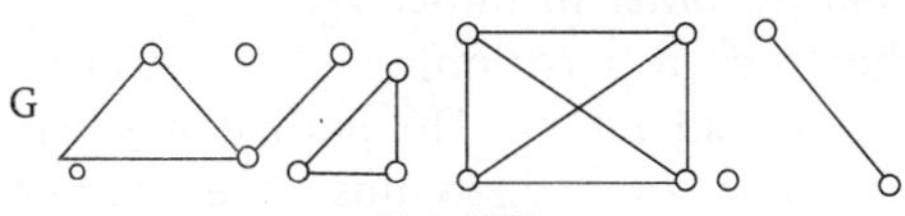

Fig. (51)

In another way, consider a vertex v_i in a disconnected graph G, not all vertices of G are joined by paths to v_i. Vertex v_i and all the vertices of G that have paths to v_i together with all the edges incident on them form a component. A component is itself a graph.

☛ A single vertex is itself a connected graph.

9.12.1 Partitions and Decompositions

Let $G = (V, E)$ be a graph.

A partition of the vertex set V of a graph G is a collection $\{V_i\}_{i \le j \le k}$ of non-empty subset of V such that

$$V = V_1 \cup V_2 \cup \ldots \cup V_k,\ k \ne 1,\ \ V_i \cap V_j = \phi, \text{ whenever } i \ne j$$

A partition of the edge set E of G is a collection $\{E_i\}_{1 \le i \le k}$ of non-empty subsets of E such that $E = E_1 \cup E_2 \cup \ldots \cup E_k$, $k \ne 1$ and $E_i \cap E_j = \phi$, whenever $i \ne j$.

☛ The partition of the edge set is called **decomposition** of G.

Theorem 1. *If G is a graph with n points and* $\delta\,(G) \ge \dfrac{n-1}{2}$, *then G is connected.*

Proof : Let us suppose G is not connected. Then graph G has more than one component. Let us consider a component $G_1 = (V_1, E_1)$ of graph G.

Let $v_1 \in V_1$ since $\delta\,(G) \ge \dfrac{n-1}{2}$ there exists at least $\dfrac{n-1}{2}$ points in G_1 which are adjacent to v_1 in G_1.

So, $\quad |V| \ge \dfrac{n-1}{2} + 1 \ \Rightarrow\ |V| \ge \dfrac{n+1}{2}$

Therefore, each component of graph G has at least $\dfrac{n+1}{2}$ points and graph G at least two components. Hence, the number of points in

$$G \ge \left(\frac{n+1}{2}\right), i.e., V\,(G) \ge n + 1$$

which is a contradiction.

Hence, graph G is a connected graph.

Theorem 2. *A graph G is connected if and only if for any partition of V into subsets* V_1 *and* V_2, *there is an edge joining a vertex of* V_1 *to a vertex of* V_2.

Proof : Let $G = (V, E)$ be a graph which is connected and $V = V_1 \cup V_2$ be a partition of V into two subsets.

Let $u \in V_1$ and $v \in V_2$

Now, since G is connected, there exists a $u - v$ path in G, say

$$u = v_0\ v_1\ v_2 \ldots v_n = V$$

Let i be the least positive integer such that $v_i \in V_2$, then $v_{i-1} \in V_1$ and the vertices $v_{i-1} \in V_1$ and $v_i \in V_2$. Conversely, let G is not connected, *i.e.*, G is disconnected, then G contains at least two components. Let V_1 be the set of all vertices of one component and V_2 be the set of remaining vertices of G. Clearly $V_1 \cup V_2 = V$ and $V_1 \cap V_2 = \phi$.

The collection $\{v_1, v_2\}$ is partition of V and there is no edge joining any vertex of V_1 to any vertex of V_2.

Theorem 3. *A graph G is disconnected if and only if its vertex set V can be partitioned into two non-empty disjoint subset* V_1 *and* V_2, *such that there exists no edge in G whose one end vertex is in subset* V_1 *and the other in subset* V_2.

Proof : Suppose that such a partitioning exists. Consider two arbitrary vertices a and b of G such that $a \in v_1$ and $b \in v_2$. No path can exist between vertices a and b, otherwise there would be at least one edge whose one end vertex would be in v_1 and the other in v_2. Hence, if a partition exists, G is not connected.

Conversly, let G be a disconnected graph. Consider a vertex a in G. Let v_1 be the set of all vertices that the joined by paths to a. Since G is disconnected, v_1 doesn't include all vertices of G. The remaining vertices will form a non-empty set v_2. No vertex in v_1 is joined to any in v_2 by an edge. Hence, the partition exists.

Theorem 4. *If a graph has exactly two vertices of odd degree, there must be a path joining these two vertices.*

Proof : Let G be a graph with all even vertices except vertices v_1 and v_2 which are odd. For every component of a disconnected graph no graph can have an odd number of odd vertices. Therefore, in graph G, v_1 and v_2 must belong to the same component and hence must have a path between them.

Theorem 5. *A simple graph with n vertices and k component can have at most $(n - k)(n - k + 1)/2$ edges.*

Proof : Let the number of vertices in each of the k components of graph G be $n_1, n_2, \ldots, n_k$.

Thus, we have $n_1 + n_2 + \ldots + n_k = n$, where $n_i \geq 1$.

Consider $\sum_{i=1}^{k} (n_i - 1) = (n_1 - 1) + (n_2 - 1) + \ldots + (n_k - 1)$

$$= (n_1 + n_2 + \ldots - n_k) - k = n - k$$

Squaring both sides, we get

$$\left[\sum_{i=1}^{k} (n_i - 1)\right]^2 = n^2 - 2nk + k^2$$

$$(n_1^2 - 2n_1 + 1) + (n_2^2 - 2n_2 + 1) + \ldots + (n_k^2 - 2n_k + 1) = n^2 - 2nk + k^2$$

$$\sum_{i=1}^{k} n_i^2 - 2(n_1 + n_2 + \ldots + n_k) + k = n^2 - 2nk + k^2$$

(terms consisting multiple of non-negative factors)

$$\sum_{i=1}^{k} n_i^2 - 2n + k \leq n^2 - 2nk + k^2$$

$$\sum_{i=1}^{k} n_i^2 \leq n^2 - 2nk + k^2 + 2n - k = n^2 - 2n(k - 1) + k(k - 1)$$

$$= n^2 + (k - 2n)(k - 1)$$

Now, the maximum number of edges in the ith component of G is ${}^nC_2$, *i.e.*, $\frac{n_i(n_i - 1)}{2}$. Therefore, the maximum number of edges in G is given by

$$\frac{1}{2}\sum_{i=1}^{k} (n_i - 1)\, n_i = \frac{1}{2}\left(\sum_{i=1}^{k} n_i^2\right) - \frac{n}{2} \leq \frac{1}{2}[n^2 - (k - 1)(2n - k)]$$

$$- \frac{n}{2} = \frac{1}{2}(n - k)(n - k + 1).$$

9.13 EULER GRAPH : KÖNIGSBERG BRIDGE PROBLEM

Two islands M and N, termed by the Preyel river in Königsberg were connected to each other and to the banks P and Q with seven bridges, as shown in Figure (a). The problem was to start at any of the four land areas of the city, M, N, P or Q, walk over each of the seven bridges exactly once, and then return to the starting point (without swimming across the river).

Euler represented this situation by means of a graph as shown in (b). Vertices represent the land area and edges represent the bridges.

The problem is same as the problem of drawing figure without lifting the pen from the paper and without retracing a line.

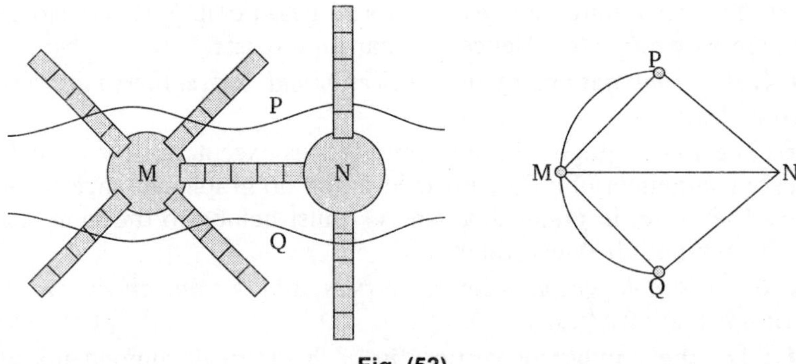

Fig. (52)

☛ Utilities problem, seating problems and electrical engineering problems are the other applications of graph theory.

9.13.1 Eulerian Path

An Eulerian path is a path which exists between any pair of vertices such that starting from one vertex reaching back to the same after travelling through all the edges once and only once.

In any connected graph, if degree of each vertex is even, then it always possess an Eulerian path.

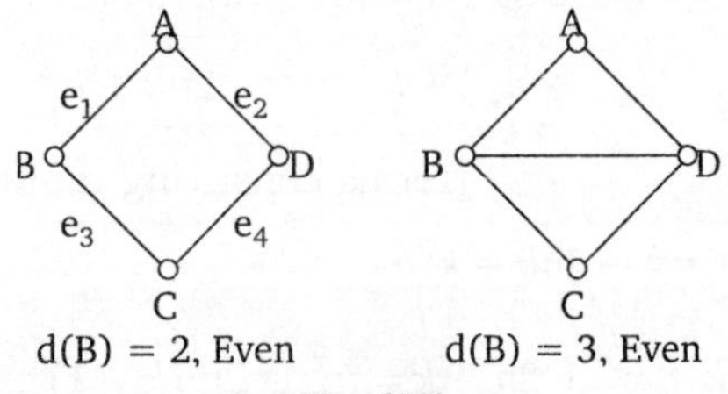

Fig. (53)

☛ Problem of Euler, is not possible because two vertex has odd degree 3 and 5.

9.13.2 Eulerian Circuit

An Eulerian circuit is a path through a graph in which the initial vertex appears second time as the terminal vertex.

9.13.3 Traversable Graph

A graph G is said to be traversable, if it has a path. A Euler path uses every edge exactly once but vertices may be repeated.

Theorem 1. *A given connected graph G is an Euler graph iff all vertices of G are of even degree.*

Proof : Suppose that G is an Euler graph. It therefore contains an Euler line (which is closed walk). In tracking this walk, we observe that every time the walk meets a vertex v, it goes through two "new" edges incident on v with one we "entered" v and with other we "exited". This is true not only of all intermediate vertices of the walk but also of the terminal vertex, because we exited and entered the same vertex at the beginning and end of the walk respectively. Thus, if G is an Euler graph, the degree of vertex is even.

Conversely, assume that all vertices of G are of even degree. Now, we construct a walk starting at an arbitrary vertex v and going through the edge of G such that no edge is traced more than once. We continue tracking as far as possible. Since vertex is of even degree, we can exit from every vertex we enter. The tracing can not stop at any vertex but v. Since v is also of even degree, we shall eventually reach v when the tracing comes to an end. If this closed walk h we first traced includes all the edges of G, G is an Euler graph. If not, we remove from G all the edges in h and obtain a subgraph h' of G formed by the remaining edges. Since both G and h have all their vertices of even degree the degree of the vertices of h' are also even. Moreover, h' must touch h at least at one vertex a, because a is connected. Starting from a we can again construct a new walk in graph h'. Since all the vertices of h' are of even degree. This walk in h' must terminate at vertex a, but this walk in h' can be combined with h to form a new walk, which start and ends at vertex v and has more edges than h. This process can be repeated until we obtain a closed walk that traverses all the edges of G. Thus, G is an Euler graph.

9.13.4 Open Eulerian or Unicursal Line

If an open walk in a graph is such that all the edges are traversed once and only once.

A graph containing a universal line is called universal graph. We start from a but not end at a and end at b so it is an unicursal graph.

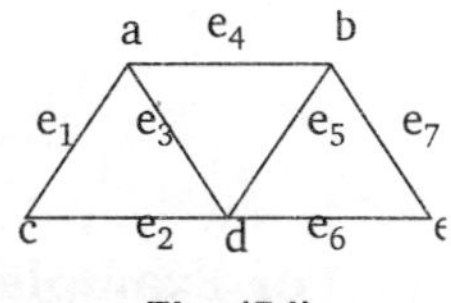

Fig. (54)

$$a\ e_1\ c\ e_2\ de_3\ ae_4\ be_5\ de_6\ e\ e_7\ b$$

Theorem 2. *In a connected graph G with exactly 2k odd vertices, there exist k edge-disjoint subgraphs such that they together contain all edges of G and that each is a unicursal graph.*

Proof : Let the odd vertices of the given graph G be named $v_1, v_2 \ldots, v_k, w_1, w_2, \ldots, w_k$ in any arbitrary order. Add k edges to G between the vertex points $(v_1, w_1)\ (v_2\ w_2) \ldots (v_k, w_k)$ to form a new graph G'.

Since every vertex of G' is of even degree, G' consists of an Euler line ρ. Now, if we remove from ρ the k edges, we just added, ρ will be split into k walks, each of which is a unicursal line. The first removal will leave a single unicursal line. The second removal will split that into two unicursal lines and each successive removal will split a unicursal line into two unicursal line, until there are k of them.

9.14 HAMILTONIAN PATH AND CIRCUITS

9.14.1 Hamiltonian Circuit

In a graph $G = (V, E)$, a Hamiltonian circuit is defined to be a closed walk which traverse every vertex of G exactly once except the starting vertex.

Definition : *A closed walk is said to be a Hamiltonian circuit if the closed walk contains every vertex of G and the degree of every vertex is 2.*

9.14.2 Hamiltonian Path

A path that passes through each of the vertices in a graph G exactly once, is called a hamiltonian path, *i.e.*, by removing any edge from the Hamiltonian circuit, we obtain a Hamiltonian path.

9.14.3 Hamiltonian Graph

A graph $G\ (V, E)$ is called Hamiltonian graph if it contains Hamiltonian circuits.

- Hamiltonian circuit is a subgraph of a graph G and Hamiltonian path is a subgraph of the Hamiltonian circuit. Thus, every graph that has a Hamiltonian circuit also has a Hamiltonian path.

- There exists many graphs that have Hamiltonian paths but do not have Hamiltonian circuits.
- In Hamiltonian circuits, the number of edges is exactly equal to the number of vertices.

9.15 LABELLED GRAPH

A graph G in which each vertex or edge is assumed a unique name or label (no two vertex have same name), then it is called labelled graph. In particular of each edge e of G is assigned a non-negative $l(e)$. The $l(e)$ is called the weight of length of (e).

For Example : Consider the graph

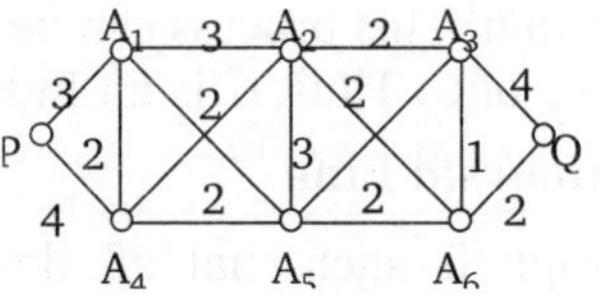

Fig. (55)

It is weighted as well as labeled graph.

For Example : consider the graph

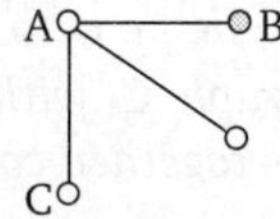

Fig. (56)

It is a labeled graph w.r.t. vertices.

ILLUSTRATIVE EXAMPLES

1. *Construct labeled graph with four vertices.*

Solution. We can construct sixteen labeled graph with four vertices given below :

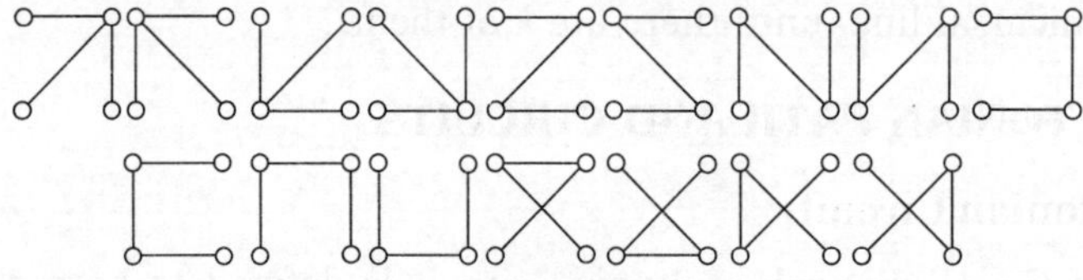

Fig. (57)

2. *Which of the following undirected graph have Eulerian circuit and path.*

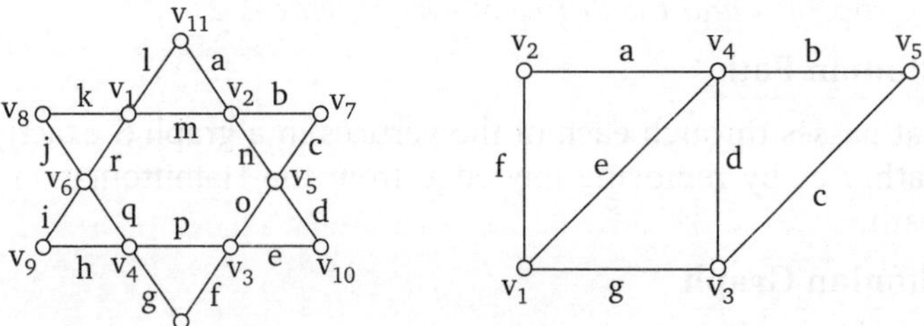

Fig. (58)

Solution. (i) This graph has an Eulerian circuit $v_1\, m\, v_2\, n\, v_5\, o\, v_3\, p\, v_4\, q\, v_6\, r\, v_1$ $l\, v_{11}\, a\, v_2\, b\, v_7\, c\, v_5\, d\, v_{10}\, e\, v_3 f\, v_{12}\, g\, v_4\, h\, v_9\, i\, v_6\, j\, v_8\, k\, v_1$

(ii) This graph has an Eulerian path $v_1\, f\, v_2\, a\, v_4\, b\, v_5\, c\, v_3\, d\, v_4\, e\, v_1\, g\, v_3$, but not Eulerian circuit.

3. *Whether the following graphs are Eulerian circuits or Eulerian path.*

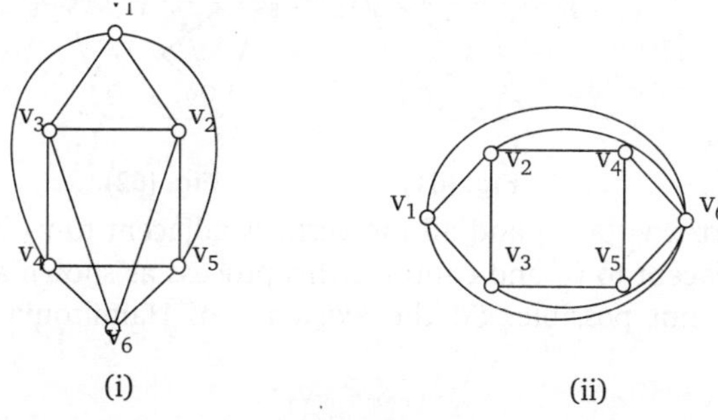

Fig. (59)

Solution. We see that given graph (i) and (ii) are connected but in graph (i) all the vertices are of even degree. Therefore, graph (i) has an Eulerian circuit. But in graph (ii) degree of v_4 and v_5 are of odd degree. Therefore, graph (ii) has an Eulerian path not Eulerian circuit.

4. *Draw a Hamiltonian circuit from the following graphs.*

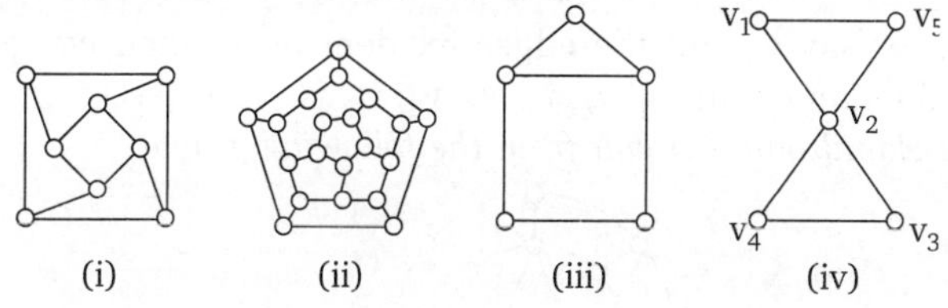

Fig. (60)

Solution. (i) This graph is a Hamiltonian circuit. We can start from any vertex and then move by covering all the vertices of the graph, hence we obtain the following Hamiltonian circuit.

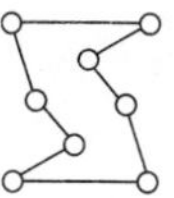

Fig. (61)

(ii) This graph is Hamiltonian circuit as shown below :

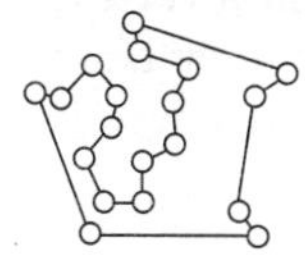

Fig. (62)

(iii) This graph is a Hamiltonian circuit as shown below :

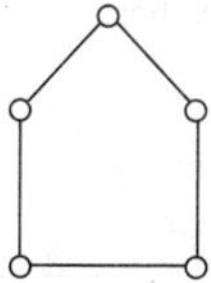

Fig. (63)

(iv) This graph is not a Hamiltonian circuit as there are four edges incident at v_2.

5. *Show that Peterson graph shown below does not have a Hamiltonian circuit but the subgraph obtained by deleting a vertex and all edges incident on it does have a Hamiltonian circuit.*

Solution.

Fig. (61) Fig. (62)

We label the vertex v_1 by v_A and all the vertices adjacent to v_A by v_B. Then, we label all the vertices by v_A adjacent to v_B and continue this process as shown above. Thus, we see that alternate labelling is not possible. So the existence of Hamiltonian path or circuit is not possible.

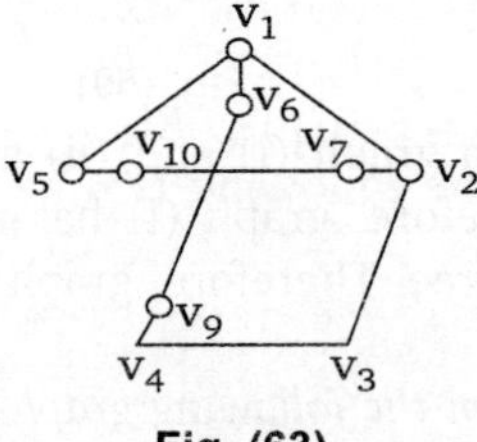

Fig. (63)

If we delete one vertex, say v_8 and the edges incident on it, then we get the adjacent graph. This graph is Hamiltonian circuit $v_1, v_6, v_9, v_4, v_3, v_2, v_7, v_6, v_{10}, v_5, v_1$.

6. *Is it possible to draw a Hamiltonian circuit from the following graph.*

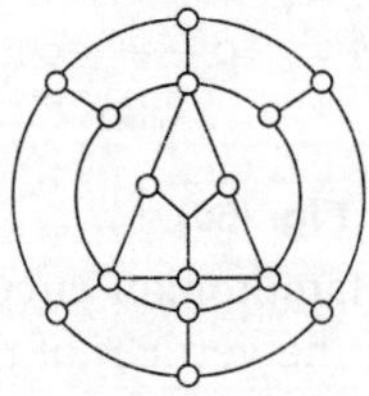
Fig. (64)

Solution. All the edges incident at a vertex. Only two can be included in a Hamiltonian circuit. Count the number of edges that must be excluded. We find 13 edges must be excluded from the graph. The number of remaining edges is insufficient to form a Hamiltonian circuit.

Exercise 9.2

1. Prove that in a complete graph G with n (odd integer greater than or equal to 3) vertices there are $(n-1)/2$ edge disjoint Hamiltonian circuits.
2. Which of the following graphs have an Eulerian circuit.

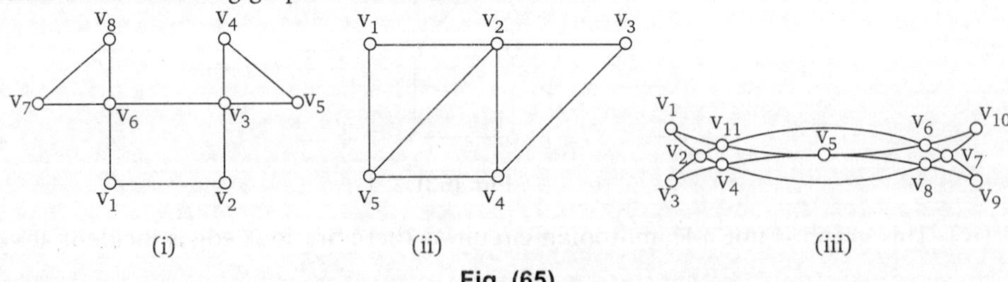

Fig. (65)

3. Check whether the adjoining graph has a Hamiltonian circuit or not.

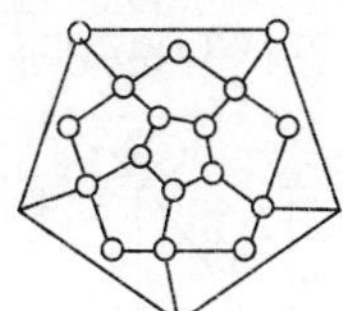

Fig. (66)

4. Draw a connected graph that becomes disconnected when any edge is removed from it.

9.16 OPERATIONS ON GRAPHS

9.16.1 Union of Two Graphs

The union of two graphs $G_1 = (V_1, E_1)$ and $G_2 = (V_2, E_2)$ is a graph denoted by $G_1 \cup G_2$ is defined by

$$G_1 \cup G_2 = (V_1 \cup V_2, E_1 \cup E_2)$$

i.e., $V(G_1 \cup G_2) = V(G_1) \cup V(G_2) = V(G)$

and $E(G_1 \cup G_2) = E(G_1) \cup E(G_2) = E(G)$

For example : Consider the following two graphs :

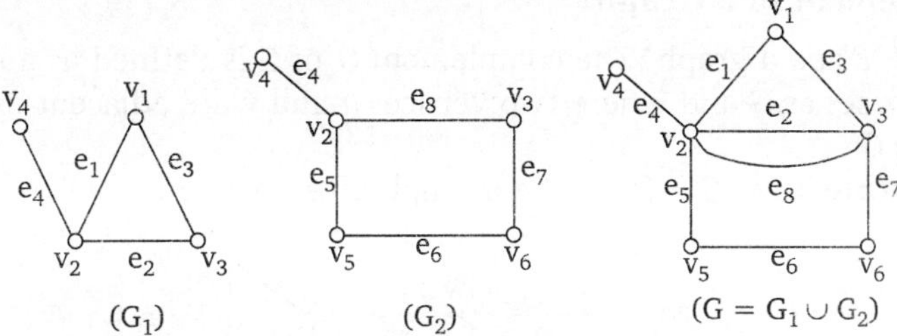

Fig. (67)

9.16.2 Intersection of Two Graphs

The intersection of two graphs $G_1 = (V_1, E_1)$ and $G_2 = (V_2, E_2)$ is a graph denoted by $G_1 \cap G_2$ defined by

$$G_1 \cap G_2 = (V_1 \cap V_2, E_1 \cap E_2)$$

i.e., $V(G_1 \cap G_2) = V(G_1) \cap V(G_2) = V(G)$

and $E(G_1 \cap G_2) = E(G_1) \cap E(G_2) = E(G)$

For example :

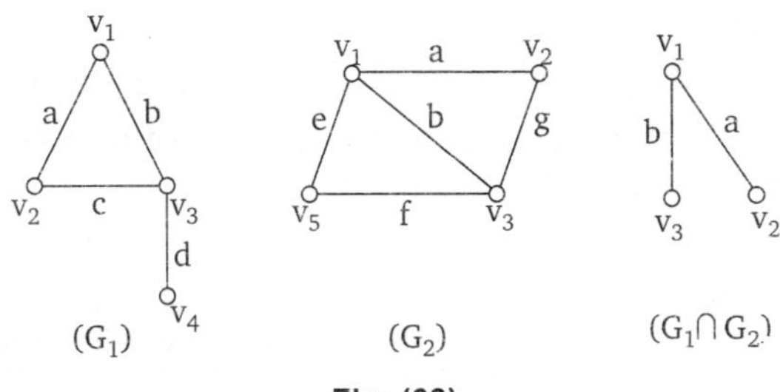

Fig. (68)

9.16.3 Ring Sum of Two Graphs

Let $G_1 = (V_1, E_1)$ and $G_2 = (V_2, E_2)$ be two graphs. The ring sum of G_1 and G_2 is a graph denoted by $G_1 \oplus G_2$ and is defined by

$$G_1 \oplus G_2 = (V_1 \cup V_2, E_1 \cup E_2 - E_1 \cap E_2) \quad i.e., \quad V(G) = V(G_1) \cup V(G_2)$$

and $E(G) = E(G_1) \cup E(G_2) - E(G_1) \cap E(G_2) = E(G_1) \Delta E(G_2)$

For example : The ring sum of two graphs defined above is given below

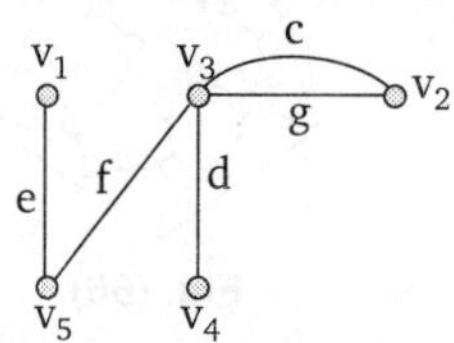

Fig. (69)

- Union, intersection and ring sums of graphs hold commutative law.
- If G_1 and G_2 are graphs such that $V(G_1) \cap V(G_2) = \phi$, then $G_1 \cup G_2$ is denoted by $G_1 + G_2$ and is called **sum of graphs.**
- If $G_1, G_2, ..., G_r$ are pairwise vertex disjoint graphs, each of which is isomorphic to G, then $G_1 + G_2 + ... + G_r$ is denoted by rG.
- If G_2 is a subgraph of G_1, then $G_1 \cup G_2 = G_1$, $G_1 \cap G_2 = G_2$ and $G_1 \oplus G_2$ is a graph obtained by G_1 by deleting all the edges of G_2.
- Every disconnected graph is the sum graph of its components.
- If G_2 is a null graph, then $G_1 \oplus G_2 = G_1 \cup G_2 = G_1$, $G_1 \cap G_2$ = null graph.

9.16.4 Complement of a Graph

Let $G = (V, E)$ be a graph. The complement $\overline{G}$ of G is defined as a simple graph with the same vertex set as G and where two vertices u and v are adjacent only when they are not adjacent in G.

For example : Let $G = (V, E)$ be a graph given below :

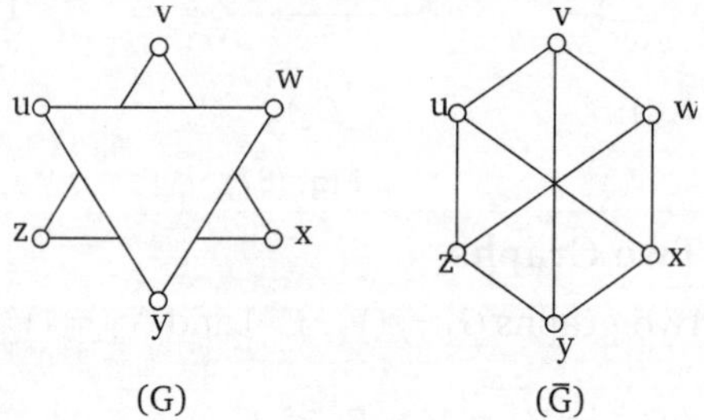

Fig. (70)

- The graph G is said to be self complementary if it is isomorphic to its complements, *i.e.*, $\overline{G} \cong G$.

9.16.5 Product of Graphs

Let $G_1 = (V_1, E_1)$ and $G_2 = (V_2, E_2)$ be two graphs. Consider any point $U = (u_1, u_2)$ and $V = (v_1, v_2)$ in $V = V_1 \times V_2$. Then U and V are adjacent in $G_1 \times G_2$ wherever $u_1 = v_1$ and u_2 adjacent v_2 or $u_2 = v_2$ and u_1 adjacent v_1.

For example :

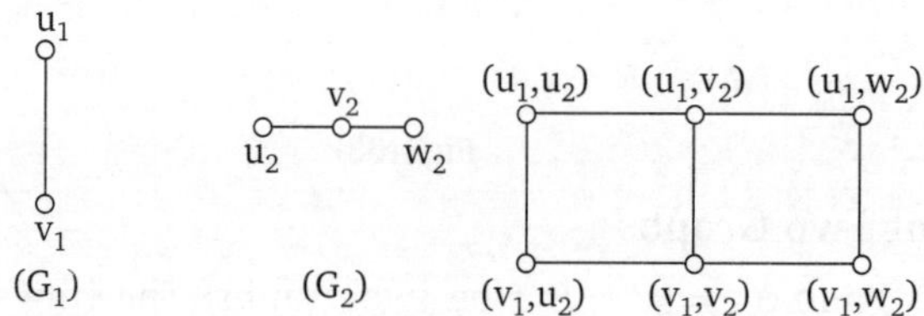

Fig. (71)

☛ The product of graphs holds the commutative law.

9.16.6 Composition or Lexigraphic Product of Graphs

Let $G_1 = (V_1, E_1)$ and $G_2 = (V_2, E_2)$ be two given graphs. The composition $G = G_1 [G_2]$ has $v = v_1 \neq v_2$ as its point set and $U = (u_1, u_2)$ is adjacent with $V = (v_1, v_2)$ whenever u_1 adjacent v_1 or $u_1 = v_1$ and u_2 adjacent v_2.

For example :

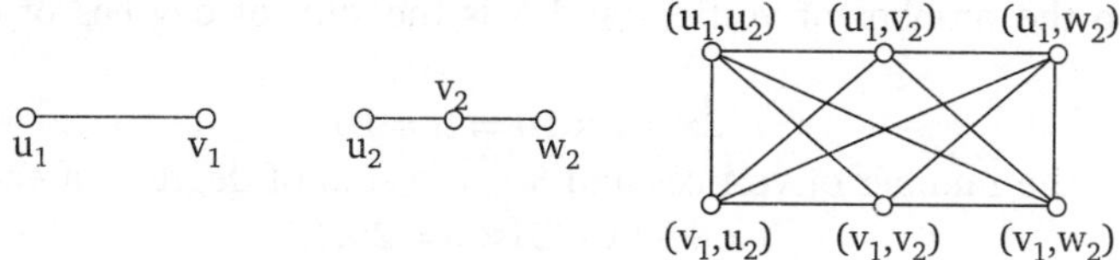

Fig. (72)

☛ The composition of graphs does not hold the commutative law.

9.16.7 Wedge Union

Let $G_1 = (V_1, E_1)$ and $G_2 = (V_2, E_2)$ be two given graphs. The join (wedge union) of two vertex disjoint graphs G_1 and G_2 denoted by $G_1 \cup G_2$ is a graph G such that

(i) $V(G) = V(G_1) \cup V(G_2)$

(ii) $E(G) = E(G_1) \cup E(G_2) \cup \{xy : x \in v(G_1), y \in v(G_2)\}$

For example :

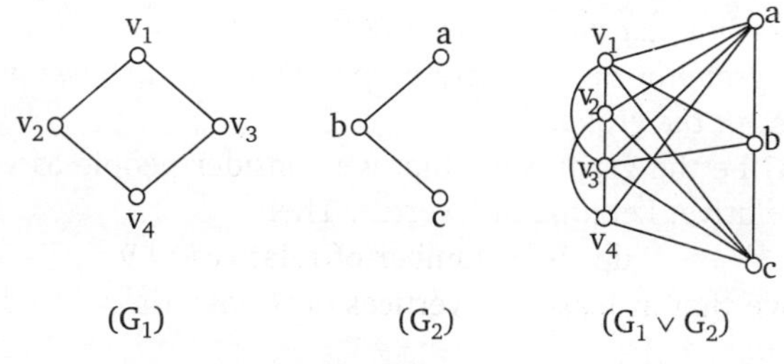

Fig. (73)

9.16.8 Normal Product or Strong Product

Let $G_1 = (V_1, E_1)$ and $G_2 = (V_2, E_2)$ be two given graphs. Then normal product of G_1 and G_2 denoted by $G_1 \bullet G_2$ is a graph G such that

(i) $V(G) = \{(x, y) : x \in (V(G_1), y \in V(G_2)\}$

(ii) $E(G) = [\{(a, b), (c, d)\}: a = c$ and $bd \in E(G_1)]$

or $[ac \in E(G_1)$ and $b = d]$ or $[ac \in E(G_1)$ and $bd \in E(G_1)]$

For example :

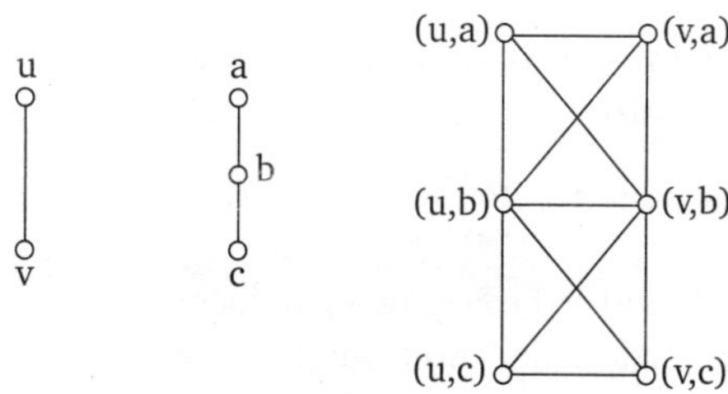

Fig. (74)

ILLUSTRATIVE EXAMPLES

1. *How many vertices do the following Graph have if they contain :*

(i) 16 edges and all vertices of degree 2 ?

(ii) 21 edges, 3 vertices of degree 4 and other each of degree 3 ?

(iii) 24 edges and all vertices of same degree.

Solution.

(i) Let x be the number of vertices and S is the sum of degrees of each vertex of the graph G, then

$$2x = 2 \times 16 \Rightarrow x = 16$$

(ii) Let x be the number of vertices and S is the sum of degrees of each vertex

$$3 \times 4 + (x - 3) \times 3 = 2 \times 21$$

$$\Rightarrow \quad 3x - 9 = 30 \qquad \Rightarrow \quad x = 13$$

(iii) Let x be the number of vertices and k be the degree of each vertices, then k satisfying

$$kx = 48$$

$$\Rightarrow \quad \text{If } x = 1, k = 48 \qquad \text{and if } x = 2, k = 24 \ldots \text{ etc.}$$

2. *What is the largest number of vertices in a graph with 35 edges if all vertices are of degree at least 2 ?*

Solution. Let x be the number of vertices, then

$$2 \times \frac{35}{x} \geq 3 \qquad \Rightarrow \quad 3x \leq 70, \text{ i.e., } x \leq \frac{70}{3}$$

Hence, the largest integer value satisfying this is 23.

3. *Show that in any group of two or more people, there are always two with exactly the same number of relatives in the group.*

Solution. Let G be the graph such that we consider people as vertices and joining two of them if they are relative. Let u be any vertex. Then

$$\text{deg. } u = \text{number of relatives of } u$$

We have to prove that at least two vertices of G have the same degree.

Let $V(G) = \{u_1, u_2, \ldots u_n\}$

Clearly, $0 \leq \text{deg. } u_i \leq n - 1$ for every value of i.

Let if possible no two vertices of G have the same degree. Then the degrees of $u_1, u_2, \ldots u_n$ are the integers $0, 1, 2, \ldots, n - 1$ in same order. However, a vertex of degree $n - 1$ is joined to every other vertex of G and therefore, no vertex can have degree zero, which is a contradiction. Hence, there exists two vertices of G with same degree.

4. *Show that the maximum number of edges all n vertices graphs with no triangles is* $\left[\frac{n^2}{4}\right]$.

Solution. Let G be a graph with n vertices. We can easily verify the required result for $n \leq 4$.

For $n > 4$, we prove the result by induction separately for odd n and even n.

Case 1 : When *n* is odd : Let us suppose result is true for all odd $n \leq 2m + 1$.

Let G be (n, q) graphs with $n = 2m + 3$ and no triangles. If $q = 0$, then $q \leq \left[\frac{n^2}{4}\right]$.

Hence, let $q > 0$. Let x and y be a pair of adjacent vertices in graph G. The subgraph $G' = G - \{x, y\}$ has $2m + 1$ vertices and no triangles. Thus, by induction hypothesis

$$q(G') \leq \left[\frac{(2m+1)^2}{4}\right] = \left[\frac{4m^2 + 4m + 1}{4}\right] = \left[m^2 + m + \frac{1}{4}\right] < m^2 + m \qquad \ldots(1)$$

Since G has no triangle, no vertices of G' can be adjacent to both x and y in G. ...(2)

Now, edges in graph G are of three types :

(i) edges of G' [$\leq m^2 + m$ in number by (1)]

(ii) edge between G' and $\{x, y\} \leq (2m + 1)$ in number (by (2))

(iii) edge xy

$$q \leq (m^2 + m) + (2m + 1) + 1 = m^2 + 3m + 2 = \frac{1}{4}[4m^2 + 12m + 8]$$

$$= \left(\frac{4m^2 + 12m + 9}{4} - \frac{1}{4}\right) = \left(\frac{(2m + 3)^2}{4}\right) = \left[\frac{n^2}{4}\right] \text{ for } n = 2m + 3$$

Also, for $n = 2m + 3$, the graph $k_{m+1, m+2}$ has no triangles and has

$$(m + 1)(m + 2) = m^2 + 3m + 2 = \left[\frac{n^2}{4}\right] \text{ edges.}$$

Hence, this maximum q is attached.

Case 2 : When *n* is even : Let us suppose result is valid for all even $n \leq 2m$.

Let the graph G be a (n, q) graph with $n = 2m + 2$ and no triangles. Let x and y be a pair of adjacent vertices in graph G and let $G' = G - \{x, y\}$

Now, G' has $2m$ vertices and no triangles.

Thus, by hypothesis $q(G') \leq \left[\frac{(2m)^2}{4}\right] = m^2$ ...(3)

Now, edges in G are of following three types :

(i) edges of G' ($\leq m^2$ in number by (3))

(ii) edges between G' and $\{x, y\}$ ($\leq 2m$ in number by an argument similar to (2))

(iii) edge xy

Therefore, $q \leq m^2 + 2m + 1 = (m + 1)^2 = \dfrac{(2m + 2)^2}{4} = \left[\dfrac{n^2}{4}\right]$

Hence, the result holds for even n. Further, we observe that for $n = 2m + 2$, $k_{m+1, m+1}$ is a $\left(n\left[\frac{n^2}{4}\right]\right)$ graph without triangles.

9.17 DIRECTED GRAPH

Definition : *A directed graph (digraph) $G = (V, E)$ consists of a set of vertices $V = \{v_1, v_2, \ldots\}$ and a set of edges $E = \{e_1, e_2, \ldots\}$ such that each edge is identified (or associated) by some ordered pair of vertices $\{v_i, v_j\}$.*

If $e_k = \{v_i, v_j\}$ is an edge then this edge in digraph is represented by a line segment between the vertices v_i and v_j along with an arrow from v_i to v_j, where v_i is called initial vertex and v_j is called the terminal vertex of the edge e_k. Also the edge e_k is incident out of the vertex v_i and is incident into the vertex v_j.

In a directed graph, we also used the following graph :

(i) the vertex v_i is the origin of the edge e_k

(ii) v_i and v_j are adjacent to each other.

(iii) v_j is a successor of v_i.

For example : Following graph is a directed graph with four vertices and seven edges.

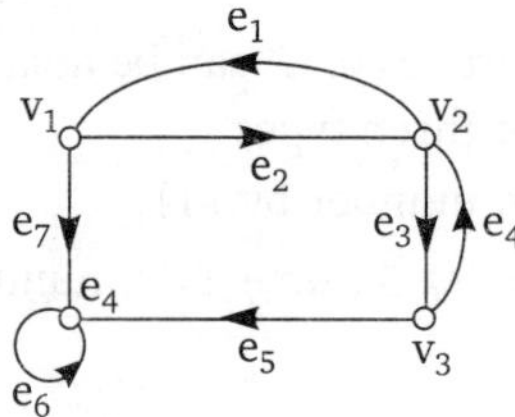

Fig. (75)

In this directed graph, the edge e_2 is incident out of the vertex v_1 and is incident into the vertex v_2. The edge e_1 is incident out of v_2 and is incident into v_1 and so on.

☛ A directed graph or digraph is also known as oriented graph.

9.17.1 Out degree and Indegree of a Vertex

The number of edges incident out of a vertex v_i is said to be the out degree (also called out valance or outward demidegree) of the vertex v_i and is denoted by $d^+(v_i)$. In a similar way, the number of edges incident into a vertex v_i is said to be the in degree (also called in-valance or inward demidegree) of the vertex v_i and is denoted by $d^-(v_i)$.

For Example : In the above directed graph

$d^+(v_1) = 2$ $\quad$ $d^-(v_1) = 1$

$d^+(v_2) = 3$ $\quad$ $d^-(v_2) = 1$

$d^+(v_3) = 1$ $\quad$ $d^-(v_3) = 2$

$d^+(v_4) = 1$ $\quad$ $d^-(v_4) = 3$

We observe that $d(v) = d^+(v) + d^-(v)$ for every vertex v in digraph.

9.17.2 Simple Digraph

A digraph is said to be simple if it has no self loop or parallel edge.

9.17.3 Symmetric Digraph

A digraph is said to be symmetric if for every edge (u, v), there is also an edge (v, u) in it.

9.17.4 Complete Symmetric Directed Digraph

A simple directed graph in which there is exactly an edge directed from every vertex to every other vertex is said to be complete symmetric directed graph. The complete symmetric graph contains $n(n-1)$ edges.

For example :

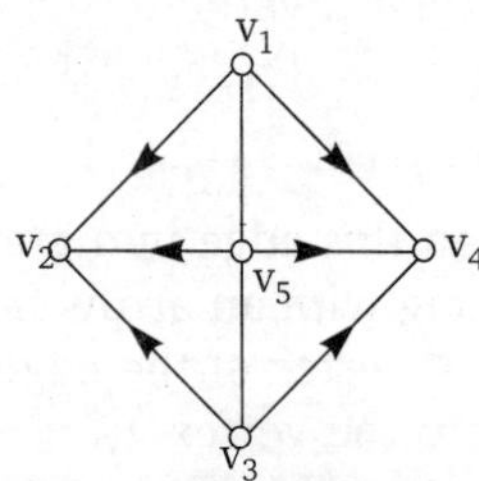

Fig. (76)

9.17.5 Asymmetric or Antisymmetric Digraph

A digraph which has almost one directed edge between a pair of vertices (self loops are allowed) is called its antisymmetric digraph.

9.17.6 Complete Asymmetric Directed Digraph

If a directed graph has exactly one edge directed between every pair of vertices then it is called complete asymmetric directed graph. It contains $^nC_2$ number of edges.

For example :

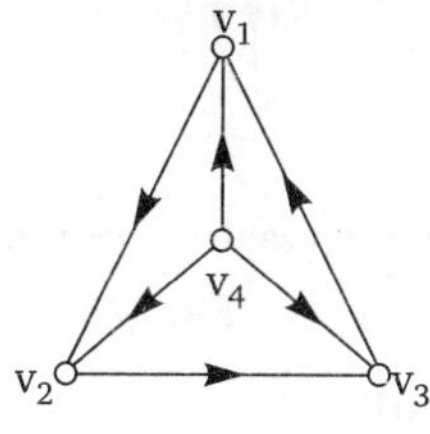

Fig. (77)

- A complete asymmetric directed graph is also called a tournament or a complete tournament.

9.17.7 Directed Walk

Let G be a directed graph. A directed walk in G from a vertex v_i to v_j is an alternating sequence of vertices and directed edges, beginning with the vertex v_i and ending with v_j such that each edge is directed from the vertex preceding it to the vertex following it. In a directed walk, no edge will appear more than once whereas a vertex can appear more than once.

9.17.8 Semi Walk

A walk in a directed graph means either a directed walk or a semi walk.

9.17.9 Directed Path

A directed path in a digraph is an open directed walk in which no edge appears more than once. In other words, a directed path P is an alternating sequence of vertices and edges, say $P = (v_0\ e_1\ v_1\ e_2\ v_2\ e_3\ v_3 \ldots e_n\ v_n)$ such that each edge e_i begins at the vertex v_{i-1} and ends at the vertex v_i.

- In a directed path, all vertices and edges are distinct.
- The length of the path is defined as the number of edges in path.

9.17.10 Semidirected Path

In a directed path, a semi directed path is defined as a path in the corresponding undirected graph.

- A semidirected path is not a directed path.

9.17.11 Connected Directed Graph

In directed path, there are two types of paths namely, directed path and semidirected paths. Thus, we define the following :

(i) Strongly Connected Directed Graph

A directed graph G is said to be strongly connected if for every two vertices u and v in G, there is path from u to v as well as path from v to u.

Definition : *A diagraph G is called strongly connected if there exists at least one directed path from every vertex to every other vertex.*

(ii) Weakly Connected Directed Graph.

A directed graph G is called weakly connected if its corresponding undirected graph is connected, but G is not strongly connected.

- A digraph which is not connected is known as disconnected.

For example :

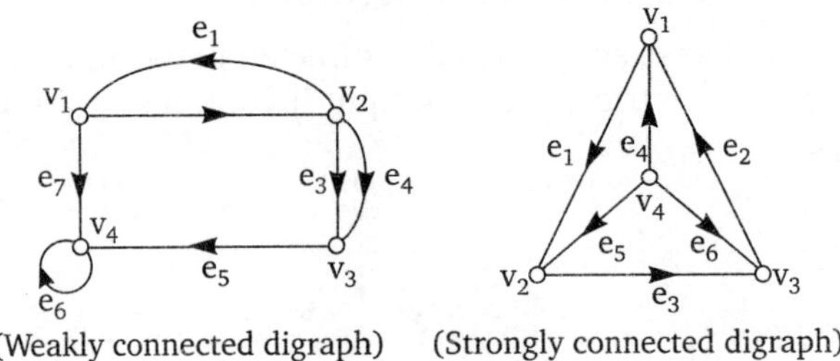

(Weakly connected digraph) (Strongly connected digraph)

Fig. (78)

9.17.12 Directed Complete Graph

A directed complete graph $G = (V, E)$ of n vertices is a graph in which each vertex is connected to every other vertex by an arrow. It is denoted by k_n. A simple directed graph in which there is exactly an edge directed from every vertex to every other vertex is said to be complete symmetric digraph.

For example :

k_3 :

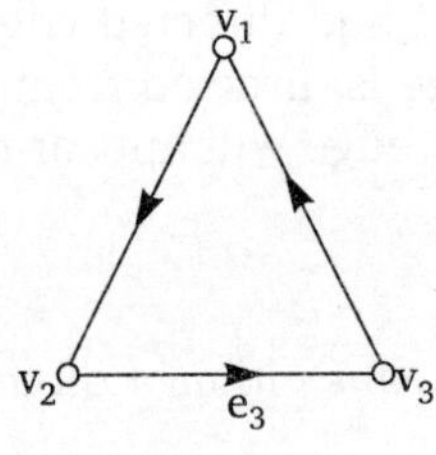

Fig. (79)

9.17.13 Directed Labeled Graph

A graph $G = (V, E)$ is said to be labeled if its edges are labeled with some name or data. We can write these labels in place of an ordered pair in the edge set.

For Example :

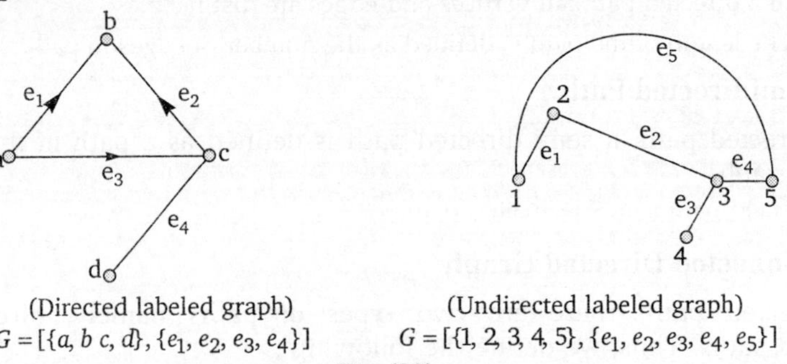

(Directed labeled graph) $G = [\{a, b\ c, d\}, \{e_1, e_2, e_3, e_4\}]$ (Undirected labeled graph) $G = [\{1, 2, 3, 4, 5\}, \{e_1, e_2, e_3, e_4, e_5\}]$

Fig. (80)

9.17.14 Balance Directed Graph or Pseudosymmetric Directed

A directed graph in which for every vertex of it, the in-degree is equal to the out-degree is called a balance directed graph or pseudosymmetric directed graph or isograph.

9.17.15 *k*-regular Directed Graph

If the in degree of all the vertices of a balance directed graph is equal to k, then it is known as a k-regular directed graph.

9.17.16 Condensation

The condensation G_c of a directed graph is a graph whose vertices are the fragments of the directed graph G and there is a directed edge from one vertex to the other vertex in G_c whenever there exists at least one directed edge from one fragment to the other corresponding fragment.

- The condensation of a strongly connected directed graph is an isolated vertex and no condensation consists a directed circuit.

9.17.17 Isomorphic Directed Graph

The directed graphs are said to be isomorphic to each other if there exists an isomorphism from the vertex set of one graph to other, which preserve adjacency and orientation as well as non-adjacency.

9.17.18 Accessibility

In a directed graph a vertex b is said to be accessible (or reachable) from vertex a if there is a directed path from a to b. Thus, a directed graph G is strongly connected if and only if every vertex in G is accessible from every other vertex.

9.17.19 Euler Directed Graph

A directed graph D is a closed directed walk (a directed walk that starts and ends at the same vertex) which traverses every edge of D exactly once is called a directed Euler line.

For :

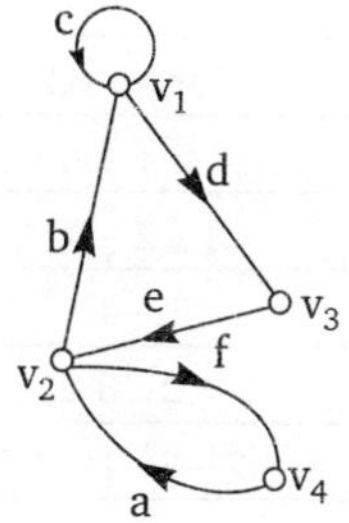

Fig. (81)

- A directed graph containing a directed Euler line is called an Euler directed graph.
- An Euler directed graph must be strongly connected.
- A directed graph is an Euler digraph if an only if it is connected and balanced.

ILLUSTRATIVE EXAMPLES

1. *Find the condensation of the graph*

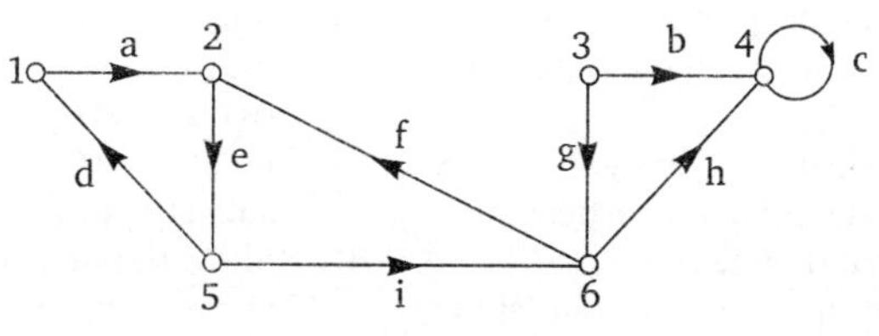

Fig. (82)

Solution. The fragments of the given graph are

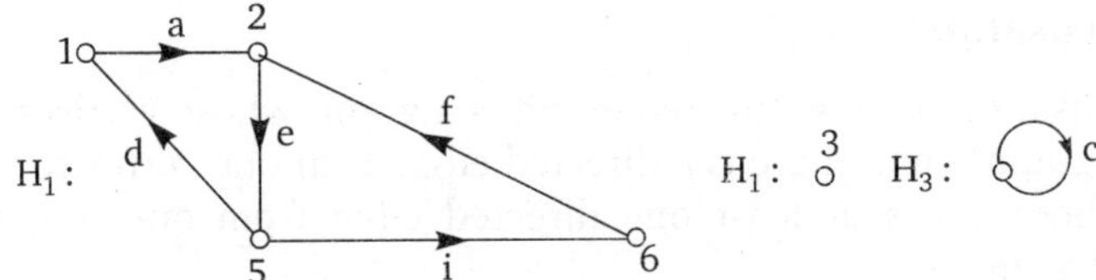

Hence, the required condensation of the graph is given by

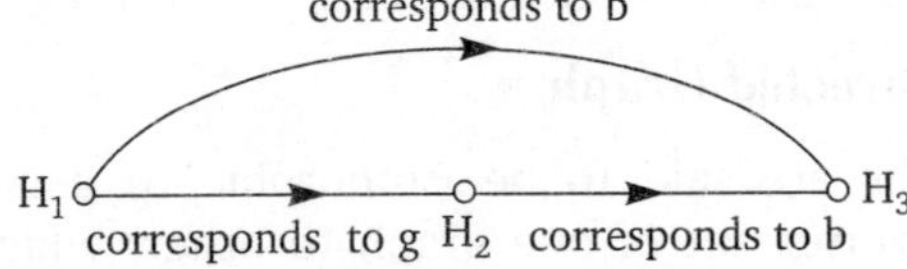

2. *Is the following directed graph strongly connected ?*

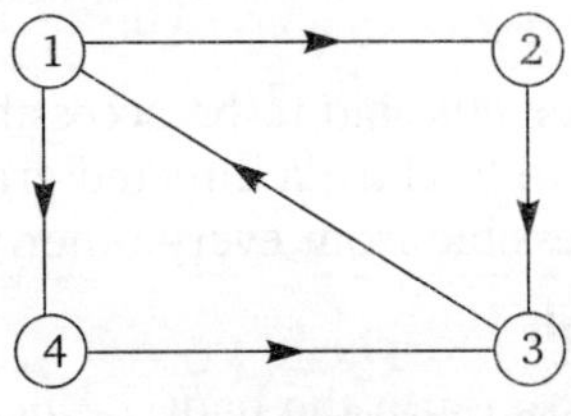

Fig. (83)

Solution. Here, the possible pair of vertices and the forward and backward paths between them are given below :

Pair of vertices	Path	
	Forward	**Backward**
(1, 2)	1 − 2	2 − 3 − 4
(1, 3)	1 − 2 − 3	3 − 1
(1, 4)	1 − 4	4 − 3 − 1
(2, 3)	2 − 3	3 − 1 − 2
(2, 4)	2 − 3 − 1 − 4	4 − 3 − 1 − 2
(3, 4)	4 − 3	4 − 3

We observe that between every pair of distinct vertices of the given graph, there exists a forward as well as backward path. hence, it is strongly connected.

Miscellaneous Exercise 9.3

1. Show that the degree of a vertex of a simple graph G on n vertices can not exceed $n-1$.
2. Show that for every even integer $n \geq 4$, there exists a 3-regular graph of order n.
3. Show that every cubic graph contains even number of points.
4. Show that for every odd integers $n \geq 5$, there exists a graph with $n + 1$ vertices such that n vertices have degree 3.
5. If a simple (p, q) graph is connected, show that $p \leq q + 1$.
6. If G_1 and G_2 are decomposition of a connected graph G, then show that $V(G_1) \cap V(G_2) \neq \phi$
7. Show that a graph G is disconnected if and only if its vertex set v can be partitioned into two subsets v_1 and v_2 such that there exists no edge in G whose one end vertex is in the subset v_1 and other in the subset v_2.
8. Show that if a and b are the only two degree vertices of a graph G then a and b are connected in G.

9. Show that the largest number of vertices in a graph with 35 edges of all vertices are degree at least 3 is 11.
10. A non-directed graph has 8 edges. Show that the number of vertices, if the degree of each vertex is 2 is given by 8.
11. A graph has 21 edges, 3 vertices of degree 4 and other vertices are on degree 3. Show that the number of vertices of G is 13.
12. Show that any two simple connected graph with n vertices, all of degree two are isomorphic.
13. Show that there can be no path longer than Hamiltonian path if it exists in a graph.
14. Show that a simple graph G is self-complimentary if G and G' are isomorphic.
15. In any graph G on n vertices, show that there exists no path between the vertices of length more than $n - 1$
16. Show that in a single graph with n vertices must be connected if it has more than $\frac{(n-1)(n-2)}{2}$ edges.

9.18 WEIGHTED GRAPH

Definition : *A graph in which a non-negative real number w (e) is assigned to its each edge e, is said to be weighted graph. Here, the number w (e) is called the weight or length of the edge e.*

9.18.1 Weight of a Path

Let G be a weighted graph. The weight of a path in the weighted graph G is equal to the sum of the weights of the edges in the path.

For example : Consider the following weighted graph :

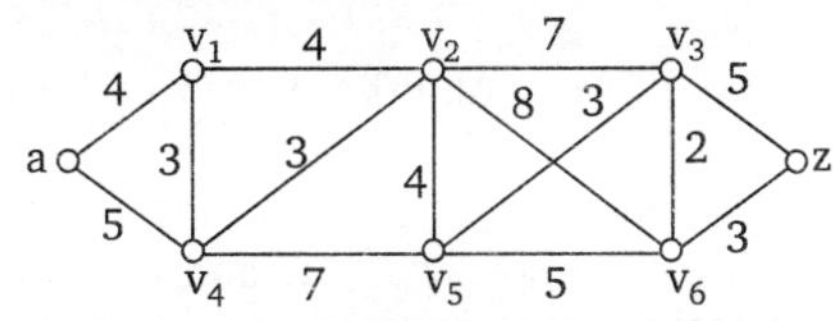

Fig. (84)

In the above weighted graph, some path from a to z and their weights are given below :

(i) Path: $(a\, v_1\, v_2\, v_5\, v_3\, v_6\, z)$; Weight : $4 + 4 + 4 + 3 + 2 + 3 = 20$
(ii) Path: $(a\, v_1\, v_4\, v_2\, v_6\, z)$; Weight : $4 + 3 + 3 + 8 + 3 = 21$
(ii) Path: $(a\, v_4\, v_5\, v_6\, v_3\, z)$; Weight : $5 + 7 + 5 + 2 + 5 = 24$ and so on.

9.18.2 Labeled Graph

A graph G is said to be labeled if its edges and/or vertices are assigned data of one kind or another.

9.19 SHORTEST PATH PROBLESMS

9.19.1 Shortest Path in a Graph without Weights

We know that the length of a path in a graph without weights denote the number of edges in a path and the shortest path is the path between two vertices in u and v that uses the least number of edges.

Breath First Search (BFS) Algorithm. Here, first we process the starting vertex s. Then we process all the neighbours of s. Then we process all the neighbours of neighbours of s and so on.

WORKING RULES

Step 1 : Label vertex s with 0, set $i = 0$.

Step 2 : Find all unlabeled vertices in G which are adjacent to vertices labeled t. If there are no such vertices, then t is not connected to s. If there are such vertices, label them $i + 1$.

Step 3 : If t is labeled go to Step 4. If not increase i to $i + 1$. Go to Step 2.

Step 4 : The length of the shortest path for s to $i + 1$. Stop.

Conclusion. Once the length of the shortest path is found from the previous algorithm. To find the actual shortest path, we use back-tracking algorithm. This algorithm uses the label λ (v) which are generated in BFS algorithm.

Back Tracking Algorithm for a shortest Path.

WORKING RULES

Step 1 : Set $i = \lambda\ (t)$ and assign $v_i = t$.

Step 2 : Obtain a vertex u adjacent to v_i and with $\lambda\ (u) = i - 1$, assign $v_{i-1} = u$.

Step 3 : If $i = 1$, stop. If not decrease i to $i - 1$ go to Step 2.

The Back Tracking Algorithm for the Number of Shortest Path

WORKING RULES

Step 1 : Set $i = \lambda\ (t)$ and $\mu\ (t) = 1$. All other vertices v for which $\lambda\ (v) = \lambda\ (t)$ are assigned $\mu\ (v) = 0$.

Step 2 : Fro each vertex v which satisfies $\lambda\ (v) = i - 1$, find the sum $\Sigma\ \mu\ (u)$ over all u's which satisfy the following conditions $\lambda\ (u) = i$ and v is adjacent to u, if there are parallel edges, $\mu\ (u)$ is repeated in this summation as many times as there are parallel edges. For each set v, set $\mu\ (v)$ equal to this sum.

Step 3 : If $i = 1$, stop. If $i \neq 1$, decrease i to $i - 1$ and go to Step 2.

ILLUSTRATIVE EXAMPLES

1. *Find the shortest path from vertex s to t and its length from the graph given below :*

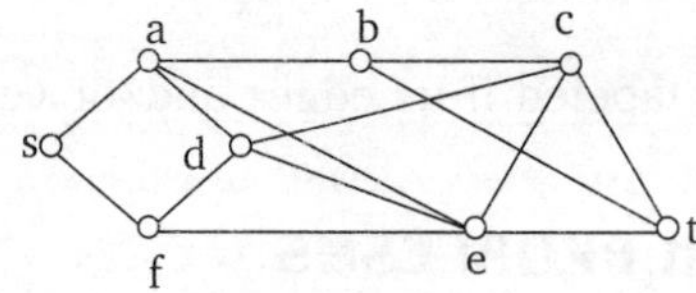

Fig. (85)

Solution. Using Breadth First Search, label $s = 0$.
Then, we labelled a and f by $0 + 1 = 1$ and b, d, e are labelled $1 + 1 = 2$.
Also c and t are labelled $2 + 1 = 3$.
Now, since t is labelled 3, so length of a shortest path from s to t is 3.

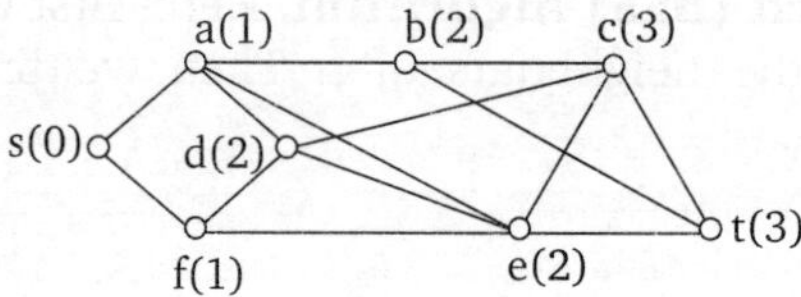

Fig. (86)

Now, using the second algorithm, since λ $(t) = 3$, we start with $i = 3$ and $v_i = t$.

Choose e (or b) adjacent to $v_3 = t$ with λ $(e) = 2$ and assign $v_2 = e$. Further, we choose f adjacent to $v_2 = e$ with λ $(f) = 1$ and assign $v_1 = f$. Finally, we take s adjacent to f with λ $(s) = 0$ and assign $v_0 = s$ which gives the shortest path v_0, v_1, v_2, v_3 = sfet from s to t.

9.19.2 Shortest Path in a Weighted Graph

We know that the length of a path in a weighted graph is the sum of the weights of the edges of this path and the shortest path between the two vertices is the minimum length of the path.

There are several different algorithm to find the shortest path between two vertices in a weight graph. Here, we shall discuss an algorithm given by Dutch mathematician Dijkstra.

9.19.3 Dijkstra Algorithm

This algorithm is used to obtain the shortest path from a specified vertex to another specified vertex. In this algorithm the vertices of the given graph are labelled. First of all the starting vertex A is assigned a permanent label O and the remaining vertices are assigned temporary label ∞. Then, at each iteration another vertex gets a permanent label according to following procedure.

WORKING RULES

Step 1. Every vertices j which are not permanently labelled yet gets a new temporary label whose value is given by min [(old value of j), (old value of i + ij)] where, i is the vertex permanently labelled in the previous iteration and d_{ij} is the length of the edge between i and j vertices. If i and j are not connected by an edge then $d_{ij} = \infty$.

Step 2. The smallest value among all the temporary labels is obtained which is assigned as the permanent label to the corresponding vertex. In case of a tie any one of the vertices may be selected for permanent labelling.

We repeat these two steps alternately till the destination vertex L gets a permanent label. In this procedure the second vertex to get a permanent label is the vertex nearest to the vertex A. the next vertex which is permanently labelled will be the next nearest vertex to A. In this way, each permanently labelled vertex will be nearest to A. Thus we will get the shortest distance and path.

ILLUSTRATIVE EXAMPLES

1. *Find the shortest path from v_1 to v_7 in the following weighted graph.*

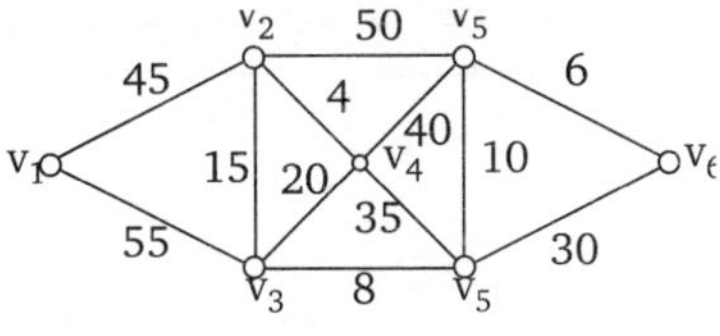

Fig. (87)

Solution. We construct table of temporary and permanent labels of vertices. The permanent labels are enclosed in squares and latest permanent label is marked as °. The lebelling procedure is given below :

i	v_1	v_2	v_3	v_4	v_5	v_6	v_7
0	[0]	[45]°	∞	∞	∞	∞	∞
1	[0]	[45]	55	∞	∞	∞	∞
2	[0]	[45]	55	[49]°	95	∞	∞
3	[0]	[45]	[55]°	[49]	89	84	∞
4	[0]	[45]	[55]	[49]	89	[63]°	∞
5	[0]	[45]	[55]	[49]	[73]°	[63]	93
6	[0]	[45]	[55]	[49]	[73]	[63]	[93]°

This shows that the shortest distance from v_1 to v_7 is 93. This method gives the shortest distance.

The shortest path can be obtained by going in backward from the terminal vertex such that we go to the predecessor vertex whose label differs by the length of the connecting edge. In case of tie, there are more than one shortest paths. The shortest path can also be obtained by keeping a record of vertices from which each vertex was labelled permanently. Here, the shortest path is $v_1 \rightarrow v_3 \rightarrow v_6 \rightarrow v_7$.

2. *Find the shortest path from v_1 to v_{12} in the following weighted graph.*

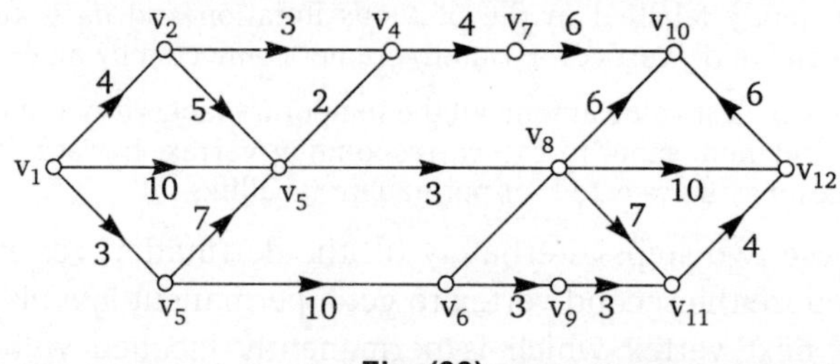

Fig. (88)

Solution. We shall use an array of length eleven (number of vertices), to show the temporary and permanent labels of the vertices as we go through the solution. The permanent labels will be shown enclosed in a square and the latest vertex assigned permanent label in the array is indicated by a mark □ . The labelling proceeds as following :

i	v_1	v_2	v_3	v_4	v_5	v_6	v_7	v_8	v_9	v_{10}	v_{11}	v_{12}
0	[0]	∞	∞	∞	∞	∞	∞	∞	∞	∞	∞	∞
1	[0]	4	[3]	∞	10	∞	∞	∞	∞	∞	∞	∞
2	[0]	[4]	[3]	∞	10	13	∞	∞	∞	∞	∞	∞
3	[0]	[4]	[3]	[7]	9	13	∞	∞	∞	∞	∞	∞
4	[0]	[4]	[3]	[7]	[9]	13	11	∞	∞	∞	∞	∞
5	[0]	[4]	[3]	[7]	[9]	13	[11]	12	∞	∞	∞	∞

6	0	4	3	7	9	13	11	12	∞	17	∞	∞
7	0	4	3	7	9	13	11	12	∞	17	19	22
8	0	4	3	7	9	13	11	12	16	17	19	22
9	0	4	3	7	9	13	11	12	16	17	19	22
10	0	4	3	7	9	13	11	12	16	17	19	22
11	0	4	3	7	9	13	11	12	16	17	19	22

Thus the shortest distance from v_1 to v_{12} is 22. Note that this method gives only the shortest distance. The shortest path can be easily obtained by going backward from the terminal vertex such that we go to that predecessor (vertex) whose label differs exactly by the length of the connecting edge. A tie indicates more than one shortest path. We can also determine the shortest path by keeping a record of the vertices from which each vertex was labelled permanently. This record can be stored in another array of length n, such that whenever a new permanent label is assigned to vertex j, the vertex from which j directly reached is recorded in the jth position to this array. In the above example, the shortest path is

$$v_1 \to v_2 \to v_5 \to v_8 \to v_{12}.$$

9.19.4 Alternative Form of Dijkstra Algorithm

Let $G = (V, E)$ be a weighted graph with w as a weight function from the set E to the set of positive real numbers. We have to determine the shortest path from the vertex a to z.

WORKING RULES

Step 1. Consider two subsets P_1 and T_1 of v such that $P_1 = \{a\}$ and $T_1 = V - P_1$. Now a shortest path from $a \in P_1$ to one of the vertices in T_1 is determined as follows :
For each vertex $x \in T_1$, let $l(x)$ be the length of a shortest path among all paths from a to x such that these path do not include any other vertex in T_1. If a and x are joined by the edge, then we take $l(x) = \infty$. Here $l(x)$ is known as index of x with regard to P_1. Let $t_1 \in T_1$ has the smallest index.

Step 2. Take $P_2 = \{a, t_1\}$, $T_2 = v - P_2$
Now, we find the index $l(x)$ of every vertex $x \in T_2$ w.r.t. P_2 as follows :
$l(x) = \min$ {index of x w.r.t. P_1, sum of the lengths of joining a to x through t_1}
Let the vertex $t_2 \in T_2$ has the minimum index $l(t_2)$ with respect to P_2.

Step 3. Let t_2 be the required vertex. We want to reach a, then process is stopped. If t_2 is not the required vertex, then take $P_3 = \{a, t_1, t_2\}$, $T_3 = v - P_3$.
Now, we find the index $l(x)$ of every vertex $x \in T_3$ w.r.t. P_3 as follows :
$l(x) = \min$ {index of x w.r.t. P_2, sum of the lengths of joining a to x through t_1 and t_2}
Let $t_3 \in T_3$ has the minimum index $l(t_3)$ w.r.t. P_3.

Step 4. Take $P_4 = \{a, t_1, t_2, t_3\}$, $T_4 = v - P_4$ and repeat the process till we get the required length of the shortest path from the vertex a to z.

ILLUSTRATIVE EXAMPLES

1. *Find the shortest path between* a and z *for the graph given below where numbers associated with edges are the weights.*

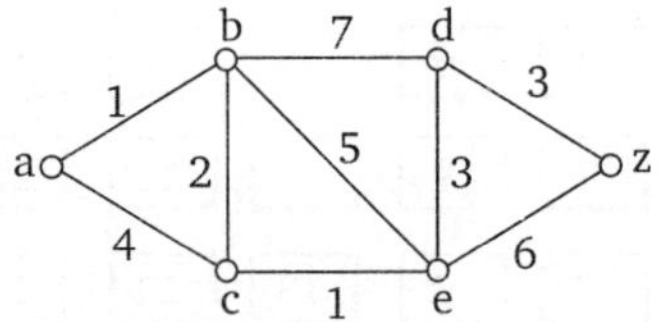

Fig. (89)

Solution. Let $G = (V, E)$ be the graph given above such that $V = \{a, b, c, d, e, z\}$

Then we use Dijkstra algorithm as follows :

Step (1) : Let $P_1 = \{a\}, T_1 = \{b, c, d, e, z\}$

$l(b) = 1, l(c) = 4, l(d) = \infty, \quad l(e) = \infty, l(z) = \infty$

$\Rightarrow \quad b \in T_1$ has the minimum index 1.

Step (2) : Taking $P_2 = \{a, b\}, T_2 = \{c, d, e, z\}$

Then, $\quad l(c) = \min\{4, 1, + 2\} = \min\{4, 3\} = 3,$

$l(d) = \min\{\infty, 1 + 7\} = 8,$

$l(e) = \min\{\infty, 1 + 5\} = 6$

$l(z) = \min\{\infty, 1 + \infty\} = \infty$

$\Rightarrow \quad c \in T_2$ has the minimum index 3.

Step (3). Taking $P_3 = \{a, b, c\}, T_3 = \{d, e, z\}$

Then, $\quad l(d) = \min\{8, 3 + \infty\} = 8$

$l(e) = \min\{6, 3 + 1\} = 4$

$l(z) = \min\{\infty, 3 + \infty\} = \infty$

$\Rightarrow \quad e \in T_3$ has the minimum index 4.

Step (4) : Taking $P_4 = \{a, b, c, e\}, T_4 = \{d, z\}$

$l(d) = \min\{8, 4 + 3\} = 7$

$l(z) = \min\{\infty, 4 + 6\} = 10$

$\Rightarrow \quad d \in T_4$ has the minimum index 7.

Step (5) : Taking $P_5 = \{a, b, c, e, d\}, T_5 = \{z\}$

$l(z) = \min\{10, 7 + 3\} = 10.$

Hence, the length of the shortest path from a to z is 10 with shortest path $a \to b \to c \to e \to d \to z$.

2. *Find the shortest path from a to z in the following weighted graph.*

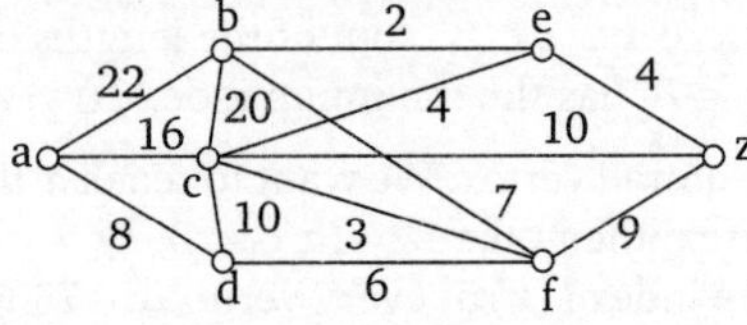

Fig. (90)

Solution. To find the shortest path from a to z, we proceed as follows :

Let $G = (V, E)$, where $V = \{a, b, c, d, e, f, z\}$

Step (1) : Let $P_1 = \{a\}, T_1 = \{b, c, d, e, f, z\}$

Then, $\quad l(b) = 22$

$l(c) = 16$

$l(d) = 8$

$l(e) = l(f) = l(z) = \infty$

$\Rightarrow \quad d \in T_1$ has the minimum index, *i.e.*, 8.

Step (2) : Taking $P_2 = \{a, d\}, T_2 = (b, c, e, f, z\}$

Then, $l(b) = \min(22, 8 + \infty) = 22$

$l(c) = \min(16, 8 + 10) = 16$

$l(e) = \min(\infty, 8 + \infty) = \infty$

$l(f) = \min(\infty, 8 + 6) = 14$

$l(z) = \min(\infty, 8 + \infty) = \infty$

$\Rightarrow$ $f \in T_2$ has the minimum index, *i.e.*, 14.

Step (3) : Taking $P_3 = \{a, d, f\}, T_3 = \{b, c, e, z\}$

Then, $l(b) = \min(22, 8 + 6 + 7) = 21$

$l(c) = \min(16, 8 + 6 + 3) = 16$

$l(e) = \min(\infty, 8 + 6 + \infty) = \infty$

$l(z) = \min(\infty, 8 + 6 + 9) = 23$

$\Rightarrow c \in T_3$ has the minimum index, *i.e.*, 16.

Step (4) : Taking $P_4 = \{a, d, f, c\}, T_4 = \{b, e, z\}$

Then, $l(b) = \min(21, 17 + 20) = 21$

$l(e) = \min(\infty, 17 + 4) = 21$

$l(z) = \min(23, 17 + 10) = 23$

$\Rightarrow$ $b \in T_4$ has the minimum index, *i.e.*, 21 (we can also take that $e \in T_4$ has the minimum index 21).

Step (5) : Taking $P_5 = \{a, d, f, c, b\}, T_5 = \{e, z\}$

Then, $l(e) = \min(21, 37 + 2) = 21$

$l(z) = \min(23, 37 + \infty) = 23$

$\Rightarrow$ $e \in T_5$ has the minimum index 21.

Step (6) : Let $P_6 = \{a, d, f, c, b, e\}, T_6 = \{z\}$

Then, $l(z) = \min(23, 39 + 4) = 23$

Hence, the length of the shortest path from a to z is 23. Also, the shortest path for this graph is $a \rightarrow d \rightarrow f \rightarrow z$.

3. *Find the shortest path between a and z in the following graph :*

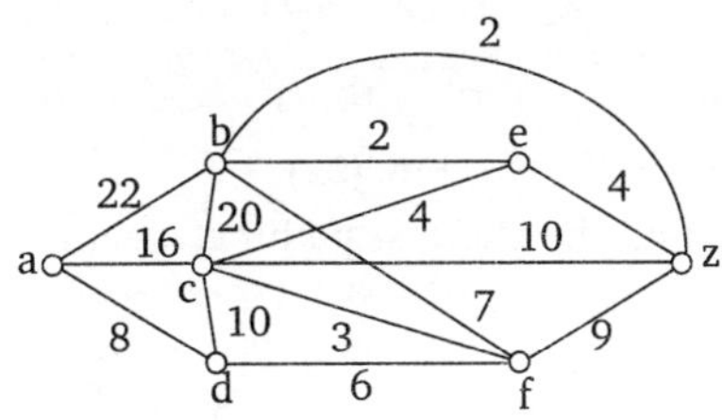

Fig. (91)

Solution. Let $G = (V, E)$ be the graph such that $V = \{a, b, c, d, e, f, z\}$

To find the shortest path and length, we proceed as follows :

Step (1) : Let $P_1 = \{a\}, T_1 = \{b, c, d, e, f, z\}$

Then, $l(b) = 22$

$l(c) = 16$

$l(d) = 8$

$l(e) = \infty = l(f) = l(z)$

$\Rightarrow$ $d \in T_1$ has the minimum index 8.

Step (2) : Let $P_2 = \{a, d\}, T_2 = \{b, c, e, f, z\}$

Then, $l(b) = \min(22, 8 + \infty) = 22$

$l(c) = \min(16, 8 + 10) = 16$

$l(e) = \min(\infty, 8 + \infty) = \infty$

$l(f) = \min(\infty, 8+6) = 14$
$l(z) = \min(\infty, 8+\infty) = \infty$

$\Rightarrow$ $f \in T_2$ has the minimum index, *i.e.*, 14.

Step (3) : Let $P_3 = \{a, d, f\}, T_3 = \{b, c, e, z\}$

Then, $l(b) = \min(22, 14+7) = 21$
$l(c) = \min(16, 14+3) = 16$
$l(e) = \min(\infty, 14+\infty) = \infty$
$l(z) = \min(\infty, 14+9) = 23$

$\Rightarrow$ $e \in T_4$ has the minimum index, *i.e.*, 16.

Step (4) : Let $P_4 = \{a, d, f, c\}, T_4 = \{b, e, z\}$

Then, $l(b) = \min(21, 17+20) = 21$
$l(e) = \min(\infty, 17+4) = 21$
$l(z) = \min(23, 17+10) = 23$

$\Rightarrow$ $e \in T_4$ has the minimum index 21.

Step (5) : Let $P_5 = \{a, d, f, c, e\}, T_5 = \{b, z\}$

Then, $l(b) = \min(21, 21+2) = 21$
$l(z) = \min(23, 21+4) = 23$

$\Rightarrow$ $b \in T_5$ has the minimum index.

Step (6) : Let $P_6 = \{a, d, f, c, e, b\}, T_6 = \{z\}$

Then, $l(z) = \min(23, 23+2) = 23$

Hence, the length of the shortest path from a to z is 23 and the shortest path for this graph is $a \to d \to f \to z$.

Exercise 9.4

1. Use the shortest path algorithm to find the shortest path from A to G in the weighted graph.

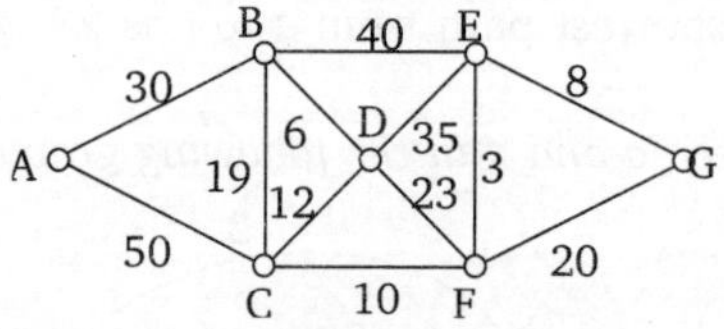

Fig. (92)

2. Use Dijkstra's algorithm to find the shortest path between the indicated vertices in the given weighted graphs.

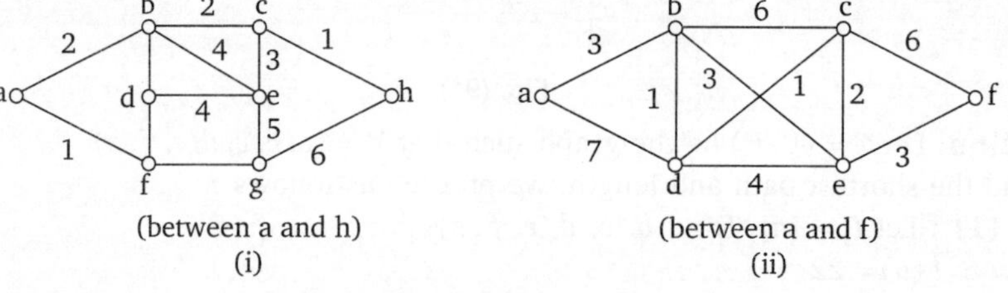

Fig. (93)

ANSWERS

1. $A \to B \to C \to F \to E \to G$, with length 77.
2. (i) $a \to b \to c \to h$; (ii) $a \to b \to d \to c \to e \to f$.

9.20 MATRIX REPRESENTATION OF GRAPHS

A graph can be represented inside a computer by using the adjacency matrix or incidence matrix of a graph matrices and graphs have many important applications in electrical network analysis and operation research.

9.20.1 Binary Matrix

A matrix whose entries are either 0 or 1 is known as binary matrix.

9.20.2 Incidence Matrix

Let G be a graph with n vertices, e edges and no self loops. Then we define the incidence matrix $A = [a_{ij}]$ as follows :

$$a_{ij} = \begin{cases} 1, & \text{if the edge } e_j \text{ incident on the vertex } v_i \\ 0, & \text{otherwise} \end{cases}$$

For example : Then, incidence matrix

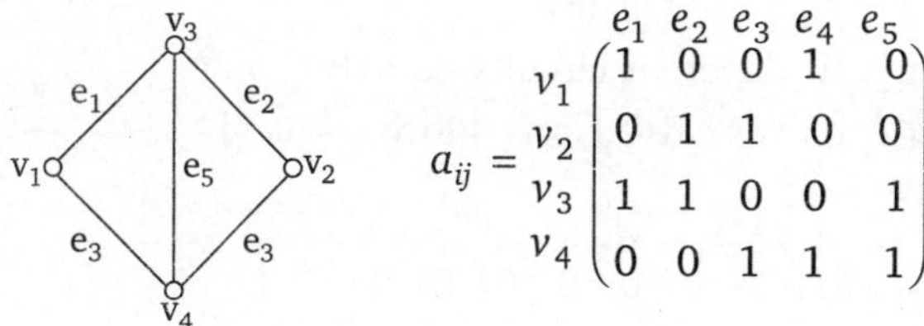

Fig. (94)

- There is a row for every vertex and a column for every edge in the incidence matrix.
- The incidence matrix of a graph is also called vertex edge incidence matrix.
- The incidence matrix is a binary matrix, *i.e.*, it contains only two elements 0 and 1.
- The incidence matrix of a graph G depends upon the ordering of the vertices and edges in G. Therefore, for different ordering we shall get different incidence matrix of the same graph G. But by interchanging some of the rows and columns of the matrix, any one of the incidence matrix of G can be obtained from another incidence matrix of G.

9.20.3 Some Important Observations

(1) The sum of each row is equal to the degree of the corresponding vertex.

(2) Every row is incident on exactly two vertices, thus each column of the incidence matrix A has exactly two 1's.

(3) If a row has all zeros, then the corresponding vertex is an isolated vertex.

(4) If graph has parallel edges, then the corresponding columns in $A = [a_{ij}]$ are identical

(5) If G is disconnected graph and G_1 and G_2 are its components, then incidence matrix can be written as

$$A\ (G) = \left[\begin{array}{c:c} A\ (G_2) & 0 \\ \hdashline 0 & A\ (G_2) \end{array}\right]$$

(6) Two graphs G_1 and G_2 are isomorphic if and only if their incidence matrix differ only by permutations of rows and column of $A\ (G)$.

(7) Permutation of any two columns represents the relabelling of the edges of graph G.

(8) Permutation of any two rows corresponds to the rearrangement of the vertices of graph G.

9.20.4 Circuit Matrix

Let $G = (V, E)$ be a graph with k different circuit and edge e. Then we define a circuit matrix $C(G) = [C_{ij}]$ of order $k \times e$ as follows

$$C_{ij} = \begin{cases} 1 & \text{if } i^{\text{th}} \text{ circuit consists of } j^{\text{th}} \text{ edge} \\ 0 & \text{otherwise} \end{cases}$$

For example : Consider the following graph

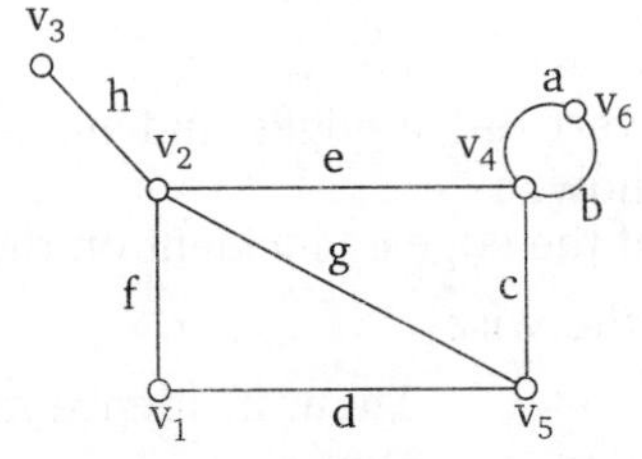

Fig. (95)

Here, we have following four different circuits namely

$$\{a, b\}, \{c, e, g\}, \{d, f, g\}, \text{ and } \{c, d, f, e\}$$

Then circuit matrix is given by

$$C[G] = \begin{array}{c} \\ C_1 \\ C_2 \\ C_3 \\ C_4 \end{array} \begin{array}{c} \begin{matrix} a & b & c & d & e & f & g & h \end{matrix} \\ \begin{pmatrix} 1 & 1 & 0 & 0 & 0 & 0 & 0 & 0 \\ 0 & 0 & 1 & 0 & 1 & 0 & 1 & 0 \\ 0 & 0 & 0 & 1 & 0 & 1 & 1 & 0 \\ 0 & 0 & 1 & 1 & 1 & 1 & 0 & 0 \end{pmatrix} \end{array}$$

☛ The circuit matrix is defined when we have at least one circuit in the given graph.

9.20.5 Adjacency Matrix

Let G be a graph with n vertices and no parallel edge (self loops are allowed). Then the adjacency matrix of G is a $n \times n$ symmetric matrix $A = [a_{ij}]$ defined by

$$a_{ij} = \begin{cases} 1 & \text{if there is an edge between } i^{\text{th}} \text{ and } j^{\text{th}} \text{ vertices} \\ 0 & \text{if there is no edge between them} \end{cases}$$

For example : Consider the following graph.

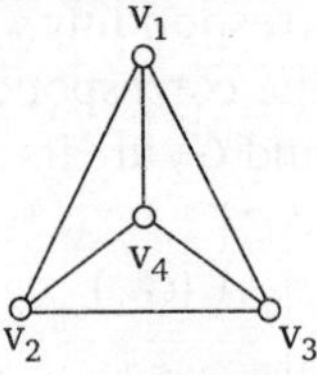

Fig. (96)

The adjacency matrix of this graph is given by

$$A = [a_{ij}] = \begin{array}{c} \\ V_1 \\ V_2 \\ V_3 \\ V_4 \end{array} \begin{array}{c} \begin{matrix} V_1 & V_2 & V_3 & V_4 \end{matrix} \\ \begin{pmatrix} 0 & 1 & 1 & 1 \\ 1 & 0 & 1 & 1 \\ 1 & 1 & 0 & 1 \\ 1 & 1 & 1 & 0 \end{pmatrix} \end{array}$$

- Since elements of adjacency matrix is either 0 or 1, therefore it is also known as bit matrix, binary matrix or boolean matrix.

9.20.6 Some Important Observations

(1) The principal diagonal of A has all zeros if and only if G has no self loop.

(2) If G is without self loops, then sum of any row (or column) is equal to the degree of the corresponding vertex, since matrix A is symmetric.

(3) Permutations of columns and of the corresponding rows imply the reordering of vertices.

(4) If G is a disconnected graph and G_1 and G_2 are its components. Then, adjoining matrix $A(G)$ can be written as

$$A(G) = \left[\begin{array}{c|c} A(G_2) & 0 \\ \hline 0 & A(G_2) \end{array}\right]$$

(5) If two graphs G_1 and G_2 are given, then they are isomorphic if and only if the adjacency of matrix of one can be derived from the adjacency matrix of the other by simply interchanging some of the columns and the corresponding rows.

(6) Number of 1′ s in a row (or a column) gives the degree of vertex corresponds to the row (or column) counting diagonal element twice. In general, deg (v) = number of 1′ s in an off diagonal row + 2 × diagonal entry.

9.20.7 Counting Path between Vertices

Let G be a graph with adjacency matrix $X(G)$ with respect to the ordering $v_1, v_2, \ldots, v_n$ (with directed or undirected edges, with multiple edges and loops allowed). The number of different path of length r from v_i to v_j, where r is a positive integer equal to the $(i, j)^{th}$ entry of X^r.

For example : Find the number of paths of length 4 from a to d in the graph G.

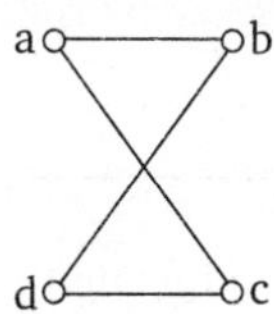

Fig. (97)

Solution. The adjacency matrix of G is

$$x(G) = \begin{array}{c} \\ a \\ b \\ c \\ d \end{array} \begin{array}{c} \begin{array}{cccc} a & b & c & d \end{array} \\ \begin{pmatrix} 0 & 1 & 1 & 0 \\ 1 & 0 & 0 & 1 \\ 1 & 0 & 0 & 1 \\ 0 & 1 & 1 & 0 \end{pmatrix} \end{array}$$

Thus, the number of paths of length 4 from a to d is the (1, 4) entry of $X^4(G)$. Since

$$X^4(G) = \begin{bmatrix} 8 & 0 & 0 & 8 \\ 0 & 8 & 8 & 0 \\ 0 & 8 & 8 & 0 \\ 8 & 0 & 0 & 8 \end{bmatrix}$$

There are exactly eight paths of length 4 from a to d. By the graph, we see that a, b, a, b, d; a, b, a, c, d; a, b, d, b, d; a, b, d, c, d; a, c, a, b, d; a, c, a, c, d; a, c, d, b, d and a, c, d, c, d are the eight paths from a to d.

ILLUSTRATIVE EXAMPLES

1. *Draw a graph with the following adjacency matrices :*

(i) $\begin{bmatrix} 0 & 1 & 1 & 1 \\ 1 & 0 & 1 & 0 \\ 1 & 1 & 0 & 1 \\ 1 & 0 & 1 & 0 \end{bmatrix}$ *(ii)* $\begin{bmatrix} 0 & 1 & 0 \\ 1 & 0 & 1 \\ 0 & 1 & 0 \end{bmatrix}$ *(iii)* $\begin{bmatrix} 0 & 1 & 1 \\ 1 & 0 & 1 \\ 1 & 1 & 0 \end{bmatrix}$

Solution. (i) A graph with the given adjacency matrix is shown in the figure.

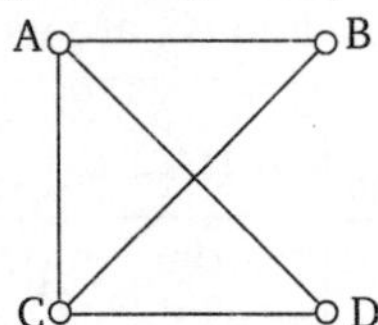

Fig. (98)

(ii) A graph with the given adjacency matix is shown below

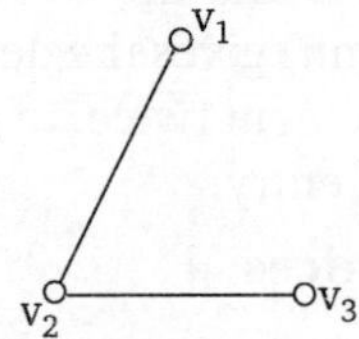

Fig. (99)

(iii) A graph with the given adjacency matrix is as given below

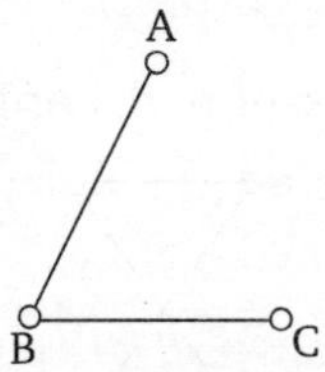

Fig. (100)

2. *Use an adjacency matrix to represent the following graphs*

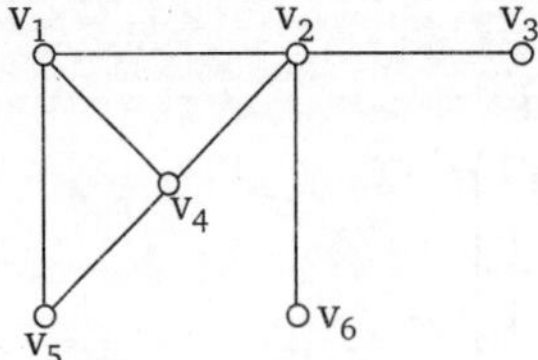

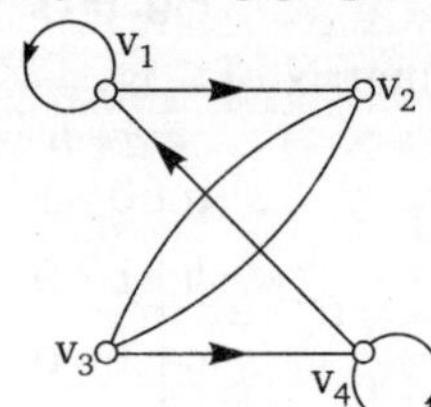

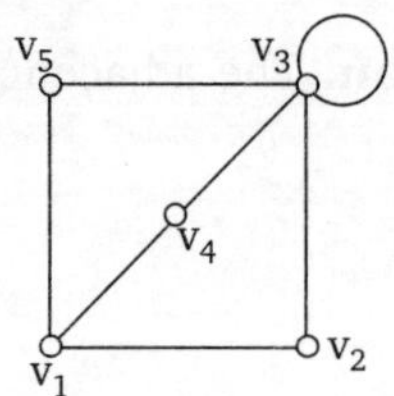

Fig. (101)

Solution. The matrix representation of the given graphs are shown below :

(i)
$$\begin{array}{c} \\ v_1 \\ v_2 \\ v_3 \\ v_4 \\ v_5 \\ v_6 \end{array} \begin{array}{c} \begin{array}{cccccc} v_1 & v_2 & v_3 & v_4 & v_5 & v_6 \end{array} \\ \begin{bmatrix} 0 & 1 & 0 & 1 & 1 & 0 \\ 1 & 0 & 1 & 1 & 0 & 1 \\ 0 & 1 & 0 & 0 & 0 & 0 \\ 1 & 1 & 0 & 0 & 1 & 0 \\ 1 & 0 & 0 & 1 & 0 & 0 \\ 0 & 1 & 0 & 0 & 0 & 0 \end{bmatrix} \end{array}$$

(ii)
$$\begin{array}{c} \\ v_1 \\ v_2 \\ v_3 \\ v_4 \end{array} \begin{array}{c} \begin{array}{cccc} v_1 & v_2 & v_3 & v_4 \end{array} \\ \begin{pmatrix} 1 & 1 & 0 & 0 \\ 0 & 0 & 1 & 0 \\ 0 & 1 & 0 & 1 \\ 1 & 0 & 0 & 1 \end{pmatrix} \end{array}$$

$$\text{(iii)}\quad \begin{array}{c} \\ v_1 \\ v_2 \\ v_3 \\ v_4 \\ v_5 \end{array}\begin{array}{c} \begin{array}{ccccc} v_1 & v_2 & v_3 & v_4 & v_5 \end{array} \\ \begin{pmatrix} 0 & 1 & 0 & 1 & 1 \\ 1 & 0 & 1 & 0 & 0 \\ 0 & 1 & 1 & 1 & 1 \\ 1 & 0 & 1 & 0 & 0 \\ 1 & 0 & 1 & 0 & 0 \end{pmatrix} \end{array}$$

3. *Find the incidence matrices of the following graphs :*

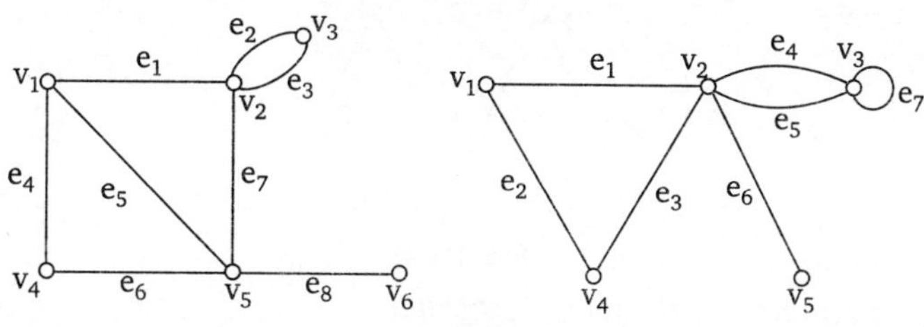

Fig. (102)

Solution : (i) The incidence matrix representing the given graph is shown below :

$$\begin{array}{c} \\ v_1 \\ v_2 \\ v_3 \\ v_4 \\ v_5 \\ v_6 \end{array}\begin{array}{c} \begin{array}{cccccccc} e_1 & e_2 & e_3 & e_4 & e_5 & e_6 & e_7 & e_8 \end{array} \\ \begin{pmatrix} 1 & 0 & 0 & 1 & 1 & 0 & 0 & 0 \\ 1 & 1 & 1 & 0 & 0 & 0 & 1 & 0 \\ 0 & 1 & 1 & 0 & 0 & 0 & 0 & 0 \\ 0 & 0 & 0 & 1 & 0 & 1 & 0 & 0 \\ 0 & 0 & 0 & 0 & 1 & 1 & 1 & 1 \\ 0 & 0 & 0 & 0 & 0 & 0 & 0 & 1 \end{pmatrix} \end{array}$$

(ii) The incidence matrix representing the given graph is

$$\begin{array}{c} \\ v_1 \\ v_2 \\ v_3 \\ v_4 \\ v_5 \end{array}\begin{array}{c} \begin{array}{ccccccc} e_1 & e_2 & e_3 & e_4 & e_5 & e_6 & e_7 \end{array} \\ \begin{pmatrix} 1 & 1 & 0 & 0 & 0 & 0 & 0 \\ 1 & 0 & 1 & 1 & 1 & 1 & 0 \\ 0 & 0 & 0 & 1 & 1 & 0 & 1 \\ 0 & 1 & 1 & 0 & 0 & 0 & 0 \\ 0 & 0 & 0 & 0 & 0 & 1 & 0 \end{pmatrix} \end{array}$$

4. *Find the incidence matrix of the adjoining graph*

Solution. The incidence matrix of the given graph is shown below :

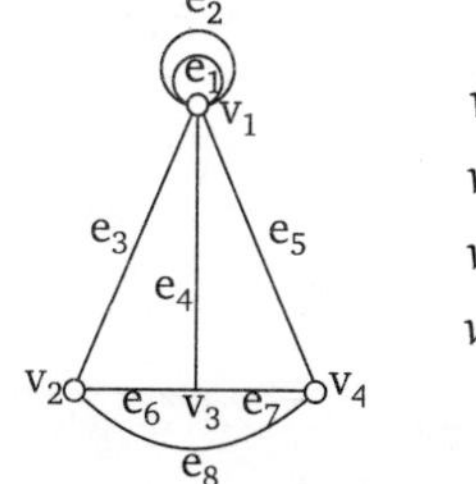

$$\begin{array}{c} \\ v_1 \\ v_2 \\ v_3 \\ v_4 \end{array}\begin{array}{c} \begin{array}{cccccccc} e_1 & e_2 & e_3 & e_4 & e_5 & e_6 & e_7 & e_8 \end{array} \\ \begin{pmatrix} 1 & 1 & 1 & 1 & 1 & 0 & 0 & 0 \\ 0 & 0 & 1 & 0 & 0 & 1 & 0 & 1 \\ 0 & 0 & 0 & 1 & 0 & 1 & 1 & 0 \\ 0 & 0 & 0 & 0 & 1 & 0 & 1 & 1 \end{pmatrix} \end{array}$$

Fig. (103)

5. *Draw the undirected graph represented by the following adjacency matrix*

$$A = \begin{bmatrix} 0 & 1 & 1 & 0 & 0 \\ 1 & 0 & 1 & 0 & 0 \\ 1 & 1 & 0 & 1 & 0 \\ 0 & 0 & 1 & 0 & 1 \\ 0 & 0 & 0 & 1 & 0 \end{bmatrix}$$

Solution. Since the given matrix is a square of order 5, therefore graph has five vertices $v_1, v_2, \ldots, v_5$. Draw an edge from v_i to v_j where $a_{ij} = 1$. Hence, required graph is given by

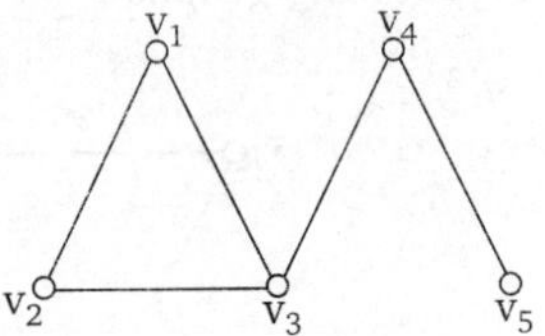

Fig. (104)

6. *Show the incidence matrix of the followir g graph :*

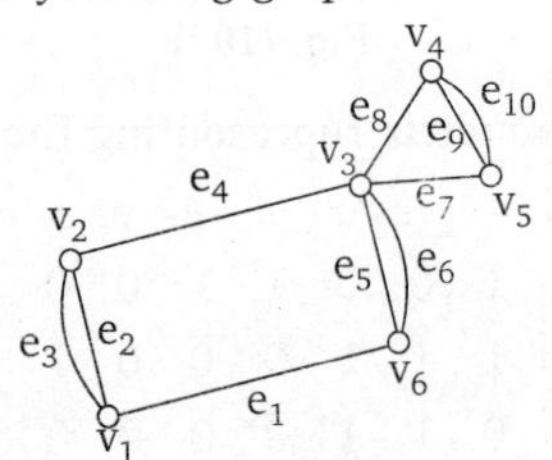

Fig. (105)

Solution. The graph has 6 vertices and 10 edges. The incidence matrix $A\,(G)$ graph is given by

$$A\,(G) = \begin{array}{c} \\ v_1 \\ v_2 \\ v_3 \\ v_4 \\ v_5 \\ v_6 \end{array} \begin{array}{c} \begin{array}{cccccccccc} e_1 & e_2 & e_3 & e_4 & e_5 & e_6 & e_7 & e_8 & e_9 & e_{10} \end{array} \\ \begin{pmatrix} 1 & 1 & 1 & 0 & 0 & 0 & 0 & 0 & 0 & 0 \\ 0 & 1 & 1 & 1 & 0 & 0 & 0 & 0 & 0 & 0 \\ 0 & 0 & 0 & 1 & 1 & 1 & 1 & 1 & 0 & 0 \\ 0 & 0 & 0 & 0 & 0 & 0 & 0 & 1 & 1 & 1 \\ 0 & 0 & 0 & 0 & 0 & 0 & 1 & 0 & 1 & 1 \\ 1 & 0 & 0 & 0 & 1 & 1 & 0 & 0 & 0 & 0 \end{pmatrix} \end{array}$$

7. *Draw the graph represented by adjacency matrix A given by*

$$A = \begin{bmatrix} 0 & 1 & 1 & 0 & 0 \\ 1 & 0 & 1 & 0 & 0 \\ 1 & 1 & 0 & 1 & 0 \\ 0 & 0 & 1 & 0 & 1 \\ 0 & 0 & 0 & 1 & 1 \end{bmatrix}$$

Solution. The adjacency matrix

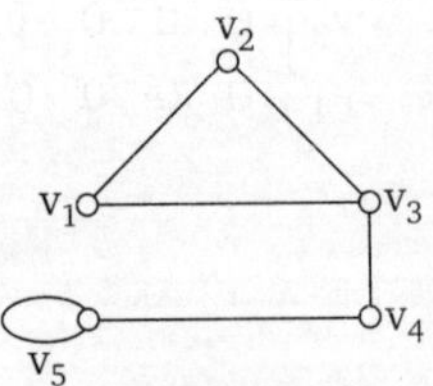

Fig. (106)

$$A = \begin{array}{c} \\ v_1 \\ v_2 \\ v_3 \\ v_4 \\ v_5 \end{array} \begin{array}{c} \begin{array}{ccccc} v_1 & v_2 & v_3 & v_4 & v_5 \end{array} \\ \begin{pmatrix} 0 & 1 & 1 & 0 & 0 \\ 1 & 0 & 1 & 0 & 0 \\ 1 & 1 & 0 & 1 & 0 \\ 0 & 0 & 1 & 0 & 1 \\ 0 & 0 & 0 & 1 & 1 \end{pmatrix} \end{array}$$

8. *Draw the undirected graph corresponding to incidence matrix*

$$A = \begin{bmatrix} 1 & 0 & 0 & 1 & 1 \\ 1 & 1 & 0 & 0 & 0 \\ 0 & 1 & 1 & 0 & 1 \\ 0 & 0 & 1 & 1 & 0 \end{bmatrix}$$

Solution. Incidency matrix is given by

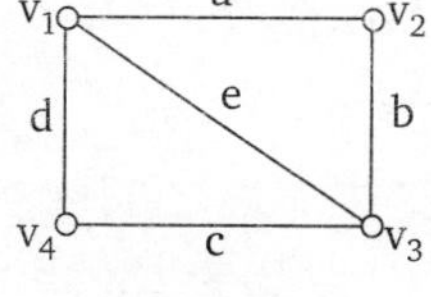

Fig. (107)

$$\begin{array}{c} \\ v_1 \\ v_2 \\ v_3 \\ v_4 \end{array} \begin{array}{c} \begin{array}{ccccc} a & b & c & d & e \end{array} \\ \begin{pmatrix} 1 & 0 & 0 & 1 & 1 \\ 1 & 1 & 0 & 0 & 0 \\ 0 & 1 & 1 & 0 & 1 \\ 0 & 0 & 1 & 1 & 0 \end{pmatrix} \end{array}$$

The required undirected graph is given above.

9. *Write the adjacency matrix for the graph :*

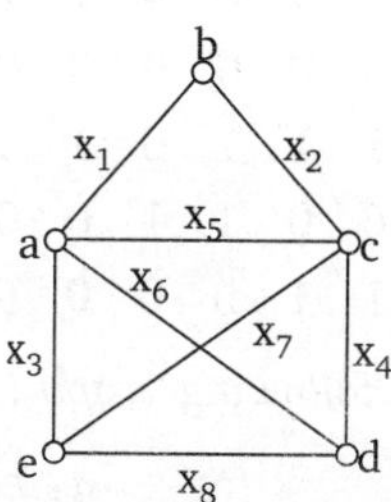

Fig. (108)

Solution. Adjacency matrix (A) is given below :

$$A = \begin{array}{c} \\ a \\ b \\ c \\ d \\ e \end{array} \begin{array}{c} \begin{array}{ccccc} a & b & c & d & e \end{array} \\ \begin{pmatrix} 0 & 1 & 1 & 1 & 1 \\ 1 & 0 & 1 & 0 & 0 \\ 1 & 1 & 0 & 1 & 1 \\ 1 & 0 & 1 & 0 & 1 \\ 1 & 0 & 1 & 1 & 0 \end{pmatrix} \end{array}$$

10. *Does there exist a graph G that corresponds to the following matrix ? Justify your answer*

$$\begin{bmatrix} 1 & 0 & 1 & 1 & 1 & 0 \\ 0 & 1 & 1 & 0 & 0 & 0 \\ 0 & 0 & 0 & 0 & 1 & 0 \\ 1 & 1 & 0 & 0 & 0 & 1 \\ 0 & 0 & 0 & 1 & 0 & 1 \end{bmatrix}$$

Solution. Yes, there exists a graph G that corresponds to the following incidence matrix

Some Important Observations

1. The sum of each row is equal to the degree of the corresponding vertex.
2. We know that every edge is incident on exactly two vertices, therefore each columns of the incidence matrix A has exactly two 1′ s.
3. If a row has all zero's, then the corresponding vertex is an isolated vertex.
4. In case the graph has parallel edges, then the corresponding column in *A* are idenctical.
5. If the incidence matrix is given, then the corresponding graph can be drawn.

11. *Find the incidence matrix for the following graph :*

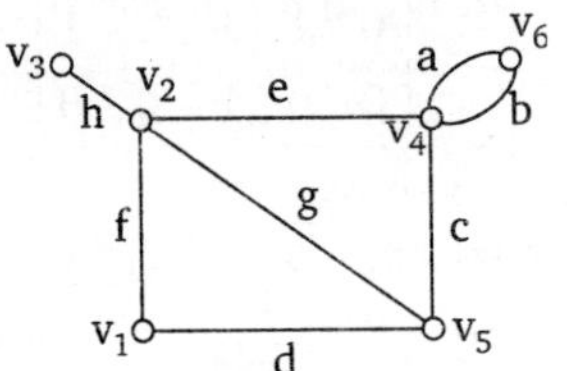

Fig. (109)

Solution. The incidence matrix *A* for the given graph is as follows :

$$A = \begin{array}{c} \\ v_1 \\ v_2 \\ v_3 \\ v_4 \\ v_5 \\ v_6 \end{array} \begin{array}{c} \begin{array}{cccccccc} a & b & c & d & e & f & g & h \end{array} \\ \begin{pmatrix} 0 & 0 & 0 & 1 & 0 & 1 & 0 & 0 \\ 0 & 0 & 0 & 0 & 1 & 1 & 1 & 1 \\ 0 & 0 & 0 & 0 & 0 & 0 & 0 & 1 \\ 1 & 1 & 1 & 0 & 1 & 0 & 0 & 0 \\ 0 & 0 & 1 & 1 & 0 & 0 & 1 & 0 \\ 1 & 1 & 0 & 0 & 0 & 0 & 0 & 0 \end{pmatrix} \end{array}$$

12. *Find the adjacency matrix for the following graph :*

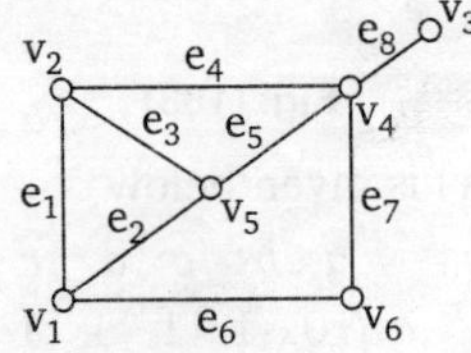

Fig. (110)

Solution. The adjacency matrix of the given graph is as follows

$$A\,(G) = \begin{array}{c} \\ v_1 \\ v_2 \\ v_3 \\ v_4 \\ v_5 \\ v_6 \end{array} \begin{array}{c} \begin{array}{cccccc} v_1 & v_2 & v_3 & v_4 & v_5 & v_6 \end{array} \\ \begin{pmatrix} 0 & 1 & 0 & 0 & 1 & 1 \\ 1 & 0 & 0 & 1 & 1 & 0 \\ 0 & 0 & 0 & 1 & 0 & 0 \\ 0 & 1 & 1 & 0 & 1 & 1 \\ 1 & 1 & 0 & 1 & 0 & 0 \\ 1 & 0 & 0 & 1 & 0 & 0 \end{pmatrix} \end{array}$$

13. *Is it possible to make the incidence matrix of the following graph ? If it is not possible, then why ?*

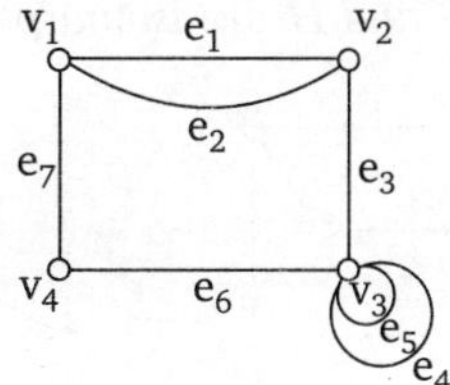

Fig. (111)

Solution. No, it is possible to make the incidence matrix of the given graph, because the graph contains two self loops.

14. *Show the incidence matrix of the following given graph.*

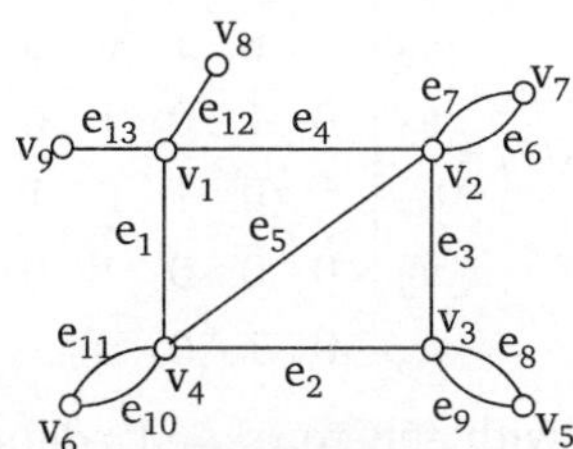

Fig. (112)

Solution. The incidence matrix of given graph is as follows :

$$\begin{array}{c} \\ v_1 \\ v_2 \\ v_3 \\ v_4 \\ v_5 \\ v_6 \\ v_7 \\ v_8 \\ v_9 \end{array}
\begin{array}{c}
\begin{array}{ccccccccccccc} e_1 & e_2 & e_3 & e_4 & e_5 & e_6 & e_7 & e_8 & e_9 & e_{10} & e_{11} & e_{12} & e_{13} \end{array} \\
\begin{pmatrix}
1 & 0 & 0 & 1 & 0 & 0 & 0 & 0 & 0 & 0 & 0 & 1 & 1 \\
0 & 0 & 1 & 1 & 1 & 1 & 1 & 0 & 0 & 0 & 0 & 0 & 0 \\
0 & 1 & 1 & 0 & 0 & 0 & 0 & 1 & 1 & 0 & 0 & 0 & 0 \\
1 & 1 & 0 & 0 & 1 & 0 & 0 & 0 & 0 & 1 & 1 & 0 & 0 \\
0 & 0 & 0 & 0 & 0 & 0 & 0 & 1 & 1 & 0 & 0 & 0 & 0 \\
0 & 0 & 0 & 0 & 0 & 0 & 0 & 0 & 0 & 1 & 1 & 0 & 0 \\
0 & 0 & 0 & 0 & 0 & 1 & 1 & 0 & 0 & 0 & 0 & 0 & 0 \\
0 & 0 & 0 & 0 & 0 & 0 & 0 & 0 & 0 & 0 & 0 & 1 & 0 \\
0 & 0 & 0 & 0 & 0 & 0 & 0 & 0 & 0 & 0 & 0 & 0 & 1
\end{pmatrix}
\end{array}$$

15. *Draw the graph whose incidence matrix A is given by*

$$A = \begin{array}{c} \\ v_1 \\ v_2 \\ v_3 \\ v_4 \\ v_5 \end{array}
\begin{array}{c}
\begin{array}{cccccc} e_1 & e_2 & e_3 & e_4 & e_5 & e_6 \end{array} \\
\begin{pmatrix}
1 & 0 & 0 & 0 & 0 & 0 \\
1 & 1 & 1 & 0 & 0 & 0 \\
0 & 1 & 0 & 1 & 1 & 1 \\
0 & 0 & 0 & 0 & 1 & 1 \\
0 & 0 & 1 & 1 & 0 & 0
\end{pmatrix}
\end{array}$$

Solution. The required graph is given by

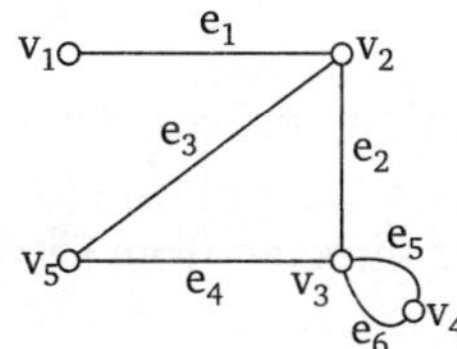

Fig. (113)

16. *Determine whether the graph G and H are isomorphic*

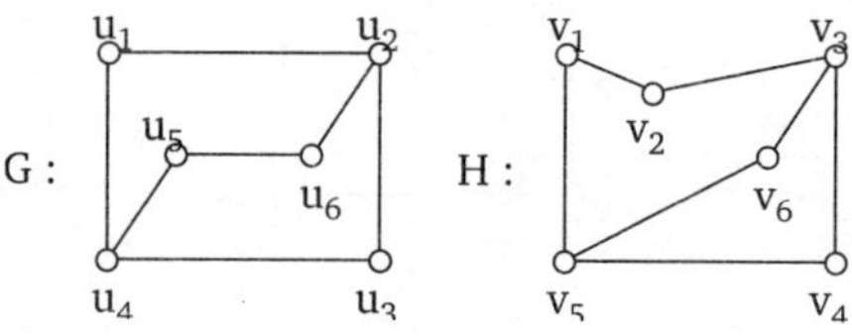

Fig. (114)

Solution : Adjacency matrix of G

$$X(G) = \begin{array}{c} \\ u_1 \\ u_2 \\ u_3 \\ u_4 \\ u_5 \\ u_6 \end{array} \begin{array}{c} \begin{array}{cccccc} u_1 & u_2 & u_3 & u_4 & u_5 & u_6 \end{array} \\ \begin{pmatrix} 0 & 1 & 0 & 1 & 0 & 0 \\ 1 & 0 & 1 & 0 & 0 & 1 \\ 0 & 1 & 0 & 1 & 0 & 0 \\ 1 & 0 & 1 & 0 & 1 & 0 \\ 0 & 0 & 0 & 1 & 0 & 1 \\ 0 & 1 & 0 & 0 & 1 & 0 \end{pmatrix} \end{array}$$

and the adjacency matrix of H with the rows and columns labeled by the images of the corresponding G

$$X(H) = \begin{array}{c} \\ v_6 \\ v_3 \\ v_4 \\ v_5 \\ v_1 \\ v_2 \end{array} \begin{array}{c} \begin{array}{cccccc} v_6 & v_3 & v_4 & v_5 & v_1 & v_2 \end{array} \\ \begin{pmatrix} 0 & 1 & 0 & 1 & 0 & 0 \\ 1 & 0 & 1 & 0 & 0 & 1 \\ 0 & 1 & 0 & 1 & 0 & 0 \\ 1 & 0 & 1 & 0 & 1 & 0 \\ 0 & 0 & 0 & 1 & 0 & 1 \\ 0 & 1 & 0 & 0 & 1 & 0 \end{pmatrix} \end{array}$$

Since $X(G) = X(H)$, it follows that f preserves edges. Hence, we conclude that G and H are isomorphic.

9.21 MATRIX REPRESENTATION OF DIGRAPHS

9.21.1 Incidence Matrix

Let $G = (V, E)$, where $v = \{v_1, \ldots v_n\}$ is the set of vertices and $E = \{e_1, e_2, \ldots e_m\}$ is the set of m edges be a digraph having no self loop. Then, the incidence matrix $A = [a_{ij}]$ of G is a $n \times m$ matrix defined by

$$a_{ij} = \begin{cases} 1 & \text{if the } j^{th} \text{ edge } e_j \text{ is incident out of the } i^{th} \text{ vertex } v_i \\ -1 & \text{if the } j^{th} \text{ edge } e_j \text{ is incident into the } i^{th} \text{ vertex } v_i \\ 0 & \text{if } e_j \text{ is not incident on } v_i \end{cases}$$

9.21.2 Adjacency Matrix

Let $G = (V, E)$ be a digraph with n vertices. Then the adjacency matrix of G is $n \times n$ matrix $A = [a_{ij}]$ defined by

$$a_{ij} = \begin{cases} r & \text{if } r \text{ edges are directed from } i^{\text{th}} \text{ vertex } v_i \text{ to } j^{\text{th}} \text{ vertex } v_j \\ 0 & \text{otherwise} \end{cases}$$

☛ If digraph has no parallel edges, then $r = 1$.

ILLUSTRATIVE EXAMPLES

1. *Find the incidence matrix of the following digraph*

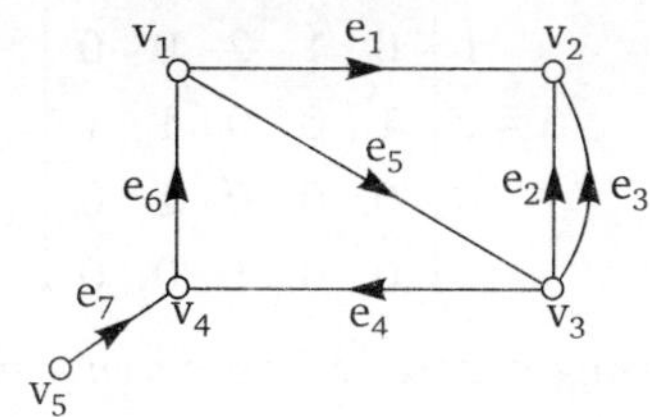

Fig. (115)

Solution. The given digraph has five vertices and seven edges. Thus, the requirement incidence matrix A is a 5×7 matrix given by

$$A = \begin{array}{c} \\ v_1 \\ v_2 \\ v_3 \\ v_4 \\ v_5 \end{array} \begin{array}{c} \begin{array}{ccccccc} e_1 & e_2 & e_3 & e_4 & e_5 & e_6 & e_7 \end{array} \\ \begin{pmatrix} 1 & 0 & 0 & 0 & 1 & -1 & 0 \\ -1 & -1 & -1 & 0 & 0 & 0 & 0 \\ 0 & 1 & 1 & 1 & -1 & 0 & 0 \\ 0 & 0 & 0 & -1 & 0 & 1 & -1 \\ 0 & 0 & 0 & 0 & 0 & 0 & 1 \end{pmatrix} \end{array}$$

2. *Draw the digraph whose incidence matrix A is given by*

$$A = \begin{array}{c} \\ a \\ b \\ c \\ d \\ e \\ f \end{array} \begin{array}{c} \begin{array}{ccccccccc} e_1 & e_2 & e_3 & e_4 & e_5 & e_6 & e_7 & e_8 & e_9 \end{array} \\ \begin{pmatrix} -1 & -1 & 0 & 1 & 0 & 0 & 0 & 0 & 0 \\ 1 & 0 & 1 & 0 & 0 & 0 & 0 & -1 & 0 \\ 0 & 1 & -1 & 0 & -1 & -1 & 0 & 0 & 0 \\ 0 & 0 & 0 & -1 & 0 & 0 & -1 & 1 & 1 \\ 0 & 0 & 0 & 0 & 1 & 1 & 1 & 0 & 0 \\ 0 & 0 & 0 & 0 & 0 & 0 & 0 & 0 & -1 \end{pmatrix} \end{array}$$

Solution. The required digraph corresponding to given incidence matrix is given below :

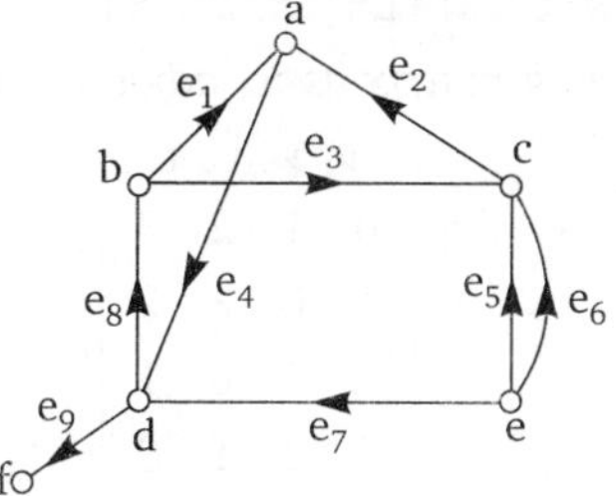

Fig. (116)

3. *Find the adjacency matrix of the following digraph*

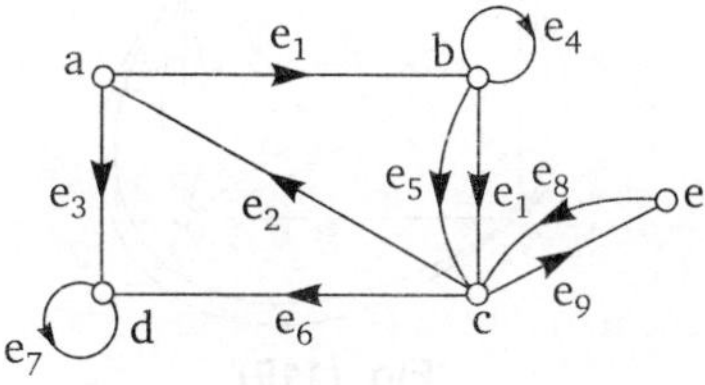

Fig. (117)

Solution. The adjacency matrix of the given digraph is

$$A = \begin{array}{c} \\ a \\ b \\ c \\ d \\ e \end{array} \begin{array}{c} \begin{array}{ccccc} a & b & c & d & e \end{array} \\ \begin{pmatrix} 0 & 1 & 0 & 1 & 0 \\ 0 & 1 & 2 & 0 & 0 \\ 1 & 0 & 0 & 1 & 1 \\ 0 & 0 & 0 & 1 & 0 \\ 0 & 0 & 1 & 0 & 0 \end{pmatrix} \end{array}$$

4. *Draw the graph having following matrix as its adjacency matrix*

$$\begin{bmatrix} 0 & 3 & 2 & 2 \\ 3 & 0 & 1 & 2 \\ 2 & 1 & 0 & 1 \\ 2 & 2 & 1 & 0 \end{bmatrix}$$

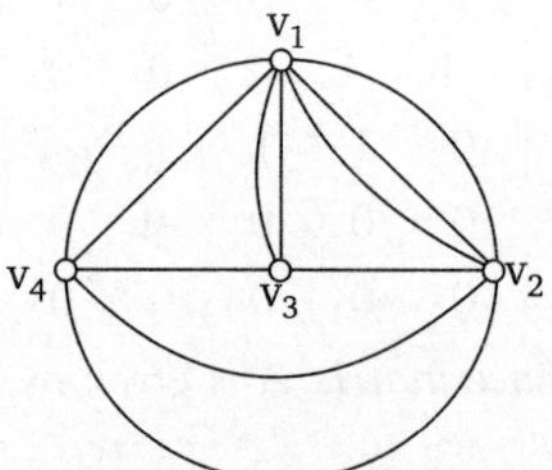

Fig. (118)

Solution. The graph corresponding adjacency matrix is

$$\begin{array}{c} \\ v_1 \\ v_2 \\ v_3 \\ v_4 \end{array} \begin{array}{c} \begin{array}{cccc} v_1 & v_2 & v_3 & v_4 \end{array} \\ \begin{bmatrix} 0 & 3 & 2 & 2 \\ 3 & 0 & 1 & 2 \\ 2 & 1 & 0 & 1 \\ 2 & 2 & 1 & 0 \end{bmatrix} \end{array}$$

5. *Draw the graph having following matrix as its adjacency matrix*

$$\begin{array}{c} \\ v_1 \\ v_2 \\ v_3 \\ v_4 \end{array} \begin{array}{c} \begin{array}{cccc} v_1 & v_2 & v_3 & v_4 \end{array} \\ \begin{pmatrix} 0 & 1 & 2 & 3 \\ 1 & 0 & 3 & 2 \\ 2 & 3 & 0 & 1 \\ 3 & 2 & 1 & 0 \end{pmatrix} \end{array}$$

Solution. The graph corresponding to the given matrix is

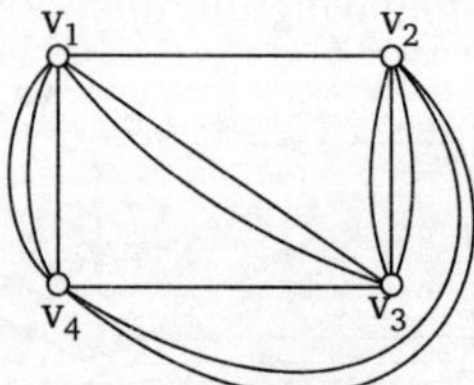

Fig. (119)

6. *Show the adjacency matrix of the following graph :*

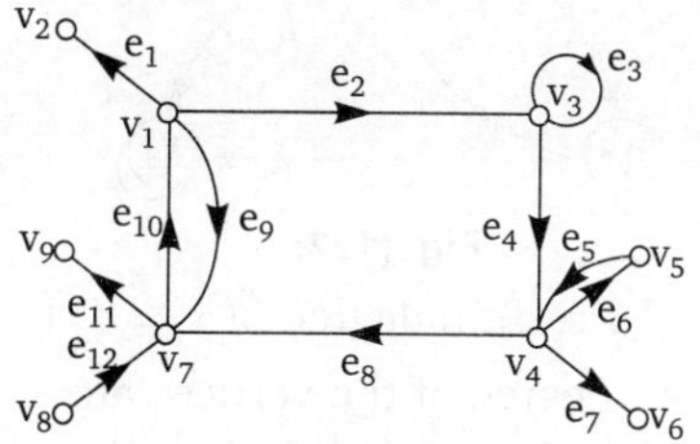

Fig. (120)

Here $(v_1, v_2,..., v_9)$ denote the vertices and $(e_1, e_2, ... e_{12})$ denote the edges in the graph.

Solution. The adjacent matrix of the given graph is

$$A = \begin{array}{c} \\ v_1 \\ v_2 \\ v_3 \\ v_4 \\ v_5 \\ v_6 \\ v_7 \\ v_8 \\ v_9 \end{array} \begin{array}{c} \begin{array}{ccccccccc} v_1 & v_2 & v_3 & v_4 & v_5 & v_6 & v_7 & v_8 & v_9 \end{array} \\ \begin{pmatrix} 0 & 1 & 1 & 0 & 0 & 0 & 1 & 0 & 0 \\ 0 & 0 & 0 & 0 & 0 & 0 & 0 & 0 & 0 \\ 0 & 0 & 1 & 1 & 0 & 0 & 0 & 0 & 0 \\ 0 & 0 & 0 & 0 & 1 & 1 & 1 & 0 & 0 \\ 0 & 0 & 0 & 1 & 0 & 0 & 0 & 0 & 0 \\ 0 & 0 & 0 & 0 & 0 & 0 & 0 & 0 & 0 \\ 1 & 0 & 0 & 0 & 0 & 0 & 0 & 0 & 1 \\ 0 & 0 & 0 & 0 & 0 & 0 & 1 & 0 & 0 \\ 0 & 0 & 0 & 0 & 0 & 0 & 0 & 0 & 0 \end{pmatrix} \end{array}$$

7. *Determine the adjacency matrix of the graph.*

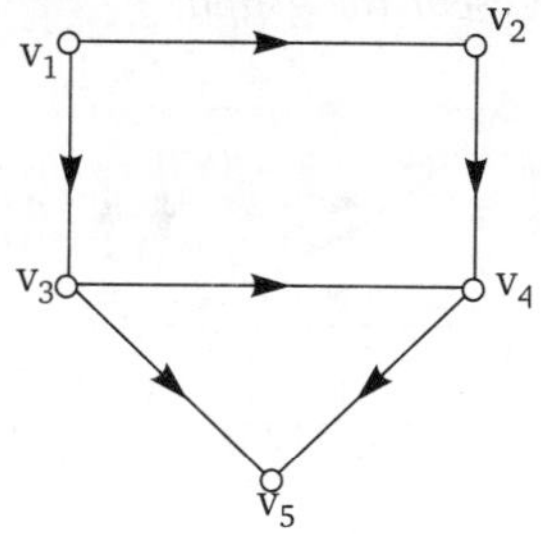

Fig. (121)

Solution. The adjacency matrix of the given graph is

$$\begin{array}{c} \\ v_1 \\ v_2 \\ v_3 \\ v_4 \\ v_5 \end{array} \begin{array}{c} \begin{array}{ccccc} v_1 & v_2 & v_3 & v_4 & v_5 \end{array} \\ \begin{pmatrix} 0 & 1 & 1 & 0 & 0 \\ 0 & 0 & 0 & 1 & 0 \\ 0 & 0 & 0 & 1 & 1 \\ 0 & 0 & 0 & 0 & 1 \\ 0 & 0 & 0 & 0 & 0 \end{pmatrix} \end{array}$$

8. *Consider the directed graph, determine the in-degree and out-degree of each of vertices of the graph.*

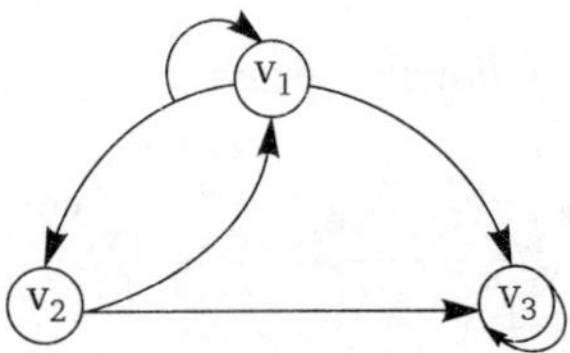

Fig. (122)

Solution. Out degree of $v_i = d^+(v_i)$. Indegree of $v_i = d^-(v_i)$

So in the given graph, the out degree of the vertices are

$$d^+(v_1) = 2;\ d^+(v_2) = 2;\ d^+(v_3) = 1$$

The in-degree of vertices are $d^-(v_1) = 2$; $d^-(v_2) = 1$; $d^-(v_3) = 3$.

9. *Find the incidence matrix to represent the graph.*

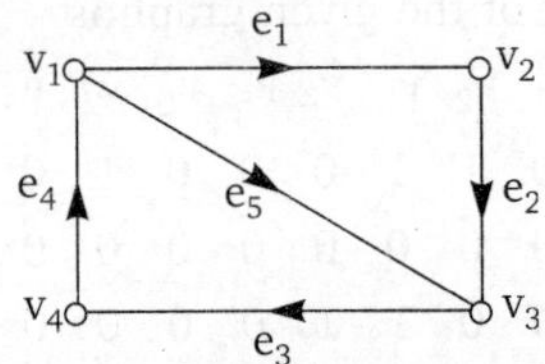

Fig. (123)

Solution. The incidence matrix corresponding to graph is given by

$$\begin{array}{c} \\ v_1 \\ v_2 \\ v_3 \\ v_4 \end{array}\begin{array}{c} \begin{array}{ccccc} e_1 & e_2 & e_3 & e_4 & e_5 \end{array} \\ \begin{pmatrix} 1 & 0 & 0 & 1 & 1 \\ 1 & 1 & 0 & 0 & 0 \\ 0 & 1 & 1 & 0 & 1 \\ 0 & 0 & 1 & 1 & 0 \end{pmatrix} \end{array}$$

10. *Find the adjacency matrix to represent the graph.*

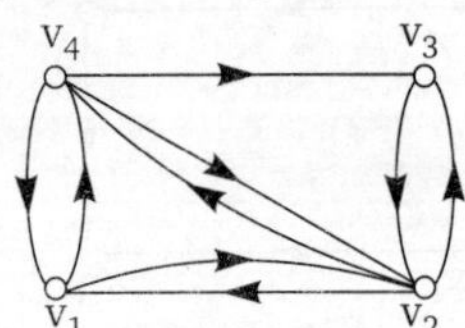

Fig. (124)

Solution. Adjcency matrix

$$\begin{array}{c} \\ v_1 \\ v_2 \\ v_3 \\ v_4 \end{array}\begin{array}{c} \begin{array}{cccc} v_1 & v_2 & v_3 & v_4 \end{array} \\ \begin{pmatrix} 0 & 1 & 0 & 1 \\ 1 & 0 & 1 & 1 \\ 0 & 1 & 0 & 0 \\ 1 & 1 & 1 & 0 \end{pmatrix} \end{array}$$

11. *Draw the graph with following incidence matrix.*

$$\begin{array}{c} \\ v_1 \\ v_2 \\ v_3 \\ v_4 \\ v_5 \end{array}\begin{array}{c} \begin{array}{cccccc} e_1 & e_2 & e_3 & e_4 & e_5 & e_6 \end{array} \\ \begin{pmatrix} 1 & 1 & 1 & 0 & 0 & 0 \\ 0 & 0 & 1 & 1 & 0 & 1 \\ 0 & 0 & 0 & 0 & 1 & 0 \\ 1 & 1 & 0 & 1 & 0 & 0 \\ 0 & 0 & 0 & 0 & 1 & 0 \end{pmatrix} \end{array}$$

Solution. Required graph is given below :

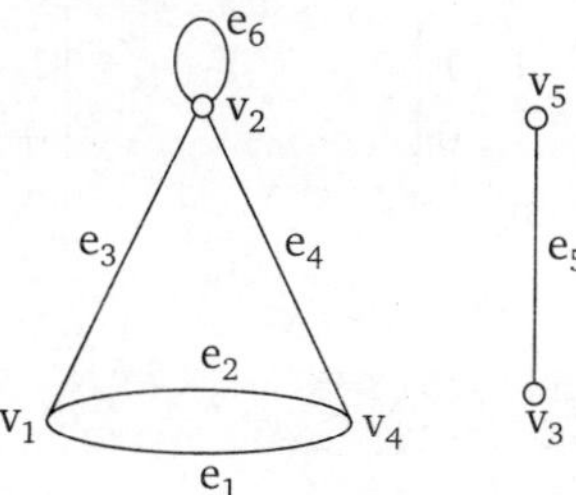

Fig. (125)

12. *Find the adjacency matrix for the following graphs.*

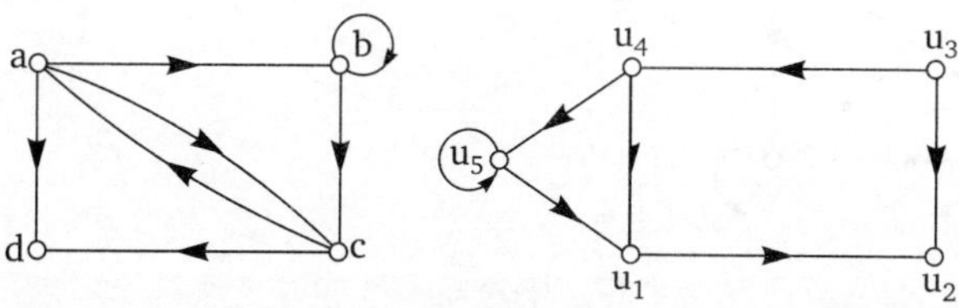

Fig. (126)

Solution. (a) The required adjacency matrix is given by

$$A = \begin{array}{c} \\ a \\ b \\ c \\ d \end{array} \begin{array}{c} \begin{array}{cccc} a & b & c & d \end{array} \\ \begin{pmatrix} 0 & 1 & 1 & 1 \\ 0 & 1 & 1 & 0 \\ 1 & 0 & 0 & 1 \\ 0 & 0 & 0 & 0 \end{pmatrix} \end{array}$$

(b) The adjacency matrix

$$\begin{array}{c} \\ u_1 \\ u_2 \\ u_3 \\ u_4 \\ u_5 \end{array} \begin{array}{c} \begin{array}{ccccc} u_1 & u_2 & u_3 & u_4 & u_5 \end{array} \\ \begin{pmatrix} 0 & 1 & 0 & 0 & 0 \\ 0 & 0 & 0 & 0 & 0 \\ 0 & 1 & 0 & 1 & 0 \\ 1 & 0 & 0 & 0 & 1 \\ 1 & 0 & 0 & 0 & 1 \end{pmatrix} \end{array}$$

Exercise 9.5

1. Draw the graph represented by the adjacency matrix

$$\begin{array}{c} \\ a \\ b \\ c \\ d \\ e \end{array} \begin{array}{c} \begin{array}{ccccc} a & b & c & d & e \end{array} \\ \begin{pmatrix} 1 & 0 & 0 & 1 & 0 \\ 0 & 1 & 1 & 0 & 1 \\ 1 & 0 & 1 & 1 & 0 \\ 0 & 1 & 0 & 1 & 0 \\ 0 & 0 & 1 & 0 & 1 \end{pmatrix} \end{array}$$

2. Find the adjacency matrix of each of the following graphs.

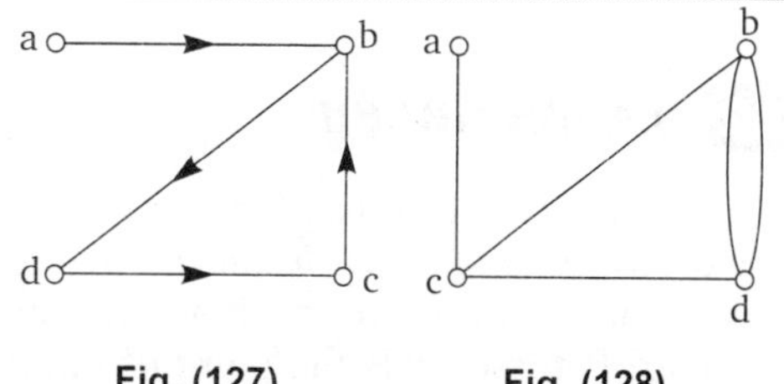

Fig. (127) Fig. (128)

3. Draw the graph represented by the given adjacency matrix

(i) $\begin{bmatrix} 0 & 1 & 0 & 1 \\ 1 & 0 & 1 & 0 \\ 0 & 1 & 0 & 1 \\ 1 & 0 & 1 & 0 \end{bmatrix}$ (ii) $\begin{bmatrix} 1 & 2 & 0 & 1 \\ 2 & 0 & 3 & 0 \\ 0 & 3 & 1 & 1 \\ 1 & 0 & 1 & 0 \end{bmatrix}$

4. Draw the directed graph using the following adjacency matrix

$$\begin{bmatrix} 0 & 0 & 0 & 0 & 1 \\ 0 & 0 & 1 & 0 & 0 \\ 1 & 0 & 0 & 0 & 0 \\ 0 & 0 & 1 & 0 & 0 \\ 0 & 1 & 0 & 0 & 0 \end{bmatrix}$$

ANSWERS

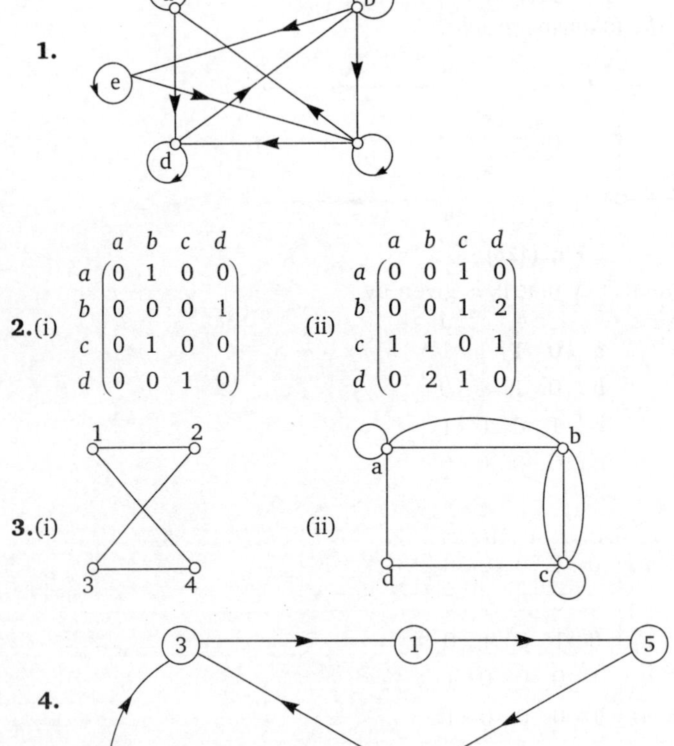

2. (i) $\begin{array}{c} \\ a \\ b \\ c \\ d \end{array}\begin{array}{c} a\ b\ c\ d \\ \begin{pmatrix} 0 & 1 & 0 & 0 \\ 0 & 0 & 0 & 1 \\ 0 & 1 & 0 & 0 \\ 0 & 0 & 1 & 0 \end{pmatrix}\end{array}$ (ii) $\begin{array}{c} \\ a \\ b \\ c \\ d \end{array}\begin{array}{c} a\ b\ c\ d \\ \begin{pmatrix} 0 & 0 & 1 & 0 \\ 0 & 0 & 1 & 2 \\ 1 & 1 & 0 & 1 \\ 0 & 2 & 1 & 0 \end{pmatrix}\end{array}$

9.22 PLANAR GRAPH

A graph G is said to be planar if there exists some geometric representation of G which can be drawn on a plane such that no two of its edge intersect. On the other hand, a graph that can not be drawn on a plane without a cross over between its edge is known as non-planar. [Crossover is the point of intersection].

A drawing of a geometric representation of a graph on any surface such that no edges intersect is known as embedding. Therefore, to declare that a graph G is non-planar, we have to show that of all possible geometric representation of G, none can be embedded by a plane.

An embedding of a planar graph G on a plane is called a plane representation of graph G.

For example : Consider the following graphs

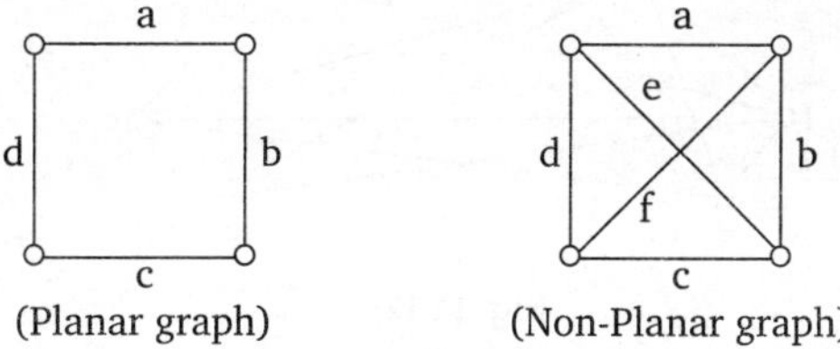

(Planar graph) (Non-Planar graph)

Fig. (129)

In the given graph, the edge e and f are intersecting, so this graph is non-planar. But if we draw edge f outside the quadrilateral, leaving the other edges unchanged. We have embedded the new geometric graph in the plane.

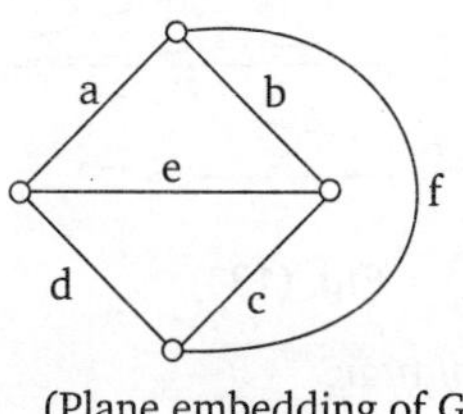

(Plane embedding of G)

Fig. (130)

- A disconnected graph is planar if and only if every component of this graph is planar.
- In order to say whether a given graph is planar or non-planar, we have to try all the possibilities to embed in a graph.
- If a graph G is non-planar but its subgraph is planar, then graph G is known as a **critical planar graph.**

9.22.1 Region of a Graph

Let G be a planar graph. Embed the graph in a plane paper and cut along its edges. The paper cut into number of pieces. The corner of each piece are the vertices of G and sides are edge of G. Then each piece is called a region of G. The total number of region of G is denoted by r. The side (edges) of the region are called boundary of the region. The number of edges in a boundary or the number of edges by which a region is bounded, is known as degree of a region.

ILLUSTRATIVE EXAMPLES

1. *Show that K_4 is planar.*

Solution. The graph K_4 can be drawn as given below :

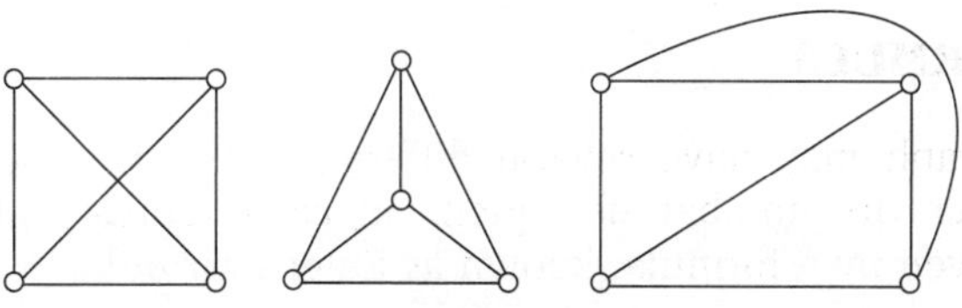

Fig. (131)

Second figure is without crossing. Therefore, K_4 is a planar graph.

2. *Whether the following graph is planar. If so, draw its planar representation.*

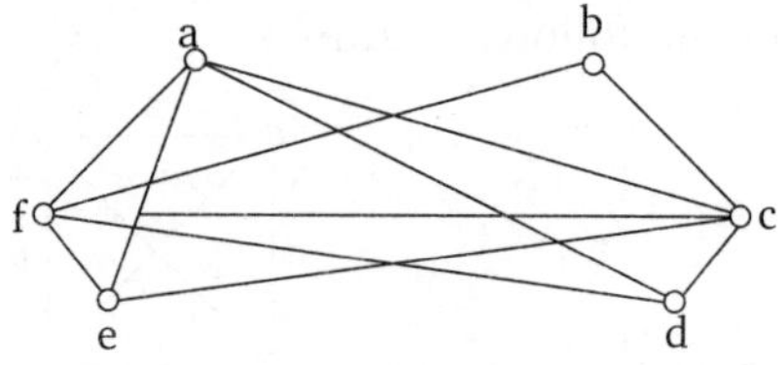

Fig. (132)

Solution. The given graph is planar. The planar representation of this graph is given below :

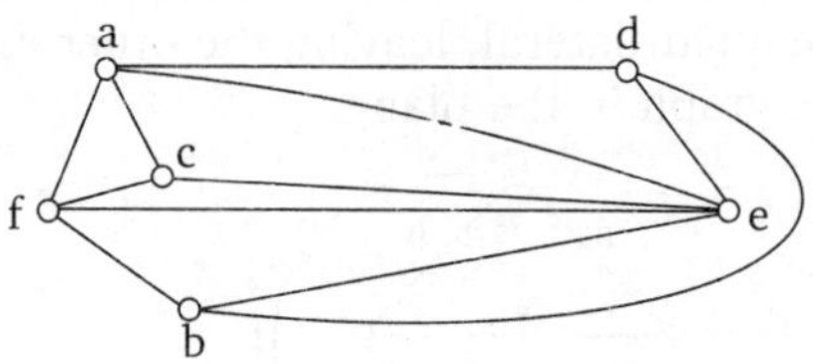

Fig. (133)

3. *Show that $K_{3,3}$ shown below is not planar.*

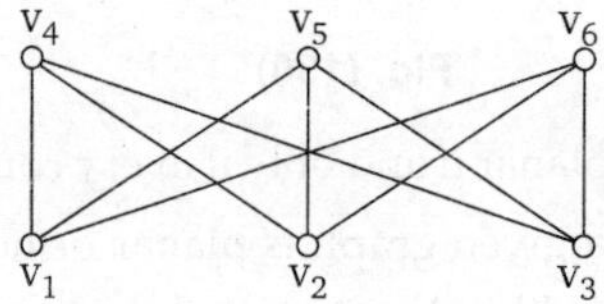

Fig. (134)

Solution. In any plane representation of the given graph, the vertices v_1 and v_2 must be connected to both v_4 and v_5. Thus, the four edges form a close curve splitting the plane into two regions R_1 and R_2 as shown below :

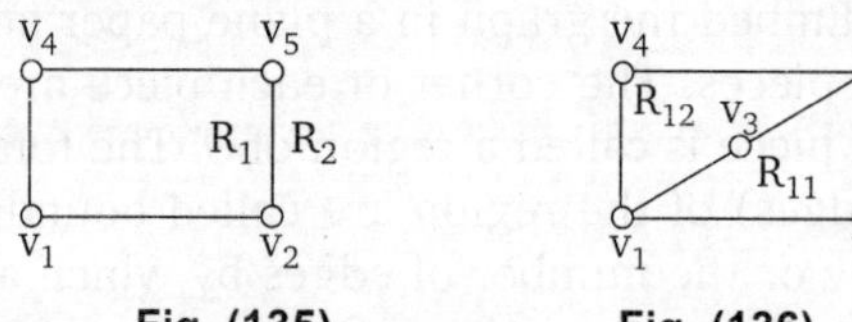

Fig. (135) Fig. (136)

The vertex v_3 will either be in R_1or in R_2. Suppose v_3 is in R_1. The edges between v_1 and v_3 and v_3 and v_5 divide R_1 into two parts say R_{11} and R_{12}.

Now, there is no way to place the vertex v_6 without facing a crossing. If v_6 is in R_2, then the edge between v_6 and v_3 can not be drawn without a crossing. Similar argument can be given to other cases. Hence $K_{3,3}$ is not planar.

9.23 EULER'S FORMULA

Since a planar graph may have several different plane representation, a natural question arises whether the number of regions in each representation is same. The answer is yes and is given by a formula, known as Euler's formula.

Theorem 1. *A connected planar graph with n vertices and e edges has $e - n + 2$ regions.*

First Proof : Let G be a connected planar graph. We shall prove the result by induction on the number of edges of G. If $e = 1$, then $n = 1$ or 2 as shown in following figure.

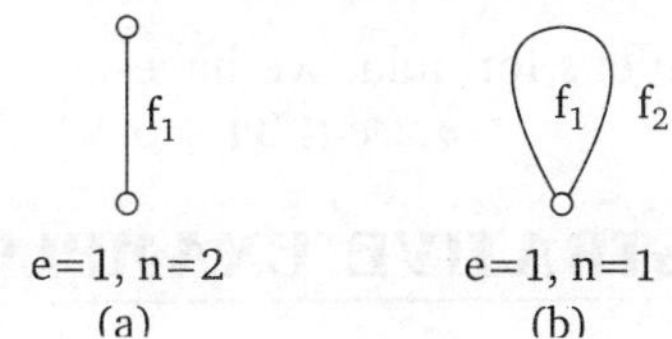

Fig. (137)

When $e = 1$ and $n = 2$, we have $e - n + 2 = 1$. Clearly graph in figure (a) above has one region.

When $e = 1$ and $n = 1$, then we have $e - n + 2 = 2$. Also, graph in figure (b) above has two regions.

Thus the result is true for $e = 1$.

We assume that the result is true for all graphs with at most $e - 1$ edges. Let G be a connected graph with e edges and f regions. If G is a tree then $e = n - 1$ and number of faces is 1 (only infinite region). Also, by the theorem, number of faces $= e - n + 2 = 1$. Hence theorem is true in this case. If G is not a tree, then it has some circuits. Let a be an edge in some circuit. Removal of the edge a from the plane representation of G will merge the two regions into one new region. Thus $G - a$ is a connected graph with n vertices, $e - 1$ edges and $f - 1$ regions (where f is the number of regions in G). By induction hypothesis, we have

$$f - 1 = (e - 1) - n + 2 \Rightarrow f = e - n + 2$$

This completes the proof.

Second Proof : It is sufficient to prove the theorem for simple graph because adding a self loop or a parallel edge simply adds one region to the graph and also increases the value of e by one. We can also disregard (*i.e.*, eliminate) all edges that do not form boundaries of any region because removal of any such edge decreases the value of e by one and decreases the value of n by one keeping the quantity $e - n + 2$ unchanged.

Since, any simple planar graph can have a plane representation such that each edge is drawn as straight line, we draw the given planar graph in such a fashion that each region is a polygon. Let the polygonal graph representing the given graph consists of f regions and let k_p be the number of p-sided regions. Since each edge lies on the boundary of exactly two regions,

$$3.k_3 + 4.k_4 + 5.k_5 + \ldots + r.k_r = 2e \qquad \ldots(1)$$

where, k_r is the number of polygons with maximum edges.

Also, we have $$k_3 + k_4 + \ldots + k_r = f \qquad \ldots(2)$$

From geometry, we know that the sum of all interior angles of a p –sided polygon is $(p - 2)\,\pi$ and the sum of the exterior angles is $(p + 2)\,\pi$. Moreover, only one region is infinite and remaining $f - 1$ regions are finite. Without loss of generality, we may assume that one region out of k_r regions of $r-$ sided polygon is infinite. Thus, the sum of all angles subtended at each vertex is

$$(3 - 2)\,\pi\, k_3 + (4 - 2)\,\pi\, k_4 + \ldots + (r - 2)\,\pi\,(k_r - 1) + (r + 2)\,\pi = 2n\pi$$

$$\Rightarrow (3k_3 + 4k_4 + \ldots + r\,k_r)\,\pi - 2\,(k_3 + k_4 + \ldots + k_r)\,\pi + 4\pi = 2n\pi$$

$$\Rightarrow 2\,e\,\pi - 2\,f\pi + 4\pi = 2\,n\,\pi \qquad \text{[Using (1) and (2)]}$$

$\Rightarrow f = e - n + 2$ this completes the proof.

Corollary 1. If a connected simple planar graph G has $n \geq 3$ vertices, e edges and f regions, then

$$3f \leq 2e \qquad \ldots(1)$$

and $$e \leq 3n - 6 \qquad \ldots(2)$$

Proof : Since each region in a simple connected graph is bounded by at least three edges and each edge belongs to exactly two regions, we have

$$3f \leq 2e$$

Now, substituting f from Euler's formula, we have

$$3(e - n + 2) \leq 2e \quad \Rightarrow \quad e \leq 3n - 6$$

ILLUSTRATIVE EXAMPLES

1. *Show that the given graph G is non-planar.*

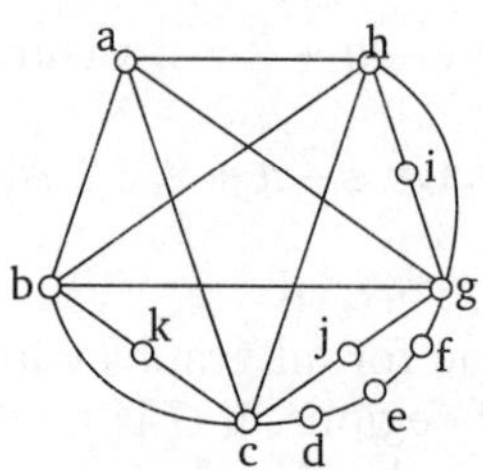

Fig. (138)

Solution. The given graph G has a subgraph H obtained by deleting the vertices i, j and k and all edges incident with these vertices. H is homeomorphic to K_5 as it can be obtained from K_5 (with vertices a, b, c, g and h) by a sequence of elementary subdivisions, adding the vertices d, e, f. Therefore G is non-planar.

2. *Prove that there exists no polyhedral planar graph with exactly seven edges.*

Solution. We know that a polyhedral graph is a regular connected planar graph in which each region is bounded by three or more edges and degree of each vertex is ≥ 3.

Since each region is bounded by 3 or more edges

$$3f \leq 2e = 2 \times 7 = 14 \quad \Rightarrow \quad f \leq 14/3$$

Since f must be an integer, therefore $f = 4$.

Again, if the graph has n vertices then

$$3n \leq 2e = 14 \quad \Rightarrow \quad n \leq 14/3$$

$\Rightarrow$ n must be less than or equal to 4.

From Euler's formula, we have

$$f + n = e + 2$$

Substituting the values of e, n and f, we get

$$4 + 4 \leq f + n = 7 + 2 = 9, \text{ a contradiction.}$$

Hence, no polyhedral graph with exactly 7 edges exists.

3. *Show that in a simple planar graph G of n vertices ($n \geq 3$), there is at least one vertex of degree ≤ 5.*

Solution. Let G be a simple planar graph with n vertices and e edges. If possible, suppose that degree of each vertex of G is greater than 5. Then

$$2e \geq 6n \quad \text{(because sum of degrees of } G \text{ is at least } 6n).$$

or $\quad e \geq 3n$

But by corollary of Euler's formula $e \leq 3n - 6$

These two inequalities contradict one another. Hence, vertex of G has degree ≤ 5.

4. *Show that in a connected planar simple graph with 6 vertices and 12 edges, each of the regions is bounded by 3 edges.*

Solution. In the given graph, we have

$$e = 12, n = 6$$

Therefore, by Euler's formula $f = e - n + 2 = 12 - 6 + 2 = 8$

Hence, the given graph has eight regions. Since the graph is simple, each region is bounded by at least three edges, we have, $2e \leq 3f$. Since $e = 12$ and $f = 8$, equality holds and hence each region is bounded by 3 edges.

5. *Count the number of vertices, number of edges and number of regions of the following :*

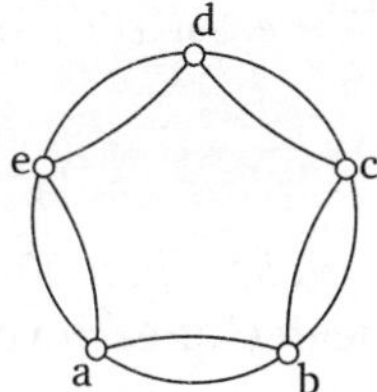

Fig. (139)

Solution. Using Euler's formula, we have $|v|-|e|+|r|=2$ where, $|v|=$ no. of vertices in G

$$|e|= \text{no. of edges in } G;$$
$$|r|= \text{no. of regions in } G.$$

Clearly, no. of vertices,

$$|v|=5$$
$$|e|=10$$

Then, by Euler's formula

$$5-10+|r|=2 \quad \Rightarrow \quad |r|=7.$$

6. *Show that a complete graph of n vertices is planar if $n \leq 4$.*

Solution. Let us take $n = 1, 2, 3, 4$. Then all graphs will be planar if these graphs are complete. If we take $n = 5$ and draw a complete graph.

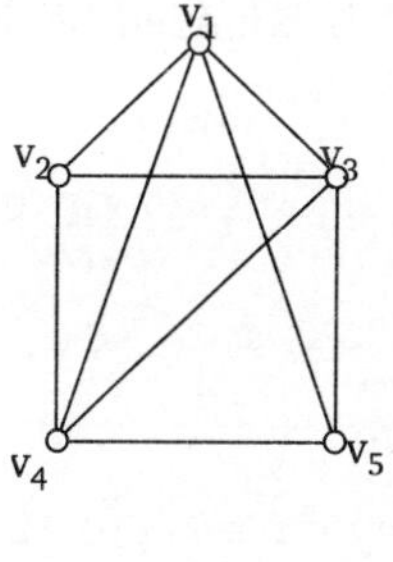

(G)

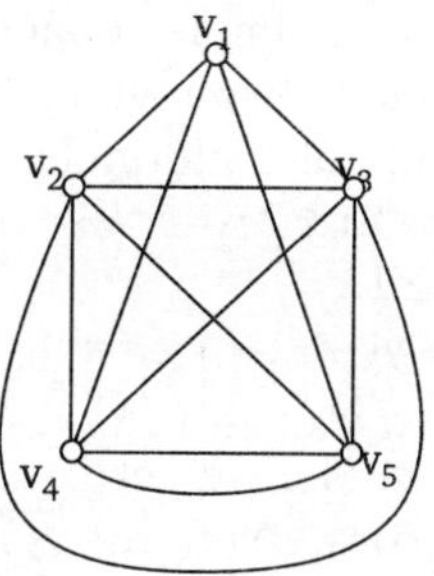

(Plane embedding of G)

Fig. (140)

Here, $e = 10$, $n = 5$

Thus, $|r| = 2 + 10 - 5 = 7$

But for a planar graph, r (no. of regions) $\leq \dfrac{2e}{3}$

But $7 \leq \dfrac{20}{3}$ is not true.

Hence, a complete graph of 5 vertices is non-planar.

Similarly, we can prove that for $n = 6, 7, \ldots$, complete graph G is non-planar.

7. *If every region of a simple planar graph with n vertices and e edges is bounded by k-edges, show that*

$$e = \frac{k\,(n-2)}{(k-2)}$$

Solution. Since, it is given that every region is bounded by $k-$ edges, therefore, we have

$$kf = 2e, \text{ where } f \text{ is the number of regions.}$$

Putting this value of f in Euler's formula, we get

$$f = e - n + 2 \Rightarrow \frac{2e}{k} = e - n + 2 \Rightarrow (n-2)k = (k-2)e$$

Hence, $e = \frac{k(n-2)}{k-2}$.

8. *Show that a simple planar graph with less than 30 edges has a vertex of degree 4 or less.*

Solution. Let us suppose that degree of each vertex is ≥ 5. Then

$$2e \geq 5n \quad \text{...(1)}$$

Since graph is simple planar, each region is bounded by at least 3 edges. Thus

$$2e \geq 3f \quad \text{...(2)}$$

By Euler's formula, we have $e = f + n - 2$

Using (1) and (2) in (3), we get

$$e \leq \frac{2e}{3} + \frac{2e}{5} - 2 = \frac{(16e - 30)}{15}$$

$\Rightarrow \quad e \geq 3$, which is a contradiction.

Because the given graph has less than 30 edges.

Thus, it follows that there is at least one vertex of degree ≥ 4.

9. *For a connected planar graph with $n \geq 3$ vertices, show that*

$$5v_1 + 4v_2 + 3v_3 + 2v_4 + v_5 \geq v_7 + 2v_8 + \ldots + (k-6)v_k + 12$$

where v_r denotes the number of vertices with degree r and k is the largest degree of a vertex in G.

Solution. If a connected simple planar graph G has $n \geq 3$ vertices, e edges and r regions.

Then, each region in a simple connected graph is bounded by at least three edges and each edge belongs to exactly two regions, we have $3r \leq 2e$

Putting this value in Euler's formula, we see

$$3(e - n + 2) \leq 2e \Rightarrow e \leq 3n - 6 \quad \text{...(1)}$$

$$\Rightarrow 6n - 2e \geq 12$$

Also, $v_1 + v_2 + v_3 + \ldots + v_k = n$ ...(2)

and $v_1 + 2v_2 + 3v_3 + \ldots + kv_k = 2e$ ...(3)

Using (2) and (3) in (1), we get

$$6(v_1 + v_2 + \ldots + v_k) - (v_1 + 2v_2 + \ldots + kv_k) \geq 12$$

$$\Rightarrow 5v_1 + 4v_2 + 3v_3 + 2v_4 + v_5 \geq v_7 + 2v_8 + \ldots + (k-6)v_k + 12.$$

9.23.1 Detection of Planarity

To check the planarity of a graph without embedding in a plane, we use the following steps :

WORKING RULES

Step 1. We know that a graph is planar if and only if its component is planar. Therefore, to check the planarity, it is sufficient to check for each component. For each component of G go to Step 2.

Step 2. Each graph is planar if and only if each of its block is planar. For each block of G go to Step 3.

Step 3. Between two vertices, if it is possible to write an edge without crossing with each other, then it is also possible to write more edges between the same pair of vertices without crossing one another. Therefore, parallel edge or self-loops does not affect the planarity. Replace all parallel edge by an edge, go to Step 4.

Step 4. Remove all self-loops.

Step 5. Two edges are said to be in series if they have exactly one vertex of degree two in common. Merging two series edges does not affect the planarity. Hence, replace all edges in series by an edge.

For Example :
Reduce the graph without affecting its planarity

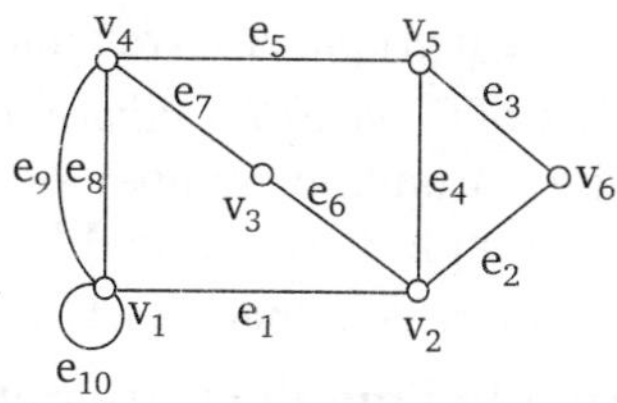

Fig. (141)

Solution. We can use above algorithm as follows (Series parallel reduction)
By step (1), we remove self loop at the vertex v_1

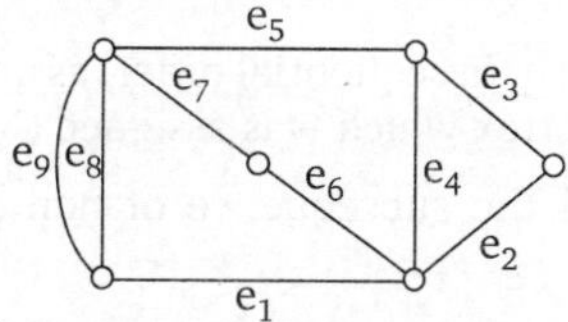

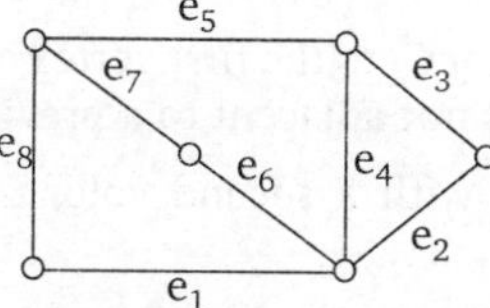

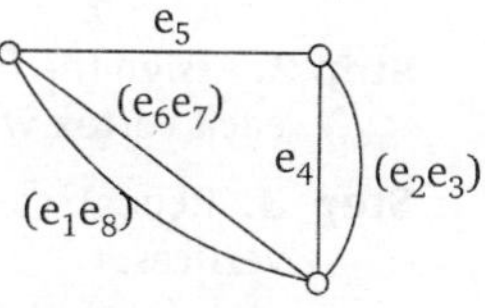

(1) (Self loop removal) (2) (Parallel edge reduced) (3) (series reduced)

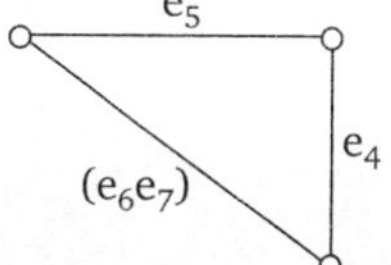

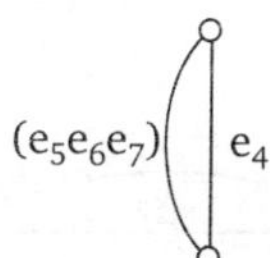

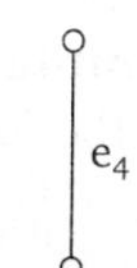

(4) (parallel reduced) (5) (series reduced) (6) (Parallel reduced)

9.23.2. Kurtowaski's Theorem.

A necessary and sufficient condition for a graph to be planar is that it does not contain the graph K_5 or $K_{3,3}$ or graph which are homeomorphic to K_5 or $K_{3,3}$.

Exercise 9.6

1. Find the degree of each region in the given graph

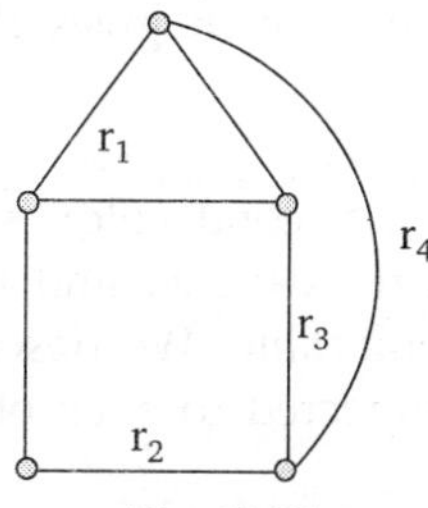

Fig. (142)

2. Show that $K_{3,3}$ is not planar.

3. Show that the graph $K_{2,2}$ and $K_{2,3}$ are planar.

4. Show that the complete graph of five vertices is non planar.

5. Show that a self-loop free planar graph is 2-connected if and only if its dual is also 2-connected.

6. Show that a connected planar graph with n vertices and e edges has $e - n + 2$ regions.

7. Show that a connected planar graph with 7 vertices and degree of each vertex equal to 4 must have 8 regions of degree 3 and 1 region of degree 4.

8. Show that every planar graph of less than 12 vertices has a vertex of degree less than or equal to 4.

ANSWERS

1. (3, 4, 3, 4)

9.24 GRAPH COLORING

Definition : *Let $G = (V, E)$ be a graph. A vertex coloring or simply coloring of G is an assignment of colors to the vertices of G such that adjacent vertices have different colors.*

Chromatic Number. The chromatic number of a graph is the least number of colors needed for coloring of the graph. It is denoted by $\chi(G)$.

☛ A graph is said to be n-colorable if there exists a coloring of G which uses n colors.

9.24.1 Wetch-Powell Algorithm to Find the Chromatic Number

WORKING RULES Let G be a graph, then, we proceed as follows :

Step 1. Order the vertices of G according to decreasing degrees.

Step 2. Assign the first color C_1 to the first vertex and then, in sequential order, assign C_1 to each vertex which is not adjacent to a previous vertex which was assigned C_1.

Step 3. Repeat step (2) with a second color C_2 and the subsequence of non-colored vertices.

Step 4. Repeat step (3) with a color C_3, then a fourth color C_4 and so on until all vertices are colored.

For Example :

Consider the following graph

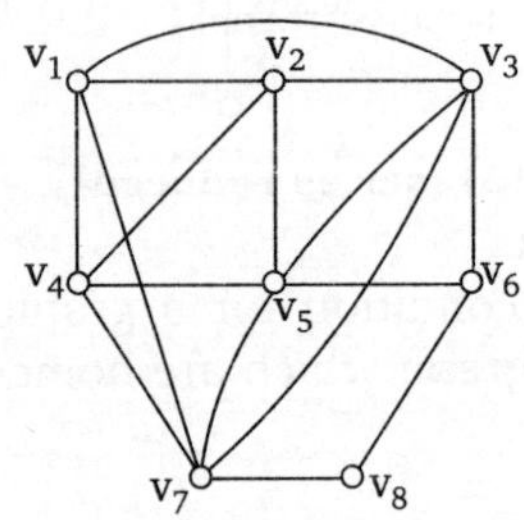

Fig. (143)

Now, to color the graph, we proceed as follows

Ordering the vertices according to decreasing degrees, which gives the following sequence

$$v_5 \quad v_3 \quad v_7 \quad v_1 \quad v_2 \quad v_4 \quad v_6 \quad v_8$$

Therefore, first color is assigned to vertex v_5 and v_1. Second color is assigned to vertices v_3, v_4 and v_8. The third color is assigned to vertices v_7, v_2 and v_6. All other vertices have been assigned a color and so graph G is 3-colorable. We observe that G is not- 2-colorable since vertices v_1, v_2 and v_3 which are connected to each other must be assigned different colors. Hence, $\chi(G) = 3$.

9.24.2 Important Observations from the Definition of Chromatic Number

(1) Chromatic number of null graph is 1.
(2) Chromatic number of complete graph K_n of n vertices is n.
(3) Chromatic number of a graph having one or more edges is at least two.
(4) Chromatic number of a graph having a triangle is at least three.

(5) If a graph is a circuit with n vertices, then
  (i) It is 2-chromatic if n is even.
  (ii) It is 3-chromatic if n is odd.
(6) If a graph is a circuit and it is given that its chromatic number is even, then it has even number of vertices (edges).
(7) If a circuit graph is 3-chromatic then number of vertices in the graph is odd.

Theorem 1. *A graph with edge $e \geq 1$ is 2-chromatic if and only if it has no circuit of odd length.*

Proof : Consider a connected graph G with circuit of even length only. Now, select any vertex and paint it with color 1. Then paint all vertices adjacent to it by color 2. If a vertex is painted with color 2, then any vertex adjacent to it will be painted with color 1. Since every circuit has even length, no adjacent vertices will have the same color. Hence, G is 2-chromatic. Conversely, let if possible, G has a circuit of odd length, then we would need at least three colors just for that circuit, then chromatic number of $G \geq 3$, which is a contradiction. Hence, G has no circuit of odd length.

Theorem 2. *If Δ is the maximum degree of the vertex in a graph G, then chromatic number of $G \leq 1 + \Delta$.*

Proof : Consider a graph G with n vertices. We shall prove the required result by induction on n. If G has 1 or 2 vertices, then result is obviously true. Further assume that result is true for all graphs having less than n vertices. Remove any vertex v and all edges incident on v. Then $G - v$ is a graph with $n - 1$ vertices and maximum degree of any vertex in $G - v$ is at most Δ. Thus, result is true for a graph with $n - 1$ vertices. Hence, by induction hypothesis, chromatic number of $G - v$ is less than or equal to $1 + \Delta$.

9.24.3 Five-color Theorem

The vertices of a planar graph can be properly colored with five colors.

Proof. We prove this theorem by mathematical induction.

Let $P(n)$ is the statement that every planar graph of n vertices can be properly coloured with five or fewer colours.

Step 1. If a graph has only one vertex, it can be coloured with fewer colours than five. **(Base step)**

Step 2. We assume that theorem hold for every planar graph with $(n - 1)$ vertices.

Step 3. We shall prove that it is true for a planar graph with n vertices. Since G is a planar and is connected it must have a vertex v such that deg $(v) \leq 5$. Let G' be the graph obtained by deleting v from G. By induction hypothesis, G' requires not more than 5 colours. When v has degree 1, 2, 3, 4 there is no difficulty. Since we can give to v one more colour. Therefore, it is sufficient to consider the case when G' has 5 colours with which its vertices are properly colours as shown in the following figures.

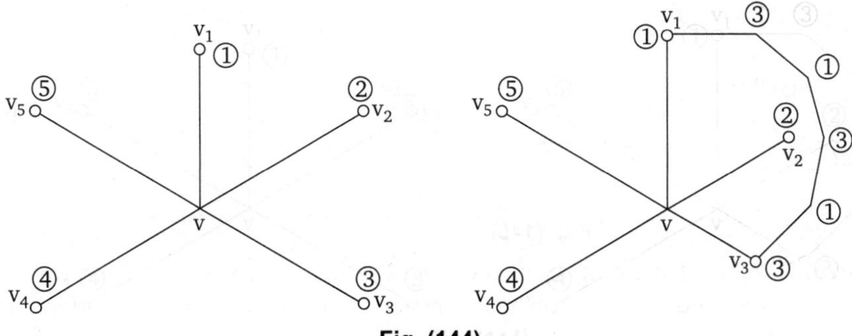

Fig. (144)

Let us suppose that there is a path between the vertices v_1 and v_3 colour alternately with colours 1 and 3, then a similar path between v_2 and v_4 can not exist, since this path, if it exists will intersect the path between v_1 and v_3, which is not possible, because G is planar. So we can interchange the colours of all vertices connected to v_2 giving colour 4 to v_2, while v_4 has still colour 4. Therefore, we can colour v with colour 2 and hence the graph $G'+v = G$ is 5 colourabe. In a similar manner we have assumed that no path exists between v_1 and v_3, we coloured v_3 with colour 1 and v with colour 3.

Hence, a planar graph with 5 vertices is colourable.

9.24.4 Four-color Theorem

Statement : This result was one of the most famous unsolved problems in graph theory and was finally proved by *k*. Appel and *w*. Haken is 1976.

Every planar graph can be properly colored using 4 colors only.

9.25 EDGE COLORING

Definition 1 : *An edge coloring of a graph G is an assignment of colors to the edge of G so that no two edges with a common vertex receive the same color.*

Definition 2 : *The minimum number of colors of G is called the chromatic index of G. It is denoted by* $\chi'(G)$.

Definition 3 : *To edge color a graph is to partition the edge set into subsets such that no two edges in the same subset have a vertex in common so that all edges in any part of the partition are disjoint. Then, the set of disjoint edges in a graph is called matching.*

- If n is odd then $\chi'(K_n) = n, (n \neq 1)$ and $\chi'(k_n) = n - 1$ if n is even.
- If G is a simple graph with maximum vertex degree Δ, then $\Delta \leq \chi'(G) \leq \Delta + 1$
- For all bipartite graph G, $\chi'(G) = \Delta$.

ILLUSTRATIVE EXAMPLES

1. *What is the chromatic number of the following graph :*

Solution : Since the vertices G, R and Y must be assigned different colors, the chromatic number of the given graph must be at least 3. We have to check whether three colors will be sufficient to color the graph. Let us suppose we assign green to G, red to R and yellow to Y. Then G can be colored green as it is adjoint to R and Y . Further c can be colored yellow as it is adjacent to R and G. Similarly e be colored yellow as it is adjoint to R and G. Similarly e can be colored red and d can be colored green. This complete the coloring of the given graph with exactly 3 colors.

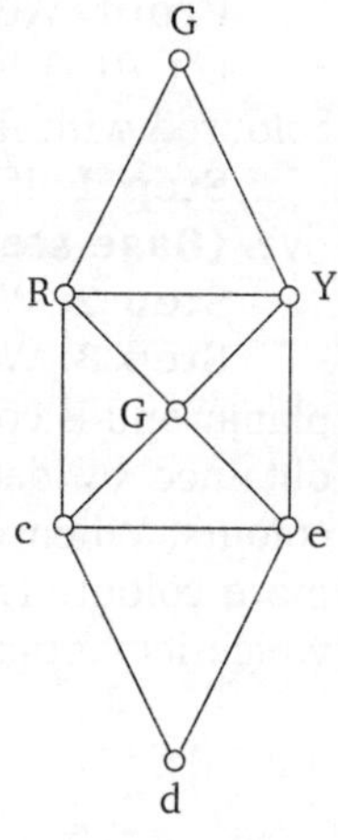

Fig. (144)

2. *Determine the chromatic number of the graph* k_6 *given below*

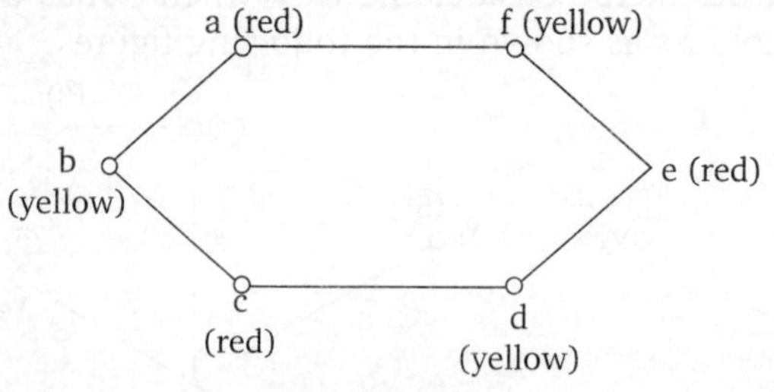

Fig. (145)

Solution. We start coloring from the vertex a and assign it red. Then we assign b as yellow, e and c red and f yellow. Thus, we need only two colors. Therefore, chromatic number of k_6 is 2.

☛ In general the chromatic number of k_n is 2 when n is even.

3. *What is the chromatic number of the complete bipartite graph $k_{3,4}$?*

Solution. The graph $k_{3,4}$ is given below :

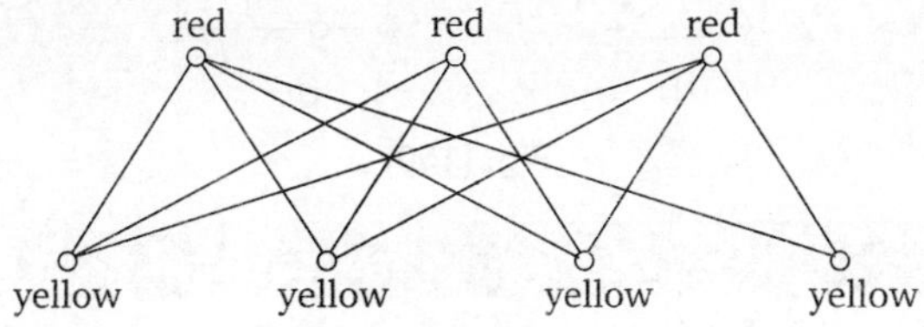

Fig. (146)

Obviously, two colors are needed for this graph. Hence, the chromatic number of $k_{3,4}$ is two.

☛ In general, the chromatic number of $k_{m,n}$ is 2.

4. *Find out chromatic number of the following graph*

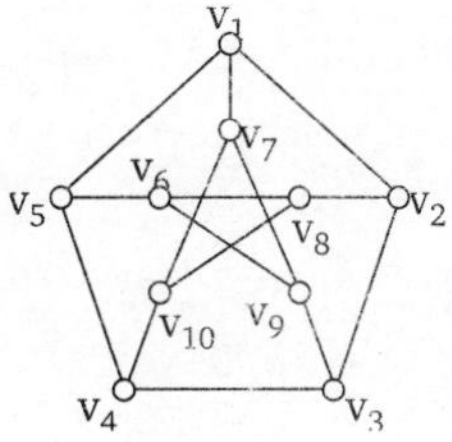

Fig. (147)

Solution. Firstly we assign a color (say red) to v_1. Then v_2, v_5 and v_7 can be assigned another color (say blue), v_4, v_6, v_8 and v_9 can be assigned again red color. v_3 and v_{10} can neither be assigned red nor blue colors. Hence, we assign v_3 and v_{10} third color (say green). Hence, chromatic number of the given graph is 3.

Exercise 9.7

1. Find chromatic number of the graph

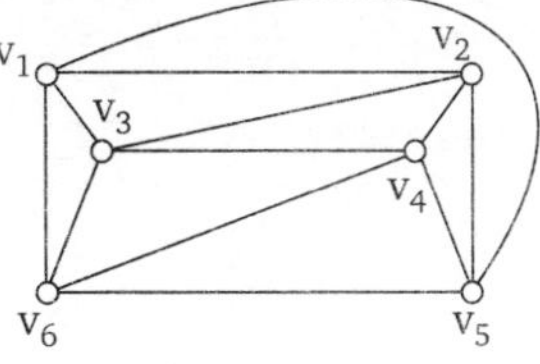

Fig. (148)

2. Find chromatic number of the graph

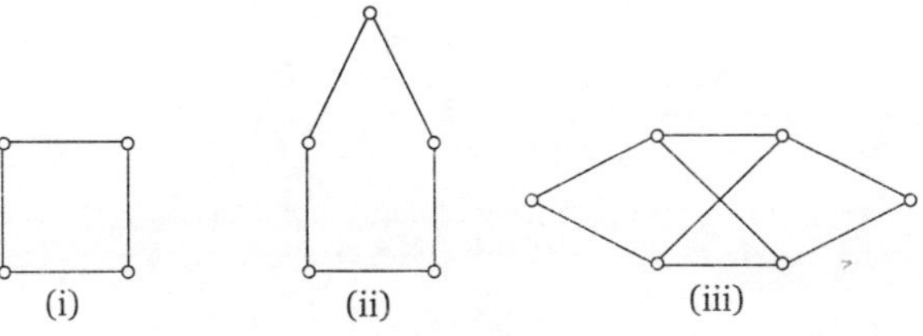

Fig. (149)

3. Find the minimum number of colors required to paint each map

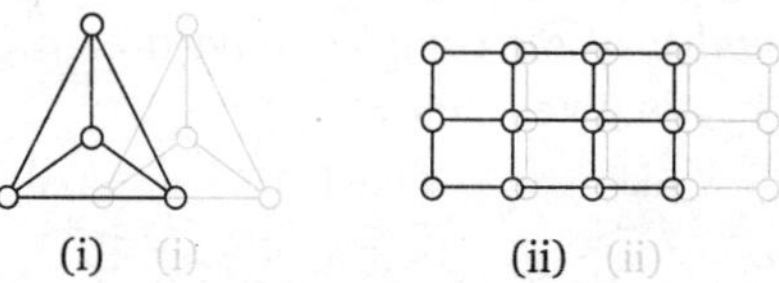

Fig. (150)

ANSWERS

1. 4 **2.** (i) 2 (ii) 3 (iii) 2 **3.** (i) 4 (ii) 3

❑❑❑

Chapter

10

Trees and Cutsets

- Trees
- Minimally connected graph
- Distance and centre
- Rooted tree
- Binary tree
- Spanning tree
- Minimal spanning trees
- Tree Traversal
- Polish notation
- Binary search tree
- Cut sets
- Connectivity and separability
- Network flows

10.1 INTRODUCTION

The graphs that we come across in most of the applications are connected, trees form one of the simplest structures encountered in all such connected graphs. The word tree suggests branching out from a root and never completing a cycle. The undirected graph that corresponds to a say genealogical chart could be considered as one of these structures and forms a subclass of a connected graph called tree. The concept of tree as we shall see subsequently is very useful in the organization of a data and also relating data in a database. In computer science specifically trees occur most often in constructing codes for data transmission and storage, data compression and optimal merging, constructing efficient algorithms to locate or search an item in a list, compiler design, sorting methods, etc.

10.2 TREES

Definition 1 : *A connected graph is called a tree if it contains no cycles. We can say a tree is a connected acyclic graph.*

Definition 2 : *A connected graph having no circuit is called a tree.*

Definition 3 : *Any set of branches in the original graph, just sufficient in number to connect all the vertices (nodes) is called a tree.*

Its edges are called branches.

For Example :

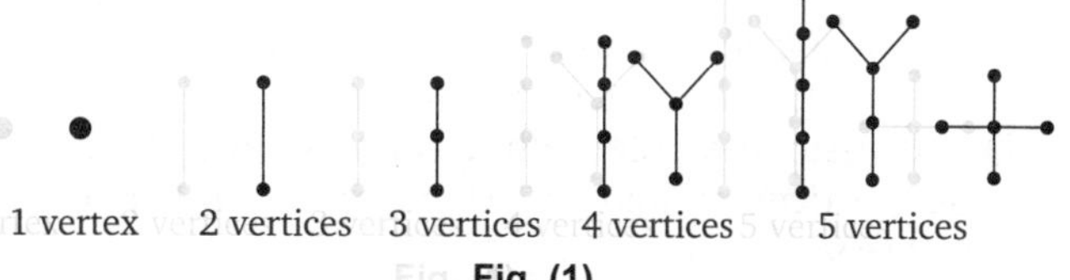

Fig. (1)

☛ Since we are considering only finite graph having at least one vertex, therefore a tree is finite and has at least one vertex.

- Also from the definition of the tree it follows that a tree is a simple graph without self loops or parallel edges.
- A tree with only one vertex is called a trivial tree or degenerate tree, otherwise it is a non trivial tree.
- A leaf (or terminal node) in a tree is a vertex of degree 1.
- A vertex of degree one is called a pendant vertex and a non pendant vertex in a tree is called an internal vertex.

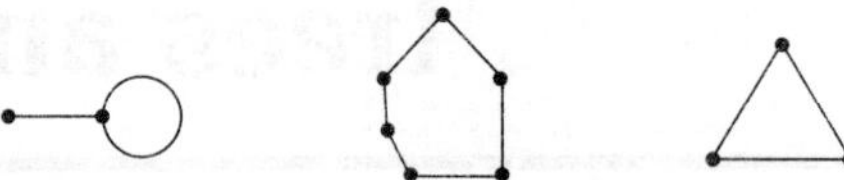

Fig. (2) : Few graphs which are not a tree

10.2.1 Forest

Let G be an acyclic graph. Then any subgraph of G must also contain no cycle. The connected components of G are also acyclic and so they are trees. This is why an acyclic graph is also called a forest.

For Example :

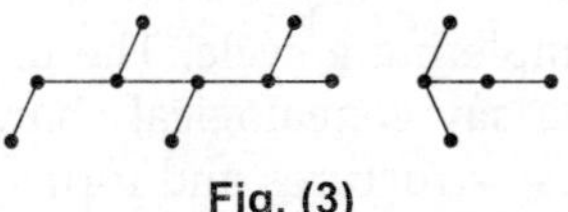

Fig. (3)

1. *Which of the following graphs are trees ?*

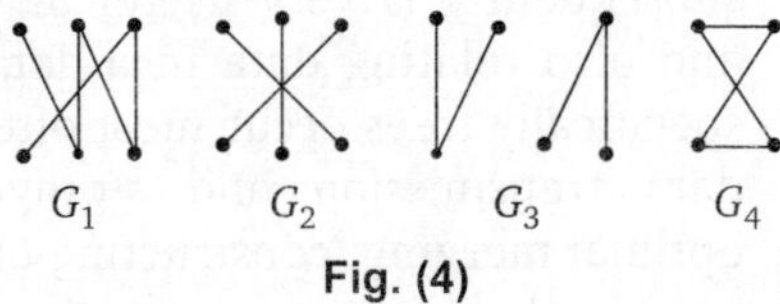

Fig. (4)

Solution. G_1 is a connected graph with no circuit and hence it is a tree. G_2 is also a connected graph with no circuit hence G_2 is also a tree. G_3 is not connected so it is not a tree. G_4 is not a tree because it has a cycle or a circuit.

Theorem 1. *There is one and only one path between every pair of vertices in a tree T.*

Proof : Since T is a connected graph, there must exist at least one path between every pair of vertices in T. Now suppose that between two vertices a and b of T, there are two distinct paths. The union of these two paths will contain a circuit and T can not be a tree.

Theorem 2. *If in a graph G, there is one and only one path between every pair of vertices, then G is a tree.*

Proof : For existence of a path between every pair of vertices, assume that G is connected. A circuit in a graph implies that there is at least one pair of vertices a, b such that there are two distinct path between a and b. Since G has one and only one path between every pair of vertices, G can have no circuit. Therefore G is a tree.

Theorem 3. *A tree of n vertices has n – 1 edges.*

Proof : The theorem will be proved by induction on the number of vertices. Theorem is true for $n = 1, 2, 3$. Assume that theorem holds for all trees with fewer than n vertices.

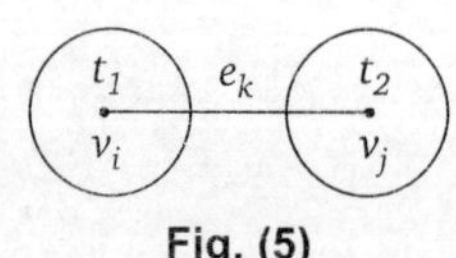

Fig. (5)

Let us assume a tree with n vertices. In T, let e_k be an edge with end vertices v_i and v_j. There is no other path between v_i and v_j except e_k.

Therefore, detection of e_k from T will disconnect the graph. Furthermore $T - e_k$ consists of exactly two components and since we have no circuit in T to begin with each of these components is a tree. Both these trees t_1 and t_2 have fewer vertices each. Therefore by induction each contain one less edge than the number of vertices in it. These $T - e_k$ have $n - 2$ edges. Hence, T has exactly $n - 1$ edges.

Theorem 4. *Every connected graph with n vertices and n – 1 edges is a tree.*

Proof : Let G be a connected graph with n vertices and $n - 1$ edges. The theorem will be proved if we show that G has no circuit. Suppose that G contains at least one circuit. Since removing an edge from a circuit does not disconnect a graph, we may remove edges, but no vertices from circuits in G until the resulting graph G^* is circuit free. Now G^* is a connected graph with n vertices and contains no circuit. Thus G^* is a tree with n vertices. Hence, G^* has $n - 1$ edges (by Theorem 3). But now the graph G has more than $n - 1$ edges, a contradiction. Hence, G has no circuit.

Theorem 5. *A graph with n vertices, (n – 1) edges and having no circuits is connected (i. e., the graph is a tree).*

Proof : If possible, let G be a graph which is disconnected and has n vertices, $n - 1$ edges and without circuits. In this case, the graph G will have more than two components. Without loss of generality, we can assume that G has two components G_1 and G_2, say. Thus there is no path between the vertices of G_1 and of G_2. Now, if we add an edge between a vertex of G_1 and a vertex of G_2, then the graph $G \cup e$ so formed will also have no circuit. Therefore, the graph $G \cup e$ is a circuitless connected graph (*i.e.*, a tree) with n vertices and n edges, which is a contradiction since a graph with n vertices and n edges must have circuits.

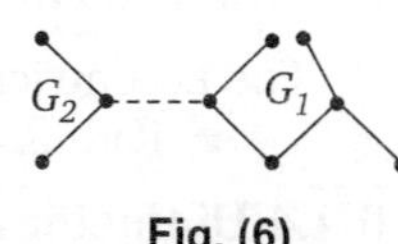

Fig. (6)

Hence, our assumption is wrong and thus the theorem is proved.

10.3 MINIMALLY CONNECTED GRAPH

A connected graph G is called minimally connected if removal of any one edge from G disconnects the graph G.

Theorem 1. *A graph G is a tree iff it is minimally connected.*

Proof : Suppose that G is tree. We show G is minimally connected. Since G is a tree, it is connected. If G is not minimally connected, then there must exist an edge e in G such that $G - e$ is connected. Therefore, e is in some circuit, which implies that G is not a tree, a contradiction. Thus G is minimally connected.

Conversely, suppose that G is a minimally connected graph. Then G is connected and can not have a circuit, otherwise, we could remove one of the edges in the circuit and still leave the graph connected. Thus a minimally connected graph is a tree.

Theorem 1. *In any tree (with two or more vertices), there are at least two pendant vertices.*

Proof : Let G be any tree having n vertices. Then G has $n - 1$ edges. Since each edge contributes two degrees, the sum of the degrees of all vertices in G is $2(n-1)$. Now $2(n - 1)$ degrees are to be divided among n vertices in G. Let the number of vertices of degree one in G be x. Since no vertex in a tree can be of zero degree, we have

$$\frac{2(n-1) - x}{n - x} \geq 2 \quad \Rightarrow \quad x \geq 2.$$

Thus, we must have at least two vertices of degree one in a tree.

10.4 DISTANCE AND CENTRE OF A TREE

Let G be a connected graph. The distance between two vertices v_i and v_j of a connected graph G is denoted by $d(v_i, v_j)$ and is defined as the length of shortest path between v_i and v_j (*i.e.*, the number of edges in the shortest path between v_i and v_j). This definition of distance also holds in case of tree, since tree is a connected graph.

☛ If graph is not a tree then there may exist several paths between a pair of vertices. We can find all the paths and select the length of the shortest path.

10.4.1 Eccentricity of a Vertex

The eccentricity $E(v)$ of a vertex v in a graph G is the distance between v and the vertex v_i farthest from v in G. Symbolically,

$$E(v) = \max_{v_i \in G} d(v, V_i)$$

10.4.2 Centre of a Graph

A center of a graph G is a vertex with minimum eccentricity in G.

☛ A tree may have more than one centre.

10.4.3 Radius of a Tree

The eccentricity of a centre in a tree is called the radius of the tree.

☛ Radius of a tree is the distance from the centre of the tree to the farthest vertex.

10.4.4 Diameter of a Tree

The diameter of a tree is defined as the length of the longest path in the tree T. In other words, the diameter of T is the maximum eccentricity of the vertex in T.

☛ Diameter of a tree is not necessarily the double of its radius.

ILLUSTRATIVE EXAMPLES

1. *Suppose a tree has n_1 vertices of degree 1, 2 vertices of degree 2, 4 vertices of degree 3 and 3 vertices of degree 4. Find n_1.*

Solution. Sum of degree of all vertices $S = 2e$, where e is the number of edges.

For a tree, number of edges $= n - 1$, where n is the number of vertices.

Sum of degrees, $s = 2e = 2(n - 1)$

$$s = n_1 \times 1 + 2 \times 2 + 4 \times 3 + 3 \times 4 = n_1 + 4 + 12 + 12 = n_1 + 28 \qquad ...(1)$$

Total vertices, $n = n_1 + 9$

We have, $s = 2(n - 1)$

$$S = 2(n_1 + 9 - 1) = 2(n_1 + 8) \qquad ...(2)$$

From (1) and (2)

$$n_1 + 28 = 2(n_1 + 8)$$

$$n_1 + 28 = 2n_1 + 16$$

$$n_1 = 12$$

Tree has 12 vertices of degree 1.

2. *Find the distance between the vertices v_1 and v_2 in figure.*

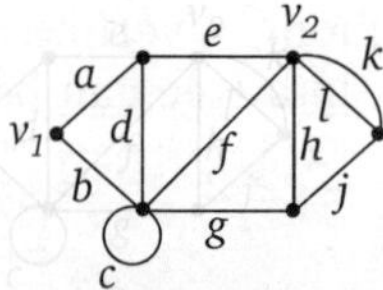

Fig. (7)

Solution. Some of the paths between vertices v_1 and v_2 in figure are

$$(a, e), (a, d, g, h), (b, f), (a, d, f), (b, g, h)$$

There are two shortest paths (a, e) and (b, f), each of length 2.

Hence, $d(v_1, v_2) = 2$.

3. *Find the eccentricities of each of the vertices of the tree T shown in following figures.*

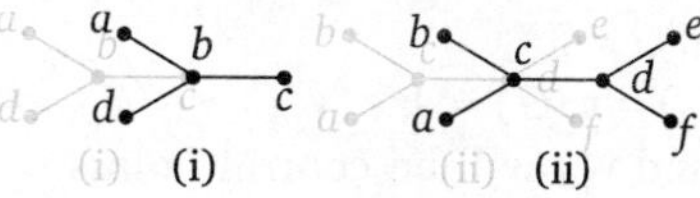

Fig. (8)

Also, find their centre.

Solution. (i) $E(a) = \max\{d(a, b), d(a, c), d(a, d)\}$

$= \max\{1, 2, 2\} = 2$

Similarly, $E(b) = 1, E(c) = 2, E(d) = 2$

Therefore, b has the minimum eccentricity, namely 1. Hence vertex b is the centre of the tree.

(ii) $E(a) = \max\{d(a, b), d(a, c), d(a, d), d(d, e), d(a, f)\}$

$= \max\{2, 1, 2, 3, 3\} = 3$

Similarly, $E(b) = 3, E(c) = 2, E(d) = 2, E(e) = 3, E(f) = 3.$

Now, there are two vertices c and d having the same minimum eccentricity, namely 2.

Hence, this graph has two centres c and d.

4. *Find the eccentricity of all vertices, radius and centre of the graph given below :*

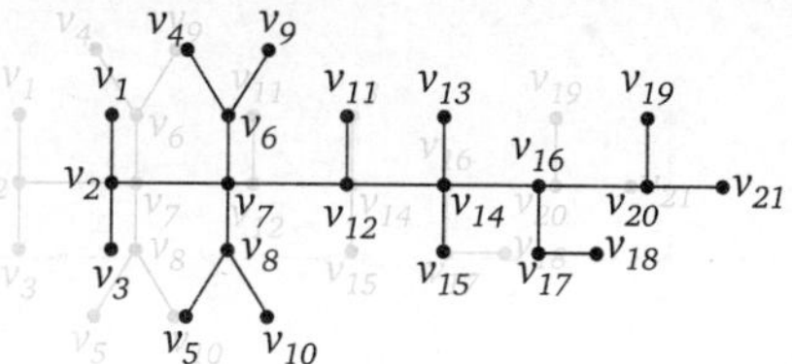

Fig. (9)

Solution : In the above graph all vertices are having equal size edges.

Now, we calculate the eccentricity for one vertex v_1

Vertex set $v = (v_2, v_3, v_4, \ldots, v_{21})$

Eccentricity $e(v_1,) = \max\{d(v_1, u)\}$ where, $u \in v$ and $v_1 \neq u$

$d(v_1, v_2) = 1$	$d(v_1, v_9) = 4$	$d(v_1, v_{16}) = 5$	$d(v_1, v_3) = 2$
$d(v_1, v_{10}) = 4$	$d(v_1, v_{17}) = 6$	$d(v_1, v_4) = 4$	$d(v_1, v_{11}) = 4$
$d(v_1, v_{18}) = 7$	$d(v_1, v_5) = 4$	$d(v_1, v_{12}) = 3$	$d(v_1, v_{19}) = 7$
$d(v_1, v_6) = 3$	$d(v_1, v_{13}) = 5$	$d(v_1, v_{20}) = 6$	$d(v_1, v_7) = 2$
$d(v_1, v_{14}) = 4$	$d(v_1, v_{21}) = 7$	$d(v_1, v_8) = 3$	$d(v_1, v_{15}) = 5$

Now, we select maximum distance for eccentricity

$$e(v_1) = \max\{1, 2, 3, 4, 5, 6, 7\} = 7$$

Similarly we can calculate eccentricity for other vertices means for $v_2, v_3, \ldots, v_{21}$

$e(v_2) = 6$	$e(v_9) = 7$	$e(v_{16}) = 5$	$e(v_3) = 7$
$e(v_{10}) = 7$	$e(v_{17}) = 6$	$e(v_4) = 7$	$e(v_{11}) = 5$
$e(v_{18}) = 7$	$e(v_5) = 7$	$e(v_{12}) = 4$	$e(v_{19}) = 7$
$e(v_6) = 6$	$e(v_{13}) = 5$	$e(v_{20}) = 6$	$e(v_7) = 5$
$e(v_{14}) = 4$	$e(v_{21}) = 7$	$e(v_8) = 6$	$e(v_{15}) = 5$

Now, we have eccentricity for all vertices. According to the definition of radius of a connected graph G.

$$\text{rad. } G = \min\{e(v)\,; v \in V\}$$

Radius of the given graph, $G = \min\{e(v)\}$

$= \min\{4, 5, 6, 7\} = 4$

Diameter of the given graph, $g = \max\{e(v)\}$

$= \max\{4, 5, 6, 7) = 7$

Radius of the given graph, $G = 4$

Diameter of the given graph, $G = 7$

$$e(v_{12}) = r(G) = 4; e(v_{14}) = r(G) = 4$$

According to definition v_{12} and v_{14} will be central points.

Centre of G = set of all central points = $\{v_{12}, v_{14}\}$

Centre of $G = \{v_{12}, v_{14}\}$

Theorem 1. *Every tree has either one or two centres.*

Proof : We know that the maximum distance, max $d(v, v_i)$ from a given vertex v to any other vertex v_i in a graph occurs only when the vertex v_i is pendant.

Consider a tree T that has more than two vertices and therefore T has at least two pendant vertices. Now we delete all pendant vertices from T and obtain another graph T_1 which again is a tree. The eccentricities of vertices in T_1 will be reduced by 1 and therefore all vertices that are centres in T_1. Again we delete all pendant vertices from T_1 and obtain another tree T_2. We contribute this process of deleting pendant vertices until there is left either a single vertex which is centre of T or a single edge whose end vertices are the two centres of T. This whole process is explained in following figure, the eccentricities of each vertex is written next to vertex.

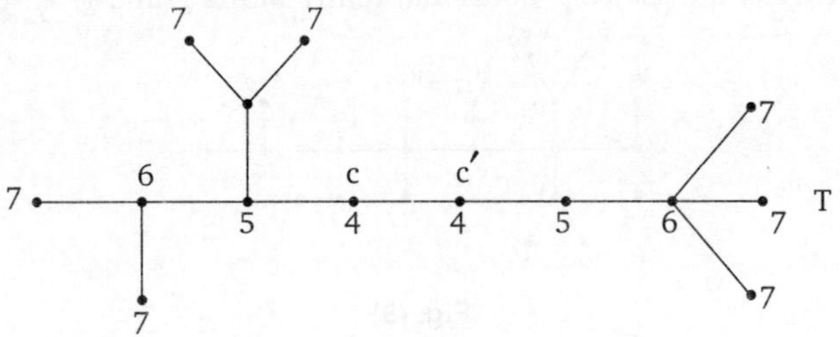

Fig. (10)

Deleting all pendant vertices from T, we get T_1

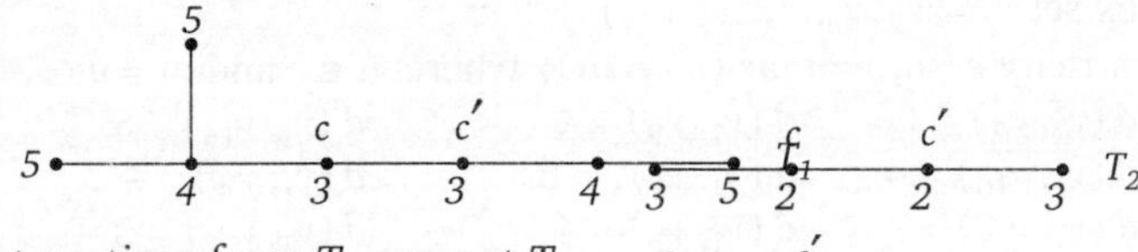

Deleting all pendant vertices from T_1, we get T_2

c c'
1 1 T_3

Deleting all pendant vertices from T_2, we get T_3

All c and c' are centres of T.

10.4.5 Directed Tree

A directed tree is an acyclic digraph which has one node called its root with indegree 0 while all other nodes have indegree 1. Every directed tree must have at least one node. An isolated node is also a directed tree.

In a directed tree any node which has outdegree 0 is called a terminal node or a leaf.

The level of any node is the length of its path from the root. The level of the root node of a directed tree is 0 while the level of any node is equal to its distance from the root.

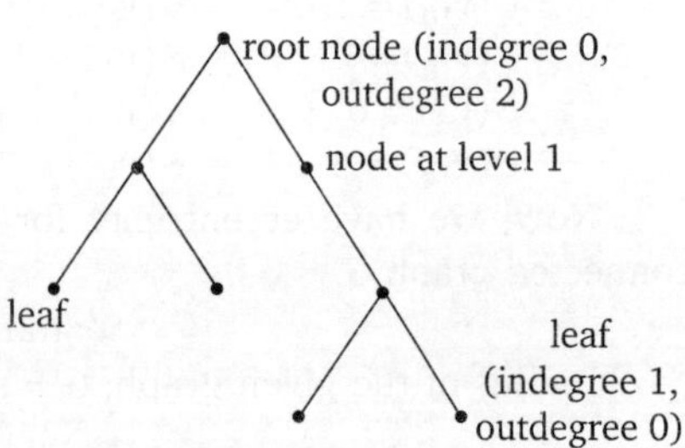

Fig. (11) : Directed Tree

10.5 ROOTED TREES

A rooted tree is a tree in which a particular vertex is distinguished from the others and is called the root.

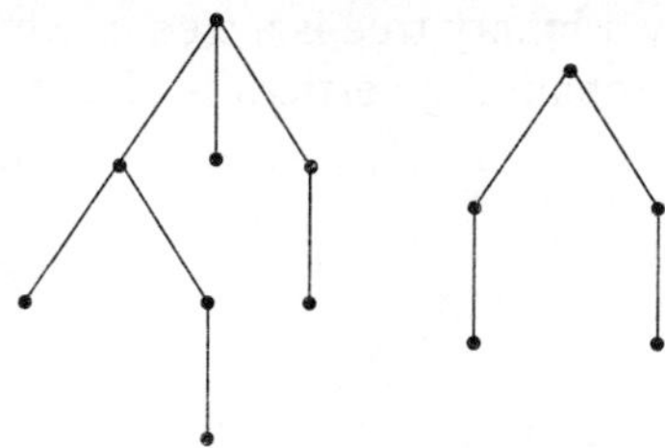

Fig. (12) Rooted Trees

- The level of a vertex is the number of edges along the unique path between it and the root.
- The level of the root is defined as 0.
- The height of a rooted tree is the maximum level to any vertex of the tree.
- The depth of a vertex v in a tree is the length of the path from the root to v
- If any vertex has no children, then it is called a leaf (or a terminal vertex).
- If any vertex has either one or two children, then it is called an internal vertex.
- The decendents of the vertex is the set consisting of all the children together with the descents of those children. Given vertices v and w, if v lies on the unique path between w and the root, then v is an ancestor of w and w is a descendent of v.

10.5.1 Subtree

Let a be a branch node in a rooted tree T. The subtree with a as root is a subgraph T' of T such that vertex set of T' is the set of sub vertices in T containing a and all of its descendants and the edge set of T' is the set of edge is all directed paths starting from a.

10.5.2 Decision Trees or Sorting Trees

Decision trees or sorting trees are labelled rooted trees which occur in applications, especially in computer programming and computer algorithms, the roots represent a starting point, later vertices represent later decision points and one proceeds downward through the tree, choosing edges at each step according to observe data.

ILLUSTRATIVE EXAMPLES

1. *Consider the rooted tree in following figure.*

(a) What is the root of T ?

(b) Find the leaves and the internal vertices of T.

(c) What are the levels of c *and* e.

(d) Find the children of c *and* e.

(e) Find the descendants of the vertices a *and* c.

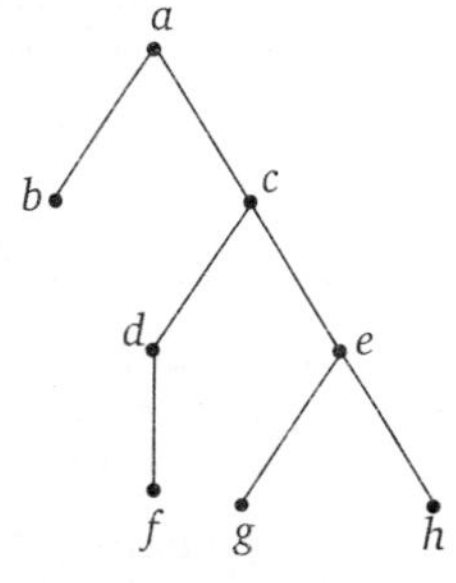

Fig. (13)

Solution. (a) Vertex a is distinguished as the only vertex located at the top of the tree, therefore a is the root.

(b) The leaves are those vertices that have no children. These are b, f, g and h. The internal vertices are c, d and e.

(c) The level of c and e are 1 and 2 respectively.

(d) The children of c are d and e and for e are g and h.

(e) the descendants of a are b, c, d, e, f, g, h. The descendants of c are d, e, f, g, h.

- Every vertex in a directed tree different from the root has a unique parent.

10.6 BINARY TREE

A binary tree is a rooted labelled tree in which any vertex has at most two children. If every vertex of a binary tree has either two children or no children, then we say the tree is a full binary tree. We can say a binary tree is a tree in which there is exactly one vertex of degree two and each of the remaining vertices is of degree one or three.

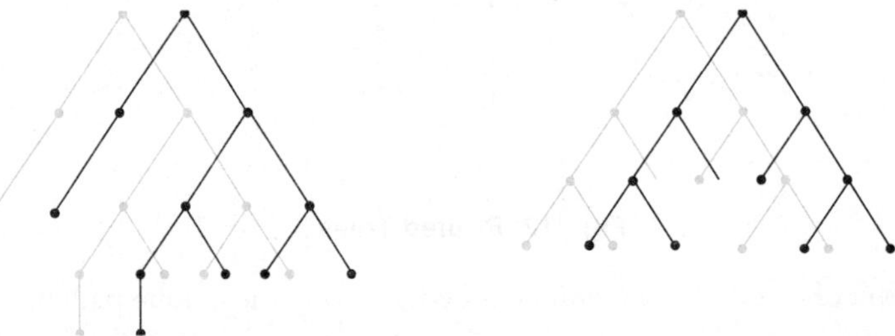

Fig. (14) : Rooted Binary Tree **Fig. (15)** : Full (Complete) Binary Tree

The vertex of degree two is distinct from all other vertices. This vertex serves as a root so we can say every binary tree is a rooted tree.

The maximum level l of any vertex in a binary tree is called the height of the tree. In a labelled tree of root r will constitute level 0. The neighbours of r will constitute level 1 and the neighbours of vertices in level 1 will have level 2 and so on. The neighbours on next level are called descendant (children) and previous level are called predecessor (father).

- ☛ Full binary tree is also known as complete binary tree.
- ☛ In a binary tree, if the heights of the left and the right sub-tree of every vertex differ by at most one is called height balanced binary tree.

10.6.1 Properties of Complete Binary Tree

(1) The number of vertices n in a complete binary tree is always odd. This is because there is exactly one vertex of even degree and the remaining $(n-1)$ vertices are of odd degrees and we know that the number of vertices of odd degrees is even. Thus $(n-1)$ is even and hence n is odd.

(2) Number of pendant vertices in any complete binary tree with n vertices is $\frac{n+1}{2}$.

Let p be the number of pendant vertices in a complete binary tree T. Then $n - p - 1$ is the number of vertices of degree three.

Therefore the number of edges in T equals

$$= \frac{1}{2}[p + 3(n - p - 1) + 2] = n - 1 \quad \Rightarrow \quad p = \frac{n+1}{2}$$

10.6.2 Height of a Binary Tree

The height of a binary tree is defined as the maximum level of any vertex in the tree and is denoted by l_{max}. The minimum possible height of an n-vertex binary tree

$$= \min l_{max} = \lceil \log_2 (n+1) - 1 \rceil$$

where $\lceil\ \rceil$ means the smallest integer function.

The construct a binary tree for a given n such that the vertex farthest from the root is as close to the root as possible, we must take two vertices at each level, except at 0-level.

Hence, $\max l_m = \frac{n-1}{2}$

10.6.3 m-ray Tree

A rooted tree is an m-ary if every internal vertex has at most m-children.

An m-ary tree is a full m-ary tree if every internal vertex has exactly m-children.

The relationship between i, the number of internal vertices and l, the number of leaves of a full m-ary can be proved by using the following theorem.

Theorem 1. *A full m-ary tree with i internal vertex has $n = mi + 1$ vertices.*

Proof : Since the tree is a full m-ray, each internal vertex has m children and the number of internal vertex is i, the total number of vertex except the root is mi.

Therefore, the tree has $n = mi + 1$ vertices.

Since l is the number of leaves, we have $n = l + i$. Using the two equalities $n = mi + 1$ and $n = l + i$, the following results can easily be deduced. A full m-ray tree with

(i) n vertices has $i = \dfrac{n-1}{m}$ internal vertices and $l = \dfrac{[(m-1)\,n + 1]}{n}$ leaves.

(ii) i internal vertices has $n = mi + 1$ vertices and $l = (m-1)\,i + 1$ leaves.

(iii) l leaves has $n = \dfrac{ml-1}{(m-1)}$ vertices and $i = \dfrac{l-1}{m-1}$ internal vertices.

Theorem 2. *There are at most m^h leaves in an m-ary tree of height h.*

Proof : We prove the theorem by mathematical induction.

Basis of Induction : For $h = 1$, the tree consists of a root with no more than m-children, each of which is a leaf. Hence, there are no more than $m^1 = m$ leaves in an m-ary of height-1.

Induction hypothesis : We assume that the result is true for all m-ary trees of height less than h.

Induction step : Let T be any m-ary of height h. The leaves of T are the leaves of subtrees of T obtained by deleting the edges from the roots to each of the vertices of level 1. Each of these subtrees has height less than or equal to $h - 1$. So by the inductive hypothesis, each of these rooted trees has at most m^{h-1} leaves. So by the inductive hypothesis, each of these rooted trees has at most m^{h-1} leaves. Since there are at most m such subtrees, each with a maximum of m^{h-1} leaves, there are at most $m.m^{h-1} = m^h$.

10.6.4 Almost Complete Binary Tree

A binary trees of depth d is said to be almost complete binary tree if

(i) Each mode in the tree is either at level d or $d - 1$.

(ii) For any node in the tree with a right descendant at level d, all the left descendants of this node are also at level d.

The following figure shows an almost complete binary tree.

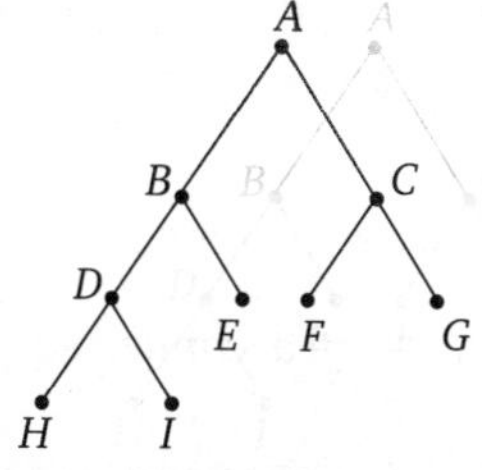

Fig. (16) :
An almost complete binary tree

ILLUSTRATIVE EXAMPLES

1. *Draw a balanced binary tree of height 4 with minimum number of vertices.*

Solution : The minimum number of vertices of a balanced tree of height

$$(h) = 2^{h+1} - 1 = 2^5 - 1 = 32 - 1 = 31.$$

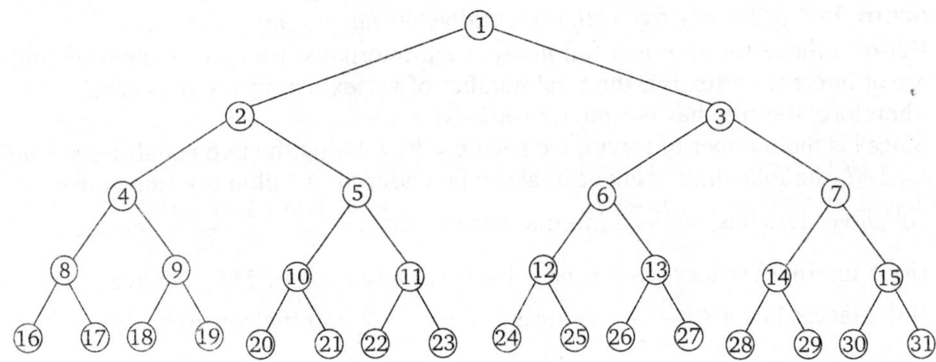

Fig. (17) : Labelled binary tree

2. *What are the left and right children of b shown in the following figure ? What are the left and right subtree of a?*

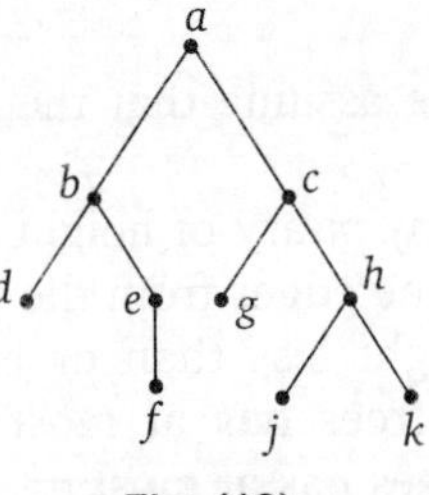

Fig. (18)

Solution. The left child of b is d and the right child is e. The left subtree of the vertex a consists of the vertices b, d, e and f and the right subtree of a consists of the vertices c, g, h, j and k whose figures are shown.

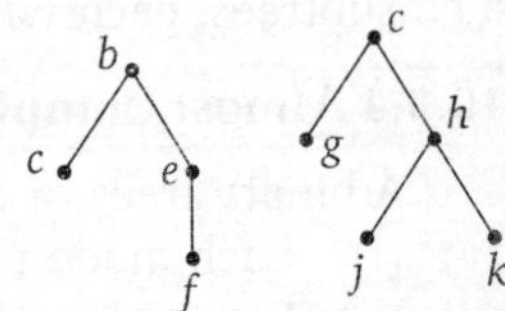

3. *Draw all the distinct binary trees of 4 vertices.*

Solution.

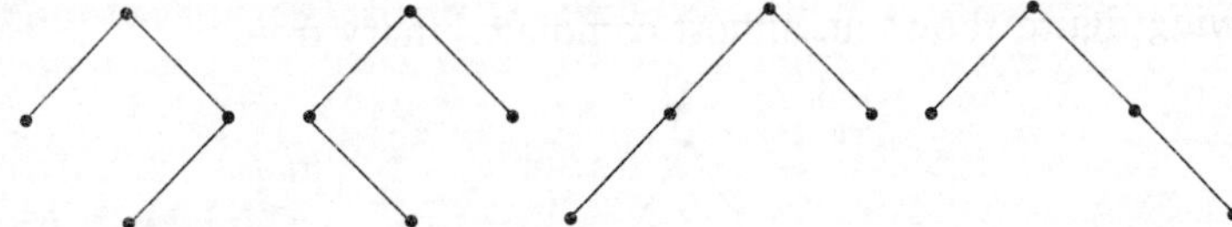

10.7 SPANNING TREE

A tree T is said to be a spanning tree of a connected graph G if T is a subgraph of G and T contains all vertices of G.

Thus for T (a subgraph of G) to be a spanning tree, we have the following requirements :

(i) T has the same set V of vertices as does G

(ii) T is a tree, and

(iii) T is a subgraph of G (so $E' \subseteq E$)

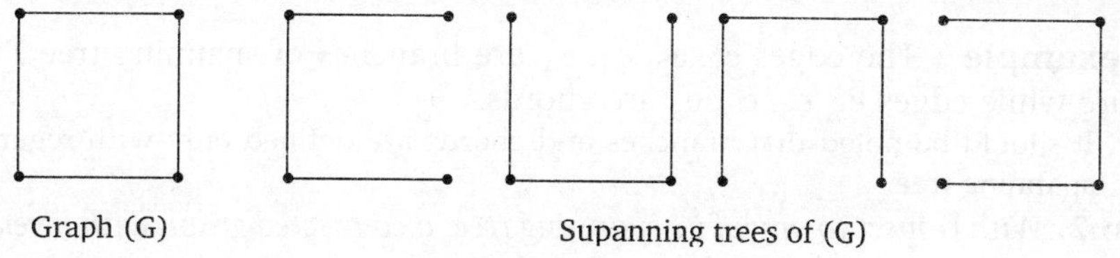

Fig. (19)

- A given graph G can have number of spanning trees.
- A disconnected graph with k components has k spanning trees which is called spanning forest.
- Spanning trees have the maximum number of edges among all trees in G, so they are the largest trees among all trees in G, therefore a spanning tree is a maximal tree or it is a maximal tree subgraphs of G.

10.7.1 Spanning subgraph

A subgraph of G which contains all the vertices of the original graph G is called spanning subgraph.

10.7.2 Co-spanning Tree

The co-spanning tree T^* of a spanning tree T of a connected graph G is the subgraph of G having all the vertices of G and exactly those edges of G which are not in T.

- A co-spanning tree may not be connected.

Theorem 1. *A simple graph G has a spanning tree if and only if G is connected.*

Proof : First, suppose that a simple graph G has a spanning tree T. T contains every vertex of G. Let a and b be vertices of G. Since a and b are also vertices of T and T is a tree, there is a path P between a and b. Since T is a subgraph, P also serves as path between a and b in G. Hence, G is connected.

Conversely, suppose that G is connected. If G is not a tree, it must contain a simple circuit. Remove an edge from one of these simple circuits. The resulting subgraph has one fewer edge but still contains all the vertices of G and is connected. If this subgraph is not a tree, it has a simple circuit : so as before, remove an edge that is in a simple circuit. Repeat this process until no simple circuit remains. This is possible because there are only a finite number of edges in the graph. The process terminates when no simple circuit remains. Thus, we eventually produce an acyclic subgraph T which is a tree. The tree is a spanning tree since it contains every vertex of G.

10.7.3 Branch and Chord of a Spanning Tree

Let G be a connected graph and T be a spanning tree of G. An edge in T is called a branch of T. An edge of G that is not in T is called a chord.

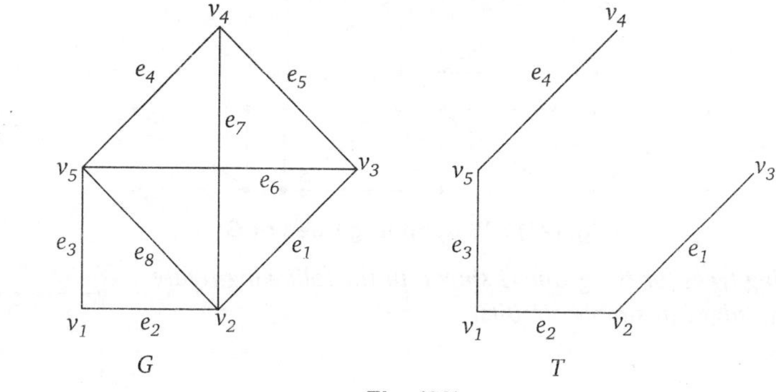

Fig. (20)

For example : The edges e_1, e_2, e_3, e_4 are branches of spanning tree T shown in above figure while edges e_5, e_6, e_7, e_8 are chords.

- It should be noted that branches and chords are defined only with regard to given spanning tree.

Theorem 2. *With respect to any of its spanning tree, a connected graph with n vertices and e edges has n – 1 tree branches and e – n + 1 chords.*

Proof : Let G be a graph with n vertices and e edges and let T be any spanning tree of G. Then T has n vertices and $n-1$ edges. The remaining $e-n+1$ edges in G are chords. Thus the graph G has $n-1$ tree branches and $e-n+1$ chords.

10.7.4 Rank and Nullity

Consider a graph G with n vertices, e edges and k components. The rank of graph G is defined as

$$\text{rank } r = n - k$$

and the nullity of the graph G is defined as nullity $\mu = e - n + k = e - r$

We note that rank + nullity = number of edges in a graph

The nullity of a graph is also called cyclomalic number of first Betti Number

- If a graph G is connected then $k = 1$ and therefore rank of a connected graph is $n - 1$ and the nullity is $e - n + 1$.
- Rank of a connected graph G = Number of branches in any spanning tree of G.
- Nullity of connected graph G = Number of chords in G.

ILLUSTRATIVE EXAMPLES

1. *Find all the spanning trees of the given graph G.*

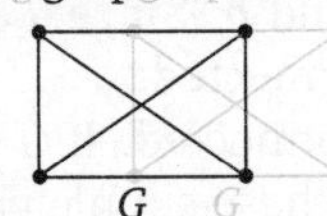

Solution. The given graph will have $(4)^{4-2} = 16$ different spanning trees.

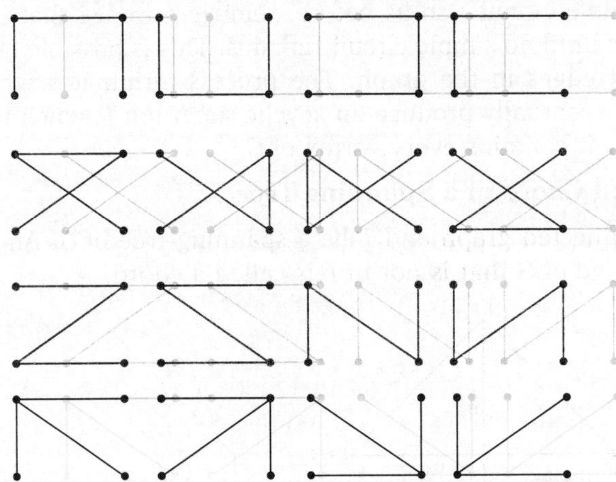

Fig. (21) : 16 spanning trees of G

2. *Find all spanning trees for the graph G shown in the following figure by removing the edges in simple circuits.*

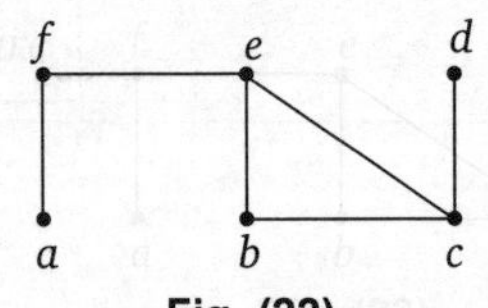

Fig. (22)

Solution. The graph G has one cycle $c\,b\,e\,c$ and removal of any edge of the cycle gives a tree. There are three trees which contain all the vertices of G and hence spanning trees.

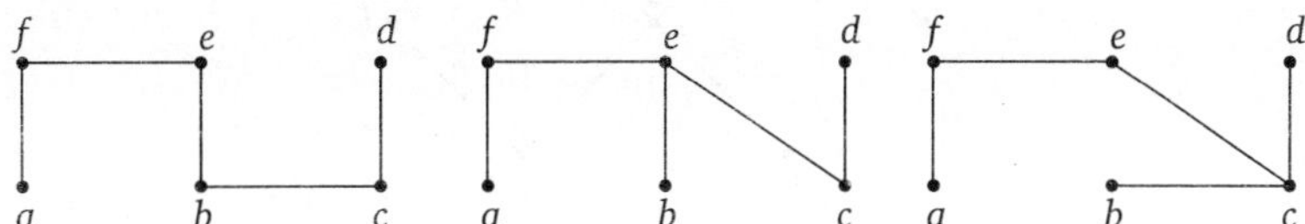

3. *Find the rank and nullity of the graph G in the following figure. Also prove that rank of G + Nullity of G = Number of edges in G.*

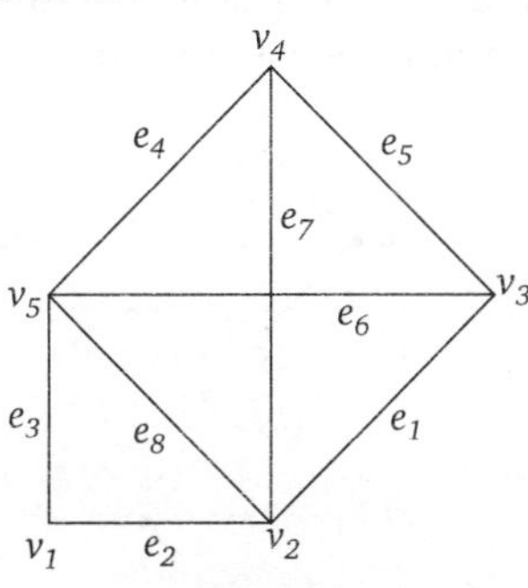

Fig. (23)

Solution. We have number of vertices in G, $n = 5$

Number of edges in G, $e = 8$

Since G is a connected graph, $\quad \therefore \quad k = 1$

$\therefore \quad$ Rank of G, $r = n - k = 5 - 1 = 4$ and nullity of G,

$$v = e - n + k = 8 - 5 + 1 = 4$$

Then, $r + \mu = 4 + 4 = 8$ = Number of edges in G.

10.7.5 Algorithm for Constructing Spanning Trees

Instead of constructing spanning trees by removing edges, spanning tree can be built up by successively adding edges. Two algorithm are based tree on this principle for finding a spanning tree are Breath-First Search (BFS) and Depth-First Search (DFS).

(a) BFS Algorithm

In this algorithm, a rooted tree will be constructed, and the underlying undirected graph of this rooted tree forms the spanning tree. The idea of BFS is to visit all vertices on a given level before going into the next level.

Procedure : Arbitrarily choose a vertex and designate it as the root. Then add all edges incident to this vertex, such that the addition of edges does not produce any cycle. The new vertices added to this stage become the vertices at level 1 in the spanning trees, arbitrarily order them. Next, for each vertex at level 1, visited in order, add each edge incident to this vertex to the tree as long as it does not produce any cycle. Arbitrarily order the children of each vertex at level 1. This produces the vertices at level 2 in the tree. Continue the same procedure until all the vertices in the tree have been added. The procedure ends, since there are only a finite number of edges in the graph. A spanning tree is produced since we have produced a tree without cycle containing every vertex of the graph.

4. *Use BFS algorithm to find a spanning tree of graph G in the following figure.*

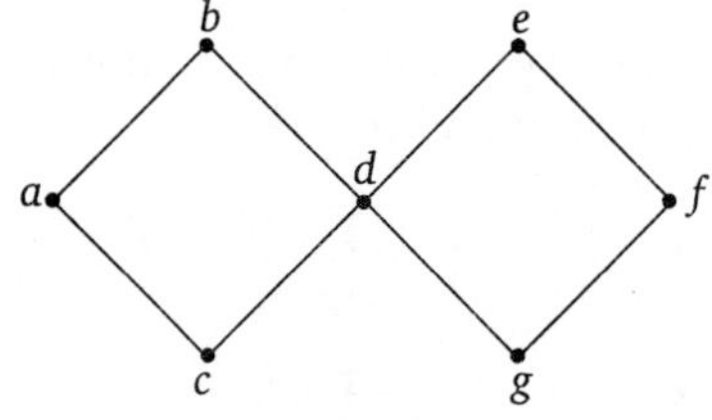

Fig. (24)

Solution. (i) Choose the vertex a to be the root.

(ii) Add edges incident with all vertices adjacent to a, so that edges $\{a, b\}$, $\{a, c\}$ are added. The two vertices b and c are in level 1 in the tree.

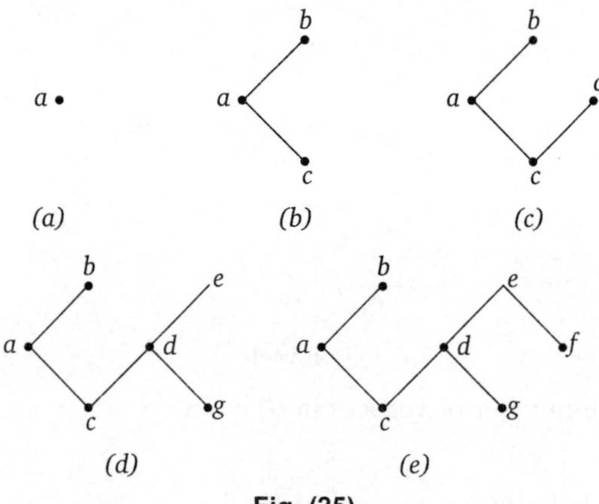

Fig. (25)

(iii) Add edges from these vertices at level 1 to adjacent vertices not already in the tree. Hence the edge $\{c, d\}$ is added. The vertex d is in level 2.

(iv) Add edge from d in level 2 to adjacent vertices not already in the tree. The edge $\{d, e\}$ and $\{d, g\}$ are added. Hence e and g are in level 3.

(v) Add edge from e at level 3 to adjacent vertices not already in the tree and hence $\{e, f\}$ is added.

The steps of BFS are shown in following figure.

(b) DFS Algorithm

An alternative to Breath-First Search is Depth-First Search which proceeds to successive levels in a tree at the earliest possible opportunity. DFS is also called back tracking.

Procedure : Arbitrarily choose a vertex from the vertices of the graph and designate it as the root. From a path starting at this vertex by successively adding edges as long as possible where each new edge is incident with the last vertex in the path without producing any cycle. If the path goes through all vertices of the graph, the tree consisting of the path is a spanning tree. Otherwise, move back to the next to the last vertex in the path, and if possible, form a new path starting at this vertex passing through vertices that were not already visited. If this can not be done, move back anther vertex in the path that is two vertices back in the path, and repeat. Repeat this procedure beginning at the last vertex visited, moving back up the path one vertex at a time, forming new paths that are as long as possible until no more edges can be added. This process ends since the graph has a finite number of edges and is connected. A spanning tree is produced.

5. *Find a spanning tree of the graph in the following figure using Depth-First Search algorithm.*

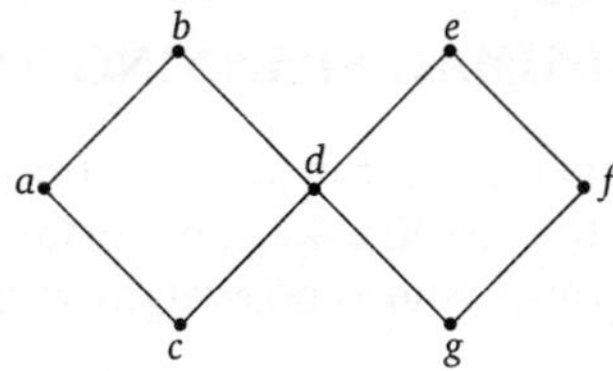

Fig. (26)

Solution. Choose the vertex a. Form a path by successively adding edges incident with vertices not already in the path as long as possible. This procedure the path $a - c - d - e - f - g$. Now back track to f. There is no path beginning at f containing vertices not already visited. Similarly, after back track at e, there is no path. So move back track at d and from the path $d - b$. This produces the required spanning tree which is shown in figure.

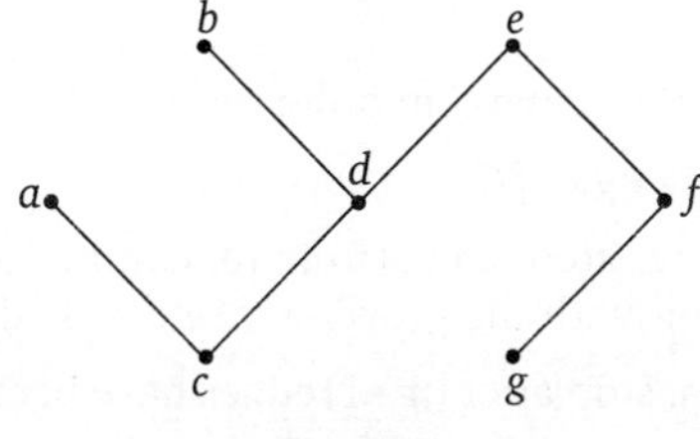

Fig. (27)

Cayley's Theorem. The complete graph k_n has n^{n-2} different spanning trees.

10.8 MINIMAL (SHORTEST) SPANNING TREES

Let G be a weighted graph. The weight of a spanning tree T of G is defined as the sum of the weights of all the branches in T. In general, different spanning trees of a weighted graph will have different weight. *A spanning tree with the smallest weight in a weighted graph is called a minimal spanning tree or a shortest spanning tree.*

- In a graph G, there may exist more than one spanning trees with the smallest weight. For example, if G is a graph with n vertices and every edge in G has unit weight then all spanning trees of G have a weight equal to $(n - 1)$ units.

10.8.1 Distance between Two Spanning Trees

Let G be a connected graph and T_1 and T_2 be its two spanning trees. The distance between T_1 and T_2 is defined as the number of edges of G present in one but not present in the other. The distance between T_1 and T_2 is denoted by $d(T_1, T_2)$.

For example : In the following figure T_1 and T_2 are two spanning trees of a graph G, then $d(T_1, T_2) = 2$.

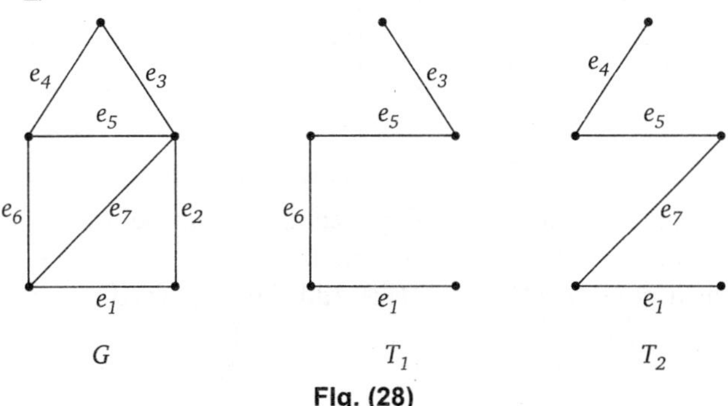

Fig. (28)

10.9 ALGORITHM FOR MINIMAL SPANNING TREES

There are several methods available for actually finding a minimal spanning tree in a given graph. Two algorithms due to Kruskal and Prim of finding a minimal spanning tree for a connected weighted graph where no weight is negative.

10.9.1 Kruskal's Algorithm

In this algorithm we first select an edge of G which has smallest weight among the edges in G which are not loops. Then for each successive step select (from all remaining edges of G) another edge of smallest weight that marks no cycle with the previously selected edges. Continue until $n-1$ edges have been selected and these edges form acyclic subgraph T of G which is a minimal spanning tree of G.

WORKING RULES

Step 1. List all edges of the graph G in order of non-decreasing weight.

Step 2. Select a smallest edge of G.

Step 3. For each successive step select (from remaining edges of G) another smallest edge that makes no circuit with the previously selected edges.

Step 4. If G has n vertices, stop after $(n-1)$ edges have been chosen. Otherwise repeat step 3.

ILLUSTRATIVE EXAMPLES

1. *Using Kruskal algorithm find a minimal spanning tree for the graph of the following figure.*

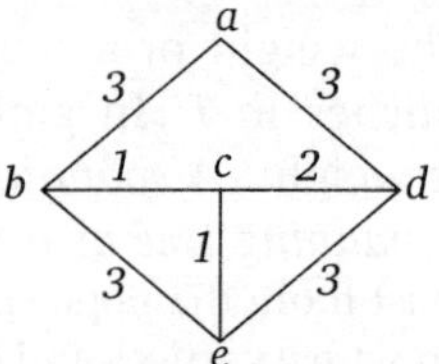

Fig. (29)

Solution. We collect the edges with their weight into a table

Edge	Weight
(b, c)	1
(c, e)	1
(c, d)	2
(a, b)	3
(e, d)	3
(a, d)	3
(b, e)	3

The steps of finding a minimal spanning tree are shown below :

(i) Choose the edge (b, c) as it has a minimal weight. b ——— c

(ii) Add the next edge (c, e) (iii) Add the edge (c, d)

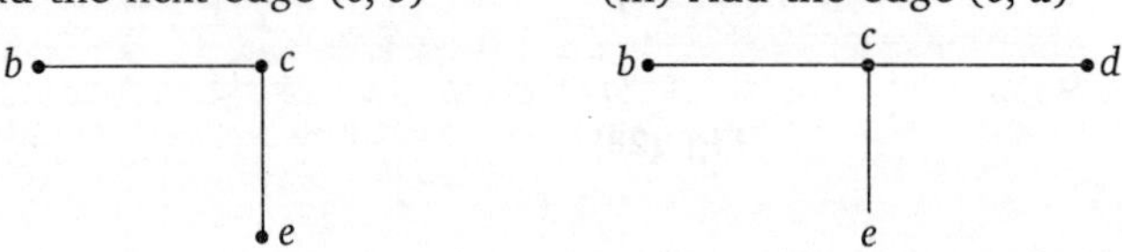

(iv) Add the edge (b, a)

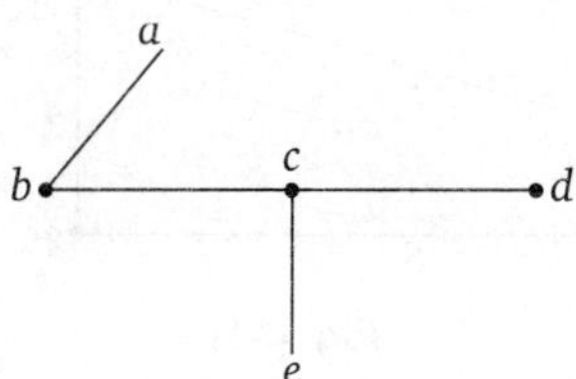

Since the vertices are 5 and we have chosen 4 edges, we stop the algorithm and the minimal spanning tree is produced

2. *Find the minimal spanning tree for the graph shown in the following figure.*

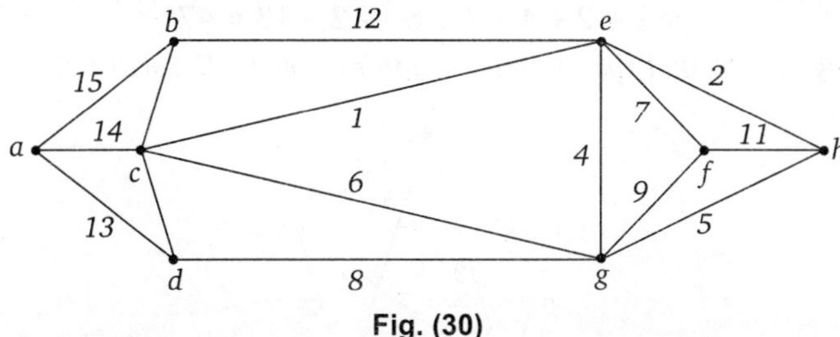

Fig. (30)

Solution. The given graph G has 8 vertices, therefore any spanning tree of G will have 7 edges. Here, we shall find the minimum spanning tree. Listing all edges of G in order of non-decreasing weights, we have

Edges	*ce*	*eh*	*eg*	*gh*	*cg*	*ef*	*dg*	*fg*	*fh*	*be*	*ad*	*ac*	*ab*
Weight	1	2	4	5	6	7	8	9	11	12	13	14	15

Firstly, we choose an edge *ce* since it is least weighted edge in G.

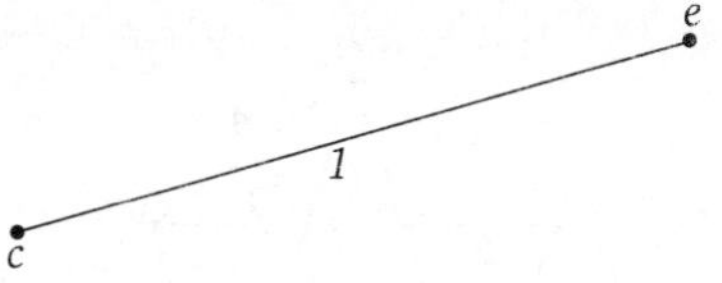

Then we chose the subsequent edges *eh* and *eg* and have

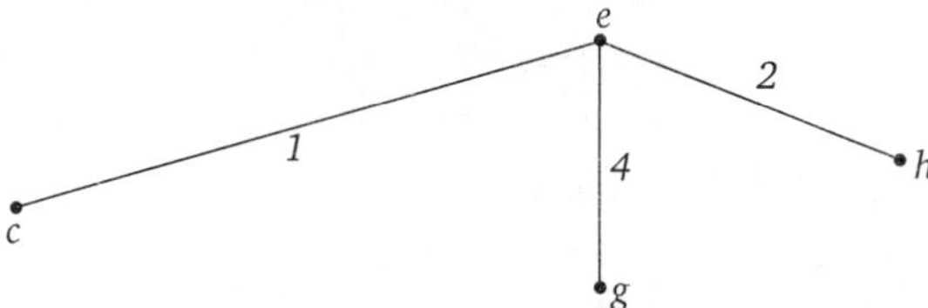

We shall delete the edges *gh* and *cg* of next higher weights since they form circuits. Next we choose the subsequent edges *ef* and *dg* and have

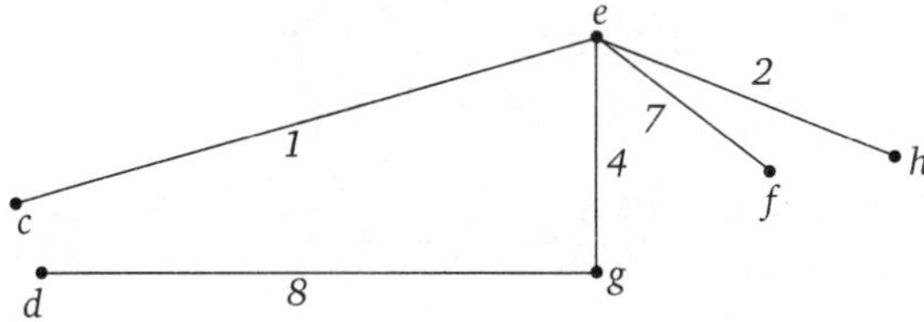

We shall delete the edges *fg* and *fh* of next higher weights since they form circuit. Next we choose the subsequent edges *be* and *ad*. We have

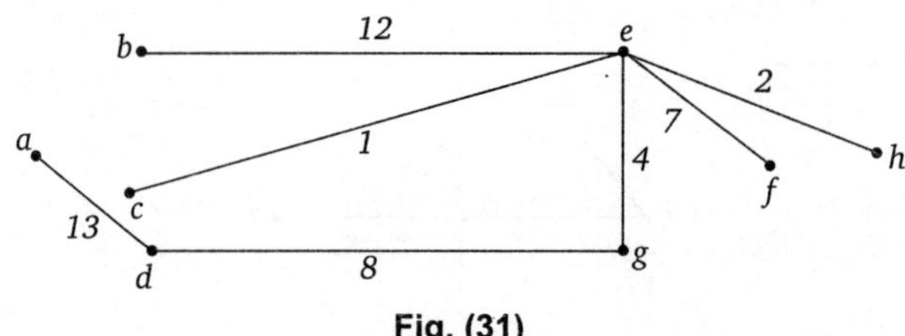

Fig. (31)

Thus, we have minimum spanning tree shown in figure 31 having 7 edges and connecting all the 8 vertices in G.

Weight of the minimum spanning tree = sum of weights

$$= 1 + 2 + 4 + 7 + 8 + 12 + 13 = 47$$

3. *For the following graph, find the minimal spanning tree by Kruskal's algorithm*

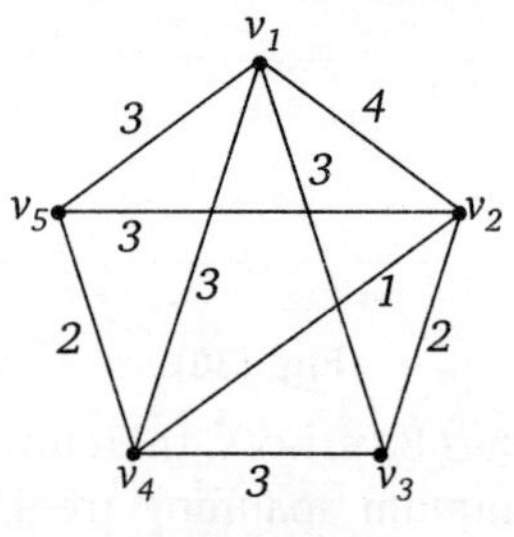

Fig. (32)

Solution. According to step 1

$$G\,[e_1, e_2, e_3, e_4, e_5, e_6, e_7, e_8, e_9] = G\,[1, 2, 2, 3, 3, 3, 3, 3, 4]$$

According to Step 2, we select

$$e = v_2\, v_4$$

Remaining edges, G [2, 2, 3, 3, 3, 3, 3, 4]

We select edge $e = v_2\, v_3$

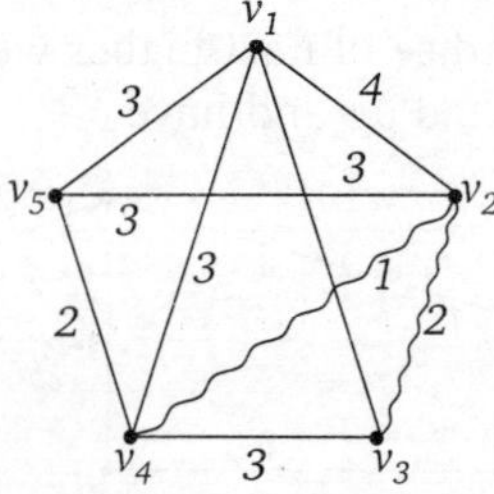

Remaining edges, G [2, 3, 3, 3, 3, 3, 4]

We select edge, $w = v_4\, v_5$

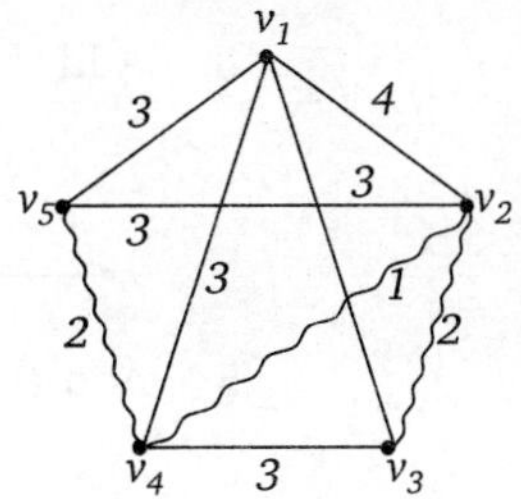

Remaining edges, G [3, 3, 3, 3, 3, 4]
We select edge, $e = v_4\, v_1$

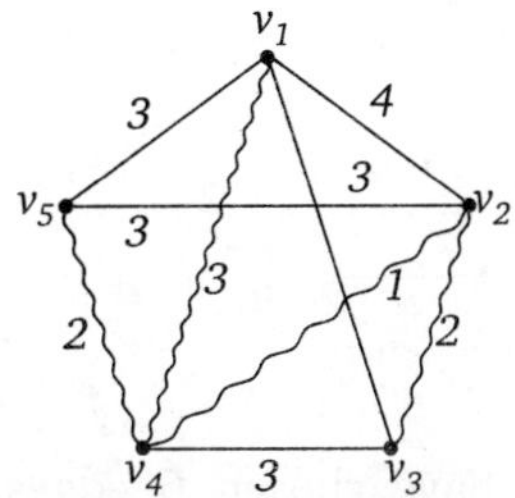

According to Step 4 of Kruskal algorithm, we have selected $n - 1$ edges, where n is the number of vertices, so we stop at this stage because in the given example, there are 5 vertices and we have selected 4 edges. So, minimal spanning tree will be

Total weight of the minimum spanning tree = 2 + 3 + 1 + 2 = 8

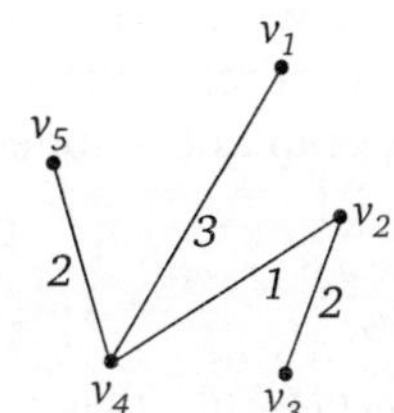

4. *Show how Kruskal's algorithm finds a minimal spanning tree of the graph in the following figure :*

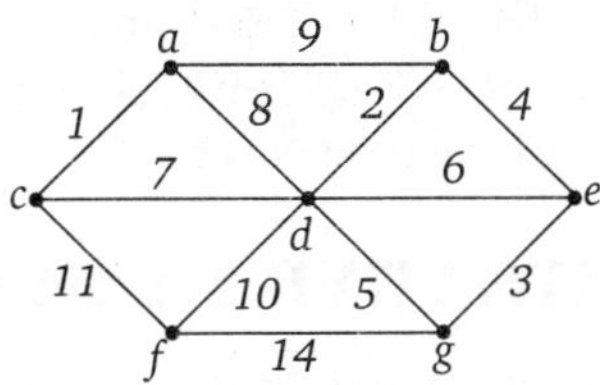

Fig. (33)

Solution. : We collect the edges with their weights into a table.

Edges	(a,c)	(b,d)	(e,g)	(b,e)	(d,g)	(d,e)	(d,c)	(a,d)	(a,b)	(d,f)	(c,f)	(f,g)
Weight	1	2	3	4	5	6	7	8	9	10	11	14

The steps of finding a minimal spanning tree are shown below :

1. Choose the edge (a, c) as it has minimal weight

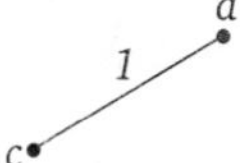

2. Add the next edge (c, d)

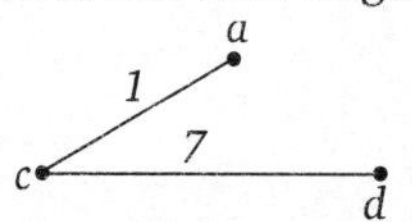

3. Add the edge (d, b)

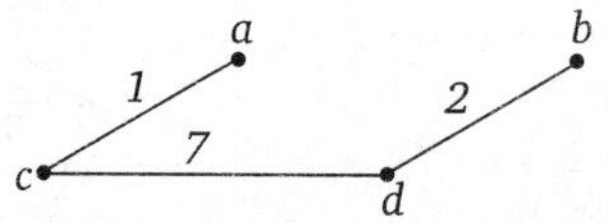

4. Add the edge (b, e)

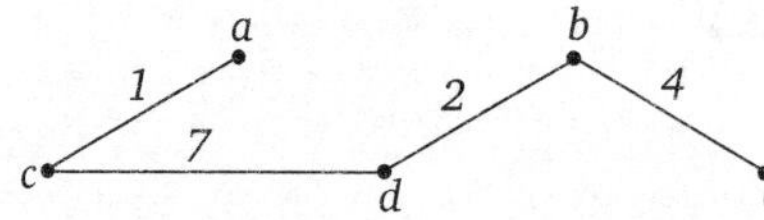

5. Add the edge (e, g)

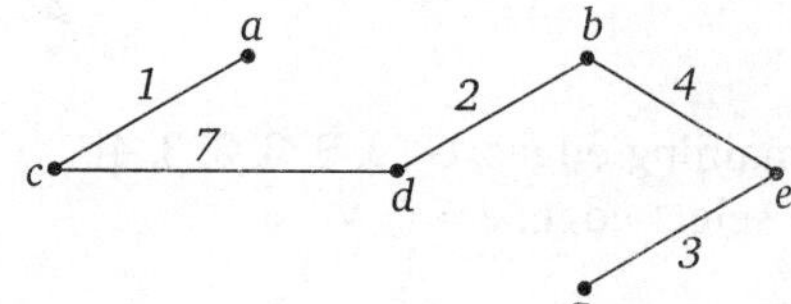

6. Add the edge (d, f)

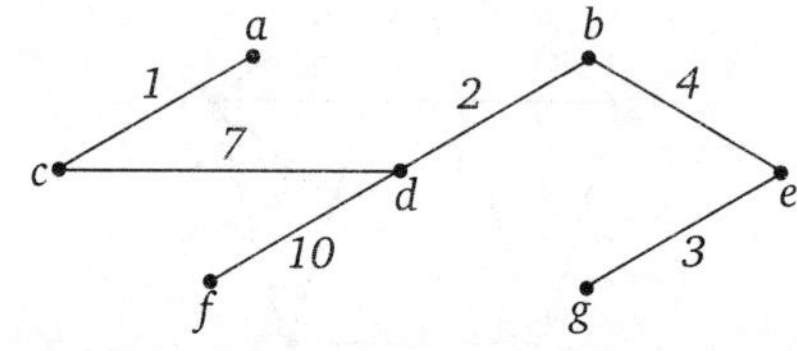

Since vertices are 7 and we have chosen 6 edges, we stop the algorithm and the minimum spanning tree is produced.

10.9.2 Prim's Algorithm

Initially the algorithm starting at a designated vertex chooses an edge with minimum weight and consider this edge and associated vertices as part of the desired tree. Then iterate, looking for an edge with minimum weight not yet selected that has one of its nodes in the tree, while the other node is not. The process terminates when $n - 1$ edges have been selected from a graph of n nodes to form a minimal spanning tree.

WORKING RULES

Step 1. Choose any vertex v_1 of G.

Step 2. Choose an edge $e_1 = v_1 v_2$ of G such that $v_2 \neq v_1$ and e_1 has smallest weight among the edges of G incident with v_1.

Step 3. If edges $e_1, e_2, \ldots, e_i$ have been choosen involving end points $v_1, v_2, \ldots, v_{i+1}$. Choose an edge $e_{i+1} = v_j v_k$ with $v_j \in \{v_1 \ldots v_{i+1}\}$ and $v_k \in \{v_1, \ldots, v_{i+1}\}$. Such that e_{i+1} has smallest weight among the edges of G with precisely one end in $[v_1, \ldots, v_{i+1}]$.

Step 4. Stop after $n-1$ edges have been chosen. Otherwise go to Step 3.

ILLUSTRATIVE EXAMPLES

1. *Find the minimal spanning tree of the weighted graph of the following figure using Prim's algorithm.*

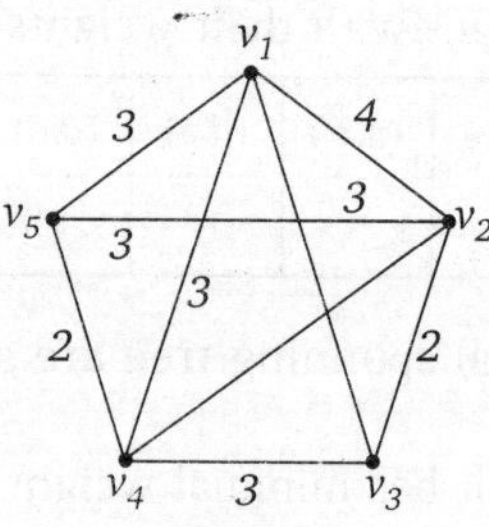

Fig. (34)

Solution. 1. We choose the vertex v_1. Now edge with smallest weight incident on v_1 is (v_1, v_3). So we choose the edge [or (v_1, v_5)].

2. Now $w = (v_1, v_2) = 4$, $w(v_1, v_4) = 3$, $w(v_1, v_5) = 3$, $w(v_3, v_2) = 2$ and $w(v_3, v_4) = 3$. We choose the edge (v_3, v_2) since it is minimum.

3. Again $w(v_1, v_5) = 3$, $w(v_2, v_4) = 1$ and $w(v_3, v_4) = 3$. We choose the edge (v_2, v_4).

4. Now we choose the edge (v_4, v_5). The minimal spanning tree is

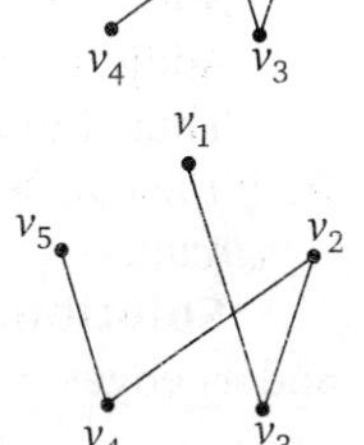

10.10 FUNDAMENTAL CIRCUITS

If we add an edge between any two vertices of a tree a circuit is created. This is because there already exists one path between any two vertices of a tree adding an edge between them creates an additional path and hence creates a circuit.

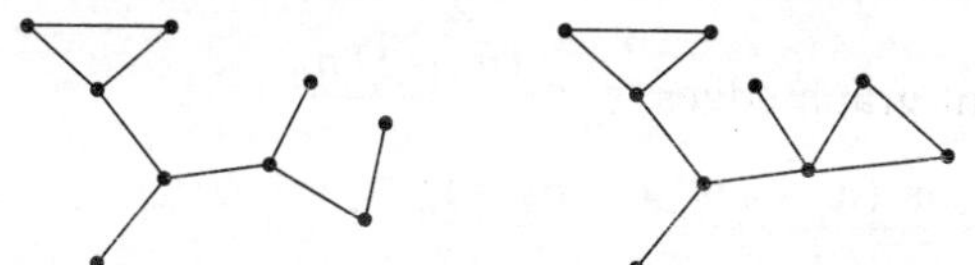

(a) Tree with one circuit (b) Tree with two circuits

Fig. (35)

If we combine any two vertices of the tree it becomes a circuit.

Definition : *Any circuit of graph G which can be obtained by adding an edge to the tree is called a fundamental circuit of G.*

Theorem 1. *A connected graph G is a tree iff adding an edge between any two vertices in G creates exactly one circuit.*

Proof : Let G be a connected graph which is tree. Let a and b be any two vertices in G. Since G is a tree, there exists one path between vertices a and b. If we add an edge between a and b, then an additional path between a and b will be created and hence it creates exactly one circuit.

Conversely, suppose that G is a connected graph such that adding an edge between any two vertices in G creates exactly one circuit. To show that G is a tree, it is sufficient to show that G has no circuit (because G is already connected). If possible, suppose that G has a circuit. Then adding an edge between two vertices of the circuit will create two circuits (because there are two paths between these vertices in the circuit). But it contradicts our hypothesis. This completes the proof.

- Let G be a connected graph, with n vertices and e edges, then the graph G has $e - n + 1$ fundamental circuits.

ILLUSTRATIVE EXAMPLES

1. *Find the fundamental circuit in the given graph G.*

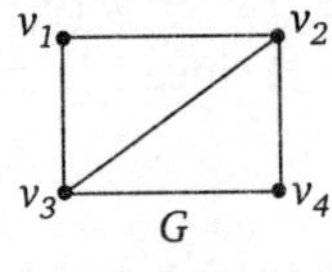

Fig. (36)

Solution. The following figure shows the spanning tree of graph G

The tree has branches $[v_1, v_2]$, $[v_1, v_3]$ and $[v_3, v_4]$

The remaining edges of G are $[v_2, v_4]$ and $[v_3, v_2]$

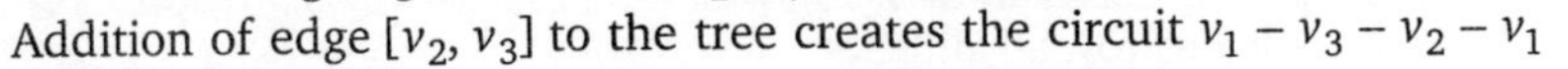

Addition of edge $[v_2, v_3]$ to the tree creates the circuit $v_1 - v_3 - v_2 - v_1$

Addition of edge $[v_2, v_4]$ to the tree creates the circuit $v_1 - v_3 - v_4 - v_2 - v_1$

Thus, fundamental circuits are $v_1 - v_3 - v_2 - v_1$ and $v_1 - v_3 - v_4 - v_2 - v_1$

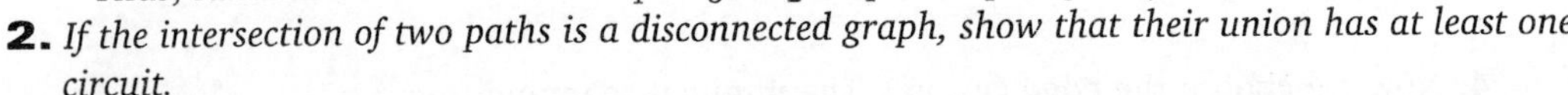

2. *If the intersection of two paths is a disconnected graph, show that their union has at least one circuit.*

Solution. Let G_1 be a path of n_1 vertices and e_1 edges and G_2 be the path of n_2 vertices and e_2 edges.

$G = G_1 \cup G_2$ and G has $n_1 + n_2$ vertices and $e_1 + e_2$ edges.

G_1 is 1-component graph of G.

We know k-components of a graph with n vertices can not have more than

$$\frac{(n-k)(n-k+1)}{2} \text{ edges}$$

For 1-component graph G_1 edges, $e_1 \leq \frac{(n_1 - 1)\, n_1}{2}$

and For 2-component graph edges, $e_2 \leq \frac{(n_2 - 1)\, n_2}{2}$

$$e_1 + e_2 = e \leq \frac{n_1(n_1 - 1)}{2} + \frac{n_2(n_2 - 1)}{2}$$

$$= \frac{n_1^2}{2} - \frac{n_1}{2} + \frac{n_2^2}{2} - \frac{n_2}{2} \leq \frac{n_1^2 + n_2^2 - (n_1 + n_2)}{2} + \frac{2n_1n_2}{2}$$

$$= \frac{[(n_1 + n_2)^2 - (n_1 + n_2)]}{2} = \frac{(n_1 + n_2)(n_1 + n_2 - 1)}{2} = \frac{n(n-1)}{2}$$

which shows that G is connected since it has only one component and there must be cut vertex which separates it into the components. Let u_1 and u_2 be two vertices in G_1 and u_{21} and u_{22} be two vertices in G_2. Since G is connected the path connecting u_1 and u_{21} as well as the path connecting u_2 and u_{22} must both pass through the cut vertex, creating the circuits. Then G has at least one circuit.

10.11 TREE WITH DIRECTED EDGES

A tree is a connected directed graph that has no circuit, neither a directed circuit nor a semicircuit. A tree of n vertices contains $n - 1$ directed edges. Tree with directed edges are of great importance in many applications, such as electrical network analysis, computer programming, etc.

10.11.1 Acyclic Directed Graph

A graph having directed circuit is called a cyclic directed graph and a directed graph having no directed circuits is called acyclic directed graph. A connected directed graph having neither a directed nor a semi-directed circuit is called a directed tree.

The process of removing cycles is called decyclization. A set of minimum number of edges of directed graph whose removal gives an acyclic directed graph is called minimal decyclization.

WORKING RULES

Step 1. Trace out all directed circuit in the given directed graph D.

Step 2. Write each directed circuit as a Boolean sum of its edges.

Step 3. Take the Boolean product of all directed circuit expressions obtained in Step 2.

Step 4. Each of the resulting terms in the sum of the products represents a set of edges whose removal will destroy all directed circuits. Select a term consisting of the smallest number of edges. This is a minimum decyclization or minimum feedback set.

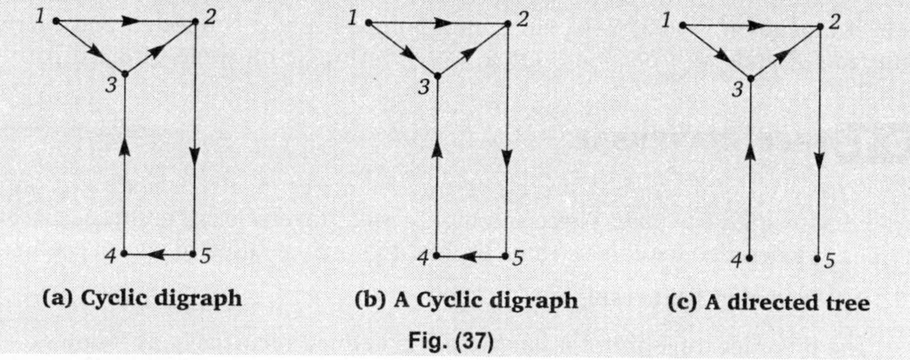

(a) Cyclic digraph (b) A Cyclic digraph (c) A directed tree

Fig. (37)

10.12 ARBORESCENCES

A digraph G is to be an arborescences if G contains no circuit (neither directed nor semi-directed) and G has exactly one vertex v of zero indegree.

The vertex v with zero indegree is called root of the arborescence.

For example :

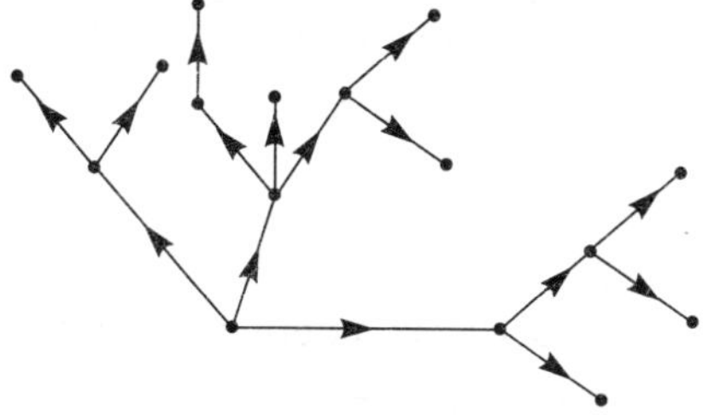

Fig. (38)

- Arborescence is also known as out-tree.
- If we assign opposite directions to each edge of an arborescence, we get a tree and this tree is known as intree.
- In an arborescence, every vertex of it is reachable from the root and the root is not accessible from any other vertex.
- A vertex of out degree zero in an arborescence is necessarily a pendant vertex.
- In a connected directed graph has a spanning tree than is an arborescence and called a spanning arborescence.

Theorem 1. *An arborescence is a tree in which every vertex of it other than one (root) has an indegree of exactly one.*

Proof : Let the number of vertices in an arborescence by n. We know that the graph is a directed tree, then it can have at most $n - 1$ edges. Now, we consider the sum of all in degree of the vertices of the graph D

$$d^-(v_1) + d^-(v_2) + \ldots + d^-(v_n) \leq n - 1$$

Since there are n vertices in the graph of which one is of indegree zero, the remaining $n - 1$ vertices is of degree at least one. By this inequality, left hand side contains only one zero term (that is for the root vertex). If a vertex has indegree at least 2, then the sum of left hand side of the inequality is greater than $n - 1$ (since there are exactly $n - 1$ non-zero term). This indegree of each vertex of D should be exactly one.

Theorem 2. *In an arborescence there is a directed path from the root R to every other vertex.*

Proof : Consider a path P which starts from the root R and traverses as far as possible in an arborescence. P can end only at a pendant vertex because otherwise we get a vertex whose indegree is two or more. But then it would be a contradiction. Hence, P ends at a pendant vertex. Now, since an arborescence is a connected and every directed path from R ends at a pendant vertex, every vertex lie on some directed path from the root R.

10.13 TREE TRAVERSAL

A traversal of a tree is a process to traverse a tree in a systematic way so that each vertex is visited exactly once. Three commonly used traversals are preorder, postorder and inorder. We describe here these three process that may be used to traverse a binary tree.

10.13.1 Preorder Traversal

The preorder traversal of a binary tree is defined recursively as follows :

(i) Visit the root

(ii) Traverse the left subtree in preorder

(iii) Traverse the right subtree in preorder

10.13.2 Postorder Traversal

The post order traversal of a binary tree is defined recursively as follows :

(i) Traverse the left subtree in post order

(ii) Traverse the right subtree in postorder

(iii) Visit the root

10.13.3 Inorder Traversal

The inorder traversal of a binary tree is defined recursively as follows :

(i) Traverse in inorder the left subtree.

(ii) Visit the root.

(iii) Traverse in inorder the right subtree.

ILLUSTRATIVE EXAMPLES

1. *Determine the order in which a preorder traversal visits the vertices of an ordered rooted tree T in the following figure.*

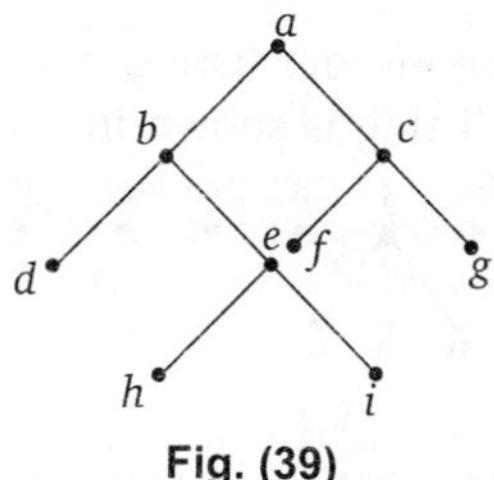

Fig. (39)

Solution. In preorder traversal of an ordered rooted tree T, first the root a is visited and the preorder list of subtree (T_1) with b as the root followed by the preorder list of subtree (T_2) with c as the root. This step could be shown as in figure (a). Next step involve starting at the root b and then visiting the root d of the subtree in preorder followed by the root e of the subtree in preorder and continuing to do so at the root f of the subtree and g of the subtree as shown in figure (b). The preorder list of the subtree with e as the root start by listing the vertex e followed by the preorder listing of the root h of the subtree and subsequently followed by preorder listing of the root i.

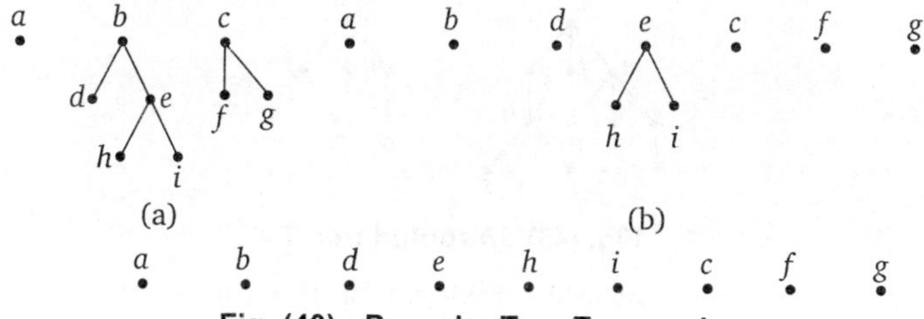

Fig. (40) : Preorder Tree Traversal

2. *In which order the vertices of an ordered rooted tree T is visited using an inorder traversal method ?*

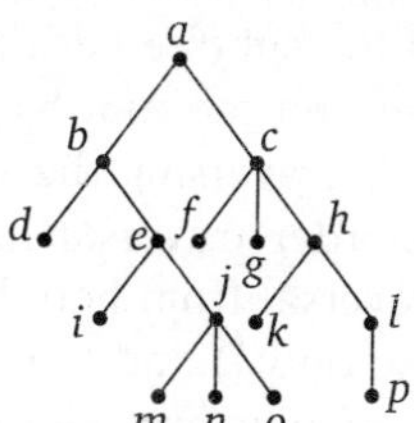

Fig. (41) : An ordered rooted tree T

Solution. For the tree T as shown above, the inorder traversal begins with the subtree with root b then root a of T followed by subtree with root at c. The inorder traversal of the subtree with root b now starts with the inorder listing of the subtree of d (here it is just d) followed by the root b and then the subtree with root e. Similar procedure is being adopted for other subtrees and this step results in figure (b). Next the inorder listing of the subtree with

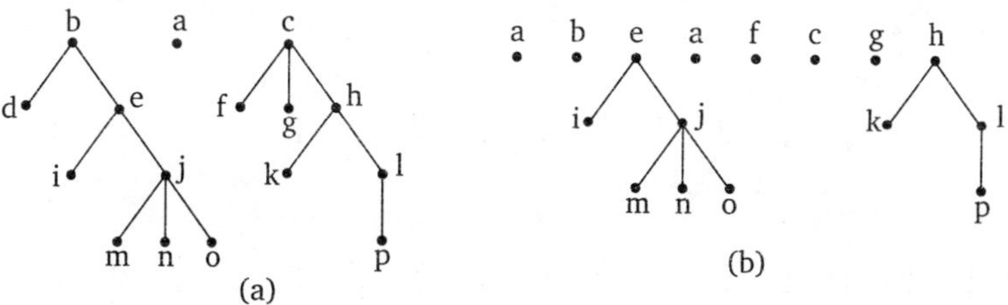

root e could be obtained by listing the subtree of i (which is just i), the root e and then the subtree of the vertex *j*. We do list the inorder listing of the subtree with root *h* similarly. The configurations obtained for inorder listing is shown in figure (*c*).

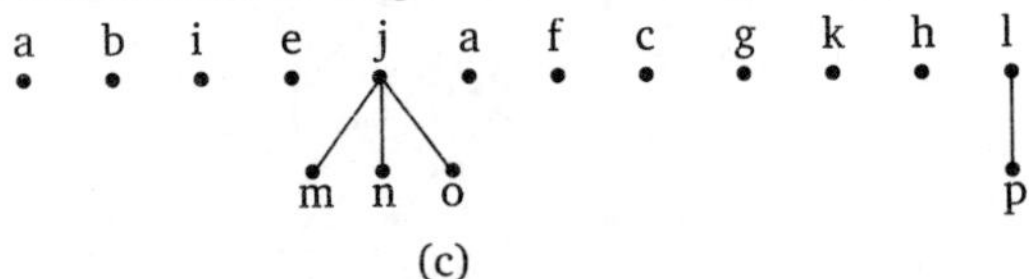

(c)

The final configuration of inorder listing of vertices in the tree could be easily obtained in figure (d).

a b i e m j n a f c g k h p l

(d)

Fig. (42)

3. *Find the order in which vertices of the following tree T are traversed in a postorder traversal.*

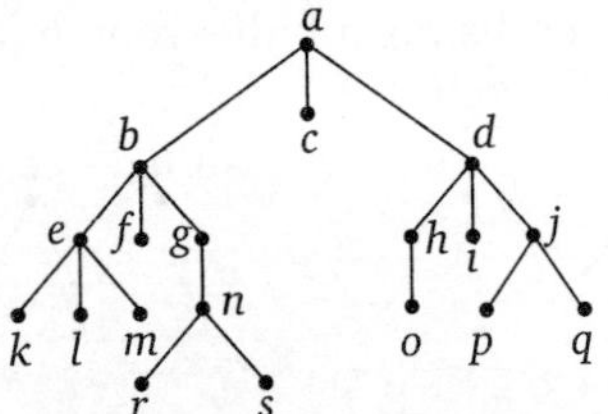

Fig. (43) : A rooted tree *T*

Solution. We begin the postorder traversal of the rooted tree T by traversing the subtree T_1 in post order with root b then subtree of c (which is just c) followed by postorder traversal of the subtree with root d and the root a at the end. The configuration that result following this step is shown in figure (a). As a next step, we start with now by traversing the subtree with root e in postorder, then the subtree with root f (which is just f only) followed by the subtree with root g in postorder and the root b at the end.

Similar procedure is adopted while traversing the subtree of h, i and j in post order. The root d is visited at the end. The postorder traversal listing at the end of this step may be shown in figure (b). The postorder traversal from here begins with the subtree of k (which is only k), then subtree of l (which is also only l) and followed by subtree of m (which is also m only) and then the root e at the end. Similar procedure is followed in postorder traversal of the tree with root g, h and j. This step results in the configuration shown in figure (c). The foregoing procedure for the postorder tree traversal is reported again for the tree with n culminating in the ordering of vertices as shown in figure (d).

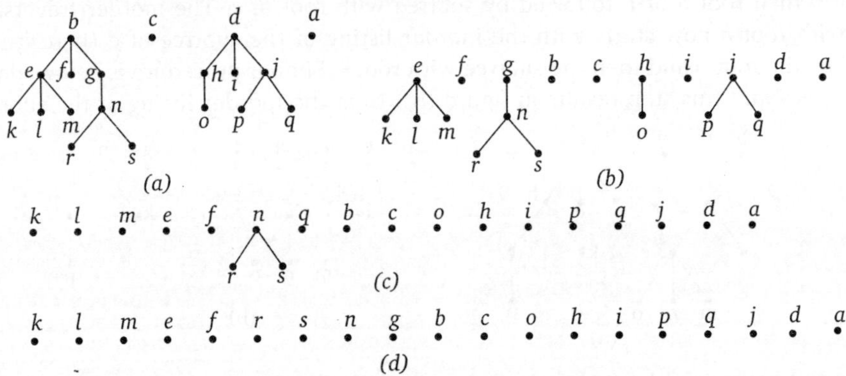

10.13.4 To Draw a Unique Binary Tree when Inorder and Preorder Traversal of the Tree is Given

WORKING RULES

Step 1. The root of T is obtained by choosing the first vertex in its preorder.

Step 2. The left child of the root vertex is obtained as follows : First use the inorder traversal to find the vertices in the left subtree of the binary tree (all the vertices to the left of this vertex in the inorder traversal are the part of the left subtree). The child of the root is obtained by selecting the first vertex in the preorder traversal of the left subtree. Draw the left child.

Step 3. Use the inorder traversal to find the vertices in the right subtree of the binary tree (all the vertices to the right of the first vertex are the part of the right subtree.) Then the right child is obtained by selecting the first vertex in the preorder traversal of the right subtree. Draw the right child.

Step 4. The procedure is repeated recursively until every vertex is not visited in preorder.

10.13.5 To Draw a Unique Binary Tree When Inorder and Postorder Traversal of the Tree is Given

WORKING RULES

Step 1. The root of the binary tree is obtained by choosing the last vertex in the postorder traversal.

Step 2. The right child of the root vertex is obtained as follows : First use the inorder traversal to find the vertices in the right subtree (all the vertices right to the root vertex in the inorder traversal are the vertices of the right subtree. The right child of the root is obtained by selecting the last vertex in the postorder traversal. Draw the right child.

Step 3. Use the inorder traversal to find the vertices in the left subtree of the binary tree. Then the left child is obtained by selecting the last vertex in the postorder traversal of the left subtree. Draw the left child.

Step 4. The process is repeated recursively until every vertex is not visited in post order.

ILLUSTRATIVE EXAMPLES

1. *Given the preorder and inorder traversal of a binary tree, draw the unique tree :*

Preorder : g b q a c p d e r

Inorder : q b c a g p e d r

Solution. Here g is the first vertex in preorder traversal, thus it is the root of the tree. Using inorder traversal, left subtree of g consists of the vertices q, b, c and a. Then the left child g is b since b is the first vertex in the preorder traversal in the left subtree. Similarly, right subtree of g consists of the vertices p, e, d are r, then the right child of g is p since p is the vertex in the preorder traversal in the right subtree. Repeating the above process with each node, we finally obtain the required tree as shown in figure.

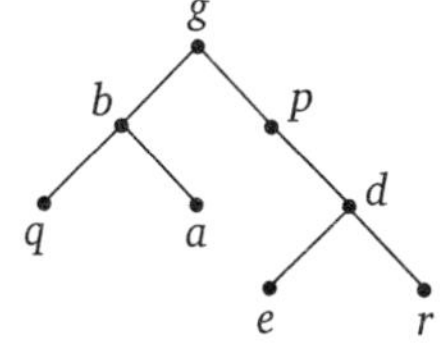

Fig. (44)

2. *Given the postorder and inorder traversal of a binary tree, draw the unique tree :*

Postorder : d e c f b h i g a

Inorder : d c e b f a h g i

Solution. Here a is the last vertex in the post order traversal, thus a is the root of the tree. Using inorder traversal, right subtree of root vertex a consists of the vertices h, g and i.

The right child of a is g. Since g is the last vertex in the post order traversal in the right subtree. Similarly, left subtree of a consists of the vertices d, c, e, b and f, then the left child of a in b is the last vertex in the post order traversal in the left subtree. Repeating the above process with each vertex, we finally obtain the required tree as shown in figure.

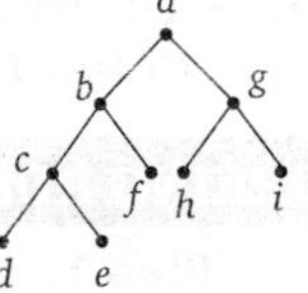

Fig. (45)

10.14 BINARY TREE REPRESENTATION OF AN EXPRESSION

Binary trees are used to represent algebraic expressions, the vertices of the tree are labelled with the numbers, variables or operations that make up the expression. The leaves of the tree can be labelled with numbers or variables. Operations such as addition, subtraction, multiplication, division or exponentiation can only be assigned to internal vertices. The operations at each vertex operate on its left and right subtree from left to right.

ILLUSTRATIVE EXAMPLES

1. *Construct the tree for the expression :* $(a + b - c \,.\, d) \div (g^3 - f)$

Solution. The tree corresponding to the given expression is given below :

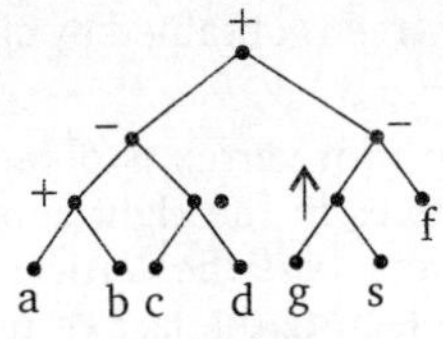

Fig. (46)

Here ↑ denotes exponentiation.

2. *Draw the trees for the following algebraic expression*

(i) $(2a + 5b)\,(8c + 9d)$ *(ii)* $d * (e - f) + p / q$

(ii) $(x - 2y)^3\,(a + 3b)^4$ *(iv)* $(a + b) * c - d / e - f$

Solution. (i) The algebraic expression for $(2a + 5b)\,(8c + 9d)$ as binary tree is shown in the following figure :

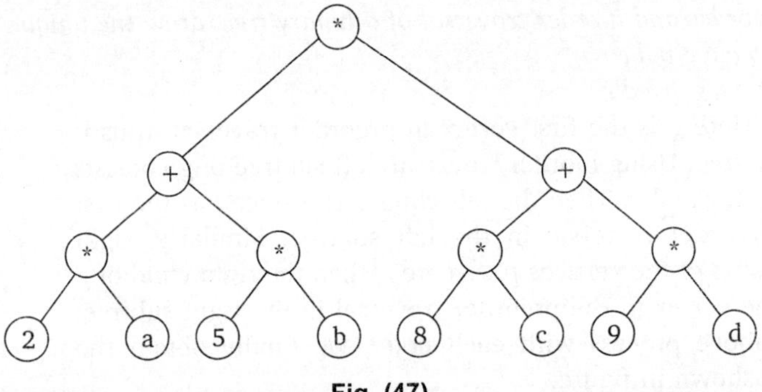

Fig. (47)

The representation of the algebraic expression for (ii), (iii) and (iv) as binary tree are shown below :

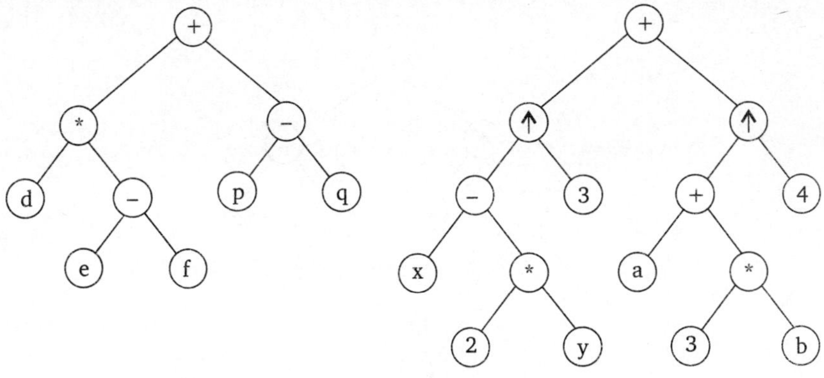

Fig. (48)

Fig. (49)

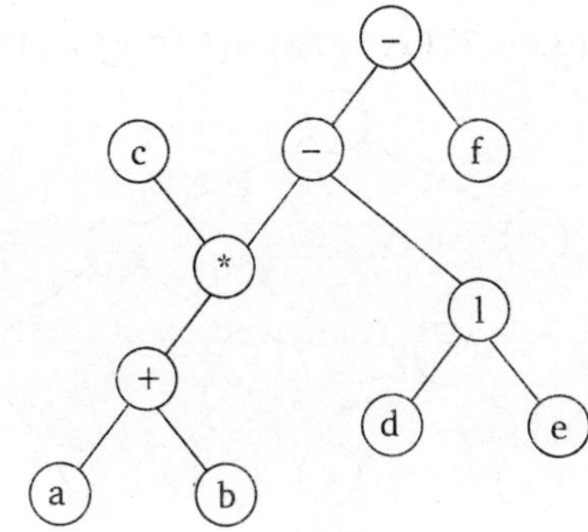

Fig. (50)

3. *Draw tree for the following algebraic expressions*

(i) $(2x + y)(5a - b)^3$ *(ii)* $(a - 3b)^2 (2x - y)$

(ii) $(3x + 7)(xy - 7z)$ *(iv)* $(a * b) - c \uparrow d - \{(e * f) + g\}$

Solution. (i) The tree for expression $(2x + y)(5a - b)^3$

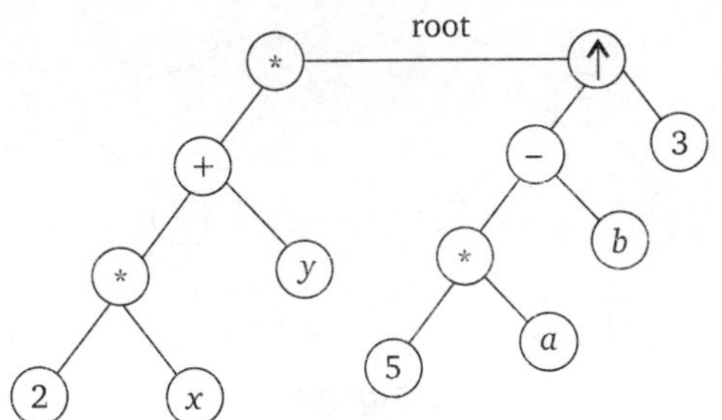

Fig. (51)

(ii) Tree for expression $(a - 3b)^2 (2x - y)$

$$= (a - 3 * b) \uparrow 2 * (2 * x - y)$$

$$= [a - (3 * b) \uparrow 2] * [(2 * x) - y)$$

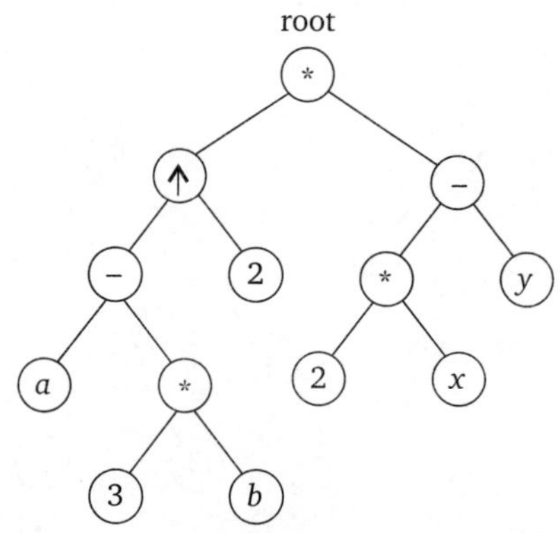

Fig. (52)

(iii) Tree for expression $(3x + 7)(xy - 7z) = [(3 * x) + 7] * [(x * y) - (7 * z)]$

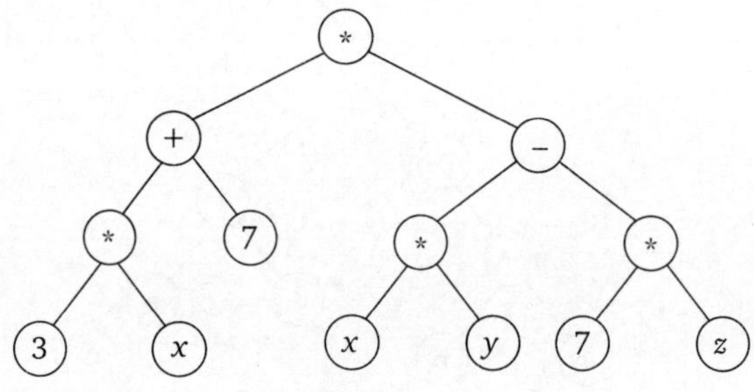

Fig. (53)

(iv) Tree for expression

$$(a * b) - c \uparrow d - \{(e * f) + g\} = ((a * b) - c \uparrow d)) - (((c * f) + g)$$
$$= (a * b) - (c \uparrow d) - ((e * f) + g)$$

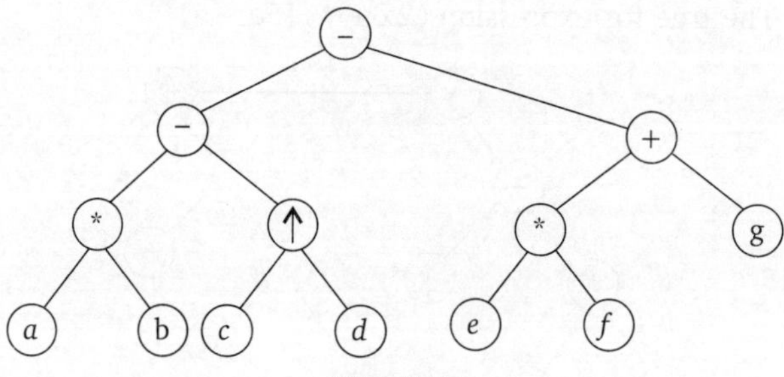

Fig. (54)

4. *Construct the binary expression tree for the expression*

(i) $(a + b) * \left(\frac{d}{c}\right)$ *(ii)* $(a - 3b)(2x - y)^3$

Solution. (i) The tree for expression $(a + b) * \left(\frac{d}{c}\right)$

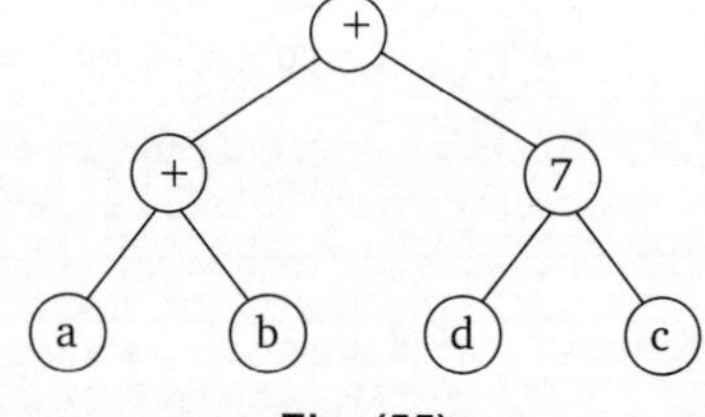

Fig. (55)

(ii) Tree for expression $(a - 3b)(2x - y)^3$

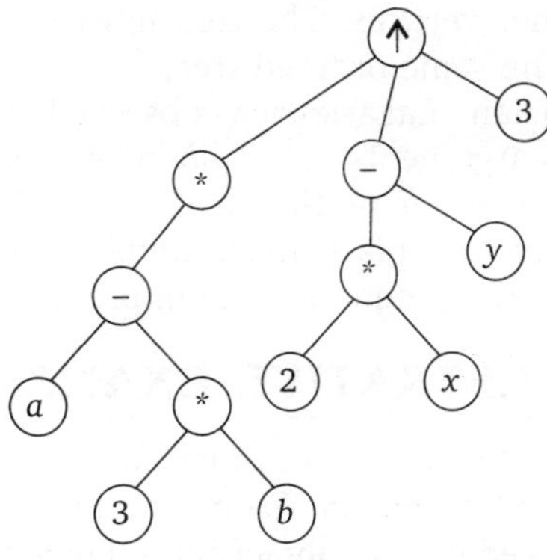

Fig. (56)

5. *Determine the value of the expression represented in a binary tree shown in the following figure.*

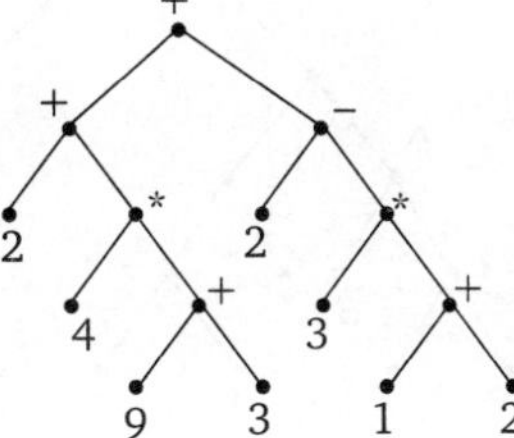

Fig. (57)

Solution. The expression represented by the binary tree is

$$(9 \div 3 \times 4 + 2) \div (((1 + 2) * 3) - 2) \text{ and the value is}$$

$$(3 * 4 + 2) \div ((3 * 3) * 2) = (12 + 2) \div (9 - 2) = 14/7 = 2.$$

6. *Prove that the maximum number of nodes in a binary tree of depth d is* $2^d - 1$ *where* $d \geq 1$.

Solution. In the binary tree, there can be only one vertex at level 0 at most two vertices at level 1, at most four vertices at level 2 and so on. Therefore the maximum number of nodes (vertices) in a binary tree of depth is $2^0 + 2^1 + 2^2 + \ldots + 2^d = 2^d - 1$

10.15 ALGEBRAIC EXPRESSIONS AND POLISH NOTATION

Any algebraic expression involving binary operations. For example, addition, subtraction, multiplication and division, can be represented by an ordered rooted tree.

For example, the following figure (a) represents the algebraic expression $\dfrac{(a - b)}{((c \times d) + e)}$

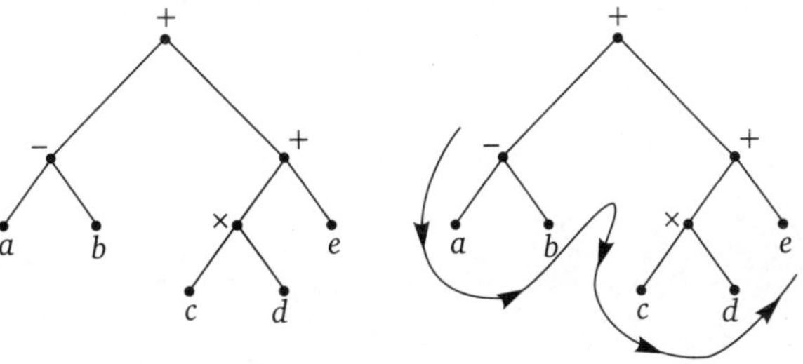

Fig. (58)

Observe that the variables in the expression a, b, c, d and e appear as leaves, and the operations appear as the other vertices. The tree must be ordered since $a - b$ and $b - a$ yield the same tree but not the same ordered tree.

The Polish mathematician Lukasiewicz observed that by placing the binary operational symbol before its arguments, *e.g.* $+ ab$ instead of $a + b$ and $/cd$ instead of c / d

One does not need to use any parenthesis. This notation is called Polish notation in prefix form (Analogously, one can place the symbol after its arguments, called Polish notation in postfix form). Rewriting equation (1) in prefix form, we obtain $/ - ab + \times cde$.

ILLUSTRATIVE EXAMPLES

1. *Obtain the binary tree representation of the expression* $((x + 2) \uparrow 3) * (y - (3 + x)) - 5$ *and write the expression in (i) prefix notation, (ii) postfix notation, (iii) infix notation*

Solution. Using the procedure, we obtain the following binary tree representation of the given expression

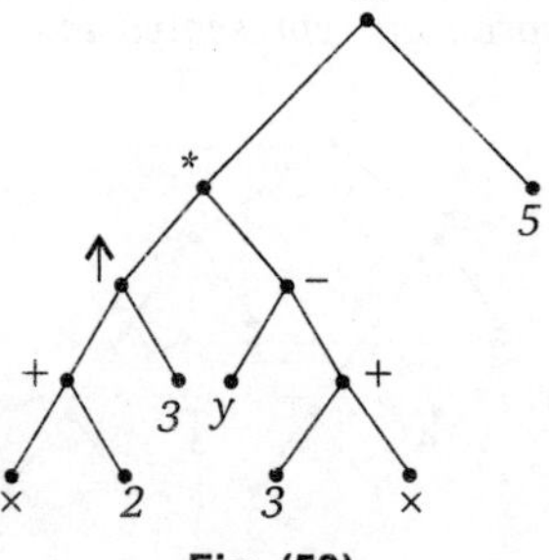

Fig. (59)

The method of preorder, postorder tree traversal gives the following form of the expression

Prefix (polish) notation : $- * \uparrow + x\,23 - y + 3x\,5$

Postfix notation : $x2 + 3 \uparrow y3x + - * 5 -$

The infix notation of the expression may be written as

$$(((x + 2) \uparrow 3) * ((y - (3 + x)) - 5) .$$

2. *Draw the ordered rooted tree for the following prefix expressions :*

(i) $* / 93 + * 24 - 76$ *(ii)* $+ - \uparrow 32 \uparrow 23 / 6 - 42$

What is the value of each of these expressions ? Write the infix expressions for each of the expressions.

Solution. (i) The ordered rooted tree is shown in figure (a). To evaluate the prefix expression, we start from right to left and perform operations using operands on the right. The various steps in evaluation are as follows :

1. $* / 93 + * 24 - 76$ **2.** $* / 93 + * 241$ **3.** $* / 93 + 81$ **4.** $* / 939$

5. $* 39$ **6.** 27

The value of the prefix expression in therefore 27. The infix expression is $((9 / 3) * ((2 * 4) + (7 - 6)))$

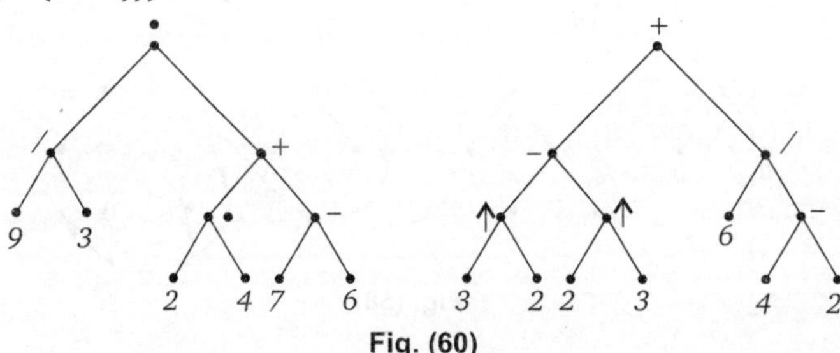

Fig. (60)

(b) The rooted tree corresponding to the prefix expression is shown in figure (b)

The steps involved in the evaluation of the prefix expression are :

1. $+ - \uparrow 3 2 \uparrow 2 3 / 6 - 4 2$ 2. $+ - \uparrow 3 2 \uparrow 2 3 / 6 2$
3. $+ - \uparrow 3 2 \uparrow 2 3 3$ 4. $+ - \uparrow 3 2 8 3$ 5. $+ - 9 8 3$ 6. $+ - 1 3$
7. 4

The value of the prefix expression is 4. The infix notation for the expression is readily obtained as :

$$((3 \uparrow 2) - (2 \uparrow 3)) + (6 / (4 - 27)).$$

10.16 BINARY SEARCH TREES

A binary search tree is a binary tree T in which data are associated with the vertices. The data are arranged so that for each vertex v in T, each data item in the left subtree of v is less than the data item in v and each data item in the right subtree of v is greater than the data item in v. Thus a binary search tree for a set S is a labelled binary tree in which each vertex v is labelled by an element $l(v) \in S$ such that

1. for each vertex u in the left subtree of v, $l(u) < l(v)$.
2. for each vertex u in the right subtree of v, $l(u) > l(v)$, and
3. for each element $a \in S$, there is exactly one vertex v such that $l(v) = a$.

The binary tree T in the following figure is a binary search tree since every vertex in T exceeds every number in its left subtree and is less than every number in its right subtree.

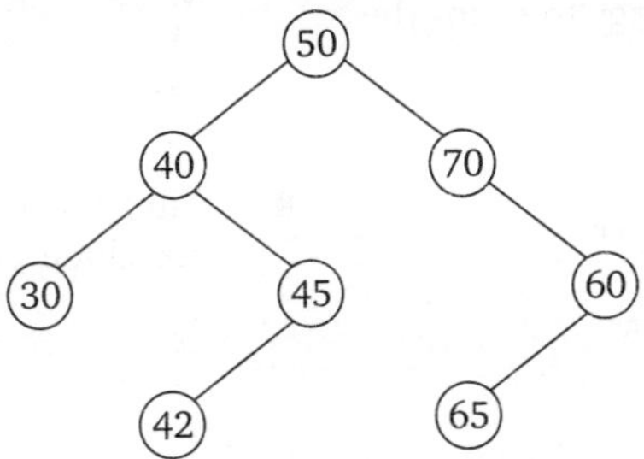

Fig. (61) : A binary search tree

Example : Build a BST for the words given below using alphabetical order :

Banana, Peach, Apple, Pear, Coconut, Mango and Papaya

Solution. Here the word banana is the root of the tree. Since the word peach comes after banana, we draw an edge to the right starting at vertex labelled banana. The next word is apple which appears before peach and also before the root banana, so an edge is drawn to the left of the vertex labelled banana and its end point is labelled apple. Next add an edge from the vertex peach or with key as peach to the right of it terminating at a vertex labelled pear since pear appear after peach. Similarly add a left child coconut to the left of the key peach, as it comes before peach. The word mango and papaya could be similarly added as shown in figure.

Banana
Apple
Peach
Coconut
Pear
Mango
Papays

Fig. (62) : A binary search tree

Exercise 10.1

1. Sketch all spanning trees of the following graph.

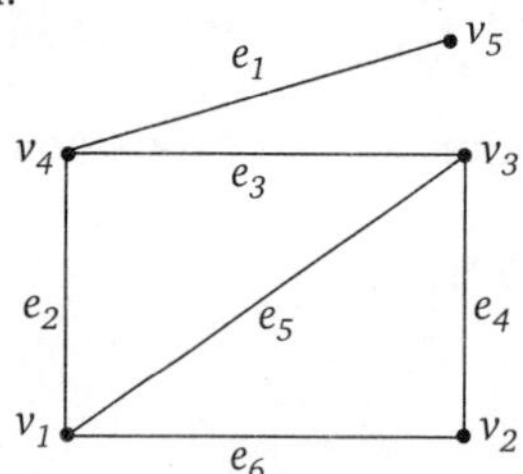

2. Consider the following graph to find the minimum cost spanning tree and also write Kruskal's algorithm to find out the minimum cost-spanning tree.

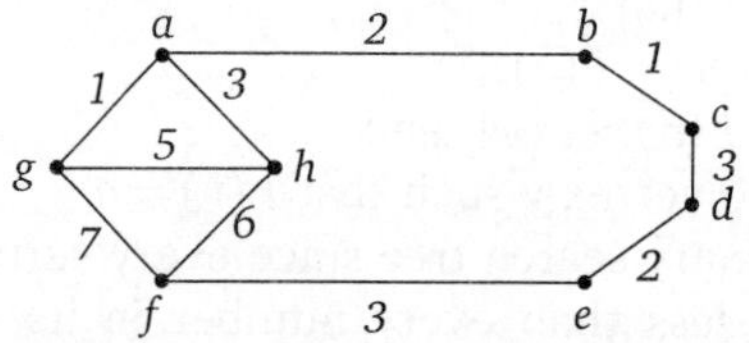

3. Use the algorithm of Kruskal to find a minimum weight spanning tree in the graph.

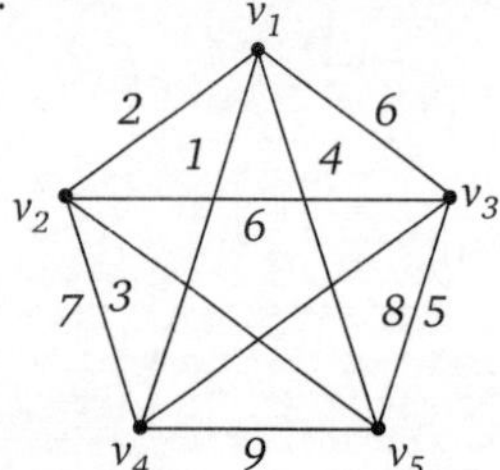

4. Sketch all the spanning trees of the following graph.

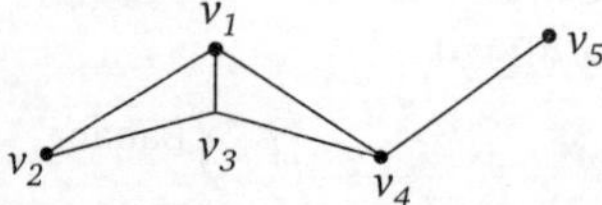

5. Find the minimal spanning tree for the connected weighted graph given below using both Kruskal's and Prim's algorithm.

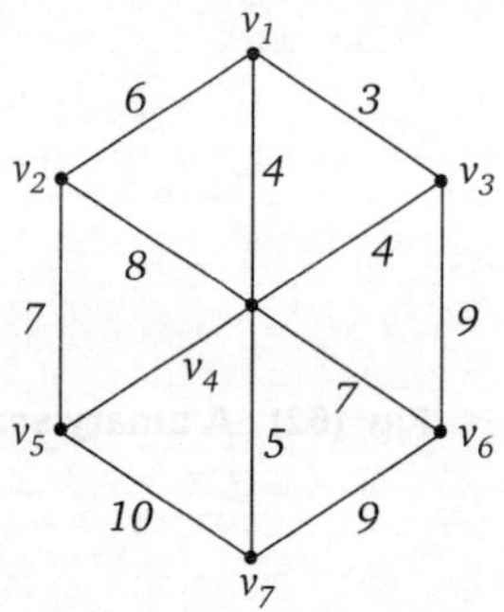

6. Determine the minimal spanning tree for the graph in the following figure using (i) Kruskal's algorithm, (ii) Prim's algorithm.

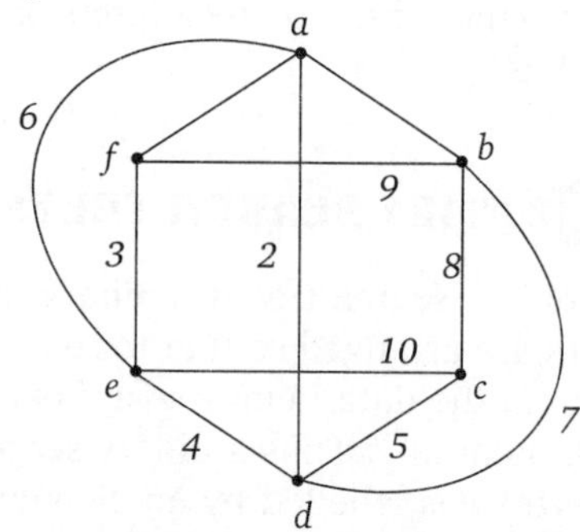

7. Using Kruskal's algorithm, determine a maximal spanning tree for the graph in the following figure.

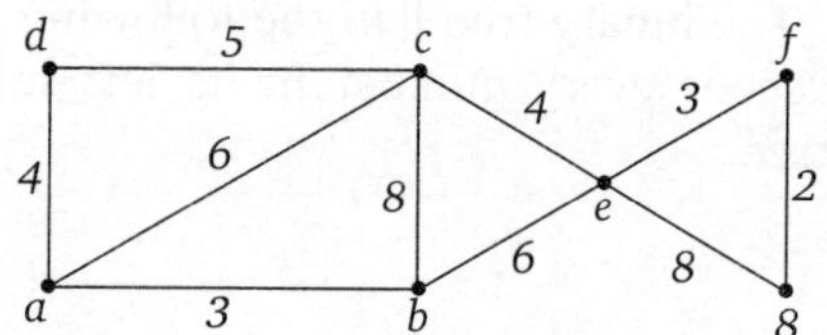

8. Find the minimum spanning tree for the weighted graph in the following figure.

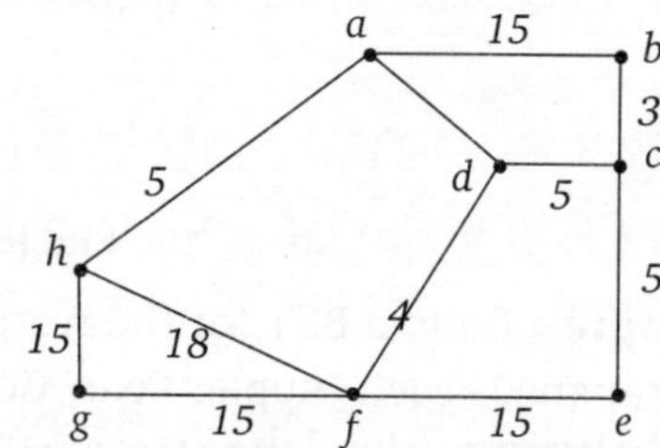

9. Find the minimum spanning tree for the connected weighted graph in the following figure :

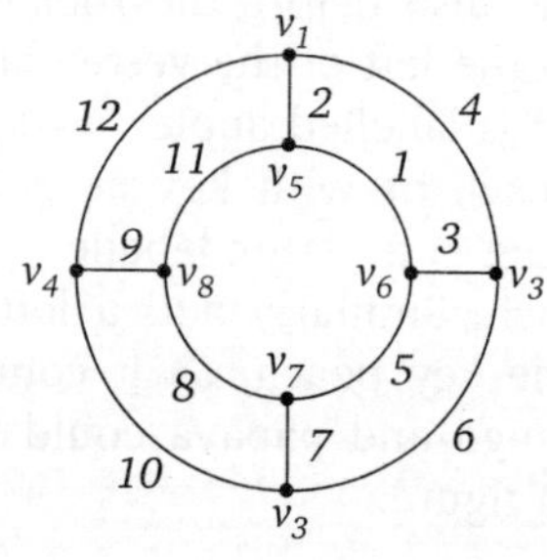

10. Find the minimum spanning tree for the weighted graph in the following figure :

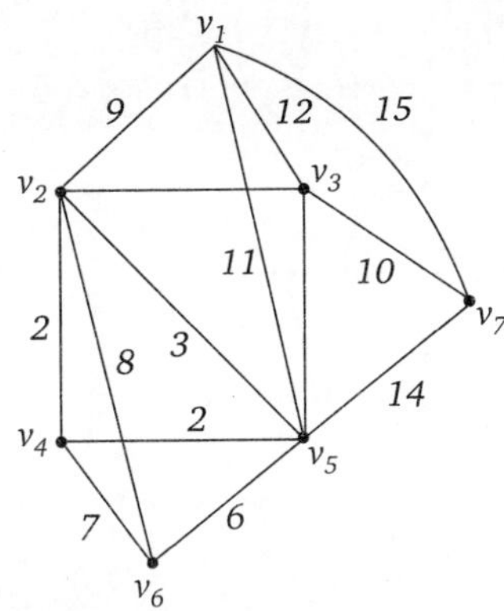

11. Find a spanning tree for each of the graph shown by removing edges :

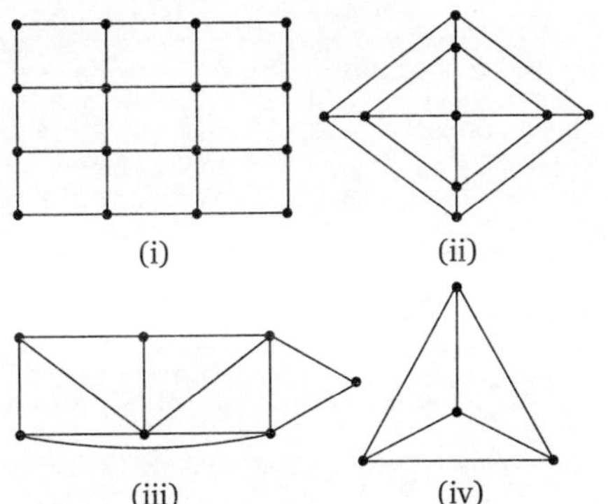

12. Draw all the spanning trees of the following graph shown below :

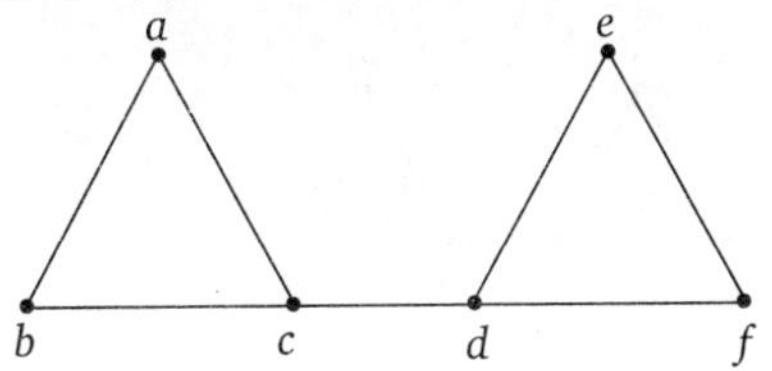

13. Obtain a spanning tree for each graph given below by using breadth first approach : With vertex ordering *a b e f c g d*.

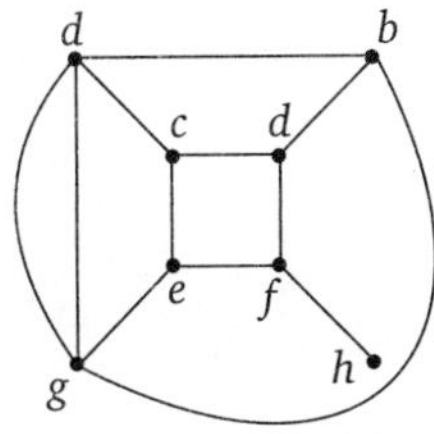

14. Obtain a spanning tree for each graph given below by using a depth-first search approach.

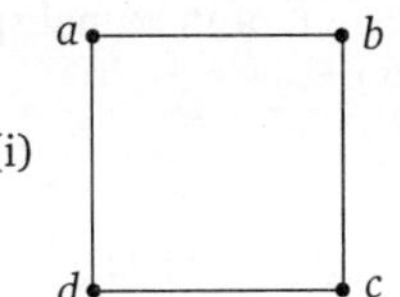

With vertex ordering *b a d c*.

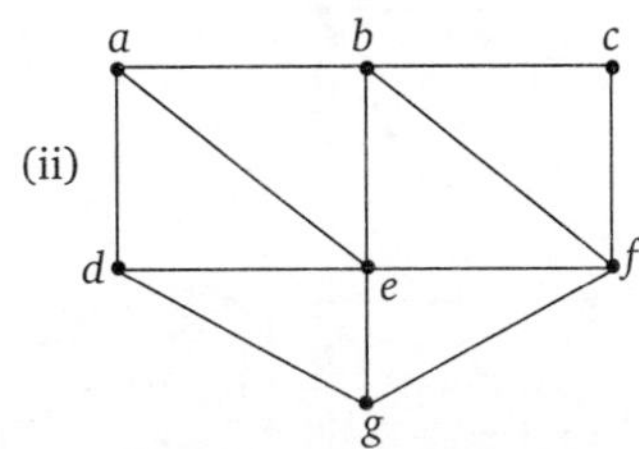

With vertex ordering *a b c f e d g*.

15. Use Kruskal's algorithm to find a minimum spanning tree for the given weighted graphs :

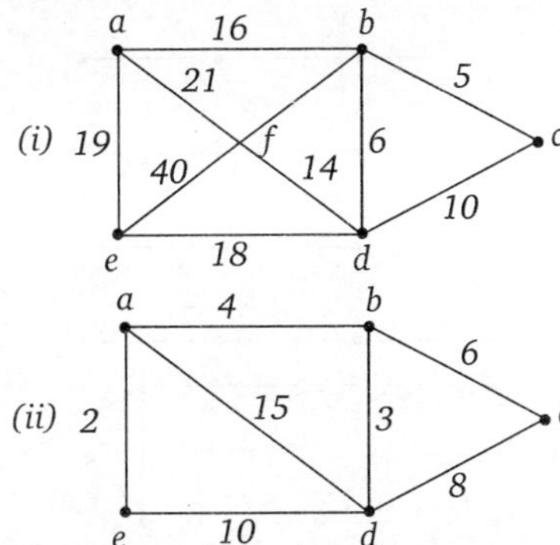

16. Use Prim's algorithm to find a minimum spanning tree for the given weighted graph :

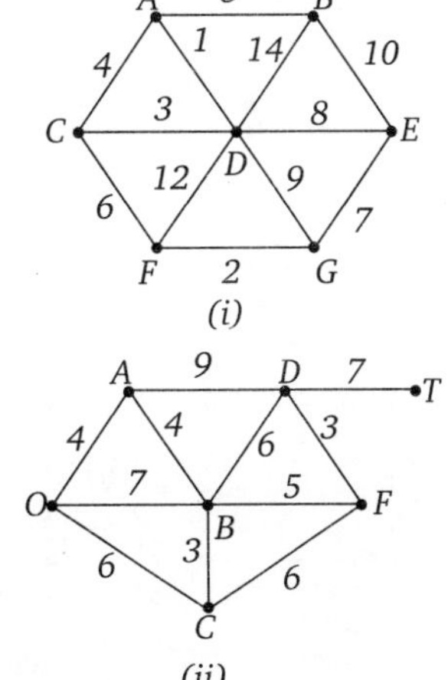

ANSWERS

6. Total weight of minimal spanning tree = 18 units
7. Total weight of maximal spanning tree = 36 units

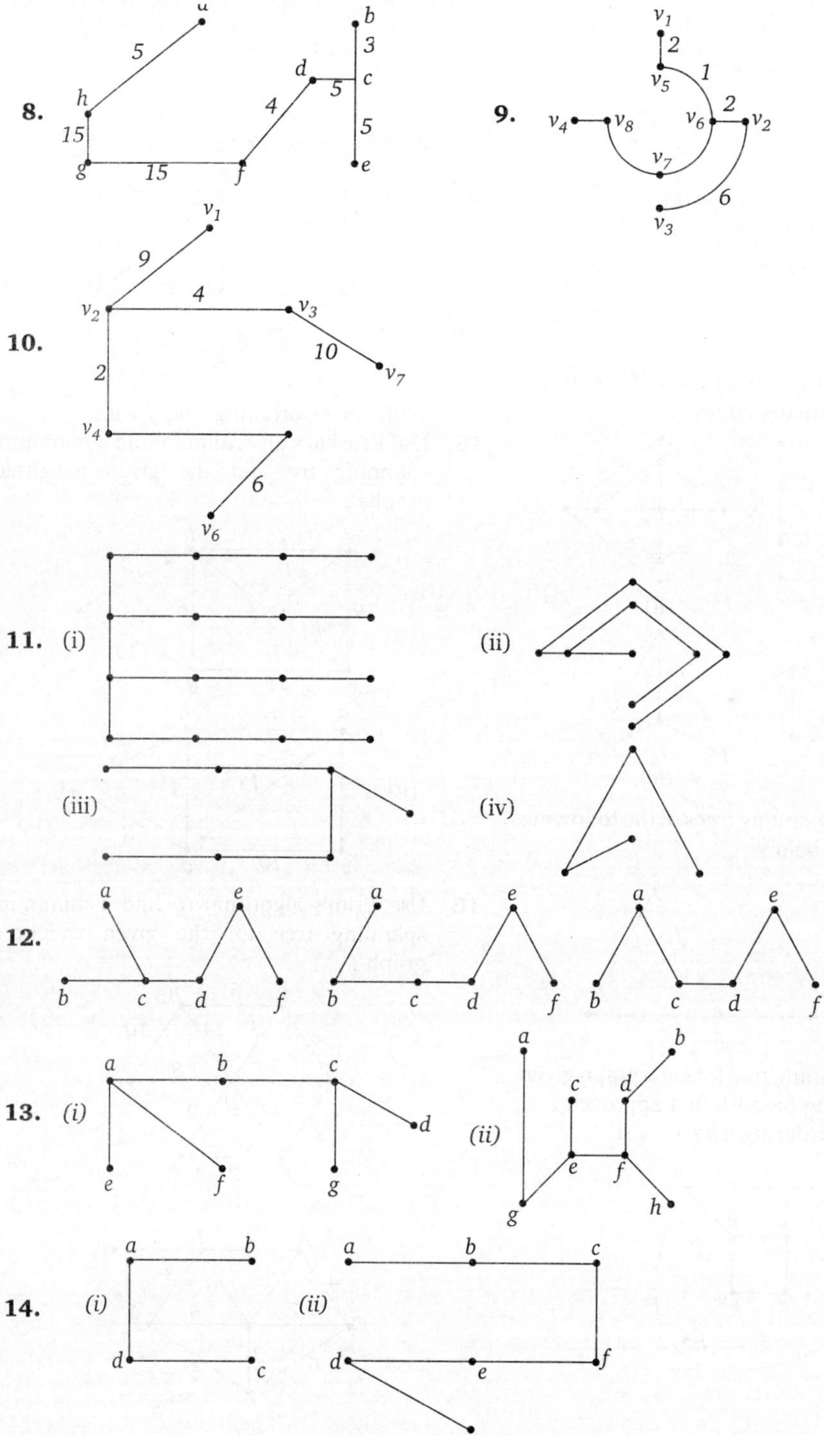

15. (i) Choose *ab, bf, bc, bd, de* (ii) Chose *ab, aa, bc, bd*

16. (i) Choose *BA, AD, DC, LF, FG, GE* (ii) Choose *OA, AB, BC, BF, FD, DT*

10.17 CUT SETS

A cut set is a set of edges in a connected graph whose removal leaves the graph disconnected and in such a manner that this set is smallest in the sense that if any edge is restored back in the graph, the graph is connected.

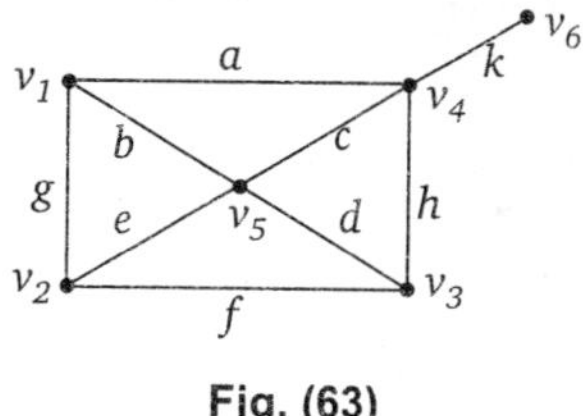

Fig. (63)

By removal of {*a, c, d, f*} these edges, then we find this graph is disconnected. If we take any edge in the set, then the graph will be connected. So, {*a, c, d, f*} is called out set. {*k*} is also a cut set. By removal of *k*, we find v_6 and isolated vertex. So graph is disconnected.

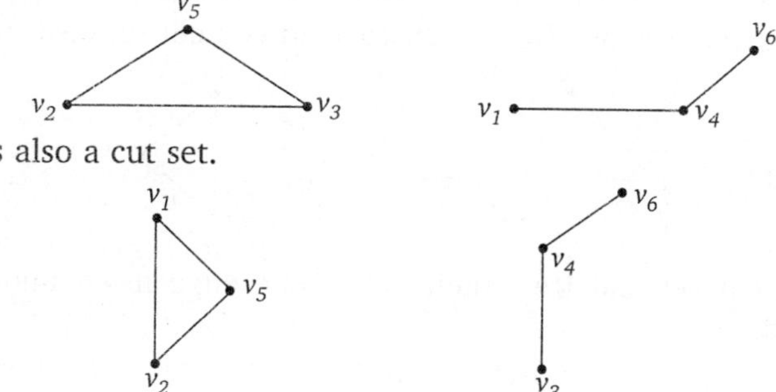

{*g, b, c, h*} is also a cut set.

☛ Cut set is also called a smallest set of minimal set.

10.17.1 Properties of a Cut Set

(1) Every cut set in a connected graph *G* must contain at least one branch of every spanning tree of *G*.

(2) In a connected graph *G*, any minimal set of edges containing at least one branch of every spanning tree of *G* is a cut set.

(3) Every circuit has an even number of edges in common with any cut set.

Theorem 1. *Every circuit has an even number of edges in common with any cut set.*

Proof : Consider a cut set *S* in graph *G*. Let the removal of *S* partition the vertices of *G* in two subsets v_1 and v_2. Consider a circuit Γ in *G*. If all the vertices in Γ are entirely within in vertex set v_1 (or v_2), the number of edges common of *S* and Γ is zero, that is $N(S \cap \Gamma) = 0$ and even number.

If, on the other hand, some vertices in Γ are in v_1 and some in v_2, we traverse back and forth between the sets v_1 and v_2 as we traverse the circuit. Because of the closed nature of a circuit, the number of edges, we traverse between v_1 and v_2 must be even and no other edge in *G* has this property. The number of edges common to *S* and Γ is even.

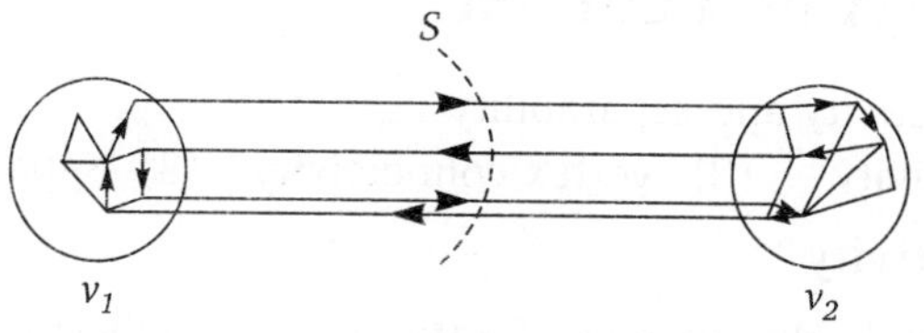

Fig. (64)

10.17.2 Fundamental Cut Set

Consider a spanning tree T of a connected graph G. Take any branch b in T. Since $\{b\}$ is a cut set in T, $\{b\}$ partitions all vertices of T into two disjoint sets. The removal of $\{b\}$ disconnect the graph, so we get a partition $V_1 = \{v_1\}$ and $V_2 = \{v_2, v_3, v_4, v_5\}$

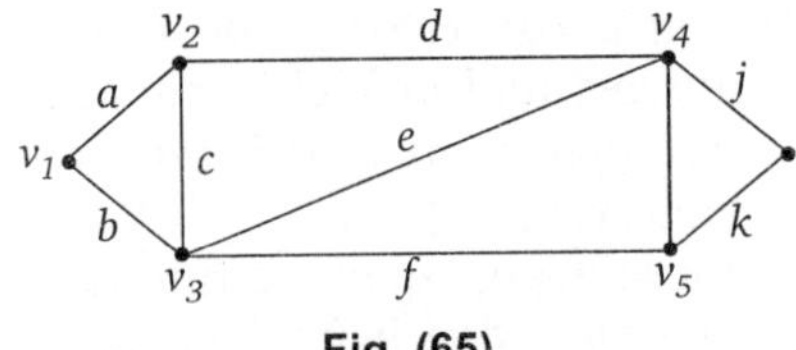

Fig. (65)

Remove $\{a, b, c\}$ from G and it will also contain V_1 only and form a cut set. In this case (a) is a chord and b is branch. A cut set contain only one branch of spanning tree and is called fundamental cut set. It is also known as 'back cut set'.

In the spanning tree, remove $\{b, c, e, d\}$, $\{b, c, d\}$ is a cut set disconnects the graph.

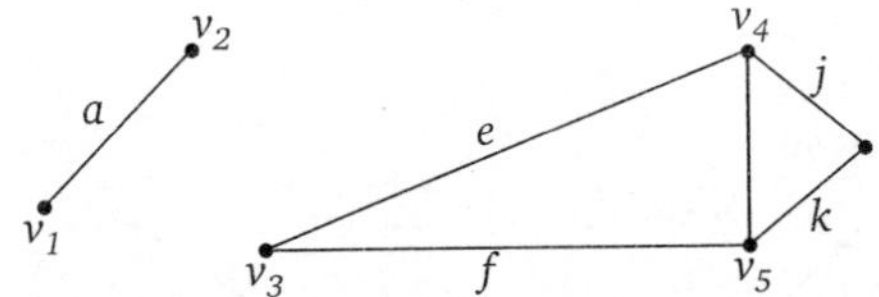

It is intersects cut set and tree, then it has one edge in common and the graph should be connected.

Theorem 2. *Every cut set in a connected graph G must contain at least one branch of every spanning tree of G.*

Proof : Any minimal set of edges containing at least one branch of every spanning tree be a cut set.

In a given connected set G, let Q be a minimal set of edges containing at least one branch of every spanning tree of G. Consider $G - Q$ the subgraph that remains after removing Q from G. Since the subgraph $G - Q$ contains no spanning tree of G, $G - Q$ is disconnected. Also, since Q is a minimal set of edges with this property and edge e from Q returned to $G - Q$, will create at least one spanning tree. Thus the subgraph $G - Q + e$ will be connected graph. Therefore, Q is minimal set of edges, whose removal from G, disconnects G.

Theorem 3. *In a connected graph G, any minimal set of edges containing at least one branch of every spanning tree of G is a cut set.*

Proof : Let S be the set of edges of G such that S contains at least one branch of every spanning tree, having minimum number of edges. If $G - S$ is connected, it means $G - S$ has spanning tree and no edge of spanning tree present in S. So, it must contain an edge of spanning tree, however, it may contain more than one branch of spanning tree. S is a minimal cut-set containing branch of the tree.

10.18 CONNECTIVITY OF A CUT SET

Cut set has connectivity and separability

(1) edge connectivity (2) vertex connectivity (3) separability

10.18.1 Edge Connectivity

In a connected graph, the number of edges in the smallest cut set is defined as the edge connectivity of G. It is usually defined by $k(G)$ or $k(e)$. Equivalently edge connectivity of a connected graph can be defined as the minimal number of edges whose removal reduces the rank of the graph by one.

10.18.2 Vertex Connectivity

The vertex connectivity of a connected graph is defined as the minimal number of vertices whose removal from G leaves the remaining subgraph disconnected. This is denoted by $k\ (v)$.

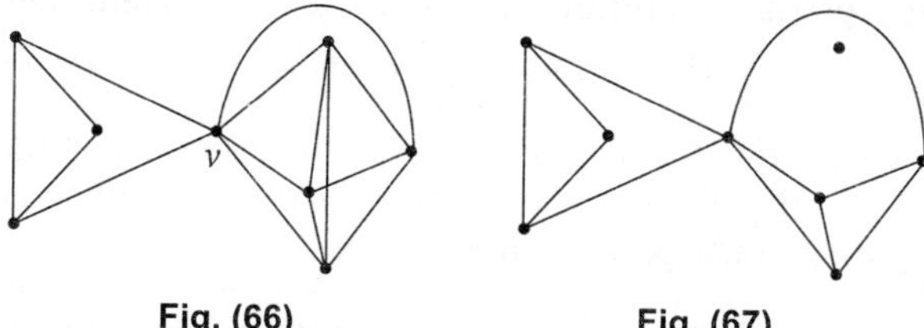

Fig. (66) Fig. (67)

$k\ (G) = 2$ The minimum number of edges by which the graph will be disconnected.

$k\ (v) = 1$ The number of vertex by the removal of this vertex the graph will be disconnected. It is minimum by removal of v.

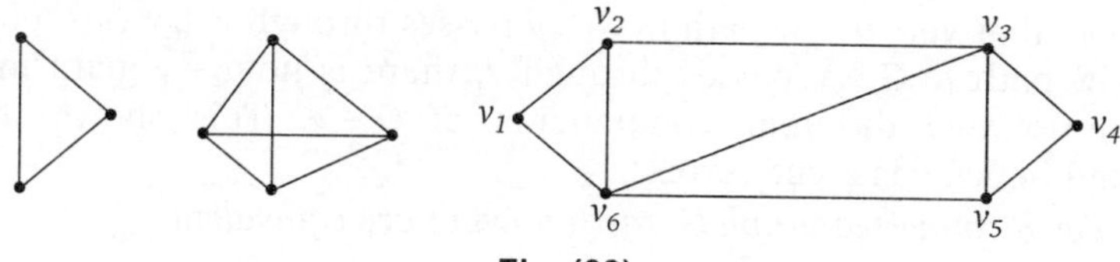

Fig. (68)

$k\ (v) = 2$. Because by the removal of v_3 and v_6, we will find a disconnected graph.

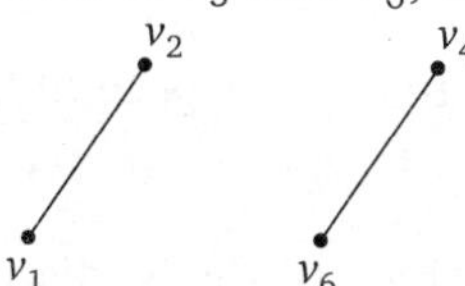

10.18.3 Separability

A connected graph is said to be separable if its vertex connectivity is one. A connected graph which is not separable is termed as non-separable graph.

Theorem 1. *The edge connectivity of a graph G can not exceed the minimum degree of a vertex in G.*

Proof : Let v be a vertex of minimum degree in G. Then removal of all the edges incident with v disconnects the vertex v in a cut set of G. But edge connectivity in the minimum cardinality of the cut sets of G, which implies that the edges connectivity is always less than the number of chords incident with the vertex v. The chord connectivity need not lie equal to the minimum degree of a vertex of G. For example in the given graph the edge connectivity is one but the minimum degree of a vertex of the graph is two.

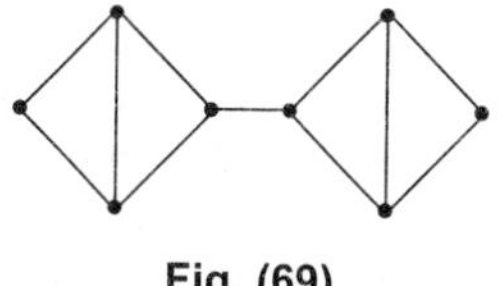

Fig. (69)

Theorem 2. *The edge connectivity of a graph G is always greater than the vertex connectivity of G.*

Proof : Let the edge connectivity of a graph be c. Then there exists a cut set S in G with c edges. Let v_1 and v_2 be the partition of the vertex set of G with respect to S. Then the edge in S are the edges of G between v_1 and v_2. If no two edges in S have the same and vertex in the set v_1 or v_2. Then removal of all the end vertices of the edges in S disconnects G, otherwise the number of vertices required to disconnect G is less than the number of edges in S.

Theorem 3. *For a connected graph G, the following are equivalent*

(i) *v is a cut vertex.*

(ii) *The vertex subset $V - v$ can be partitioned as $U \cup W$ such that for each $u \in U$ and $w \in W$, $u - w$ path passes through v.*

(iii) *There exist vertices $u, w \in V - v$ such that $u - w$ path in G passes through v.*

Proof : (i) $\Rightarrow$ (ii)

Let v be a cut vertex. Therefore, $G - v$ is disconnected. Let $G_1, G_2, \ldots, G_k$ be components of $G - v$. Let $u = V(G)$ and $W = \sum_{i=1}^{k} V(G_i) \Rightarrow w \in V(G_i)$ for some $2 \le i \le k$.

If there is a $u - w$ path P in G not passing through v, then P connects u and w in $G - v$ also. But u and w are the components and $G - v$ are connected which shows a contradiction. Thus, path P must pass through v.

(ii) $\Rightarrow$ (iii)

If $u, w \in G_1$ and G is connected, then $u - w$ path in G passes through v. If $u \in G_2$ and $w \in G_3$. Also G_2, G_3 are the components and $G - v$ is connected so we can find $u - w$ path in G passes through v.

(iii) $\Rightarrow$ (i)

We are given that very $u - w$ path in $G - v$ passes through v. Let $G - v$ be connected. Since each $u - w$ path in $G - v$ passes through v, there is no $u - w$ path in $G - v$. Thus u and w belong to two different components of $G - v$. This shows that $G - v$ is disconnected and hence v is a cut vertex.

Theorem 4. *For a connected graph G, the following are equivalent.*

(i) *e is a cut edge of G.*

(ii) *If $e = (a, b)$, then is a partition of the edges subset $E - e$ as $E_1 \cup E_2$ with $a \in V(E_1)$, $b \in V(E_2)$ such that for any $u \in V(E_1)$ and $w \in V(E_2)$ each $u - w$ path contains e.*

(iii) *Then exist vertices u, w such that $u - w$ path in G contains e.*

(iv) *e is not a cycle edge of G.*

Proof : (i) $\Rightarrow$ (ii)

Let e be a cut edge of G. Then $G - e$ is disconnected. Let G_1 and G_2 be two components of $G - e$. Let $E_1 = E(G_1)$ and $E_2 = E(G_2)$. If $u \in V(G_1)$ and $w \in V(G_2)$, such that there is a $u - w$ path P in G which does not contain e, then u, w are connected in $G - e$ by P which shows that $G - e$ is connected is a contradiction. Hence, each $u - w$ path in G must contain e.

(ii) $\Rightarrow$ (iii); (iii) $\Rightarrow$ (iv)

Suppose e lies on a cycle c. Then $c - e$ gives an $a - b$ path not containing e with vertices u and w following condition (iii). Let P be the $u - w$ path without any loss of generality. We may assume that a, b occur in the natural order in P. Let u_0 and w_0 be the first and the last vertices that P has in common with G. Then $P_{u, u_0} \cup P_{u_0, w_0} \cup P_{w_0, w}$ is a $u - w$ path P of G which does not contain e is a contradiction because $u - w$ path contains e. Hence, e is not a cycle edge of G.

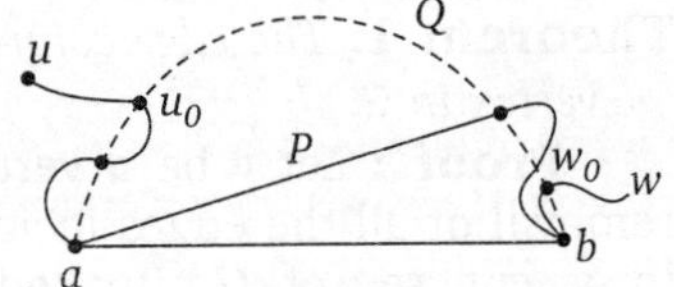

Fig. (70)

(iv) $\Rightarrow$ (i)

Let, if possible, $G - e$ be connected then there is an $a - b$ path P in $G - e$. But then $P \cup e$ is a cycle containing e which contradicts condition (iv). This shows that $G - e$ is disconnected. Hence e is cut edge of G.

Fig. (71)

10.18.4 Block

A block in a connected graph is connected subgraph which doesn't have any cut vertex.

☛ Normally a block doesn't have any cut edge.

Only in $k_2 = (\{a, b\}, e)$, the block has cut edge. This block is known as trivial block and all other blocks are non-trivial blocks.

Definition : *A block of a graph G is a maximal subgraph H of G such that H is a block, i. e., H does not have a cut vertex. But for any $v \in V(G) - V(H)$. Then $V(H) \cup \{v\}$ is either disconnected graph or a separable graph.*

Theorem 5. *For a connected graph G, the following are equivalent :*

(i) G is a non-trivial block.

(ii) Any two vertices of G lie on a cycle.

(iii) Given any vertex v and any edges vw, there is a cycle of G containing both.

(iv) Given any pair of edges $e = uv$ and $e' = u'v'$, there is a cycle containing both.

(v) Given any pair of vertices u and u' and any edge $e = vw$, there is a uu' path of G containing e.

Proof : (i) ⇒ (ii)

Let G be a non-trivial block and let u and v be any two vertices of G. We prove the result by the principal of induction on distance between u and v, *i.e.*, $d(u, v)$. If $d(u, v) = 1$, then uv is an edges and since G is a non-trivial block, uv is not a cut edge. Hence, uv is a cycle edge and u and v lies in it.

Now, let us assume that if u is any vertex, then any vertex v' at a distance at most $k - 1$ from u lies on cycle with u. Let v be a vertex at distance k from u. We show that uv lies on a cycle.

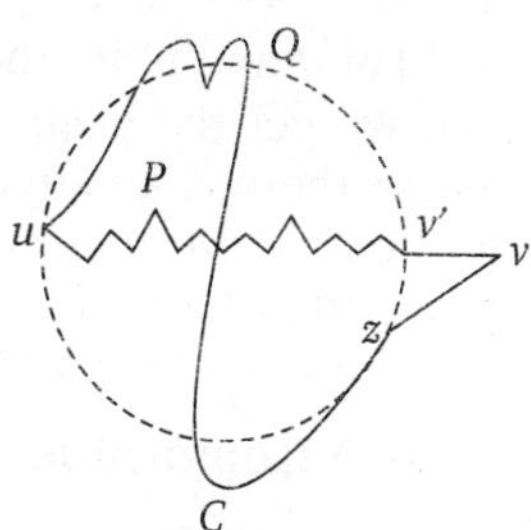

Fig. (72)

Let P be a shortest uv path and v' be the penultimate vertex on P from u. By induction, there is a cycle c containing u and v'. Since v' is not a cut vertex, because it is a block. There is a uv path Q not passing through v'. Let z be the last vertex from u that Q has in common with c. Then c_{uv}, $U(vv')\,U\,Q_{uz}\ Uc_{zu}$ is a cycle of G containing both u and v.

(ii) ⇒ (i)

Suppose any two vertices of G lies on a cycle in G. We show that G is a block. Let, if possible, G have a cut vertex u, then there are vertices v and w such that vw path passes through u. Therefore there doesn't exist any cycle in G containing v and w, which is a contradiction. So, G is a non-trivial block.

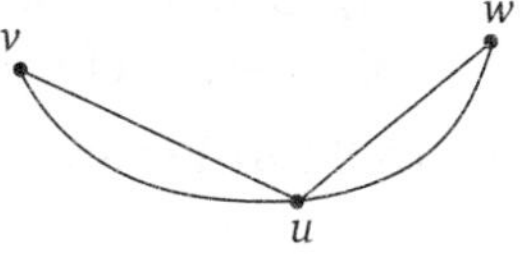

Fig. (73)

(ii) ⇒ (iii)

Let us suppose that any two vertices of G lies on a cycle in G. Let vertex u and edge vw be given. Since G is a block vw is not a cut edge. Let c be a cycle containing vw. If c contains u also, then we have done, the theorem is proved. If not, there is a cycle z containing u and v. Taking any orientation of z, let x and y be the first and the last vertices from u that z has in common with c. Then the $u - x$ segment of $z\ [z(u, x)]$ and $x - y$ segment of c containing vw and the $y - u$ segment of z constitute a cycle of G containing u and vw. If the cycle passes through u. then it contains both u and uw.

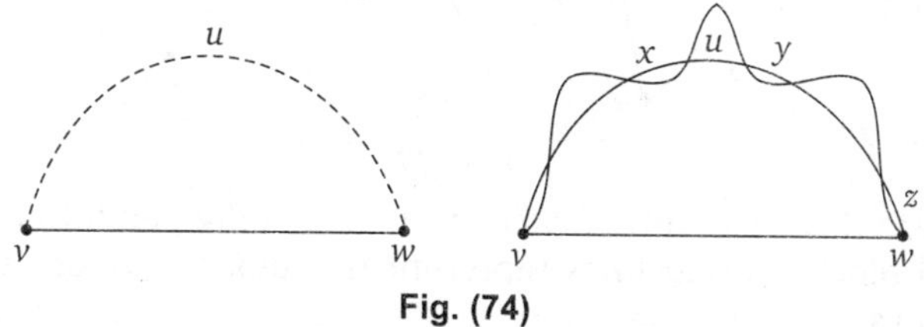

Fig. (74)

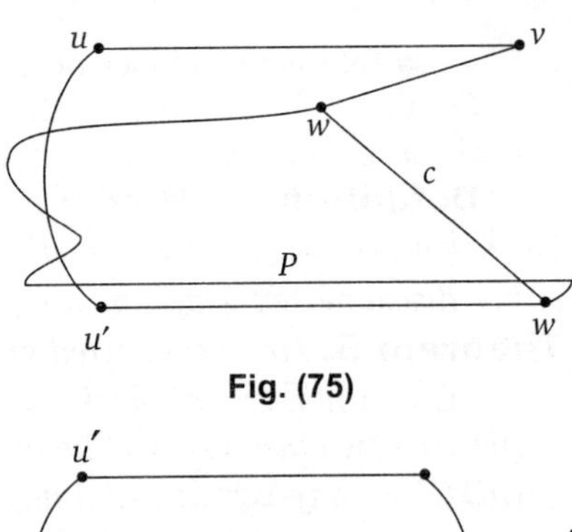

Fig. (75)

(iii) ⇒ (ii)

Let u and v be any two vertices. Hence, v cannot be an isolated vertex, then there is an edge vw in G. Hence, by hypothesis, there exists a cycle c containing u and edge vw, *i.e.*, there exists a cycle c countaining u and v.

(iii) ⇒ (iv)

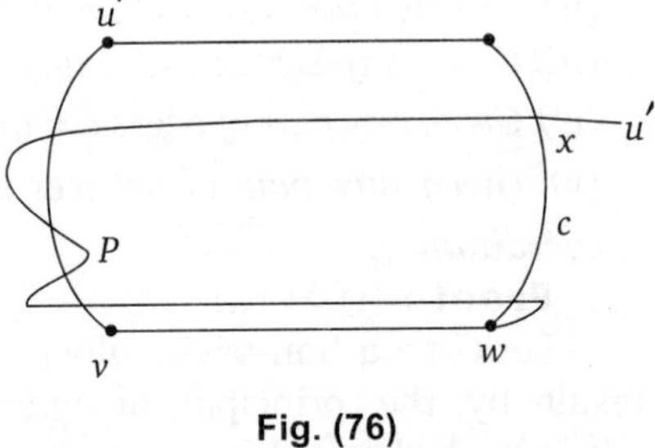

Fig. (76)

Let $e - uv$ and $e' = u'v'$ be the given edge. By (iii), there ia a cycle c containing u and e'. If it passes through u also, then we are done. If not, since G is a block, u is not a cut vertex and hence, there is a $v - v'$ path P in G, not passing through u. Let w be the first vertex from v, then P has in common with c. Then $v - w$ segment of P, $w - u'$ segment vertex from v, then P has in common with c. Then $v - w$ segment of P, $w - u'$ segment of c containing $u'v'$ and the edge uv constitute a cycle of G containing both e and e'.

(iv) ⇒ (v)

Let u and u' be the vertices and vw be the given edge if $u u'$ is also an edge then by (iv), we get the desired result. If not, then since G is a block u is not an isolated vertex and so there is an edge $u v_1$ in G by (iv) there is a cycle containing vu_1 and vw. Further u not be a cut vertex, there is a u' w path P not passing through u. Let x be the first vertex from u' that P has in common with c. Then, the $u' - x$ segment of P and $x - u$ segment of c containing $v - w$ constitute the desired path.

10.18.5 Isomorphic

If G and G' are two graphs we will say they are isomorphic if

(i) both have same number of edges,

(ii) both have same number of vertices, and

(iii) degree of each vertex of both graphs are same.

10.18.6 Cut Vertex

A vertex v is said to be cut vertex if removal of the vertex disconnects the graph.

10.18.7 Separable

A connected graph having one cut vertex is called as a separable graph.

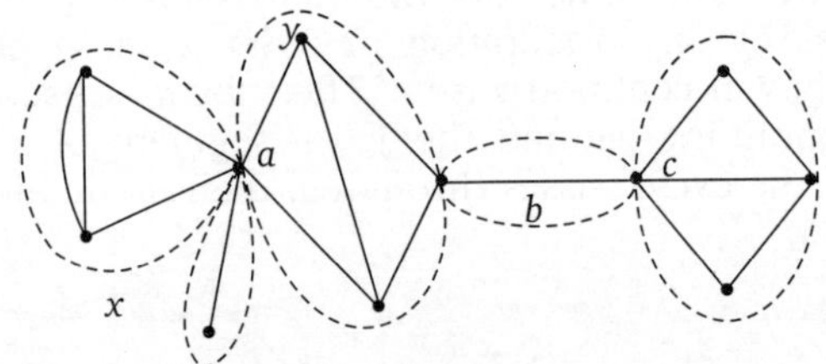

Fig. (77)

Here, a, b, c are three cut vertex each cut vertex is divided into two subgraphs. Each subgraph is called block. Dotted lines represent five blocks. We are writing these blocks separately as follows :

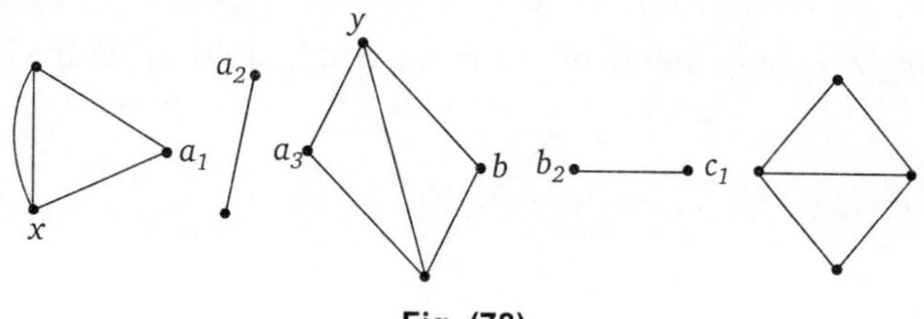

Fig. (78)

First cut vertex is divided into three cut vertices, say a_1, a_2, a_3, second and third into two b_1, b_2 and c_1, c_2, respectively.

10.18.8 1-Isomorphic

Two graph G_1 and G_2 are said to be 1-isomorphic if they become isomorphic to each other under repeated application of the following :

(i) split a cut vertex into two vertices to produce two disjoint subgraphs.
There graphs are 1-isomorphic but not isomorphic.
1 isomorphic $\nRightarrow$ isomorphism
i.e., Every 1-isomorphic is not necessarily isomorphism
However, isomorphism $\Rightarrow$ 1-isomorphism
i.e., Every isomorphism is 1-isomorphism
If we join x to y, then
This graph is 1-isomorphic to the (i) and (ii).

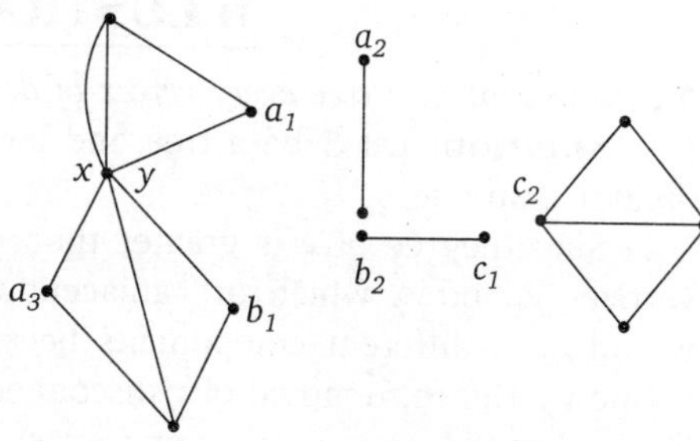

Fig. (79)

10.18.9 2-Isomorphic

A graph is said to be 2-isomorphic, if

(i) split a cut vertex into two vertices to produce two disjoint subgraphs.

(ii) split the vertex x into x_1, x_2 and the vertex y into y_1 and y_2 such that G is split into $g_1\ \bar{g}_1$.

Let vertices x_1, y_1 go with g_1 and x_2, y_2 with $\bar{g}_1$. Now, rejoin x_1 with y_2 and x_2 with y_1.

The graph will satisfy both conditions.

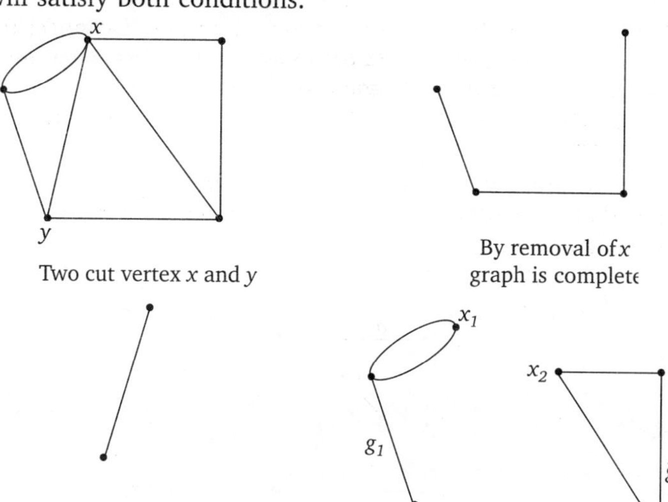

g_1 and $\bar{g}_1$ are two components of the main graph. Join x_1 with y_2 and x_2 with y_1.

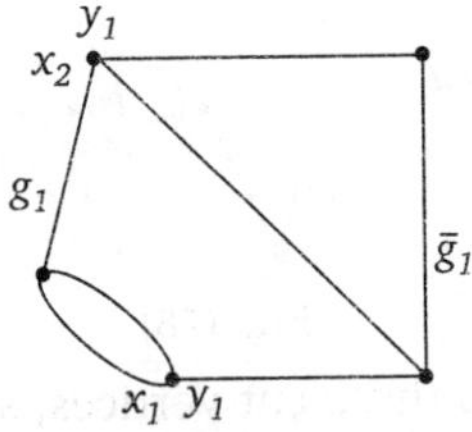

Fig. (80)

The graph is 2-isomorphic. But is not isomorphic. In (i), x has degree y but in (ii), no one vertex has degree y.

ILLUSTRATIVE EXAMPLES

1. *Prove that in a tree every vertex of degree greater than one is a cut-vertex.*

Solution. Let G be a tree and let v be a vertex in G, whose degree is greater than one.

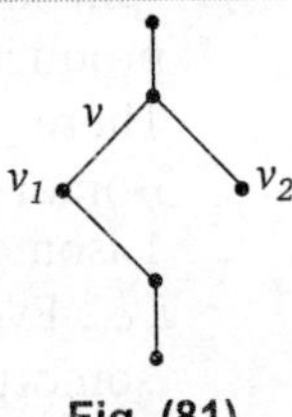

Fig. (81)

Since degree of v is greater than one, it means there are at least two vertices v_1 and v_2 which are adjacent to v. Clearly removal of v will put v_1 and v_2 in different components because there is no other path between v_1 and v_2. Hence, removal of v disconnects the graph. Thus v is a cut-vertex.

2. *Find the edge connectivity and vertex connectivity of the graph in the given diagram.*

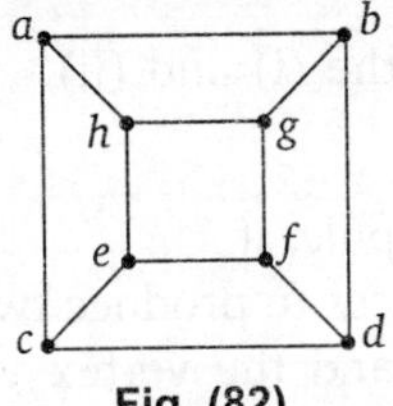

Fig. (82)

Solution. It can be seen that removal of any two vertices does not disconnect the graph However, removal of three vertices a, b and f disconnects the graph. Hence, vertex connectivity is three.

Similarly, we see that removal of any two edges does not disconnect the graph but the set of any three edges incident on a vertex forms a cut set. Hence, edge connectivity is three.

- In the graph of the above diagram, we have
  Edge connectivity = Vertex connectivity = degree of any vertex

3. *What is minimum cut ?*

Solution. In a weighted graph, the minimum cut or minimum cut set represents that cut set of edges such that the sum of the weights associated with these edges is the smallest of all the cuts sets of the graph.

4. *Find the minimal cut set of the given graph.*

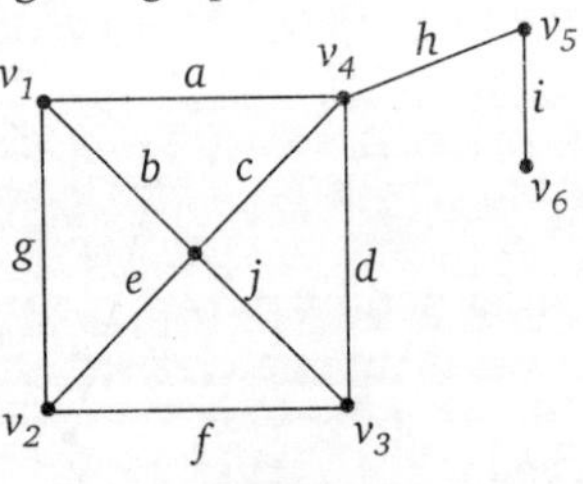

Fig. (83)

Solution. A cut set means a set of edges whose removal from G leaves the graph G disconnected.

$[h]$ will be minimal cut set.

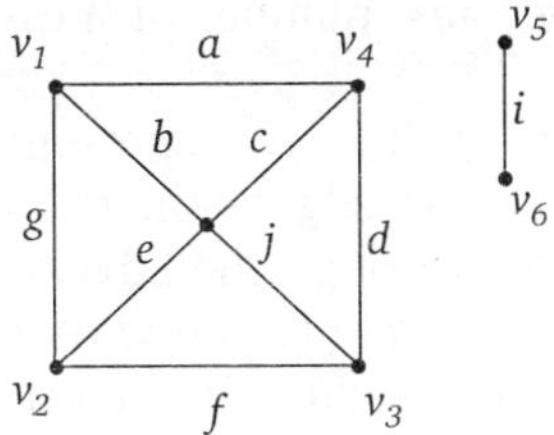

5. *Find the edge connectivity and the vertex connectivity of the following graph.*

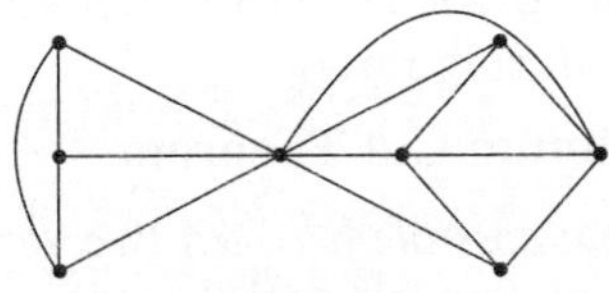

Fig. (84)

Solution. The minimum number of edges removal disconnects the graph is 3 hence edge connectivity $\lambda(G) = 3$.

The minimum number of vertices required to disconnect the graph is 1, hence vertex connectivity = 1

6. *Find the edge connectivity and vertex connectivity of the graph given below :*

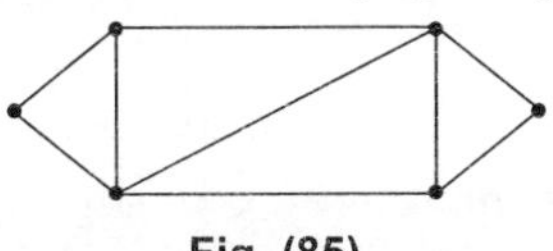

Fig. (85)

Solution. Edge connectivity = 2

Vertex connectivity = 2

7. *Give an example for a graph whose vertex connectivity is equal to edge connectivity.*

Solution.

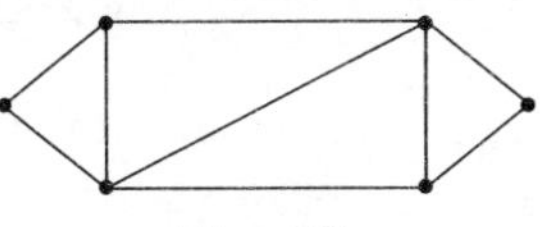

Fig. (86)

8. *The edge connectivity of a graph is less than or equal to the degree of the vertex with the smallest degree in G.*

Solution. If v is the vertex with smallest degree, v can be separated from G by removing at least $d(v)$ edges. Hence, edge connectivity $\leq d(u)$.

v is having degree one, *i.e.*, minimum degree vertex. If we disconnect e which gives the edge connectivity one. So edge connectivity is equal to the degree of the vertex with the smallest degree in G.

10.19 NETWORK FLOWS

Nowadays, we are largely governed by networks, such as network of telephone lines, network of transportation (highways, rails, etc.). Thus the mathematical analysis of such networks has become very important. In this section, we will see that network analysis is essentially a study of graphs.

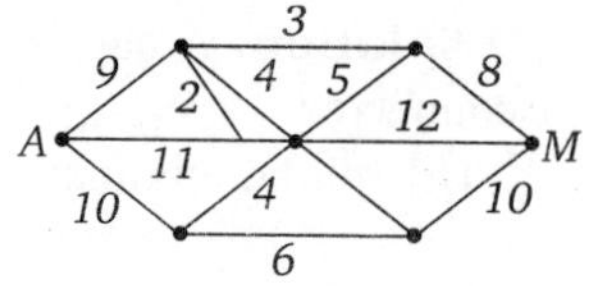

Fig. (87)
A netwok of flow

A network is represented by a weighted connected graph that contains no self loops. The vertices denote the stations (or places) and the edges are links through which the given commodity (such as oil, gas, number of messages, etc.) flows.

The weight associated with each edge represents the capacity of the edge. The capacity of an edge is the maximum amount of material which can flow through the edge.

The following assumptions are made for network problems :

(i) At any intermediate vertex, the amount of material entering is equal to the amount of material leaving.

(ii) The flow in any edge can not exceed the capacity of that edge.

(iii) There is no loss of material during flows.

10.19.1 Maximum Flow Minimum Cut Theorem

Theorem 1. *The maximum flow possible between two vertices in a network is equal to the minimum of the capacities of all cut sets with respect to two vertices.*

Proof : Consider any cut set S with respect to the vertices v_1 and v_2. In the subgraph $G - S$ a graph after removing S from G. There is no path between v_1 and v_2. Hence, every path between v_1 and v_2 must contain at least one edge of S. Hence, the total flow rate between v_1 and v_2 can not exceed the capacity of S. Since this holds for all cut sets with respect to v_1 and v_2, the flow rate can not exceed the minimum of their capacities.

Exercise 10.2

1. List all cut sets with respect to the vertex v_1, v_2 in the graph

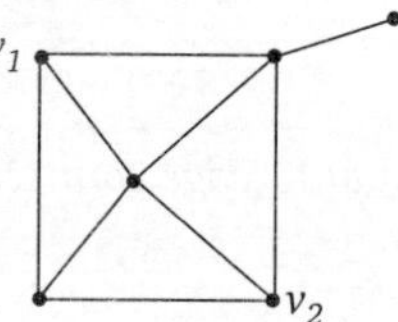

2. Give an example of a simple regular graph of degree three that is separable.

3. Prove that every connected graph with three or more vertices has at least two vertices which are not cut vertices.

4. What is the vertex connectivity of the complete graph of n vertices ?

5. Find the edge connectivity and vertex connectivity of the following graphs.

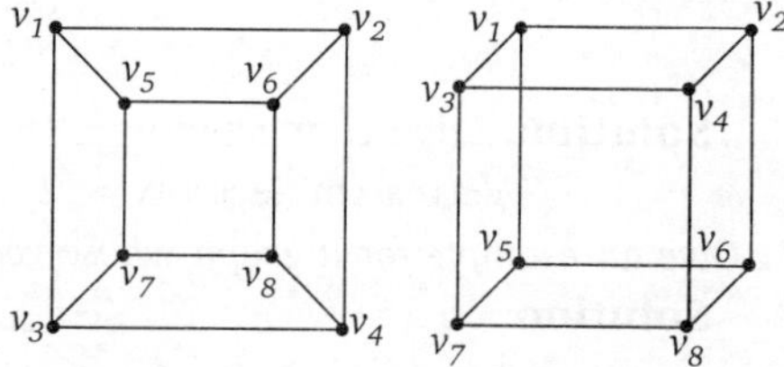

6. (a) Give an example of a graph whose edge connectivity is one.

(b) Give an example of a graph whose vertex connectivity is one.

(c) Give an example of a graph whose edge connectivity and vertex connectivity both are equal to one.

7. Obtain the maximum flow in the network.

☐☐☐

Chapter

11

Automata Theory

- Finite automata
- Transition function
- Deterministic and non-deterministic automata
- Mealy and moore machine
- Formal language
- Chomasky Hierarchy
- Regular sets and regular grammer
- Pumping lemma
- Context free language
- Normal forms
- Turing machine

11.1 THE CONCEPT OF AUTOMATION

An automation is an abstract model of a digital computer. Every automation includes several features. It has a mechanism for reading input. It will be assumed that the input is a string over a given alphabet, written on an input file which the automation can read but not change. The input file is divided into cells each of which can hold one symbol. The input mechanism can read the input file left to right one symbol at a time. The input mechanism can also detect the end of the input string. The automation produce output in some form. It may have a temporary storage device consisting an unlimited number of cells, each is capable of holding a single symbol from an alphabet. The automation has a control unit which can be in any one of a finite number of internal states and which can change state in some defined manner.

The internal state of control unit at the next time step is determined by next step or transition function. The transition function gives the next state in terms of current state and the current input symbol.

Definition : *A finite automation* $M = (Q, \Sigma, \delta, q_0, F)$ *is a collection of following five things :*

(1) *Q is a finite set of states.*

(2) *It has an initial state or start state, i.e.,* $q_0 \in Q$.

(3) *It have some (may be none) final states, i.e.,* $F \subseteq Q$.

(4) *An alphabet* Σ *of possible input letters.*

(5) *A finite set of transition that tells for each state and for each letter of input alphabet which state to go to the next, i.e.,* $\delta : Q \times \Sigma \to Q$ *where* δ *is a transition function.*

- **Finite automation :** Finite because the number of possible states and number of letters in alphabet both are finite.
- **Automation :** because the change of states is totally governed by the input.
- Final state is also called accepting state.
- The transition from one state to another are governed by transition function δ, for example if $\delta(q_i, a) = q_j$ it means that if finite automata in state q_i and current input signal is a then the finite automata will go to state q_j.

- A string ω is accepted by a finite automata $\delta(q_o, \omega) = q$, if $q \in F$. This is the acceptability of a string by final state.

11.2 TRANSITION DIAGRAM

We represent a finite automata by a transition diagram or transition graphs in which each state is represented by a small circle and directed edges represent transitions. The label in the circles are names of the states while names on the edges are current value of the input symbol. The initial state will be identified by incoming unlabelled arrow not outgoing at any circle. Final state is drawn with double circle.

A transition system accepts a string w in Σ^* if

(a) there exists a path which originates from some initial state goes along the arrow and terminates at some final state and

(b) the path value obtained by concatenation of all edge labels of the path is equal to w.

Every finite automation $(Q, \Sigma, \delta, q_o, F)$ can be viewed as a transition system $(Q, \Sigma, \delta', Q_o, F)$. If we take $Q_o = \{q_o\}$ and $\delta' = \{q, w, \delta(q, w) : q \in Q, w \in \Sigma^*\}$. But a transition system need not be a finite automation. A transition system may contain more than one initial state.

11.2.1 Properties of Transition Functions

Property 1 : $\delta(q, \wedge) = q$ in a finite automation. This means the state of the system can be changed only by an input symbol.

Property 2 : For all strings w and input symbol a

$$\delta(q, aw) = \delta(\delta(q, a), w)$$
$$\delta(q, wa) = \delta(\delta(q, w), a)$$

This property gives the state after the automation consumes or reads the first symbol of a string aw and the state after automation consumes a prefix of the string wa.

ILLUSTRATIVE EXAMPLES

1. *Prove that for any transition function δ and for any two input strings x and y*

$$\delta(q, xy) = \delta(\delta(q, x), y) \quad \text{...(1)}$$

Proof : We prove it by method of induction $|y|$, *i.e.*, length of y.

Basis : When $|y| = 1$ and let $y = a \in \Sigma$

L.H.S. (1) $\delta(q, xy) = \delta(q, xa)$
$= \delta(\delta(q, x), a)$ by property 2
$=$ R.H.S. (1)

Induction Step : Let us assume the result is true of (1) for all strings x and strings y with $|y| = n$. Let y be a string of length $n + 1$. To prove that equation (1) is true for strings of length $n + 1$. Let $y = y_1a$, where $|y_1| = n$.

L.H.S. of (1) $\delta(q, xy) = \delta(q, xy_1, a)$ $\quad \because y = y_1a$
$= \delta(q, x_1a)$ $\quad$ Let $xy_1 = x_1$
$= \delta(\delta(q, x_1), a)$ $\quad$ by property 2
$= \delta(\delta(q, xy_1), a)$
$= \delta(\delta(\delta(q, x) y_1), a)$ $\quad$ by induction hypothesis.

R.H.S. of (1)
$= \delta(\delta(q, x), y) = \delta(\delta(q, x), y_1a)$
$= \delta(\delta(\delta(q, x), y_1), a)$ $\quad$ by property 2.

Hence, L.H.S. = R.H.S. This proves equation (1) is true for all strings of length $n + 1$ so by induction hypothesis it is true for all strings.

2. *Prove that if $\delta(q, x) = \delta(q, y)$, then $\delta(q, xz) = \delta(q, yz)$ for all strings z in Σ^*.*

Solution : $\delta(q, xz) = \delta(\delta(q, x), z)$

$= \delta(\delta(q, y), z)$ given that $\delta(q, x) = \delta(q, y)$

Now, $\delta(q, yz) = \delta(\delta(q, y), z) = \delta(q, xz)$.

ILLUSTRATIVE EXAMPLE

1. *If $(M = \{q_0, q_1, q_2\}, \{a, b\}, \delta, q_0, \{q_2\})$ finite automata where transition function δ is given by*

$\delta(q_0, a) = q_1$ $\quad \delta(q_0, b) = q_2;$ $\quad \delta(q_1, a) = q_0$ $\quad \delta(q_1, b) = q_2;$

$\delta(q_2, a) = q_2$ $\quad \delta(q_2, b) = q_2$

Solution : The graph in following figure represents the D.F.A.

This finite automata accepts the string *ab*, string in state q_0. The symbol *a* is read first. It checks the edges of the graph automation goes into state q_1. Next *b* is read and automation goes into state q_2. We are now at the end of the string and at the same time in a final state q_2. Therefore string *ab* is accepted.

The *FA* does not accept the string *aa*. After reading two consecutive *a*'s, it will be in state q_0 which is not a final state. Therefore, Finite automata rejects this string.

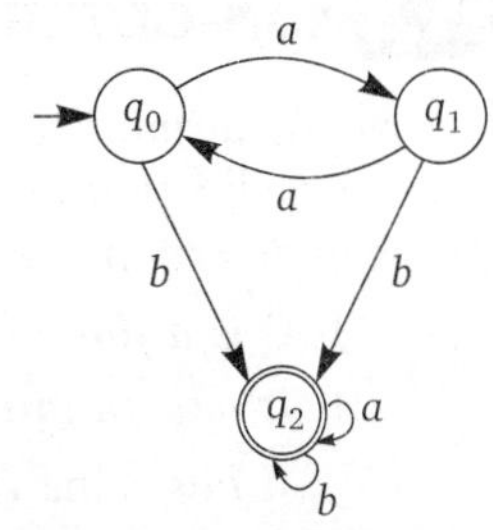

Fig. (1)

11.3 ACCEPTABILITY OF STRING BY A FINITE AUTOMATA

Let us consider a finite automata whose transition function δ is given in the form of a transition table $M = (\{q_0, q_1, q_2\}, \{a, b\}, \delta, q_0, \{q_2\})$.

Given the sequence of states for input string *aaab*.

Transition Table

States	**Inputs**	
	a	**b**
$\rightarrow q_0$	q_1	q_2
q_1	q_0	q_2
(q_2)	q_2	q_2

$$\delta(q_0, \overset{\downarrow}{a}aab) = \delta(q_1, \overset{\downarrow}{a}ab)$$
$$= \delta(q_0, \overset{\downarrow}{a}b)$$
$$= \delta(q_1\ \overset{\downarrow}{b})$$
$$= \delta(q_2\ \wedge)$$
$$= q_2 \text{ (Final state)}$$

The symbol $\downarrow$ indicates current input symbol.

11.4 EXTENDED TRANSITION FUNCTION

We denote extended function by δ^*, *i.e.*, $Q \times \Sigma^* \rightarrow Q \cdot \Sigma^*$ is a string rather than a single letter and its value gives the state automation will be in after reading the string.

If $\quad \delta(q_0, a) = q_1$ and $\delta(q_1, b) = q_2$

Then $\quad \delta^*(q_0, ab) = q_2$

We can define δ^* recursively by

1. $\delta^*(q, \wedge) = q$, *i.e.*, finite automata changed state after reading an input symbol.

2. $\delta^*(q, wa) = \delta(\delta^*(q, w), a)$. For all $q \in Q$, $w \in \Sigma^*$, $a \in \Sigma$

ILLUSTRATIVE EXAMPLES

1. *If* $\delta(q_0, a) = q_1$ *and* $\delta(q_1, b) = q_2$, *then* $\delta^*(q_0, ab) = \delta(q_1, b) = q_2$.

Solution : $\delta^*(q_0, ab) = \delta(\delta^*(q_0, a), b)$

$= \delta(\delta(\delta^*(q_0, \wedge), a), b)$

$= \delta(\delta(q_0, a), b)$ $\quad \because \delta^*(q_0, \wedge) = q_0$

$= \delta(q_1, b)$ $\quad \because \delta(q_0, a) = q_1$

$= q_2$

11.5 NON-DETERMINISTIC FINITE AUTOMATA

Non-deterministic means a choice of moves for an automation rather than a unique move in each situation we allow a set of possible moves.

Non-deterministic automata $M = (Q, \Sigma, \delta, q_0, F)$ *is a collection of following five things*

(1) Q is a finite set of states.

(2) It has an initial state or start state, i.e., $q_0 \in Q$.

(3) It has some (may be none) final states, i.e., $F \subseteq Q$.

(4) An alphabet Σ *of possible input letter.*

(5) A transition function $\delta : Q \times \Sigma \to 2^Q$, 2^Q *is the power set of Q.*

11.6 DIFFERENCE BETWEEN DETERMINISTIC FINITE AUTOMATA AND NON-DETERMINISTIC FINITE AUTOMATA

1. In non-deterministic automata, the range of δ is in the power set of Q, so that its value is not a single element of Q but a subset of it. This subset defines the set of all possible states that can be reached by transition. For example if current state is q_1 and the symbol is a. Then, $\delta(q_1, a) = \{q_0, q_2\}$. Then, either q_0 or q_2 could be the next state of the non-deterministic finite automata. But in Deterministic finite automata, it is not possible. In DFA $\delta(q_1, a) = q_0$. In DFA, for a single input letter there is only one move possible from a state to next state.

2. We can allow in NDFA $\delta(q_0, \wedge) = q$. This means that NDFA can make a transition without consuming an input symbol. In DFA, the system changes state if $\delta(q_i, a) = q_j$. DFA does not change state without consuming any letter.

3. In an NDFA the set $\delta(q_i, a)$ may be empty, *i.e.*, there is no transition defined for this specific situation.

For Example : Consider the transition graph in following figure :

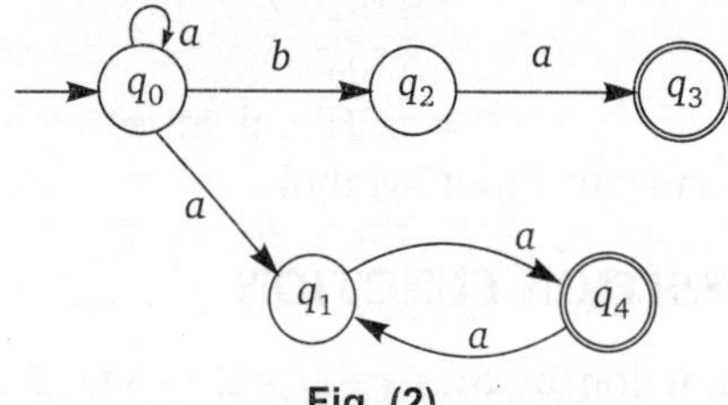

Fig. (2)

It describes a non-deterministic finite automata (NDFA) since there are two labeled a out of q_0.

Definition : *The language L accepted by an NDFA* $M = (Q, \Sigma, \delta, q_0, F)$ *is defined as the set of all input strings accepted by M.*

$$L(M) = \{w \in \Sigma^* : \delta^*(q_0, w) \cap F \neq \phi\}$$

This means the language consists of all strings w for which there is a walk labeled w from initial state of transition diagram to some final states.

Definition : *A string $w \in \Sigma^*$ is accepted by NDFA M if $\delta(q_0, w)$ contains some final states.*

11.7 FINITE AUTOMATA WITH OUTPUT

Mealy and Moore Machine : The finite automata which we considered in this section have binary output, *i.e.*, they accept the string or do not accept the string. The acceptability of any string decided on the basis of reachability of the final state by the initial state. If we remove this restriction and consider the model where output can be chosen from some other alphabet. The value of output function $X(t)$ is a function of present state $q(t)$ and the present input $x(t)$, *i.e.*, $X(t) = \lambda(q(t), x(t))$

This model is called Mealy machine where λ is output function. If the output function depends on present state and is independent of current input. The output function will be $X(t) = \lambda(q(t))$.

This model is called Moore machine.

11.7.1 Moore Machine

The Moore machine is a six-tuple $(Q, \Sigma, \Delta, \delta, \lambda, q_0)$, in which

1. *Q is a finite non-empty set of states.*
2. *Σ is a set of input letters called input alphabets.*
3. *Δ is an alphabet of possible output.*
4. *δ is a transition function $\Sigma \times Q \to Q$, i.e., a transition table that shows for each state and each input letter, what state is reached next.*
5. *λ is the output function mapping Q into Δ.*
6. *q_0 is the initial state.*

- A Moore machine does not define a language of accepted words because every possible input string creates an output string and there is no such things as final state. The processing is terminated when the last input letter is read and last output character is printed.
- Moore machine have pictorial representation very similar to FA. The difference is that instead of having only the name of state inside the little circle, we also specify the output character printed by that state. The two symbol inside the circle are separated by a slash "/". On the left side is the name of the state and on the right side is output from that state.

Example : Let us consider an example defined by following transition table of Moore machine.

Input alphabet $\Sigma = \{a, b\}$
Output alphabet $\Delta = \{0, 1\}$
States $Q = \{q_0, q_1, q_2, q_3\}$
q_0 is the initial state.

$$M = (Q, \Sigma, \Delta, \delta, \lambda, q_0)$$

Present State	Next State S		Output
	a	*b*	
q_0	q_1	q_3	1
q_1	q_3	q_1	0
q_2	q_0	q_3	0
q_3	q_3	q_2	1

The pictorial representation of the Moore machine is shown in the Figure 3.

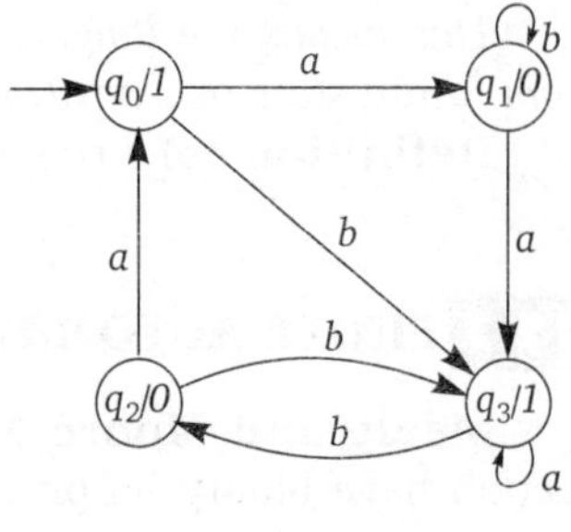

Fig. (3)

Let us trace the operation of this machine on input string *abab*. We start at state q_0 which automatically prints out the character 1. When we read the first letter of the input string which is an *a* and which sends us to state q_1. This state prints 0. The next input letter is *b* and loop shows that we return to state q_1. Being in q_1 again we print another 0. Then we read *a* go to q_3 and print 1. Next we read *b* go to q_2 and print 0. This is end of the run. The output sequence has been 10010.

11.7.2 Mealy Machine

Mealy machine is another variation of FA. A Mealy machine is like a Moore machine except that now output letter printing while we are travelling along the edge, not in state themselves. What we print output depends on edge we take. If there are two different edges from one state (*i.e.*, one *a*-edge, one *b*-edge) to another state it is possible that they will have different printing instructions for us. We take no printing instruction from state itself.

Definition : A Mealy machine is a six tuple $(Q, \Sigma, \Delta, \delta, \lambda, q_0)$

1. *Q is a finite non-empty set of states.*
2. *Σ is a set of input letters called input alphabets.*
3. *Δ is an alphabet of possible output character.*
4. *δ is a transition function $\Sigma \times Q \to Q$, i.e., a transition table that shows for each state and each input letter, what state is reached next.*
5. *λ is the output function mapping $Q \times \Sigma \to \Delta$.*
6. *q_0 is the initial state.*

☛ As with Moore machine the Mealy machine does not define a language by accepting and rejecting input string so it has no final state.

Example : The adjoining Fig. 4 represents a Mealy machine.

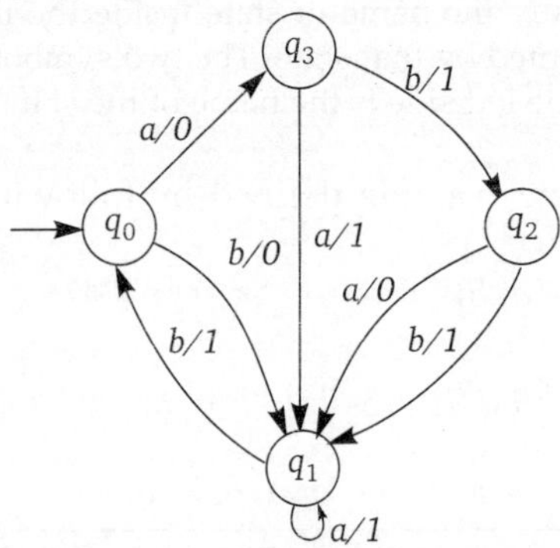

Fig. (4)

Transition table for this Mealy machine

Present state	Next state			
	Input a		Input b	
	State	Output	State	Output
$\rightarrow q_0$	q_3	0	q_1	0
q_1	q_1	1	q_0	1
q_2	q_1	0	q_1	1
q_3	q_1	1	q_2	1

- For a Moore machine if one input string is of length n, the output string is of length $n + 1$. Since the first output is $\lambda(q_0)$ for all output strings. In the case of Mealy machine if the input string is of length n, the output string is also of same length n.

11.8 PROCEDURE FOR CONVERTING THE MOORE MACHINE INTO MEALY MACHINE

We can modify the acceptability of input string by Moore machine by neglecting the response of the Moore machine to input $\wedge$. Let us consider a Mealy machine M and Moore machine M' are equivalent if for all input strings w, $bX_M(w) = X_{M'}(w)$, where b is the output of Moore machine for its initial state. The following method help to construct an equivalent Mealy machine.

WORKING RULES

Step (i). First we define the output function λ' for Mealy machine as a function of present state and input symbol. We define λ' by $\lambda'(q, a) = \lambda(\delta(q, a))$ for all states q and input symbol a.

Step (ii). The transition function is the same as that of the given Moore machine.

ILLUSTRATIVE EXAMPLES

(Based on the Conversion of mealy to more meeting and Vice-Versa)

1. *Construct a Mealy machine equivalent to following Moore machine.*

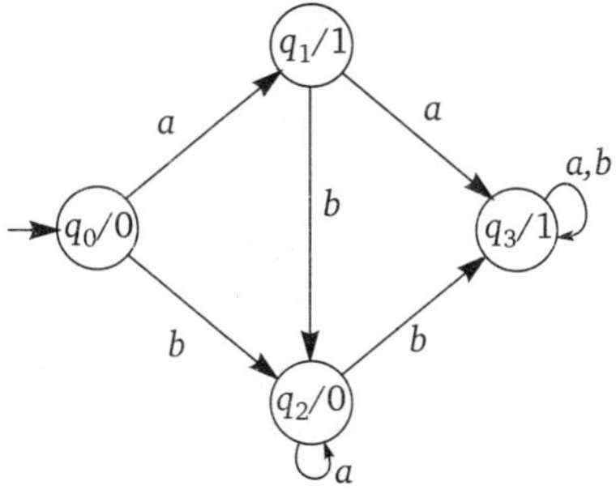

Fig. (5)

Solution : Transition table for the given Moore machine

Present state	Next state		Output
	a	b	
q_0	q_1	q_2	0
q_1	q_3	q_2	1
q_2	q_2	q_3	0
q_3	q_3	q_3	1

In the case of Moore machine, we construct pair consisting of the state and corresponding output and reconstruct the table for Mealy machine.

If we compare the present state column and output column, then

Output 0 is associated with state q_0; Output 1 is associated with state q_1

Output 0 is associated with state q_2; Output 1 is associated with state q_3

In Mealy machine, the state q_0, q_1, q_2, q_3 in next state column should be associated with output 0, 1, 0, 1 respectively.

The transition table for the Mealy machine

Present state	Next state			
	Input a		Input b	
	State	Output	State	Output
$\rightarrow q_0$	q_1	1	q_2	0
q_1	q_3	1	q_2	0
q_2	q_2	0	q_3	1
q_3	q_3	1	q_3	1

Transition diagram for Mealy machine

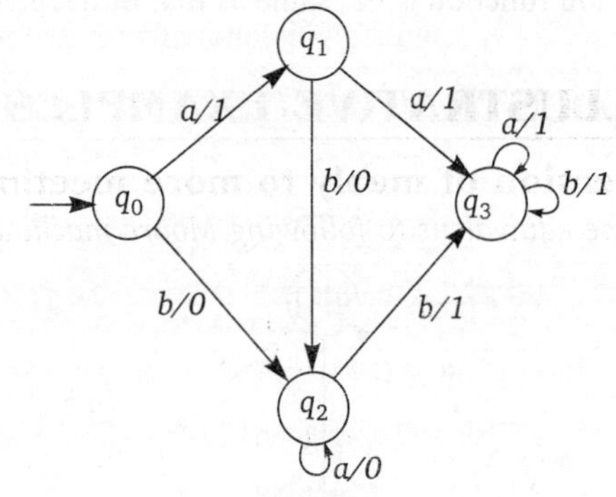

Fig. (6)

2. *Construct a Mealy machine which is equivalent to Machine given in the table.*

Present state	Next state		Output
	a	b	
$\rightarrow q_0$	q_2	q_1	0
q_1	q_1	q_3	0
q_2	q_2	q_3	1
q_3	q_3	q_2	1

Solution : Now, compare the present state column and output column. Then, output 0, 0, 1, 1 are associated with state q_0, q_1, q_2, q_3 respectively.

In Mealy machine, the state q_0, q_1, q_2, q_3 in the next state column should be associated with output 0, 0, 1, 1 respectively.

Transition table for Mealy machine

Present state	Next state			
	Input *a*		Input *b*	
	State	Output	State	Output
$\rightarrow q_0$	q_2	1	q_1	0
q_1	q_1	0	q_3	1
q_2	q_2	1	q_3	1
q_3	q_3	1	q_2	1

☛ We can reduce the number of states in any machine (*i.e.*, Moore machine or Mealy machine) by considering states with identical transition. If two states have identical transition (*i.e.*, rows corresponding to these two states are identical), then we can remove one of them.

3. *Consider the Moore machine given by the following transition table. Construct the corresponding Mealy machine.*

Present state	Next state		Output
	a	*b*	
$\rightarrow q_1$	q_1	q_2	0
q_2	q_2	q_4	0
q_3	q_1	q_3	1
q_4	q_1	q_3	1

Solution : Now, construct the transition table for the Mealy Machine.

Output 0, 0, 1, 1 are associated with states q_1, q_2, q_3, q_4 respectively.

Now, the transition table for Mealy machine :

Present state	Next state			
	Input *a*		Input *b*	
	State	Output	State	Output
$\rightarrow q_1$	q_1	0	q_2	0
q_2	q_2	0	q_4	1
q_3	q_1	0	q_3	1
q_4	q_1	0	q_3	1

In the table rows corresponding to q_3 and q_4 are identical so we can remove one of two states, *i. e.*, q_3 or q_4. Now, we remove q_4 by q_3. The reconstructed table is

Present State	Next State			
	Input a		Input b	
	State	Output	State	Output
$\rightarrow q_1$	q_1	0	q_2	0
q_2	q_2	0	q_4	1
q_3	q_1	0	q_3	1

11.9 PROCEDURE FOR TRANSFORMING A MEALY MACHINE INTO A MOORE MACHINE

Now, we develop a procedure for transforming a Mealy machine into Moore machine, so that for a given input string, the output string is the same (except the first symbol) in both the machine.

ILLUSTRATIVE EXAMPLES

1. *Consider the Mealy machine represented by the following figure. Construct a Moore machine equivalent to this machine.*

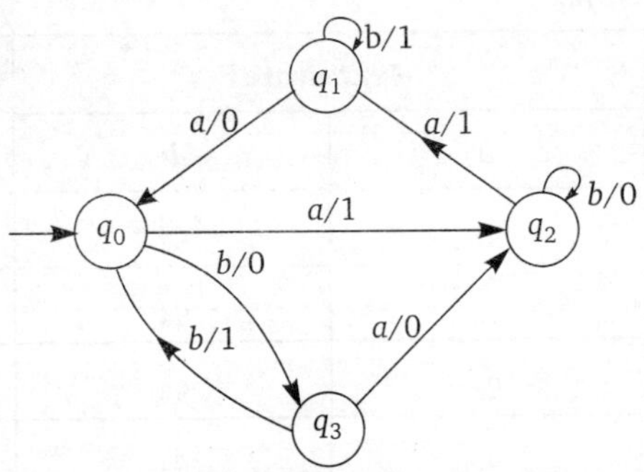

Fig. (7)

Solution : We convert the transition diagram into the following transition table

Present State	Next State			
	Input a		Input b	
	State	Output	State	Output
$\rightarrow q_0$	q_2	1	q_3	0
q_1	q_0	0	q_1	1
q_2	q_1	1	q_2	0
q_3	q_2	0	q_0	1

In the first step we develop the procedure so that both machines accept exactly the same set of input sequence. We first look into the next state column for any state say q_i and determine the number of different outputs associated with q_i in that column.

We split the q_i into several different states. The number of such states being equal to the number of different outputs associated with q_i.

Now, we apply this method in given problem from the next state column.

q_0 is associated with two outputs 0 and 1. So, we split it into two states q_{00} associated with output 0 and q_{01} associated with output 1. We must select the initial state for new machine so let us arbitrarily select q_{00}.

q_1 is associated with output 1.

q_2 is associated with output 0 and 1 so we split it into two states q_{20} associated with output 0 and q_{21} associated with output 1.

q_3 is associated with output 0.

Now, reconstructed table for new states.

Present State	Next State			
	Input a		Input b	
	State	Output	State	Output
$\rightarrow q_{00}$	q_{21}	1	q_3	0
q_{01}	q_{21}	1	q_3	0
q_1	q_{00}	0	q_1	1
q_{20}	q_1	1	q_{20}	0
q_{21}	q_1	1	q_{20}	0
q_3	q_{20}	0	q_{01}	1

The pair of states and outputs in next state column can be rearranged as in the following table.

The Revised Table :

Present State	Next State		
	Input a	Input b	Output
$\rightarrow q_{00}$	q_{21}	q_3	0
q_{01}	q_{21}	q_3	1
q_1	q_{00}	q_1	1
q_{20}	q_1	q_{20}	0
q_{21}	q_1	q_{20}	1
q_3	q_{20}	q_{01}	0

Now, the transition diagram for the required Moore machine is given in Figure 8.

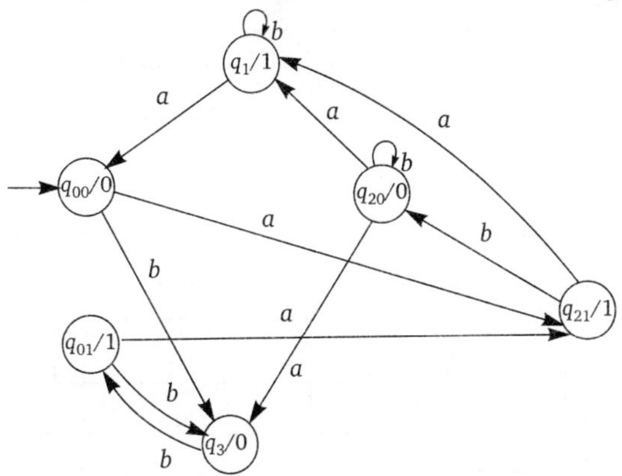

Fig. (8)

2. *Construct a Moore machine equivalent to Mealy machine M given by the table*

Present State	Next State			
	Input *a*		Input *b*	
	State	Output	State	Output
$\to q_1$	q_1	1	q_2	0
q_2	q_4	1	q_4	1
q_3	q_2	1	q_3	1
q_4	q_3	0	q_1	1

Solution : We see the next state column

q_1 is associated with output 1.

q_2 is associated with output 0 and 1 so we split it into two states, q_{20} is associated with output 0 and q_{21} is associated with output 1.

q_3 is associated with two different outputs 0 and 1, so we split it into two states, q_{30} associated with output 0 and q_{31} associated with output 1.

q_4 is associated with output 1.

Now, we reconstruct the table for new states.

Present State	Next State			
	Input *a*		Input *b*	
	State	Output	State	Output
$\to q_1$	q_1	1	q_{20}	0
q_{20}	q_4	1	q_4	1
q_{21}	q_4	1	q_4	1
q_{30}	q_{21}	1	q_{31}	1
q_{31}	q_{21}	1	q_{31}	1
q_4	q_{30}	0	q_1	1

The pair of states and outputs in next state column can be rearranged as in the following table.

The revised Table :

Present State	Next State		
	Input *a*	Input *b*	Output
$\to q_1$	q_1	q_{20}	1
q_{20}	q_4	q_4	0
q_{21}	q_4	q_4	1
q_{30}	q_{21}	q_{31}	0
q_{31}	q_{21}	q_{31}	1
q_4	q_{30}	q_1	1

This table is for required Moore machine. But here we observe that initial state q_1 is associated with output 1. This means that if we give input $\wedge$ then we get output 1 if machine starts at state q_1.

Thus, this Moore machine accepts zero length sequence which is not accepted by the given Mealy machine. To overcome this problem we must neglect the response of Moore machine to input $\wedge$. We add a new initial state q_0 whose transition are same with q_1 but output 0.

The new table is

Present State	Next State		
	Input *a*	Input *b*	Output
$\to q_0$	q_1	q_{20}	0
q_1	q_1	q_{20}	1
q_{20}	q_4	q_4	0
q_{21}	q_4	q_4	1
q_{30}	q_{21}	q_{31}	0
q_{31}	q_{21}	q_{31}	1
q_4	q_{30}	q_1	1

11.10 CONCEPTS OF FORMAL LANGUAGE

We are familiar with the notation of language such as Hindi, English and French still most of us would probably find it difficult to say what the word "language" means. Dictionaries define the term informally as a system suitable for expression of certain ideas, facts or concepts including a set of symbols and rules for their manipulation.

The theory of formal languages is an area with a number of applications in computer science. The concept of CFG (Context Free Grammar) was invented by Linguist Noam Chomesky in 1956. Chomesky gave several mathematical model for languages.

Before giving the definition of grammar, let us present an example to motivate the formal definition. Suppose we limit ourselves to a very restrictive subset of the sentence in English. We may begin by asking what a sentence in language is. Suppose a sentence can be one of two forms

1. <Noun> <Verb> <Adverb>
2. <Noun> <Verb>

<Noun> <Verb> <Adverb> or <Noun> <Verb> is not a sentence of English language but only the description of a particular sentence.

Now, we must specify what is a <noun>, <verb>, <adverb>

1. A <Noun> is dog, cat, Ram, Shyam
2. A <Verb> is chases, meets, runs, ate, ran
3. An <Adverb> is slowly, rapidly

If we replace <noun>, <verb> and <adverb> by suitable words, we get actual grammatically correct sentence. If we replace <noun> by cat, <verb> by runs and <adverb> by slowly in <noun> <verb> <adverb>, we get the sentence *Cat runs slowly*, similarly the sentence *Ram ran* can be given in the form <noun> <verb>.

Let us call <noun>, <verb>, <adverb> as **variables** or **non-terminals**. Words like dog, cat, Ram, Shyam, chases, meets, runs, ate, ran, slowly, rapidly which form the sentences are called **terminals**. Let S be a variable denoting a sentence.

We can form the following rules to generate two types of sentences :

$S \to$ <Noun> <verb> <adverb>

$S \to$ <Noun> <verb>

<Noun> → dog
<Noun> → cat
<Noun> → Ram
<Noun> → Shyam
<Verb> → chases
<Verb> → meets
<Verb> → runs
<Verb> → ate
<Verb> → ran
<Adverb> → slowly
<Adverb> → rapidly

The arrow in each rule indicates that the word on the right side of arrow can replace the word on left side of the arrow. We denote the collection of rules by P. We can describe the grammar by 4 tuple (V_N, Σ, P, S)

1. V is a finite set of objects called variables or non-terminals.
2. Σ is a finite set of objects called terminals.
3. S is a special variable, *i.e.*, $S \in V$ called start symbol.
4. P is a finite set of productions, whose elements $\alpha \to \beta$ where α and β are strings on $V_N \cup \Sigma$, where α has at least one symbol from V.
5. $V_N \cap \Sigma = \phi$

Example : $G = (V_N, \Sigma, P, S)$ is a grammar where

$V = \{S, \text{noun, verb, adverb}\}$

$\Sigma = $ {dog, cat, Ram, Shyam, chases, meets, runs, ate, ran, slowly, rapidly}

P (Productions) which denote collection of rules.

The sentence are obtained starting with S replacing words using productions and terminating when a string of terminals is obtained.

☛ Reverse substitution is not permitted. For example if $A \to a$ is a production then we can replace A by a but we can not replace a by A.

11.10.1 Alphabet

A finite nonempty set Σ of symbols called the alphabet.

11.10.2 Strings

Strings are finite sequences of symbols of the alphabet. For example if the alphabet $\Sigma = \{a, b\}$, then *abab* and *aaabbba* are strings on Σ. If $w = abaa$ indicate that the string named w has the specific value *abaa*.

11.10.3 Language

The language generated by the grammar G denoted by $L(G)$ as a subset of Σ^* defined as $\{w \in \Sigma^* : S \overset{*}{\underset{G}{\Rightarrow}} w\}$.

11.10.4 Sentences or Words

A string in language L is called the sentence of $L(G)$.

11.10.5 Production

The productions are grammatical rules that specify how sentences in the language can be made up. A production is of the form $\alpha \to \beta$ where α and β are strings of terminals and non-terminals.

11.10.6 Derivation

The sequence of application of the rules that produce the finished string of terminals from the starting symbol is called a derivation or a generation of the word. The derivation may or may not be unique which means that by applying productions to the start symbol in two different ways we may still produce the same finished product. Any derivation involves the application of productions when the number of times we apply production is one we write $\alpha \Rightarrow \beta$ when it is more than one we write $\alpha \overset{*}{\Rightarrow} \beta$

Example : Consider the grammar

$G = (\{S\}, \{a, b\}, P, S)$ where P is given by

$S \rightarrow a\, S\, b$

$S \rightarrow \wedge$

Then derivation of string *aabb*

$S \Rightarrow a\, S\, b$ By applying $S \rightarrow a \quad S \quad b$

$\Rightarrow aa\, S\, bb$ By applying $S \rightarrow a \quad S \quad b$

$\Rightarrow aa \wedge bb$ By applying $S \rightarrow \wedge$

$= aabb$

So, we can write $S \overset{*}{\Rightarrow} aabb$

11.10.7 Working String or Sentential From

If $w \in L(G)$, then the sequence $S \Rightarrow w_1 \Rightarrow w_2 \Rightarrow \ldots \Rightarrow w_n \Rightarrow w$ is a derivation of sentence w. The string $S, w_1, w_2, \ldots, w_n$ which contain variables as well as terminals are called sentential form of the derivation.

In above example *aabb* is sentence in the language generated by G while $aa\, S\, bb$ is a sentential form. The derivation of a string is complete when working string can not be modified if the final string does not contain any variable.

Notation :

(1) If A is any set, then A^* is a set of all string over A.

(2) A^+ denotes $A^* - \{\wedge\}$ where $\wedge$ is empty string. Empty string which is a string with no symbol at all. It is denoted by $\wedge$.

(3) We denote variables or non-terminals by capital letters, *i.e.*, A, B, C, S, A_1, A_2 are all non-terminals.

(4) We denote terminals by small letters, *i.e.*, $a, b, c, d, \ldots$ are all terminals, arithmetic operator, braces, etc. are all terminals.

(5) $x, y, z, w, \ldots$ denote strings of terminals

(6) $\alpha, \beta, \gamma, \ldots$ denote elements of $(V_N \cup \Sigma)^*$

(7) $X^0 = \wedge$ for any symbol X in $V_N \cup \Sigma$.

☛ If $B \rightarrow \alpha$ is production where $B \in V_N$, then this production is called B-production. If $B \rightarrow \alpha_1 / \alpha_2 / \alpha_3, \ldots$ are B-productions these productions are written as $B \rightarrow \alpha_1, B \rightarrow \alpha_2, B \rightarrow \alpha_3, \ldots$.

Now, we give some examples of grammar and language generated by these grammar.

ILLUSTRATIVE EXAMPLES

1. *If* $G = (\{S\}, \{a\}, \{S \rightarrow aS, S \rightarrow \wedge\}, S)$, *find* $L(G)$.

Solution : Given productions are $S \rightarrow aS$; $S \rightarrow \wedge$

as $S \rightarrow \wedge$ is a production

$S \Rightarrow \wedge$ so $\wedge \in L(G)$

For $n \geq 1$

$$S \Rightarrow aS \Rightarrow aaS \Rightarrow aaaS \overset{*}{\Rightarrow} a^nS \Rightarrow a^n$$

Therefore, $\{a^n : n \geq 0\} \subseteq L(G)$

To show that $L(G) \subseteq \{a^n : n \geq 0\}$. We start $w \in L(G)$.

The derivation of w starts with start symbol S. If we apply $S \to \wedge$, we get $w = \wedge$. If we apply first production $S \to aS$ once at several times and at any stage if we apply $S \to \wedge$, we get a terminal string. Thus, the derivation of w is of the form

$$S \overset{*}{\Rightarrow} a^nS \Rightarrow a^n \wedge = a^n \quad i.e., L(G) \subseteq \{a^n : n \geq 0\}$$

Therefore, $L(G) = \{a^n : n \geq 0\}$

2. *$G = (\{S\}, \{a, b\}, \{S \to a\,S\,a, S \to b\,S\,b, S \to \wedge\}, S)$ find language generated by grammar G.*

Solution : The productions are $S \to a\,S\,a;\ S \to b\,S\,b;\ S \to \wedge$

We know $L(G) = \{a, b\}$ *

All productions are S productions

If we apply $S \to \wedge$

Then

$$S \Rightarrow \wedge \qquad \wedge \in L(G)$$

Let us consider a string $x = a_1a_2 \ldots a_m a_{m'} \ldots a_1$ where each a_i is a or b.

The first production in the derivation $x = a_1a_2 \ldots a_m a_{m'} \ldots a_1$ is $S \to a\,S\,a$ or $S \to b\,S\,b$ according as $a_1 = a$ or b. The subsequent productions are obtained in similar way. The last production $S \to \wedge$. So, $x = a_1a_2 \ldots a_m a_{m'} \ldots a_1 \in L(G)$. $L(G)$ is set of even palindrome over $\Sigma = \{a, b\}$.

3. *Find the language generated by the following grammar*

$$S \to 0\,S\,1\,0\,B\,1;\ B \to 1\,B\,1\,1$$

Solution : In given productions, we observe that all words start with 0 and end with 1. So, we apply any S production they generate equal number of 0's and 1's. If we apply first S production $S \to 0\,S\,1$ several times

So, $\quad S \overset{*}{\Rightarrow} 0^nS\,1^n$

Now, if we apply $S \to 0\,B\,1$ several times, because B production has terminal string $B \to 1$.

$$S \overset{*}{\Rightarrow} 0^nS\,1^n \overset{*}{\Rightarrow} 0^n\,0^m\,B\,1^m1^n$$

But B generates string of 1's only.

So, $\quad S \overset{*}{\Rightarrow} 0^nS\,1^n \overset{*}{\Rightarrow} 0^n\,0^m\,1^k\,1^m\,1^n = 0^{m+n}\,1^{k+m+n} = 0^k\,1^l$

Let $m + n = k,\ k + m + n = l$

So, $\quad L(G) = \{0^k\,1^l : l > k \geq 1\}$

4. *Let $G = (\{S, A\}, \{a, b\}, \{S \to a\,A\,a, A \to a\,A\,a \mid b, S)$, find $L(G)$.*

Solution : Given productions

$$S \to a\,A\,a;\ A \to a\,A\,a;\ A \to b$$

Only S production.

$$S \Rightarrow a\,A\,a \Rightarrow a\,b\,a \quad a\,b\,a \in L(G)$$

Now, $\quad S \Rightarrow a\,A\,a \Rightarrow aa\,A\,aa \Rightarrow a^n\,A\,a^n \Rightarrow a^n\,b\,a^n$

Therefore, $a^nb\,a^n \in L\,(G)$ where $n \geq 1$.

Therefore, $\{a^n\,b\,a^n : n \geq 1\} \subseteq L(G)$

To show $L(G) \subseteq \{a^n\,b\,a^n\}]$

We have only S-production $S \to a\,A\,a$ so we apply this production first. If we apply $A \to b$ we get a terminal string, otherwise we apply $A \to a\,A\,a$ either once or several times. So, we get $a^n\,A\,a^n$. To get a terminal string we replace A by b by applying $A \to b$, so any derivation of the form $S \overset{*}{\Rightarrow} a^n\,A\,a^n \Rightarrow a^n\,b\,a^n \quad n \geq 1$.

Therefore, $\quad L(G) \subseteq \{a^n\,b\,a^n : n \geq 1\}$

Thus, $\quad L(G) = \{a^n\,b\,a^n : n \geq 1\}$

5. *Construct a grammar for the language* $L = \{0000, 0011, 1100, 1111\}$.

Solution : Since, L has a finite number of strings, we can simply list all strings in the language. Thus, $\Sigma = \{0, 1\}$ be the set of terminals, let $V_N = \{S\}$ be the set of variable and S be the start symbol. We have the set of productions.

$$S \to 0000;\ S \to 0011;\ S \to 1100;\ S \to 1111$$

We can have a slightly simple grammar. Let $V_N = \{S, A\}$ be the set of variables and S is the starting symbol. The following set of productions will also specify the language L

$$S \to AA$$
$$A \to 00$$
$$A \to 11$$

So, $G = (\{S, A\}, \{0, 1\}, \{S \to AA, A \to 00, A \to 11\}, S)$

6. *Construct a grammar for the language* $L = \{0^i\,1^{2i} : i \geq 1]$.

Solution : Let $\Sigma = \{0, 1\}$ and $V_N = \{S\}$ with S being the starting symbol. Let the set of productions $S \to 0\,S\,11;\ S \to 011$

To obtain the string 000011111111 as follows

$S \Rightarrow 0\,S\,11$ By applying $S \to 0\,S\,11$
$\Rightarrow 00\,S\,1111$ By applying $S \to 0\,S\,11$
$\Rightarrow 000\,S\,111111$ By applying $S \to 0\,S\,11$
$\Rightarrow 000011111111$ By applying $S \to 011$

7. *Construct a grammar for the language*

$$L = \{a^m b^m c^n : m \geq 1, n \geq 0\}.$$

Solution : $L = L_1 \cup L_2$ where $L_1 = \{a^m b^m : m \geq 1\};\ L_2 = \{a^m b^m c^n : m \geq 1, n \geq 1\}$

Now, we construct L_1 by recursion and L_2 by concatenating elements of L_1 and $c^n \quad n \geq 1$.

Now, we define set of productions

$$S \to X;\ X \to ab;\ X \to a\,X\,b;\ S \to Sc$$

So, $\quad G = (\{S, X\}, \{a, b, c\}, P, S)$ for $m \geq 1, n \geq 0$.

Thus, for example we obtain the string $a^4\,b^4\,c^2$ as follows

$$S \Rightarrow Sc \Rightarrow Scc \Rightarrow a\,X\,b\,c^2 \Rightarrow aaXbbc^2 \Rightarrow aaa \times bbbc^2 \Rightarrow aaaabbbbc^2.$$

8. *Construct a grammar for the language* $L = \{w : w \in \{a, b\}^*$, *the number of a's in w is multiple of 3*$\}$.

Solution : Let $\Sigma = \{a, b\}$ and $V_N = \{S, X, Y\}$ with S being starting symbol. The set of productions

$$S \to bS;\ S \to b;\ S \to aX;\ X \to bX;\ X \to aY;\ Y \to bY;\ Y \to aS;\ Y \to a$$

Now, we derive a string *bbababbab*

$$S \Rightarrow bS \Rightarrow bbS \Rightarrow bbaX \Rightarrow bbabX \Rightarrow bbabaY \Rightarrow bbababY \Rightarrow bbababbY$$
$$\Rightarrow bbababbaS \Rightarrow bbababbab$$

9. *Construct a grammar for language* $L = \{a^m b^n : m, n \geq 1, m \neq n\}$.

Solution : $L_1 \cup L_2$ where $L_1 = \{a^m b^n : m > n\};\ L_2 = \{a^m b^n : m < n\}$

Now, for language L_1 productions are

$$X \to aX;\ X \to aY;\ Y \to a\,Y\,b;\ Y \to ab \qquad \ldots(1)$$

is a set of productions in grammar for L_1 where $\Sigma = \{a, b\}$ is the set of terminals and $V_N = \{A, B\}$ is the set of variables with X being starting symbol.

Now, for language L_2 the productions are :

$$A \rightarrow Ab;\ A \rightarrow Bb;\ B \rightarrow a\,B\,b;\ B \rightarrow ab \qquad \ldots(2)$$

is the set of productions in grammar for L_2, where $\{a, b\}$ is the set of terminals and $\{A, B\}$ is the set of variables, A being the starting symbol.

Now, we add the two productions

$$S \rightarrow X;\ S \rightarrow A$$

to productions in (1) and (2) we have a grammar for L with S being the starting symbol.

Thus, the complete grammar is

$$S \rightarrow X;\ S \rightarrow A;\ X \rightarrow aX;\ X \rightarrow aY;\ Y \rightarrow a\,Y\,b;\ Y \rightarrow ab;$$
$$A \rightarrow Ab;\ A \rightarrow Bb;\ B \rightarrow a\,B\,b;\ B \rightarrow ab$$

11.11 THE CHOMSKY HIERARCHY OF GRAMMARS

Noam Chomsky a founder of formal language theory provided an initial classification into four language type 0 to type 3 in terms of production.

11.11.1 Type 0 Grammar

A type 0 grammar is any phrase structure grammar without any restriction. All grammars we have considered are type 0 grammars. Type 0 includes all productions of the form.

any string $\rightarrow$ any string

$$i.e., \alpha \rightarrow \beta$$

where α is in $(V_N \cup \Sigma)^+$ and β is in $(V_N \cup \Sigma)^*$.

11.11.2 Type 0 Production

A production without any restrictions is called a type 0 production.

11.11.3 Type 1 Grammar

A grammar G is said type 1 grammar in which the left side of each production is not larger than the right side is called context sensitive grammar denoted as CSG or type 1, *i.e.*, a grammar G is said to be type 1 if every production $\alpha \rightarrow \beta$ has the property $|\alpha| \leq |\beta|$ where $|\alpha|$ and $|\beta|$ are length of left side and length of right side respectively.

11.11.4 Type 1 Language

A language generated by type 1 grammar is called context sensitive or type 1 language. The production $A \rightarrow \wedge$ is also allowed in type 1 grammar but in the case S does not appear on right side of any production.

In a context sensitive grammar the production are of the form $\alpha\, A\, \alpha' \rightarrow \alpha\, \beta\, \alpha'$, where α is left context and α' is right context.

- The name "context sensitive" comes from the fact that we can replace the variable A by □ in a word only when A lies between α and α'.

11.11.5 Type 2 Grammar

A grammar G is said to be of type 2 if every production is of the form $A \rightarrow \beta$, *i.e.*, the left side is a variable and right side is a word in one or more symbols, or $A \in V_N$ and $\beta \in (V_N \cup \Sigma)$.

- Type 2 grammar is also called context free grammar.

- A grammar G is said to be context free if the production are of the form $A \rightarrow \beta$. The name context free comes from the fact that we can replace the variable A by □ regardless of where A appears.

11.11.6 Type 3 Grammar

A grammar G is said to be type 3 or regular grammar if all the productions are of the form $A \rightarrow a$ or $A \rightarrow aB$, *i.e.*, the left side is a single variable and the right side is either a single terminal or a terminal followed by variable. $A \rightarrow \wedge$ is allowed in type 3 grammar but in this case S does not appear on the right side of any production.

For Example :

(a) Determine the type of the grammar G which consist of the productions
$S \rightarrow aA,\ A \rightarrow a\ A\ B,\ B \rightarrow b,\ A \rightarrow a$.

Solution : (a) $S \rightarrow aA$ is type 3 production, $A \rightarrow a\ AB$ is type 2 production, $B \rightarrow b$ and $A \rightarrow a$ are type 3 production. Highest type number is type 2. So, given grammar is Type 2 grammar.

(b) $A \rightarrow aB,\ B \rightarrow aA,\ B \rightarrow a,\ B \rightarrow b,\ A \rightarrow aB,\ A \rightarrow a$

Solution : In the given grammar each production is of type 3 so given grammar is type 3 or regular grammar.

(c) $S \rightarrow a\ AB,\ AB \rightarrow bB,\ B \rightarrow b,\ A \rightarrow aB$

Solution : $S \rightarrow aA\ B$ is type 2 production, $AB \rightarrow bB$ is type 1 production, *i.e.*, length of right side equal to length of left side, $B \rightarrow b$ and $A \rightarrow aB$ are of type 3 production. In given grammar, the highest type number 1 so given grammar is type 1 grammar.

(d) $S \rightarrow a\ A\ B,\ AB \rightarrow C,\ A \rightarrow b,\ B \rightarrow AB$

Solution : $S \rightarrow a\ A\ B$ is type 2 production, $AB \rightarrow C$ is type 0 production, $A \rightarrow b$ is type 3 production, $B \rightarrow AB$ is type 2 production. Highest type number is 0. So, given grammar is type 0 grammar.

11.12 RELATION BETWEEN LANGUAGES

The recursively enumerable language (L_{RE}), *i.e.*, type 0, the context sensitive languages (L_{CS}), *i.e.*, type 1, the context free language (L_{CF}), *i.e.*, type 2 and regular language (L_{REG}), *i.e.*, type 3 one way of exhibiting the relationship between these families is by Chomesky hierarchy. Each language family of type i is a proper subset of the family of type $i - 1$. The following diagram exhibits the relationship clearly.

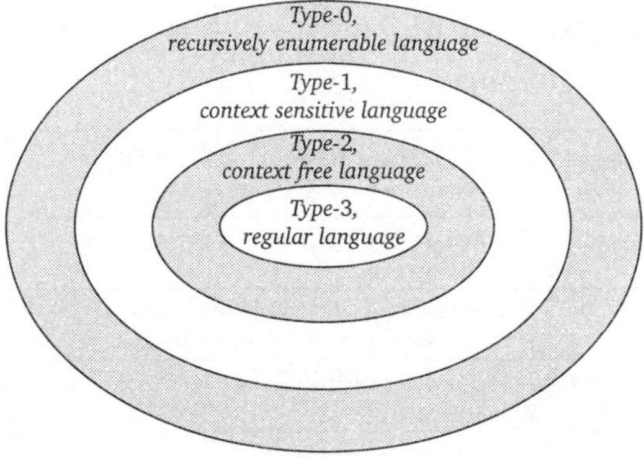

Fig. (9)

- $L_{REG} \subseteq L_{CF} \subseteq L_{CS} \subseteq L_{RE}$, *i.e.*, type 3 $\subseteq$ type 2 $\subseteq$ type 1 $\subseteq$ type 0

Theorem 1 : *Every context sensitive language L recursive.*

Proof : Let us consider context-sensitive language L with associated context-sensitive grammar G. Then, derivation of string w

$$S \Rightarrow \alpha_1 \Rightarrow \alpha_2 \Rightarrow \alpha_3 \Rightarrow \ldots \Rightarrow \alpha_n = w$$

We assume that without loss of generality that all sentential form in a single derivation are different, *i.e.*, $\alpha_i \neq \alpha_j$ The number of steps in any derivation is bounded function of $|w|$ and we know that

$$|x_j| \leq |x_{j+1}|$$

because G is context sensitive or non-contracting in the sense that the length of successive sentential forms can never decrease. Now, we need to add is that there exist some m, depending only on G and w such that

$$|\alpha_j| < |\alpha_{j+m}|$$

for all j with $m = m(|w|)$ a bounded function of $|V_N \cup \Sigma|$ and $|w|$. This follows because the finiteness of $|V_N \cup \Sigma|$ implies that there are only a finite number of strings of a given length. Therefore, the length of a derivation of $w \in L$ is at most $|w| m(|w|)$

11.12.1 Algorithm to check whether $w \in L$

We check all derivation of length up to $|w| m(|w|)$ since set of productions of G is finite. There are only a finite number of these. If any of them gives w, then $w \in L$ otherwise it is not.

Theorem 2 : *There exists a recursive set which is not a context-sensitive language over* $\Sigma = \{a, b\}$.

Proof : Consider the set of all context-sensitive grammar on $\Sigma = \{a, b\}$. We can write the elements of Σ^* as a sequence, *i.e.*, *a, b, aa, ab, ba, bb, aaa, ...* in this case *aab* is 9th element.

Every grammar is defined in terms of a finite set of alphabet and a finite set of productions. We can write all context sensitive grammars over Σ as a sequence, *i, e,.* G_1, G_2, G_3.

We define $X = \{w_i \in \Sigma^* : w_i \notin L(G_i)\}$. X is recursive if $w \in \Sigma^*$. Then we can find i such that $w = w_i$. This can be done in a finite number of steps. For example if $w = aba$, then $w = w_{20}$. As G_{20} is context sensitive, we have an algorithm to test whether $w = w_{20} \in L(G_{20})$.

We prove by contradiction that X is not a context sensitive language. Let $X = L(G_n)$ for some n. Consider w_n the nth element of Σ^*. By definition of X, $w_n \in X$ implies $w_n \notin L(G_n)$. This contradicts $X = L(G_n)$. $w_n \notin X$ implies $w_n \in L(G_n)$ and once again this contradicts $X = L(G_n)$. Thus, $X \neq L(G_n)$ for any n. X is not context sensitive language.

11.12.2 Language and Automata

Type	Name of language generated	Production restriction $X \to Y$	Acceptor
0	Recursively enumerable	X = any string with non-terminals Y = any string	Turning Machine (TM)
1	Context sensitive	X = any string with non-terminals Y = any string as long as or longer than X	Linear Bounded Automata (LBA)
2	Context free	X = one non-terminal Y = any string	Push Down Automata (PDA)
3	Regular	X = one non-terminals $Y = a\,B$ or $y = a$ where a is terminal and B is non-terminal	Finite Automata

Exercise 11.1

1. Test whether (i) 110011, (ii) 10110, (iii) 101011, (iv) 11010 are accepted by the following transition system

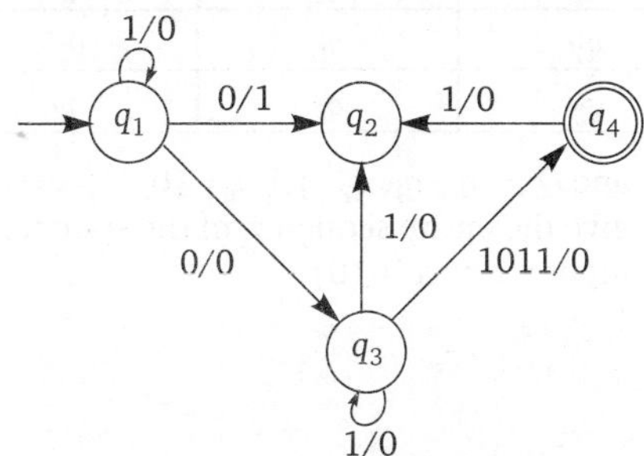

2. Construct a Mealy Machine which can output EVEN, ODD according as the total number of 1's encountered is even or odd. The input symbols are 0 and 1.
3. Construct a Mealy machine which is equivalent to the Moore machine given in the table :

Present State	Next State		
	Input a	Input b	Output
$\to q_1$	q_2	q_3	0
q_2	q_4	q_3	1
q_3	q_3	q_2	1
q_4	q_1	q_4	1

4. Construct a Moore Machine equivalent to the Mealy Machine given in the table

Present State	Next State			
	Input 0		Input 1	
	State	Output	State	Output
$\to q_1$	q_1	1	q_2	0
q_2	q_4	1	q_4	0
q_3	q_2	0	q_3	1
q_4	q_4	1	q_1	1

5. Find the language generated by following grammars :
   (i) $S \to aSb \mid aAb,\ A \to bA \mid b$
   (ii) $S \to 0S1 \mid 0A \mid 0 \mid 1B \mid 1,\ A \to 0A \mid 0,\ B \to \mid B \mid 1$
   (iii) $S \to AA$
   (iv) $S \to 0S \mid 1S \mid 0 \mid 1$
   (v) $S \to Xa,\ X \to aX \mid bX \mid \wedge$
   (vi) $S \to XaaX,\ X \to aX \mid bX \mid \wedge$
   (vii) $S \to aAa,\ A \to aAa \mid b$
6. Construct the grammar accepting each of the following sets :
   (i) The set of all palindrome over {0, 1}
   (ii) $L = \{w \subset w^T : w \in \{0, 1\}\}$
   (iii) $L = \{0^m 1^n 0^n 1^m : m, n \geq 1\}$
   (iv) The set of all strings over {0, 1} ending in 0.
   (v) The set of all strings over {0, 1} beginning with 0.
7. Find the highest type number which can be applied to the following grammar :
   (i) $S \to A0,\ A \to 0 \mid B0,\ B \to 011$
   (ii) $S \to ASB \mid 0,\ A \to 0A \mid 1$
   (iii) $S \to 0A \mid 0$
   (iv) $S \to 0AA \mid 1,\ AA \to 0 \mid 1,\ AA \to 0A$
   (v) $AS \to B,\ B \to 0B \mid 1$
   (vi) $S \to ASB \mid D,\ A \to aA$
8. State whether the following statements are true or false. Justify your answer with a proof.
   (i) If G_1 and G_2 are equivalent, then they are of same type.
   (ii) If L is a finite subset of Σ^*, then L is context free language.
   (iii) If L is a finite subset of Σ^*, then L is a regular language.
9. If each production in grammar G has same variable on its right side, what can you say about $L(G)$?
10. Construct the Moore Machine equivalent to the following Mealy Machine

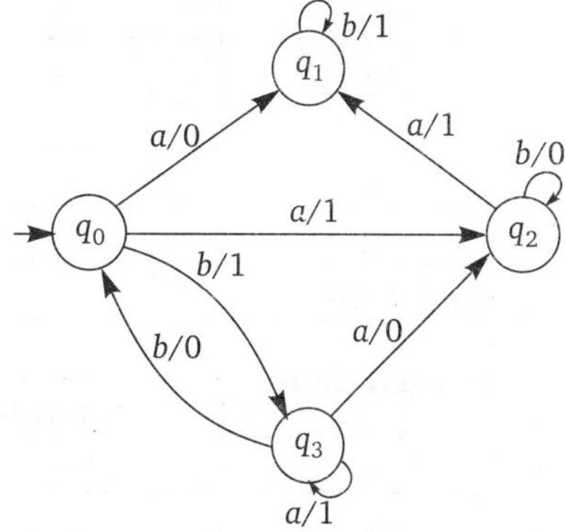

11. Construct the context free grammar generating following language $\{a^n b^n : n \geq 1\} \cup \{a^m b^{2m} : m \geq 1\}$
12. If $G = (\{S\}, \{0, 1\}, \ \{S \to 0S1, S \to \wedge\}, S)$, find $L(G)$.

13. Find the language generated by the grammar
$S \to AB, A \to A1/0, B \to 2B \mid 3.$
Can the above language be generated by grammar of higher type?

14. Show that a grammar consisting of productions from the form $A \to xB \mid y$ where x, y are in Σ^* and $A, B \in V_N$ is equivalent to a regular grammar.

15. For the finite state machine whose transition function δ is given in the table

State	Inputs	
	0	1
$\to q_0$	q_2	q_1
q_1	q_3	q_0
q_2	q_0	q_3
q_3	q_1	q_2

and $Q = \{q_0, q_1, q_2, q_3\}$, $q = \{0, 1\}$, $F = \{q_0\}$, give the entire sequence of the state for the input string 110101.

ANSWERS

1. (i) No (ii) No (iii) Yes (iv) No

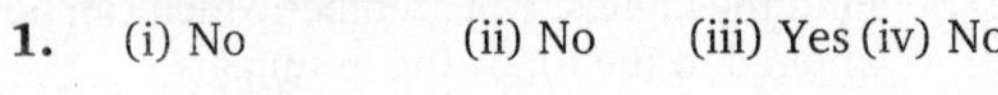

2.

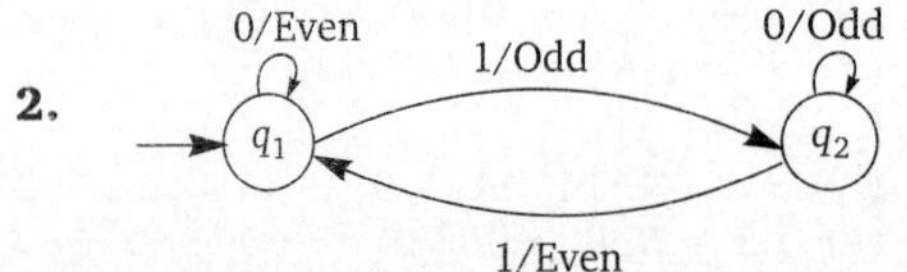

3.

Present State	Next State			
	Input 0		Input 1	
	State	Output	State	Output
$\to q_1$	q_2	1	q_3	1
q_2	q_4	1	q_3	1
q_3	q_3	1	q_2	1
q_4	q_1	0	q_4	1

4.

Present State	Next State		
	Input 0	Input 1	Output
$\to q_1$	q_1	q_2	1
q_2	q_{41}	q_{40}	0
q_3	q_2	q_3	1
q_{40}	q_3	q_1	0
q_{41}	q_3	q_1	1

Revised table

Present State	Next State		
	Input 0	Input 1	Output
$\to q_0$	q_1	q_2	0
q_1	q_1	q_2	1
q_2	q_{41}	q_{40}	0
q_3	q_2	q_3	1
q_{40}	q_3	q_1	0
q_{41}	q_3	q_1	1

5. (i) $L(G) = \{a^m b^n : n > m \geq 1\}$

(ii) $L(G) = \{0^m 1^n : m \neq n$ and at least one of m and $n \geq 1\}$ clearly $0^m \in L(G)$ and $1^n \in L(G)$ where $m, n \geq 1$

(iii) $L(G) = \phi$ since there is no terminal in grammar.

(iv) $L(G) = \{0, 1\}^+$

(v) $L(G) = (a + b)^* a$

(vi) $L(G) = (a + b)^* aa (a + b)^*$

(vii) $L(G) = a^n b a^n$

6. (i) $S \to 0S0$, $S \to 1S1$, $S \to a$, $S \to b$, $S \to \wedge$

(ii) $S \to 0S0 \mid 1S1 \mid C$

(iii) $S \to 0S1$, $S \to 0A1$, $A \to 1A0$, $A \to 10$

(iv) $S \to 0S$, $S \to 1S$, $S \to 0$

(v) $S \to 0A$, $A \to 0A$, $A \to 1A$, $A \to a$, $A \to b$

7. (i) $S \to A0$ is type 2, $A \to 0$ is type 3, $A \to B0$ is type 2, $B \to 011$ is type 2. So, type 2 is the highest type number.

(ii) $S \to ASB$ is type 2, $S \to 0$ is type 3, $A \to 0A$ is type 3, $A \to 1$ is type 3. So, type 2 is highest type number.

(iii) $S \to 0A$ is type 3, $S \to 0$ is type 3. So, highest type number is 3.

(iv) $S \to 0AA$ is type 2, $S \to 1$ is type 3, $AA \to 0$ is type 0, $AA \to 1$ is type 0, $AA \to 0A$ is type 1. So, highest type number is type 0.

(v) $A\,S \to B$ is type 0, $B \to 0B$ is type 1, $B \to 1$ is type 3. Highest type number is type 0.

(vi) $S \to ASB$ is type 2, $S \to D$ is type 2, $A \to aA$ is type 3. Highest type number is type 2.

8. (i) False. Two grammars of different type may generate the same language. For example $S \to aS \mid bS \mid a \mid b$. It generate language $L(G) = \{a, b\}^+$. All productions in the grammar are type 3, so grammar is type 3.

Let us consider another grammar $S \to SS \mid aS \mid bS \mid a \mid b$ which is generated also language $L(G) = \{a, b\}^+$. But grammar is type 2 since $S \to SS$ is type 2 production. Both grammar are of different types but generate same language.

(ii) True. Since, L is a finite subset of Σ^* so $L = \{w_1, w_2, \ldots, w_n\}$. Then, $G = (\{S\}, \Sigma, P, S)$ where P consist of $S \to w, \mid w_2 \mid, \ldots \mid w_n$.

(iii) True. It is enough to show that $\{w\}$ is regular where $w \in \Sigma^*$. Let $w = a_1 a_2, \ldots, a_n$. Then, the grammar whose productions are $S \to a_1 A_1$, $A_1 \to a_2 A_2, \ldots, A_{n-1} \to a_n$ generate $\{w\}$.

9. If each production in grammar G has some variable on its right side, then $L(G) = \phi$. We get a variable on application of each production and so no terminal string result.

10. Fist convert the Mealy Machine into transition table. Then use procedure to convert Mealy Machine into Moore Machine. The table of the Moore Machine

Present State	Next State		
	Input a	**Input b**	**Output**
$\to q_{00}$	q_{21}	q_3	0
q_{01}	q_{21}	q_3	1
q_1	q_{00}	q_1	1
q_{20}	q_1	q_{20}	0
q_{21}	q_1	q_{20}	1
q_3	q_{20}	q_{01}	0

12. $L(G) = 0^n 1^n$

13. The language generated by given grammar is $L(G) = \{01^m2^n3 : m, n \geq 1\}$. This language can be generated by type 3 grammar.

14. Let the given grammar be G_1. The production $A \to x\,B$, where $x = a_1a_2\ldots a_n$ is replaced by $A \to a_1A_1, A_1 \to a_2A_2, \ldots, A_{n-1} \to a_nB$. The production $A \to y$, where $y = b_1b_2\ldots b_m$ is replaced by $A \to b_1B_1, B_1 \to b_2B_2\ldots B_{m-1} \to b_m$. The grammar G_2 whose productions are the new productions is regular and equivalent to G_1.

15. $\delta(q_0, \overset{\downarrow}{1}10101) = \delta(q_1, \overset{\downarrow}{1}0101) = \delta(q_0, \overset{\downarrow}{0}101) = \delta(q_2, \overset{\downarrow}{1}01) = \delta(q_3, \overset{\downarrow}{0}1)$

$$= \delta(q_1, \overset{\downarrow}{1}) = \delta(q_0, \wedge) = q_0$$

Hence, $q_0 \xrightarrow{1} q_1 \xrightarrow{1} q_0 \xrightarrow{0} q_2 \xrightarrow{1} q_3 \xrightarrow{0} q_1 \xrightarrow{1} q_0$

11.13 CONSTRUCTION OF DETERMINISTIC FINITE AUTOMATA FROM NONDETERMINISTIC (NFA)

Here, we come across to a fundamental question. In what sense are DFA and NFA's different? There is a difference in their definition but this does not imply that there is any essential distinction between them. To explore this question, we introduce the notion of equivalence between automata.

Definition : *Two finite acceptors M and M_1 are said to be equivalent if $L(M) = L(M_1)$, i.e., if both accept the same language.*

Definition : ***T(M)*** **:** *The set accepted by an automation M (deterministic or non-deterministic) is the set of input strings accepted by M. It is denoted by $T(M)$.*

Theorem 1. *Let L be the language accepted by non-deterministic finite automata $M = (Q, \Sigma, \delta, q_0, F)$. Then there exists a deterministic finite automata $M' = (Q', \Sigma, \delta', q_0', F')$ which also accepts L.*

Proof : Let $M = (Q, \Sigma, \delta, q_0, F)$ be NDFA accepts L we can construct a DFA M' as

$$M' = (Q', \Sigma, \delta, q_0', F')$$

where

(i) $Q' = 2^Q$ any state in Q' is denoted by $[q_1, q_2, \ldots, q_k]$
where $q_1, q_2, \ldots, q_k \in Q$

(ii) $q_0' = [q_0]$

(iii) F' is the set of all subsets of Q containing an element of F.

11.13.1 Construction of Q', q_0' and F'

M is initially at q_0. But by applying any input symbol say a, M can reach any of the states in $\delta(q_0, a)$. Just after the application of the input symbol a we require all the possible state that M can reach after the application of a. Hence, M' has to remember all these possible states at any time. Hence, the states of M' are defined as subsets of Q.

As M starts with start state q_0, q_0' defined as $[q_0]$. A string w belongs to $T(M)$. If a final state is one of the possible state M reaches on processing w. So, a final state in M', *i.e.*, an element of F'.

Construction of δ'

(iv) $\delta'([q_1, q_2, \ldots, q_k], a) = \delta(q_1, a) \cup \delta(q_2, a) \cup \ldots \cup \delta(q_k, a)$

Then, $\delta'([q_1, q_2, \ldots, q_k], a) = [p_1, \ldots, p_k]$ If and only if

$$\delta'\{q_1, q_2, \ldots, q_k\}, a) = \{p_1, p_2, \ldots, p_k\}$$

Before proving $L = T(M')$ we prove an auxiliary result.

$$\delta'(q_0', y) = [q_1, \ldots, q_k]$$

if and only if $\qquad \delta(q_0, y) = \{q_1, \ldots, q_k\}$ for all $x \in \Sigma^*$ $\qquad$ ...(i)

We prove this by induction on $|y|$

$$\delta'(q'_0, x) = [q_1, q_2, ..., q_k] \text{ if } \delta(q_0, x) = \{q_1, ..., q_k\} \quad ...(ii)$$

When $|y| = 0$, then $\delta(q_0, \wedge) = \{q_0\}$ and by definition of δ'

$$\delta'(q'_0, \wedge) = q'_0 = [q_0]$$

Equation (i) is true for y with $|y| = 0$. This is the basis for induction.

Now, we assume equation (ii) is true for all steps z with $|z| \leq n$.

Let y be a string of length $n + 1$. We can write y as za where $|y| = n$ and $a \in \Sigma$.

11.14 REGULAR SETS AND REGULAR GRAMMAR

Regular expressions are useful for representing certain sets of strings in an algebraic fashion. A language is regular if there exists a finite automata for it. Therefore, every regular language can be described by some DFA or some NDFA, such a description can be very useful for example if we want to show the logic by which we decide if a given string is in a certain language. One way of describing regular languages is via notation of regular expression. This notation involves a combination of strings of symbols from some alphabet Σ, parentheses and the operators +,. and *.

Definition : Recursive definition of regular expression over Σ follows the following rules :

Rule 1 : Every letter of Σ can be made into regular expression by writing it in bold face, ^ itself is a regular expression.

Rule 2 : If R_1 and R_2 are two regular expressions, then

(i) (R_1) is a regular expression.

(ii) R_1R_2 is also a regular expression, *i.e.*, the concatenation of two regular expressions R_1 and R_2 is a regular expression.

(iii) $R_1 + R_2$ is also a regular expression, *i.e.*, the union of two regular expressions R_1 and R_2 is a regular expression.

(iv) R^* is also a regular expression, *i.e.*, the iteration (closure) of regular expression R is also a regular expression.

Rule 3 : A string is a regular expression over Σ and is obtained recursively by rule 2 once or several times.

- We use *a* for regular expression just to distinguish it from the symbol (string) *x*.
- The parentheses used in rule 2 (i) influence the order of evaluation of a regular expression.
- In the absence of parentheses, hierarchy of operations as iteration (closure), concatenation and union. This hierarchy is similar to that followed for arithmetic expressions (exponentiation, multiplication and addition).

11.14.1 Regular Set

Any set represented by a regular expression is called a regular set. For example, If $\{0, 1\}, \in \Sigma$, then

(a) $\{0\}$ is represented by regular expression $\mathbf{R} = 0$.

(b) $\{0, 1\}$ is represented by regular expression $\mathbf{R} = 0 + 1$

(c) $\{01\}$ is represented by regular expression $\mathbf{R} = \mathbf{a}^*$.

(d) $\{^\wedge, a, aa, aaa, aaaa, ...\}$ is represented by regular expression a^*.

ILLUSTRATIVE EXAMPLES

1. *Find a regular expression R over* $\Sigma = \{a, b\}$

(i) $L_1 = \{b^m a b^n, m > 0, n > 0\}$ *(ii)* $L_2 = \{a, ab, ab^2, ...\}$

(iii) $L_3 = \{a^m b^n, m > 0, n > 0\}$

Solution : (i) L_1 consists of all words with exactly one a which is neither the first nor last letter of the word, *i.e., there is one or more b's* before and after a.

Thus, regular expression for $R = bb * abb *$.

(ii) L_2 consists of all words beginning with an a followed by one or more b's. Thus regular expression for L_2 is $R = ab *$.

(iii) L_3 consists of all words beginning with one or more a's followed by one or more b's. Thus regular expression for L_3 is $R = aa * bb *$ or $R = a * ab * b$ is also a solution.

2. *Describe the following sets by regular expression R over* $\Sigma = \{a, b\}$

(a) L_1 = *the set of all strings which consist of all words with exactly two a's.*

(b) L_2 = *the set of all strings which consist of all words beginning with 'a' and followed by zero or more b's.*

(c) L_3 = *will consist of all words beginning and ending in 'a' and enclosing one or more b's.*

(d) L_4 = *will consist of all words with exactly two a's.*

(e) $L_5 = \{\wedge, a, aa, aaa, aaaa, \ldots\}$

(f) $L_6 = \{\wedge, ab\}$

(g) $L_7 = \{abb, a, b, bba\}$

(h) $L_8 = \{ab\ ba\}$

(i) L_9 = *A language of all words without a double a*

(j) L_{10} = *The set of all strings of a's and b's that at some point contain a double letter.*

(k) $L_{11} = \{\wedge, aa, bb, aaaa, aabb, abab, abba, baab, baba, bbaa, bbbb, aaaaaa, aaaabb, aaabab, \ldots\}$

Solution : (a) The regular expression for L_1 is $\mathbf{R = b * ab * ab *}$

(b) The L_2 is obtained by concatenating a and b^*. Therefore regular expression for L_1 is $R = ab *$.

(c) L_3 is represented by regular expression $\mathbf{R = abb * a}$

(d) L_4 is represented by regular expression $\mathbf{R = a * ba * ba *}$

(e) Any element of L_5 is either $\wedge$ or concatenation of 0's so the regular expression for L_5 is $\mathbf{R = a *}$.

(f) The set $L_6 = \{\wedge, ab\}$ is represented by $\mathbf{R = \wedge + ab}$

(g) The set $L_7 = \{abb, a, b, bba\}$ is represented by regular expression $\mathbf{R = abb + a + b + bba}$

(h) The set $L_8 = \{ab\ ba\}$ is represented by the regular expression $\mathbf{R = ab + ba.}$

(i) L_9 is a language without double a and is represented by regular expression $\mathbf{R = b * (abb^*) * (\wedge + a)}$. All words start here some b's. Then come repeated factors of the form $abb *$, *i.e.*, an a followed by at least one b. Then we finish up with a final a or we leave the last b's.

(j) The regular expression for the language L_{10} is $R = (a + b) * (aa + bb) (a + b) *$. This is the set of strings of a's and b's that at some point contain a double letter.

(k) $L_{11} = \{\wedge, aa, bb, aaaa, aabb, abab, abba, \ldots\}$ not thast L_{11} do not have the same number of a's and b's just an even number quantity of each., $R = [aa + bb + (ab + ba) (aa + bb) * (ab + ba)] *$.

11.15 PUMPING LEMMA FOR REGULAR SETS

Pumping Lemma gives a necessary condition for an input string to belong to a regular set. The result is called pumping lemma since it gives a method of pumping many input strings from a given string. Pumping lemma is used to show that certain sets are not regular. The pumping lemma is based on the observation that in a transition graph with n vertices any walk of length n or longer must repeat some vertex that contain a cycle.

Theorem 1 : *Let $L(G)$ be an infinite regular language. Then there exists some positive integer n such that any $w \in L(G)$ with $|w| \geq n$ can be decomposed as*

$$w = xyz$$

with $|xy| \leq n$ and $|y| \geq 1$

such that $w_i = xy^iz$ is also in $L(G)$ for all $i = 0, 1, 2, \ldots$

A long string in $L(G)$ can be broken into three parts in such a way that an arbitrary number of repetition of middle part produces another string in L. We can say that the middle string is "pumped" hence the result is called pumping lemma.

Proof : If $L(G)$ is regular, then there exists a finite automata for it that recognizes it. Let such DFA have states $q_0, q_1, q_2, q_3, \ldots, q_n$. Now we take a string w in $L(G)$ such that $|w| \geq n = m + 1$. Since $L(G)$ is assumed as infinite language.

Let us consider the set of states. The automation goes through as it processes w. Let $q_0, q_a, q_b, \ldots, q_f$.

Since this sequence has exactly $|w| + 1$ entries, at least one state must be repeated and such a repetition must start no later than the nth move. Thus the sequence must look like $q_0, q_a, q_b, \ldots, q_s, \ldots, q_s, \ldots, q_f$.

Now consider the substring x, y, z of w such that

$$\delta^* (q_0, x) = q_s$$
$$\delta^* (q_s, y) = q_s$$
$$\delta^* (q_s, z) = q_f$$

with $|xy| \leq m + 1 = n$ and $|y| \geq 1$

Then $\delta^* (q_0, xz) = q_f$ as well as $\delta^* (q_0, xy^2z) = q_f$

$\delta^* (q_0, xy^3z) = q_f$ so on completing the proof of the theorem.

11.16 APPLICATION OF PUMPING LEMMA

Pumping Lemma is used to prove that certain sets are not regular. The following steps are needed for proving that a given set is not regular.

WORKING RULES

Step 1 : Assume that given language is regular and n be the states in corresponding finite automata.

Step 2 : Choose a string w such that $|w| \geq n$ breaking the string into three substring x, y, z, i.e., $w = xyz$ with $|xy| \leq n$, $|y| \geq 1$.

Step 3 : Find a suitable integer i such that $xy^iz \notin L$. This is a contradiction. Hence L is not regular.

ILLUSTRATIVE EXAMPLES

1. *Show that $L = \{a^nb^n : n \geq 1\}$ is not regular.*

Solution : Step 1 : Let L is regular and n be the number of states in corresponding FA.

Step 2 : Let $w = a^nb^n$. Then $|w| = 2n > n$ and

Let $w = xyz$ with $|xy| \leq n$ and $|y| \geq 1$

Step 3 : To find $xy^iz \notin L$, substring y can be in any of the following forms

Case (i) : y has a's, i.e., $y = a^m$ for some $m \geq 1$

Case (ii) : y has b's, *i.e.*, $y = b^k$ for some $k \geq 1$

Case (iii) : y has both a's and b's, *i.e.*, $y = a^m b^k$

An in case (i) if $i = 0$, as $xyz = a^n b^n \Rightarrow xz = a^{n-m} b^n$ as $m \geq 1$, $n - m \neq n$, so $xy^i z \neq L$.

As in case (ii) if $i = 0$, as $xyz = a^n b^n \Rightarrow xz = a^n b^{n-k}$ as $k \geq 1$, $n - k \neq n$, then $xy^i z \notin L$

As in case (iii) if $xyz = a^n b^n$, then $xyz = a^{n-m} a^m b^k b^{n-k}$

$$xy^2 z = a^{n-m} a^m b^k a^m b^k b^{n-k}$$

$xy^2 z$ is not of the form $a^n b^n$, so $xy^2 z \neq L$

2. *Show that set $L = \{a^{n^2} : n \geq 1\}$ is not regular.*

Solution : Step 1 : Suppose L is regular. Let n be the number of states in FA. Accepting L.

Step 2 : Let $w = a^{m^2}$. Then $|w| = m^2 > n$ and let $w = xyz$ where $|xy| \leq n$ and $|y| \geq 1$.

Step 3 : Let us consider $xy^2 z$.

Now $|xyz| = m^2$

$$m^2 = |x| + |y| + |z| \leq |xy^2 z|$$

As $|xy| \leq m$, $|y| \leq m$, therefore $|xy^2 z|$

$$|xy^2 z| = |x| + 2|y| + |z| \leq m^2 + m$$

$$m^2 < |xy^2 z| \leq m^2 + m$$

$$m^2 < |xy^2 z| < m^2 + m + m + 1$$

$$m^2 < |xy^2 z| < (m + 1)^2$$

Hence, $|xy^2 z|$ lies between m^2 and $(m + 1)^2$ but not equal any one of them so $xy^2 z \in L$.

3. *Show that set $L = \{a^p : p$ is prime number$\}$ is not regular.*

Solution : Step 1 : Suppose that L is regular and we get a contradiction. Let n be the number of states in FA accepting L.

Step 2 : Let p be a prime number and greater than n. Let $w = a^p$. By pumping lemma $w = xyz$ and $|w| = p$. With $|xy| \leq n$ and $|y| > 0$ and x, y, z are simply strings of a's. Let $y = a^k$ for same $k \geq 1$ and $k \leq n$.

Step 3 : Let $i = p + 1$. Then $|xy^i z| = |xyz| + |y^{i-1}| = p + (i - 1)k = p + pk = p(1 + k)$.

We know that prime number is not divisible. $p(1 + k)$ is not a prime number. So $xy^i z \notin L$. This is a contradiction. Thus L is not regular.

11.17 CONTEXT FREE LANGUAGE

Context free language is applied in parser design. It is also useful for describing block structure in programming language. Context free grammars are capable of describing most, but not all of syntax of programming language. A context-free grammar CFG is a collection of three things.

1. An alphabet Σ of letters called terminal from which we are going to make strings that will be the word of a language.
2. A set of symbols called nonterminals. One of which is the symbol S called start symbol.
3. A finite set of productions of the form

   One nonterminal $\rightarrow$ finite string of terminals and/or nonterminal

i.e., a grammar G is context free. If every production is of the form $A \rightarrow \alpha$ where $\alpha \in (V_N \cup \Sigma)^*$ and $A \in V_N$

ILLUSTRATIVE EXAMPLES

1. *The grammar $G = (\{S\}, \{a, b\}, P, S)$ with productions*

$$S \to a\,S\,a$$
$$S \to b\,S\,b$$
$$S \to \wedge$$

is context free. A typical derivation in this grammar is

$S \Rightarrow a\,S\,a$	*By production 1*
$S \Rightarrow aa\,S\,aa$	*By production 1*
$S \Rightarrow aab\,S\,baa$	*By production 2*
$S \Rightarrow aab \wedge baa$	*By production 3*
$S \Rightarrow aabbaa$	

It is clear that

$$L(G) = \{ww^R : w \in (a, b)^*\}$$

2. *Let the terminal be a and b. The nonterminals be S, X and Y and productions be $G = (\{S, X, Y\}, \{a, b\}, P, S)$*

$$S \to X$$
$$S \to Y$$
$$X \to \wedge$$
$$Y \to aY$$
$$Y \to bY$$
$$Y \to a$$
$$Y \to b$$

All the words in this language are either of type X, if the first production in their derivation is

$$S \to X$$

or type Y. If the first production in their derivation is

$$S \to Y$$

The only possible continuation for words of type X is one production

$$X \to \wedge$$

Therefore, $\wedge$ is the only word of type X.

Any string of a's and b's except the null string can be produced from Y. Putting together the type X and the type Y words, we see that the total language generated by this CFG is all strings of a's and b's, null or otherwise. The language generated is $(a + b)^*$.

3. *The grammer G with productions*

$$S \to abY$$
$$X \to aaYb$$
$$Y \to bbXa$$
$$X \to \wedge$$

is context free language. Give the derivation of the word *abbbaabba*.

$$\begin{aligned} S &\Rightarrow abY \\ &\Rightarrow abbbXa \\ &\Rightarrow abbbaaYb \\ &\Rightarrow abbbaabbXa \\ &\Rightarrow abbbaabb\wedge a \\ &\Rightarrow abbbaabba \end{aligned}$$

11.18 CHOMESKY NORMAL FORM

A context free grammar which has only productions of the form

variable $\rightarrow$ *string of exactly two variables*

or, of the form *variable* $\rightarrow$ *one terminal*

is said to be in Chomesky normal form or CNF.

Reduction to Chomesky Normal Form : For any context free language L, the non $\wedge$ words of L can be generated by a grammar in which all productions are in CNF.

Step 1 : Eliminate all unit and null production

Step 2 : Elimination of terminals on RHS. Then

$$G_1 = (V'_N, \Sigma, P', S)$$

where P' and V'_N are constructed as follows

(i) All the productions of the form $X \rightarrow b$ or $X \rightarrow AB$ included in P'. All the variables are included in V'_N.

(ii) Consider $A \rightarrow Y_1Y_2....Y_n$ with some terminals on RHS. If $Y_1 = a_i$ (say) add a new variable X_a to V'_N and $X \rightarrow a_i$ to P'.

In production $A \rightarrow Y_1Y_2....Y_n$ every terminal on RHS is replaced by the corresponding new variable and resulting production is added to P'. Thus new CFG will be $G_1 = (V'_N, \Sigma, P', S)$.

Step 3 : Reduce the number of variables on RHS : For any production in P' the RHS consists of either a single terminal or two or more variables. Then new CFG will be $G_2 = (V''_N, \Sigma, P'', S)$ as follows

(i) All productions in P' are added to P''. If they are in Chomesky normal from. All variable in V'_N added to V''_N.

(ii) Consider $A \rightarrow Y_1Y_2...Y_k$ is a new production where $k \geq 3$ and all Y_i are variables. Then we reduce this production as

$$A \rightarrow Y_1X_1,\ X_1 \rightarrow Y_2X_2, ..., X_{k-2} = X_{k-1}Y_k$$

and new variable and production $X_1, X_2, ... X_{k-2}$ are added to V''_N and P'' respectively. Then the new CFG G_2 in Chomesky normal form.

ILLUSTRATIVE EXAMPLES

1. *Reduce the following grammar in Chomesky normal form*

$$S \rightarrow aSa \mid bSb \mid a \mid b \mid a\ a \mid b\ b$$

Solution : Step 1: There is no unit and null production in the grammar.

Step 2 : Let $G = (V'_N, \{a, b\}, P', S)$ where P' and V'_N are constructed as

(i) $S \rightarrow a$ and $S \rightarrow b$ are in Chomesky normal form included in P'

(ii) $S \rightarrow aSa$ gives $S \rightarrow ASA$ and $A \rightarrow a$

$S \rightarrow bSb$ gives $S \rightarrow B\,S\,B$ and $B \rightarrow b$

$S \rightarrow aa$ gives $S \rightarrow AA$

$S \rightarrow bb$ gives $S \rightarrow BB$

$V'_N = \{S, A, B\}$

P' consists of

$S \rightarrow ASA$

$S \rightarrow BSB$

$A \rightarrow a$

$B \rightarrow b$

$S \rightarrow AA$

$S \rightarrow BB$

$S \rightarrow a$
$S \rightarrow b$

Step 3 : Now we convert first two productions into Chomesky normal form. By introducing new variable.

$S \rightarrow ASA$ is replaced by
$S \rightarrow AS_1$ and $S_1 \rightarrow SA$
$S \rightarrow BSB$ is replaced by $S \rightarrow BS_2$ and $S_2 \rightarrow SB$

Now P'' is $S \rightarrow AS_1$
$S_1 \rightarrow SA$
$S \rightarrow BS_2$
$S_2 \rightarrow SB$
$A \rightarrow a$
$B \rightarrow b$
$S \rightarrow AA$
$S \rightarrow BB$
$S \rightarrow a$
$S \rightarrow b$ $\quad V_N'' = \{S, S_1, S_2, A, B\}$

Then, $G_2 = (S, S_1, S_2, A, B\}, \{a, b\}, P'', S)$

In new CFG, G_2 consists of all productions in Chomesky normal form.

11.19 GOLDBACH NORMAL FORM

Another useful grammatical form is the Goldbach normal form. In this form we put restrictions not on the length of right sides of a production but on the position in which terminals and variables can appear. Goldbach normal form is a little complicated and not very transparent. Constructing a grammar in GNF equivalent to a given context free grammar is tedious. Goldbach normal form has many theoretical and practical consequences. A context free grammar is said to be Goldbach normal form if all productions have the form

$$A \rightarrow ax, \text{ where } a \in \Sigma \text{ and } x \in V_N''$$

ILLUSTRATIVE EXAMPLES

1. *Convert the given grammar into GNF.*

$$X \rightarrow XY;\ X \rightarrow aX \mid bY \mid b;\ Y \rightarrow b$$

Solution : The given grammer is not in GNF.

In G, $X \rightarrow XY$ is not in the form $A \rightarrow ax$

So we replace X in right side of $X \rightarrow XY$ by the X-productions $X \rightarrow aX \mid bY \mid b$, we get

$$X \rightarrow aXY,\ X \rightarrow bYY,\ X \rightarrow bY$$

These X-productions are in GNF

So new grammar $G' = (V, \Sigma, P', X)$

P' is $X \rightarrow aXY \mid bYY \mid by$
$X \rightarrow aX \mid bY \mid b$
$Y \rightarrow b$

all productions in G' are in GNF.

2. *Convert the grammar into GNF $A \rightarrow abAb \mid aa$.*

Solution : Production $A \rightarrow abAb$ and $A \rightarrow aa$ are not in GNF

Now, we introduce new variable

$A \rightarrow aBAB$
$A \rightarrow aC$

$$B \to b$$
$$C \to a$$

New grammar in GNF.

11.20 PUMPING LEMMA FOR CONTEXT FREE LANGUAGE

Theorem 1. *If G is any CFG in Chomesky normal form, then any $w \in L(G)$ with $|w| \geq m$, then we break up w into five substrings.*

$$w = uvxyz \quad \ldots(1)$$

with $$|vxy| \leq m \quad \ldots(2)$$

and $$|vy| \geq 1 \quad \ldots(3)$$

such that $uv^n xy^n z \in L(G)$ for all $n = 0, 1, 2, \ldots$. This is known as pumping lemma for context free language.

Proof : Let us consider language $L(G) - \wedge$ and we assume that G is without null and unit productions.

Since the length of the string on the right side of any production is bounded by k, then length of the derivation of any word w must be $|w| / k$. Since $L(G)$ is infinite, then there exist long derivations and corresponding tree of arbitrary height.

Now, we consider a high derivation tree and a long path from root to leaf. Since the number of variables in G is finite, then obviously some variables repeat on this path as shown in figure. Now a derivation corresponding to derivation tree

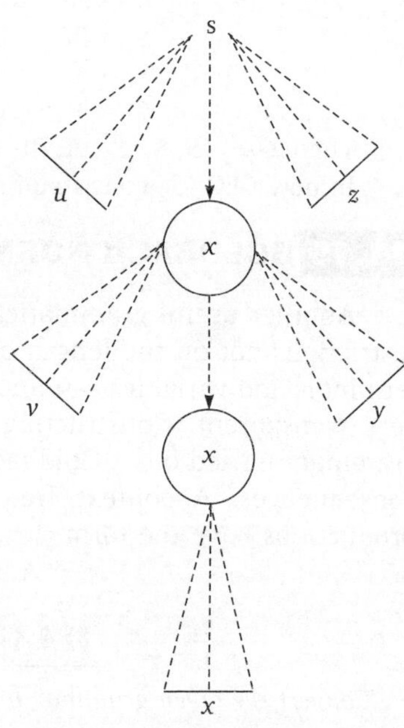

Fig. (10)

$$S \overset{*}{\Rightarrow} uXz$$
$$\overset{*}{\Rightarrow} uv\ Xyz$$
$$\overset{*}{\Rightarrow} uv\ Xyz$$

where u, v, x, y, z are all strings of terminals from the above derivation. We see that $X \overset{*}{\Rightarrow} vXy$ and $X \overset{*}{\Rightarrow} x$. So all strips $uv^n xy^n z$, $n = 0, 1, 2, 3, \ldots$ can be generated by the grammar G and therefore in $L(G)$. Now in derivation $X \overset{*}{\Rightarrow} vXy$ and $X \overset{*}{\Rightarrow} x$, if we assume no variable repeats, therefore the length of strips v, x and y depends only on the productions of grammar and can be bounded independently of w so that condition (2) holds and there is no unit and null productions in the grammar so v and y can not both empty strings so condition (3) holds. This completes the argument that condition (1) and (4) holds.

11.21 PROCEDURE TO SHOW THAT LANGUAGE *L(G)* IS NOT CONTEXT FREE

We use pumping lemma to show that language $L(G)$ is not a context free language. We assume that $L(G)$ is context free. By applying we get a contradiction. For this we use following steps.

WORKING RULES

Step 1 : Assume L is context free. Let n be the natural number obtained by using the pumping lemma.

Step 2 : Choose $w \in L(G)$ so that $w \geq n$. Write $w = uvxyz$ using pumping lemma.

Step 3 : Find a suitable integer i such that $uv^ixy^iz \notin L$. This is a contradiction and so $L(G)$ is not context free.

ILLUSTRATIVE EXAMPLES

1. *Show that $L(G) = \{a^nb^nc^n : n \geq 1\}$ is not context free.*

Solution : Step 1 : Assume that $L(G)$ is context free. Let n be the natural number obtained by using pumping lemma.

Step 2 : Let $w = a^nb^nc^n$. Then $|w| = 3n > m$ taking $w = uvxyz$ where $|vy| \geq 1$, *i.e.*, at least one of v or y is not $\wedge$.

Step 3 : $uvxyz = a^nb^nc^n$ as $1 \leq |vy| \leq n$, v or x can not contain all the tree symbols a, b, c

(i) v or x is of the form a^ib^j or b^ic^j for some i, j such that $i + j \leq n$

(ii) v or y is a string constructed by repetition of only one symbol among a, b, c

When v or y is of the form a^ib^j, then $v^2 = a^ib^ja^ib^j$ or $y^2 = a^ib^ja^ib^j$.

As v^2 is substring of uv^2xy^2z, then uv^2xy^2z can not be of the form $a^mb^mc^m$ so $uv^2wx^2y \notin L$.

When both v and y are formed by repetition of single symbol, *i.e.*, $u = a^i$, $v = b^j$ for some i and j, $i \leq n$, $j \geq n$, then substring uxz contain remaining symbol. a_1^n will be substring of uxz as a_1 does not occur in v or x. The number of occurrence of one of two other symbols in uwy is less than n. So, $uv^0xy^0z = uxz \neq L$.

2. *Show that $L = \{a^p : p \text{ is prime}\}$ is not context free language.*

Solution : Step 1 : Suppose $L(G)$ is context free. Let n be the natural number obtained by using pumping lemma.

Step 2 : Let $w = a^p \in L$ where p is the prime number and $p > n$ and let $w = uvxyz$.

Step 3 : By pumping lemma $uv^0xy^0z = uxz \in L$ so $|uxz|$ is a prime number say q. Let $|vy| = r$. Then $|uv^2xy^2z| = q + qr = q(1 + r)$ is not a prime number, since $q + qr$ has two factors q and $(1 + r)$. $uv^qxy^qz \notin L$. This is a contradiction. Therefore, L is not context free.

11.22 TURING MACHINE

The turing machine is a simple mathematical model of general purpose computer. Turing machine is capable of performing any calculation which can be performed by any computing machine.

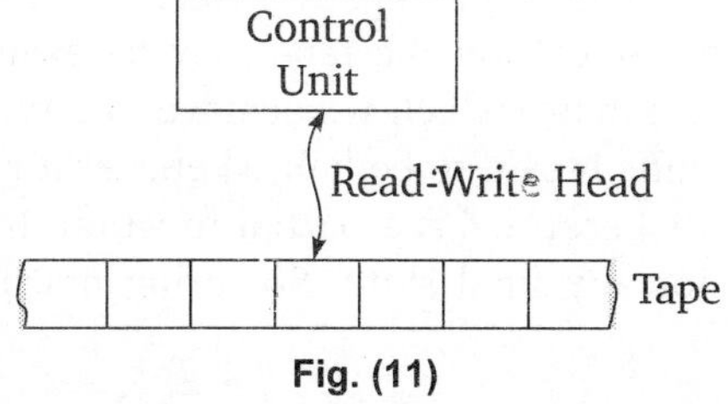

Fig. (11)

Turing machine is an automation whose temporary storage is a tape. This tape is divided into cells. Each cell is capable of holding one input. A read and write head is associated with the tape that can move right or left on the tape and that can read and write a single symbol on each move. On each move the machine examines the present symbol under Read-Write head on the tape.

A turing machine is defined by

$$M = (Q, \Sigma, \Gamma, \delta, q_0, b, F)$$

where

1. Q is the set of states
2. Σ is a set of input symbols and $b \notin \Sigma$.
3. Γ is a set of tape symbols called tape alphabet.
4. δ is the transition function such that δ is
$$Q \times \Gamma \to Q \times \Gamma \times \{L, R\}$$
Current state of the machine and current tape symbol being read. The result is a new state of the machine, a new tape symbol, which replaces old one and moves symbol L or R.
5. $q_0 \in Q$ is the initial state.
6. $b \in \Gamma$ is a special symbol called blank.
7. $F \subseteq Q$ is the set of final states.

11.22.1 Moves in a Turing Machine

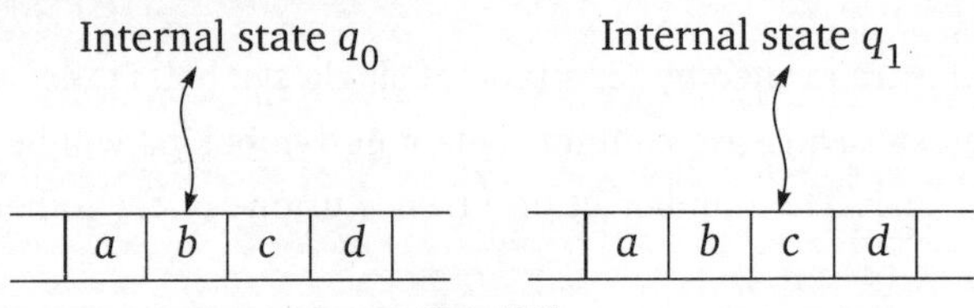

Fig. (12)

The figure shows the situation before and after moves caused by the transition.

$$\delta(q_0, b) = (q_1, b, R)$$

In one move, the machine examines the present symbol under the Read-Write head on the tape and present state of automation to determine

(i) A new symbol to be written on tape in the cell under R/W head.

(ii) A motion of the Read-Write head along the tape either the head moves one cell left (L) or one cell right (R).

(iii) The next state of automation

(iv) Whether halt or not

(v) The acceptability of a string is decided by reachability from initial state to some final state.

Automation always starts in the given initial state with some information on the tape. It then goes through a sequence of steps controlled by transition function δ. During this process the symbols on any cell on the tape may be examined and changed many times. Whole process may terminate, which we achieve in a turing machine by putting it into halt state. A turing machine is said to be halt whenever it reaches a configuration for which δ is not defined. This is because δ is a partial function. In fact we will assume that no transitions are defined for any final state. So turing machine will halt whenever it enters a final state.

11.23 REPRESENTATION OF TURING MACHINE

We can represent turing machine by (i) transition table, (ii) transition diagram, (iii) instantaneous description using move relation.

11.23.1 Transition Table Representation of TM

We can give the definition of δ in the form of a table called transition table if $\delta(q_0, a) = (q_1, d, R)$. We write $d\,R\,q_0$ under d column and q_0 row. So if we get $d\,R\,q_0$ in the table d written in the current cell, R (right) gives the moment of the head (L or R) and q_1 denotes the new state into which turing machine enters.

ILLUSTRATIVE EXAMPLES

1. *Let us consider a Turing Machine with four states $Q = \{q_0, q_1, q_2, q_3, q_4\}$, $F = \{q_4\}$, q_0 is the initial state. The tape symbols are $\Gamma = \{0, 1, b\}$*

Transition Table

Present state	Tape symbol		
	0	1	b
$\rightarrow q_0$	$1\,R\,q_1$	$1\,R\,q_0$	$b\,L\,q_2$
q_1		$1\,R\,q_1$	
q_2		$0\,L\,q_3$	$b\,R\,q_4$
q_3		$1\,L\,q_3$	
q_4			

Draw the computation sequence of the input string 111011.

Solution : For the string 111011b we get the following sequence

$$q_0111011b \vdash 1q_011011b \vdash 1q_01011b \vdash 111q_0011b \vdash 1111q_111b \vdash 11111q_11b$$
$$\vdash 111111q_1b \vdash 11111q_21b \vdash 1111q_310b \vdash 111q_3110b \vdash 11q_31110b$$
$$\vdash 11q_31110b \vdash 1q_311110b \vdash q_3111110b \vdash q_3b111110b \vdash q_4111110b$$

q_4 is the final state string that is accepted in turing machine.

11.23.2 Representation by Instantaneous Description (ID)

Snapshots of turing machine in action can be used to describe a turing machine. These give instantaneous description of a turing machine

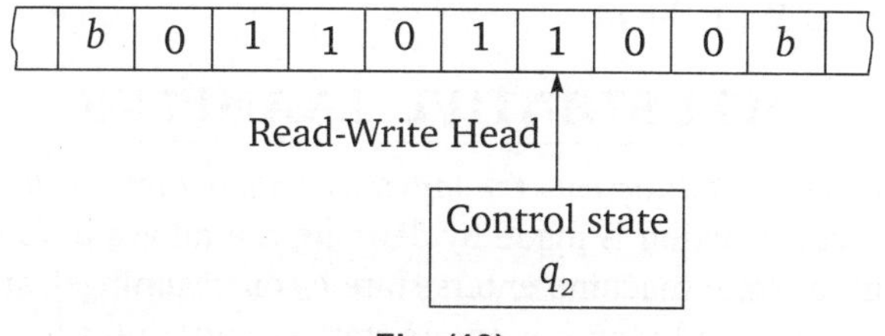

Fig. (13)

The string 01101, written to the left q_2. The sequence of non blank symbols to the right of current input symbol 1 is 00. Thus the Instantaneous Description as given

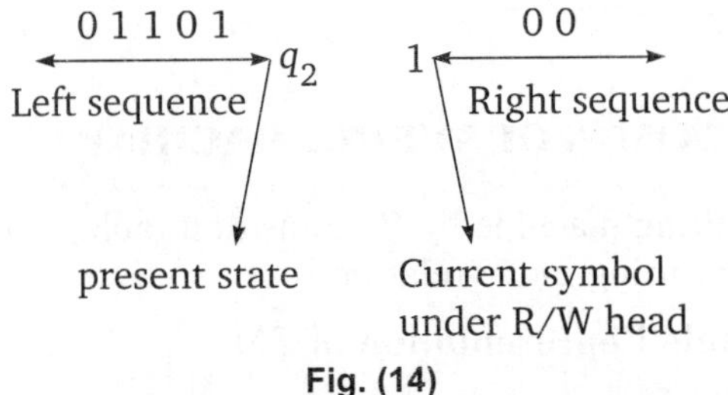

Fig. (14)

11.24 REPRESENTATION BY TRANSITION DIAGRAM

We can use transition system to represent the turing machine. The states are represented by vertices, directed edges are used to represent transition of states. The tables are triple of the form (α, β, γ), where $\alpha, \beta \in \Gamma$ and $\gamma \in \{L, R\}$

It means $\delta(q_i, \alpha) = (q_j, \beta, \gamma)$

Each edge in transition system can be represented by 5-tuple $(q_i, \alpha, \beta, \gamma, q_j)$, each turing machine can be described by sequence of 5-tuple representing all the directed edges. The initial state indicated by $\rightarrow$ and any final state is marked with circle (O).

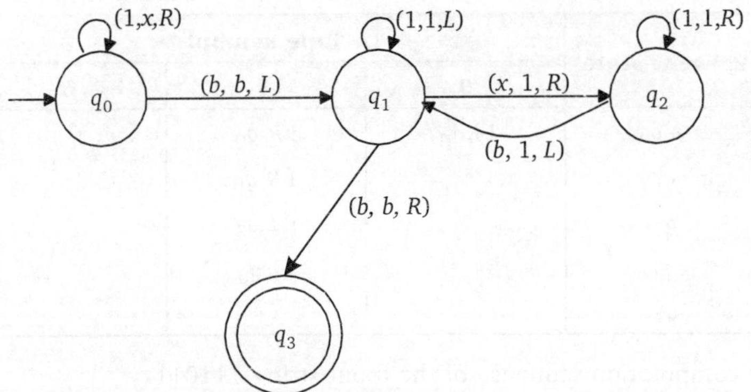

Fig. (15) : Transition System for Turing Machine

11.25 DESIGN OF THE TURING MACHINE

(i) The main objective in scanning a symbol by Read-Write head is to know what to do in the future. The turing machine remembers the past symbols scanned. The turing machine can remember this by going to the next unique state.

(ii) The number of states must be minimized. This can be achieved by changing the state only when there is a change in written symbol or when there is a change in the movement of Read-Write head.

ILLUSTRATIVE EXAMPLES

1. *Design a Turing Machine that accepts the language denoted by regular expression 11*.*

Solution : The construction is made by defining the moves in following manner :

(i) q_0 is the initial state machine enters state q_0 on scanning 1 and writes 1.

(ii) If M is in state q_0 and scan b it enters state q_1 and writes b.

(iii) q_1 is only accepting state.

$$M = (\{q_0, q_1\}, \{1\}, \{1, b\}, \delta, q_0, b, q_1)$$

δ is defined as

Present state	b	1
$\rightarrow q_0$	$b\,R\,q_1$	$1\,R\,q_0$
(q_1)		

Let us obtain computation sequence of 11 and 110

$$q_0 11 \vdash 1q_0 1 \vdash 11q_0 b \vdash 11bq_1$$

As q_1 is final state, string 11 is accepted by the machine

Now $\quad q_0 110 \vdash 1q_0 10 \vdash 11q_0 0$

The machine halt and q_0 is not final sate. So 110 is rejected by the machine.

2. *Design a Turing Machine that accepts odd number of 1's.*

Solution : The construction is made by defining move in following manner

(i) q_0 is the initial state machine enter state q_1 on scanning 1 and writes b.

(ii) If machine is in state q_2 scans 1, it enters q_1 and writes b.

(iii) q_1 is the final state.

Transition table

Present state	1
$\rightarrow q_0$	$bR\,q_1$
(q_1)	$bR\,q_0$

Let us obtain sequence of 111 and 1111

$$q_0 111 \vdash bq_1 11 \vdash bbq_0 1 \vdash bbbq_1$$

q_1 is the final state. So string 111 is accepted by turing machine.

Now string $\quad q_0 1111 \vdash bq_1 111 \vdash bbq_0 11 \vdash bbbq_1 1 \vdash bbbbq_0$

Machine halt and q_0 is non final state so the string is rejected by the machine.

3. *Design a turing machine over* $\{1, b\}$ *which can compute concatenation function over* $\Sigma = \{1\}$ *if pair of words* (x_1, x_2) *is the input, the output has to be* x_1x_2.

Solution : Assume that the two words x_1 and x_2 written initially on that tape separated by the symbol b.

If $x_1 = 111$ and $x_2 = 11$, then

initially input tape

b	1	1	1	b	1	1	b	

output tape

b	1	1	1	1	b	

The construction is made by following manner.

(i) First when separating symbol is found, replace it by 1.

(ii) When right most 1 is found, replace it by b.

(iii) Read-write head, return to starting position.

Transition table

Present state	Tape symbol	
	1	b
$\rightarrow q_1$	$1R\,q_1$	$1R\,q_2$
q_2	$1R\,q_2$	$bL\,q_3$
q_3	$bL\,q_4$	—
q_4	$1L\,q_4$	$bR\,q_5$
(q_5)	—	—

The input string $11b11$ computation sequence

$$q_1 11b11 \vdash 1q_1 1b11 \vdash 11q_1 b11 \vdash 111q_2 11 \vdash 1111q_2 1 \vdash 11111q_2 b \vdash 1111q_3 1b \vdash 111q_4 1b \vdash 11q_4 11b \vdash 1q_4 111b \vdash q_4 1111b \vdash q_4 b1111b \vdash bq_5 1111$$

q_5 is the final state.

4. *Design a Turing Machine that accepts all strings consisting of even number of 1's.*

Solution : The construction is made by defining move in following manner

(i) q_0 is the initial state M enter state q_1 on scanning 1.

(ii) If M is in state q_2 and scans 1, it enters q_1.

(iii) q_0 state is used for remembering that even number of 1's are encountered and q_1 is used to remember that odd number of 1's are encountered.

(iv) q_f is the accepting state.

Transition table

Present state	0	1	b
$\rightarrow q_0$	$0Rq_0$	$1Rq_1$	bRq_f
q_1	$0Rq_1$	$1Rq_0$	—
q_f	—	—	—

For the input string 101101, the computation sequence is given as

$$q_0 101101 \vdash q_1 01101 \vdash 10q_1 1101 \vdash 101q_0 101 \vdash 1011q_1 01 \vdash 10110q_1 \vdash 101101q_0 b \vdash 101101bq_f$$

q_f is the final state. String is accepted by the machine.

Exercise 11.2

1. Construct a non-deterministic finite automation accepting $\{ab, ba\}$ and use it to find a deterministic automation accepting the same set.
2. Construct the DFA for the NDFA given in the following figure.

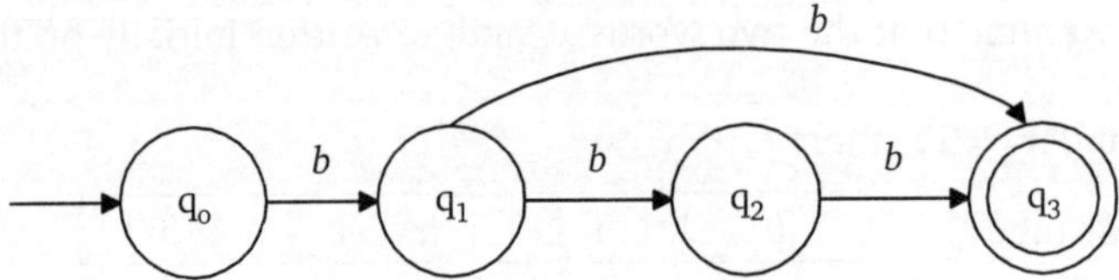

3. Construct the DFA for the NDFA given in the following figure.

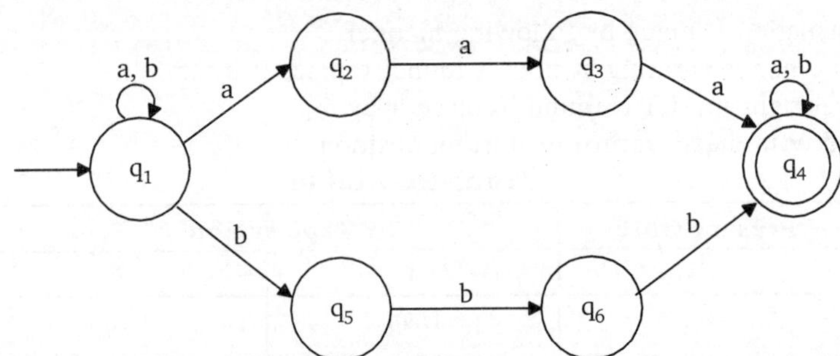

4. Construct a non-deterministic automation accepting all input with *bb* in them and use it to find a deterministic automation accepting the same set.
5. Construct a minimum state automation equivalent to a given automation whose transition table is given as

State	Inputs	
	0	1
$\rightarrow q_1$	q_1	q_4
q_2	q_3	q_6
q_3	q_4	q_5
q_4	q_1	q_6
q_5	q_1	q_7
q_6	q_2	q_5
q_7	q_2	q_4

6. Construct a minimum state automation equivalent to the finite automation given in following figure.

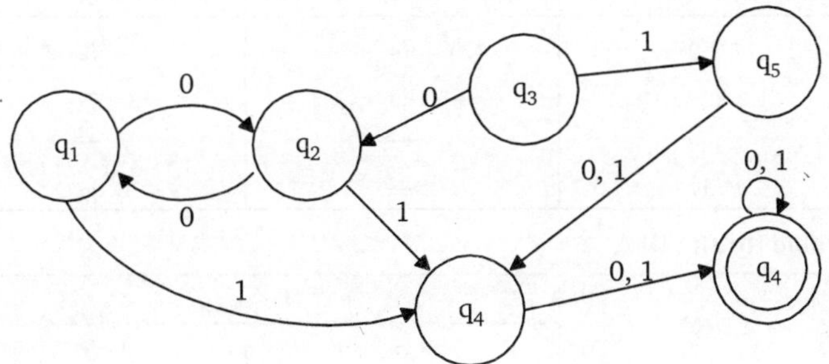

7. Consider the transition system given in the figure. Prove that the strings recognised are $(a + a(b + aa)^* b)^* a(b + aa)^* a$.

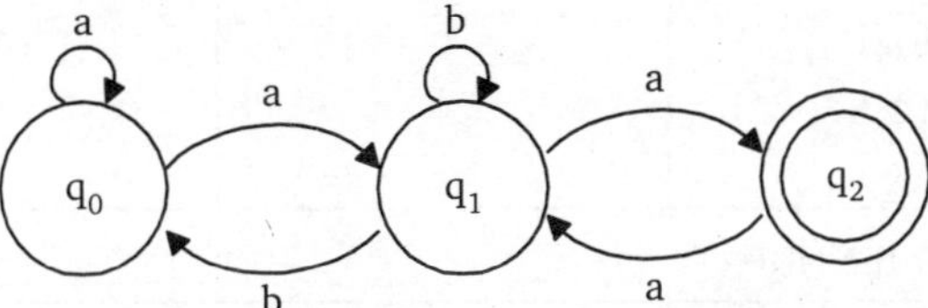

8. Prove that the FA whose transition diagram is given in figure accepts the set of all strings over the alphabet $\{a, b\}$ with an equal number of $a's$ and $b's$ such that each prefix has at most one more a than $b's$ and at most one more b than $a's$..

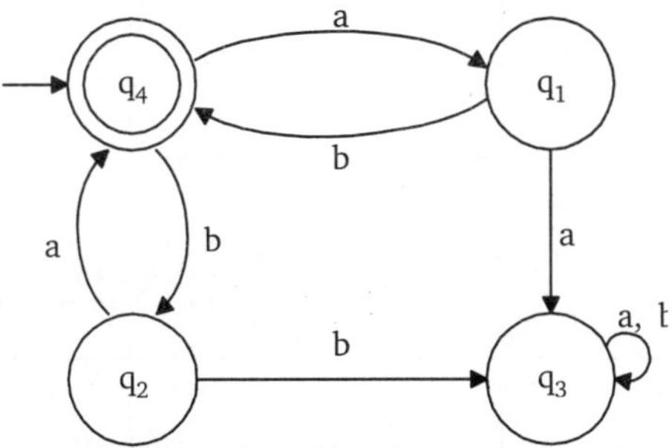

9. Design a Turing machine that can compute proper subtraction *i.e.*, $m - n$ where m and n are positive integers $m - n$ is defined as $m - n$ if $m > n$ and $m - n$ is 0 if $m \leq n$.

10. Design a Turing machine that copies q is to find a machine that perform the computation $q_0w \vdash^* q_iww$ for any $w \in \{1\}^*$.

ANSWERS

1. NDFA for {*ab*, *ba*}

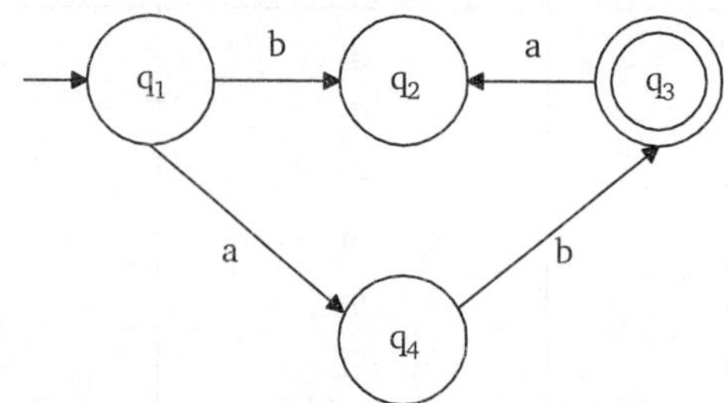

Transition table for the NDFA

State	Inputs	
	a	***b***
$\rightarrow q_1$	q_4	q_2
q_2	q_3	—
q_3	—	—
(${q_4}$)	—	q_3

Transition table for the DFA

State	Inputs	
	a	***b***
$\rightarrow [q_1]$	$[q_4]$	$[q_2]$
$[q_2]$	$[q_3]$	ϕ
$[q_4]$	ϕ	$[q_3]$
$[q_3]$	ϕ	ϕ
(ϕ)	ϕ	ϕ

2. Transition table for the NDFA

State	Inputs	
	a	***b***
$\rightarrow q_0$	—	q_1
q_1	—	q_2, q_3
(${q_2}$)	—	q_3
q_3		—

Transition table for the DFA

State	Inputs	
	a	***b***
$\rightarrow [q_0]$	ϕ	$[q_1]$
$[q_1]$	ϕ	$[q_2, q_3]$
$[q_2, q_3]$	ϕ	$[q_3]$
$[q_3]$	ϕ	ϕ
ϕ	ϕ	ϕ

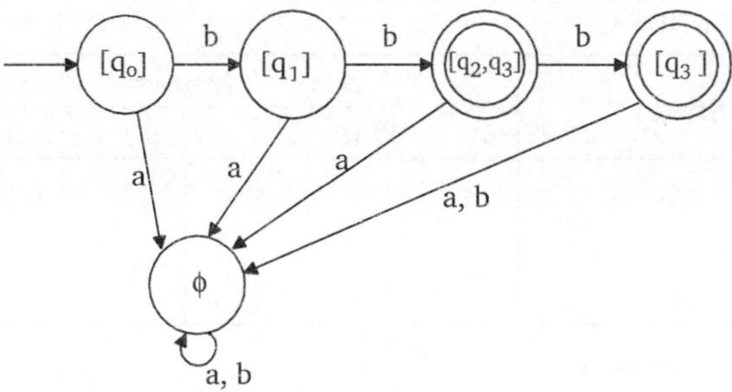

3. Transition table for the NDFA

State	Inputs	
	a	*b*
$\rightarrow q_1$	q_1, q_2	q_1, q_5
q_2	q_3	—
q_3	q_4	—
q_5	—	q_6
q_6	—	q_4
q_4	q_4	q_4

Transition table for the DFA

State	Inputs	
	a	*b*
$\rightarrow [q_1]$	$[q_1,q_2]$	$[q_1,q_5]$
$[q_1,q_2]$	$[q_1,q_2,q_3]$	$[q_1,q_5]$
$[q_1,q_5]$	$[q_1,q_2]$	$[q_1,q_5,q_6]$
$[q_1,q_2,q_3]$	$[q_1,q_2,q_3,q_4]$	$[q_1, q_5]$
$[q_1,q_5,q_6]$	$[q_1,q_2]$	$[q_1,q_4,q_5,q_6]$
$[q_1,q_2,q_3,q_4]$	$[q_1,q_2,q_3,q_4]$	$[q_1,q_4,q_5]$
$[q_1,q_4,q_5,q_6]$	$[q_1,q_2,q_4]$	$[q_1,q_4,q_5,q_6]$
$[q_1,q_4,q_5]$	$[q_1,q_2,q_4]$	$[q_1,q_5,q_6,q_4]$
$[q_1,q_2,q_4]$	$[q_1,q_2,q_3,q_4]$	$[q_1,q_4,q_5]$

4. NDFA that accepts all inputs with *bb* in them is

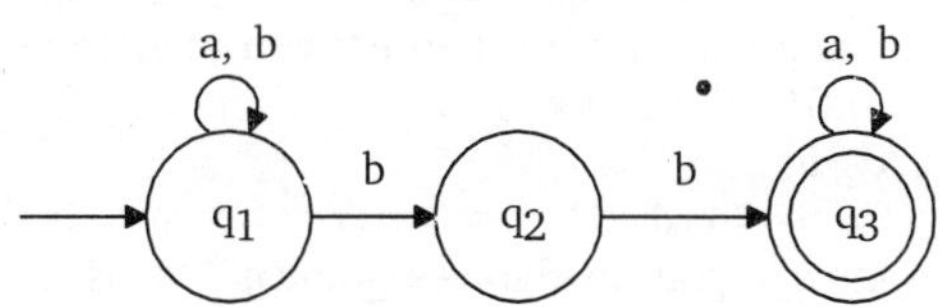

Transition table for NDFA

State	Inputs	
	a	b
$\rightarrow q_1$	q_1	q_1, q_2
q_2	—	q_3
q_3	q_3	q_3

Transition table for the DFA

State	Inputs	
	a	b
$\rightarrow [q_1]$	$[q_1]$	$[q_1, q_2]$
$[q_1, q_2]$	$[q_1]$	$[q_1, q_2, q_3]$
$[q_1, q_2, q_3]$	$[q_1, q_3]$	$[q_1, q_2, q_3]$
$[q_1, q_3]$	$[q_1, q_3]$	$[q_1, q_2, q_3]$

Last two rows are identical so we reduce $[q_1, q_2, q_3]$ row

State	Inputs	
	a	b
$\rightarrow [q_1]$	$[q_1]$	$[q_1, q_2]$
$[q_1, q_2]$	$[q_1]$	$[q_1, q_2, q_3]$
$[q_1, q_2, q_3]$	$[q_1, q_3]$	$[q_1, q_2, q_3]$

Transition diagram for DFA

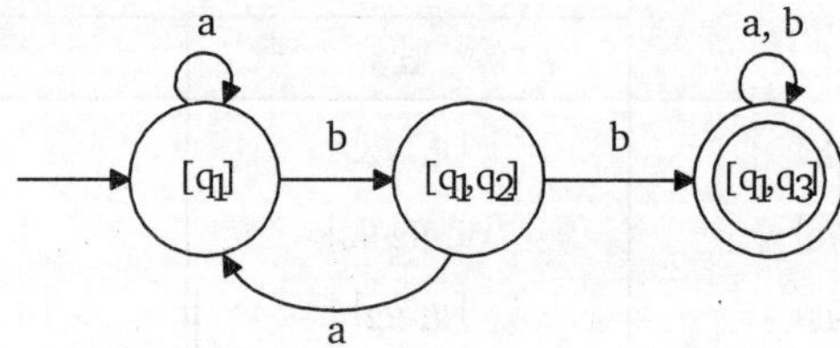

5. π's are given below

$$\pi_0 = \{\{q_7\}, \{q_1, q_2, q_3, q_4, q_5, q_6\}\}$$
$$\pi_1 = \{\{q_7\}, \{q_1, q_2, q_3, q_4, q_5, q_6\}, \{q_5\}\}$$
$$\pi_2 = \{\{q_7\}, \{q_1, q_2, q_4\}, \{q_3, q_6\}, \{q_5\}\}$$
$$\pi_3 = \{\{q_7\}, \{q_1\}, \{q_2\}, \{q_4\}, \{q_3, q_6\}, \{q_5\}\}$$
$$\pi_4 = \{\{q_7\}, \{q_1\}, \{q_2\}, \{q_4\}, \{q_3\}, \{q_6\}, \{q_5\}\}$$

or $$\pi_4 = \{\{q_1\}, \{q_2\}, \{q_3), \{q_4\}, \{q_5\}, \{q_6\}, (q_7\}\}$$

Here $\pi = Q$. The minimum state automation is simply the given automation.

6. $\pi_0 = \{\{q_6\}, \{q_1, q_2, q_3, q_4, q_5\}\}$; $\pi_1 = \{\{q_6\}, \{q_4\}, \{q_1, q_2, q_3, q_5\}\}$
$\pi_2 = [\{q_6\}, \{q_4\}, \{q_1, q_2\}, \{q_3\}, \{q_5\}\}$

7. Three equations for q_0, q_1, q_2
$q_0 = q_0 a + q_1 b + \wedge$; $q_1 = q_1 b + q_0 a + q_2 a$; $q_2 = q_1 a$
q_2 is the final state so by applying Arfen's theorem you solve the equations.
Value of $q_2 = (a + a(b + aa)^* b)^* a(b + aa)^* a$

8. The equation for q_0, q_1, q_2, q_3
$q_0 = q_2 a + q_1 b + \wedge$; $q_1 = q_0 a$; $q_2 = q_0 b$; $q_3 = q_1 a + q_2 b + a_3 a + q_3 b$
q_0 is the only final state and q_0 equation involves q_2 and q_1. We use only q_2, q_1 equations. So we get $q_1 = (ab + ba)^*$

❑❑❑

Chapter

12

Semigroups and Monoids

- Algebraic Structure
- Groupoid
- Semigroups
- Monoids (Semigroups with Identity)
- Sub Semigroups
- Commutative Monoid
- Congruence Relation and Quotient Semigroups

12.1 INTRODUCTION

It is known that a binary operation on a set X is a mapping from $X \times X$ to X. It is usually denoted by means of a symbol such as $+$, $\times$, $.$, $*$, o, $\oplus$. If we denote a binary operation on a set X by $*$, then the result of $*$ on the elements x, y of X is expressed by $x * y$. If $f : X \times X \to X$ be a binary composition in X and $x, y \in X$. Then $f(x, y)$ is called the composite of x and y under the composition f.

For example :

(1) Addition and multiplication are binary composition in the following sets.

(i) The set $\boldsymbol{N}$ of natural numbers.

(ii) The set $\boldsymbol{Z}$ of all numbers.

(iii) The set $\boldsymbol{Q}$ of all rational numbers.

(iv) The set $\boldsymbol{R}$ of real numbers.

(2) Subtraction is not a binary composition in $\boldsymbol{N}$ because for given $m, n, \in \boldsymbol{N}$, $m - n$ may not be an element of $\boldsymbol{N}$, *e.g.*, $2 - 3$

☛ The adjective binary is used because our rule combines two elements at a time.

12.1.1 *n*-Ray Operations

Let $\boldsymbol{X}$ be any non empty set $\boldsymbol{A}$ mapping $f : X^n \to X$ is called an n-ray operations for $n = 1, 2, \ldots$ on the set X.

For $n = 1$, such an operation is called a unary operation. Therefore any function from X to it self is a unary operation on X.

For $n = 2$, n-ary operation is called binary operation.

For $n = 3$, i.e., a function from $X \times X \times X$ into is X called 3-ary (or ternary) operation.

12.2 ALGEBRAIC STRUCTURE

Definition : *A non-empty set with one or more operations, each of which is an n-ary operation for some $n \in \mathbf{N}$ is called algebraic structure.* An algebraic structure is denoted by $(X, +, \times \ldots)$ where X is a non-empty set and $+, \times \ldots$ are operations on X. In this chapter, we shall study some of those algebraic structure which involves binary operations only.

12.3 GROUPOID

Let S be any non-empty set and $*$ be any binary operation. Then the pair $(S, *)$ is said to be groupoid. As defined above if $(S, *)$ is a grouped. Then S is the set underlying this grouped. We shall however, use the phrases 'S is a groupoid with respect to $*$' and 'S is a groupoid' in case $\times$ is taken for granted. This convection of using S for the structure as well as the set underlying the structure will be followed throughout and should cause no confusion.

For Example :

$(N, +)$, where N is the set of natural numbers and $+$ is the operation on N, is a groupoid. It is also noted that $(N, -)$ is not a groupoid.

12.3.1 Commutative groupoid

Definition : *A groupoid $(S, *)$ is said to be commutative groupoid if $*$ is a commutative operation.*

For Example :

(i) $(\mathbf{N}, +)$, $(\mathbf{Z}, \times)$ are both commutative groupoids.

(ii) $(\mathbf{Q}, -)$ and $(\mathbf{R} \sim \{0\}, \div)$ are not commutative groupoids because neither subtraction is in $\mathbf{Q}$ nor division in the set of non-zero real nuber is a commutative operations.

(iii) Let S be the set of all square matrices of order 3 over $\mathbf{R}$ and let $\oplus$ and $\otimes$ denote respectively the matrix addition and matrix multiplication. Then $(S, \oplus)$ is commutative but $(S, \otimes)$ is not a commutative groupoid.

12.4 SEMIGROUPS

Definition : *A groupoid $(S, *)$ is said to be semigroup if $*$ is associative, i.e., if S is any non-empty set and $*$ is a binary composition on S such that $a * (b * c) = (a * b) * c$, $\forall a, b, c \in S$. Then $(S, *)$ a semigroup.*

For Example :

(i) $(\mathbf{N}, +)$, $(\mathbf{N}, .)$ $(\mathbf{Z}, \times)$ and $(\mathbf{R}, +)$ are the examples of semigroup.

(ii) Let $*$ be defined by $a * b = |a - b|$, $\forall a, b \in \mathbf{R}$

Then clearly, $(\mathbf{R}, *)$ is a commutative groupoid but not a semigroup.

12.4.1 Relation between Groupoid, Commutative Groupoid and Semigroup

Let us denote the set of all semigroups by S and C denote the set of all commutative groupoid. Then following diagram shows the relation between the set of groupoids, commutative groupoid and semigroup.

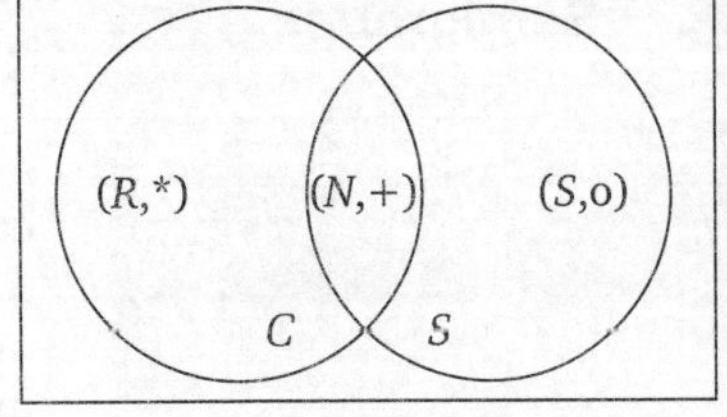

Fig. (1)

12.5 MONOIDS (SEMIGROUPS WITH IDENTITY)

Definition : *A semigroup $(S, *)$ is said to be a* ***monoid*** *if S has an identity element with respect to $*$, i.e., $(S, *)$ is a monoid if S is non-empty set and $*$ is a binary composition on S such that*

(i) $a * (b * c) = (a * b) * c$, $\forall a, b, c \in S$ **(Associativity)**

and (ii) $\exists e \in S$ such that $a * e = e * a = a \, \forall \, a \in S$.

For Example :

(i) The semigroup $(\mathbf{N}, \times)$, $(\mathbf{Q}, \times)$, $(\mathbf{R}, \times)$ are all monoids. 1 being the identity element for each.

(ii) (**N**, +) is a semigroup which is not a monoid.

(iii) (**Q**, +), (**R**, +) are both monoids, 0 being the identity element for each.

(iv) ($P(X)$, ∩) is a monoid, X being the identity element.

(v) ($P(X)$, ∪) is a monoid, ϕ being the identity element.

(vi) Let S be the set of all square matrices of order 4 over **R** and let ⊕ and ⊗ denote respectively the operations of matrix addition and matrix multiplication. Then both (S, ⊕) and (S, ⊗) are monoids. The identity of (S, ⊕) is the null matrix of under 4 and that of (S, ⊗) the unit matrix of order 4.

12.5.1 Relation between the Groupoid, Commutative Groupoid, Semigroup and Monoids

Let M denote the set of all monoids. Then following diagram shows the relation between these algebraic structures

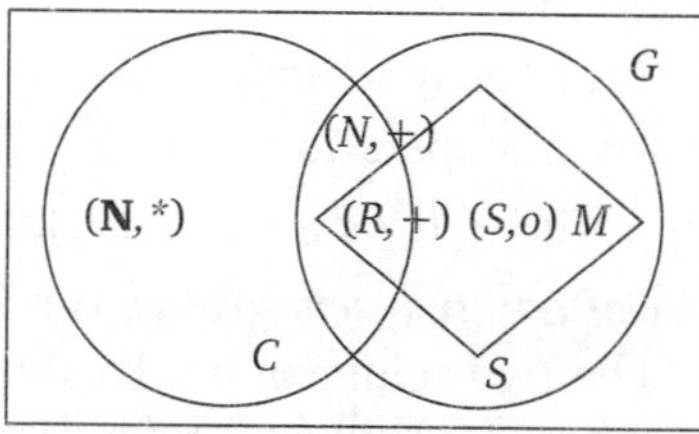

Fig. (2)

12.5.2 Monoids with Inverse

Definition : *A monoid (S, *) in which every element has an inverse is known as a group.*

For Example :

(i) (**R**, +) is a group

(ii) Let S denotes the set of real functions and ∘ is the operation of composite of functions. Then (S, ∘) is not a group because only the bijective functions have an inverse.

Illustrations :

- Algebraic structure $(G, .)$ is a monoid in each of the following cases, where '.' denote multiplication of numbers.
  (a) $G = \{2^n : n \in \mathbf{Z}\}$
  (b) G = {the set of all positive rational numbers}.
  (c) $G = \{1, -1, i, -i\}$
  (d) $G = \{2 : z \in \mathbf{C}.|z| = 1\}$
  (e) $G = \{a + b\sqrt{2} : a, b \in \mathbf{Q}, a^2 + b^2 \neq 0\}$
- Algebraic structure $(G, +)$ is a monoid in each of the following cases, where + denote addition of numbers.
  (a) $G = \{a + b\sqrt{2} : a \in \mathbf{Q}, b \in \mathbf{Q}\}$
  (b) $G = \{3n : n \in \mathbf{Z}\}$
  (c) $G = \{a + ib : a \in \mathbf{Z}, b \in \mathbf{Z}\}$
  (d) $G = \{a + ib, a \in \mathbf{Q}, b \in \mathbf{Q}\}$
- Following are the examples of monoids.
  (a) The set of all real numbers of the form $x + y\sqrt{3}$ where $x, y \in \mathbf{Q}$, with respect to addition.
  (b) The set of all non-zero complex numbers with respect to the multiplication.
  (c) The set of real numbers R with respect to * where $a * b = a + b - ab, \forall a, b \in \mathbf{R}$
  (d) The set **Z** with respect to *, where $a * b = a + b + 1, \forall a, b \in \mathbf{Z}$.
  (e) The set **Z** with respect to * where $a * b = a + b - 1, \forall a, b \in \mathbf{Z}$.
  (f) The set of all matrices of the form $\begin{pmatrix} a & 0 \\ 0 & 0 \end{pmatrix}$ where $a \in R$ is a monoid with respect to the matrix multiplication
  (g) The set of square symmetric matrices of order n over **C**.

12.6 SUB SEMIGROUPS

Definition : Let $(S, *)$ be a semigroup and let T be any subset of S. If the set T is closed under the operation $*$ (*i.e.*, $a * b \in T, \forall\ a, b \times T$) defined on S. Then $(T, *)$ is called a sub semigroup of $(S, *)$.

For Example.

Let E denote the set of even positive integers. Then $(E, .)$ where '.' is the multiplication of integers is a subsemigroup of semigroup $(\mathbf{N}, .)$.

- It is known that the associative property holds in any subset of a semigroup Thus a sub semigroup $(T, *)$ of a semigroup $(S, *)$ is itself a semigroup.
- If $(S, *)$ is a semigroup then $(S, *)$ itself is a subsemigroup of $(S, *)$. Such a subsemigroup is called trivial sub semigroup.

12.6.1 Direct Product of Subsemigroup

Definition : *Let $(S, *)$ and (T, o) be two semigroups. Then direct product of $(S, *)$ and (T, o) is the system $(S \times T, \oplus)$ where the operation $\oplus$ on $S \times T$ is defined as follows*

$$(a, b) \oplus (c, d) = (a * c, b \circ d)\ ;\ \forall\ (a, b), (c, d) \in S \times T$$

Theorem 1. *The direct product of any two semigroups is a semigroup.*

Proof : Let $(S, *)$ and $(T, \circ)$ be two semigroups. To show that $(S \times T, \oplus)$ where $\oplus$ is defined by $(a, b) \oplus (c, d) = (a * c, b \circ d)$ for all $(a, b), (c, d) \in S \times T$ is a semigroup.

Let $\quad a, c \in S \Rightarrow a * c \in S \quad [\because S$ is a semigroup$]$

Similarly, $b, d \in T \Rightarrow b \circ d \in T \quad [\because T$ is a semigroup$]$

We have to show that $\oplus$ is associative on $S \times T$.

Let (a, b), (c, d) and (e, f) be any three elements in $S \times T$.

Then $((a, b) \oplus (c, d) \oplus (e, f)) = (a * c, b \circ d) \oplus (e. f)$

$$= ((a * c) * e, (b \circ d) \circ f)$$
$$= (a * (c * e), b \circ (d \circ f))$$

$[\because *$ and $\circ$ both are associative$]$

$$= (a, b) \oplus ((c * e, d \circ f))$$
$$= (a, b) \oplus ((c, d) \oplus (e, f))$$

$\Rightarrow \quad \oplus$ is associative. Hence, $(S \times T, \oplus)$ is a semigroup.

Theorem 2. *The product of two commutative semigroups is again commutative semigroup.*

Proof : In Theorem 1, we have already proved that if $(S, *)$ and (T, o) are semigroups. Then their product is again a semigroup. Therefore, here it remains to prove that $S \times T$ is commutative.

Let (a, b) and (c, d) be any two elements in $S \times T$. Then

$$(a, b) \oplus (c, d) = (a \times c, b \circ d)$$
$$= (c \times a, d \circ b) \qquad [\because * \text{ and } \circ \text{ both are commutative}]$$
$$= (c, d) \oplus (a, b)$$

$\Rightarrow \quad \oplus$ is commutative operator on $S \times T$.

Hence, $(S \times T, \oplus)$ is a commutative semigroup.

12.6.2 Morphisms of Semigroups

Definition 1 : *Let $(S, *)$ and $(T, \circ)$ be any two semigroups. A mapping $f : S \to T$ is called semigroup* **homomorphism** *if $f(a * b) = f(a)\ o\ f(b), \forall a, b \in S$ (structure preserving property).*

- A homomorphism of a semigroup $(S, *)$ into itself is called **endomorphism** *i.e.*, endomorphism is a mapping $f : S \to S$ such that $f(a * b) = f(a) * f(b)\ ; \forall\ a, b \in S$.

Definition 2 : *Let (S, *) and (T, o) be two semigroups. A semigroup homomorphism $f : S \to T$ is called*

(a) **epimorphism** if f is onto also.
(b) **monomorphism** if f is one-one also.
(c) **isomorphism** if f is one-one and onto (*i.e.*, bijective)

Definition 3 : *A semigroup isomopphism from a semigroup (S, ∗) onto itself is called semigroup automorphism, i. e., a semigroup automorphism is a mapping from a semigroup S,∗) onto itself which is homomorphism and bijective.*

Definition 4 : Two semigroups $(S, *)$ and $(T, \circ)$ are said to be isomorphic if there exists a semigroup isomorphism from S to T (or from T to S).

☛ If two semigroups and $(T, \circ)$ are isomorphic then they are structurally identical in the sense that they can differ only in the nature of their elements and the involved operations.

Definition 5 : *An element a in a semigroup (S, *) is called an idempotent if* $a^2 = a$.

Theorem 3. *Let $f : S \to T$ be an onto mapping from a semigroup (S, *) to an algebraic structure $(T, \circ)$ where $\circ$ is a binary operation on T. If f is a semigroup* ***homomorphism*** *then $(T,\circ)$ is a semigroup.*

Proof : We have to prove that $(T, \circ)$ is a semigroup. For this we shall prove that $\circ$ is an associative operation on T.

Let x, y, z be any three elements in T. Now, since f is onto mapping, there exist a, b, c in S such that $x = f(a)$

$$y = f(b) \text{ and } z = f(c)$$

Consider $(x \circ y) \circ z = (f(a) \circ f(b) \circ f(c))$

$= f(a * b) \circ f(c)$ [∵ f is a homomorphism]
$= f((a * b) * c)$ [∵ f is a homomorphism]
$= f(a \times (3 \times c))$ [∵ f is associative]
$= f(a) \circ f(b * c)$ [∵ f is a homomorphism]
$= f(a) \circ (f(b) \circ f(c))$ [∵ f is a homomorphism]
$= x \circ (y \circ z)$

⇒ $\circ$ is associative. Hence $(T, \circ)$ is a semigroup.

Theorem 4. *Let f be a semigroup homomorphism from semigroup (S, *) to semigroup $(T, \circ)$. If $a \in S$ is an idempotent element then f (a) is idempotent in $(T, \circ)$.*

Proof : Let $(S, *)$ and $(T, \circ)$ be two semigroups and let $f : S \to T$ be a homomorphism.

Let $a \in S$ be idempotent element.

Then, by definition $a^2 = a$

Now $[f(a)]^2 = f(a) \circ f(a) = f(a * a)$ [∵ f is a homomorphism]
$= f(a^2) = f(a)$

⇒ $f(a)$ is an idempotent element in $(T, \circ)$. [∵ $a^2 = a$]

Theorem 5. *Every finite semigroup has an idempotent element.*

Proof : Let S be a finite semigroup on S and x be any element in S.

Consider positive integral powers of x.

$$x, x^2, x^3 \ldots$$

Since S is finite, therefore all these powers of x can not be distinct. Thus, there exists positive integers r and s such that $x^r = x^s$.

Let, $k = r - s$. Then $k \geq 1$.

Also, we can write $x^r = x^s$ as

Now, $x^{s+k} = x^s$ ...(1)

$x^{2s+k} = x^s . x^{s+k} = x^s . x^s = x^{2s}$ [using (1)]

Similarly $x^{3s+k} = x^{3s}$.

Then we can write by the principle of mathematical induction

$x^{ms+k} = x^{ms}$ for any $m \in \mathbf{N}$ ...(2)

Further $x^{ms+2k} = x^{ms+k} . x^k = x^{ms} . x^k$ [using (2)]

$= x^{ms+k}$

$= x^{ms}$

and $x^{ms+3k} = x^{ms+2k} . x^k$ ...(3)

$= x^{ms} . x^k = x^{ms+k}$

$= x^{ms}$

Again using principle of induction, we have $x^{ms+nk} = x^{ms}$ for $n \in \mathbf{N}$

For $m = k$ and $n = s$, we get

$x^{ks+sk} = x^{ks} \Rightarrow x^{2ks} = x^{ks} \Rightarrow \mathbf{a^2 = a}$ [for $a = x^{ks}$]

Hence, every finite semigroup has an idempotent element.

Theorem 6. *If f is a homomorphism from a commutative semigroup $(S, *)$ onto a semigroup $(T, \circ)$ then $(T, \circ)$ is also commutative.*

Proof : Let $f : S \to T$ be a homomorphism.

Let x and y be any two elements of T. Now since f is onto, therefore there exist elements a and b in S such that $x = f(a)$ and $y = f(b)$

Consider $x \circ y = f(a) \circ f(b)$

$= f(a * b)$ [$\because f$ is a homomorphism]

$= f(b * a)$ [$\because (S, *)$ is commutative]

$= f(b) \circ f(a)$

$= y \circ x$

Hence, $(T, \circ)$ is commutative.

Theorem 7. *Let f be a homomorphism from a semigroup $(S, *)$ to semigroup (T, o). If $\mathbf{H}$ is a subsemigroup of $(S, *)$. Then $f(\mathbf{H}) = \{t \in T : t = f(h)$, for some $h \in \mathbf{H}\}$ is a subsemigroup of $(T, \circ)$.*

Proof : Let x and y be any two elements of $f(H)$. By definition of $f(H)$, there must exist a and b in H such that $x = f(a)$ and $y = f(b)$

Now, $x \circ y = f(a) \circ f(b) = f(a * b)$

$= f(c)$, where $c = a * b \in H$

$\Rightarrow f(c) \in f(H) \Rightarrow f(H)$ is closed under the operator $\circ$ of T.

Hence, $f(H)$ is a subsemigroup of $(T, \circ)$.

12.7 COMMUTATIVE MONOID

Definition : *A monoid $(M, *)$ is said to be commutative if the operator $*$ is commutative on M.*

12.7.1 Cyclic Monoid

A monoid $(M, *)$ is said to be cyclic if there exists an element $a \in M$ such that every element of M can be written as some integral powers of a. In such a case a is called the generator of the cyclic monoid M.

Theorem 1. *Every cyclic monoid is a commutative monoid.*

Proof : Let $(M, *)$ be a cyclic monoid generated by an element a of M. Further let x, y be any two elements of M.

Then $x = a^m$ and $y = a^n$ for some $m, n \in \mathbf{N}$.

Consider $x * y = a^m * a^n$

$$= a^{m+n} = a^{n+m}$$

$$= a^n * a^m = y * a$$

Hence, $(M, *)$ is commutative.

Theorem 2. *The identity element of a monoid is unique.*

Proof : Let $(M, *)$ be a monoid. Let if possible e and e' be two identity elements in $(M, *)$.

Now, since e is the identity element therefore $e * e' = e' * e = e'$ ...(1)

Further, since e' is the identity element, therefore $e' * e = e * e' = e$ ...(2)

From (1) and (2) we conclude that $e = e'$

Hence, the identity element of a monoid is unique.

12.7.2 Submonoids

Let $(M, *)$ be a monoid with identity element e and let T be a non-empty subset of M. If T is closed under the operation $*$ and identity $e \in T$, then $(T, *)$ is known as submonoid of $(M, *)$.

☛ Since associativity holds in a monoid $(M, *)$, therefore it holds in any subset of monoid M. Hence, a submonoid of a monoid is also a monoid.

Theorem 3. *The intersection of two submonoids of a monoid $(M, *)$ is again a submonoid of $(M, *)$.*

Proof : Let $(T_1, *)$ and $(T_2, *)$ be two submonoids of a monoid $(M, *)$. Again e be the identity element in M. Using the definition of monoid, we have

$$e \in T_1, e \in T_2 \text{ which implies } T_1 \cap T_2 \neq \phi$$

Firstly, we shall prove that $T_1 \cap T_2$ is closed with respect to $*$ let $x, y \in T_1 \cap T_2$

$\Rightarrow$ $x, y \in T_1$ and $x, y \in T_2$ $\Rightarrow$ $x * y \in T_1$ and $x * y \in T_2$

$\Rightarrow$ $x * y \in T_1 \cap T_2$

$\Rightarrow$ $T_1 \cap T_2$ is closed with respect to $*$. Finally, we show that $e \in T_1 \cap T_2$.

Since T_1 and T_2 are submonoids, thus T_1 and T_2 both contains identity elements.

Now $e \in T_1$ and $e \in T_2$ $\Rightarrow$ $e \in T_1 \cap T_2$

$\Rightarrow$ $T_1 \cap T_2$ contains identity element e of monoid $(M, *)$.

Hence, $(T_1 \cap T_2, *)$ is a submonoid.

Theorem 4. *If $(M, *)$ is a commutative monoid, then the set of all idempotent elements of M forms a submonoid.*

Proof : Let S be the set of all idempotent elements of M such that

$$S = \{x \in M : x^2 = x\}$$

It is known that the identity element $e \in M$ is always idempotent. Thus, we have $e \in S$.

Now to show that S is closed with respect to $*$.

Let $a \in S$, $b \in S$. Then $a^2 = a$, $b^2 = b$ ...(1)

Consider $(a * b)^2 = (a * b) * (a * b)$

$$= a * (b * a) * b \quad [\because * \text{ is associative}]$$

$$= a * (a * b) * b \quad [\because * \text{ is commutative}]$$

$$= (a * a) * (b * b)$$

$$= a^2 * b^2 = a * b \quad [\text{Using (1)}]$$

$\Rightarrow$ $a * b$ is idempotent element of M. Hence, $a * b \in S$.

Hence, S is a submonoid.

12.7.3 Morphisms of Monoids

Definition 1. *Let $(M, *)$ and $(T, \circ)$ be two monoids with e and e′ as identity elements respectively. Then a mapping $f : M \to T$ is said to be a monoid homomorphism if for any two elements $a, b \in M$*

$$f(a * b) = f(a) \circ f(b). \quad \text{[Structure preserving property]}$$

and $f(e) = e'$

Definition 2 : *A homomorphism of a monoid into itself is called a monoid endomorphism.*

Definition 3 : *A homomorphism $f : M \to T$ is called*

(i) monoid epimorphism if f is onto also.

(ii) monoid monomorphism if f is one-one also.

(iii) monoid isomorphism if f is one-one and onto (bijective) also.

Definition 4 : *Two monoids $(M, *)$ and $(T, \circ)$ are said to be isomorphic monoids if there exists a monoid isomorphism from M to T.*

Definition 5 : *An isomorphism of a monoid onto itself is called a monoid automorphism. Thus a mapping $f : (M, *) \to (M, *)$ is said to be a monoid automorphism of M if*

(i) f is bijective *(ii) $f(a * b) = f(a) * f(b)$; $\forall\, a, b \in M$* *(iii) $f(e) = e$*

Theorem 5. *Let $(M, *)$ and $(T, \circ)$ be two monoids with identity elements e and e′ respectively. If f is onto from M onto T such that*

$$f(a * b) = f(a) \circ f(b) \; ; \forall\, a, b \in M$$

Then $f(e) = e'$

Proof : Define a map $f : M \xrightarrow{\text{onto}} T$

Let y be any element of T. Since, f is onto therefore, there exists an element $x \in M$ such that $f(x) = y$

Now $y = f(x) = f(x * e)$

$= f(x) \circ f(e) = y \circ f(e)$ [$\because f$ is onto]

Similarly $y = f(x) = f(e * x) = f(e) \circ f(x) = f(e) \circ y$

$\Rightarrow$ $f(e) \circ y = y \circ f(e) = y$ $\Rightarrow$ $f(e)$ is the identity for T.

$\Rightarrow$ $e' = f(e)$ [$\because$ Identity element of a monoid is always unique]

Theorem 6 : *Let $x = x \circ f(e) = f(e) \circ x$ be a monoid with identity element $f(e)$ and $(T, \circ)$ be any algebraic structure. If a mapping $(T, \circ)$ is onto and satisfying.*

$$f(a * b) = f(a) \circ f(b) \; \forall\, a, b \in M$$

then $(T, \circ)$ is a monoid with $a, b \in M$ as its identity element.

Proof : Firstly, we shall prove that $\circ$ is associative. Let x, y and z be any three elements in T. Now, since f is onto then there exist a, b and c in M such that

$$x = f(a),\ y = f(b),\ z = f(c)$$

Consider $x \circ (y \circ z) = f(a) \circ (f(b) \circ f(c))$

$= f(a) \circ f(b * c)$ [$\because f(b * c) = f(b) \circ f(c)$]

$= f(a * (b * c)) = f((a * b) * c)$ [by associativity]

$= f(a * b) \circ f(c)$

$= (f(a) \circ f(b)) \circ f(c) = (x \circ y) \circ z$

$\Rightarrow$ $\circ$ is associative in T.

Further, we shall show that $f(e)$ is the identity for T. Let x be any element of T.

Now, since f is onto, there exists an element $a \in M$ such $f(a) = x$

Consider, $x = f(a)$

$= f(a * e)$

$= f(a) \circ f(e) = x \circ f(e)$

Similarly, we can prove that $x = f(a) = f(e * a) = f(e) \circ f(a) = f(e) \circ x$

Therefore, $x = x \circ f(e) = f(e) \circ x$

Thus, we conclude that $f(e)$ is the identity element for $(T, \circ)$.

Hence $(T, \circ)$ is a monoid with $f(e)$ as identity element.

- In view of the Theorem-1, an onto mapping f from a monoid $(M, *)$ onto monoid $(T, \circ)$ is called homomorphism if for any two elements $a, b \in M$

$$f(a * b) = f(a) \circ f(b)$$

This is because $f(e) = e'$ is a consequence of f being onto and satisfying

$$f(a * b) = f(a) \circ f(b) \ ; \forall\ a, b \in M.$$

- If f is a homomorphism from a commutative monoid $(M, *)$ onto a monoid $(T, \circ)$, then T is also commutative.

12.8 CONGRUENCE RELATION AND QUOTIENT SEMIGROUPS

Definition : *An equivalence relation R on the semigroup $(S, *)$ is called a congruence relation if $a\ R\ a'$ and $b\ R\ b' \Rightarrow (a * b)\ R\ (a' * b')$.*

Theorem 1. *Let $(S, *)$ be a congruence relation on a semigroup R. Then quotient set S/R is a semigroup with respect to the operation defined by*

$$a \in S, [a]$$

where $[a]$ denotes the equivalence class of element a in S corresponding to the relation R.

Proof : Let $(S, *)$ be a semigroup and R be a congruence relation on S. Then, for any $a \in S$, the equivalence class $[a]$ is a set containing all those elements of S which are related to a under the relation R. Now, for $a, b \in S$, define an operation $\oplus$ on S/R such that $[a] \oplus [b] = [a * b]$

Firstly, we shall prove that operation $\oplus$ is well defined on S/R.

Now, suppose that $[a] = [a']$ and $[b] = [b']$

Since $[a] = [a']$ and $[b] = [b']$ therefore $a\ R\ a'$ and $b\ R\ b'$

Now, since R is a congruence relation, therefore

$$a\ R\ a' \text{ and } b\ R\ b' \Rightarrow a * b\ R\ a' * b'$$

$\Rightarrow \quad (a * b) = (a' * b') \qquad \Rightarrow \quad \oplus$ is well defined on S/R

$\Rightarrow \quad \oplus$ is binary operation on S/R.

Further, we shall show that $\oplus$ is associative.

For $a, b, c \in S$, we have

$$\begin{aligned}[a] \oplus [b] \oplus [c] &= [a] \oplus [b * c] \\ &= [a * (b * c)] \\ &= [(a * b) * c] \\ &= [a * b] \oplus [c] \\ &= ([a] \oplus [b]) \oplus [c] \qquad [\because * \text{ is associative}]\end{aligned}$$

Therefore, $\oplus$ is associative. Hence S/R is a semigroup.

Definition : *The semigroup defined in above theorem $(S/R, \oplus)$ is called quotient semigroup or factor semigroup of semigroup $(S, *)$ by the congruence relation R.*

- The operator $\oplus$ on S/R is called quotient binary relation.

Theorem 2. *Let R be a congruence relation on the monoid $(M, *)$ Then $(M/R, \oplus)$ is a monoid, where $\oplus$ on M/R is defined by $[a] + [b] = [a * b]$.*

Proof : Let $(M, *)$ be a monoid with e as identity. Using Theorem-1, we can say that $(M/R, \oplus)$ is a semigroup. We have to show that $(M/R, \oplus)$ has identity element. For this we shall prove that $[e]$ is the identity element in $(M/R, \oplus)$. Let $[a]$ be any element in M/R where $a \in M$. Then we have

$$\begin{aligned}[a] \oplus [e] &= [a * e] \\ &= [a] \\ &= [e * a] \\ &= [e] \oplus [a]\end{aligned}$$

Hence, $[e]$ is the identity in M/R. Therefore $(M/R, \oplus)$ is a monoid.

Theorem 3. *Let $a \in S$ be a semigroup and let $f(a) = [a]$ be a congruence relation on $\Rightarrow f$ Then there exists a homomorphism from f onto quotient semigroup $(S/R, a \in S)$.*

Proof. Define a function $f : S \to S/R$ such that

$$f(a) = [a], \text{ the equivalence class of } a \in S \text{ under } R.$$

To show, f is onto and homomorphism.

(i) **f** is onto

Let $[a]$ be any arbitrary element in S/R. Then $a \in S$ and $f\,(a) = [a]$

$\Rightarrow$ f is onto.

(ii) **f** is homomorphism.

Let $a \in S$, $b \in S$, Then we have $f\,(a * b) = [a * b], = [a] \oplus [b]$, by definition of $\oplus$

$$= f(a) \oplus f(b)$$

$\Rightarrow$ Structure preserving property is satisfied by f.

$\Rightarrow$ f is a homomorphism.

☛ The mapping $f : S \to S/R$ defined by $f(a) = [a]$ is called **natural** homomorphism.

Theorem 4. *Let $f : S \to T$ be a semigroup homomorphism from the semigroup $(S, *)$ to the semigroup $(T, \circ)$. Then the relation R on $(S, *)$ defined by $a\,R\,b$ is and only if $f(a) = f(b)$ for $a, b \in S$ is a congruence relation on $(S, *)$ is an equivalence relation.*

Proof : It is known that a relation is said to be equivalence if it is reflexive, symmetric and transitive.

(i) **R is symmetric**. Since $f(a) = f(a)$ $\Rightarrow$ $a\,R\,a : \forall\, a \in S$

(ii) **R is symmetric.** Let $a\,R\,b$ then $f(a) = f(b)$ which can easily be written as $f(b) = f(a)$

$\Rightarrow$ $b\,R\,a$

(iii) **R is transitive.** Let us suppose that $a\,R\,a$ and $b\,R\,c$

Now $a\,R\,b \Rightarrow f(a) = f(b)$, $b\,R\,c \Rightarrow f(b) = f(c)$.

Therefore, $a\,R\,b$ and $b\,R\,c$ $\Rightarrow$ $f(a) = f(c)$ $\Rightarrow$ $a\,R\,c$.

Therefore, R is an equivalence relation.

Further, it remains to prove that R is a congruence relation.

Suppose that $a\,R\,a'$ and $b\,R\,b'$

Then $a\,R\,a' \Rightarrow f(a) = f(a')$ and $b\,R\,b' \Rightarrow f(b) = f(b')$

Now, $f(a * b) = f(a) \circ f(b)$ [$\because f$ is a homomorphism]

$$\begin{aligned} &= f(a') \circ f(b') \\ &= f(a' * b') \Rightarrow (a * b)\,R\,(a' * b')\end{aligned}$$

Therefore, $a\,R\,a$ and $b\,R\,b' \Rightarrow (a * b)\,R\,(a' * b')$

Hence R is a congruence relation on $(S, *)$.

Theorem 5. *(Fundamental Theorem of Homomorphism of Semigroups)*

Let $(S, *)$ and $(T, \circ)$ be two semigroups. If $f : S \to T$ is a semigroup homomorphism, then semigroup $(T, \circ)$ is isomorphic to some quotient semigroup of $(S, *)$.

Proof : Let $f: S \to T$ be a homomorphism of the semigroup $(S, *)$ onto the semigroup $(T, \circ)$. Let R be the congruence relation on $(S, *)$ corresponding to the homomorphism f. Then, R is defined by aRb iff $f(a) = f(b); \forall a, b \in S$.

To show that $(T, \circ)$ is isomorphic to quotient semigroup $(S/R, \oplus)$ of $(S, *)$

If $a \in S$, then $[a] \in S/R$ and $f(a) \in T$

Now, define a mapping $\psi : S/R \to T$ by setting $\psi([a]) = f[a]$ for all $a \in S$.

Firstly, we shall prove that ψ is semigroup isomorphism. For this first of all we shall prove that ψ is well defined.

Let us suppose $a, a' \in S$ and $[a] = [a']$.

Now $[a] = [a'] \Rightarrow aRa' \quad \Rightarrow \quad f[a] = f[a']$

$\Rightarrow \quad \psi([a]) = \psi([a']) \quad \Rightarrow \quad \psi$ is well defined.

ψ is one-one. Let us suppose that $\psi([a]) = \psi([b])$ Then, we have $f[a] = f[b]$

$\Rightarrow \quad aRa \quad \Rightarrow \quad [a] = [b] \quad \Rightarrow \quad \psi$ is one-one

ψ is onto. Let b be any arbitrary element of T. Now, since f is onto, then there exists an element a in S such that $b = f[a]$.

Now, $\psi([a]) = f[a] = b \quad \Rightarrow \quad \psi$ is onto.

ψ is a homomorphism.

Let $[a]$, $[b]$ be any two elements in S/R. Then

$$\begin{aligned} \psi([a] \oplus [b]) &= \psi([a * b]) \\ &= f(a * b) \\ &= f[a] \circ f[b] \quad [\because f \text{ is a homomorphism}] \\ &= \psi([a]) \circ \psi([b]) \end{aligned}$$

Hence, we conclude that ψ is an isomorphism of S/R onto T. Therefore, $(S/R, \oplus)$ and $(T, \circ)$ are isomorphic.

ILLUSTRATIVE EXAMPLES

1. *Let **R** be the set of real numbers and operation $*$ on **R** be defined as follows*

$$a * b = |a - b|; \forall a, b \in \mathbf{R}$$

*Then $*$ is commutative binary operation on **R** but $(\mathbf{R}, *)$ is not a semigroup.*

Solution. By definition of an absolute value of numbers we have

$|a - b|$ is always positive real number and $a * b \in \mathbf{R}; \forall a, b \in \mathbf{R}$

$\Rightarrow \quad *$ is binary operation on **R**.

Now, since $|a - b| = |b - a|; \forall a, b \in \mathbf{R}$, we have $a * b = b * a; \forall a, b \in \mathbf{R}$

Therefore, $*$ is a commutative operation. Also, $*$ is not associative because

$$1 * (2 * 3) = 1 * |2 - 3| = 1 * 1 = |1 - 1| = 0$$

But $(1 * 2) * 3 = |1 - 2| * 3 = 1 * 3 = |1 - 3| = 2 \quad \Rightarrow \quad 1 * (2 * 3) \neq (1 * 2) * 3$

2. *Show that there exists a semigroup homomorphism from the semigroup $(\mathbf{N}, +)$ of natural numbers under addition to the semigroup $(\{0, 1, 2, 3\}, +_4)$ where $+_4$ denotes the operation of addition modulo 4 on the set $\{0, 1, 2, 3\}$.*

Solution. Define a mapping $f: \mathbf{N} \to \{0, 1, 2, 3\}$ such that

$$\begin{aligned} f(a) &= a \pmod 4; \forall a \in \mathbf{N} \\ &= \text{The remainder } r, 0 \le r < 4 \text{ when } a \text{ is divided by } 4. \end{aligned}$$

Now, for any $a, b \in \mathbf{N}$, let $f(a) = i$ and $f(b) = j$. Then

$$\begin{aligned} f(a + b) &= (a + b) \pmod 4 \\ &= (i + j) \pmod 4 \\ &= i +_4 j = f(a) +_4 f(b) \end{aligned}$$

Hence, f is a homomorphism.

3. *Show that monoids $(\mathbf{Z}, +)$ and $(E, +)$ are Isomorphic where **Z** is the set of integers and E is the set of even integers.*

Solution. Define a mapping $f: \mathbf{Z} \to E$ such that $f(x) = 2x; \forall x \in \mathbf{Z}$

We shall prove that f is an isomorphism.

(i) f is one-one

Let $x, y \in \mathbf{Z}$

Now, suppose that

$\Rightarrow \quad f(x) = f(y) \qquad \Rightarrow \quad 2x = 2y \qquad \Rightarrow \quad x = y$

Therefore, $f(x) = f(y)$

$\Rightarrow \quad x = y \qquad \Rightarrow \quad f$ is one-one

(ii) f is onto

Let y be any arbitrary element of E. Then y is an even integer. Thus $\frac{y}{2}$ is an integer.

Let $x = \frac{y}{2}$, then $x \in \mathbf{Z}$ such that $f(x) = 2x = 2 \cdot \frac{y}{2} = y \quad \Rightarrow \quad f$ is onto

(iii) f is a homomorphism.

Let $x, y \in \mathbf{Z}$ then $f(x + y) = 2(x + y) = 2x + 2y = f(x) + f(y)$

Therefore, f is a homomorphism. Hence, monoids $(\mathbf{Z}, +)$ and $(E, +)$ are isomorphic.

4. *Let A be a non-empty set of symbols and let A^* be the free semigroup generated by A under the operation of concatenation. Show that the functions $f : A^* \to \mathbf{N}$ defined by $f(\alpha) = l(\alpha)$, the length of α, where $l(\alpha)$ denotes the number of symbols in α (each symbol is counted as many times as it appears in α) is a homomorphism and if R is the relation induced by f, then show that $(A^*/R, \oplus)$ is isomorphic to $(\mathbf{N}, +)$.*

Solution. Let α, β be any two elements (words) in A^*

Then, we have $f(\alpha, \beta) = l(\alpha \cdot \beta) = l(\alpha) + l(\beta) = f(\alpha) + f(\beta)$

$\Rightarrow \quad f$ is a homomorphism.

By Theorem 18 (just before the solved examples), f induces a congruence relation on A^* defined by $\alpha R \beta$ if and only if $l(\alpha) = l(\beta)$

Now, since f is onto, then by fundamental theorem of semigroup homomorphism $(A^*/R, \oplus)$ is isomorphic to $(\mathbf{N}, +)$

Exercise 12.1

1. Let $S = \{1, 2, 3, 6, 12\}$ and the operation $*$ be defined as follows
   $a * b$ = greatest common divisor of a and b.
   Show that $(S, *)$ is a semigroup.
2. Show that the set $\mathbf{N}$ of natural numbers is a semigroup under operation $*$ defined by $x * y = \max\{x, y\}$
3. Show that every finite semigroup has an idempotent element.
4. Show that the non-empty interesection of two sub-semigroups of a semigroup is again a sub-semigroup.
5. Show that the union of two sub semigroups of a semigroup need not be a sub semigroup.
6. Show that the set of all semigroups endomorphism of a semigroup is a semigroup under the operation of composition of functions.
7. Let $(S, *)$ and $(T, *)$ be two semigroups. Show that $S \times T$ and $T \times S$ are isomorphic semigroups
8. Show that the monoids $(\{0, 1, 2, 3\}, +_4)$ and $(\{1, 3, 7, 9\}, \times_{10})$ are isomorphic.
9. Show that the set of all invertible elements of a monoid forms a monoid under the same operation as that of the monoid.
10. Show that the intersection of two congruence relation on a semigroup is a congruence relation.

❑❑❑

1.	**Aho, A.V., Hopocroft, J.E. and Ullman, J.D.**	*The Design and analysis of computer Algorithm;* **Addison-Wesley.**
2.	**Brand, L**	*Differential and difference equation;* **Wiley.**
3.	**Conway, J.H.**	*Regular Algebra and finite machines;* **Chapman and Hall.**
4.	**Hall, P**	*On representation of subsets;* **J. Hondon, Math. Soc.**
5.	**Halmas, P.R.**	*Lecturers on Boolean Algebra;* **Springer-Verlag.**
6.	**Liu, C.L.**	*Elements of Discrete Mathematics;* **McGraw-Hill.**
7.	**Pundir, S.K.**	*Introductory Discrete Mathematics;* **JPNCo.**
8.	**Ramsey, F.P.**	*On a problem in formal logic;* **Proc. London, Math. Soc.**
9.	**Truss, J**	*Discrete Mathematics for computer Scientists;* **Pearson Education Asia.**
10.	**Klison, R.J.**	*Introduction to graph Theory;* **Longman.**

INDEX

A

B

C

D

E

P

Q

R

S

T

U

V

W

Z